ANALYTICAL DYNAMICS

ANALYTICAL DYNAMICS

Haim Baruh
Rutgers University

Boston   Burr Ridge, IL   Dubuque, IA   Madison, WI   New York   San Francisco   St. Louis
Bangkok   Bogotá   Caracas   Lisbon   London   Madrid
Mexico City   Milan   New Delhi   Seoul   Singapore   Sydney   Taipei   Toronto

WCB/McGraw-Hill

A Division of The **McGraw-Hill** Companies

ANALYTICAL DYNAMICS

Copyright © 1999 by The McGraw-Hill Companies, Inc. All rights reserved. Printed in the United States of America. Except as permitted under the United States Copyright Act of 1976, no part of this publication may be reproduced or distributed in any form or by any means, or stored in a data base or retrieval system, without the prior written permission of the publisher.

This book is printed on acid-free paper.

1 2 3 4 5 6 7 8 9 0 DOC/DOC 9 3 2 1 0 9 8

ISBN 0-07-365977-0

Vice president and editorial director: *Kevin T. Kane*
Publisher: *Thomas Casson*
Senior sponsoring editor: *Debra Riegert*
Marketing manager: *John T. Wannemacher*
Project manager: *Kari Geltemeyer*
Production supervisor: *Scott M. Hamilton*
Designer: *JoAnne Schopler*
Supplement coordinator: *Kimberly D. Stack*
Compositor: *Publication Services, Inc.*
Typeface: *10/12 Times Roman*
Printer: *R. R. Donnelley & Sons Company*

Library of Congress Cataloging-in-Publication Data
Baruh, Haim.
    Analytical dynamics / Haim Baruh.
      p.   cm.
    Includes bibliographical references and index.
    ISBN 0-07-365977-0
    1. Dynamics.   I. Title.
QA845.B38   1999
531'.11—dc21   98-34940

http://www.mhhe.com

Dedicated in memory of my dear mother, Ester Baruh, ע״ה

Contents

Preface xiii

Chapter 1 Basic Principles 1

1.1 Introduction 1
1.2 Systems of Units 1
1.3 Review of Vector Analysis 4
    1.3.1 Rectilinear (Cartesian) Coordinates 5
    1.3.2 Curvilinear Coordinates 8
1.4 Newtonian Particle Mechanics 26
1.5 Degrees of Freedom and Constraints 33
1.6 Impulse and Momentum 36
1.7 Work and Energy 43
    1.7.1 Gravitational Potential Energy 44
    1.7.2 Potential Energy of Springs 46
    1.7.3 Elastic Strain Energy 48
    1.7.4 Work-Energy Relations 48
1.8 Equilibrium and Stability 54
1.9 Free Response of Linear Systems 58
1.10 Response to Harmonic Excitation 62
1.11 Forced Response of Linear Systems 68
1.12 First Integrals 71
1.13 Numerical Integration of Equations of Motion 75
    References 77
    Homework Exercises 77

Chapter 2 Relative Motion 87

2.1 Introduction 87
2.2 Moving Coordinate Frames 88
2.3 Representation of Vectors 91
2.4 Transformation of Coordinates, Finite Rotations 97
2.5 Infinitesimal Rotations, Angular Velocity 107
2.6 Rate of Change of a Vector, Angular Acceleration 117
2.7 Relative Velocity and Acceleration 124
2.8 Observations from a Moving Frame 133
    References 144
    Homework Exercises 144

Chapter 3   Dynamics of a System of Particles   153

- 3.1 Introduction 153
- 3.2 Equations of Motion 153
- 3.3 Linear and Angular Momentum 157
- 3.4 Work and Energy 161
- 3.5 Impact of Particles 164
- 3.6 Variable Mass and Mass Flow Systems 170
- 3.7 Concepts from Orbital Mechanics: The Two Body Problem 173
- 3.8 The Nature of the Orbit 177
- 3.9 Orbital Elements 187
- 3.10 Plane Kinetics of Rigid Bodies 193
- 3.11 Instant Centers and Rolling 200
- 3.12 Energy and Momentum 203
- References 206
- Homework Exercises 207

Chapter 4   Analytical Mechanics: Basic Concepts   215

- 4.1 Introduction 215
- 4.2 Generalized Coordinates 216
- 4.3 Constraints 219
  - 4.3.1 Holonomic Constraints 220
  - 4.3.2 Nonholonomic Constraints 223
- 4.4 Virtual Displacements and Virtual Work 230
- 4.5 Generalized Forces 237
- 4.6 Principle of Virtual Work for Static Equilibrium 241
- 4.7 D'Alembert's Principle 245
- 4.8 Hamilton's Principles 249
- 4.9 Lagrange's Equations 253
- 4.10 Lagrange's Equations for Constrained Systems 260
- References 265
- Homework Exercises 265

Chapter 5   Analytical Mechanics: Additional Topics   273

- 5.1 Introduction 273
- 5.2 Natural and Nonnatural Systems, Equilibrium 273
- 5.3 Small Motions about Equilibrium 279
- 5.4 Rayleigh's Dissipation Function 286

- 5.5 Eigenvalue Problem for Linearized Systems 289
- 5.6 Orthogonality and Normalization 293
- 5.7 Modal Equations of Motion and Response 296
- 5.8 Generalized Momentum, First Integrals 300
- 5.9 Routh's Method for Ignorable Coordinates 302
- 5.10 Impulsive Motion 304
- 5.11 Hamilton's Equations 308
- 5.12 Computational Considerations: Alternate Descriptions of the Motion Equations 311
- 5.13 Additional Differential Variational Principles 314
  - References 317
  - Homework Exercises 317

Chapter 6 RIGID BODY GEOMETRY 323

- 6.1 Introduction 323
- 6.2 Center of Mass 323
- 6.3 Mass Moments of Inertia 326
- 6.4 Transformation Properties 334
  - 6.4.1 Translation of Coordinates 335
  - 6.4.2 Rotation of Coordinate Axes 337
- 6.5 Principal Moments of Inertia 343
- 6.6 Inertia Ellipsoid 347
  - References 351
  - Homework Exercises 351

Chapter 7 RIGID BODY KINEMATICS 355

- 7.1 Introduction 355
- 7.2 Basic Kinematics of Rigid Bodies 355
  - 7.2.1 Pure Translation 356
  - 7.2.2 Pure Rotation 356
  - 7.2.3 Combined Translation and Rotation 359
- 7.3 Euler's and Chasles's Theorems 360
- 7.4 Relation between Direction Cosines and Angular Velocity 365
- 7.5 Euler Angles 367
  - 7.5.1 Euler Angle Sequences 368
  - 7.5.2 Angular Velocity and Acceleration 369
  - 7.5.3 Axisymmetric Bodies 372
- 7.6 Angular Velocities as Quasi-Velocities (Generalized Speeds) 378
- 7.7 Euler Parameters 380
  - 7.7.1 Relating the Direction Cosines to the Euler Paramaters 382

7.7.2 Relating the Euler Parameters to Angular Velocities 384
7.7.3 Relating the Euler Parameters to the Direction Cosines 386
7.7.4 Relating the Euler Parameters to the Euler Angles and Vice Versa 387
7.7.5 Other Considerations 388
7.8 Constrained Motion of Rigid Bodies, Interconnections 391
7.8.1 Rotational Contact, Joints 391
7.8.2 Translational Motion and Sliding Contact 395
7.8.3 Combined Sliding and Rotation 396
7.9 Rolling 400
References 412
Homework Exercises 412

Chapter 8 RIGID BODY DYNAMICS: BASIC CONCEPTS 421

8.1 Introduction 421
8.2 Linear and Angular Momentum 422
8.3 Transformation Properties of Angular Momentum 425
8.3.1 Translation of Coordinates 425
8.3.2 Rotation of Coordinates 426
8.4 Resultant Force and Moment 429
8.5 General Equations of Motion 431
8.6 Rotation about a Fixed Axis 443
8.7 Equations of Motion in State Form 445
8.8 Impulse-Momentum Relationships 448
8.9 Energy and Work 451
8.10 Lagrange's Equations for Rigid Bodies 456
8.10.1 Virtual Work and Generalized Forces 456
8.10.2 Generalized Coordinates 458
8.10.3 Lagrange's Equations in Terms of the Euler Angles 462
8.10.4 Lagrange's Equations in Terms of the Euler Parameters 463
8.10.5 Lagrange's Equations and Constraints 464
8.11 D'Alembert's Principle for Rigid Bodies 474
8.12 Impact of Rigid Bodies 477
References 481
Homework Exercises 481

Chapter 9 DYNAMICS OF RIGID BODIES: ADVANCED CONCEPTS 489

9.1 Introduction 489
9.2 Modified Euler's Equations 490

- 9.3 Moment Equations about an Arbitrary Point   498
- 9.4 Classification of the Moment Equations   503
- 9.5 Quasi-Coordinates and Quasi-Velocities (Generalized Speeds)   504
- 9.6 Generalized Speeds and Constraints   510
- 9.7 Gibbs-Appell Equations   517
- 9.8 Kane's Equations   524
- 9.9 The Fundamental Equations and Constraints   531
- 9.10 Relationships between the Fundamental Equations and Lagrange's Equations   532
- 9.11 Impulse-Momentum Relationships for Generalized Speeds   539
  - References   542
  - Homework Exercises   542

Chapter 10 QUALITATIVE ANALYSIS OF RIGID BODY MOTION   547

- 10.1 Introduction   547
- 10.2 Torque-Free Motion of Inertially Symmetric Bodies   548
- 10.3 General Case of Torque-Free Motion   556
- 10.4 Polhodes   557
- 10.5 Motion of a Spinning Top   563
- 10.6 Rolling Disk   573
- 10.7 The Gyroscope   580
  - 10.7.1 Free Gyroscope   580
  - 10.7.2 The Gyrocompass   582
  - 10.7.3 Single-Axis Gyroscope   585
  - References   587
  - Homework Exercises   587

Chapter 11 DYNAMICS OF LIGHTLY FLEXIBLE BODIES   591

- 11.1 Introduction   591
- 11.2 Kinetic and Potential Energy: Small or No Rigid Body Motion   592
  - 11.2.1 Kinematics and Geometry   592
  - 11.2.2 Kinetic and Potential Energy for Beams   597
  - 11.2.3 Kinetic and Potential Energy for Torsion   599
  - 11.2.4 Operator Notation   601
- 11.3 Equations of Motion   604
  - 11.3.1 Extended Hamilton's Principle   604

11.3.2 Boundary Conditions 606
 11.3.3 Simplification 608
 11.3.4 Operator Notation 611
11.4 Eigensolution and Response 616
11.5 Approximate Solutions: The Assumed Modes Method 626
 11.5.1 Approximation Techniques and Trial Functions 626
 11.5.2 Method of Assumed Modes 628
 11.5.3 Convergence Issues 632
11.6 Kinematics of Combined Elastic and Large Angle Rigid Body Motion 637
11.7 Dynamics of Combined Elastic and Large Angle Rigid Body Motion 645
 11.7.1 Equations of Motion in Hybrid Form 646
 11.7.2 Equations of Motions in Discretized Form 649
11.8 Analysis of the Equations of Motion 651
11.9 Case 3, Rotating Shafts 659
11.10 Hybrid Problems 663
 References 664
 Homework Problems 664

Appendix A A Brief History of Dynamics 669

A.1 Introduction 669
A.2 Historical Evolution of Dynamics 669
A.3 Biographies of Key Contributors to Dynamics 678
 References 685

Appendix B Concepts from the Calculus of Variations 687

B.1 Introduction 687
B.2 Stationary Values of a Function 687
B.3 Stationary Values of a Definite Integral 693
B.4 The Variational Notation 696
B.5 Application of the Variational Notation to Dynamics Problems 699
 References 700
 Homework Exercises 700

Appendix C Common Inertia Properties 701

Index 707

PREFACE

APPROACH

Two occurrences in the second part of the 20th century have radically changed the nature of the field of dynamics. The first is the increased need to model and analyze complex, multibodied and often elastic-bodied structures, such as satellites, robot manipulators, and vehicles. The second is the proliferation of the digital computer, which has led to the development of numerical techniques to derive the describing equations of a system, integrate the equations of motion, and obtain the response. This new computational capability has encouraged scientists and engineers to model and numerically analyze complex dynamical systems which in the past either could not be analyzed, or were analyzed using gross simplifications.

The prospect of using computational techniques to model a dynamical system has also led dynamicists to reconsider existing methods of obtaining equations of motion. When evaluated in terms of systematic application, ease of implementation by computers, and computational effort, some of the traditional approaches lose part of their appeal. For example, to obtain Lagrange's equations, one is traditionally taught first to generate a scalar function called the Lagrangian and then to perform a series of differentiations. This approach is computationally inefficient. Moreover, certain terms in the differentiation of the kinetic energy cancel each other, resulting in wasted manipulations.

As a result of the reevaluation of the methods used in dynamics, new approaches have been proposed and certain older approaches that were not commonly used in the past have been brought back into the limelight. What has followed in the literature is a series of papers and books containing claims by proponents of certain methods, each extolling the virtues of one approach over the other without a fair and balanced analysis. This, at least in the opinion of this author, has not led to a healthy environment and fruitful exchange of ideas. It is now possible for basic graduate level courses in dynamics to be taught at different schools with entirely different subject material.

These developments of recent years have inspired me to compile my lecture notes into a textbook. Realizing that much of the research done lately in the field of dynamics has been reported very subjectively, I have tried to present in this book a fair and balanced description of dynamics problems and formulations, from the classical methods to the newer techniques used in today's multibody environments. I have emphasized the need to know both the classical methods as well as the newer techniques and have shown that these approaches are really complementary. Having the knowledge and experience to look at a problem in a number of ways not only facilitates the solution but also provides a better perspective. For example, the book discusses Euler parameters, which lead to fewer singularities in the solution and which lend themselves to more efficient computer implementation.

The focus of this book is primarily the kinematics and derivation of the describing equations of dynamics. We also consider the qualitative analysis of the response. We discuss a special case of quantitative analysis, namely the response of motion linearized about equilibrium.

We discuss means to analyze the kinematics and to describe the equations of motion. We study force and moment balances, as well as analytical methods. In most dynamics problems, the resulting equations of motion are nonlinear and lengthy, so that closed-form solutions are generally not available. We discuss analytical solutions, motion integrals and basic stability concepts. We make use of integrals of the motion, which are derived quantities that give qualitative information about the system without having to solve for the exact solution.

For linearized systems, we discuss the closed-form response. We outline concepts from vibration theory and eigenvector expansions. This also is done for continuous systems, in the last chapter of the book. We discuss the importance of numerical solutions.

CONTENTS

The book is organized into eleven chapters and three appendixes. The first eight chapters are intended for an introductory level graduate or advanced undergraduate course. The later chapters of the book can be used as part of a second, more advanced graduate level course. The book follows a classical approach, in which one first deals with particle mechanics and then extends the concepts into rigid bodies. The Lagrange's equations are initially discussed for a system of particles and plane motion of rigid bodies.

We follow in this book this school of thought for a number of reasons. First, graduate students come from a variety of backgrounds. Many times, students have not considered dynamics since the sophomore dynamics, or the freshman physics course. Also, this organization presents a more natural flow of the concepts used in dynamics. Nevertheless, the book is suitable also for instructors who prefer to teach three-dimensional rigid body dynamics before introducing analytical methods. Following is a description of the chapters:

In Chapter 1 we study fundamental concepts of dynamics and see their applications to particle mechanics problems. We discuss Newton's laws and energy and momentum principles. We look at integrals of motion and basic ideas from stability theory. The chapter outlines the response of linearized systems, which forms an introduction to vibration theory. This chapter should be covered in detail if the course is an undergraduate one. Less time should be spent on it for a graduate course or if the students taking the course are familiar with the basic ideas.

Chapter 2 discusses relative motion. Coordinate frames, rotation sequences, angular velocities, and angular accelerations are introduced. The significance of taking time derivatives in different coordinate systems is emphasized. We derive the relative motion equations and consider motion with respect to the rotating earth.

Chapter 3 is a chapter on systems of particles and plane kinetics of rigid bodies. It is primarily included for pedagogical considerations. I recommend its use for an undergraduate course; for a graduate level course, it should serve as independent reading. A number of sections in this chapter are devoted to an introduction to celestial mechanics problems, namely the two-body problem. The sections on plane kinetics of rigid bodies basically review the sophomore level material. This review is included here mainly because the approaches in the next chapter are described in terms of particles and plane motion of rigid bodies.

The subject of classical analytical mechanics is discussed in Chapters 4 and 5. Chapter 4 introduces the basic concepts, covering generalized coordinates, constraints, and degrees of freedom. We derive the principle of virtual work, D'Alembert's principle, and Hamilton's principle and then we develop Lagrange's equations. Analytical mechanics makes use of the calculus of variations, a subject covered separately in Appendix B. While Chapters 4 and 5 are written so that one does not absolutely need to learn the calculus of variations as a separate subject, it has been the experience of this author that some initial exposure of students to the calculus of variations is very helpful.

Chapter 5 revisits the concept of equilibrium and outlines the distinction between natural and nonnatural systems. We derive the linearized equations about equilibrium. The response of linearized systems is analyzed, which in essence is vibration theory for multidegree of freedom systems. Generalized momenta and motion integrals are considered.

In Chapter 6, we discuss the internal properties of a rigid body. In a departure from traditional approaches, we discuss moments of inertia independent of the kinetic energy and angular momentum.

Chapter 7 is devoted to a detailed analysis of the kinematics of a rigid body, where we learn of methods of quantifying the angular velocity vector. We present a discussion of Euler angles and Euler parameters. We then discuss constraints acting on the motion and quantify these constraints and the resulting kinematic relations.

Chapter 8 explores basic ideas associated with the kinetics of rigid bodies. We first begin with the application of force and moment balances. We express the equations of motion in terms of both the Euler angles and angular velocities. We discuss the relative merits of deriving the equations of motion in terms of generalized coordinates as well as in terms of angular velocity components. We analyze impulse-momentum and work-energy principles. We discuss the physical interpretation of Lagrange's equations and integrals of the motion associated with the Lagrangian.

Chapter 9 introduces more advanced concepts in the analysis of rigid body motion. We analyze the modified Euler's equations and then consider the moment equations about an arbitrary point. We discuss quasi-velocities, also known as generalized speeds, and their applications. We demonstrate that such coordinates are desirable when dealing with nonholonomic systems. We demonstrate the equivalence of the Gibbs-Appell and Kane's equations and discuss momentum balances in terms of the generalized speeds.

Many of the analytical methods described in this chapter could have been introduced in Chapters 4 or 5. However, the power of these methods, which are equally applicable to both particles and rigid bodies, is better appreciated when we consider applications to complex rigid body problems.

Chapter 10 covers the qualitative analysis of rigid body motion, and in particular gyroscopic effects. The chapter initially goes into a qualitative study of torque-free motion and the differences in the response between axisymmetric and arbitrary bodies. We then discuss interesting classical applications of gyroscopic motion, such as a spinning top, a rolling disk, and gyroscopes.

Chapter 11 investigates the subject of dynamics of lightly flexible bodies. Recent problems in dynamics have demonstrated the importance of including the elasticity of a body in the describing equations. The emphasis is the analysis of bodies that undergo combined rigid and elastic motion, typical examples being robot manipulators and spacecraft with appendages. We derive the classical boundary value problem and examine a shortcoming in the traditional formulation. The combined large-angle rigid and elastic motions are modeled in terms of the superposition of a primary motion, as the motion of a moving reference frame, and a secondary motion, the motion of the body as observed from the moving reference frame.

Appendix A is a historical survey of dynamics and a synopsis of the work of the many people who contributed to this field.

Appendix B presents an introduction to the calculus of variations. It is recommended that at least part of this appendix be studied before Chapter 4. However, Chapter 4 is written such that a brief introduction to virtual displacements should be sufficient to understand basic concepts from analytical mechanics.

Appendix C gives the mass moments of inertia of common shapes.

Pedagogical Tools

The book contains several examples and homework problems. It has been my experience that students understand a subject best when they see many examples. I encourage anyone teaching dynamics, either at the undergraduate or the graduate level, to use as many examples as possible.

Another pedagogical tool emphasized in the book is computational techniques. While we do not go into details of numerical integration, we discuss the numerical integration of equations of motion. Many of the examples and homework problems in the book can be assigned as computer projects. I encourage every student to keep pace with new advances in scientific software, such as symbolic manipulators, because, as discussed earlier, the availability of computational tools has changed the nature of dynamics.

Supplement

The book is supplemented by an Instructor's Solutions Manual which includes detailed solutions to all of the problems in the book.

Acknowledgments

Writing a textbook is a lengthy and difficult task and I would like to acknowledge the support of many people. First, I am richly indebted to all the teachers I have had in my life, from my elementary school teacher in Istanbul, Mrs. Cahide Şen, to my advisor in graduate school, Dr. Leonard Meirovitch, who has always been a source of advice and inspiration. I am very grateful to my employer, Rutgers University, and especially to the leadership of the Mechanical and Aerospace Engineering Department, the current chair, Professor Abdelfattah Zebib, and the previous chair, Professor Samuel Temkin, for giving their faculty the time and resources to conduct scholarly work. I thank Verica Radisavljevic emphatically, who read the entire manuscript, helped with figures and made very valuable suggestions. Professor Leon Bahar gave valuable suggestions on variational mechanics. Many graduate students who used earlier versions of the manuscript as their course notes helped correct errors and improve the text. Many thanks are due to McGraw-Hill, and especially to Senior Editor Debra Riegert, who gave me the opportunity to publish with them. Victoria St. Ambrogio was an outstanding copy editor and the competent staff of McGraw-Hill and Publication Services facilitated the production of the book.

I am grateful to the reviewers of the manuscript, Professors Maurice Adams of Case Western Reserve University, Vincent K. Choo of New Mexico State University, William Clausen of Ohio State University, John E. Prussing of the University of Illinois at Urbana-Champaign, and Thomas W. Strganac of Texas A&M University; all provided very meaningful comments and suggestions.

I thank my beloved wife Rahel for all her support and encouragement, as well as for reading parts of the manuscript and improving the text. I am forever thankful to my dear parents, Ester and Yasef Baruh, who worked tirelessly so their children could get a good education and also be good people. They taught me, by teaching and by example, what the most important things in life really are. I am very grateful to my dear uncle, Selim Baruh, who has always been a source of inspiration and who made it possible for me to attend graduate school.

I dedicate this book to the memory of my mother, who passed away a year before the publication of this book. Ester Baruh was a symbol of honesty, integrity, decency, kindness, compassion, and common sense. She was unfailingly devoted to her family and to her community. May her memory be a blessing.

While both the author and the publisher have made every effort to produce an error-free book, there may nevertheless be errors. I would be most appreciative if a reader who spots an error or has a comment about the book would bring it to my attention. My current e-mail addresses are baruh@jove.rutgers.edu or baruh@rci.rutgers.edu. To inform the reader of any discovered errors I am establishing a page on my professional Web site (www-srac.rutgers.edu/~baruh). Just go to my Web site and you will find the link. If my web address at any time has changed, try www.rutgers.edu and find me from there.

chapter 1

BASIC PRINCIPLES

1.1 INTRODUCTION

This chapter discusses basic principles and concepts used in the field of dynamics. It describes these principles and concepts within the context of particles. First, we look at the systems of units commonly used in dynamics. Then, we analyze the kinematics of a particle and outline the coordinate systems used to describe the motion. We distinguish between coordinate systems that deal with components of the motion in fixed directions and coordinate systems based on the properties of the path followed by the particle. The kinematic analysis is followed by the kinetics of a particle, and Newton's laws are given. We move on to analyze the concept of force and discuss the integration of the equations of motion. The distinction is drawn between the qualitative and quantitative analysis of the motion. Integration of the equations of motion lead to energy and momentum expressions and in some cases to other integrals of the motion, which are useful in analyzing the nature of the motion. There is an introduction to the concepts of equilibrium and stability, and the closed-form integration of linearized equations of motion. This subject forms the basis of vibration analysis.

This chapter is a collection of the fundamental principles that one uses in obtaining the equations of motion and in analyzing these equations of motion. The developments in subsequent chapters in this book are built on these principles. The reader is encouraged to understand all the concepts discussed in this chapter thoroughly before continuing with the rest of the text.

1.2 SYSTEMS OF UNITS

While anyone can develop a set of units to describe the evolution of a dynamical system, in classical mechanics two basic sets are widely used: Système International

(SI), or metric, and U.S. Customary. The primary difference between these systems is that the SI system is universal and absolute, and the U.S. system is local (or gravitational), that is, valid on earth. Today, most countries in the world have adopted the SI system as their standard.

To describe the evolution of a body, three fundamental quantities are needed. In all commonly used systems there is agreement on two: *length* and *time,* whose dimensions are denoted by L and T, respectively. The U.S. and SI systems differ on the nature of the third quantity. The SI system uses *mass* (M), defined as the amount of matter (absolute) contained in a body. By contrast, the U.S. system uses *force* (F). The justification for this system is that the weight of a body, as the force with which the body is pulled toward the center of mass of the coordinate system (in our case, earth), is easier to visualize. Weight is a relative quantity. It changes depending on the amount of gravitational attraction.

Corresponding to each fundamental quantity there is a *base unit*[1] that describes standardized amounts of a fundamental quantity. Evolution of the base units has been in many cases based on convenience (e.g., *foot*), but in some cases a rational explanation is not available. The base units are usually abbreviated by symbols. The fundamental quantities and corresponding base units in the U.S. and SI systems are as follows (the dimensions and symbols are denoted in the parentheses):

SI Units		U.S. Customary Units	
Mass (M):	kilogram (kg)	Force (F):	pound (lb)
Length (L):	meter (m)	Length (L):	foot (ft)
Time (T):	second (s)	Time (T):	second (sec or s)

In the SI system, force is a derived quantity. The base unit of force ($F = ML/T^2$) is denoted by a *newton* (N), where $1 \text{ N} = 1 \text{ kg} \cdot \text{m/s}^2$. In the U.S. system, mass is a derived quantity, and the unit of mass ($M = FT^2/L$) is denoted by a *slug,* with 1 slug = 1 lb·sec²/ft. When using the SI system many people commonly, and at the same time erroneously, refer to mass as a unit of weight. This is due to the perpetuation of the original definition of *kilogram* as a unit of weight. The kilogram was formally redefined as a unit of mass in the year 1859.

The selection of base units in a given system of units is not absolute. In many cases one switches to a different system of units to describe the motion. For example, the speed of an automobile is usually described in kilometers per hour (or miles per hour), and the speed of a ship is described by knots (1 knot = 1 nautical mile/hr, where 1 nautical mile = 1.852 km).

To describe rotational displacements one may use *degrees* (°) or *radians* (rad). Going around a full circle takes 360 degrees or 2π radians and is referred to as a *revolution.* A radian is a dimensionless unit. The angle of 1 radian = 57.2958° is depicted in Fig. 1.1.

---

[1] Some refer to the fundamental quantities as base units and to the base units as dimensions.

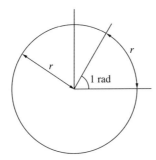

Figure 1.1 One radian

An interesting source of debate has been the need to reexamine and to recalibrate the standards used in defining the base units. Many precision calculations require formal definitions for several quantities. For example, one meter was originally defined as one 10-millionth of the distance from one of the earth's poles to the equator. The current formal definition of a meter is the distance that light travels in one 299,792,458th of a second. One inch is defined as being exactly equal to 2.54 cm. The current formal definition of a second is the time it takes for 9,192,631,770 vibrations to occur in cesium atoms excited by microwave. A second was originally defined as 1/60 of a minute, which is 1/60 of an hour, which is 1/24 of a day. A kilogram is defined as the mass of water contained in a volume of one liter and at a temperature of 4° Celsius. A block with a mass of 1 kg is maintained in a vault in Paris as the standard kilogram. At the time of writing of this text, there was ongoing debate on revising the formal definition of the kilogram.

To relate the mass of an object to its weight, we make use of the *gravitational constant*. The gravitational constant is denoted by g and it has the units of acceleration. The general form for this constant will be given in Section 1.4. On earth at sea level the value of g is approximated as $g = 9.81$ m/s^2 or $g = 32.2$ ft/sec^2. On earth an object of mass 1 kg weighs 9.81 N, and an object of weight 1 lb has a mass of 1/32.2 slugs. It is interesting that the weight of a medium-sized apple is about one newton.

We have so far treated the three fundamental quantities as independent of each other. This is a correct assumption as long as relativistic effects are ignored (when the speeds involved are much less than the speed of light). As the speeds involved approach the speed of light, mass, length, and time become interrelated. This is the foundation of relativistic mechanics and it will not be pursued in this text.

It is always important to check that the dimensions of the quantities being manipulated match. Equations of motion and equilibrium equations must have all of their terms with the same dimension. Checking dimensional homogeneity is a good way of spotting errors.

Analogous to the preceding discussion regarding standards, one must also be careful in the accuracy of the solution and rounding off numbers when solving a problem. Many engineering problems require an accuracy of about one part in a thousand or better. Accordingly, in most problems in this text we will retain four and sometimes five significant digits in the solution (e.g., 0.7655, 8.975, 12.34 or 0.76549, 8.9762, 12.345). Following this line of thought, we note that the standard assumptions for the gravitational constant given above are not extremely accurate. In celestial mechanics problems, which we will discuss in Chapter 3, one needs to go to much higher levels of accuracy.

Example 1.1

Suppose a new coordinate system is developed such that the density of water and the acceleration of gravity are both of unit magnitude. If the pound is taken as the unit of force, how do the units of length and time in the new system compare with the U.S. Customary system?

Solution

Denote the base units in the old system by lb, ft, and sec and the units in the new system by lb, ft*, and s*. We are given that

$$\text{Acceleration of gravity } g = 32.2 \text{ ft/sec}^2 = 1 \text{ ft*}/(s^*)^2 \quad [a]$$

$$\text{Density of water } \gamma = \text{Specific wt/gravity} = (62.8 \text{ lb/ft}^3)/(32.2 \text{ ft/sec}^2)$$

$$= 1 \text{ lb/ft*}^2/(1 \text{ ft*}/s^{*2})$$

$$\rightarrow 1.950 \text{ lb sec}^2/\text{ft}^4 = 1 \text{ lb } (s^*)^2/(ft^*)^4 \quad [b]$$

These two relations are two equations that can be solved for the two unknowns ft* and s*. From the first relation, $1 \text{ ft*} = 32.2 \text{ ft } (s^*)^2/\text{sec}^2$. Substituting this into the second relation, we obtain

$$1.950 \text{ lb sec}^2/\text{ft}^4 = 1 \text{ lb } (s^*)^2/[32.2 \text{ ft } (s^*)^2/\text{sec}^2]^4 \quad [c]$$

which can be solved to give $1 \text{ sec}/s^* = 11.31$. Substitution of this result into the first relation yields $1 \text{ ft}/\text{ft}^* = 3.973$.

1.3 REVIEW OF VECTOR ANALYSIS

In this section we review the sets of coordinates and associated unit vectors commonly used in dynamics. As with systems of units, one can develop a specific set of vectors to describe the orientation of a body. The study of dynamics uses two basic types of coordinates: rectilinear and curvilinear. Rectilinear coordinates describe the components of the motion in fixed directions. Curvilinear coordinate systems incorporate the properties of the path that the particle follows.

In a coordinate system, one defines a set of three *principal directions*. The *coordinate axes* and associated *unit vectors* are directed along the principal directions. When a set of unit vectors e_1, e_2, e_3 obeys the cross product rule $e_i \times e_j = \varepsilon_{ijk} e_k$ where $\varepsilon_{ijk} = 1$ if i, j, k are in order ($i = 1, j = 2, k = 3$, or $i = 2, j = 3, k = 1$, or $i = 3, j = 1, k = 2$), $\varepsilon_{ijk} = -1$ if i, j, k are not in order ($i = 1, j = 3, k = 2$, or $i = 2, j = 1, k = 3$, or $i = 3, j = 2, k = 1$), and $\varepsilon_{ijk} = 0$ if any two indices are repeated, the set is referred to as a *mutually orthogonal triad*. The coordinate system that uses these unit vectors is called a *right-handed coordinate system*.

1.3.1 RECTILINEAR (CARTESIAN) COORDINATES

In this coordinate system, the origin and the principal directions remain fixed. These directions are usually referred to as x, y, and z axes or X, Y, and Z axes, with associated unit vectors $\mathbf{i}$, $\mathbf{j}$, and $\mathbf{k}$ or $\mathbf{I}$, $\mathbf{J}$, and $\mathbf{K}$, respectively. These unit vectors are time invariant. Because the unit vectors are independent of the path followed, this coordinate system is called *extrinsic*.

Consider Fig. 1.2, where a particle P is moving along a curve. The displacement vector $\mathbf{r}(t)$ that describes the location of point P is

$$\mathbf{r}(t) = x\mathbf{i} + y\mathbf{j} + z\mathbf{k} \qquad [1.3.1]$$

The distance r from the origin O to point P then becomes

$$r = \sqrt{\mathbf{r}(t) \cdot \mathbf{r}(t)} = \sqrt{x^2 + y^2 + z^2} \qquad [1.3.2]$$

Denoting the position of the particle at time $t + \Delta t$ by $\mathbf{r}(t + \Delta t)$, the velocity of the particle is defined by

$$\mathbf{v}(t) = \frac{d\mathbf{r}(t)}{dt} = \dot{\mathbf{r}}(t) = \lim_{\Delta t \to 0} \frac{\mathbf{r}(t + \Delta t) - \mathbf{r}(t)}{\Delta t} \qquad [1.3.3]$$

Introduction of Eq. [1.3.1] into Eq. [1.3.3] yields

$$\mathbf{v}(t) = \dot{x}\mathbf{i} + \dot{y}\mathbf{j} + \dot{z}\mathbf{k} = v_x\mathbf{i} + v_y\mathbf{j} + v_z\mathbf{k} \qquad [1.3.4]$$

Because the unit vectors are time invariant, their derivatives vanish. In a similar fashion, one can obtain expressions for the acceleration

$$\mathbf{a}(t) = \ddot{\mathbf{r}}(t) = \ddot{x}\mathbf{i} + \ddot{y}\mathbf{j} + \ddot{z}\mathbf{k} = a_x\mathbf{i} + a_y\mathbf{j} + a_z\mathbf{k} \qquad [1.3.5]$$

This set of coordinates is useful when the components of the motion can be analyzed separately from each other. Projectile motion problems provide a classical example.

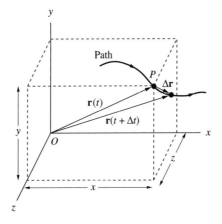

Figure 1.2

Let us investigate a few interesting cases for motion in one direction, also known as *rectilinear motion*. Taking the displacement variable as x, we can express the acceleration in the most general case as $a = a(x, \dot{x}, t)$. If the acceleration is constant, $a = c$, then we can obtain the velocity and displacement by direct integration as $v(t) = v_0 + ct$, $x(t) = x_0 + v_0 t + ct^2/2$, with x_0 and v_0 denoting the initial displacement and initial velocity. Direct integration can again be used if the acceleration is only a function of time, $a = f(t)$. If the acceleration is a function of the displacement only, $a = f(x)$, one uses the transformation $a = dv/dt = f(x)$. Multiplying both sides by dx, we obtain

$$f(x)\,dx = \frac{dv}{dt}dx = v\,dv \qquad [1.3.6]$$

where each side is a function of x or v only. It follows that each side can be integrated separately from the other. If the acceleration is a function of velocity alone, $a = f(v)$, we write $a = dv/dt = f(v)$. Carrying the dt term to the right side and the $f(v)$ term to the left, we obtain

$$\frac{dv}{f(v)} = dt \qquad [1.3.7]$$

where we have again separated the variables so that each side of the above equation can be integrated independently of the other.

The case when $a = f(x, \dot{x})$ is left as a homework exercise. Any other combination lends itself to more complicated problems. The situation gets worse when there is two- or three-dimensional motion, and the acceleration in one direction is related to motion in more than one direction, such as $a_x = f(x, y)$. One has to then go into coordinate systems that take advantage of the properties of the path followed by the particle, as we will see next.

Example 1.2

A projectile is thrown from an inclined surface with an initial velocity v_0 and an angle of $\theta = 30°$ from the incline, as shown in Fig. 1.3. The plane of the incline makes an angle of $\phi = 15°$ with the horizontal. Find the time elapsed before the particle falls to the ground and the distance traveled along the incline.

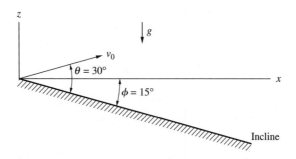

Figure 1.3

Solution

The equations of motion in the horizontal and vertical directions are independent of each other and have the form

$$ma_x = 0 \qquad ma_z = -mg \qquad \text{[a]}$$

which can be integrated directly to yield the velocity and displacement relations

$$v_x(t) = v_{x0} = \text{constant} \qquad x(t) = v_{x0}t$$

$$v_z(t) = v_{z0} - gt \qquad z(t) = v_{z0}t - \frac{1}{2}gt^2 \qquad \text{[b]}$$

where v_{x0} and v_{z0} are the components of the initial velocity. From Fig. 1.3, the initial velocities are

$$v_{x0} = v_0 \cos(\theta - \phi) = v_0 \cos 15° \qquad v_{z0} = v_0 \sin(\theta - \phi) = v_0 \sin 15° \qquad \text{[c]}$$

Denote the time it takes for the projectile to touch the ground by t_f and the coordinates of that point by $x_f = x(t_f)$ and $z_f = z(t_f)$. From Fig. 1.4 (not drawn to scale) we can relate z_f and x_f by

$$-z_f = x_f \tan \phi \qquad \text{[d]}$$

Substituting the values for x_f and z_f from Eq. [b], we obtain

$$-(v_{z0}t_f - 0.5gt_f^2) = v_{x0}t_f \tan \phi \qquad \text{[e]}$$

which leads to the following equation for t_f:

$$0.5gt_f^2 - (v_{z0} + v_{x0} \tan \phi)t_f = 0 \qquad \text{[f]}$$

Solving for t_f for the final time, we obtain

$$t_f = \frac{v_{z0} + v_{x0} \tan \phi}{\frac{1}{2}g}$$

$$= \frac{v_0 \sin 15° + v_0 \cos 15° \tan 15°}{\frac{1}{2}g} = \frac{4v_0 \sin 15°}{g} = 1.035 \frac{v_0}{g} \qquad \text{[g]}$$

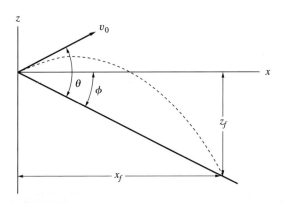

Figure 1.4

The final distance traveled is calculated by substituting the value for t_f into Eqs. [b] and [d], and we have

$$x(t_f) = v_{x0}t_f = \frac{4v_0^2 \sin 15° \cos 15°}{g} = \frac{v_0^2}{g} \quad \text{[h]}$$

$$z(t_f) = -x_f \tan \phi = -\frac{v_0^2}{g}\tan 15° = -0.2680\frac{v_0^2}{g} \quad \text{[i]}$$

1.3.2 CURVILINEAR COORDINATES

Properties of the path followed by a particle turn out to be extremely useful quantities for understanding the nature of the motion. This observation has led to the development of curvilinear coordinate systems. There are several such coordinate systems that one can develop. We will begin with a coordinate system that is entirely based on the properties of the path. We will then study two commonly used coordinate systems that make use of the properties of the path as well as the position of the particle. Curvilinear coordinate systems are referred to as *intrinsic*, as these coordinate systems depend on the path followed by the particle.

Normal-Tangential Coordinates This coordinate system is attached to the particle whose motion is considered (see Fig. 1.5). The distance traversed along the path is designated by s, measured from a reference position. This coordinate system is primarily used to describe the nature of the path followed by the particle.[2] The variables used are referred to as *path variables*.

While it may appear to be more desirable from a mathematical perspective to just look at the distance between the initial and final positions, considering the path

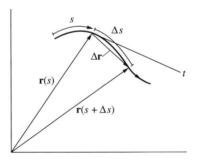

Figure 1.5

---

[2] In this regard, the normal-tangential coordinates do not constitute a coordinate system that can describe displacement, velocity, or acceleration. Rather, they are a means of obtaining information about the path followed by the particle. The other curvilinear coordinates that we will study in this section can, in essence, be derived from the path variables.

followed and the total distance traversed along the path is very important. Someone traveling from one location to another, say by driving, usually selects a route that minimizes the distance traveled along the road (the path).

We define two principal directions to describe the motion. The first is along the tangent to the curve and along the direction of the motion. This direction is called the *tangential direction.* We denote the unit vector associated with this direction as $\mathbf{e}_t$. Consider Fig. 1.5 and the position of the particle after it has traveled distances of s and $s + \Delta s$. The associated position vectors, measured from a fixed location, are denoted as $\mathbf{r}(s)$ and $\mathbf{r}(s + \Delta s)$, respectively. Define by $\Delta \mathbf{r}$ the difference between $\mathbf{r}(s)$ and $\mathbf{r}(s + \Delta s)$, thus

$$\Delta \mathbf{r} = \mathbf{r}(s + \Delta s) - \mathbf{r}(s) \qquad [1.3.8]$$

From Fig. 1.5, as Δs becomes small $\Delta \mathbf{r}$ and Δs have the same length and become parallel to each other. Further, $\Delta \mathbf{r}$ becomes aligned with the tangential direction. We hence define the unit vector in the tangential direction as

$$\mathbf{e}_t = \lim_{\Delta s \to 0} \frac{\Delta \mathbf{r}}{\Delta s} = \frac{d\mathbf{r}}{ds} \qquad [1.3.9]$$

The unit vector $\mathbf{e}_t$ changes direction as the particle moves. We can obtain the velocity of the particle by differentiating the displacement vector with respect to time. Using the chain rule for differentiation,

$$\mathbf{v}(t) = \frac{d\mathbf{r}}{dt} = \frac{d\mathbf{r}}{ds} \frac{ds}{dt} \qquad [1.3.10]$$

Now using the definition of $\mathbf{e}_t$ from Eq. [1.3.9] and noting that the speed v is the rate of change of the distance traveled along the path, $v = ds/dt$, we obtain

$$\mathbf{v}(t) = v\mathbf{e}_t \qquad [1.3.11]$$

The second principal direction is defined as being normal to the curve and directed toward the *center of curvature* of the path, as shown in Fig. 1.6. This direction is defined as the *normal direction.* The associated unit vector is denoted by $\mathbf{e}_n$. The center of curvature of a path associated with a certain point on the path lies along a line perpendicular to the path at that point. An infinitesimal arc in the vicinity of that point can be viewed as a circular path, with the center of curvature as the center of the circle. The radius of the circle is called the *radius of curvature.* Because the normal to the curve is perpendicular to the tangential direction, the two unit vectors are orthogonal, that is, $\mathbf{e}_t \cdot \mathbf{e}_n = 0$.

Differentiation of Eq. [1.3.11] with respect to time gives the acceleration of the particle $\mathbf{a}(t)$ as

$$\mathbf{a}(t) = \dot{\mathbf{v}}(t) = \dot{v}\mathbf{e}_t + v\dot{\mathbf{e}}_t \qquad [1.3.12]$$

To obtain the derivative of $\mathbf{e}_t$ we displace the particle by an infinitesimal distance ds, and refer to the unit vectors associated with the new location by $\mathbf{e}_t(s + ds)$, as shown in Fig. 1.6. The center of curvature associated with all points on the arc is the same. The arc length can be expressed as $ds = \rho\, d\phi$; $d\phi$ is the infinitesimal angle

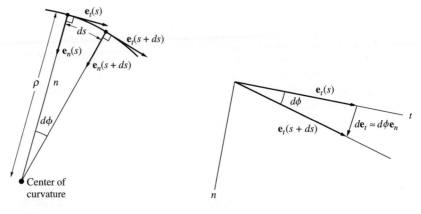

Figure 1.6

Figure 1.7

traversed as the particle moves by a distance ds. Defining the vector connecting $\mathbf{e}_t(s)$ and $\mathbf{e}_t(s+ds)$ by $d\mathbf{e}_t$, we can write $d\mathbf{e}_t = \mathbf{e}_t(s+ds) - \mathbf{e}_t(s)$. From Fig. 1.7, the angle between $\mathbf{e}_t(s+ds)$ and $\mathbf{e}_t(s)$ is very small, so that

$$|d\mathbf{e}_t| \approx \sin d\phi |\mathbf{e}_t(s)| \approx d\phi |\mathbf{e}_t| = \frac{ds}{\rho} \qquad [1.3.13]$$

or

$$\left|\frac{d\mathbf{e}_t}{ds}\right| = \frac{1}{\rho} \qquad [1.3.14]$$

The radius of curvature is a measure of how much the curve bends. For motion along a straight line, the curve does not bend and the radius of curvature has the value of infinity. For plane motion, using the coordinates x and y such that the curve is described by $y = y(x)$, the expression for the radius of curvature can be shown to be

$$\frac{1}{\rho} = \frac{|d^2y/dx^2|}{(1+(dy/dx)^2)^{3/2}} \qquad [1.3.15]$$

The absolute value sign in this equation is necessary because we defined the radius of curvature as a positive quantity. Considering the sign convention that we adopted above, we have

$$\frac{d\mathbf{e}_t}{ds} = \frac{\mathbf{e}_n}{\rho} \qquad [1.3.16]$$

Using the chain rule, we obtain the time derivative of $\mathbf{e}_t$ as

$$\dot{\mathbf{e}}_t = \frac{d\mathbf{e}_t}{ds}\frac{ds}{dt} = v\frac{\mathbf{e}_n}{\rho} \qquad [1.3.17]$$

Introduction of this relation into Eq. [1.3.12] yields

$$\mathbf{a}(t) = \dot{\mathbf{v}}(t) = \dot{v}\mathbf{e}_t + \frac{v^2}{\rho}\mathbf{e}_n \qquad [1.3.18]$$

The first term on the right in this equation is the component of the acceleration due to a change in speed, referred to as *tangential acceleration* (a_t). The second term is the contribution due to a change in direction, referred to as the *normal acceleration* or *centripetal acceleration* (a_n). The acceleration expression can be written as

$$\mathbf{a}(t) = a_t\mathbf{e}_t + a_n\mathbf{e}_n \qquad [1.3.19]$$

with

$$a_t = \dot{v} \qquad a_n = \frac{v^2}{\rho} \qquad [1.3.20a,b]$$

It is a common misconception to treat an object moving along a curved path with constant speed as having no acceleration. Note that the normal component of the acceleration is always directed toward the center of curvature.

The normal and tangential directions define the plane of the motion for that particular instant. This instantaneous plane of motion is called the *osculating plane* (after the Italian word *osculari,* which means *to kiss*). The orientation of the osculating plane changes as the particle moves. This is referred to as the *twisting of the osculating plane*. The unit vector $\mathbf{e}_b$, referred to as the unit vector in the *binormal direction* and defined as $\mathbf{e}_b = \mathbf{e}_t \times \mathbf{e}_n$, is perpendicular to the osculating plane. Consider, for the sake of illustration, plane motion as shown in Fig. 1.8. The direction of $\mathbf{e}_b$ and hence the osculating plane alternates as the curvature of the path switches from convex to concave.

The unit vector in the binormal direction is used to obtain the derivative of the unit vector in the normal direction. While $\mathbf{e}_b$ is not needed to describe the velocity or acceleration, it is used to describe the evolution of the plane of motion. The twisting of the osculating plane is a path parameter.

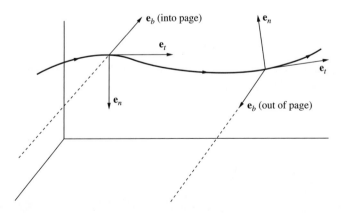

Figure 1.8

To analyze the movement of the plane of motion, we investigate the derivatives of $\mathbf{e}_n$ and $\mathbf{e}_b$ with respect to s. To this end, we take the derivative of $\mathbf{e}_b$ using the definition of the cross product as

$$\frac{d\mathbf{e}_b}{ds} = \mathbf{e}_t \times \frac{d\mathbf{e}_n}{ds} + \frac{d\mathbf{e}_t}{ds} \times \mathbf{e}_n \qquad [1.3.21]$$

Using Eq. [1.3.16], $d\mathbf{e}_t/ds = \mathbf{e}_n/\rho$, so that the second term in the right side of the above equation vanishes. From the first term on the right side we conclude that the rate of change of $\mathbf{e}_b$ along the path has no tangential component. To show that $d\mathbf{e}_b/ds$ has no component in the binormal direction either, we differentiate the dot product $\mathbf{e}_b \cdot \mathbf{e}_b = 1$, with the obvious result

$$\frac{d}{ds}(\mathbf{e}_b \cdot \mathbf{e}_b) = 2\mathbf{e}_b \cdot \frac{d\mathbf{e}_b}{ds} = 0 \qquad [1.3.22]$$

It follows that $d\mathbf{e}_b/ds$ can only have a component in the normal direction. This can be explained physically the same way we explained Eqs. [1.3.13] through [1.3.16]. We define a quantity denoted by τ, which is called the *torsion of the curve*, as

$$\frac{d\mathbf{e}_b}{ds} = -\frac{1}{\tau}\mathbf{e}_n \qquad [1.3.23]$$

The torsion of the curve is a measure of how much the plane of motion twists, or how the osculating plane changes direction. For plane motion, while the curve may change from convex to concave, the binormal direction does not change and hence τ is equal to infinity.

We next obtain the derivative of $\mathbf{e}_n$ with respect to s. From the above discussion we know that $d\mathbf{e}_n/ds$ cannot have a component in the normal direction. We proceed to write this as

$$\frac{d\mathbf{e}_n}{ds} = c_1\mathbf{e}_t + c_2\mathbf{e}_b \qquad [1.3.24]$$

in which c_1 and c_2 are coefficients to be determined. We make use of the property that the unit vectors are orthogonal to each other and write

$$\mathbf{e}_n \cdot \mathbf{e}_t = 0 \qquad \mathbf{e}_n \cdot \mathbf{e}_b = 0 \qquad [1.3.25]$$

Differentiation of these equations with respect to s and using Eqs. [1.3.16] and [1.3.23] yields

$$\mathbf{e}_n \cdot \frac{d\mathbf{e}_b}{ds} + \mathbf{e}_b \cdot \frac{d\mathbf{e}_n}{ds} = -\frac{1}{\tau} + \mathbf{e}_b \cdot \frac{d\mathbf{e}_n}{ds} = 0$$

$$\mathbf{e}_n \cdot \frac{d\mathbf{e}_t}{ds} + \mathbf{e}_t \cdot \frac{d\mathbf{e}_n}{ds} = \frac{1}{\rho} + \mathbf{e}_t \cdot \frac{d\mathbf{e}_n}{ds} = 0 \qquad [1.3.26]$$

Taking the dot product of Eq. [1.3.24] with $\mathbf{e}_t$ and $\mathbf{e}_b$ and comparing with Eqs. [1.3.26], we obtain

$$c_1 = -\frac{1}{\rho} \qquad c_2 = \frac{1}{\tau} \qquad [1.3.27]$$

We are now in a position to write the rate of change of the unit vectors in the normal-tangential coordinate system as

$$\frac{d\mathbf{e}_t}{ds} = \frac{1}{\rho}\mathbf{e}_n \qquad \frac{d\mathbf{e}_n}{ds} = -\frac{1}{\rho}\mathbf{e}_t + \frac{1}{\tau}\mathbf{e}_b \qquad \frac{d\mathbf{e}_b}{ds} = -\frac{1}{\tau}\mathbf{e}_n \qquad [1.3.28a,b,c]$$

Eqs. [1.3.28a,b,c] are known as *Frenet's formulas*. They enable us to describe how the curve and the associated unit vectors change.

Note that up to now, we derived the unit vectors and their derivatives as a function of s, the distance traversed along the path. But we did not find an expression for s itself. To accomplish this, we need to express position in terms of a path variable. Denoting this variable by α, we can write the position vector as

$$\mathbf{r} = \mathbf{r}(\alpha) = x(\alpha)\mathbf{i} + y(\alpha)\mathbf{j} + z(\alpha)\mathbf{k} \qquad [1.3.29]$$

Using the definition of $\mathbf{e}_t$ in Eq. [1.3.29], we obtain

$$\mathbf{e}_t = \frac{d\mathbf{r}}{ds} = \frac{d\mathbf{r}}{d\alpha}\frac{d\alpha}{ds} = \left(\frac{dx}{d\alpha}\mathbf{i} + \frac{dy}{d\alpha}\mathbf{j} + \frac{dz}{d\alpha}\mathbf{k}\right)\frac{d\alpha}{ds} \qquad [1.3.30]$$

We then take the dot product of $\mathbf{e}_t$ with itself:

$$\mathbf{e}_t \cdot \mathbf{e}_t = 1 = \left(\frac{d\alpha}{ds}\right)^2 \left[\left(\frac{dx}{d\alpha}\right)^2 + \left(\frac{dy}{d\alpha}\right)^2 + \left(\frac{dz}{d\alpha}\right)^2\right] \qquad [1.3.31]$$

which can be solved to yield

$$ds = \sqrt{\left(\frac{dx}{d\alpha}\right)^2 + \left(\frac{dy}{d\alpha}\right)^2 + \left(\frac{dz}{d\alpha}\right)^2}\, d\alpha \qquad [1.3.32]$$

Integrating this expression, we obtain

$$s = \int_{\alpha_0}^{\alpha} \sqrt{\left(\frac{dx}{d\alpha}\right)^2 + \left(\frac{dy}{d\alpha}\right)^2 + \left(\frac{dz}{d\alpha}\right)^2}\, d\alpha \qquad [1.3.33]$$

where α_0 is the initial value of the path parameter. When the path parameter is time, we get the familiar expression of displacement being the integral of the velocity over time.

We next express the radius of curvature, torsion, and the unit vectors in terms of the path variables. We first write the unit vector in the tangential direction and the derivative of the position vector as

$$\mathbf{e}_t = \frac{d\mathbf{r}}{ds} = \frac{d\mathbf{r}}{d\alpha}\frac{d\alpha}{ds} = \frac{\mathbf{r}'}{s'} \qquad \mathbf{r}' = s'\mathbf{e}_t \qquad [1.3.34a,b]$$

Primes denote differentiation with respect to the path variable α. Differentiation of Eq. [1.3.34b] with respect to the path parameter α, and using Eq. [1.3.16], yields

$$\mathbf{r}'' = s''\mathbf{e}_t + s'\frac{d\mathbf{e}_t}{ds}\frac{ds}{d\alpha} = s''\mathbf{e}_t + \frac{(s')^2}{\rho}\mathbf{e}_n \qquad [1.3.35]$$

As one would expect, if the path variable is selected as time, Eqs. [1.3.34b] and [1.3.35] yield the expressions for velocity and acceleration. To express the unit vector in the normal direction in terms of the path variables, we multiply Eq. [1.3.35] by s' and subtract from it Eq. [1.3.34b] multiplied by s'', with the result

$$\mathbf{e}_n = \frac{\rho}{(s')^3}(\mathbf{r}''s' - \mathbf{r}'s'') \qquad [1.3.36]$$

It is customary to express $\mathbf{e}_n$ in terms of the first derivative of s. Noting from Eqs. [1.3.34b] and [1.3.35] that $\mathbf{r}'' \cdot \mathbf{r}' = s's''$ and substituting this relationship in the above equation, we obtain

$$\mathbf{e}_n = \frac{\rho}{(s')^4}(\mathbf{r}''(s')^2 - \mathbf{r}'(\mathbf{r}'' \cdot \mathbf{r}')) \qquad [1.3.37]$$

To find the unit vector in the binormal direction, we use the definition $\mathbf{e}_b = \mathbf{e}_t \times \mathbf{e}_n$ and substitute the values for $\mathbf{e}_t$ and $\mathbf{e}_n$ from Eqs. [1.3.34a] and [1.3.36], with the result

$$\mathbf{e}_b = \frac{\mathbf{r}'}{s'} \times \frac{\rho}{(s')^3}(\mathbf{r}''s' - \mathbf{r}'s'') = \frac{\rho}{(s')^3}(\mathbf{r}' \times \mathbf{r}'') \qquad [1.3.38]$$

Finding the radius of curvature and the torsion requires manipulation of the above equations. The radius of curvature is found by making use of Eq. [1.3.37] and the property that the norm of $\mathbf{e}_n$ is 1. To find the torsion τ we invoke Eq. [1.3.28c] and carry out the algebra. The results are

$$\frac{1}{\rho} = \frac{1}{(s')^3}\sqrt{(\mathbf{r}'' \cdot \mathbf{r}'')s'^2 - (\mathbf{r}' \cdot \mathbf{r}'')^2} \qquad \frac{1}{\tau} = -\frac{\rho^2}{(s')^6}[\mathbf{r}'' \cdot (\mathbf{r}' \times \mathbf{r}''')] \qquad [1.3.39a,b]$$

The derivations are left as an exercise. Note that to find ρ one needs the second derivative of $\mathbf{r}$ with respect to the path variable, whereas to find the torsion τ the third derivative of $\mathbf{r}$ is required. This can be explained by the following observation: Let time be the path variable. Given the velocity and acceleration of a particle at a point in time, one can determine what the plane of the motion is and the radius of curvature, but not how the plane of motion is twisting.

Finally, in terms of the derivatives with respect to s, we can write

$$\frac{d\mathbf{r}}{ds} = \mathbf{e}_t \qquad \frac{d^2\mathbf{r}}{ds^2} = \frac{1}{\rho}\mathbf{e}_n \qquad [1.3.40]$$

We now compare the two coordinate systems that we have seen so far. With rectilinear coordinates, one describes velocity and acceleration as the rate of change of absolute distance from an origin. With normal-tangential coordinates, one describes velocity and acceleration using properties of the path that the particle follows. The distance from an origin does not come into the picture. The two descriptions can be used together to give a better understanding of the nature of the motion.

Example 1.3

A particle moves on a path on the xy plane defined by the curve $y = 3x^2$, where x varies with the relation $x = \sin\alpha$. Find the radius of curvature of the path and the unit vectors in the normal and tangential directions when $\alpha = \pi/6$.

Solution

The position vector can be written as

$$\mathbf{r} = \sin\alpha \mathbf{i} + 3\sin^2\alpha \mathbf{j} \qquad [a]$$

so that the unit vector in the tangential direction is

$$\mathbf{e}_t = \frac{d\mathbf{r}}{ds} = \frac{d\mathbf{r}}{d\alpha}\frac{1}{s'} = \frac{1}{s'}[\cos\alpha\mathbf{i} + 6\sin\alpha\cos\alpha\mathbf{j}] \qquad [b]$$

where $s' = ds/d\alpha$. Noting that the magnitude of $\mathbf{e}_t$ is 1, we write

$$s' = \sqrt{\cos^2\alpha + 36\sin^2\alpha\cos^2\alpha} \qquad [c]$$

To find the radius of curvature we use Eq. [1.3.39a], which requires expressions for $\mathbf{r}'$ and $\mathbf{r}''$. These derivatives are

$$\mathbf{r}' = \frac{d\mathbf{r}}{d\alpha} = \cos\alpha\mathbf{i} + 6\sin\alpha\cos\alpha\mathbf{j} \qquad \mathbf{r}'' = \frac{d^2\mathbf{r}}{d\alpha^2} = -\sin\alpha\mathbf{i} + 6(\cos^2\alpha - \sin^2\alpha)\mathbf{j} \qquad [d]$$

Evaluating these expressions at $\alpha = \pi/6$, we obtain

$$s'\left(\frac{\pi}{6}\right) = \sqrt{7.5} \qquad \mathbf{r}'\left(\frac{\pi}{6}\right) = \frac{\sqrt{3}}{2}(\mathbf{i} + 3\mathbf{j}) \qquad \mathbf{r}''\left(\frac{\pi}{6}\right) = -\frac{\mathbf{i}}{2} + 3\mathbf{j} \qquad [e]$$

which, when substituted into Eq. [1.3.39a] yields

$$\frac{1}{\rho} = \frac{1}{(s')^3}\sqrt{(\mathbf{r}''\cdot\mathbf{r}'')s'^2 - (\mathbf{r}'\cdot\mathbf{r}'')^2} = 0.1897 \qquad [f]$$

When $\alpha = \pi/6$, the unit vector in the tangential direction has the value

$$\mathbf{e}_t = \frac{1}{\sqrt{7.5}}\left(\frac{\sqrt{3}}{2}\mathbf{i} + \frac{3\sqrt{3}}{2}\mathbf{j}\right) = \frac{1}{\sqrt{10}}(\mathbf{i} + 3\mathbf{j}) = 0.3162\mathbf{i} + 0.9487\mathbf{j} \qquad [g]$$

To find the unit vector in the normal direction we need to use Eq. [1.3.37], which yields

$$\mathbf{e}_n = \frac{\rho}{(s')^4}(\mathbf{r}''(s')^2 - \mathbf{r}'(\mathbf{r}''\cdot\mathbf{r}')) = -0.9487\mathbf{i} + 0.3162\mathbf{j} \qquad [h]$$

Note that we determined all the path parameters using their given formulas. We could have solved this problem by assuming that the parameter α is time and by using the expressions for the normal and tangential accelerations.

Example 1.4

The motion of a particle is described in Cartesian coordinates as

$$x(t) = 2t^2 + 4t \qquad y(t) = 0.1t^3 + \cos t \qquad z(t) = 3t \qquad [a]$$

Find the radius of curvature and the torsion of the path at time $t = 0$.

Solution

From Eq. [1.3.20b], the radius of curvature is related to the speed of the particle and its normal acceleration as

$$\rho = \frac{v^2}{a_n} \qquad [b]$$

and the components of the acceleration in the normal-tangential and Cartesian coordinates are related by

$$a^2 = a_n^2 + a_t^2 = a_x^2 + a_y^2 + a_z^2 \quad \text{[c]}$$

We can obtain the components of the velocity and acceleration by differentiating Eq. [a]:

$$\dot{x}(t) = 4t + 4 \qquad \dot{y}(t) = 0.3t^2 - \sin t \qquad \dot{z}(t) = 3$$
$$\ddot{x}(t) = 4 \qquad \ddot{y}(t) = 0.6t - \cos t \qquad \ddot{z}(t) = 0 \quad \text{[d]}$$

The square of the speed of the particle at $t = 0$ is

$$v^2(0) = [\dot{x}^2(0) + \dot{y}^2(0) + \dot{z}^2(0)] = [4^2 + 0 + 3^2] = 25 \quad \text{[e]}$$

The tangential acceleration can be found by differentiating the expression for the speed

$$a_t = \dot{v} = \frac{d}{dt}\sqrt{\dot{x}^2(t) + \dot{y}^2(t) + \dot{z}^2(t)} = \frac{1}{2}\frac{1}{\sqrt{\dot{x}^2 + \dot{y}^2 + \dot{z}^2}}(2\dot{x}\ddot{x} + 2\dot{y}\ddot{y} + 2\dot{z}\ddot{z}) \quad \text{[f]}$$

At $t = 0$, we have

$$a_t(0) = \frac{4(4) + 0(-1) + 3(0)}{\{4^2 + 0^2 + 3^2\}^{1/2}} = \frac{16}{5} \quad \text{[g]}$$

Substituting the numerical values into Eq. [c], we obtain

$$\left(\frac{v^2}{\rho}\right)^2 + a_t^2 = a_x^2 + a_y^2 + a_z^2 \Rightarrow \left(\frac{25}{\rho}\right)^2 + \left(\frac{16}{5}\right)^2 = 4^2 + 1^2 + 0 \quad \text{[h]}$$

which can be solved for ρ to yield

$$\rho = \sqrt{\frac{25^2}{17 - (16/5)^2}} = \frac{125}{13} = 9.615 \quad \text{[i]}$$

Next, we find the radius of curvature using Eq. [1.3.39a]. Noting that the path variable is time, at $t = 0$ we have

$$\dot{\mathbf{r}}(0) = 4\mathbf{i} + 3\mathbf{k} \qquad \ddot{\mathbf{r}}(0) = 4\mathbf{i} - \mathbf{j} \qquad \dot{s}(0) = \sqrt{\dot{x}^2(0) + \dot{y}^2(0) + \dot{z}^2(0)} = 5 \quad \text{[j]}$$

Substituting these values into Eq. [1.3.39a] we obtain

$$\frac{1}{\rho} = \frac{1}{125}\sqrt{17(5)^2 - 16^2} = \frac{13}{125} \quad \text{[k]}$$

which gives the same result as the one obtained in Eq. [i].

To find the torsion of the curve we make use of Eq. [1.3.39b]. We note that the third derivative of the position vector is required. From Eq. [d] we obtain

$$\mathbf{r}''' = \dddot{\mathbf{r}}(t) = (0.6 + \sin t)\mathbf{j} \quad \text{[l]}$$

so that at $t = 0$, $\dddot{\mathbf{r}}(t) = 0.6\mathbf{j}$. Introducing this and the expressions for $\mathbf{r}'$ and $\mathbf{r}''$ into Eq. [1.3.39b] we obtain

$$\frac{1}{\tau} = -\frac{9.615^2}{5^6}\{(4\mathbf{i} - \mathbf{j}) \cdot [(4\mathbf{i} + 3\mathbf{k}) \times 0.6\mathbf{j}]\} = 0.04260 \quad \text{[m]}$$

so that the torsion of the curve at $\tau = 0$ is $\tau = 1/0.04260 = 23.47$. Comparing with the value of the radius of curvature, we see that the curve is bending more than it is twisting at this instant.

The above analysis could also be carried out using a vector approach from the beginning. We can write the velocity as $\mathbf{v} = 4\mathbf{i} + 3\mathbf{k}$, from which we obtain the unit vector in the tangential direction as

$$\mathbf{e}_t = \frac{\mathbf{v}}{|\mathbf{v}|} = \frac{4\mathbf{i} + 3\mathbf{k}}{5} \qquad [\mathbf{n}]$$

The component of the acceleration in the tangential and normal directions can then be computed as

$$a_t = \mathbf{a} \cdot \mathbf{e}_t \qquad a_n = |\mathbf{a} - (\mathbf{a} \cdot \mathbf{e}_t)\mathbf{e}_t| = \sqrt{a^2 - a_t^2} \qquad [\mathbf{o}]$$

For the problem at hand, $\mathbf{a} = 4\mathbf{i} - \mathbf{j}$, so that substituting it into the above equation yields the results that we obtained earlier. This latter approach is in many cases more efficient in obtaining the path parameters.

Cylindrical Coordinates   This set of coordinates is particularly useful if the particle is moving along a curved path, the position of the particle is of interest, and one component of the motion can be separated from the other two.

Fig. 1.9 describes this coordinate system. The inertial coordinate system xyz is chosen such that the component of the motion that can be separated from the other two is defined as the z direction. We take the path of the particle and project it onto the xy plane. The parameters describing the motion are:

1. The height z.
2. The absolute distance from the origin of the coordinate system to the projection of the path of the particle on the xy plane, denoted by R or r.

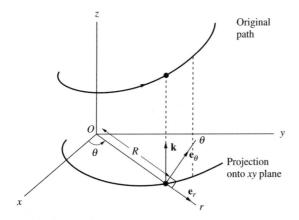

Figure 1.9

3. On the xy plane, the amount of counterclockwise rotation from a starting line (usually selected as the x axis) to reach the line joining the origin and projection of the particle on the xy plane. Denoted by θ and measured in radians.

The selection of the origin of the coordinate system is very important when using cylindrical coordinates, as R and θ change by changing the location of the origin. The first principal direction is fixed, and it is the z direction, with associated unit vector **k**. To describe the component of the motion in the xy plane, we define two perpendicular directions. The *radial direction* is defined as outward from the origin of the coordinate system to the projection of the particle on the xy plane. The associated unit vector is referred to as $\mathbf{e}_r$. The *transverse direction* is perpendicular to the radial direction and it is denoted by θ, with unit vector $\mathbf{e}_\theta$. Essentially, the radial and transverse directions are obtained by rotating the x and y axes counterclockwise by θ about the z axis. The unit vectors $\mathbf{e}_r$, $\mathbf{e}_\theta$, and **k** form a mutually orthogonal triad, with $\mathbf{e}_r \times \mathbf{e}_\theta = \mathbf{k}$. For plane motion, these coordinates are referred to as *polar coordinates*. The unit vectors can be expressed in terms of the unit vectors in Cartesian coordinates as

$$\mathbf{e}_r = \cos\theta \mathbf{i} + \sin\theta \mathbf{j} \qquad \mathbf{e}_\theta = -\sin\theta \mathbf{i} + \cos\theta \mathbf{j} \qquad [\mathbf{1.3.41}]$$

The position of a particle is expressed in cylindrical coordinates as

$$\mathbf{r}(t) = R\mathbf{e}_r + z\mathbf{k} = R\cos\theta \mathbf{i} + R\sin\theta \mathbf{j} + z\mathbf{k} = x\mathbf{i} + y\mathbf{j} + z\mathbf{k} \quad [\mathbf{1.3.42}]$$

To obtain the velocity we differentiate this equation, to get

$$\mathbf{v}(t) = \dot{\mathbf{r}}(t) = \dot{R}\mathbf{e}_r + R\dot{\mathbf{e}}_r + \dot{z}\mathbf{k} \qquad [\mathbf{1.3.43}]$$

which requires the derivative of the unit vector in the radial direction. To calculate this derivative, consider the particle at a point P', as shown in Fig. 1.10. The coordinate system has rotated by an angle of $\Delta\theta$. Denoting the unit vectors associated with the new position as $\mathbf{e}_r(\theta + \Delta\theta)$ and $\mathbf{e}_\theta(\theta + \Delta\theta)$, we relate them to $\mathbf{e}_r(\theta)$ and $\mathbf{e}_\theta(\theta)$ by

$$\mathbf{e}_r(\theta + \Delta\theta) = \mathbf{e}_r(\theta)\cos\Delta\theta + \mathbf{e}_\theta(\theta)\sin\Delta\theta$$

$$\mathbf{e}_\theta(\theta + \Delta\theta) = -\mathbf{e}_r(\theta)\sin\Delta\theta + \mathbf{e}_\theta(\theta)\cos\Delta\theta \qquad [\mathbf{1.3.44}]$$

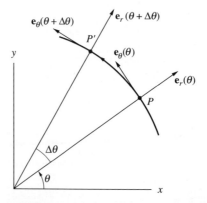

Figure 1.10

Using a small angles assumption of $\sin \Delta\theta \approx \Delta\theta$, $\cos \Delta\theta \approx 1$, and taking the limit as $\Delta\theta$ approaches zero, we obtain

$$\lim_{\Delta\theta \to 0} \frac{[\mathbf{e}_r(\theta + \Delta\theta) - \mathbf{e}_r(\theta)]}{\Delta\theta} = \frac{d\mathbf{e}_r}{d\theta} = \mathbf{e}_\theta$$

$$\lim_{\Delta\theta \to 0} \frac{[\mathbf{e}_\theta(\theta + \Delta\theta) - \mathbf{e}_\theta(\theta)]}{\Delta\theta} = \frac{d\mathbf{e}_\theta}{d\theta} = -\mathbf{e}_r \qquad [1.3.45]$$

Using the chain rule of differentiation and Eq. [1.3.44], we obtain

$$\dot{\mathbf{e}}_r = \frac{d\mathbf{e}_r}{d\theta}\frac{d\theta}{dt} = \dot{\theta}\mathbf{e}_\theta \qquad \dot{\mathbf{e}}_\theta = \frac{d\mathbf{e}_\theta}{d\theta}\frac{d\theta}{dt} = -\dot{\theta}\mathbf{e}_r \qquad [1.3.46]$$

which, when substituted in the expression for the velocity, yield

$$\mathbf{v}(t) = \dot{R}\mathbf{e}_r + R\dot{\theta}\mathbf{e}_\theta + \dot{z}\mathbf{k} \qquad [1.3.47]$$

The first term on the right side of this expression corresponds to a change in the radial direction and the second term to a change in angle. The third term is the component of the velocity in the z direction.

In a similar fashion we can find the expression for acceleration. Differentiation of Eq. [1.3.47] yields

$$\mathbf{a}(t) = \ddot{\mathbf{r}}(t) = \ddot{R}\mathbf{e}_r + \dot{R}\dot{\mathbf{e}}_r + \dot{R}\dot{\theta}\mathbf{e}_\theta + R\ddot{\theta}\mathbf{e}_\theta + R\dot{\theta}\dot{\mathbf{e}}_\theta + \ddot{z}\mathbf{k} \qquad [1.3.48]$$

Substituting in the values for the derivatives of the unit vectors and combining terms, we obtain

$$\mathbf{a}(t) = (\ddot{R} - R\dot{\theta}^2)\mathbf{e}_r + (R\ddot{\theta} + 2\dot{R}\dot{\theta})\mathbf{e}_\theta + \ddot{z}\mathbf{k} \qquad [1.3.49]$$

We can attribute a physical meaning to the acceleration terms. The first term, $\ddot{R}$, describes the rate of the change of the component of the velocity in the radial direction. The second term, $R\dot{\theta}^2$, is the centripetal acceleration. It is always in the negative radial direction as R is a positive quantity. The term $R\ddot{\theta}$ describes the acceleration due to the change in the angle θ. The last term, $2\dot{R}\dot{\theta}$, is known as the *Coriolis acceleration*, named after the French military engineer Gustave Coriolis, who first explained the significance and existence of this component of the acceleration. It results from two effects: The first effect is due to the differentiation of the $\mathbf{e}_r$ term in $\dot{R}\mathbf{e}_r$, and it is associated with a change in direction. The second effect is due to the differentiation of R in the expression $R\dot{\theta}\mathbf{e}_\theta$, and it is associated with a change in magnitude.

Example 1.5

A car travels with constant speed v on a spiraling path along a mountain. The shape of the mountain is approximated as a paraboloid with base radius a and height h, as shown in Fig. 1.11. It takes the car exactly six full turns around the mountain to reach the top. Find the velocity of the car as a function of its radial distance R from the center of the mountain.

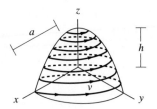

Figure 1.11

Solution

The height reached by the car is related to the radial distance by the relation

$$h - z = \frac{h}{a^2} R^2 \qquad [a]$$

The height is also related to the angle traversed by

$$z = \frac{h}{6} \frac{\theta}{2\pi} \qquad [b]$$

One can express the height and the angle traversed as a function of the radial distance R as

$$z = h - \frac{h}{a^2} R^2 \qquad \theta = \frac{12\pi}{h} z = 12\pi \left(1 - \frac{R^2}{a^2}\right) \qquad [c]$$

Differentiating Eqs. [c] and introducing into the expression for velocity, we obtain

$$\dot{z} = -2 \frac{h}{a^2} R\dot{R} \qquad \dot{\theta} = -24\pi \frac{R\dot{R}}{a^2} \qquad [d]$$

and

$$\mathbf{v} = \dot{R} \mathbf{e}_r + R\dot{\theta} \mathbf{e}_\theta + \dot{z} \mathbf{k} = \dot{R} \mathbf{e}_r - 24\pi \frac{R^2 \dot{R}}{a^2} \mathbf{e}_\theta - 2h \frac{R\dot{R}}{a^2} \mathbf{k} = v \mathbf{e}_t \qquad [e]$$

We are given that the speed is constant, so that we take the magnitude of the velocity as

$$v = \sqrt{\mathbf{v} \cdot \mathbf{v}} = \dot{R} \sqrt{1 + 576\pi^2 \frac{R^4}{a^4} + 4h^2 \frac{R^2}{a^4}} \qquad [f]$$

Defining the variable G as

$$G = \sqrt{1 + 576\pi^2 \frac{R^4}{a^4} + 4h^2 \frac{R^2}{a^4}} \qquad [g]$$

we can write $\dot{R} = v/G$. Equation [f] gives the relation for the rate of change of the radial distance as a function of the radial distance itself.

In a similar fashion, we can come up with an expression for the magnitude of the acceleration.

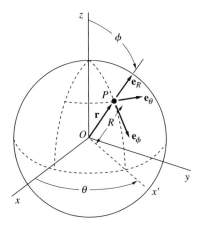

Figure 1.12

Spherical Coordinates When neither one of the components of the motion is separable from the other and a path-related coordinate system is needed, *spherical coordinates* are suitable. The configuration of this coordinate system is shown in Fig. 1.12. The parameters describing the path are the absolute distance from the origin of the coordinate system to the particle, denoted by R, and two angles θ and ϕ, referred to as the *polar* and *azimuthal* angles, respectively. Note that the parameter R used here is different from the parameter R used in cylindrical coordinates.

The principal directions are referred to as the *radial, polar,* and *azimuthal*. The radial direction is defined as outward from the origin to the particle. The corresponding unit vector is denoted by $\mathbf{e}_R$. The other directions depend on the polar angle θ and the azimuthal angle ϕ. The azimuthal angle is the angle between $\mathbf{r}$ and the z axis, and the polar angle θ is the angle between the x axis and the projection of $\mathbf{r}$ on the xy plane. Note the similarity between the polar angle θ and the transverse angle when using cylindrical coordinates. It follows that the polar direction, with the unit vector $\mathbf{e}_\theta$, is tangent to the circle obtained by traversing the component of $\mathbf{r}(t)$ in the xy plane. The unit vector associated with the azimuthal direction is denoted by $\mathbf{e}_\phi$, and it is tangent to the circle on the $x'z$ plane, which is obtained by rotating the xz plane about the z axis by an angle of θ. The three unit vectors are mutually orthogonal, with the relation

$$\mathbf{e}_R \times \mathbf{e}_\phi = \mathbf{e}_\theta \qquad [1.3.50]$$

We express the coordinates of a particle as

$$x = R\sin\phi\cos\theta \qquad y = R\sin\phi\sin\theta \qquad z = R\cos\phi \qquad [1.3.51]$$

and the displacement vector has the form

$$\mathbf{r} = R\mathbf{e}_R \qquad [1.3.52]$$

To obtain the velocity, we differentiate the above equation,

$$\mathbf{v}(t) = \dot{\mathbf{r}} = \dot{R}\mathbf{e}_R + R\dot{\mathbf{e}}_R = \dot{R}\mathbf{e}_R + R\frac{\partial \mathbf{e}_R}{\partial \theta}\frac{d\theta}{dt} + R\frac{\partial \mathbf{e}_R}{\partial \phi}\frac{d\phi}{dt} \qquad [1.3.53]$$

indicating that the derivatives of $\mathbf{e}_R$ must be found with respect to both angles. To accomplish this, we will obtain the derivatives of $\mathbf{e}_R$ using their expressions in Cartesian coordinates. From Eqs. [1.3.51] and [1.3.52], we can write

$$\mathbf{e}_R = \sin\phi\cos\theta\mathbf{i} + \sin\phi\sin\theta\mathbf{j} + \cos\phi\mathbf{k} \qquad [1.3.54]$$

Differentiation of this equation with respect to θ and ϕ yields

$$\frac{\partial \mathbf{e}_R}{\partial \theta} = -\sin\phi\sin\theta\mathbf{i} + \sin\phi\cos\theta\mathbf{j} = \sin\phi(-\sin\theta\mathbf{i} + \cos\theta\mathbf{j}) \qquad [1.3.55]$$

$$\frac{\partial \mathbf{e}_R}{\partial \phi} = \cos\phi\cos\theta\mathbf{i} + \cos\phi\sin\theta\mathbf{j} - \sin\phi\mathbf{k} = \cos\phi(\cos\theta\mathbf{i} + \sin\theta\mathbf{j}) - \sin\phi\mathbf{k}$$

From Fig. 1.12, we can express the unit vectors in the polar and azimuthal directions as

$$\mathbf{e}_\theta = -\sin\theta\mathbf{i} + \cos\theta\mathbf{j} \qquad \mathbf{e}_\phi = \cos\phi(\cos\theta\mathbf{i} + \sin\theta\mathbf{j}) - \sin\phi\mathbf{k} \qquad [1.3.56]$$

Note that when the azimuthal angle $\phi = 90°$, spherical and cylindrical coordinates coincide. Indeed, we have

$$\mathbf{e}_R(\phi = 90°) = \mathbf{e}_r \qquad \mathbf{e}_\theta(\phi = 90°) = \mathbf{e}_\theta \qquad \mathbf{e}_\phi(\phi = 90°) = -\mathbf{k} \qquad [1.3.57]$$

Introducing Eqs. [1.3.56] into [1.3.55] we obtain

$$\frac{\partial \mathbf{e}_R}{\partial \theta} = \sin\phi\mathbf{e}_\theta \qquad \frac{\partial \mathbf{e}_R}{\partial \phi} = \mathbf{e}_\phi \qquad [1.3.58]$$

The above relations can be used to obtain the time derivative of $\mathbf{e}_R$ as

$$\frac{d\mathbf{e}_R}{dt} = \dot{\theta}\sin\phi\mathbf{e}_\theta + \dot{\phi}\mathbf{e}_\phi \qquad [1.3.59]$$

As a result, one expresses the velocity as

$$\mathbf{v}(t) = \dot{R}\mathbf{e}_R + R\dot{\theta}\sin\phi\mathbf{e}_\theta + R\dot{\phi}\mathbf{e}_\phi \qquad [1.3.60]$$

To obtain the acceleration, we need to generate the time derivatives of the unit vectors in the polar and azimuthal directions. Using a procedure similar to the above, we obtain

$$\frac{d\mathbf{e}_\theta}{dt} = -\dot{\theta}\sin\phi\mathbf{e}_R - \dot{\theta}\cos\phi\mathbf{e}_\phi$$

$$\frac{d\mathbf{e}_\phi}{dt} = -\dot{\phi}\mathbf{e}_R + \dot{\theta}\cos\phi\mathbf{e}_\theta \qquad [1.3.61]$$

Using this equation and manipulating the algebra, the expression for the acceleration has the form

$$\mathbf{a}(t) = \ddot{\mathbf{r}}(t) = (\ddot{R} - R\dot{\theta}^2\sin^2\phi - R\dot{\phi}^2)\mathbf{e}_R + (R\ddot{\theta}\sin\phi + 2\dot{R}\dot{\theta}\sin\phi + 2R\dot{\theta}\dot{\phi}\cos\phi)\mathbf{e}_\theta$$
$$+ (R\ddot{\phi} + 2\dot{R}\dot{\phi} - R\dot{\theta}^2\sin\phi\cos\phi)\mathbf{e}_\phi \qquad [1.3.62]$$

In Eq. [1.3.62], the terms having the squares of the first derivatives correspond to the centripetal accelerations, and the mixed first derivative terms correspond to the Coriolis accelerations. The second derivatives indicate accelerations in the radial direction and in the polar and azimuthal angles.

Example 1.6

The length of the spherical pendulum shown in Fig. 1.13 varies by the relation $L = 2 + \sin \pi t$ m. The pendulum spins with the constant rate of $\dot\theta = 2$ rad/s. We are given that the angle β is related to the length of the pendulum and θ by

$$L^2 \dot\theta \sin^2 \beta = 8 \text{ m}^2/\text{s} \qquad [a]$$

Assuming that β is always in the range $-\pi/2 \leq \beta \leq \pi/2$, find the velocity of the tip of the pendulum at $t = 0.1$ s.

Solution

We attach a set of spherical coordinates to the pendulum. Note that according to our definition of the polar and azimuthal angles, ϕ and β are related by $\beta = \pi - \phi$. We observe that $\sin \beta = \sin \phi$. Fig. 1.14 shows the coordinate system in side view using the z and radial axes.

Considering the range of β, we can obtain the expression for $\sin \beta$ from Eq. [a] as

$$\sin \beta = \sqrt{\frac{8}{L^2 \dot\theta}} = \frac{2}{2 + \sin \pi t} \qquad [b]$$

To find $\dot\beta$, we differentiate this equation, with the result

$$\cos \beta \dot\beta = -\frac{2\pi \cos \pi t}{(2 + \sin \pi t)^2} \quad \to \quad \dot\beta = \frac{1}{\sqrt{1 - \sin^2 \beta}} \frac{-2\pi \cos \pi t}{(2 + \sin \pi t)^2} \qquad [c]$$

The expression for the velocity is given by Eq. [1.3.60]. We identify the individual values as

$$\dot R = \dot L = \pi \cos \pi t \qquad \dot\phi = -\dot\beta \qquad [d]$$

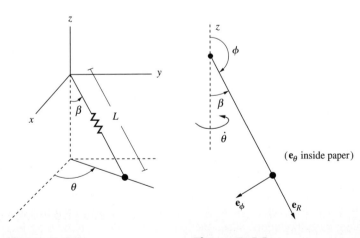

Figure 1.13   **Figure 1.14**

At $t = 0.1$ s, we have

$$R = 2 + \sin(0.1\pi) = 2.309 \text{ m} \qquad \dot{R} = \pi\cos(0.1\pi) = 2.988 \text{ m} \qquad \sin\phi = \sin\beta \quad \textbf{[e]}$$

$$\sin\phi = \frac{2}{2 + \sin(0.1\pi)} = 0.8662 \qquad \dot{\phi} = \frac{1}{\sqrt{1 - \sin^2\beta}}\frac{2\pi\cos(0.1\pi)}{(2 + \sin(0.1\pi))^2} = 2.243 \text{ rad/s}$$

which, when substituted in Eq. [1.3.60], yields

$$\mathbf{v}(t) = \dot{R}\mathbf{e}_R + R\dot{\theta}\sin\phi\,\mathbf{e}_\theta + R\dot{\phi}\mathbf{e}_\phi = 2.988\mathbf{e}_R + 4.000\mathbf{e}_\theta + 5.179\mathbf{e}_\phi \text{ m/s} \qquad \textbf{[f]}$$

Equation [a] is an angular momentum conservation relation. This relationship arises because the speed of the pendulum can be written as

$$v = \sqrt{\dot{R}^2 + R^2(\dot{\theta}^2\sin^2\phi + \dot{\phi}^2)} \qquad \textbf{[g]}$$

The speed v is independent of θ. Many systems in dynamics have this property.

Mixed Descriptions As discussed earlier, in certain cases it may not be sufficient to use a single set of coordinates to solve a problem; one may also need to take advantage of another coordinate system. The different coordinate systems provide different types of information about the motion, and the added information arising from the use of more than one coordinate system facilitates our understanding of the nature of that motion. Usually, one exploits the relationships between the two coordinate systems by first writing the unit vectors of the coordinate systems in terms of the motion variables. Then, one considers the relationship between the unit vectors in the different coordinate systems.

One question that arises is whether one can use any set of parameters to define a coordinate system. The answer to this question is positive, provided that three parameters p_1, p_2, p_3 can be found such that there exists a unique transformation between the components of the motion in rectilinear coordinates x, y, and z (or any other proper coordinate system) and p_1, p_2, and p_3. For example, for cylindrical coordinates, $p_1 = R$, $p_2 = \theta$, and $p_3 = z$. The transformations from x, y, and z to R, θ, and z are given in Eq. [1.3.42]. Considering these equations and Fig. 1.9, we can relate R, θ, and z to x, y, and z as

$$R = \sqrt{x^2 + y^2} \qquad \theta = \tan^{-1}\left(\frac{y}{x}\right) \qquad z = z \qquad \textbf{[1.3.63]}$$

where the range of interest in the inverse tangent is $(-\pi/2, \pi/2)$. A proper selection of the coordinate frame is crucial to the understanding of the problem and to the solution.

There are several other parabolic and hyperbolic coordinate systems that facilitate derivation of the governing equations, as well as the solution, for a specific problem. Examples can be found not only from motion analysis but from other problems such as heat transfer. To develop a coordinate system with orthogonality properties among its components, additional requirements need to be introduced. These requirements are described in the text by Ginsberg listed in the reference section at the end of the chapter.

1.3 REVIEW OF VECTOR ANALYSIS

The reader is always encouraged to explore the possibility of using more than one coordinate system when tackling a dynamics problem. One note of caution is in order, though. When selecting coordinate systems and the variables associated with them, be careful to avoid the ambiguities that can result from an improper selection of the variables. A good way to avoid this problem is to make sure that the transformation from one set of variables to another is indeed unique.

Example 1.7

In Fig. 1.15, the pin attached to the circle of 6-inch radius is sliding in the slot with the constant speed of $v = 2$ in/sec. Find the values of $\dot\theta$ and $\ddot\theta$ at the instant when $\phi = 90°$.

Solution

We will make use of polar as well as normal-tangential coordinates. The geometry of the system is shown in Fig 1.16. The expressions for R and θ are

$$R = \sqrt{6^2 + 9^2} = 10.82 \text{ in} \qquad \theta = \tan^{-1}\left(\frac{6}{9}\right) = 33.69° \qquad \text{[a]}$$

The relationship between the two sets of unit vectors can be written as

$$\mathbf{e}_t = \cos\theta \mathbf{e}_r - \sin\theta \mathbf{e}_\theta \qquad \mathbf{e}_n = -\cos\theta \mathbf{e}_\theta - \sin\theta \mathbf{e}_R \qquad \text{[b]}$$

where $\sin\theta = 0.5547$, $\cos\theta = 0.8321$. Writing the expression for velocity in the two coordinate systems,

$$\mathbf{v} = v\mathbf{e}_t = \dot{R}\mathbf{e}_r + R\dot\theta \mathbf{e}_\theta \qquad \text{[c]}$$

so that to find $\dot R$ and $\dot\theta$ we take the inner product of $\mathbf{v}$ with $\mathbf{e}_r$ and $\mathbf{e}_\theta$, thus

$$\mathbf{v}\cdot\mathbf{e}_r = \dot R = v\mathbf{e}_t\cdot\mathbf{e}_r = v\cos\theta = 2(0.8321) = 1.664 \text{ in/sec}$$

$$\mathbf{v}\cdot\mathbf{e}_\theta = R\dot\theta = v\mathbf{e}_t\cdot\mathbf{e}_\theta = -v\sin\theta = -2(0.5547) = -1.109 \text{ in/sec} \qquad \text{[d]}$$

with the conclusion

$$\dot R = 1.664 \text{ in/sec} \qquad \dot\theta = -\frac{1.109}{R} = -\frac{1.109}{10.82} = -0.1025 \text{ rad/sec} \qquad \text{[e]}$$

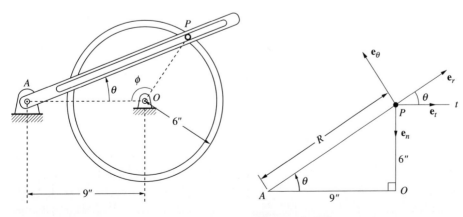

Figure 1.15   **Figure 1.16**

To find the acceleration, we make use of the fact that the speed is constant, so that the acceleration has the form

$$\mathbf{a} = \frac{v^2}{\rho}\mathbf{e}_n = (\ddot{R} - R\dot{\theta}^2)\mathbf{e}_r + (R\ddot{\theta} + 2\dot{R}\dot{\theta})\mathbf{e}_\theta \qquad [\mathbf{f}]$$

with the radius of curvature being $\rho = 6$ in. Taking the inner product of $\mathbf{a}$ and $\mathbf{e}_\theta$ we obtain

$$\mathbf{a} \cdot \mathbf{e}_\theta = R\ddot{\theta} + 2\dot{R}\dot{\theta} = \frac{v^2}{\rho}(-\cos\theta) \qquad [\mathbf{g}]$$

Solving for $\ddot{\theta}$ we obtain

$$\ddot{\theta} = -\frac{(v^2 \cos\theta \div \rho + 2\dot{R}\dot{\theta})}{R} = -\frac{(4 \cdot 0.8321 \div 6 - 2 \cdot 1.664 \cdot 0.1025)}{10.82} = -0.01974 \text{ rad/sec}^2 \qquad [\mathbf{h}]$$

1.4 NEWTONIAN PARTICLE MECHANICS

Newtonian mechanics is based upon the three laws of Newton, and the counterpart of Newton's second law for rotational motion. It is valid for reference systems at rest or moving with uniform velocity with respect to each other. Such reference systems are known as *inertial*. The best approximation to an inertial reference frame is to consider the distant stars fixed. A less accurate approximation is to consider the sun fixed. In the vicinity of the earth, if the velocities involved are much less than the *escape velocity*,[3] or if the distances traversed are much smaller than the radius of the earth, and if the duration of the motion is not long, the earth is used as an inertial reference frame.

Isaac Newton developed his three laws while he was working on the motion of rigid bodies. While the laws govern the motion of the center of mass of a body, they do not account for the rotational motion. The rotational equations relate the moments applied to a body to the change in angular momentum; these were developed by Euler around 1750. (Newton published his three laws in his *Principia* in 1687, but he had developed them about 20 years before, while he was at Cambridge University.) Here, we will consider Newton's equations within the context of particles. To this end, we first describe what we mean by a *particle*.

A particle is defined as *a body with no physical dimensions,* implying that the entire mass of the particle is concentrated at one point. This, of course, is an idealization and it is a valid assumption only when the physical dimensions of the body are small compared to its weight. Neglecting the physical dimensions implies that the rotational motion of the body is ignored.

Consider a particle of mass m and let $\mathbf{F}(t)$ denote the sum of all forces acting on the particle. Define the *linear momentum* of the particle by $\mathbf{p}(t) = m\mathbf{v}(t)$. The

---

[3]The escape velocity of a spacecraft is the speed that the spacecraft must attain in order to escape from the gravitational attraction of the celestial body it is orbiting and to assume a parabolic orbit.

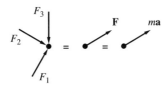

Figure 1.17

physical explanation of linear momentum is the tendency of the body to continue to translate. The mass of the body is its resistance to translation. Newton's laws can be stated as follows.

First law: If no forces act on a particle, the particle retains its linear momentum. If the particle is at rest it remains at rest, and if it is moving, it moves with constant linear momentum. A particle of constant mass moves along a straight line with constant velocity. In other words, $\mathbf{F} = \mathbf{0} \rightarrow \mathbf{v}(t) = $ constant.

Second law: The rate of change of the linear momentum of a particle is equal to the sum of all forces acting on it.

$$\frac{d\mathbf{p}(t)}{dt} = \frac{d(m\mathbf{v}(t))}{dt} = \mathbf{F} \qquad [1.4.1]$$

For a particle whose mass remains constant, the above relation reduces to

$$m\mathbf{a}(t) = \mathbf{F} \qquad [1.4.2]$$

Newton's first law is a special case of his second law. Fig. 1.17 illustrates this law. We will treat variable mass systems in Chapter 3.

Third law: When two particles exert forces upon each other, these forces are equal in magnitude and opposite in direction, and they lie along the line joining the two particles.

Denoting by $\mathbf{F}_{ij}$ the force exerted on particle i by particle j, we arrive at the conclusion that $\mathbf{F}_{ij} = -\mathbf{F}_{ji}$. Consider, for example, Fig. 1.18, a dog pulling a cart. The free-body diagrams of the dog and cart are given in Fig. 1.19. The friction between the dog's legs and the ground provides the forward thrust. The dog and cart exert forces on each other that are equal in magnitude and opposite in direction.

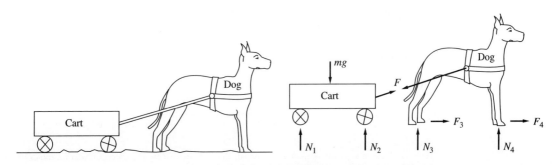

Figure 1.18   **Figure 1.19**

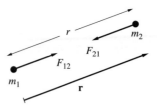

Figure 1.20

A very important application of Newton's third law is in celestial mechanics. The forces that two bodies exert on each other, as shown in Fig. 1.20, are governed by Newton's law of gravitation as

$$F_{12} = -F_{21} = \frac{Gm_1m_2}{r^2} \qquad [1.4.3]$$

in which G is the universal constant of gravitation, and r is the distance between the particles. The value of G is $G = 6.673(10^{-11})$ m³/kg·s² in SI units, and $G = 3.439(10^{-8})$ ft⁴/lb·sec⁴ in U.S. Customary units. Because G is such a small quantity, for any two small bodies the gravitational attraction is extremely small. The gravitational force becomes significant when at least one of the bodies involved is very large, such as in an analysis of motion in the vicinity of a celestial body. Equation [1.4.3] can be expressed in vector form as

$$\mathbf{F}_{12} = -\mathbf{F}_{21} = \frac{Gm_1m_2}{r^3}\mathbf{r} \qquad [1.4.4]$$

where $\mathbf{r}$ is the position vector between the centers of mass of the two bodies. For motion near the earth, using the values of mass of the earth as $m_e = 5.976(10^{24})$ kg and mean radius as $r_e = 6,378$ km, we assume that the distance of the body from the surface of the earth is negligible compared to the radius. Defining the gravitational constant as $g = Gm_e/r_e^2$, we obtain the force of gravity as $F = m_2 g$, where the mean value for the gravitational constant is

$$g = \frac{6.673(10^{-11}) \times 5.976(10^{24})}{[6.378(10^6)]^2} = 9.803 \text{ m/s}^2 \qquad [1.4.5]$$

Using the above equation to calculate the gravitational constant g is not accurate. Equation [1.4.5] is based on treating the earth as a particle (or as a rigid uniform sphere), and it ignores centrifugal effects due to the rotation of the earth. Furthermore, gravitational effects due to the sun and moon also affect the value of g. The radius of the earth is not constant,[4] and its density is not uniform, resulting in different values of g at different points on the earth. For dynamics problems that do not require a substantial amount of precision, average values of $g = 9.81$ (or 9.807) m/s² or $g = 32.2$ (or 32.17) ft/sec² are used at sea level.

---

[4] The actual shape of the earth is an oblate spheroid. The earth looks more like an apple, with the radius larger around the equator, smaller at the poles, and with the poles slightly pressed in.

A more accurate approximation is the 1980 International Gravity Formula. It assumes that the earth is a rigid ellipsoid and takes into consideration the rotation of the earth. The approximation for g is given as a function of the latitude as

$$g = 9.780327[1 + 0.005279 \sin^2 \lambda + 0.000023 \sin^4 \lambda] \text{ m/s}^2 \quad \text{[1.4.6]}$$

where $\lambda = 90 - \phi$ is the latitude angle, with ϕ being the azimuthal angle defined when we considered spherical coordinates in Fig. 1.12. Even this approximation is not exactly accurate, because it fails to take into consideration both the dip in the earth's shape at the poles and the nonrigidity of the earth. At a latitude of 45° and at sea level, the commonly used value of g to 5-digit accuracy is $g = 9.8066$ m/s² or 32.174 ft/sec².

We introduced above the concept of a force without rigorously defining what it is. It turns out that we know of the existence of forces and we know their effects, but we cannot rigorously define what a force is. Here is a commonly used definition:

> A force is the effect of one body on another.

We also note that no method exists to directly measure a force. We have developed theories on how forces affect systems and have validated those theories by measuring the effects of the forces in the form of deformations or accelerations. In essence, in dynamics we use cause and effect relationships. For example, people weighing themselves on a mechanical bathroom scale read an output of the spring deflection caused by their weight, multiplied by the spring constant.

We can categorize the forces acting on bodies into two general types: (1) Contact forces, such as friction, impact, spring, dashpot, and so forth. These forces are applied to a point or an area on the body. (2) Field forces, such as gravitational and electromagnetic. These forces are applied to the body uniformly.

A special class of force that is of importance in dynamics is *friction*. Friction forces are developed when a body in contact with another moving or fixed body has a tendency to slide (slip) over that other body. The contact force between the two bodies is the normal force, hence a reaction force. The amount of slippage depends on the material properties and surface characteristics (rough, smooth, etc.) of the contacting bodies. The study of friction is a very complex subject, and an accurate representation of friction forces is difficult. Often, we use the model known as *dry friction* or *Coulomb friction* to describe friction. This model approximates the friction force as a coefficient of friction multiplied by the normal force. The coefficient of friction, denoted by μ, is an approximate quantity whose value depends on the material properties of the contacting bodies, the relative speed of the contacting bodies, and to a lesser degree, the temperature. Many engineering handbooks have tables of the coefficient of friction for a variety of materials and surfaces.

Another factor that affects the value of the coefficient of friction is whether the point of contact is moving or not. If there is slip between the point of contact and the surface, one uses the coefficient of kinetic friction. Mathematically,

$$F_f = \mu_k N \qquad [1.4.7]$$

where F_f is the friction force, N is the normal force, and μ_k is the coefficient of kinetic friction. If there is no slip between the contacting bodies the friction force is a quantity less than or equal to the static coefficient of friction multiplied by the normal force, and it can be expressed as

$$F_f \leq \mu_s N \qquad [1.4.8]$$

in which μ_s is the static coefficient of friction. In general, $\mu_k \leq \mu_s$. (When no other information is given, one assumes $\mu_k = \mu_s$.) Sliding begins when the friction force reaches $\mu_s N$, and after that point μ_k is used to describe the amount of friction.

The friction force acting on a body always opposes the impending motion. For moving bodies the friction force opposes the velocity of the contacting point relative to the point of contact. Fig. 1.21 illustrates this concept. The friction force can be represented in vector form as

$$\mathbf{F}_f = -\frac{\mu_k N \mathbf{v}}{|\mathbf{v}|} \quad \text{or} \quad \mathbf{F}_f = -\mu_k N \mathbf{e}_t \qquad [1.4.9]$$

For rectilinear motion one can write

$$F_f = -\mu_k N \operatorname{sign}(v) \qquad [1.4.10]$$

in which

$$\operatorname{sign}(v) = 1 \text{ when } v > 0 \qquad \operatorname{sign}(v) = -1 \text{ when } v < 0 \qquad [1.4.11]$$

The friction force is a nonlinear function. When solving problems involving friction one must be careful in determining the direction the friction force should be acting and whether there is slipping or not. In many cases the direction of the friction force may not be obvious. To determine whether there is slipping, one can begin by assuming that there is slipping (or that there is no slipping) and then check the validity of this assumption. For example, if one assumes no slipping, one can calculate

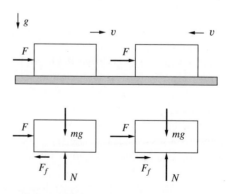

Figure 1.21

the magnitude of the friction force and compare it with the maximum magnitude the friction force can attain.

When solving a dynamics problem one should:

1. Isolate the bodies involved.
2. Select a coordinate system and positive directions, and draw free-body diagrams.
3. Write the force balances.
4. Use the kinematics of the problem to eliminate redundant variables.

If the objective is an instantaneous analysis, the accelerations and reaction forces are calculated. If one desires to solve for the response, the kinematics is used further to eliminate the reaction forces. The resulting equation(s) are differential equations in terms of the motion variables only. Such equations are called *equations of motion.*

When considering the response, one first needs to decide whether to obtain a qualitative or a quantitative solution of the equations of motion. A quantitative solution implies actual solution of the differential equations of motion and it depends on the nature of the differential equation of motion as well as the form of the forcing **F**. The tremendous evolution in the field of differential equations in the past few centuries has produced several methods of solution. Several approximate methods have also been developed that can be implemented on digital computers.

A qualitative solution gives information about the nature of the response, or it gives the response at specific points in time and space, without having to solve for the response explicitly. Such qualitative analysis includes

1. Impulse-momentum relationships
2. Work-energy relationships
3. Other motion integrals
4. Equilibrium and stability

We will discuss these approaches in the remaining sections of this chapter.

Example 1.8

A collar of mass m slides in a circular track of radius R, as shown in Fig. 1.22. The coefficient of friction between the collar and the track is μ. The collar is given an initial velocity v_0. Find the distance traveled by the collar when it comes to a rest.

Solution

We can use either normal-tangential or cylindrical coordinates to solve this problem. The free-body diagram of the collar is given in Fig. 1.23. Summing forces perpendicular to the plane of the motion yields

$$mg = N_1 \qquad [a]$$

where N_1 is the component of the normal force perpendicular to the plane of the motion. Summing forces in the normal and tangential directions, we obtain

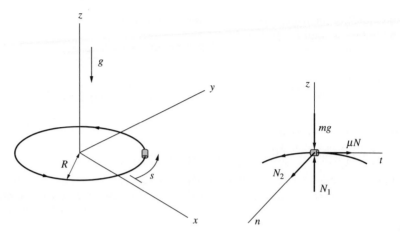

Figure 1.22 **Figure 1.23**

$$F_n = m\frac{v^2}{R} = N_2 \qquad [b]$$

$$F_t = ma_t = m\frac{dv}{dt} = -\mu N \qquad [c]$$

where N_2 is the component in the normal direction and N is the total normal force

$$N = \sqrt{N_1^2 + N_2^2} = \sqrt{m^2 g^2 + \frac{m^2 v^4}{R^2}} \qquad [d]$$

Introducing Eq. [d] into Eq. [c] and eliminating the mass term, we obtain

$$\frac{dv}{dt} = -\mu\sqrt{g^2 + \frac{v^4}{R^2}} \qquad [e]$$

which represents the equation of motion. It can be solved by moving the dt term to the right and the radical to the left. The problem, however, asks us to find the distance traveled, so that we seek to convert Eq. [e] to one in terms of the displacement, as

$$\frac{dv}{dt} = \frac{dv}{ds}\frac{ds}{dt} = v\frac{dv}{ds} = -\mu\sqrt{g^2 + \frac{v^4}{R^2}} \qquad [f]$$

or

$$\frac{v\,dv}{\sqrt{R^2 g^2 + v^4}} = -\frac{\mu}{R}\,ds \qquad [g]$$

Equation [g] can be integrated from the initial velocity v_0 to the final velocity 0 to yield

$$-\frac{\mu}{R}s = \frac{1}{2}\ln(Rg) - \frac{1}{2}\ln\left[v_0^2 + \sqrt{R^2 g^2 + v_0^4}\right] \qquad [h]$$

which can be solved for the distance traveled by the collar as

$$s = \frac{R}{2\mu} \ln\left[\frac{v_0^2 + \sqrt{R^2 g^2 + v_0^4}}{Rg}\right] \qquad [\mathrm{i}]$$

The distance traveled is inversely proportional to the coefficient of friction.

Example 1.9

The mass-spring system shown in Fig. 1.24 consists of a block of mass m attached to a wall with a spring of constant k. The coefficient of friction between the block and the surface it slides on is μ. A force $F(t)$ acts on the block. Find the equation of motion.

Solution

We draw the free-body diagrams in Figs. 1.25 and 1.26. Summing forces in the vertical direction yields

$$N = mg \qquad [\mathrm{a}]$$

To find the friction force we multiply the normal force with the friction coefficient. To find the direction of the friction force, we note that the block moves back and forth, so that the friction force is in a different direction depending on the velocity of the block. We have that

$$\text{when } \dot{x} > 0, \quad F_f = -\mu N = -\mu mg \qquad [\mathrm{b}]$$

$$\text{when } \dot{x} < 0, \quad F_f = \mu N = \mu mg \qquad [\mathrm{c}]$$

so that the equation of motion can be written considering the two regimes as

$$\text{when } \dot{x} > 0, \quad m\ddot{x} + kx = F - \mu mg \qquad [\mathrm{d}]$$

$$\text{when } \dot{x} < 0, \quad m\ddot{x} + kx = F + \mu mg \qquad [\mathrm{e}]$$

where we note that both equations of motion are linear. The two equations of motion [d] and [e] can be combined into a single nonlinear equation by

$$m\ddot{x} + kx = F - \mu mg \, \text{sign}(\dot{x}) \qquad [\mathrm{f}]$$

1.5 DEGREES OF FREEDOM AND CONSTRAINTS

In the beginning of this chapter we studied coordinate systems commonly used in dynamics. Three physical coordinates were required to specify the position of a

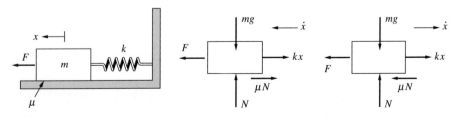

Figure 1.24   **Figure 1.25**   **Figure 1.26**

particle. In the previous section we studied Newton's second law, which related the resultant acceleration of a particle to the applied force. Equation [1.4.2] is a vector relationship, and it can be separated into three scalar components.

In many circumstances the motion of a particle is constrained to move in a subset of the three-dimensional configuration space. Examples of this include a vehicle moving along a track or on a fixed surface. In these cases, it is not necessary to use all three coordinates associated with a reference frame to describe the motion; depending on the constraints, one or two coordinates are sufficient. For example, if the particle shown in Fig. 1.27 is constrained to move along a surface whose mathematical description is

$$f(x, y, z, t) = 0 \qquad [1.5.1]$$

then two coordinates are sufficient to describe the motion. Selecting them as x and y, the z coordinate can be ascertained by solving Eq. [1.5.1]. A constraint of the form above is known as a *configuration constraint*. A more general form of a constraint is when the constraint is expressed in the form of differentials, such as

$$a_x\,dx + a_y\,dy + a_z\,dz + a_0\,dt = 0 \qquad [1.5.2]$$

or, dividing Eq. [1.5.2] by dt,

$$a_x\dot{x} + a_y\dot{y} + a_z\dot{z} + a_0 = 0 \qquad [1.5.3]$$

in which the coefficients a_0, a_x, a_y, and a_z are functions of x, y, and z. We will see a more complete analysis of constraints in Chapter 4.

A constraint equation is a geometric or kinematic representation of the restrictions on the motion of a body. What causes the body to execute such restricted motion is the constraint force associated with the constraint. When the constraint can be represented as a surface, as in Eq. [1.5.1], the associated constraint force is always normal to the surface. While we will prove this formally in Chapter 4, we can explain it physically here by noting that the particle velocity is always tangent to the surface. Hence, the particle cannot have a motion normal to the surface. This is caused by the constraint force.

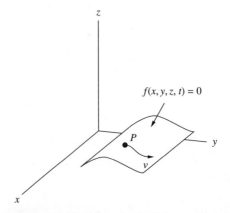

Figure 1.27

1.5 Degrees of Freedom and Constraints

We define by *degree of freedom* the minimum number of independent coordinates necessary to describe the configuration of a system. Each constraint applied to a system reduces the number of degrees of freedom by one. For example, the orientation of a vehicle moving up a spiral in Example 1.5 can be described using cylindrical coordinates. The radius is a function of the angle traversed, which constitutes one constraint. The rise of the spiral is also described in terms of the angle, which constitutes another constraint. Hence, only one of the coordinates R, θ, or z is sufficient to describe the motion. The number of degrees of freedom can be calculated using the relationship

$$\text{No. of degrees of freedom} = \text{No. of coordinates} - \text{No. of constraints} \quad \textbf{[1.5.4]}$$

Newton's second law for a particle yields three scalar equations. The number of equations of motion is determined by the number of constraints. If one (or two) constraints act on a particle there will be two (or one) equations of motion and the remaining equations will represent reactions. The reaction equations can usually be identified easily. We will see a more complete analysis of constraints in Chapter 4.

Example 1.10

The system shown in Fig. 1.28 consists of a vehicle undergoing rectilinear motion. A pendulum is attached to the vehicle. Find the equations of motion.

Solution

This is a two degree of freedom problem. If unrestricted, the vehicle has one degree of freedom and the mass has three. The wire restricts the motion of the pendulum. Denoting the displacements of the pendulum in the horizontal and vertical directions by x_P and y_P, we can express them as

$$x_P = x + L \sin\theta \qquad y_P = -L \cos\theta \qquad \textbf{[a]}$$

which constitutes two constraint equations. Hence, the combined system has two degrees of freedom.

The free-body diagrams of the vehicle and pendulum are given in Fig. 1.29. For the vehicle, summing forces along the horizontal we obtain

$$M\ddot{x} = T\sin\theta + F \qquad \textbf{[b]}$$

For the pendulum, we have the force balances in the horizontal and vertical directions as

$$m\ddot{x}_P = -T\sin\theta \quad \text{or} \quad m(\ddot{x} + L\ddot{\theta}\cos\theta - L\dot{\theta}^2\sin\theta) = -T\sin\theta \qquad \textbf{[c]}$$

$$m\ddot{y}_P = T\cos\theta - mg \quad \text{or} \quad m(L\ddot{\theta}\sin\theta + L\dot{\theta}^2\cos\theta) = T\cos\theta - mg \qquad \textbf{[d]}$$

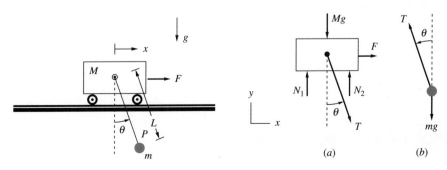

Figure 1.28

Figure 1.29

Eqs. [a], [c], and [d] are three equations that contain the constraint force (in this case the tension in the pendulum) explicitly. We can eliminate T from the above equations and obtain two equations of motion, the same number of equations as the degrees of freedom. This can be accomplished in a number of ways. One way is as follows: First, we introduce Eq. [c] into Eq [b], which yields

$$(M + m)\ddot{x} + mL\ddot{\theta}\cos\theta - mL\dot{\theta}^2\sin\theta = F \quad \text{[e]}$$

Then, we multiply Eq. [c] by $\cos\theta$ and Eq. [d] by $\sin\theta$ and add the resulting expressions. Doing so yields

$$mL\ddot{\theta} + m\ddot{x}\cos\theta + mg\sin\theta = 0 \quad \text{[f]}$$

which can be put into a more familiar form by multiplying it with L, so that

$$mL^2\ddot{\theta} + mL\ddot{x}\cos\theta + mgL\sin\theta = 0 \quad \text{[g]}$$

Equations [e] and [g] are the two equations of motion in terms of the two independent variables x and θ.

1.6 IMPULSE AND MOMENTUM

Newton's second law states that the rate of change of the linear momentum of a particle is equal to the applied force, or

$$\mathbf{F}(t) = \frac{d\mathbf{p}}{dt} \quad \text{[1.6.1]}$$

If we multiply the above equation by dt and integrate from an initial time t_1 to final time t_2, we obtain

$$\int_{t_1}^{t_2} \mathbf{F}(t)\,dt = \int_{\mathbf{p}(t_1)}^{\mathbf{p}(t_2)} d\mathbf{p} = \mathbf{p}(t_2) - \mathbf{p}(t_1) = m\mathbf{v}(t_2) - m\mathbf{v}(t_1) \quad \text{[1.6.2]}$$

Eq. [1.6.2] is the *impulse-momentum theorem* for a particle. The term on the left is called the *impulse*, which is equal to the change in linear momentum.[5] The unit of linear momentum is mass times velocity, ML/T. In the U.S. system one commonly uses lb • sec and in the SI system, N • s.

When there is more than one particle, the linear momentum of the entire system is obtained by adding up the linear momentum of each particle that comprises the system. For example, for a system of two particles of masses m_1 and m_2, the linear momentum is obtained by simply adding the linear momenta of each particle, $\mathbf{p} = m_1\mathbf{v}_1 + m_2\mathbf{v}_2$.

An interesting special case is when $\mathbf{F}(t) = \mathbf{0}$, or over the interval (t_1, t_2) the integral of $\mathbf{F}(t)$ is zero. It follows that in such cases the initial and final values of

---

[5]Note that this definition is different than the common use of the word.

the linear momentum are the same, $\mathbf{p}(t_2) = \mathbf{p}(t_1)$, which denotes the *principle of conservation of linear momentum*. The principle states that if the net effect of forces acting on a particle is zero over a time period, then the linear momentum of the particle has the same values at the beginning and end of the interval.

The principle of conservation of linear momentum is generally of more use when more than one particle is involved. Note that linear momentum of a system can be conserved in a certain direction of the motion only and not be conserved in the other directions. Defining a unit vector $\mathbf{e}$ along the direction the linear momentum is conserved and taking the dot product of Eq. [1.6.2] with $\mathbf{e}$, we obtain

$$\int_{t_1}^{t_2} \mathbf{F}(t) \cdot \mathbf{e}\, dt = \int_{\mathbf{p}(t_1)}^{\mathbf{p}(t_2)} d\mathbf{p} \cdot \mathbf{e} = \mathbf{p}(t_2) \cdot \mathbf{e} - \mathbf{p}(t_1) \cdot \mathbf{e} \qquad [\mathbf{1.6.3}]$$

An interesting application of the impulse-momentum theorem is when the duration of the applied force is very short. A large force applied through a very short time period is defined as an *impulsive force*. Denoting by ε the duration of the impulse, and taking the limit as ε approaches to zero, we have

$$\lim_{\varepsilon \to 0} \int_{t_1}^{t_1+\varepsilon} \mathbf{F}(t)\, dt = \hat{\mathbf{F}}(t_1) \qquad [\mathbf{1.6.4}]$$

where $\hat{\mathbf{F}}$ is the impulsive force and has the dimension of force × time. Theoretically, the amplitude of the impulsive force approaches infinity. To mathematically describe an impulsive force we make use of the Dirac delta function.

Consider Fig. 1.30. The Dirac delta function at point $t = a$ is denoted by $\underline{\underline{\delta}}(t-a)$ and is defined as[6]

$$\underline{\underline{\delta}}(t-a) = 0 \quad \text{when } t \neq a$$

$$\int_{-\infty}^{\infty} \underline{\underline{\delta}}(t-a)\, dt = 1 \qquad [\mathbf{1.6.5}]$$

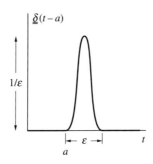

Figure 1.30   Dirac delta function

---

[6]We are using a different notation than the traditional delta to differentiate between the Dirac delta, the Kronecker delta, and the variation functions.

The quantity on the right side of this equation is recognized as the moment of the resultant of the external forces about point O, $\mathbf{M}_O = \mathbf{r} \times \mathbf{F}$, so that

$$\frac{d\mathbf{H}_O}{dt} = \mathbf{M}_O \qquad [1.6.15]$$

The rate of change of angular momentum of a particle about a point is equal to the applied moment about that point. The above equation is of considerably more use when dealing with systems of particles.

There is debate among scientists on whether Eq. [1.6.15] is a derived relationship or a stated principle, like Newton's second law. The argument in support of the stated principle viewpoint is based on the idea that shear forces are neglected in the derivation, and that every physical body has a nonzero volume. What makes the derivation above possible is the assumption that a particle has no physical dimensions.

Similar to the linear momentum case, we can integrate Eq. [1.6.15] over time, yielding the *angular impulse-momentum theorem* as

$$\int_{t_1}^{t_2} \mathbf{M}_O(t)\, dt = \int_{\mathbf{H}_O(t_1)}^{\mathbf{H}_O(t_2)} d\mathbf{H}_O = \mathbf{H}_O(t_2) - \mathbf{H}_O(t_1) \qquad [1.6.16]$$

The term on the left is defined as the *angular impulse*. As with linear momentum, if the integral over time of the moment about a point is zero, angular momentum is conserved about that point. Also, because Eq. [1.6.16] is a vector relationship, angular momentum may be conserved in a particular direction while not conserved in another direction.

If the applied moment is impulsive, that is, its duration is infinitesimally short, the angular impulse-momentum relationship for zero initial conditions becomes

$$\hat{\mathbf{M}}_O(0) = \mathbf{H}_O(0^+) \qquad [1.6.17]$$

An impulsive angular moment can be generated by a very large torque applied over a very small interval, or by an impulsive force applied through a moment arm. For particle mechanics problems, the commonly used coordinates for angular momentum problems are cylindrical coordinates. The central force problem, as demonstrated in Example 1.12, is a classic case. Remember that when we discussed cylindrical coordinates, we emphasized the importance of attentively selecting the origin of the coordinate frame. The same argument is valid when selecting the point about which to calculate the angular momentum.

Example 1.11

One of the earliest industrial applications of the principle of conservation of angular momentum is the centrifugal governor. James Watt used the centrifugal governor to control flow in steam engines. Fig. 1.32 shows a typical governor. As the speed of the governor changes, the arms of the governor move. Because the only external force acting on the governor is along the shaft, the angular momentum about a point along the axis of the shaft is conserved.

The links are of length $L = 0.2$ m and assumed to be massless. The balls are 0.6 kg each. The governor is originally rotating at 50 rpm and the arms make an angle of $\theta = 30°$ with the vertical. What is the value of θ when the governor's speed becomes 75 rpm?

1.7 WORK AND ENERGY

In the previous section, we integrated Newton's second law over time to obtain momentum relationships. Here, we integrate it over the spatial variable. To this end, we consider a particle whose position is described by the vector $\mathbf{r}$ and which is acted upon by a force $\mathbf{F}$, as shown in Fig. 1.34. We define by dW the *incremental work* that the force does on the particle as the particle moves by an incremental distance $d\mathbf{r}$ as

$$dW = \mathbf{F} \cdot d\mathbf{r} = |\mathbf{F}||d\mathbf{r}|\cos\psi = m\mathbf{a} \cdot d\mathbf{r} \qquad [1.7.1]$$

where ψ is the angle between $\mathbf{F}$ and $d\mathbf{r}$. Recalling that $\mathbf{v} = d\mathbf{r}/dt$ and $\mathbf{a} = d\mathbf{v}/dt$, and multiplying and dividing the right side of the above equation by dt, we obtain

$$dW = \mathbf{F} \cdot d\mathbf{r} = m\frac{d\mathbf{v}}{dt} \cdot \frac{d\mathbf{r}}{dt}dt = m\frac{d\mathbf{v}}{dt} \cdot \mathbf{v}\,dt = m\mathbf{v} \cdot d\mathbf{v} = \frac{1}{2}m\,d(\mathbf{v} \cdot \mathbf{v}) \qquad [1.7.2]$$

We next define by T the *kinetic energy* of the particle as

$$T = \frac{1}{2}m\mathbf{v} \cdot \mathbf{v} = \frac{1}{2}mv^2 \qquad [1.7.3]$$

where $v = \sqrt{\mathbf{v} \cdot \mathbf{v}}$ is the speed. The incremental change in kinetic energy is $dT = m\,d(\mathbf{v} \cdot \mathbf{v})/2 = m\mathbf{v} \cdot d\mathbf{v}$. The kinetic energy is an absolute quantity and, hence, dT is a perfect differential. We then write Eq. [1.7.2] as

$$dW = dT \qquad [1.7.4]$$

The work done on the system by the force $\mathbf{F}$ is denoted by $W_{1 \to 2}$, and it is obtained by integrating the motion from point 1 to point 2, hence

$$W_{1 \to 2} = \int_{\mathbf{r}_1}^{\mathbf{r}_2} \mathbf{F} \cdot d\mathbf{r} = \int_{T_1}^{T_2} dT = T_2 - T_1 \qquad [1.7.5]$$

which gives the *work-energy theorem:*

$$T_1 + W_{1 \to 2} = T_2 \qquad [1.7.6]$$

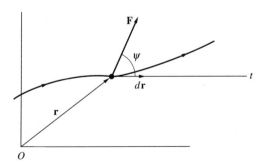

Figure 1.34

We distinguish between cases when dW is a perfect differential and when it is not. When dW is not a perfect differential, we cannot express it as the differential of a function, but merely as an infinitesimal element. The necessary condition for dW to be a perfect differential is that the force depends on the position vector alone, $\mathbf{F} = \mathbf{F}(\mathbf{r})$ (although there are cases when $\mathbf{F} = \mathbf{F}(\mathbf{r})$ and dW is *not* a perfect differential).

One can take advantage of cases when some of the forces acting on a body lead to perfect differentials. Such forces are known as *conservative forces*. Examples of conservative forces include spring forces, gravitational forces, and certain electromagnetic forces. The incremental work can be expressed as the derivative of a potential function as

$$dW = \mathbf{F}(\mathbf{r}) \cdot d\mathbf{r} = -dV(\mathbf{r}) \qquad [1.7.7]$$

where the potential function V is an explicit function of $\mathbf{r}$ only. Because dW is a perfect differential, its integral is independent of the path followed and it is dependent only on the end points of the integration. Over a closed path the value of the integral is zero, or

$$\oint \mathbf{F}(\mathbf{r}) \cdot d\mathbf{r} = 0 \qquad [1.7.8]$$

Moreover, Eq. [1.7.7] can be evaluated by integrating from a reference position $\mathbf{r}_R$ (or datum) to the location of the particle to yield

$$V(\mathbf{r}) = -\int_{\mathbf{r}_R}^{\mathbf{r}} \mathbf{F}(\mathbf{r}) \cdot d\mathbf{r} \qquad [1.7.9]$$

The potential function $V(\mathbf{r})$ is also known as the *potential energy*. Physically, potential energy is explained as the potential of a body to do work, or the stored energy.

Note that while kinetic energy is an absolute quantity, potential energy is relative: its value depends on the reference point about which it is measured. Because the interest is in increments of potential energy, selection of the reference point does not make any difference. One selects the datum to either simplify computations or to give the problem at hand a better physical interpretation.

The dimension of work, kinetic energy, and potential energy is force times distance, FL, or ML^2/T^2. In the SI system one commonly uses the units N·m. A *joule* (J) is defined as a unit of energy as $1\ \text{J} = 1\ \text{N} \cdot \text{m}$. In the U.S. Customary system the commonly used unit is ft·lb. Do not confuse the unit of energy with the unit of a moment, which has the same dimension as energy. In the U.S. system, the unit of a moment is usually denoted by lb·ft. Also, joule is never used as a unit of moment in the SI system.

We now consider three types of forces that lead to potential energy.

1.7.1 Gravitational Potential Energy

The gravitational attraction force between two bodies was given in Eq. [1.4.3] as $F = Gm_1m_2/r^2$, where G is the universal gravitational constant, m_1 and m_2 are the

In the Cartesian coordinate system, the del operator has the form

$$\nabla = \frac{\partial}{\partial x}\mathbf{i} + \frac{\partial}{\partial y}\mathbf{j} + \frac{\partial}{\partial z}\mathbf{k} \qquad [1.7.23]$$

so that, expressing the force vector as $\mathbf{F} = F_x\mathbf{i} + F_y\mathbf{j} + F_z\mathbf{k}$, one relates the components of the force to the potential energy as

$$F_x = -\frac{\partial V}{\partial x} \qquad F_y = -\frac{\partial V}{\partial y} \qquad F_z = -\frac{\partial V}{\partial z} \qquad [1.7.24]$$

When using cylindrical coordinates the del operator has the form

$$\nabla = \frac{\partial}{\partial R}\mathbf{e}_r + \frac{1}{R}\frac{\partial}{\partial \theta}\mathbf{e}_\theta + \frac{\partial}{\partial z}\mathbf{k} \qquad [1.7.25]$$

and the components of the force are related to the potential energy by

$$F_r = -\frac{\partial V}{\partial R} \qquad F_\theta = -\frac{1}{R}\frac{\partial V}{\partial \theta} \qquad F_z = -\frac{\partial V}{\partial z} \qquad [1.7.26]$$

Now consider that some of the forces acting on a particle are conservative and some are not. Forces that are not conservative are referred to as *nonconservative*. Express the force vector as $\mathbf{F} = \mathbf{F}_c + \mathbf{F}_{nc}$, the notation being obvious. It follows that the incremental work done by the force also can be divided into two parts; that is,

$$dW = dW_c + dW_{nc} = -dV + dW_{nc} \qquad [1.7.27]$$

where $dW_{nc} = \mathbf{F}_{nc} \cdot d\mathbf{r}$ is the work done by the nonconservative forces. We obtain the total work by integrating the overall work as

$$W_{1\to 2} = \int_{\mathbf{r}_1}^{\mathbf{r}_2} dW = \int_{\mathbf{r}_1}^{\mathbf{r}_2} (\mathbf{F}_c + \mathbf{F}_{nc}) \cdot d\mathbf{r} = V_1 - V_2 + \int_{\mathbf{r}_1}^{\mathbf{r}_2} \mathbf{F}_{nc} \cdot d\mathbf{r} \qquad [1.7.28]$$

where the work done by the nonconservative forces is

$$W_{nc_{1\to 2}} = \int_{\mathbf{r}_1}^{\mathbf{r}_2} \mathbf{F}_{nc} \cdot d\mathbf{r} \qquad [1.7.29]$$

Substituting Eqs. [1.7.28] and [1.7.29] into Eq. [1.7.6] yields

$$T_1 + V_1 + W_{nc_{1\to 2}} = T_2 + V_2 \qquad [1.7.30]$$

The *total energy* of the system is defined as $E = T + V$. One writes the energy balance as

$$E_1 + W_{nc_{1\to 2}} = E_2 \qquad [1.7.31]$$

The total energy of a dynamical system under the influence of nonconservative forces changes as the system moves. If all the forces acting on the body that do work are conservative, Eq. [1.7.31] indicates that its total energy remains the same. This is known as the *principle of conservation of energy*, and it explains the name *conservative force*. This principle can be written as

$$E_1 = E_2 \qquad [1.7.32]$$

It is sometimes more convenient to express work as an integral over time. We divide and multiply the integrand in the expression for work by dt, so that

$$W_{1\to 2} = \int_{\mathbf{r}_1}^{\mathbf{r}_2} \mathbf{F} \cdot d\mathbf{r} = \int_{t_1}^{t_2} \mathbf{F} \cdot \frac{d\mathbf{r}}{dt} dt = \int_{t_1}^{t_2} \mathbf{F} \cdot \mathbf{v}\, dt \qquad [1.7.33]$$

in which t_1 and t_2 denote initial and final times, respectively. The integrand on the right side of the above equation is defined as *power*, and it is denoted by P,

$$P = \mathbf{F} \cdot \mathbf{v} \qquad [1.7.34]$$

Power is basically the measure of how fast a force can do work. The dimension of power is work/time. In the SI system, the unit *watt* (W) is defined to represent power as $1\text{ W} = 1\text{ J/s} = 1\text{ N}\cdot\text{m/s}$. The U.S. system represents power using either ft·lb/min or horsepower (Hp), where $1\text{ Hp} = 550\text{ ft}\cdot\text{lb/sec}$. We have

$$\frac{dW}{dt} = P \qquad W_{1\to 2} = \int_{t_1}^{t_2} P\, dt \qquad [1.7.35]$$

Some texts define work as the integral of power over time. The above equation can also be viewed as the integral of the expression

$$\frac{dE}{dt} = \mathbf{F}_{nc} \cdot \mathbf{v} = \dot{W}_{nc} \qquad [1.7.36]$$

over time. It indicates that the rate of change of the total energy (rate of work done) is equal to the power of the nonconservative forces acting on the body.

A special category of forces comprises the forces that do no work. From Eq. [1.7.1], for a nonzero force to not do any work, either $d\mathbf{r} = \mathbf{0}$, or $\mathbf{F}$ is perpendicular to $d\mathbf{r}$. Included in this category are normal forces, other reaction forces perpendicular to the direction of motion (perpendicular to the tangential direction), and forces applied to points that have zero velocity. Obviously, if a force is applied to a stationary point, there is no work done. A very important force that does no work is the friction force in a rigid body rolling without slip.

Example 1.13

Consider the 30 m long incline of 15° shown in Fig. 1.41. Along the incline there are two paths, one a straight line and the other a semicircle. A mass of 0.5 kg is released at the bottom with an initial speed of 15 m/s, first along the straight path and next along the circular path. The coefficients of friction between the mass and the incline are $\mu_s = \mu_k = 0.1$. What will be the speed of the mass as it reaches the top of the incline following the straight path, and then, following the circular path?

Solution

This problem can be solved using the work-energy theorem. From the free-body diagram on Fig. 1.42, the forces acting on the mass are gravity, contributing to the potential energy; the normal force N, which does no work; and the friction force F, a nonconservative force. From the force balance along the incline we obtain

$$N = mg\cos 15° = 0.5(9.807)(0.9660) = 4.736\text{ N} \qquad [a]$$

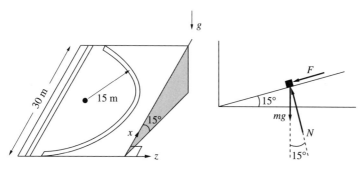

Figure 1.41   **Figure 1.42**

so that the friction force is

$$F = \mu_k N = 0.1\, mg \cos 15° = 0.1(0.5)(9.807)(0.9660) = 0.4736 \text{ N} \quad \text{[b]}$$

Note that the friction force is in the direction opposing the motion. We assume that friction acts only at the bottom of the paths and that the walls of the path are frictionless. Hence, the magnitude of the friction force is the same for both paths, even though its direction is different for the two paths. We take as a datum point for the potential energy the bottom of the incline. This way, $V_1 = 0$. The initial kinetic energy is

$$T_1 = \frac{1}{2}mv^2 = \frac{1}{2}(0.5)(15)^2 = 56.25 \text{ N} \cdot \text{m} \quad \text{[c]}$$

Let us first consider the motion of the mass along the straight path. Assuming that the mass reaches the top, the potential energy at the top of the incline is

$$V_2 = 30\,mg \sin 15° = 38.07 \text{ N} \cdot \text{m} \quad \text{[d]}$$

The work done by the nonconservative force can be expressed as

$$W_{1 \to 2} = -F(30) = -0.4736(30) = -14.21 \text{ N} \cdot \text{m} \quad \text{[e]}$$

Using the work-energy theorem, we obtain

$$T_2 = T_1 + V_1 + W_{1 \to 2} - V_2 = 56.25 + 0 - 14.21 - 38.07 = 3.970 \text{ N} \cdot \text{m} \quad \text{[f]}$$

and we calculate the velocity as

$$v_2 = \sqrt{\frac{2T_2}{m}} = \sqrt{\frac{2 \times 3.97}{0.5}} = 3.985 \text{ m/s} \quad \text{[g]}$$

Considering motion along the semicircular path, V_2 remains the same, but the nonconservative work changes, as the path is longer. To reach the top, the path length that is traversed is the circumference of a semicircle of radius 15 m, so that the nonconservative work becomes

$$W_{1 \to 2} = -F(\pi)(15) = -0.4736(3.142)(15) = -22.32 \text{ N} \cdot \text{m} \quad \text{[h]}$$

Comparing Eqs. [c], [d], and [h], we conclude that the mass does not reach the top of the incline, as the value of kinetic energy at that point would be negative. We then ask how far the mass travels along the incline before it comes to a stop. To find this value, it is preferable to work with the angle ϕ, as shown in Fig. 1.43. The mass has traveled a distance of $R\phi$, with R being the radius of 15 m. The height of the mass is $R(1 - \cos\phi)\sin 15°$. It follows that the potential energy and nonconservative work become

$$V_2 = mgR(1 - \cos\phi)\sin 15° = 0.5(9.807)(15)(0.2588) = 19.04 - 19.04\cos\phi$$
$$W_{1\to 2} = -FR\phi = -0.4736(15)\phi = -7.104\phi \qquad \text{[i]}$$

Noting that $T_2 = 0$ when the mass comes to a rest, we write the work-energy theorem as

$$T_1 + W_{1\to 2} - V_2 = 56.25 - 7.104\phi - 19.04 + 19.04\cos\phi$$
$$= 37.21 - 7.104\phi + 19.04\cos\phi = 0 \qquad \text{[j]}$$

The solution of Eq. [j] can be obtained numerically, and it can be shown to be

$$\phi = 2.756 \text{ rad} = 157.9° \qquad \text{[k]}$$

One can next ask whether the mass, after coming to a rest, slides back or not. To examine this issue, we need to perform a force balance along a line tangent to the path at $\phi = 157.9°$. Fig. 1.43 shows the configuration. We define the plane of the incline as the xz plane, with the y direction perpendicular to the incline. The normal force and the magnitude of friction force remain the same as before. However, we are now summing forces along the tangential direction, that is, the direction of impending motion. We write the gravity force as

$$\mathbf{F}_g = -mg\sin 15°\mathbf{i} - mg\cos 15°\mathbf{j} \qquad \text{[l]}$$

The mass will move if the component of the gravity force along the tangential direction is larger than the friction force. From Fig. 1.43 we express the unit vector $\mathbf{i}$ in terms of normal-tangential coordinates as

$$\mathbf{i} = -\sin\psi\mathbf{e}_t - \cos\psi\mathbf{e}_n \qquad \text{[m]}$$

in which $\psi = 180° - \phi = 22.1°$. The component of gravity along the path becomes

$$\mathbf{F}_g \cdot \mathbf{e}_t = mg\sin 15°\sin 22.1° = 0.4773 \text{ N} \qquad \text{[n]}$$

Comparing Eq. [n] with Eq. [b] we conclude that the mass moves back, as the component of gravity in the tangential direction is greater than the friction force.

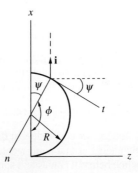

Figure 1.43

Example 1.14

Consider the system in Fig. 1.28. The pendulum is released from rest from the position $\theta = \pi/2$. Find a relation for the velocity of the base as a function of θ, for $F = 0$.

Solution

We treat the two masses as one system. The only forces external to the system are gravity and the normal force acting on the cart. We conclude that (1) there are no nonconservative forces that do work, so energy is conserved, and (2) there are no external forces in the horizontal direction, so the linear momentum in the horizontal direction is conserved.

Using as datum the horizontal position of the pendulum, we can obtain the total energy by considering the initial condition as

$$E\left(\theta = \frac{\pi}{2}\right) = T\left(\theta = \frac{\pi}{2}\right) + V\left(\theta = \frac{\pi}{2}\right) \quad \text{[a]}$$

where $T\left(\theta = \frac{\pi}{2}\right) = 0$, as there is no initial motion and $V = 0$. It follows that the total energy is $E = 0 =$ constant. We next take an arbitrary position of the pendulum and write the kinetic and potential energies

$$T = \frac{1}{2}M\dot{x}^2 + \frac{1}{2}mv_P^2, \qquad V = -mgL\cos\theta \quad \text{[b]}$$

where $v_P^2 = v_{Px}^2 + v_{Py}^2$ and

$$v_{Px} = \frac{d}{dt}(x + L\sin\theta) = \dot{x} + L\dot{\theta}\cos\theta, \qquad v_{Py} = \frac{d}{dt}(-L\cos\theta) = L\dot{\theta}\sin\theta \quad \text{[c]}$$

Substituting Eqs. [b] and [c] into Eq. [a] we obtain for the total energy

$$E = 0 = \frac{1}{2}(M + m)\dot{x}^2 + \frac{1}{2}mL^2\dot{\theta}^2 + mL\dot{x}\dot{\theta}\cos\theta - mgL\cos\theta \quad \text{[d]}$$

We next look at the conservation of linear momentum in the horizontal direction. The initial linear momentum is zero. The linear momentum for any value of θ is

$$p = M\dot{x} + mv_{Px} = M\dot{x} + m(\dot{x} + L\dot{\theta}\cos\theta) = 0 \quad \text{[e]}$$

which can be rewritten as

$$(M + m)\dot{x} = -mL\dot{\theta}\cos\theta \quad \text{[f]}$$

or

$$\dot{\theta} = -\frac{(M + m)\dot{x}}{mL\cos\theta} \quad \text{[g]}$$

We eliminate $\dot{\theta}$ from Eq. [d] by substituting Eq. [g] into Eq. [d] and simplifying, which yields

$$-\frac{1}{2}(M + m)\dot{x}^2 + \frac{1}{2m}\frac{(M + m)^2\dot{x}^2}{\cos^2\theta} - mgL\cos\theta = 0 \quad \text{[h]}$$

When $\cos\theta = 0$, the above relation cannot be used. However, from the initial conditions and Eq. [f], when $\cos\theta = 0$ then $\dot{x} = 0$ as well. Also, note that Eq. [h] is in terms of $\dot{x}^2$. The sign of the velocity of the cart can be determined from Eq. [f].

1.8 Equilibrium and Stability

Equilibrium is an essential and very useful concept in dynamics. For a system of particles or interconnected bodies, *static equilibrium* is defined as the state when all the particles and bodies comprising the system are at rest. Both the velocity $\mathbf{v}(t)$ and the acceleration $\mathbf{a}(t)$ of each body are zero at equilibrium.

When discussing the equilibrium of a system, two important questions come to mind: (1) How does one calculate the equilibrium position? (2) What happens to the system if it is displaced from equilibrium?

To calculate the equilibrium position one can use Newton's second law. Setting the acceleration to zero results in the equilibrium relation

$$\mathbf{F} = 0 \qquad [1.8.1]$$

where $\mathbf{F}$ is the sum of all the external forces. An alternate procedure that gives more of a qualitative insight is to look at the energy. Consider a particle and the case when the kinetic energy is only quadratic in terms of the velocity of the particle.[9] We take the differential form of the work-energy principle and remove from it all velocity- and time-dependent terms (including the kinetic energy as well as all time-dependent parts of the nonconservative force). From Eq. [1.7.28] we have

$$dW = (\mathbf{F}_c + \mathbf{F}_{nc}) \cdot d\mathbf{r} = -dV + dW_{nc} = 0 \qquad [1.8.2]$$

The right side of this equation is zero because at equilibrium the particle is not moving; hence, no work is done on it by the external forces. The equilibrium equation then becomes

$$dV = dW_{nc} \qquad [1.8.3]$$

When the system is conservative, the equilibrium condition is defined as

$$dV = 0 \qquad [1.8.4]$$

which can be interpreted as the potential energy having a stationary value at equilibrium. For a single degree of freedom system, Eq. [1.8.3] can be solved easily, and it is the preferred approach for systems consisting of interconnected components. In Chapter 4 we will derive the general form of Eq. [1.8.3] for multiple degree of freedom systems.

Now consider the second question: What happens to the particle when it is disturbed from equilibrium? Three scenarios are possible:

1. The particle returns to the equilibrium position and stays there. In this case, the equilibrium position is called *stable*, or *asymptotically stable*.

---

[9]The relevance of this requirement will be discussed in Chapter 5.

2. The particle hovers around equilibrium without staying at one point, but it does not return to or get away from the equilibrium position. The equilibrium position is referred to as *critically stable, merely stable,* or *neutrally stable.*

3. The particle moves away from the equilibrium point and it never returns there. The equilibrium position is called *unstable.*

There are several approaches that enable one to examine behavior in the vicinity of an equilibrium position. The simplest is to linearize the equations of motion about the equilibrium position. A theorem from stability theory states that if the linearized equations of motion in the neighborhood of equilibrium reveal *significant behavior* (continuous decaying or growing in amplitude), then the general motion (in the large) is governed by this significant behavior.

Consider a particle of mass m, and write the equation of motion as

$$\ddot{x}(t) = \frac{F(x(t), \dot{x}(t))}{m} = f(x(t), \dot{x}(t)) \qquad [1.8.5]$$

At equilibrium, all the velocity and acceleration terms are zero; therefore, the equilibrium condition is obtained by

$$f(x_e, 0) = 0 \qquad [1.8.6]$$

where x_e denotes the equilibrium position. Depending on the nature of function f, there can be more than one equilibrium position. We now define a local variable $\varepsilon = x - x_e$, and expand each term in the equation of motion about the equilibrium position $x = x_e$, $\dot{x}_e = 0$, $\ddot{x}_e = 0$. The term $\ddot{x}(t)$ simply becomes $\ddot{\varepsilon}(t)$. We expand the right side of Eq. [1.8.5] in a Taylor series and retain the linear terms:

$$f(x, \dot{x}) \approx f(x_e, 0) + \frac{\partial f}{\partial x}\bigg|_{x_e} \varepsilon + \frac{\partial f}{\partial \dot{x}}\bigg|_{x_e} \dot{\varepsilon} \qquad [1.8.7]$$

The first term on the right side of Eq. [1.8.7] is zero by definition of the equilibrium relation in Eq. [1.8.6]. Defining by γ_1 and γ_2 the partial derivatives evaluated at the equilibrium position,

$$\gamma_1 = -\frac{\partial f}{\partial x}\bigg|_{x_e} \qquad \gamma_2 = -\frac{\partial f}{\partial \dot{x}}\bigg|_{x_e} \qquad [1.8.8]$$

and using the local variable ε, we can express $f(x, \dot{x})$ as

$$f(x, \dot{x}) \approx -\gamma_1 \varepsilon(t) - \gamma_2 \dot{\varepsilon}(t) \qquad [1.8.9]$$

This results in the linearized equation of motion for small motions (local behavior) around equilibrium,

$$\ddot{\varepsilon}(t) + \gamma_2 \dot{\varepsilon}(t) + \gamma_1 \varepsilon(t) = 0 \qquad [1.8.10]$$

Note that all coefficients in the above equation are constant. Consider a time-dependent solution of $\varepsilon(t) = E e^{\lambda t}$, where λ denotes the time dependency and E is the amplitude. Substitution of this into Eq. [1.8.10], yields

$$(\lambda^2 + \gamma_2 \lambda + \gamma_1) E e^{\lambda t} = 0 \qquad [1.8.11]$$

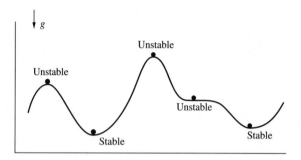

Figure 1.44

Because $Ee^{\lambda t}$ cannot be zero for a nontrivial solution, one must have

$$\lambda^2 + \gamma_2 \lambda + \gamma_1 = 0 \qquad [1.8.12]$$

This is known as the *characteristic equation,* whose roots are

$$\lambda = \frac{-\gamma_2 \pm \sqrt{\gamma_2^2 - 4\gamma_1}}{2} \qquad [1.8.13]$$

The behavior in the neighborhood of equilibrium is dictated by the roots of the characteristic equation. If the roots λ are both real negative or complex with negative real parts, $\varepsilon(t)$ decays exponentially and the equilibrium position is stable. If any one of the roots has a positive real part, then $\varepsilon(t)$ grows exponentially. The equilibrium position is unstable. If the λ roots are purely imaginary, then $\varepsilon(t)$ oscillates with constant amplitude. The linearized equations in this case do not represent *significant behavior* and they are not conclusive. One has to conduct additional analyses to determine the nature of the motion. Such analyses include qualitative approaches such as energy theorems, the Liapunov method, or quantitative analyses such as numerical integration.

More advanced concepts from stability theory are beyond the scope of this text. We will discuss one important stability theorem here, referred to as the *potential energy theorem.* The theorem states that

for conservative systems, if the potential energy has a minimum in the equilibrium position, then the equilibrium position is critically stable. Otherwise, it is unstable.

The theorem is illustrated conceptually in Fig. 1.44.

Example 1.15

Find the equilibrium position for the two links attached to a spring in Fig. 1.45. The spring is not stretched when the links are horizontal.

Solution

We have a conservative, single degree of freedom system. The displacement of every component (both links and the spring) can be expressed in terms of the angle θ. Denoting the spring deflection by x, we can write

$$x = 2L(1 - \cos\theta) \qquad [a]$$

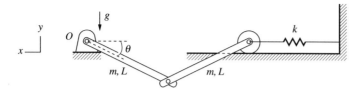

Figure 1.45

To solve for the equilibrium position by a vector approach, we need to draw a free-body diagram for each link and invoke the equilibrium relations. This is quite cumbersome, as we need to model and include in our calculations the reaction forces at the joints. It is preferable to seek a scalar solution by means of the potential energy function. The free-body diagram of the system is shown in Fig. 1.46. There are three external forces that do work acting on the system. Two of these are the force of gravity on the rods; we denote them by

$$\mathbf{F}_1 = \mathbf{F}_2 = -mg\mathbf{j} \qquad [\mathbf{c}]$$

and they act at the midpoints of the beams. The third force is the spring force and it is expressed in the form

$$\mathbf{F}_3 = -kx\mathbf{i} = -kL(1 - \cos\theta)\mathbf{i} \qquad [\mathbf{d}]$$

We write the potential energy as

$$V = \frac{1}{2}kx^2 - 2mg\left(\frac{L}{2}\sin\theta\right) = \frac{1}{2}k[2L(1-\cos\theta)]^2 - mgL\sin\theta \qquad [\mathbf{e}]$$

We obtain the equilibrium position by differentiating the potential energy and setting it to zero; thus,

$$0 = \frac{dV}{d\theta} = 4kL^2(1-\cos\theta)\sin\theta - mgL\cos\theta \qquad [\mathbf{f}]$$

Upon rearranging, this gives

$$\frac{\sin\theta}{\cos\theta}(1-\cos\theta) = \frac{mgL}{4kL^2} = \frac{mg}{4kL} \qquad [\mathbf{g}]$$

to solve for θ. Solution of Eq. [g] can be obtained numerically.

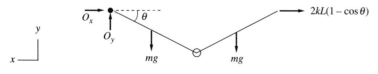

Figure 1.46

Example 1.16

A particle of mass $m = 1$ is acted upon by an excitation $F(x) = -x + x^2/4$. Find the equilibrium positions and determine their stability.

Solution

This problem is similar to a mass-spring system with a nonlinear spring constant. We can find the equilibrium position by setting $F(x) = 0$, with the result

$$f(x) = \frac{F(x)}{m} = -x\left(1 - \frac{x}{4}\right) = 0 \quad \Rightarrow \quad x_e = 0 \quad \text{or} \quad x_e = 4 \qquad \text{[a]}$$

We have two equilibrium positions. To find the linearized equations, we expand $f(x)$ about equilibrium. From Eq. [1.8.8], $\gamma_2 = 0$ for all positions, as there are no terms in f that are functions of $\dot{x}$. Differentiating Eq. [a] with respect to x we obtain $f'(x) = -1 + x/2$. We then evaluate γ_1 for each of the equilibrium positions to yield

$$\text{For } x_e = 0, \quad \gamma_1 = 1 \qquad \text{For } x_e = 4, \quad \gamma_1 = -1 \qquad \text{[b]}$$

so that the linearized equations of motion about equilibrium become

$$\text{For } x_e = 0, \quad \ddot{\varepsilon}(t) + \varepsilon(t) = 0 \qquad \text{[c]}$$

$$\text{For } x_e = 4, \quad \ddot{\varepsilon}(t) - \varepsilon(t) = 0 \qquad \text{[d]}$$

Equation [c] represents a simple sinusoid, so that the linearized equations do not exhibit significant behavior. One can further analyze this case using the potential energy theorem. The potential energy has a minimum at that point, as from Eqs. [1.7.24] and [1.8.8] the term γ_1 describes the second derivative of the potential energy. Hence, the equilibrium position is critically stable. By contrast, the equilibrium position $x_e = 4$ is unstable, as it has an increasing exponential solution. The characteristic equation for this case is

$$\lambda^2 - 1 = 0 \qquad \text{[e]}$$

which has the solution $\lambda = \pm 1$, indicating one real positive root. One can also physically explain this result: If $-f(x)$ is regarded as the spring force, once the variable x passes the equilibrium point $x_e = 4$, the spring force becomes negative, thus offering negative resistance. Of course, in an actual mass-spring system, once the point $x_e = 4$ is crossed there would no longer be a spring.

One can verify the result for $x_e = 4$ by invoking the energy theorem. The second derivative for the potential energy is negative, indicating a local maximum and an unstable equilibrium position.

1.9 FREE RESPONSE OF LINEAR SYSTEMS

In previous sections we obtained the equations of motion and equilibrium, and then linearized the equation of motion about equilibrium. We now consider the explicit response of a linear (or linearized), constant coefficient, single degree of freedom system.

In this section, we shall analyze the free response, that is, when there are no external forces. The linearized equation of motion of a system with one degree of freedom is given in Eq. [1.8.10]. Introducing the notation $\omega_n = \sqrt{\gamma_1}$ and $\zeta = \gamma_2/2\omega_n$, and the variable x to describe the motion in the vicinity of equilibrium, we rewrite Eq. [1.8.10] as

$$\ddot{x}(t) + 2\zeta\omega_n \dot{x}(t) + \omega_n^2 x(t) = 0 \qquad [1.9.1]$$

The quantities ω_n and ζ are known as *natural frequency* and *damping factor*, respectively. The natural frequency is a measure of the amount of stiffness versus mass in a system. It is related to the potential energy. The damping factor ζ is a measure of the energy dissipation in the system. Energy dissipation is caused by internal friction, as well as by dissipative forces, such as forces generated by a dashpot. When $\zeta = 0$ the motion is referred to as *undamped* and when $\zeta > 0$, the motion is called *damped*.

To observe the physical significance of these expressions, we begin by considering the mass-spring-dashpot system in Fig. 1.47. Consider a linear model for the spring and dashpot, so that the spring force is described by $F_s = -kx(t)$ and the dashpot force is in the form $F_d = -c\dot{x}(t)$, with c referred to as the *viscous damping coefficient*. This coefficient indicates the strength of the dashpot. This way of modeling dissipative force acting on a body is convenient, as it leads to equations of motion that are linear. The external force is denoted by F. Summing forces, we obtain

$$m\ddot{x}(t) = -c\dot{x}(t) - kx(t) + F \quad \rightarrow \quad m\ddot{x}(t) + c\dot{x}(t) + kx(t) = F \qquad [1.9.2]$$

In the above equation, $\omega_n = \sqrt{k/m}$ and $\zeta = c/2\sqrt{km}$. We analyze the natural frequency as m and k are varied. As the spring constant is increased, we have a stiffer system, and the natural frequency becomes larger. As the mass increases the system gets heavier, so the natural frequency decreases.

We first consider systems with no damping, $\zeta = 0$. The equation of motion reduces to

$$\ddot{x}(t) + \omega_n^2 x(t) = 0 \qquad [1.9.3]$$

To solve this equation, we introduce the general solution $x(t) = Xe^{\lambda t}$ and collect terms, so

$$(\lambda^2 + \omega_n^2)Xe^{\lambda t} = 0 \qquad [1.9.4]$$

For there to be a nontrivial solution, $Xe^{\lambda t}$ cannot be zero, from which we conclude that

$$\lambda^2 + \omega_n^2 = 0 \qquad [1.9.5]$$

which is recognized as the *characteristic equation*. The roots of the characteristic equation are $\lambda_{1,2} = \pm i\omega_n$, where $i^2 = -1$. Complex roots indicate an oscillatory

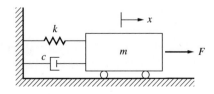

Figure 1.47

system. The response can be expressed as

$$x(t) = X_1 e^{i\omega_n t} + X_2 e^{-i\omega_n t} \qquad [1.9.6]$$

where X_1 and X_2 are complex constants of integration. Because $x(t)$ is real and $e^{i\omega_n t}$ and $e^{-i\omega_n t}$ are complex conjugates, X_1 and X_2 must be complex conjugates of each other as well. Introducing the real valued constants *amplitude A* and *phase angle ϕ* such that

$$X_1 = \frac{1}{2} A e^{-i\phi} \qquad X_2 = \frac{1}{2} A e^{i\phi} \qquad [1.9.7]$$

and substituting in Eq. [1.9.6], we obtain

$$x(t) = \frac{1}{2} A (e^{i(\omega_n t - \phi)} + e^{-i(\omega_n t - \phi)}) = A \cos(\omega_n t - \phi) \qquad [1.9.8]$$

The constants A and ϕ are determined from the initial conditions. Given initial conditions of $x(0) = x_0$ and $\dot{x}(0) = v_0$, it is easy to show that

$$A = \sqrt{x_0^2 + \left(\frac{v_0}{\omega_n}\right)^2} \qquad \phi = \tan^{-1}\left(\frac{v_0}{\omega_n x_0}\right) \qquad [1.9.9]$$

and that the response can also be expressed as

$$x(t) = x_0 \cos \omega_n t + \frac{v_0}{\omega_n} \sin \omega_n t \qquad [1.9.10]$$

The nature of the motion is harmonic, repeating itself in cycles. Using the relation

$$A \cos(\omega_n t - \phi) = A \cos(2\pi + \omega_n t - \phi) \qquad [1.9.11]$$

the amplitude of the motion attains the same value after a time of period T so that

$$T = \frac{2\pi}{\omega_n} \qquad [1.9.12]$$

The value of T is known as the *period of oscillation;* it is usually measured in seconds, and it describes the length of a cycle. The units of natural frequency are in radians/second. Another quantity commonly used to describe harmonic motion is the *frequency, f_n*, defined as

$$f_n = \frac{1}{T} = \frac{\omega_n}{2\pi} \qquad [1.9.13]$$

The unit of frequency is cycles per second (cps) or *hertz* (Hz). A frequency of 1 Hz = 2π rad/s. Fig. 1.48 shows a plot of $x(t)$. An interesting property of the natural frequency of a system is that it is a quantity dependent on the parameters of a system and not on the initial conditions.

We next consider the case when the damping is not zero. Introducing $x(t) = X e^{\lambda t}$ to Eq. [1.9.1] and using the same line of thought, we obtain the characteristic equation

$$\lambda^2 + 2\zeta \omega_n \lambda + \omega_n^2 = 0 \qquad [1.9.14]$$

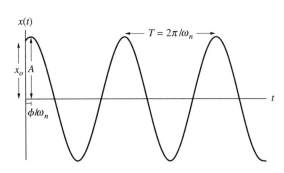

Figure 1.48 Undamped response

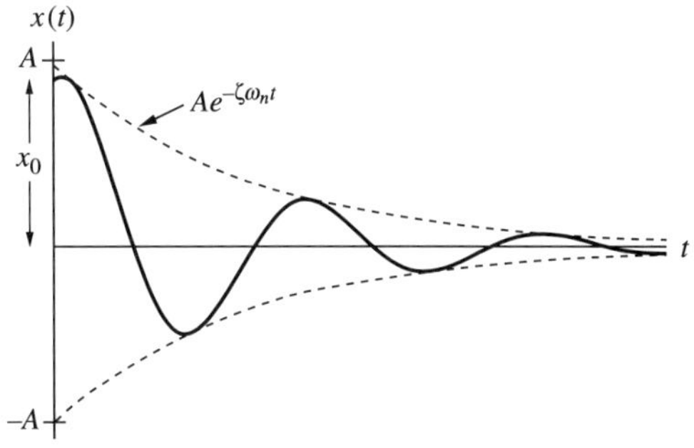

Figure 1.49 Damped response

whose roots are

$$\lambda_{1,2} = \omega_n(-\zeta \pm \sqrt{\zeta^2 - 1}) \quad [1.9.15]$$

The nature of the motion depends on the values of the damping coefficient. We identify the following five cases:

1. When $\zeta > 1$, the roots are real, negative, and distinct. The motion is in the form of a decaying exponential and it is not periodic. Such a system is called *overdamped*. The response has the general form

$$x(t) = A_1 e^{\lambda_1 t} + A_2 e^{\lambda_2 t} \quad [1.9.16]$$

with A_1 and A_2 being real quantities whose values depend on the initial conditions. The above equation can also be expressed in terms of hyperbolic sines and cosines.

2. When $0 < \zeta < 1$, the roots are complex conjugates with negative real parts in the form

$$\lambda_{1,2} = \omega_n(-\zeta \pm i\sqrt{1 - \zeta^2}) = -\zeta\omega_n \pm i\omega_d \quad [1.9.17]$$

where $\omega_d = \omega_n\sqrt{1 - \zeta^2}$ is the *damped natural frequency*. This quantity basically represents the frequency of oscillation of the damped system.

To obtain the response, we note the identity $e^{(a+b)} = e^a e^b$, so that

$$\begin{aligned} x(t) &= X_1 e^{(-\zeta\omega_n + i\omega_d)t} + X_2 e^{(-\zeta\omega_n - i\omega_d)t} \\ &= e^{-\zeta\omega_n t}(X_1 e^{i\omega_d t} + X_2 e^{-i\omega_d t}) \end{aligned} \quad [1.9.18]$$

Following the approach used for the undamped system, one can express this equation as

$$x(t) = A e^{-\zeta\omega_n t} \cos(\omega_d t - \phi) \quad [1.9.19]$$

The motion is in the form of a decaying sinusoidal, with an exponential decay envelope, as shown in Fig. 1.49. Such a system is known as *underdamped*. One can show that, in terms of the initial displacement x_0 and initial velocity v_0, the response has the form

$$x(t) = e^{-\zeta\omega_n t}\left(x_0 \cos \omega_d t + \frac{v_0 + \zeta\omega_n x_0}{\omega_d} \sin \omega_d t\right) \quad \text{[1.9.20]}$$

Note that, similar to the natural frequency, the damping factor is also not dependent on the initial conditions, but it is a function of the system parameters.

3. The case $\zeta = 1$ represents the border between underdamped and overdamped systems. It is called *critically damped*. The roots of the characteristic equation are real, negative, and equal to each other, $\lambda_1 = \lambda_2 = -\omega_n$. The motion is in the form of a decaying exponential. The response has the form

$$x(t) = (A_1 + A_2 t)e^{-\omega_n t} \quad \text{[1.9.21]}$$

in which both A_1 and A_2 are real.

4. The case $\zeta = 0$ represents the undamped case that we saw above. Here, $\omega_d = \omega_n$, and Eq. [1.9.20] reduces to the undamped response of Eq. [1.9.10].

5. When $\zeta < 0$, the roots of the characteristic equation 0ave positive real parts, and they may or may not be complex. A positive real root implies an exponentially growing solution and instability, as we saw in the previous section. Such a system is sometimes called *negatively damped*.

Example 1.17

Consider Fig. 1.47. The block weighs 20 lb and the spring is of constant $k = 5$ lb/in. Damping is negligible and $F = 0$. The block is released from rest with an initial displacement of $x_0 = 3$ in. Find its position and velocity after four seconds.

Solution

The equation of motion is

$$m\ddot{x}(t) + kx(t) = 0 \quad \text{[a]}$$

with $m = 20/g = 20/32.2 = 0.6211$ slugs $= 0.6211$ lb·sec²/ft $= 0.05176$ lb·sec²/in. The natural frequency is

$$\omega_n = \sqrt{\frac{k}{m}} = \sqrt{\frac{5}{0.05176}} = 9.829 \text{ rad/sec} \quad \text{[b]}$$

From Eq. [1.9.10] we can write the response to an initial displacement as

$$x(t) = x_0 \cos \omega_n t \qquad \dot{x}(t) = -\omega_n x_0 \sin \omega_n t \quad \text{[c]}$$

so that at $t = 4$ sec we have

$$x(4) = 3\cos(9.829 \times 4) = -0.1382 \text{ in} \qquad \dot{x}(4) = -3(9.829)\sin(9.829 \times 4) = -29.46 \text{ in/sec} \quad \text{[d]}$$

1.10 Response to Harmonic Excitation

In this section we consider single degree of freedom systems subjected to harmonic excitation. Rather than studying the general case of response to arbitrary excitation first and then considering the special case of harmonic excitation, we prefer to study harmonic excitation independently. The reason for this is twofold: First, excitations of a harmonic nature play a very important role in engineering. For example, most machines have rotating parts, which generate harmonic forces on their supports or other bodies they are in contact with. Second, one can study response to harmonic excitation using the steady state motion approach, where one is primarily concerned with the motion amplitudes and phase angles, rather than with initial conditions.

Consider a dynamical system, whose linearized equations of motion have the form

$$\ddot{x}(t) + 2\zeta\omega_n \dot{x}(t) + \omega_n^2 x(t) = f(t) \qquad [1.10.1]$$

where $f(t)$ is the external excitation. For harmonic excitation with a single driving frequency ω, $f(t)$ can be written in the general compact complex form

$$f(t) = A\omega_n^2 e^{i\omega t} \qquad [1.10.2]$$

One of the advantages in writing the excitation in this form is that the excitation amplitude A has the same units as $x(t)$. For example, for translational mechanical systems $f(t)$ has the units of force/mass and A has units of displacement. Also, because $e^{i\omega t} = \cos\omega t + i\sin\omega t$, one can solve for the response for a cosine or a sine excitation simultaneously. The response to a complex excitation will also be complex, so that the real part of the solution will be the response to a cosine excitation, and the imaginary part, response to a sine excitation.

We are interested in the steady state motion. At steady state, all effects due to transient sources, such as initial conditions and impulsive forces, have died out, and the response is only due to the external harmonic excitation. Hence, we assume that the response is of the same nature as the excitation and consider a steady state solution in the form $x(t) = X(i\omega)e^{i\omega t}$. Introducing this solution and Eq. [1.10.2] into Eq. [1.10.1] and collecting the coefficients, we obtain

$$(-\omega^2 + 2i\omega\zeta\omega_n + \omega_n^2)X(i\omega)e^{i\omega t} = A\omega_n^2 e^{i\omega t} \qquad [1.10.3]$$

which can be rearranged and solved for $X(i\omega)$ as

$$X(i\omega) = \frac{A}{1 - (\omega/\omega_n)^2 + 2i\zeta\omega/\omega_n} \qquad [1.10.4]$$

The amplitude $X(i\omega)$ can be characterized by the ratio of the excitation amplitude A multiplied by a nondimensional ratio. This ratio dictates the amplitude of the response, so we define it as

$$G(i\omega) = \frac{1}{1 - (\omega/\omega_n)^2 + 2i\zeta\omega/\omega_n} \qquad [1.10.5]$$

where $G(i\omega)$ is called the *frequency response*. Recalling that a complex number $a + ib$ can be expressed as $Me^{i\psi}$, where $M = \sqrt{a^2 + b^2}$ and $\psi = \tan^{-1}(b/a)$, we can express the frequency response as

$$G(i\omega) = |G(i\omega)|e^{-i\psi} \qquad [1.10.6]$$

where

$$|G(i\omega)| = \frac{1}{([1 - (\omega/\omega_n)^2]^2 + [2\zeta\omega/\omega_n]^2)^{1/2}} \qquad \psi = \tan^{-1}\frac{2\zeta\omega/\omega_n}{1 - (\omega/\omega_n)^2}$$

[1.10.7]

The quantity $|G(i\omega)|$ is referred to as the *magnification factor*, and the angle ψ is the *phase angle*, not to be confused with the phase angle ϕ defined in the previous section. The steady state response can then be written as

$$x_s(t) = A|G(i\omega)|e^{i(\omega t - \psi)}$$

[1.10.8]

The value of the magnification factor is affected by the amount of damping, as well as by the ratio of the driving frequency to the natural frequency. A plot of $|G(i\omega)|$ versus ω/ω_n is given in Fig. 1.50 for various values of the damping factor. The magnification factor becomes smaller with increasing amounts of damping, as one would expect. For a given amount of damping, the magnification factor becomes larger as ω/ω_n approaches unity. For an undamped system when $\omega/\omega_n = 1$ the magnification factor becomes infinity. This phenomenon, called *resonance*, manifests itself as very high amplitude vibrations when the driving frequency becomes very close to the natural frequency.

Dynamical systems subjected to harmonic excitation are designed such that resonant frequencies are avoided, or encountered as briefly as possible. Furthermore,

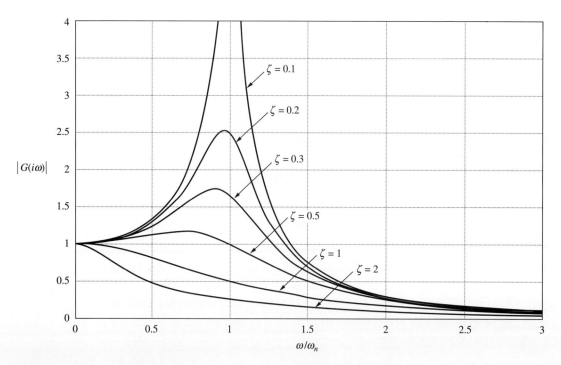

Figure 1.50 Magnification factor

such systems have a certain amount of damping to reduce vibration amplitudes. A poorly designed system can experience dangerous levels of vibration, which can cause physical damage, discomfort to occupants, and reduction in precision.

A plot of the phase angle ψ versus ω/ω_n is given in Fig. 1.51. The phase angle is always less than 90° when $\omega/\omega_n < 1$ and greater than 90° when $\omega/\omega_n > 1$. The resonance point becomes the defining factor for the phase angle, as all curves go through the point $\psi = \pi$ and $\omega/\omega_n = 1$.

When $\omega/\omega_n < 1$, the response has the same sign as the excitation. In this case the response is referred to as being *in phase* with the excitation. By contrast, when $\omega/\omega_n > 1$, the response has the opposite sign as the excitation, indicating that it is *out of phase* with the excitation.

An interesting special case to analyze is that of undamped vibration. Here, setting $\zeta = 0$ we can write the steady state response as

$$x_s(t) = \frac{A}{1 - (\omega/\omega_n)^2} e^{i\omega t} \qquad [1.10.9]$$

When $\omega/\omega_n = 1$ and $f(t) = A\omega_n^2 \cos \omega_n t$, the response can be shown to be

$$x(t) = \frac{A}{2}\omega_n t \sin \omega_n t \qquad [1.10.10]$$

indicating that the amplitude of the response increases with time in a linear amplification envelope. Fig. 1.52 shows the response. An infinite amplitude is never

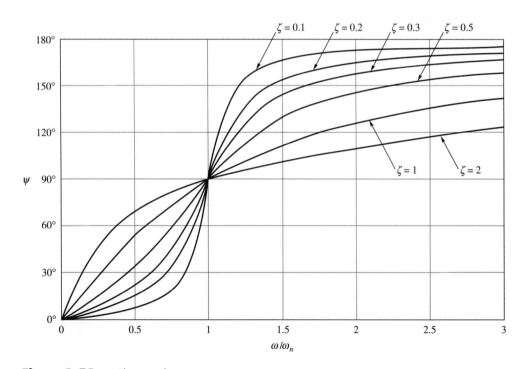

Figure 1.51   Phase angle

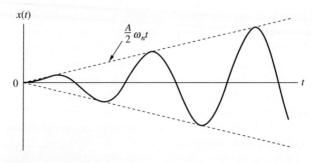

Figure 1.52 Resonance

reached in reality. As the amplitude increases, either the mathematical analysis becomes invalid and nonlinear and other effects begin to dominate the motion, or the amplitudes become high enough to damage the system. Also, all physical systems have some energy dissipation mechanism, in the form of viscous damping or in some other form, which helps reduce vibration amplitudes.

Example 1.18

A very interesting application of response to harmonic excitation is the motion of a vehicle over a wavy terrain. Consider such a vehicle, modeled as a single degree of freedom system, traveling with constant speed v over a road that has a sinusoidal shape, as shown in Fig. 1.53. Derive the magnification factor.

Solution

The road surface can be modeled as

$$y(x) = Y \sin\left(\frac{2\pi x}{L}\right) \qquad [a]$$

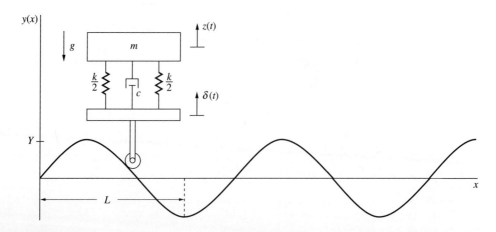

Figure 1.53

where Y and L denote the amplitude and length of the wave. Because the vehicle is traveling at constant speed v, we can express its location on the x axis as $x(t) = vt$. Substituting this into the above equation, we obtain

$$y(t) = Y \sin\left(\frac{2\pi vt}{L}\right) = Y \sin \omega t \qquad \text{[b]}$$

where $\omega = 2\pi v/L$. The problem now is reduced to one of a mass-spring-damper system whose base is undergoing a prescribed harmonic motion. The speed of the vehicle dictates the frequency of the excitation.

Consider the free-body diagram of the vehicle. The deflection of the spring is $\delta(t) = z(t) - y(t)$. Writing Newton's second law, we obtain

$$m\ddot{z}(t) = -k\delta(t) - c\dot{\delta}(t) - mg \qquad \text{[c]}$$

which leads to the equation of motion as

$$m\ddot{z}(t) + c\dot{z}(t) + kz(t) = -mg + c\dot{y}(t) + ky(t) \qquad \text{[d]}$$

We divide this equation by m and make use of the expressions for natural frequency and the damping factor. Further, we can put $y(t)$ and $\dot{y}(t)$ into exponential form as $y(t) = Y \operatorname{Im}(e^{i\omega t})$, $\dot{y}(t) = Y \operatorname{Im}(i\omega e^{i\omega t})$, and write the equation of motion as

$$\ddot{z}(t) + 2\zeta\omega_n \dot{z}(t) + \omega_n^2 z(t) = -g + Y\omega_n^2(i2\zeta\omega/\omega_n + 1)e^{i\omega t} \qquad \text{[e]}$$

The first term on the right side has the effect of lowering the mass by a height of $g/\omega_n^2 = mg/k$. It is the static deformation of the mass due to its own weight. Measuring $z(t)$ from static equilibrium, we have two terms for the excitation, so that the response also consists of two terms

$$z(t) = YH(i\omega)e^{i\omega t} \qquad \text{[f]}$$

where

$$H(i\omega) = G(i\omega)(i2\zeta\omega/\omega_n + 1) \qquad \text{[g]}$$

is the frequency response of this system. The magnification factor (also known as *transmissibility* for this problem) is the magnitude of $H(i\omega)$. Recalling from complex algebra that $|(a + ib)(c + id)| = |(a + ib)| \times |(c + id)|$, we have

$$|H(i\omega)| = \sqrt{1 + (2\zeta\omega/\omega_n)^2}\, |G(i\omega)| \qquad \text{[h]}$$

Fig. 1.54 plots the transmissibility for various values of the damping factor. As expected, as the driving frequency gets close to the natural frequency, the transmissibility becomes very large. Peak values occur at frequencies slightly lower than the natural frequencies. It is interesting to note that as ω/ω_n gets larger—that is, as the vehicle goes faster—the transmissibility becomes less than 1 and approaches zero. Also, this ratio goes to zero faster when the damping factor becomes smaller.

Anyone can observe these results when driving a vehicle. As a vehicle goes faster, the effects of uneven terrain and of potholes are felt less by the passengers. (It should be cautioned that when a vehicle goes faster over potholes, the damping forces exerted by the dashpots also become larger; hence, such driving wears out the shock absorbers.)

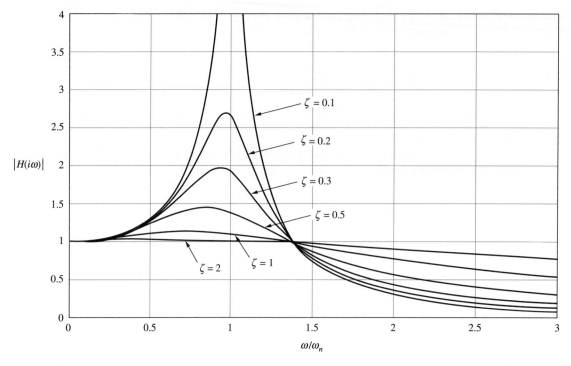

Figure 1.54   Transmissability

1.11 FORCED RESPONSE OF LINEAR SYSTEMS

We will obtain the general response of a linear system by making use of its response to impulsive loading. Consider the system in Eq. [1.10.1] subjected to an impulsive excitation $\hat{f} = \hat{F}/m$ at time $t = 0$

$$\ddot{x}(t) + 2\zeta\omega_n\dot{x}(t) + \omega_n^2 x(t) = \hat{f}\underline{\delta}(t - 0) \qquad [1.11.1]$$

in which $\hat{f}(t)$ is the impulsive force per unit mass. In Section 1.6 we learned that an impulsive force applied to a system results in a sudden change in velocity. For a system originally at rest, from Eq. [1.6.12], the position and velocity immediately after the impulse are

$$x(0^+) = x(0) \qquad v(0^+) = \hat{f} \qquad [1.11.2]$$

One can view the response of the system after the application of the impulse as the free response with the position and velocity right after the impulse as initial conditions. From Eq. [1.9.20], for zero initial displacement, the response has the form

$$x(t) = \frac{\hat{f}}{\omega_d} e^{-\zeta\omega_n t} \sin \omega_d t \qquad [1.11.3]$$

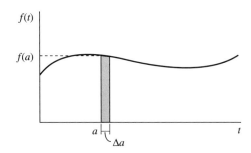

Figure 1.55

We generalize this result to the case when the amplitude of the impulsive force is unity by calling it *impulse response g(t)* as

$$g(t) = \frac{1}{\omega_d} e^{-\zeta\omega_n t} \sin \omega_d t \qquad [1.11.4]$$

We next represent an arbitrary force as a summation of impulses. Consider a force $f(t)$, as shown in Fig. 1.55. At any point $t = a$, the impulse due to the force applied over a time period of Δa has the form $\hat{f}(t - a) = f(a)\Delta a$. Considering that Δa is very small, we can treat $f(a)\Delta a$ as impulsive and, from the above equation, give the response to an impulse applied at $t = a$ the form

$$x_a(t) = \hat{f}(t - a)g(t - a) = f(a)g(t - a)\Delta a u(t - a) \qquad [1.11.5]$$

where the subscript in $x_a(t)$ signifies that it is the part of the response due to the impulse at $t = a$, and $u(t - a)$ is the unit step function, defined as

$$u(t - a) = 1 \text{ when } t \geq a$$
$$u(t - a) = 0 \text{ when } t < a$$

To find the response to the entire excitation, we sum Eq. [1.11.5] over all the impulses. As Δa becomes smaller, the summation is replaced by integration and we get

$$x(t) = \int_0^t f(a)g(t - a)\,da \qquad [1.11.6]$$

Equation [1.11.6] is known as the *convolution integral*. By a proper change of variables (say, $t - a = \tau, a = t - \tau, da = -d\tau$), one can show that

$$x(t) = \int_0^t f(t - \tau)g(\tau)\,d\tau = \int_0^t f(\tau)g(t - \tau)\,d\tau \qquad [1.11.7]$$

When solving for the response, one can use either form of Eq. [1.11.7]. One bases the selection of which form to use on the ease with which the convolution integral can be evaluated. For example, if the excitation is constant, use of the form with $f(t - \tau)$ is preferable.

Note that the response obtained from the convolution integral is the response of the system to zero initial conditions. To find the total response, we superpose this with the response to initial conditions only, given in Eq. [1.9.20], with the result

$$x(t) = e^{-\zeta\omega_n t}\left(x_0 \cos\omega_d t + \frac{v_0 + \zeta\omega_n x_0}{\omega_d}\sin\omega_d t\right) + \frac{1}{\omega_d}\int_0^t f(t-\tau)e^{-\zeta\omega_n \tau}\sin\omega_d \tau\, d\tau$$
[1.11.8]

For an undamped system, the general response reduces to

$$x(t) = x_0 \cos\omega_n t + \frac{v_0}{\omega_n}\sin\omega_n t + \frac{1}{\omega_n}\int_0^t f(t-\tau)\sin\omega_n \tau\, d\tau \quad [1.11.9]$$

As stated earlier, there are other ways of solving Eq. [1.10.1]. One way is the homogeneous plus particular solution approach (obtain the homogeneous solution, the particular solution, add them up and then impose the initial conditions), and the other is the Laplace transform solution. Actually, the Laplace transform solution is equivalent to the convolution integral. Note that with this approach we obtain the solution as the sum of two quantities, the response of an otherwise free system with the given initial conditions plus the response to the excitation for zero initial conditions.

Example 1.19

An interesting special case is the response of systems subjected to Coulomb type frictional forces, such as the mass-spring system in Fig. 1.24 in Example 1.9. While the friction force is nonlinear, the equation of motion can be split into two linear equations, depending on the value of the velocity. Ignoring the external force F, Eqs. [d] and [e] in Example 1.9 can be written as

$$\text{when } \dot{x} > 0, \quad \ddot{x} + \omega_n^2 x = -\mu g \quad \text{[a]}$$

$$\text{when } \dot{x} < 0, \quad \ddot{x} + \omega_n^2 x = \mu g \quad \text{[b]}$$

Consider as initial conditions an initial displacement x_0 and zero initial velocity. We first need to determine whether motion will take place or not. This depends on whether the spring force is larger than the friction force. At the point when the velocity is zero, there will be subsequent motion if

$$|kx| > \mu mg \quad \text{or} \quad |\omega_n^2 x| > \mu g \quad \text{[c]}$$

Assuming that at the onset of motion, the spring force is larger than the friction force, the velocity first encountered will be negative. Therefore, we use Eq. [b] as the equation of motion. From Eq. [1.11.9] we obtain

$$x(t) = x_0 \cos\omega_n t + \frac{1}{\omega_n}\int_0^t \mu g \sin\omega_n \tau\, d\tau$$

$$= x_0 \cos\omega_n t + \frac{1}{\omega_n}\mu g \frac{1}{\omega_n}(1-\cos\omega_n t) = \frac{\mu g}{\omega_n^2} + \left(x_0 - \frac{\mu g}{\omega_n^2}\right)\cos\omega_n t \quad \text{[d]}$$

The velocity becomes zero at a half cycle, $t = T/2 = \pi/\omega_n$. The displacement at this point is

$$x\left(\frac{\pi}{\omega_n}\right) = -x_0 + \frac{2\mu g}{\omega_n^2} \quad \text{[e]}$$

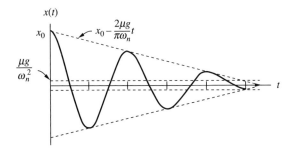

Figure 1.56 Response to Coulomb friction

We now switch to Eq. [a] as the equation of motion, as long as Eq. [c] holds. The response in the second half cycle can be shown to be

$$x(t) = -\frac{\mu g}{\omega_n^2} + \left(x_0 - 3\frac{\mu g}{\omega_n^2}\right)\cos\omega_n t \qquad \frac{\pi}{\omega_n} < t \leq \frac{2\pi}{\omega_n} \qquad [\mathbf{f}]$$

When $t = T = 2\pi/\omega_n$, the velocity is again zero, at which point the displacement has the value

$$x\left(\frac{2\pi}{\omega_n}\right) = x_0 - \frac{4\mu g}{\omega_n^2} \qquad [\mathbf{g}]$$

It turns out that at every half cycle the amplitude of vibration is reduced by $2\mu g/\omega_n^2$. Hence, the response subject to Coulomb friction is in the form of a decaying curve, with the decay envelope in the form of two straight lines of slope $2\mu g/\pi\omega_n$. A typical response curve is shown in Fig. 1.56. The motion stops when the velocity is zero and the spring force is less than the friction force.

The problem with dealing with Coulomb friction is that the solution has to be obtained for every half cycle individually, making the analysis cumbersome. Often, engineers replace the Coulomb friction model with an equivalent viscous damping model and use linear equations. While this is a gross simplification, if one considers the uncertainties in determining the friction force and the damping factor, the simplification does not look as bad. It is difficult to model the damping properties of a system accurately.

1.12 FIRST INTEGRALS

In preceding sections we learned about quantities that were derived by integrating Newton's second law. For the linear and angular momenta, Newton's second law was integrated over time, and for energy, the integration was carried over the displacement variable. We also studied the circumstances under which energy and momentum were conserved.

When applicable, the principles of conservation of momentum and energy give qualitative information about the motion without solving for the response explicitly. This is a desirable feature, especially for systems with complicated equations of motion and when one needs to know the nature of the motion but not the explicit response. When they are conserved we refer to energy, linear momentum, and angular momentum as *first integrals,* or *integrals of the motion.* The terminology is due to the equations of motion being integrated once to arrive at the first integrals. First integrals involve expressions in which the highest order derivative is one less than the highest order derivative in the equations of motion.

Energy and momentum are not the only first integrals that can be found for a dynamical system. There exist other first integrals such as the Jacobi integral and generalized momenta associated with ignorable coordinates. In most cases, integrals of the motion have physical explanations, even though one can generate a first integral that may not have an obvious physical interpretation. First integrals also come in handy when integrating the equations of motion numerically, as they can be used to check the accuracy of the numerical integration. One of the first steps when analyzing a dynamical system should be to search for the existence of first integrals in order to ascertain the qualitative description of the system behavior.

First integrals are also used for the geometric analysis of motion. A typical example of this is phase plane analysis, where one plots the dependent variable versus its time derivative. We will not go into this subject in detail but will outline some of the uses of the phase portrait when dealing with conservative systems.

Consider, for example, a particle of mass m moving in one direction and being acted upon by a force $F(x)$. From Newton's second law, we have $m\ddot{x}(t) = F(x)$. Energy is conserved. One can arrive at this conclusion by integrating the equation of motion over x, which yields

$$\dot{x}^2 - G(x) = C = \text{constant} \qquad [1.12.1]$$

where $dG(x)/dx = 2F(x)/m$. C is basically a measure of the total energy, and it depends on the initial conditions. We can rewrite Eq. [1.12.1] to show the relationship between x and $\dot{x}$ as

$$\dot{x} = \pm\sqrt{C - G(x)} \qquad [1.12.2]$$

From Eq. [1.12.2], the *phase portrait* of $\dot{x}$ versus x has a number of properties: It is symmetric about the x axis. Also, the phase portrait is continuous with continuous derivatives, so that there are no sharp corners in the phase portrait. For conservative systems, each curve of the phase portrait corresponds to a specific energy level. One can plot the phase portraits for different energy levels and analyze the motion. It can be shown that phase portrait curves below (or above) the x axis corresponding to different energy levels never cross each other.[10] If the phase portraits did cross each other, one would not know the energy level associated with the point where the curves met.

The phase portrait is also a useful tool for systems that do not admit integrals of the motion. In such systems, the phase portrait will not be symmetric about the x axis. Phase plane analysis is one of the cornerstones of nonlinear stability theory; it is widely used in mechanical and other systems (electrical, chemical, etc.) as well as in control theory.

---

[10]Except a curve that crosses the x axis, which ends up intersecting its symmetric counterpart. Such a curve defines critical points and regions on the phase space, separating stable and unstable regions of the phase space. It is called a *separatrix.*

Example 1.20

Plot $\dot{x}$ versus x, the phase portrait for a particle of unity mass subjected to the forces (1) $F(x) = -x + x^3/9$, and (2) $F(x) = -x - x^3/9$.

Solution

From the analysis in the previous section, we can write Eq. [1.12.1] for case (1) as

$$\dot{x}^2 + x^2 - \frac{x^4}{18} = E_1 = \text{constant} \quad \text{[a]}$$

The energy constant, of course, can be evaluated using the initial conditions $x(0)$ and $\dot{x}(0)$ or the values of displacement and velocity at any other time when they are known. For case (2) the integral of the motion becomes

$$\dot{x}^2 + x^2 + \frac{x^4}{18} = E_2 = \text{constant} \quad \text{[b]}$$

Before we plot the phase trajectories, we investigate the equilibrium positions and their stability. For case (1), the equilibrium points can be shown to be $x = 0$ and $x = \pm 3$. A stability analysis and plot of the potential energy indicates that the equilibrium point $x = 0$ is critically stable and the two equilibrium points $x = \pm 3$ are unstable. To see this, we note that $f(x) = -x + x^3/9$, so that its derivative is

$$f'(x) = -1 + \frac{x^2}{3} \quad \text{[c]}$$

At the equilibrium position $x = 0$, $\gamma_1 = 1$, and for $x = \pm 3$, $\gamma_1 = -2$. For case (2), there is only one equilibrium point, $x_e = 0$, and it is stable.

The phase portraits are shown in Figs. 1.57 and 1.58, where velocity is plotted versus displacement for various values of the energy constants E_1 and E_2. It should be noted from Section 1.7 that the two cases considered in this example are representative of softening and hardening springs, respectively. The quantity $F(x)$ can be treated as the spring force acting on the body. In case (1) the spring force is zero or negative in the range $-3 \leq x \leq 3$, and positive when $|x| > 3$. This implies that as the deflection of the spring gets bigger, the spring force gets smaller. When $x > 3$ (or $x < -3$), the spring force becomes less than zero, leading to an unstable equilibrium position.

By contrast, case (2) represents a spring force that keeps growing as x gets larger. Consequently, there is no equilibrium point other than $x = 0$. As x moves further away from equilibrium, the force pulling it toward equilibrium becomes larger.

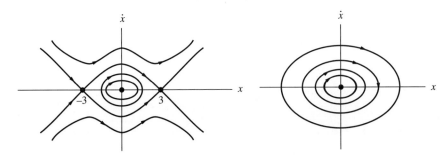

Figure 1.57 Case (1)   **Figure 1.58** Case (2)

For case (1) when disturbed from unstable equilibrium (in this case for $|x| < 3$), the particle leaves the vicinity of the unstable equilibrium point and moves toward the stable equilibrium point. This is typical of systems that have stable and unstable equilibrium positions.

Example 1.21

A particle of mass m, shown in Fig. 1.59, is acted upon by a force expressed in polar coordinates as $\mathbf{F} = k/r^2 \mathbf{e}_r$, where r is the distance from the origin to the particle. Find the integrals of the motion.

Solution

This problem is similar to the central force problem discussed in Example 1.12. The angular momentum about the origin O is conserved, so that it is an integral of the motion. To show that this indeed is the case, we write Newton's second law in polar coordinates as

$$\mathbf{F} = m\mathbf{a} \;\rightarrow\; \frac{k}{r^2}\mathbf{e}_r = m(\ddot{r} - r\dot{\theta}^2)\mathbf{e}_r + m(r\ddot{\theta} + 2\dot{r}\dot{\theta})\mathbf{e}_\theta \quad \text{[a]}$$

We resolve Eq. [a] into its radial and transverse components as

$$m(\ddot{r} - r\dot{\theta}^2) = \frac{k}{r^2} \quad \text{[b]}$$

$$m(r\ddot{\theta} + 2\dot{r}\dot{\theta}) = 0 \quad \text{[c]}$$

We first manipulate Eq. [c] and note that

$$\frac{d}{dt}(r^2\dot{\theta}) = r^2\ddot{\theta} + 2r\dot{r}\dot{\theta} = r(r\ddot{\theta} + 2\dot{r}\dot{\theta}) = 0 \quad \text{[d]}$$

which implies that $r^2\dot{\theta}$ is constant. Hence, one integral of the motion is

$$r^2\dot{\theta} = h = \text{constant} \quad \text{[e]}$$

This first integral has a physical significance, as it is the angular momentum of the particle about the origin of the coordinate system

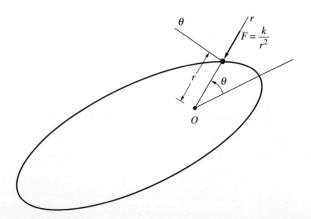

Figure 1.59

$$\mathbf{H}_O = \mathbf{r} \times m\mathbf{v} = r\mathbf{e}_r \times m(\dot{r}\mathbf{e}_r + r\dot{\theta}\mathbf{e}_\theta) = mr^2\dot{\theta}\mathbf{k} \qquad [\mathbf{f}]$$

This integral of the motion can be directly identified by noting that the applied force $\mathbf{F}$ does not create a moment around O. Next, consider Eq. [b] and eliminate the $\dot{\theta}$ term from it using Eq. [e]. Writing Eq. [e] as $\dot{\theta} = h/r^2$, and introducing it into Eq. [b], we obtain

$$m\ddot{r} = F(r) = \frac{mh^2}{r^3} + \frac{k}{r^2} \qquad [\mathbf{g}]$$

The above equation describes a conservative system, with $F(r)$ denoting the conservative force. The energy is conserved, and it has the form

$$\frac{1}{2}m\dot{r}^2 + \frac{1}{2}\frac{mh^2}{r^2} + \frac{k}{r} = E = \text{constant} \qquad [\mathbf{h}]$$

We identify the above expression as another integral of the motion. For general central force problems, $\mathbf{F} = f(r)\mathbf{e}_r$, Eq. [h] becomes $\frac{1}{2}m\dot{r}^2 + \frac{1}{2}mh^2/r^2 + V(r) = E$, where

$$V(r) = -\int_{r_0}^{r} f(r)\,dr \qquad [\mathbf{i}]$$

For orbital mechanics problems it is convenient to introduce the variable $u = 1/r$ and to analyze the equations of motion in terms of u. We will study this subject in more detail in Chapter 3.

1.13 NUMERICAL INTEGRATION OF EQUATIONS OF MOTION

In previous sections, we saw closed-form approaches for obtaining the response of a single degree of freedom system described by linear, constant coefficient equations. Such equations are almost always approximations to more complex equations, often about equilibrium. Even then, if the external excitation is a complicated function, finding a closed-form solution of the convolution integral may become prohibitive. In cases when no analytical tool is available to obtain the response, numerical integration of the equations of motion by means of a digital computer is often a viable alternative.

There are several types of computational methods for integrating the equations of motion of a system and getting its response. The different types of methods are useful in treating different types of equations. We will not go into the various approaches here, but we will discuss the general principles behind numerical integration.

Dynamical systems are governed by differential equations, where the excitation and response are usually continuous functions of time. Digital computers are designed to deal with discrete phenomena. Hence, integration of equations of motion on a digital computer needs to be discretized and carried out on an incremental basis. The basic idea is as follows: Consider the equation of motion

$$\ddot{x}(t) = h(x(t), \dot{x}(t)) + f(t) \qquad [1.13.1]$$

We select a time increment, say Δ, and initial time, say t_0. The time increment Δ is also called the *sampling period*. We feed the computer the initial conditions for position and velocity at $t = t_0$, $x(t_0)$, and $\dot{x}(t_0)$, as well as the value of the external force, $f(t_0)$, and invoke the numerical integration routine. The output of the numerical integration routine will be approximations to the position and velocity at $t = t_0 + \Delta$, $x(t_0 + \Delta)$ and $\dot{x}(t_0 + \Delta)$. This is considered as one step of the integration.

We then go to the second step of the integration. Values generated for the displacement and velocity at the end of the first step, $x(t_0 + \Delta)$ and $\dot{x}(t_0 + \Delta)$, are fed into the computer as initial conditions, together with the external force $f(t_0 + \Delta)$. The output is approximations to the position and velocity at the end of the second step, $x(t_0 + 2\Delta)$ and $\dot{x}(t_0 + 2\Delta)$. The process continues until the final time or other selected condition is reached.

Two obvious questions involve the selection of the numerical integration approach, and the selection of the time increment Δ. One selects the numerical integration approach based on the nature of $h(x(t), \dot{x}(t))$ as well as of $f(t)$. One selects the sampling period such that the results of the numerical integration are accurate. A very small value of Δ increases the computational effort quite a bit, while a large value of Δ leads to inaccuracies. Often, the nature of the equations of motion gives guidelines for the selection of Δ. For example, in a vibrating system like the one considered in Sections 1.9–1.11, the sampling period Δ should be less than $T/6$, one sixth of the period of vibration.

It turns out that almost all numerical integration methods require that the describing equations be cast into what is known as *state form*. In state form, the system is represented by first-order differential equations. The left side of the equations contain first-order derivatives of the variables and the right side of the equations do not have any time derivatives. For example, to write the vibrating system of Eq. [1.10.1] in state form, we introduce two new variables,

$$z_1(t) = x(t) \qquad z_2(t) = \dot{x}(t) \qquad [1.13.2]$$

and write the equation of motion in state form as the two equations

$$\dot{z}_1(t) = z_2(t)$$

$$\dot{z}_2(t) = -\omega_n^2 z_1(t) - 2\zeta\omega_n z_2(t) + f(t) \qquad [1.13.3]$$

For multidegree of freedom systems we have a set of variables $\{z(t)\} = [z_1(t), z_2(t) \ldots z_m(t)]^T$, and can write the state equations as

$$\{\dot{z}(t)\} = [A(\{z(t)\})] + [B(\{z(t)\})]\{f(t)\} \qquad [1.13.4]$$

in which $\{f(t)\}$ is the excitation vector. When linear, constant coefficient systems are expressed in state form, the equations have the form

$$\{\dot{z}(t)\} = [A]\{z(t)\} + [B]\{f(t)\} \qquad [1.13.5]$$

where $[A]$ and $[B]$ are constant coefficient matrices.

Example 1.22

Identify the $[A]$ and $[B]$ matrices for the system described by Eqs. [1.13.3].

Solution

We have one input, $f(t)$, so that $\{f(t)\}$ is a scalar. Thus, $[A]$ is a 2×2 matrix, and $[B]$ is a matrix of order 2×1, hence a vector of order 2. We have

$$\{z(t)\} = \begin{bmatrix} z_1(t) \\ z_2(t) \end{bmatrix} \qquad [A] = \begin{bmatrix} 0 & 1 \\ -\omega_n^2 & -2\zeta\omega_n \end{bmatrix} \qquad [B] = \begin{bmatrix} 0 \\ 1 \end{bmatrix} \qquad \{f(t)\} = f(t) \qquad [\text{a}]$$

REFERENCES

Ginsberg, J. H. *Advanced Engineering Dynamics*. New York: Harper & Row, 1988.
Goldstein, H. *Classical Mechanics*. 2nd ed. Reading, MA: Addison-Wesley, 1980.
Greenberg, M. D. *Foundations of Applied Mathematics*. Englewood Cliffs, NJ: Prentice-Hall, 1978.
Greenwood, D. T. *Principles of Dynamics*. 2nd ed. Englewood Cliffs, NJ: Prentice-Hall, 1988.
Lanczos, C. *The Variational Principles of Mechanics*. New York: Dover Publications, 1986.
Meirovitch, L. *Methods of Analytical Dynamics*. New York: McGraw-Hill, 1970.

HOMEWORK EXERCISES

SECTION 1.2

1. Given the values for $g = 32.174$ ft/sec^2 and that a block of mass 1 kg has a weight of 2.2046 lb, determine how many newtons equals a force of 1 lb.

SECTION 1.3

2. Show that the radius of curvature of a projectile's trajectory reaches a minimum at the top of the trajectory.

3. Consider normal and tangential coordinates and show that the radius of curvature and torsion are related to the path variable s by Eqs. [1.3.39]. Then, consider two-dimensional motion (x, y) and derive the expression for the radius of curvature given in Eq. [1.3.15].

4. A pin is constrained to move in a guide slot whose curve is defined by $y = -x^2 + x$, as shown in Fig. 1.60. The pin is being pushed by a motor such that it

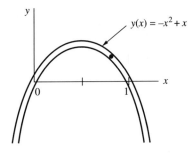

Figure 1.60

has an acceleration $\ddot{x}(t) = 1 + 0.2t$. The initial conditions are at $t = 0$, $x(0) = -0.6$, $\dot{x}(0) = 0$. Find the acceleration of the particle when $t = 0.5$ s, and find the radius of curvature at that instant.

5. A particle moves along a surface defined by $z = 2xy$, such that $x = 3\sin\alpha$ and $y = 3\cos\alpha$, in which α is a parameter. Find the unit vectors in the normal, tangential, and binormal directions, as well as the radius of curvature and torsion of the curve when $\alpha = \pi/4$.

6. At a certain instant, the velocity and acceleration of a particle are defined by $\mathbf{v} = 3\mathbf{i} + 4\mathbf{j} - 6\mathbf{k}$ m/s and $\mathbf{a} = -2\mathbf{i} + 3\mathbf{k}$ m/s^2. Find the radius of curvature and change of speed of the particle at that instant.

7. Consider Example 1.3 and find the distance traveled by the particle on the curve between the points $\alpha = 0$ and $\alpha = \pi/6$.

8. A vehicle modeled as a particle of mass m is moving up a spiraled road of constant radius R, as shown in Fig. 1.61. It takes the vehicle five full circles to reach the top, which is at a height h from the bottom.
   a. Express the position, velocity, and acceleration of the particle using cylindrical coordinates.
   b. Obtain the relationships between the unit vectors in the cylindrical coordinates and the normal-tangential coordinates. What is the radius of curvature?

9. Using a spherical coordinate system, express R, θ, and ϕ in terms of x, y, and z. In other words, find the counterpart of Eq. [1.3.63] in spherical coordinates.

10. A particle is traveling along an elliptic path, described by the equation $x^2/a^2 + y^2/b^2 = 1$, with $a \geq b$. Show—using path variables—that an approximate solution for the perimeter of the ellipse is $2\pi\sqrt{0.5(a^2 + b^2)}$, and that the solution becomes exact when $a = b$. *Hint:* You will need to use elliptic integrals and their tables to arrive at your result.

11. Equations [1.3.32] and [1.3.33] relate the distance traversed along the path in terms of the rectilinear coordinates x, y, and z as well as the path parameter α. Derive the equivalent expressions for ds in terms of spherical and cylindrical coordinates.

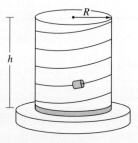

Figure 1.61

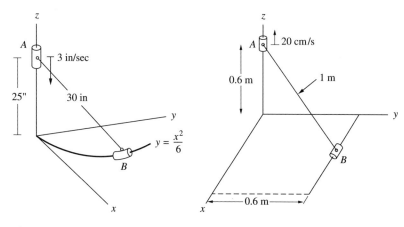

Figure 1.62 **Figure 1.63**

12. A rod is attached to two sliders moving in guide bars by universal joints as shown in Fig. 1.62. Find the velocity of slider B when slider A is at a height of 25 in. and is sliding down with a speed of 3 in/s.

13. A rod is attached to two sliders moving in guide bars by universal joints as shown in Fig. 1.63. Find the velocity of slider B when slider A is at a height of 60 cm and it is sliding up with a speed of 20 cm/s.

Section 1.4

14. Using spherical coordinates, find the equations of motion of the pendulum in Fig. 1.13.

15. Find the equations of motion for the double pendulum in Fig. 1.64.

16. Consider the vehicle moving up a spiraled road of problem 1.8. There is a coefficient of friction of μ between the road and the vehicle. Derive the equation of motion. First, use all three cylindrical coordinates as motion variables and generate three constrained equations. Then, using the traversed angle as the motion variable, reduce the equations of motion to one. Treat the vehicle as a bead sliding around a helical wire. A force pushes the vehicle up.

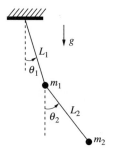

Figure 1.64

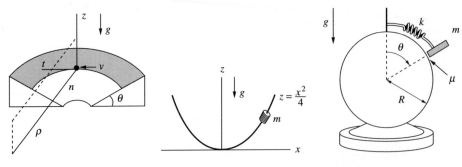

Figure 1.65 **Figure 1.66** **Figure 1.67**

17. A vehicle modeled as a particle of mass m is moving with constant speed v along a curved road with radius of curvature ρ, as shown in Fig. 1.65. The coefficient of friction between the road surface and the vehicle is μ. In order to reduce slipping, the road is banked by an angle θ. Find a relationship between μ and θ that will prevent the vehicle from slipping when the brakes are applied and the vehicle slows down with deceleration a.

18. A bead of mass m slides without friction on a wire shaped as the curve $z = x^2/4$, as shown in Fig. 1.66. Find the equation of motion for the bead.

19. Find the equation of motion for the mass sliding with friction over a disk of radius R, as shown in Fig. 1.67. A spring connects the mass with the top of the disk. The spring is unstretched when $\theta = 0$.

20. A particle of mass m is being acted upon by a force $F(x, \dot{x}, t) = x\dot{x}$. Find its response using initial conditions $x(0) = x_0$, $\dot{x}(0) = v_0$. Consider two cases: (a) $v_0 > x_0^2/2m$, and (b) $v_0 < x_0^2/2m$.

21. Find the equation of motion of the system in Fig. 1.68. The two sliders have mass m and the link is massless. Friction affects only the slider moving horizontally.

22. A steel ball of mass 0.2 kg is released into the frictionless inner surface of a cone with an apex angle of 15°, at a distance of 65 cm from the apex, as shown

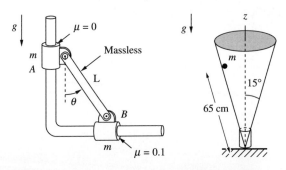

Figure 1.68 **Figure 1.69**

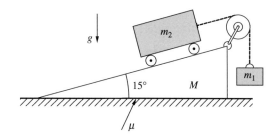

Figure 1.70

in Fig 1.69. What should be the initial speed of the ball, so that the ball does not fall to the bottom of the cone and its elevation remains the same? Use spherical coordinates.

23. The wedge (15°) in Fig. 1.70 of mass M rests on a rough platform with coefficient of friction μ. A mass m_1 is suspended by a string and is attached by a pulley to another mass m_2 which slides without friction on the wedge.
    a. Solve for the accelerations of m_1, m_2, and the tension in the string when μ is sufficient to keep the wedge from moving.
    b. Find the smallest value of μ for which the wedge remains at rest.

SECTION 1.5

24. Determine the number of degrees of freedom for the systems shown in Figs. 1.71 and 1.72.

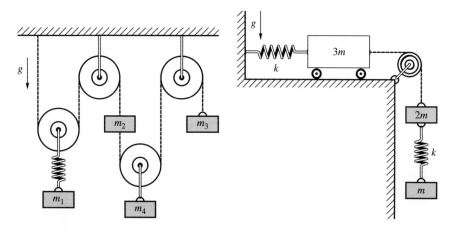

Figure 1.71   **Figure 1.72**

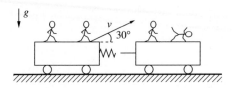

Figure 1.73

Section 1.6

25. A fugitive who weighs 180 lb is running on top of a train. The train car weighs 10,000 lb and is moving at a speed of 5 mph, as shown in Fig. 1.73. The fugitive's goal is to jump onto the next car, which has the same mass and speed and which has just separated from the car behind it. As the fugitive jumps, he resembles a projectile which has left the ground with an angle of 30° and a speed of 20 mph relative to the train. What is the speed of the second car after the fugitive jumps on it and he comes to a stop on the train?

26. Consider the double pendulum in Fig. 1.64. An impulsive force $\hat{F}$ is applied to m_2 in the horizontal direction. Find $\dot{\theta}_1$ and $\dot{\theta}_2$ immediately after the impulse.

Section 1.7

27. The forces acting on a particle, expressed in the Cartesian coordinate system, are $F_x = 2x + 3x^2y - y^3 + 4$, $F_y = 2y + x^3 - 3xy^2 + 1$, $F_z = 0$. What is the potential energy V?

28. Find the period of oscillation of a simple pendulum for arbitrary motions of the pendulum.

29. A particle slides over a circular cylinder of radius R, as shown in Fig. 1.74. The angle that the line connecting the center of the cylinder with the particle makes with the vertical is denoted by θ. The particle is slightly tilted with speed

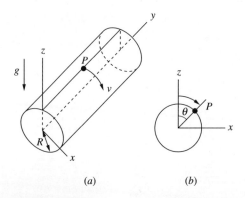

Figure 1.74

v_0 in the x direction at the top of the cylinder. Find the value of the angle θ when the particle and cylinder lose contact.

30. Consider the particle in the previous problem. The particle is released from the top with an initial velocity of $\mathbf{v}_0 = v_{xo}\mathbf{i} + v_{yo}\mathbf{j}$, and $\mu = 0.1$. Derive the equations of motion, and discuss the difficulties in solving for the value of θ when the particle loses contact with the cylinder.

SECTION 1.8

31. Find the equilibrium position for the system in Fig. 1.72.
32. Find the equilibrium position(s) of a particle of mass $m = 1$, which is acted upon by the forces:
    a. $F(x, \dot{x}) = 0.6\dot{x} + x - 0.1x^3$ b. $F(x, \dot{x}) = -0.4\dot{x} + x + x^2 + 0.05\dot{x}^2$
33. A particle of mass m is at the tip of a massless rod of length L and is being used as an inverted pendulum attached to two springs, as shown in Fig. 1.75. Find the equilibrium positions for the mass and check for their stability. Assume small motions and that the springs always deflect horizontally.
34. Assess the stability of the equilibrium points associated with the mass in Problem 1.32.

SECTION 1.9

35. Obtain the equation of motion for the system shown in Fig. 1.76. The rod is massless. Then, assume small motions and linearize.
36. Obtain the equation of motion and calculate the natural frequency of the two-pulley system shown in Fig. 1.77. Assume that the pulleys are smooth and massless.
37. Consider Example 1.17 and that a dashpot of constant $c = 0.3 \text{ lb} \cdot \text{sec/in}$ acting on the mass-spring system. Find the displacement and velocity after two seconds.

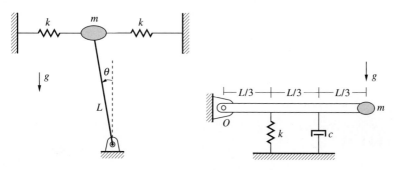

Figure 1.75   **Figure 1.76**

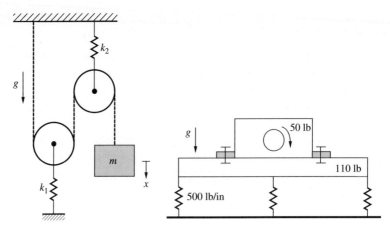

Figure 1.77 **Figure 1.78**

SECTION 1.10

38. A machine of weight 50 lb is mounted on a 110 lb table (vibration isolator), which is supported by three springs of constant 500 lb/in each, as shown in Fig. 1.78. The rotor inside the machine rotates with a speed of 1000 rpm, and it generates a force that varies harmonically between -30 and 30 lbs. Find the amplitude of the response, assuming that the table has no rotational motion. What is the force transmitted to the support?

39. A machine with rotating components is to be placed on four springs and four dashpots. The machine weighs 200 lb and it operates at 600 rpm. Design the spring and dashpot constants such that only 50 percent of the shaking force of the unit is transmitted to the supporting structure, and $\zeta = 0.3$.

40. A mass-spring-damper system has a mass of 10 kg and spring constant of 500 N/m. The viscous damping coefficient is not known and is to be determined experimentally from the frequency response. When the system is excited by a frequency $\omega = 14.5$ rad/s, the magnification factor is observed to be 0.3, and when $\omega = 11$ rad/s, the magnification factor is 0.65. Find the damping coefficient that will give values for the magnification factor closest to the two measurements.

41. In Fig. 1.28, the mass M moves according to the relation $x(t) = x_0 \cos \omega t$. Find the equation of motion for the pendulum, assuming that θ remains small at all times.

42. Consider the small motions of the double pendulum in Fig. 1.64. Let θ_1 be specified and find the resonance condition for θ_2. Then let θ_2 be specified and find the resonance condition for θ_1.

SECTION 1.11

43. The step response of a system is the response to the excitation $f(t) = 1u(t)$ in Eq. [1.10.1], with zero initial conditions. Find the step response for an undamped

system and show that the derivative of the step response is the impulse response $g(t)$.

44. An undamped mass-spring system ($m\ddot{x} + kx = F$) is subjected to the excitation $F(t) = tu(t)$, with zero initial conditions. Find the response.

45. An undamped mass-spring system is subjected to the excitation $F(t) = F_0$ ($0 \le t \le t_0$) and no excitation after $t > t_0$. The initial displacement is $x_0 = -F_0/3k$, with no initial velocity. Find the response.

46. An undamped mass-spring system is subjected to the excitation $F(t) = F_0 \sin 2\omega_n t$, with zero initial conditions. Derive an expression for the response. Compare this expression with the frequency response in Section 1.10 and comment on the difference.

SECTION 1.12

47. A particle of mass m is being acted on by a force expressed in polar coordinates as $\mathbf{F} = k\theta/r \, \mathbf{e}_\theta$, where k is a constant. Find the integrals of motion of this system.

48. A particle of mass m is being acted on by a force expressed in spherical coordinates as $\mathbf{F} = kR\mathbf{e}_R$, where k is a constant. Find the integrals of motion of this system.

49. Find the integrals of the motion for the pendulum shown in Fig. 1.13.

chapter

2

RELATIVE MOTION

2.1 INTRODUCTION

When motion is observed from an inertial frame, the expressions for velocity and acceleration have simple forms. Often it is necessary or advantageous to observe motion from a moving reference frame rather than an inertial frame. One can find examples of this from, among other cases, machine dynamics, vehicle dynamics, and motion relative to the rotating earth. In machine dynamics, one needs to relate the motion of one component to the other. Measurement of motion from a moving vehicle or platform is a common necessity. And, in an expanded sense, all motion measured on the earth is with respect to a rotating coordinate system.

Ignoring the earth's rotation is a realistic assumption in a number of problems. Motion over short distances and over short time intervals can be accurately analyzed without considering the earth's rotation. In a number of cases, though, the effect of having a noninertial reference frame must be considered. For example, for computations associated with weather patterns and ocean dynamics, and just about any type of motion over long time periods, neglecting to consider rotation of the earth gives incorrect results.

In this chapter we consider the motion of reference frames with respect to each other and establish relative motion equations. A major difference between an inertial and a noninertial reference frame arises when we calculate the derivatives of vectors. We distinguish between local derivatives, that is, derivatives calculated from moving reference frames, and global (or total) derivatives, which are calculated with respect to inertial reference frames. We discuss the differences in form between the translational velocity vector and angular velocity vector and emphasize that angular velocity is a defined vector, rather than the derivative of another vector. We consider motion involving the rotating earth. Within this context, the reader is introduced to a basic perturbation technique to obtain approximate solutions to complex problems.

2.2 Moving Coordinate Frames

Consider two coordinate frames: XYZ, a fixed frame with origin at O, and xyz, a moving frame with origin at B, as shown in Fig. 2.1. We define the unit vectors along the fixed axes X, Y, and Z by $\mathbf{I}$, $\mathbf{J}$, and $\mathbf{K}$ and along the moving axes x, y, and z by $\mathbf{i}$, $\mathbf{j}$, and $\mathbf{k}$, respectively.

We are primarily interested in the case when the moving xyz frame rotates. Consider the motion of a point P. One can observe the displacement of P from the inertial frame, using the vector $\mathbf{r}_P$, or from the relative frame, by the vector $\mathbf{r}_{P/B}$. From vector calculus we write

$$\mathbf{r}_P = \mathbf{r}_B + \mathbf{r}_{P/B} \qquad [2.2.1]$$

For the sake of discussion, assume that points O and B coincide, and drop the subscript P. One can express $\mathbf{r} = \mathbf{r}_P$ as $\mathbf{r} = X\mathbf{I} + Y\mathbf{J} + Z\mathbf{K}$ or $\mathbf{r} = x\mathbf{i} + y\mathbf{j} + z\mathbf{k}$. To investigate the relationships between the velocities as observed from the different frames, we differentiate $\mathbf{r}$ with respect to time. In terms of the inertial frame components, because the unit vectors $\mathbf{I}$, $\mathbf{J}$, and $\mathbf{K}$ are fixed in direction, we have

$$\mathbf{v} = \frac{d}{dt}\mathbf{r} = \dot{X}\mathbf{I} + \dot{Y}\mathbf{J} + \dot{Z}\mathbf{K} \qquad [2.2.2]$$

and in terms of the moving frame

$$\mathbf{v} = \dot{x}\mathbf{i} + \dot{y}\mathbf{j} + \dot{z}\mathbf{k} + x\dot{\mathbf{i}} + y\dot{\mathbf{j}} + z\dot{\mathbf{k}} \qquad [2.2.3]$$

The first three terms on the right in Eq. [2.2.3] describe the velocity as observed from the relative frame. The last three terms describe the rate of change of the unit vectors and, hence, the contribution due to the motion of the relative frame itself. The normal-tangential, cylindrical, and spherical coordinates that we studied in Chapter 1 are in essence rotating coordinate systems.

We identify two types of terms: the *local derivative* terms, taken in the relative frame, and an added set of terms that depend on the motion of the relative frame. The local derivatives together with the added terms constitute the *global derivative* terms, measured in the inertial frame. From this comes the simple but important conclusion that *derivatives of a vector are different quantities when taken in different reference frames*. One must clearly specify which reference frame the derivative is taken in.

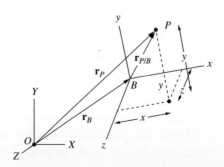

Figure 2.1

Using a similar approach, we obtain the expression for the acceleration of point P as

$$\mathbf{a} = \frac{d}{dt}\mathbf{v} = \ddot{x}\mathbf{i} + \ddot{y}\mathbf{j} + \ddot{z}\mathbf{k} + x\ddot{\mathbf{i}} + y\ddot{\mathbf{j}} + z\ddot{\mathbf{k}} + 2(\dot{x}\dot{\mathbf{i}} + \dot{y}\dot{\mathbf{j}} + \dot{z}\dot{\mathbf{k}}) \quad \text{[2.2.4]}$$

The first three terms on the right in this equation describe the acceleration as observed from the relative frame. They constitute the local second derivative. The next three terms describe the acceleration of the relative frame. The last three terms exist because there is motion with respect to a moving reference frame. These terms arise from two sources: the differentiation in time of the unit vectors in the expression $\dot{x}\mathbf{i} + \dot{y}\mathbf{j} + \dot{z}\mathbf{k}$, and the differentiation of the displacement variables in the expression $x\mathbf{i} + y\mathbf{j} + z\mathbf{k}$. These are known as the Coriolis terms.

We obtained Eqs. [2.2.3] and [2.2.4] by a straightforward differentiation of the displacement expressions. We followed this procedure in Chapter 1 when obtaining derivatives of unit vectors associated with the coordinate systems we were considering. One may ask whether there is a more general way to obtain derivatives of vectors. Indeed, there is, as we will see later on in this chapter.

When selecting a moving coordinate system, one must establish the relationship between the inertial and the moving coordinate systems. A simple illustration of a coordinate transformation is given in the example that follows.

Example 2.1

Water is flowing out of the garden sprinkler in Fig. 2.2 with the constant speed of 2 m/s. The sprinkler arm rotates counterclockwise at the constant rate of $20/\pi$ rpm. As it exits the sprinkler, the water makes an angle of 15° with the horizontal. Find the velocity and acceleration of a particle of water as it leaves the sprinkler arm, and the velocity 0.05 seconds later.

Solution

We have two convenient locations to define the origin of the relative axes: to place point B at the pivot, or to place point B at the tip of the sprinkler. Let us locate point B at the pivot, so that $\mathbf{r}_B = \mathbf{0}$, $\mathbf{v}_B = \mathbf{0}$. We define the inertial coordinate frame with the Z direction along the vertical. The orientation of the relative axes is selected such that the Z and z axes coincide and that the projection of the sprinkler on the XY plane is along the x axis. Defining the angle between the X and x axes by θ, the xyz frame is obtained by a counterclockwise rotation about the Z axis by θ, as shown in Fig 2.3. We have

$$\mathbf{i} = \cos\theta\mathbf{I} + \sin\theta\mathbf{J} \qquad \mathbf{j} = \cos\theta\mathbf{J} - \sin\theta\mathbf{I} \qquad \mathbf{k} = \mathbf{K} \qquad \text{[a]}$$

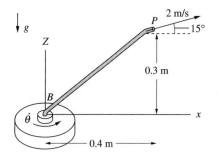

Figure 2.2

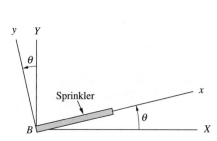

Figure 2.3

Differentiation of Eq. [a] with respect to time yields the rates of change of the unit vectors **i**, **j**, and **k** as

$$\frac{d\mathbf{i}}{dt} = -\dot{\theta}\sin\theta\mathbf{I} + \dot{\theta}\cos\theta\mathbf{J} = \dot{\theta}\mathbf{j} \qquad \frac{d\mathbf{j}}{dt} = -\dot{\theta}\cos\theta\mathbf{I} - \dot{\theta}\sin\theta\mathbf{J} = -\dot{\theta}\mathbf{i} \qquad \frac{d\mathbf{k}}{dt} = 0$$

$$\frac{d^2\mathbf{i}}{dt^2} = \ddot{\theta}\mathbf{j} + \dot{\theta}\frac{d\mathbf{j}}{dt} = \ddot{\theta}\mathbf{j} - \dot{\theta}^2\mathbf{i} \qquad \frac{d^2\mathbf{j}}{dt^2} = -\ddot{\theta}\mathbf{i} - \dot{\theta}\frac{d\mathbf{i}}{dt} = -\ddot{\theta}\mathbf{i} - \dot{\theta}^2\mathbf{j} \qquad \frac{d^2\mathbf{k}}{dt^2} = 0 \qquad [\mathbf{b}]$$

indicating that the component of the motion in the z-direction is not affected by the motion of the relative frame. The rotation of the moving frame is described by

$$\dot{\theta} = \frac{20 \text{ rev}}{\pi \text{ min}}\frac{1 \text{min}}{60 \text{s}}\frac{2\pi \text{rad}}{\text{rev}} = \frac{2}{3} \text{ rad/s} \qquad \ddot{\theta} = 0 \qquad [\mathbf{c}]$$

It is useful to write the position vector associated with a water particle inside the sprinkler as

$$\mathbf{r}_P = \mathbf{r}_{P/B} = x\mathbf{i} + z\mathbf{k} \qquad [\mathbf{d}]$$

so that the coordinate y and all its derivatives are zero. Using this information, Eqs. [2.2.3], [2.2.4], and the vector derivatives in Eqs. [b],

$$\mathbf{v}_P = \dot{x}\mathbf{i} + \dot{z}\mathbf{k} + x\dot{\mathbf{i}} = \dot{x}\mathbf{i} + \dot{z}\mathbf{k} + x\dot{\theta}\mathbf{j} \qquad [\mathbf{e}]$$

$$\mathbf{a}_P = \ddot{x}\mathbf{i} + \ddot{z}\mathbf{k} + x\ddot{\mathbf{i}} + 2\dot{x}\dot{\mathbf{i}} = \ddot{x}\mathbf{i} + \ddot{z}\mathbf{k} + x(\ddot{\theta}\mathbf{j} - \dot{\theta}^2\mathbf{i}) + 2\dot{x}\dot{\theta}\mathbf{j} \qquad [\mathbf{f}]$$

As the water particle is about to leave the sprinkler, we have from Fig. 2.2,

$$x = 0.4 \text{ m} \qquad z = 0.3 \text{ m} \qquad \dot{x} = 2\cos 15° = 1.932 \text{ m/s}$$
$$\dot{z} = 2\sin 15° = 0.5176 \text{ m/s} \qquad \ddot{x} = \ddot{z} = 0 \qquad [\mathbf{g}]$$

Substituting the above values into Eqs. [e] and [f], we obtain

$$\mathbf{v}_P = 1.932\mathbf{i} + 0.2667\mathbf{j} + 0.5176\mathbf{k} \text{ m/s} \qquad [\mathbf{h}]$$

$$\mathbf{a}_P = 0.4\left(0\mathbf{j} - \frac{4}{9}\mathbf{i}\right) + 2(1.932)\frac{2}{3}\mathbf{j} = -0.1778\mathbf{i} + 2.576\mathbf{j} \text{ m/s}^2 \qquad [\mathbf{i}]$$

Because the xyz axes are continuously changing direction, the velocity and acceleration can be expressed more meaningfully using their components in the vertical and in the horizontal plane. We can hence write

$$v_{\text{vert}} = 0.5176 \text{ m/s} \qquad v_{\text{horiz}} = \sqrt{1.932^2 + 0.2667^2} = 1.950 \text{ m/s} \qquad [\mathbf{j}]$$

As soon as the water particle leaves the nozzle, the only force that acts on it is gravity. Also, we are no longer viewing motion from a set of rotating coordinates. It is more convenient now to look at the horizontal and vertical components of the velocity. Ignoring air resistance, the horizontal component of the velocity does not change. The vertical component changes due to gravity, which we can express as $\mathbf{a} = -9.807\mathbf{K}$ m/s^2. After 0.05 seconds, the vertical component of the velocity becomes

$$v_{\text{vert}} = 0.5176 - 0.05(9.807) = 0.02725 \text{ m/s} \qquad [\mathbf{k}]$$

indicating that the water particle is about to start going down.

2.3 REPRESENTATION OF VECTORS

In this section we describe different ways of representing vectors. Consider a rectilinear coordinate system with unit vectors $\mathbf{e}_1$, $\mathbf{e}_2$, and $\mathbf{e}_3$ and two vectors $\mathbf{r}$ and $\mathbf{u}$ defined as

$$\mathbf{r} = r_1\mathbf{e}_1 + r_2\mathbf{e}_2 + r_3\mathbf{e}_3 \qquad \mathbf{u} = u_1\mathbf{e}_1 + u_2\mathbf{e}_2 + u_3\mathbf{e}_3 \qquad [\mathbf{2.3.1}]$$

We will refer a set of vectors described this way as *geometric vectors* or *spatial vectors*. The dot and cross products of these vectors yield the results

$$\mathbf{r} \cdot \mathbf{u} = r_1 u_1 + r_2 u_2 + r_3 u_3$$

$$\mathbf{r} \times \mathbf{u} = (r_2 u_3 - r_3 u_2)\mathbf{i} + (r_3 u_1 - r_1 u_3)\mathbf{j} + (r_1 u_2 - r_2 u_1)\mathbf{k} \qquad [\mathbf{2.3.2}]$$

We also express the vectors $\mathbf{r}$ and $\mathbf{u}$ in *column vector format* as

$$\{r\} = \begin{bmatrix} r_1 \\ r_2 \\ r_3 \end{bmatrix} \qquad \{u\} = \begin{bmatrix} u_1 \\ u_2 \\ u_3 \end{bmatrix} \qquad [\mathbf{2.3.3}]$$

The column vectors $\{r\}$ and $\{u\}$ are also referred to as *algebraic vectors*. Using this description, we can express the dot product of two geometric vectors in column vector format by

$$\mathbf{r} \cdot \mathbf{u} \rightarrow \{r\}^T \{u\} \qquad [\mathbf{2.3.4}]$$

where T denotes the matrix transpose. To express the cross product $\mathbf{r} \times \mathbf{u}$ using column vectors, we introduce the skew-symmetric matrix $[\tilde{r}]$ associated with the vector $\mathbf{r}$, and write

$$[\tilde{r}] = \begin{bmatrix} 0 & -r_3 & r_2 \\ r_3 & 0 & -r_1 \\ -r_2 & r_1 & 0 \end{bmatrix} \qquad [\mathbf{2.3.5}]$$

so that

$$\mathbf{r} \times \mathbf{u} \rightarrow [\tilde{r}]\{u\} = \begin{bmatrix} r_2 u_3 - r_3 u_2 \\ r_3 u_1 - r_1 u_3 \\ r_1 u_2 - r_2 u_1 \end{bmatrix} \qquad [\mathbf{2.3.6}]$$

Hence, the geometric vector operation of $\mathbf{r} \times \mathbf{u}$ and the column vector operation of $[\tilde{r}]\{u\}$ are equivalent. Note that because $\mathbf{r} \times \mathbf{u} = -\mathbf{u} \times \mathbf{r}$, we can also write $[\tilde{r}]\{u\} = -[\tilde{u}]\{r\}$.

In dynamics, one frequently encounters the vector product $\mathbf{r} \times (\mathbf{r} \times \mathbf{u})$, which is used to describe centripetal acceleration. The expression is commonly shortened to $\mathbf{r} \times \mathbf{r} \times \mathbf{u}$, with the understanding that the cross product between $\mathbf{r}$ and $\mathbf{u}$ is performed first. Using the notation introduced above,

$$\mathbf{r} \times (\mathbf{r} \times \mathbf{u}) \rightarrow [\tilde{r}][\tilde{r}]\{u\} \qquad [\mathbf{2.3.7}]$$

and we note that the matrix multiplications in $[\tilde{r}][\tilde{r}]\{u\}$ can be performed in any order.

The relations derived above are only valid when all vectors involved are expressed in the same coordinate system. If the vectors **r** and **u** are not represented in the same coordinate system, Eqs. [2.3.4] and [2.3.6] no longer hold. For example, if we have two coordinate systems with unit vectors $\mathbf{e}_1 \mathbf{e}_2 \mathbf{e}_3$ and $\mathbf{e}'_1 \mathbf{e}'_2 \mathbf{e}'_3$ and two spatial vectors **r** and **u** in the form

$$\mathbf{r} = r_1 \mathbf{e}_1 + r_2 \mathbf{e}_2 + r_3 \mathbf{e}_3 \qquad \mathbf{u} = u_1 \mathbf{e}'_1 + u_2 \mathbf{e}'_2 + u_3 \mathbf{e}'_3 \qquad [2.3.8]$$

their dot product becomes

$$\mathbf{r} \cdot \mathbf{u} = \sum_{j=1}^{3} \sum_{k=1}^{3} r_j u_k \mathbf{e}_j \cdot \mathbf{e}'_k \rightarrow \{r\}^T [c] \{u\} \qquad [2.3.9]$$

in which **r** and **u** are given in Eq. [2.3.1] and the matrix $[c]$ has the form

$$[c] = \begin{bmatrix} \mathbf{e}_1 \cdot \mathbf{e}'_1 & \mathbf{e}_1 \cdot \mathbf{e}'_2 & \mathbf{e}_1 \cdot \mathbf{e}'_3 \\ \mathbf{e}_2 \cdot \mathbf{e}'_1 & \mathbf{e}_2 \cdot \mathbf{e}'_2 & \mathbf{e}_2 \cdot \mathbf{e}'_3 \\ \mathbf{e}_3 \cdot \mathbf{e}'_1 & \mathbf{e}_3 \cdot \mathbf{e}'_2 & \mathbf{e}_3 \cdot \mathbf{e}'_3 \end{bmatrix} \qquad [2.3.10]$$

The expression $[c]\{u\}$ can be viewed as the column vector representation of **u** using the $\mathbf{e}_1 \mathbf{e}_2 \mathbf{e}_3$ triad. Conversely, the expression $\{r\}^T [c]$ can be viewed as the transpose of $[c]^T \{r\}$, the column vector representation of **r** in the $\mathbf{e}'_1 \mathbf{e}'_2 \mathbf{e}'_3$ triad. In the next section, we will formally define the entries of $[c]$ as direction cosines.

It should be noted that sets of geometric vectors can also be represented in column vector format. The elements of the column vector will be geometric vectors. For instance, the triad $\mathbf{e}_1 \mathbf{e}_2 \mathbf{e}_3$ can be written as

$$\{e\} = \begin{bmatrix} \mathbf{e}_1 \\ \mathbf{e}_2 \\ \mathbf{e}_3 \end{bmatrix} \qquad [2.3.11]$$

This form is also useful when taking dot or cross products of unit vectors. For example, the expression $\mathbf{g} = g_1 \mathbf{e}_1 + g_2 \mathbf{e}_2 + g_3 \mathbf{e}_3$, where $g_i (i = 1, 2, 3)$ are scalars, can be written using the column vector format as $\mathbf{g} = \{e\}^T \{g\}$ in which $\{g\} = [g_1\ g_2\ g_3]^T$.

The column vector notation is not restricted to describing geometric vectors. This formulation is commonly used to represent variables in a Euclidean space.

Next, consider differentiation of scalars and vectors with respect to other vectors. This procedure can be conveniently illustrated using column vectors. Consider the scalar S and a vector $\{q\} = [q_1\ q_2\ \ldots\ q_n]^T$ of dimension n, where the elements $q_1, q_2, \ldots, q_n$ are independent of each other. The derivative of S with respect to $\{q\}$ is defined as the n-dimensional row vector $dS/d\{q\}$, whose elements have the form

$$\frac{dS}{d\{q\}} = \begin{bmatrix} \frac{\partial S}{\partial q_1} & \frac{\partial S}{\partial q_2} & \cdots & \frac{\partial S}{\partial q_n} \end{bmatrix} \qquad [2.3.12]$$

In compact form, $dS/d\{q\}$ is also written as $S_{\{q\}}$. When $\{q\}$ is the column vector representation of a geometric vector in rectilinear coordinates, the above operation becomes similar to the gradient operation. We can write $\nabla S = [dS/d\{q\}]^T$.

In a similar fashion, we obtain the derivative of a column vector with respect to another. Given the column vector $\{u\}$ of order m, where $\{u\} = [u_1 \ u_2 \ \ldots \ u_m]^T$, the derivative of $\{u\}$ with respect to a vector $\{q\}$ of order n is obtained by differentiating every element of $\{u\}$ with respect to every element of $\{q\}$. The end result is the $m \times n$ matrix, referred to as the *Jacobian,* and denoted by $[\{u\}_{\{q\}}]$ or by $[u_q]$, having the form

$$\frac{d\{u\}}{d\{q\}} = [\{u\}_{\{q\}}] = [u_q] = \begin{bmatrix} \frac{\partial u_1}{\partial q_1} & \frac{\partial u_1}{\partial q_2} & \ldots & \frac{\partial u_1}{\partial q_n} \\ \frac{\partial u_2}{\partial q_1} & \frac{\partial u_2}{\partial q_2} & \ldots & \frac{\partial u_2}{\partial q_n} \\ \ldots & & & \\ \frac{\partial u_m}{\partial q_1} & \frac{\partial u_m}{\partial q_2} & \ldots & \frac{\partial u_m}{\partial q_n} \end{bmatrix} \quad [2.3.13]$$

We now investigate some of the special forms of the scalar S and its derivative with respect to a vector. Consider, for example, the n-dimensional column vectors $\{v\} = [v_1 \ v_2 \ \ldots \ v_n]^T$ and $\{q\}$ and define the scalar S as

$$S = \{v\}^T\{q\} = \{q\}^T\{v\} = \sum_{k=1}^{n} v_k q_k \quad [2.3.14]$$

in which $v_k (k = 1, 2, \ldots, n)$ are the elements of $\{v\}$ and $q_k (k = 1, 2, \ldots, n)$ are the elements of $\{q\}$. Taking the derivative of S with respect to $\{q\}$, we obtain

$$\frac{dS}{d\{q\}} = \frac{d(\{v\}^T\{q\})}{d\{q\}} \quad [2.3.15]$$

$$= \left[v_1 + \sum_{k=1}^{n} q_k \frac{\partial v_k}{\partial q_1} \quad v_2 + \sum_{k=1}^{n} q_k \frac{\partial v_k}{\partial q_2} \quad \ldots \quad v_n + \sum_{k=1}^{n} q_k \frac{\partial v_k}{\partial q_n} \right]$$

For the special case when the elements of $\{v\}$ are not functions of q_k, $v_j \neq v_j(q_k)$, $(j, k = 1, 2, \ldots, n)$ we obtain

$$\frac{dS}{d\{q\}} = \frac{\partial}{d\{q\}}\left(\{v\}^T\{q\}\right) = \{v\}^T \quad [2.3.16]$$

If we write the vector $\{v\}$ as $\{v\} = [D]\{h\}$ where $[D]$ is a square matrix of order n and $\{h\}$ is an n-dimensional vector, so that $S = \{h\}^T[D]^T\{q\} = \{q\}^T[D]\{h\}$ and consider the case where none of the elements of $[D]$ and $\{h\}$ are a function of $\{q\}$, then

$$\frac{dS}{d\{q\}} = \frac{d(\{h\}^T[D]^T\{q\})}{d\{q\}} = \{h\}^T[D]^T \quad [2.3.17]$$

It follows that if $[D]$ is not a function of $\{h\}$ the derivative of S with respect to $\{h\}$ is

$$\frac{dS}{d\{h\}} = \frac{d(\{h\}^T[D]^T\{q\})}{d\{h\}} = \frac{d(\{q\}^T[D]\{h\})}{d\{h\}} = \{q\}^T[D] \quad [2.3.18]$$

In mechanics, one commonly encounters scalar quantities that are quadratic in terms of the motion variables, such as kinetic and potential energy. Define S as

$S = \{q\}^T[D]\{q\}$. The derivative of S with respect to $\{q\}$ in this case has the form

$$\frac{dS}{d\{q\}} = \{q\}^T[D] + \{q\}^T[D]^T \qquad [2.3.19]$$

and when the matrix D is symmetric, this becomes

$$\frac{d\left(\{q\}^T[D]\{q\}\right)}{d\{q\}} = 2\{q\}^T[D] \qquad [2.3.20]$$

Consider the symmetric form above and the case when the matrix $[D]$ has elements that are functions of $\{q\}$. We frequently encounter this situation in Lagrangian mechanics. The derivative of a matrix with respect to a vector is a tensor of order three. We can avoid dealing with tensors of order three by taking the derivative of the product $[D]\{q\}$, so that we now have

$$\frac{d\left(\{q\}^T[D]\{q\}\right)}{d\{q\}} = \{q\}^T[D] + \{q\}^T\frac{d\left([D]^T\{q\}\right)}{d\{q\}} \qquad [2.3.21]$$

Next, let us obtain the derivatives of functions of several variables with respect to time. Consider the scalar x, which is a function of n variables $q_1, q_2, \ldots, q_n$ and time t, so that $x = x(q_1, q_2, \ldots, q_n, t)$. The derivative of x with respect to time is obtained using the chain rule as

$$\dot{x} = \frac{dx}{dt} = \sum_{k=1}^{n}\frac{\partial x}{\partial q_k}\dot{q}_k + \frac{\partial x}{\partial t} \qquad [2.3.22]$$

We can express this relationship in column vector format. Indeed, introducing the column vector $\{q\} = [q_1 \ q_2 \ \ldots \ q_n]^T$, we write Eq. [2.3.22] as

$$\dot{x} = \frac{dx}{dt} = \frac{\partial x}{\partial\{q\}}\{\dot{q}\} + \frac{\partial x}{\partial t} = x_{\{q\}}\{\dot{q}\} + \frac{\partial x}{\partial t} \qquad [2.3.23]$$

where we recall that the derivative of a scalar with respect to a column vector is a row vector. Extending this to the case when the time derivative of a column vector $\{r\}$ is sought, where $\{r\} = [r_1 \ r_2 \ \ldots \ r_m]^T$, where $r_j = r_j(q_1, q_2, \ldots, q_n, t)$, ($j = 1, 2, \ldots, m$) we obtain

$$\{\dot{r}\} = [\{r\}_{\{q\}}]\{\dot{q}\} + \left\{\frac{\partial r}{\partial t}\right\} = [r_q]\{\dot{q}\} + \left\{\frac{\partial r}{\partial t}\right\} \qquad [2.3.24]$$

in which $[\{r\}_{\{q\}}] = [r_q]$ is recognized to be a matrix of order $m \times n$. When $\{r\}$ is the column vector representation of a geometric vector, $m = 3$ and $[r_q]$ becomes a $3 \times n$ matrix with $[r_q]_{jk} = \partial r_j/\partial q_k$ ($j = 1, 2, 3; k = 1, 2, \ldots, n$). To visualize this better we express the vector $\{r\}$ in terms of a set of unit vectors as

$$\{r\} = [r_1 \ r_2 \ r_3]^T \qquad \mathbf{r} = r_1\mathbf{e}_1 + r_2\mathbf{e}_2 + r_3\mathbf{e}_3 \qquad [2.3.25]$$

so that

$$\dot{\mathbf{r}} = \sum_{k=1}^{n}\frac{\partial \mathbf{r}}{\partial q_k}\dot{q}_k + \frac{\partial \mathbf{r}}{\partial t} \quad \text{or} \quad \dot{\mathbf{r}} = \sum_{j=1}^{3}\frac{dr_j}{dt}\mathbf{e}_j = \sum_{j=1}^{3}\left[\sum_{k=1}^{n}\frac{\partial r_j}{\partial q_k}\dot{q}_k + \frac{\partial r_j}{\partial t}\right]\mathbf{e}_j$$

$$[2.3.26]$$

To evaluate the second derivative of $\{r\}$, we use the chain rule again. We wish to avoid taking the partial derivative of a matrix with respect to a vector, so we make use of Eq. [2.3.21] and write

$$\{\ddot{r}\} = [r_q]\{\ddot{q}\} + ([r_q]\{\dot{q}\})_{\{q\}}\{\dot{q}\} + 2\frac{\partial}{\partial t}[r_q]\{\dot{q}\} + \left\{\frac{\partial^2 r}{\partial t^2}\right\} \qquad [2.3.27]$$

The second term on the right side of this equation can also be expressed as

$$([r_q]\{\dot{q}\})_{\{q\}}\{\dot{q}\} = [\{\dot{q}\}^T[r_{1qq}]\{\dot{q}\} \quad \{\dot{q}\}^T[r_{2qq}]\{\dot{q}\} \quad \{\dot{q}\}^T[r_{3qq}]\{\dot{q}\}]^T \qquad [2.3.28]$$

in which the elements of the matrices $[r_{jqq}](j = 1, 2, 3)$ are

$$[r_{jqq}]_{ik} = \frac{\partial^2 r_j}{\partial q_i \partial q_k} \qquad j = 1, 2, 3; \; i, k = 1, 2, \ldots, n \qquad [2.3.29]$$

It is important to remember that the differentiation operation is conducted in the same reference frame as the vector **r** is measured in. In the above equations we considered an inertial frame.

Example 2.2

Consider the two coordinate systems *XYZ* and *xyz*, with unit vectors **IJK** and **ijk**. The *xyz* coordinate system is obtained by rotating the *XYZ* system first by an angle of 30° counterclockwise about the *Z* axis and then rotating the resulting intermediate $x'y'z'$ coordinate system by 45° clockwise about the y' axis. Fig. 2.4 shows the rotation sequence. Given the vectors **a** = 3**I** + 4**J** + 6**K** and **b** = 2**i** + **j** + 2**k**, find **a**·**b** and the matrix $[c]$.

Solution

We first represent the unit vectors **ijk** in terms of **IJK**. From Fig. 2.4, we can write the unit vectors of the intermediate $x'y'z'$ axes as

$$\mathbf{i}' = \cos 30°\mathbf{I} + \sin 30°\mathbf{J} \qquad \mathbf{j}' = -\sin 30°\mathbf{I} + \cos 30°\mathbf{J} \qquad \mathbf{k}' = \mathbf{K} \qquad [\mathbf{a}]$$

The xyz coordinate system is related to the intermediate axes by a clockwise rotation of 45° about the y' axis, so that—using the short notation for the sine (s) and cosine (c) of an

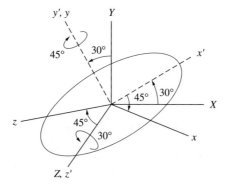

Figure 2.4

angle—the unit vectors in the coordinate systems are related by

$$\mathbf{i} = c\,45°\mathbf{i}' + s\,45°\mathbf{k}' = c\,45°(c\,30°\mathbf{I} + s\,30°\mathbf{J}) + s\,45°\mathbf{K} = \frac{\sqrt{6}}{4}\mathbf{I} + \frac{\sqrt{2}}{4}\mathbf{J} + \frac{\sqrt{2}}{2}\mathbf{K} \quad [\mathbf{b}]$$

$$\mathbf{j} = \mathbf{j}' = -s\,30°\mathbf{I} + c\,30°\mathbf{J} = -\frac{1}{2}\mathbf{I} + \frac{\sqrt{3}}{2}\mathbf{J} \quad [\mathbf{c}]$$

$$\mathbf{k} = -s\,45°\mathbf{i}' + c\,45°\mathbf{k}' = -s\,45°(c\,30°\mathbf{I} + s\,30°\mathbf{J}) + c\,45°\mathbf{K} = -\frac{\sqrt{6}}{4}\mathbf{I} - \frac{\sqrt{2}}{4}\mathbf{J} + \frac{\sqrt{2}}{2}\mathbf{K} \quad [\mathbf{d}]$$

Introducing these expressions to the vector **b**, we obtain

$$\mathbf{b} = 2\mathbf{i} + \mathbf{j} + 2\mathbf{k} = 2\left(\frac{\sqrt{6}}{4}\mathbf{I} + \frac{\sqrt{2}}{4}\mathbf{J} + \frac{\sqrt{2}}{2}\mathbf{K}\right) - \frac{1}{2}\mathbf{I} + \frac{\sqrt{3}}{2}\mathbf{J} + 2\left(-\frac{\sqrt{6}}{4}\mathbf{I}\right.$$

$$\left. - \frac{\sqrt{2}}{4}\mathbf{J} + \frac{\sqrt{2}}{2}\mathbf{K}\right) = -\frac{1}{2}\mathbf{I} + \frac{\sqrt{3}}{2}\mathbf{J} + 2\sqrt{2}\mathbf{K} \quad [\mathbf{e}]$$

We can now find the dot product of **a** and **b** as

$$\mathbf{a}\cdot\mathbf{b} = 3\left(-\frac{1}{2}\right) + 4\left(\frac{\sqrt{3}}{2}\right) + 6(2\sqrt{2}) = 18.94 \quad [\mathbf{f}]$$

Let us now generate the direction cosine matrix and write

$$[c] = \begin{bmatrix} \mathbf{I}\cdot\mathbf{i} & \mathbf{I}\cdot\mathbf{j} & \mathbf{I}\cdot\mathbf{k} \\ \mathbf{J}\cdot\mathbf{i} & \mathbf{J}\cdot\mathbf{j} & \mathbf{J}\cdot\mathbf{k} \\ \mathbf{K}\cdot\mathbf{i} & \mathbf{K}\cdot\mathbf{j} & \mathbf{K}\cdot\mathbf{k} \end{bmatrix} = \begin{bmatrix} c\,45°\,c\,30° & -s\,30° & -s\,45°\,c\,30° \\ c\,45°\,s\,30° & c\,30° & -s\,45°\,s\,30° \\ s\,45° & 0 & c\,45° \end{bmatrix} \quad [\mathbf{g}]$$

$$= \begin{bmatrix} \sqrt{6}/4 & -1/2 & -\sqrt{6}/4 \\ \sqrt{2}/4 & \sqrt{3}/2 & -\sqrt{2}/4 \\ \sqrt{2}/2 & 0 & \sqrt{2}/2 \end{bmatrix}$$

To find the dot product using column vectors we use Eq. [2.3.9], which yields

$$\mathbf{a}\cdot\mathbf{b} = \{a\}^T[c]\{b\} = [3\ 4\ 6]\begin{bmatrix} \sqrt{6}/4 & -1/2 & -\sqrt{6}/4 \\ \sqrt{2}/4 & \sqrt{3}/2 & -\sqrt{2}/4 \\ \sqrt{2}/2 & 0 & \sqrt{2}/2 \end{bmatrix}\begin{bmatrix} 2 \\ 1 \\ 2 \end{bmatrix} = 18.94 \quad [\mathbf{h}]$$

which, of course is the same as the result in Eq. [f].

Example 2.3

Figure 2.5 shows a door opened at an angle θ. On the door at point C is an ant that starts crawling upwards in a straight line. Its path makes an angle ϕ, which is fixed, with the bottom of the door. Determine the position of the ant, and obtain the velocity of the ant using Eqs. [2.3.23] through [2.3.26].

Solution

We attach the inertial coordinates XYZ to the door frame and the moving coordinates xyz to the door, as shown in Fig. 2.6. The xyz axes are obtained by a clockwise rotation of XYZ by an angle of θ about the Z axis. The position of the ant is

$$\mathbf{r} = -L\mathbf{k} + h\mathbf{i} + s(-\cos\phi\,\mathbf{i} + \sin\phi\,\mathbf{k}) = (h - s\cos\phi)\mathbf{i} + (-L + s\sin\phi)\mathbf{k} \quad [\mathbf{a}]$$

We relate the inertial and moving coordinates by

$$\mathbf{k} = \mathbf{K} \qquad \mathbf{i} = \cos(-\theta)\mathbf{I} + \sin(-\theta)\mathbf{J} = \cos\theta\,\mathbf{I} - \sin\theta\,\mathbf{J} \quad [\mathbf{b}]$$

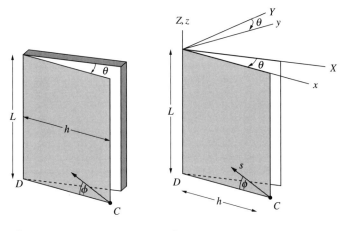

Figure 2.5   **Figure 2.6**

which, when introduced into Eq. [a], yields

$$\mathbf{r} = (h - s\cos\phi)\cos\theta\mathbf{I} - (h - s\cos\phi)\sin\theta\mathbf{J} + (-L + s\sin\phi)\mathbf{K} \quad [\mathbf{c}]$$

Two variables that can describe the motion are θ and s so that $\{q\} = [\theta \ s]^T$. To find the velocity of the ant, we write the position vector in column vector form. We then find $[r_q]$, which will be a 3×2 matrix, with its columns containing derivatives of $\{r\}$ with respect to θ and s. We thus have

$$\{r\} = \begin{bmatrix} (h - s\cos\phi)\cos\theta \\ -(h - s\cos\phi)\sin\theta \\ -L + s\sin\phi \end{bmatrix} \quad [r_q] = \begin{bmatrix} -(h - s\cos\phi)\sin\theta & -\cos\phi\cos\theta \\ -(h - s\cos\phi)\cos\theta & \cos\phi\sin\theta \\ 0 & \sin\phi \end{bmatrix} \quad [\mathbf{d}]$$

Noting that there is no explicit time dependence in the position vector, we can write the velocity as

$$\{v\} = [r_q]\{\dot{q}\} = \begin{bmatrix} -(h - s\cos\phi)\sin\theta & -\cos\phi\cos\theta \\ -(h - s\cos\phi)\cos\theta & \cos\phi\sin\theta \\ 0 & \sin\phi \end{bmatrix} \begin{bmatrix} \dot{\theta} \\ \dot{s} \end{bmatrix}$$

$$= \begin{bmatrix} -(h - s\cos\phi)\sin\theta\dot{\theta} - \cos\phi\cos\theta\dot{s} \\ -(h - s\cos\phi)\cos\theta\dot{\theta} + \cos\phi\sin\theta\dot{s} \\ \sin\phi\dot{s} \end{bmatrix} \quad [\mathbf{e}]$$

While one can obtain the above relation by direct differentiation of Eq. [c], the use of Eq. [2.3.24] lends itself to efficient computer implementation and is preferred for more complex problems.

2.4 TRANSFORMATION OF COORDINATES, FINITE ROTATIONS

Consider two coordinate frames, $x_1 x_2 x_3$ with unit vectors $\mathbf{e}_1$, $\mathbf{e}_2$, and $\mathbf{e}_3$; and $x'_1 x'_2 x'_3$ with unit vectors $\mathbf{e}'_1$, $\mathbf{e}'_2$, and $\mathbf{e}'_3$. Without loss of generality we select the frames such

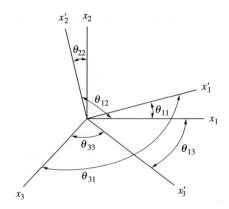

Figure 2.7

that their origins coincide, as shown in Fig. 2.7. The position vector to a point P can be expressed as

$$\mathbf{r} = x_1\mathbf{e}_1 + x_2\mathbf{e}_2 + x_3\mathbf{e}_3 = x'_1\mathbf{e}'_1 + x'_2\mathbf{e}'_2 + x'_3\mathbf{e}'_3 \qquad [2.4.1]$$

in which

$$x_1 = \mathbf{r}\cdot\mathbf{e}_1 \qquad x_2 = \mathbf{r}\cdot\mathbf{e}_2 \qquad x_3 = \mathbf{r}\cdot\mathbf{e}_3$$
$$x'_1 = \mathbf{r}\cdot\mathbf{e}'_1 \qquad x'_2 = \mathbf{r}\cdot\mathbf{e}'_2 \qquad x'_3 = \mathbf{r}\cdot\mathbf{e}'_3 \qquad [2.4.2]$$

These terms can be generalized to resolve the vector $\mathbf{r}$ into its components along a coordinate system with unit vectors $\mathbf{e}_1\mathbf{e}_2\mathbf{e}_3$ as

$$\mathbf{r} = (\mathbf{r}\cdot\mathbf{e}_1)\mathbf{e}_1 + (\mathbf{r}\cdot\mathbf{e}_2)\mathbf{e}_2 + (\mathbf{r}\cdot\mathbf{e}_3)\mathbf{e}_3 \qquad [2.4.3]$$

In a similar fashion, we can express unit vectors in the two coordinate systems in terms of each other. For example, we express the primed unit vectors as

$$\mathbf{e}'_1 = (\mathbf{e}'_1\cdot\mathbf{e}_1)\mathbf{e}_1 + (\mathbf{e}'_1\cdot\mathbf{e}_2)\mathbf{e}_2 + (\mathbf{e}'_1\cdot\mathbf{e}_3)\mathbf{e}_3$$
$$\mathbf{e}'_2 = (\mathbf{e}'_2\cdot\mathbf{e}_1)\mathbf{e}_1 + (\mathbf{e}'_2\cdot\mathbf{e}_2)\mathbf{e}_2 + (\mathbf{e}'_2\cdot\mathbf{e}_3)\mathbf{e}_3$$
$$\mathbf{e}'_3 = (\mathbf{e}'_3\cdot\mathbf{e}_1)\mathbf{e}_1 + (\mathbf{e}'_3\cdot\mathbf{e}_2)\mathbf{e}_2 + (\mathbf{e}'_3\cdot\mathbf{e}_3)\mathbf{e}_3 \qquad [2.4.4]$$

Let us now examine the nature of the dot product terms in Eqs. [2.4.4]. Take, for example, $\mathbf{e}'_1\cdot\mathbf{e}_2 = \mathbf{e}_2\cdot\mathbf{e}'_1$. Evaluating that expression, we obtain

$$\mathbf{e}'_1\cdot\mathbf{e}_2 = \mathbf{e}_2\cdot\mathbf{e}'_1 = |\mathbf{e}'_1||\mathbf{e}_2|\cos\theta_{21} = \cos\theta_{21} \qquad [2.4.5]$$

where θ_{21} is the angle between the x_2 and x'_1 axes (Fig. 2.7). The dot product between two unit vectors is equal to the cosine of the angle between them. We define a quantity called *direction cosine* between the two axes x_i and x'_j by the cosine of the angle between the x_i and x'_j axes, and denote it by $c_{ij} = \mathbf{e}_i\cdot\mathbf{e}'_j = \cos\theta_{ij}$ ($i, j = 1, 2, 3$). The angles that the coordinate axes make with the axes of another coordinate system are called *direction angles*. Considering the preceding relations,

2.4 TRANSFORMATION OF COORDINATES, FINITE ROTATIONS

one can write for x_1'

$$x_1' = \mathbf{r} \cdot \mathbf{e}_1' = (x_1\mathbf{e}_1 + x_2\mathbf{e}_2 + x_3\mathbf{e}_3) \cdot [(\mathbf{e}_1' \cdot \mathbf{e}_1)\mathbf{e}_1 + (\mathbf{e}_1' \cdot \mathbf{e}_2)\mathbf{e}_2 + (\mathbf{e}_1' \cdot \mathbf{e}_3)\mathbf{e}_3]$$
$$= x_1\mathbf{e}_1' \cdot \mathbf{e}_1 + x_2\mathbf{e}_1' \cdot \mathbf{e}_2 + x_3\mathbf{e}_1' \cdot \mathbf{e}_3 = x_1 c_{11} + x_2 c_{21} + x_3 c_{31} \qquad \textbf{[2.4.6]}$$

Similar expressions are derived for x_2' and x_3'. Define the column vectors $\{r\}$ and $\{r'\}$ as

$$\{r\} = [x_1 \quad x_2 \quad x_3]^T \qquad \{r'\} = [x_1' \quad x_2' \quad x_3']^T \qquad \textbf{[2.4.7]}$$

and the *direction cosine matrix* $[c]$ as

$$[c] = \begin{bmatrix} c_{11} & c_{12} & c_{13} \\ c_{21} & c_{22} & c_{23} \\ c_{31} & c_{32} & c_{33} \end{bmatrix} \qquad \textbf{[2.4.8]}$$

which leads to the relationship between $\{r'\}$ and $\{r\}$ as

$$\{r'\} = [c]^T \{r\} \qquad \textbf{[2.4.9]}$$

Next, we express x_1, x_2, x_3 in terms of x_1', x_2', x_3'. Following the same procedure as above, we obtain

$$x_1 = c_{11} x_1' + c_{12} x_2' + c_{13} x_3'$$
$$x_2 = c_{21} x_1' + c_{22} x_2' + c_{23} x_3'$$
$$x_3 = c_{31} x_1' + c_{32} x_2' + c_{33} x_3' \qquad \textbf{[2.4.10]}$$

which can be written as

$$\{r\} = [c]\{r'\} \qquad \textbf{[2.4.11]}$$

Equation [2.4.9] can be inverted to yield $\{r\} = [c]^{-T}\{r'\}$. Comparing with Eq. [2.4.11], we conclude that the direction cosine matrix is a *unitary* (also called *orthonormal*) matrix, that is, its inverse is equal to its transpose, or

$$[c]^{-1} = [c]^T \qquad [c][c]^T = [1] \qquad \textbf{[2.4.12a,b]}$$

where [1] is the *identity matrix*.

Observe from the preceding equations that the unit vectors can also be expressed in terms of each other using the direction cosine matrix. Defining the column vectors

$$\{e\} = \begin{bmatrix} \mathbf{e}_1 \\ \mathbf{e}_2 \\ \mathbf{e}_3 \end{bmatrix} \qquad \{e'\} = \begin{bmatrix} \mathbf{e}_1' \\ \mathbf{e}_2' \\ \mathbf{e}_3' \end{bmatrix} \qquad \textbf{[2.4.13]}$$

it is easy to show that

$$\{e'\} = [c]^T \{e\} \qquad \{e\} = [c]\{e'\} \qquad \textbf{[2.4.14]}$$

Be aware that the definition of direction cosine we are using here is not universally accepted. Some texts instead define the direction cosine as $c_{ij} = \mathbf{e}_i' \cdot \mathbf{e}_j$.

We defined one direction cosine for each angle between the *i*th and *j*th coordinate axes, for a total of nine. The question arises as to how many of the direction cosines are independent of each other. Equation [2.4.12b] represents six independent equations that relate the direction cosines (six because of the symmetry of the equations), reducing the number of independent direction cosines c_{ij}, and hence independent angles θ_{ij}, to three.[1] It follows that, at most, three parameters are necessary to represent the transformation from any given configuration of coordinate axes to another one.

The important question is how to select these parameters. One possibility is to use three rotation angles. In this case, any two successive rotations need to be about nonparallel axes. Otherwise the rotation angles will not be distinguishable. Another possibility is to use a single rotation about a particular axis. We will make use of this case in Chapter 7.

Let us consider three rotation angles and analyze how one can accomplish rotations of coordinate systems, and explore means of expressing rotations of coordinate systems and the rates of change of these rotations. To this end, we identify two approaches: a body-fixed rotation sequence and a space-fixed rotation sequence.

To carry out a *body-fixed rotation sequence,* begin with an initial frame and rotate it about one of its axes. Make the next rotation about one of the axes of the rotated coordinate system, which leads to a third coordinate system. Then rotate this third coordinate system about one of its axes to obtain the final frame. This rotation sequence can be visualized by imagining a box attached to the moving reference frame. Each rotation is performed along one of the edges of the box. The position of the box with respect to the final rotated coordinate frame is the same as its position with respect to the initial frame.

Consider an initial frame $x_1 x_2 x_3$ as shown in Fig. 2.8, and rotate it by an angle of θ_1 about the x_1 axis. Denoting the resulting frame by $y_1 y_2 y_3$, we have

$$y_1 = x_1 \qquad y_2 = x_2 \cos\theta_1 + x_3 \sin\theta_1 \qquad y_3 = -x_2 \sin\theta_1 + x_3 \cos\theta_1$$
[2.4.15]

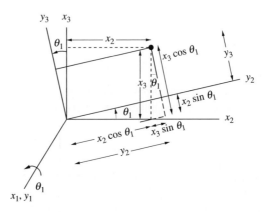

Figure 2.8 A 1 rotation

---

[1] One can demonstrate this by writing [c] as three column vectors [{a_1} {a_2} {a_3}]. These vectors are orthonormal vectors, and they represent the direction angles of the axes of the transformed coordinates. Equation [2.14.12b] represents the six possible dot products among these vectors.

or, in matrix form,

$$\{y\} = [R_1]\{x\} \quad [2.4.16]$$

in which

$$\{y\} = \begin{bmatrix} y_1 \\ y_2 \\ y_3 \end{bmatrix} \quad \{x\} = \begin{bmatrix} x_1 \\ x_2 \\ x_3 \end{bmatrix} \quad [R_1] = \begin{bmatrix} 1 & 0 & 0 \\ 0 & \cos\theta_1 & \sin\theta_1 \\ 0 & -\sin\theta_1 & \cos\theta_1 \end{bmatrix} \quad [2.4.17]$$

where $[R_1]$ is referred to as the *rotation matrix*. We recognize that $[R_1]$ is the transpose of the direction cosine matrix between the two coordinate systems, $[c_1] = [R_1]^T$. The above transformation is also known as a *1 rotation,* denoting the axis about which the rotation takes place.

Take the $y_1 y_2 y_3$ axes and rotate them by an angle θ_2 about the y_3 axis (Fig. 2.9). This type of rotation is called a *3 rotation.* Denoting the resulting frame by $z_1 z_2 z_3$, we can show that

$$\{z\} = [R_2]\{y\} \quad [2.4.18]$$

in which

$$\{z\} = \begin{bmatrix} z_1 \\ z_2 \\ z_3 \end{bmatrix} \quad [R_2] = \begin{bmatrix} \cos\theta_2 & \sin\theta_2 & 0 \\ -\sin\theta_2 & \cos\theta_2 & 0 \\ 0 & 0 & 1 \end{bmatrix} \quad [2.4.19]$$

The rotation matrix $[R_2]$ is the transpose of the direction cosine matrix between $\{z\}$ and $\{y\}$. Finally, rotate the $z_1 z_2 z_3$ axes by θ_3 about the z_2 axis (a *2 rotation*) to obtain the $x_1' x_2' x_3'$ axes (Fig. 2.10). Similar to the previous rotations, we have

$$\{x'\} = [R_3]\{z\} \quad [2.4.20]$$

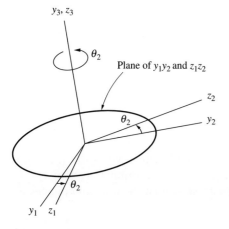

Figure 2.9   A 3 rotation

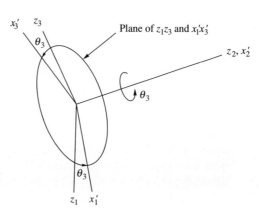

Figure 2.10   A 2 rotation

in which

$$\{x'\} = \begin{bmatrix} x'_1 \\ x'_2 \\ x'_3 \end{bmatrix} \qquad [R_3] = \begin{bmatrix} \cos\theta_3 & 0 & -\sin\theta_3 \\ 0 & 1 & 0 \\ \sin\theta_3 & 0 & \cos\theta_3 \end{bmatrix} \qquad [2.4.21]$$

To obtain the $x'_1 x'_2 x'_3$ axes from the $x_1 x_2 x_3$ axes, we combine Eqs. [2.4.16] through [2.4.21], which yields

$$\{x'\} = [R_3][R_2][R_1]\{x\} = [c_3]^T [c_2]^T [c_1]^T \{x\} = [R]\{x\} \qquad [2.4.22]$$

in which $[R]$ is the rotation matrix between the unprimed and primed coordinates. It is clear that $[R] = [c]^T$, where $[c] = [c_1][c_2][c_3]$. The transformation we have just performed is referred to as a *1-3-2 rotation sequence,* describing the order of the axes about which the coordinate systems are transformed.

We can continue to perform more rotations of the kind above, and in some cases it may be convenient to do so. However, performing more rotations than three introduces a redundancy. As an illustration, consider again the same rotation sequence. Given the direction cosine matrix between the initial and rotated frames, and what the rotation sequence is, one can uniquely determine angles θ_1, θ_2, and θ_3. This is because Eq. [2.4.12b] describes three independent equations which can be solved for the three unknowns θ_1, θ_2, θ_3. If we have a fourth rotation, with an angle θ_4, we still have three independent equations; but now we have four unknowns with no unique solution.

All rotation matrices $[R_i]$ ($i = 1, 2, 3$) have determinants equal to 1, that is, $\det[R_i] = 1$. From linear algebra,

$$\det[R] = \det([R_3][R_2][R_1]) = \det[R_3]\det[R_2]\det[R_1] = 1 \qquad [2.4.23]$$

so that the combined transformation from $\{x\}$ to $\{x'\}$ is carried out by a matrix whose determinant is equal to 1. This implies that the direction cosine matrix between $\{x\}$ and $\{x'\}$ has a determinant of unity. The determinant of a general orthogonal matrix is ± 1, so that we have in effect shown that for an orthonormal matrix to represent a direction cosine matrix, its determinant must be equal to unity. One can show independently of the preceding argument that the direction cosine matrix between any two right-handed coordinate systems has a determinant equal to 1.

The second approach mentioned for describing rotation transformations between coordinate systems is by means of the *space-fixed rotation sequence,* where the rotation transformations are carried out about the initial axes. Consider a set of initial coordinates $x_1 x_2 x_3$. We first rotate this frame about the x_1 axis by an angle θ_1 to obtain the $y_1 y_2 y_3$ axes and call this rotation matrix $[R_1]$. Then, we rotate the $y_1 y_2 y_3$ coordinates about the x_2 axis by an angle θ_2 to obtain $z_1 z_2 z_3$ coordinates. We denote this transformation matrix by $[R_2]$. In a similar fashion, we perform the third transformation about the x_3 axis by θ_3 to obtain the final coordinate system $x'_1 x'_2 x'_3$. Denoting the rotation matrix by $[R_3]$, one can show that the final coordinates are related to the original coordinates by

$$\{x'\} = [R]\{x\} = [R_1][R_2][R_3]\{x\} \qquad [2.4.24]$$

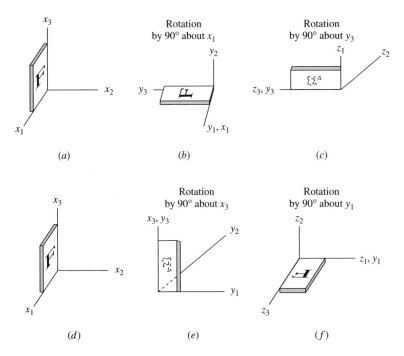

Figure 2.11 Finite rotations do *not* commute

Looking at Eq. [2.4.24], we see that the final transformation matrix is in reverse order compared with the transformation matrix for body-fixed transformations.

In general, one uses body-fixed transformations to relate one coordinate system to another. It is usually more convenient and meaningful to visualize the motion and to express angular velocities and accelerations in terms of a set of axes attached to the body. Nevertheless, space-fixed rotations provide an alternate description, and they help one to visualize the rotation angles.

We are interested in expressing transformations from one coordinate system to another as vectors. We can see from the preceding analysis that the order in which the rotations are performed makes a difference in the orientation of the transformed coordinate system. One can verify this visually, by just taking a book and rotating it about two axes in different sequences. The concept is illustrated in Fig. 2.11 for a body-fixed rotation sequence. One can illustrate this concept using a space-fixed rotation sequence as well. Therefore, it is not possible to represent rotations of coordinate systems by finite angles as vector operations, because the commutativity rule will not hold.

Example 2.4

We analyze the minimum amount of information needed to determine the direction cosine matrix uniquely. We begin with a set of axes $x_1 x_2 x_3$ and transform it into $x_1' x_2' x_3'$. We have nine direction cosines and six independent equations resulting from Eq. [2.4.12b], so that three of the direction cosines have to be specified. Consider first the case where the following

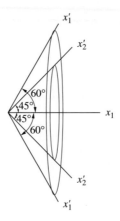

Figure 2.12

information is given:

Angle between the x_1 and x'_1 axes is 60°

Angle between the x_1 and x'_2 axes is 45°

Angle between the x_1 and x'_3 axes is 60°

We are given the task of finding the direction cosines. We need to first check to see if the information given is consistent, or, in other words, geometrically compatible. Given the angle between two axes, the locus of points that are compatible defines a right circular cone with the apex angle as the angle between the two axes, as shown in Fig. 2.12. In this example, the axis x'_1 defines a right circular cone about x_1 with apex angle 60°. Similarly, the x'_2 axis defines a cone about x_1 with apex angle 45°. It follows that the maximum angle between any two lines on the cones is 105°, making it possible for the angle between the x'_2 and x'_1 axes to be 90°. In a similar fashion, it is possible to have a 90° angle between the x'_1 and x'_3 axes and the x'_2 and x'_3 axes. Therefore, the information given is compatible with a right-handed coordinate system.

(As an illustration of a geometrically incompatible case, suppose we were given the problem above, except that the angle between the x_1 and x'_1 axes is 30°. It follows that the maximum angle that one can have between the x'_1 and x'_2 axes is 75°, making it impossible to have a right angle between x'_1 and x'_2.)

Returning to the original problem, once we determine that the information we have is consistent, we proceed with finding the direction cosines. From the above relations

$$c_{11} = 0.5 \qquad c_{12} = 0.707 \qquad c_{13} = 0.5 \qquad \text{[a]}$$

Equation [2.4.12b] written in terms of c_{ij} results in the six equations

$$c_{11}^2 + c_{12}^2 + c_{13}^2 = 1 \qquad c_{21}^2 + c_{22}^2 + c_{23}^2 = 1 \qquad c_{31}^2 + c_{32}^2 + c_{33}^2 = 1$$

$$c_{11}c_{21} + c_{12}c_{22} + c_{13}c_{23} = 0 \qquad c_{11}c_{31} + c_{12}c_{32} + c_{13}c_{33} = 0$$

$$c_{21}c_{31} + c_{22}c_{32} + c_{23}c_{33} = 0 \qquad \text{[b]}$$

The values in Eq. [a] satisfy the first of Eqs. [b] uniquely, so that we are reduced to five equations for the six unknown direction cosines. Hence, the direction cosines cannot be solved for. The physical explanation of this is that only the x_1 axis is uniquely specified with respect to the $x'_1 x'_2 x'_3$ frame. The x_2 and x_3 axes are not specified at all. A coordinate system obtained by any amount of rotation about the x_1 axes will satisfy Eq. [b].

The conclusion is that, to define the direction cosines, one needs consistent and compatible information about the direction angles of at least two different axes in any frame. Consider now the following information:

Angle between the x_3 and x'_3 axes is 15°

Angle between the x_2 and x'_2 axes is 35°

Angle between the x_1 and x'_2 axes is 125°

This information results in the direction cosines

$$c_{33} = 0.9659 \qquad c_{22} = 0.8192 \qquad c_{12} = -0.5736 \qquad \text{[c]}$$

A quick examination of c_{12} and c_{22} indicates that $c_{32} = 0$, so that the x'_2 axis lies on the plane generated by the x_1 and x_2 axes. The transformation from $x_1 x_2 x_3$ to $x'_1 x'_2 x'_3$ is accomplished in two rotations. The first is a counterclockwise rotation about the x_3 axis by an angle of 35°, resulting in the intermediate coordinate system $y_1 y_2 y_3$. The second rotation is about the y_2 axis by an angle of 15°. What we do not know at this point is whether this second rotation is clockwise or counterclockwise. Fig. 2.13 illustrates the rotations. It follows that in this case, we need one more piece of information to uniquely determine the transformed coordinates. This additional information can be in the form of a sketch.

Example 2.5

One way to visualize body-fixed rotation transformations is to attach an imaginary box to the coordinate frame and observe what happens to the box as the coordinate system is rotated. Consider the box in Fig. 2.14. Rotate the box about line OA clockwise by 30°, then about line OB counterclockwise by 105°. What are the coordinates of point D in the initial frame after these rotations?

Solution

The relative frame is attached to the box. We denote the initial frame by XYZ. The intermediate frame after rotation about OA (the X axis) is denoted by $x'y'z'$, as shown in Fig. 2.15. The final configuration xyz is obtained by a counterclockwise rotation about OB (the y' axis) by 105°, as shown in Fig. 2.16.

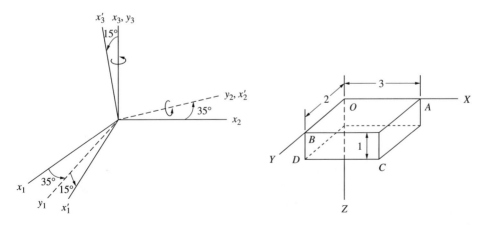

Figure 2.13   **Figure 2.14**

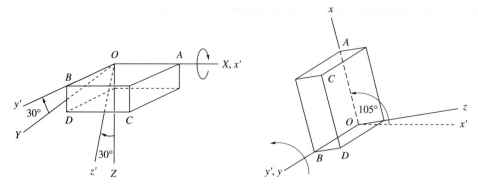

Figure 2.15 **Figure 2.16**

Using rotation transformation equations, we write the relations between the coordinate frames as

$$\begin{bmatrix} x' \\ y' \\ z' \end{bmatrix} = \begin{bmatrix} 1 & 0 & 0 \\ 0 & \cos(-30°) & \sin(-30°) \\ 0 & -\sin(-30°) & \cos(-30°) \end{bmatrix} \begin{bmatrix} X \\ Y \\ Z \end{bmatrix} = \begin{bmatrix} 1 & 0 & 0 \\ 0 & \frac{\sqrt{3}}{2} & -\frac{1}{2} \\ 0 & \frac{1}{2} & \frac{\sqrt{3}}{2} \end{bmatrix} \begin{bmatrix} X \\ Y \\ Z \end{bmatrix} = [R_1] \begin{bmatrix} X \\ Y \\ Z \end{bmatrix} \quad \text{[a]}$$

$$\begin{bmatrix} x \\ y \\ z \end{bmatrix} = \begin{bmatrix} \cos(105°) & 0 & -\sin(105°) \\ 0 & 1 & 0 \\ \sin(105°) & 0 & \cos(105°) \end{bmatrix} \begin{bmatrix} x' \\ y' \\ z' \end{bmatrix} = \begin{bmatrix} -0.2588 & 0 & -0.9659 \\ 0 & 1 & 0 \\ 0.9659 & 0 & -0.2588 \end{bmatrix} \begin{bmatrix} x' \\ y' \\ z' \end{bmatrix} = [R_2] \begin{bmatrix} x' \\ y' \\ z' \end{bmatrix} \quad \text{[b]}$$

so that the relation between the original and the rotated frames is

$$\begin{bmatrix} x \\ y \\ z \end{bmatrix} = [R_2][R_1] \begin{bmatrix} X \\ Y \\ Z \end{bmatrix} = [R] \begin{bmatrix} X \\ Y \\ Z \end{bmatrix} \quad \text{[c]}$$

in which

$$[R] = [R_2][R_1] = \begin{bmatrix} -0.2588 & -0.4830 & -0.8365 \\ 0 & 0.8660 & -0.5000 \\ 0.9659 & -0.1294 & -0.2241 \end{bmatrix} \quad \text{[d]}$$

is the final transformation matrix. Equation [c] is valid when relating the initial, as well as the final, orientations of points on the box as the box is moved. Denoting these initial values by the subscript i and the final coordinates by the subscript f, we can write

$$\begin{bmatrix} x_i \\ y_i \\ z_i \end{bmatrix} = [R] \begin{bmatrix} X_i \\ Y_i \\ Z_i \end{bmatrix} \qquad \begin{bmatrix} x_f \\ y_f \\ z_f \end{bmatrix} = [R] \begin{bmatrix} X_f \\ Y_f \\ Z_f \end{bmatrix} \quad \text{[e, f]}$$

On the other hand, because the moving frame is attached to the box, the coordinates of a point on the box before rotations in the initial frame are the same as the coordinates of that point in the rotated frame after rotations. We therefore have

$$\begin{bmatrix} x_f \\ y_f \\ z_f \end{bmatrix} = \begin{bmatrix} X_i \\ Y_i \\ Z_i \end{bmatrix} \qquad [\mathbf{g}]$$

Introducing the inverse of Eq. [f] into Eq. [g], we obtain

$$\begin{bmatrix} X_f \\ Y_f \\ Z_f \end{bmatrix} = [R]^T \begin{bmatrix} x_f \\ y_f \\ z_f \end{bmatrix} = [R]^T \begin{bmatrix} X_i \\ Y_i \\ Z_i \end{bmatrix} \qquad [\mathbf{h}]$$

To find the coordinates of point D, we denote the initial and final positions of point D as D_i and D_f. The initial coordinates of point D are $(0, 2, 1)$; thus, its final coordinates are

$$\begin{bmatrix} X_{Df} \\ Y_{Df} \\ Z_{Df} \end{bmatrix} = [R]^T \begin{bmatrix} X_{Di} \\ Y_{Di} \\ Z_{Di} \end{bmatrix} = [R]^T \begin{bmatrix} 0 \\ 2 \\ 1 \end{bmatrix} = \begin{bmatrix} 0.9659 \\ 1.603 \\ -1.224 \end{bmatrix} \qquad [\mathbf{i}]$$

2.5 INFINITESIMAL ROTATIONS, ANGULAR VELOCITY

In the previous section we saw that consecutive rotations of coordinate frames by finite angles do not lend themselves to representation as vectors. Hence, one does not have a vector to differentiate in order to represent rotation rates. To express rates of rotations, we begin by analyzing infinitesimal rotations. Consider, for example, a 1-2-3 body-fixed rotation sequence with rotation angles of θ_1, θ_2, and θ_3 (θ_1 about x_1, θ_2 about y_2, and θ_3 about z_3). The final transformation matrix can be shown to be

$$[R] = \begin{bmatrix} c\theta_2 c\theta_3 & c\theta_1 s\theta_3 + s\theta_1 s\theta_2 c\theta_3 & s\theta_1 s\theta_3 - c\theta_1 s\theta_2 c\theta_3 \\ -c\theta_2 s\theta_3 & c\theta_1 c\theta_3 - s\theta_1 s\theta_2 s\theta_3 & s\theta_1 c\theta_3 + c\theta_1 s\theta_2 s\theta_3 \\ s\theta_2 & -s\theta_1 c\theta_2 & c\theta_1 c\theta_2 \end{bmatrix} \qquad [\mathbf{2.5.1}]$$

Now, consider that all the rotation angles θ_1, θ_2, and θ_3 are very small, and replace them with $\Delta\theta_1$, $\Delta\theta_2$, and $\Delta\theta_3$. We also assume that these small rotations take place during a short time period of Δt. Invoking the small angle assumption of $\sin \Delta\theta_i \approx \Delta\theta_i$, $\cos \Delta\theta_i \approx 1$, and neglecting second- and higher-order terms in $\Delta\theta_i$, the rotation matrix becomes

$$[R] = \begin{bmatrix} 1 & \Delta\theta_3 & -\Delta\theta_2 \\ -\Delta\theta_3 & 1 & \Delta\theta_1 \\ \Delta\theta_2 & -\Delta\theta_1 & 1 \end{bmatrix} \qquad [\mathbf{2.5.2}]$$

It should be stressed that $\Delta\theta_1$, $\Delta\theta_2$, and $\Delta\theta_3$ are not the differentials of finite expressions but differential quantities themselves. This observation is critical to understanding the definition of angular velocity. Examining Eqs. [2.5.1] and [2.5.2] more closely, it becomes clear that no matter what the order of transformation is, $[R]$ in Eq. [2.5.2] has the same form, indicating that infinitesimal rotations are commutative. The orientation of the transformed coordinate system does not depend on the sequence of the infinitesimal rotations.

Let us explore ways to express infinitesimal rotations, and their rates, as vectors. We write the relation between the coordinates of a point in terms of the initial and transformed coordinates as

$$\{x'\} = [R]\{x\} \qquad \{x\} = [R]^T\{x'\} \qquad [2.5.3]$$

We express the rotation matrix as $[R] = [1] - [\Delta\theta]$, in which $[1]$ is the identity matrix, and

$$[\Delta\theta] = \begin{bmatrix} 0 & -\Delta\theta_3 & \Delta\theta_2 \\ \Delta\theta_3 & 0 & -\Delta\theta_1 \\ -\Delta\theta_2 & \Delta\theta_1 & 0 \end{bmatrix} \qquad [2.5.4]$$

is a skew-symmetric matrix, that is, $[\Delta\theta]^T = -[\Delta\theta]$. It follows that $[R]^T = [1] + [\Delta\theta]$.

We now obtain a relationship between the initial and final coordinates of a point as the reference frame is transformed. Denoting quantities pertinent to the initial and final positions by the subscripts i and f,

$$\{x'_i\} = [R]\{x_i\} \qquad \{x'_f\} = [R]\{x_f\} \qquad \{x'_f\} = \{x_i\} \qquad [2.5.5]$$

Next, we define by $\{\Delta x'\}$ the change in the coordinates by

$$\{\Delta x'\} = \{x'_f\} - \{x'_i\} = \{x_i\} - \{x'_i\} = [R]^T\{x'_i\} - \{x'_i\} \qquad [2.5.6]$$

Now, dropping the subscript i, and using the relation $[R]^T = [1] + [\Delta\theta]$, we can express $\{\Delta x'\}$ as

$$\{\Delta x'\} = [R]^T\{x'\} - \{x'\} = ([1] + [\Delta\theta])\{x'\} - \{x'\} = [\Delta\theta]\{x'\} \qquad [2.5.7]$$

Note that we could have derived the equivalent of Eq. [2.5.7] in terms of the initial coordinates $x_1 x_2 x_3$. Indeed, defining the change in coordinates as $\{\Delta x\} = \{x_f\} - \{x_i\}$ and substituting into Eqs. [2.5.5], one obtains $\{\Delta x\} = [\Delta\theta]\{x\}$.

Eq. [2.5.7] can be viewed as the column vector representation of the relation

$$\mathbf{\Delta r} = \mathbf{\Delta\theta} \times \mathbf{r} \qquad [2.5.8]$$

where

$$\mathbf{\Delta\theta} = \Delta\theta_1 \mathbf{e}'_1 + \Delta\theta_2 \mathbf{e}'_2 + \Delta\theta_3 \mathbf{e}'_3 \qquad [2.5.9]$$

is an infinitesimal rotation vector.[2] The concept is illustrated in Fig. 2.17. The rotation takes place about an axis passing through the vector $\mathbf{\Delta\theta}$. The rotation is by an amount $\Delta\theta$, which is the magnitude of $\mathbf{\Delta\theta}$. Note that the boldface in the above equation, indicating that the quantity is a vector, is over the entire expression $\mathbf{\Delta\theta}$ and not just over the θ. This signifies that $\mathbf{\Delta\theta}$ is not the infinitesimal value of a vector but a defined quantity consisting of a collection of infinitesimal rotations.

---

[2] We defined by Eq. [2.5.9] the infinitesimal rotation vector without rigorously proving that it indeed is a vector. The proof requires that certain transformation properties be satisfied. It can be found in the text by Meirovitch.

2.5 INFINITESIMAL ROTATIONS, ANGULAR VELOCITY

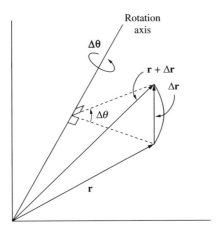

Figure 2.17

Let us divide Eq. [2.5.8] by the time increment Δt during which the rotations take place, and take the limit as Δt approaches zero. The left side leads to a simple derivative term, written

$$\lim_{\Delta t \to 0} \frac{\Delta \mathbf{r}}{\Delta t} = \frac{d\mathbf{r}}{dt} = \dot{\mathbf{r}} \qquad [2.5.10]$$

To evaluate the right side, we take Eq. [2.5.9], divide it by Δt, and take the limit as Δt approaches zero. We define the resulting expression as the *angular velocity* of the moving frame with respect to the initial frame and write it as

$$\boldsymbol{\omega} = \lim_{\Delta t \to 0} \frac{\Delta \boldsymbol{\theta}}{\Delta t} \qquad [2.5.11]$$

where

$$\boldsymbol{\omega} = \omega_1 \mathbf{e}'_1 + \omega_2 \mathbf{e}'_2 + \omega_3 \mathbf{e}'_3 \qquad [2.5.12]$$

in which

$$\omega_i = \lim_{\Delta t \to 0} \frac{\Delta \theta_i}{\Delta t} \qquad i = 1, 2, 3 \qquad [2.5.13]$$

are the components of the angular velocity, also referred to as the *instantaneous angular velocities of the rotating frame*.

We have shied away from writing the right side of Eq. [2.5.11] as a derivative. What should be emphasized is that *angular velocity is a defined quantity* and that *it is not the derivative of any vector*. For this reason, the angular velocity vector is referred to as *nonholonomic*, a term that is associated with expressions that cannot be expressed as derivatives of other terms. A nonholonomic expression cannot be integrated to another expression. The way one arrives at the angular velocity vector is completely different from the derivation of the expression for translational velocity, or the rate of change of any defined vector.

In view of this discussion, one can write the rate of change of the position vector as

$$\dot{\mathbf{r}} = \boldsymbol{\omega} \times \mathbf{r} \qquad \frac{d}{dt}\{x'\} = [\tilde{\omega}]\{x'\} \qquad [2.5.14]$$

in which $[\tilde{\omega}]$ is the matrix representation of the angular velocity vector $\boldsymbol{\omega}$,

$$[\tilde{\omega}] = \begin{bmatrix} 0 & -\omega_3 & \omega_2 \\ \omega_3 & 0 & -\omega_1 \\ -\omega_2 & \omega_1 & 0 \end{bmatrix} \qquad [2.5.15]$$

It should be reiterated that $\mathbf{r}$ is a vector whose components are fixed in the relative frame $x'_1 x'_2 x'_3$. Equation [2.5.14] is illustrated in Fig. 2.18. The change in $\mathbf{r}$ is due to a change in direction, and hence, $\dot{\mathbf{r}}$ is orthogonal to $\mathbf{r}$. It is also orthogonal to the angular velocity vector, as $\mathbf{r}$ can be visualized as rotating about the axis generated by the angular velocity vector. By definition of the cross product, $\dot{\mathbf{r}}$ is perpendicular to the plane generated by the vectors $\boldsymbol{\omega}$ and $\mathbf{r}$.

Now that we have defined angular velocity as a vector, we can obtain the angular velocity of a reference frame by adding up the angular velocities associated with the rotations that lead to the orientation of the reference frame.

Equation [2.5.14] is valid not only for a vector describing the velocity of a point, but for any vector $\mathbf{u}$ whose components are constant relative to the moving frame. We have

$$\dot{\mathbf{u}} = \boldsymbol{\omega} \times \mathbf{u} \qquad [2.5.16]$$

An example is when $\mathbf{u}$ is a unit vector. For the unit vectors considered in this section, we have

$$\begin{aligned} \dot{\mathbf{e}}'_1 &= \boldsymbol{\omega} \times \mathbf{e}'_1 = -\omega_2 \mathbf{e}'_3 + \omega_3 \mathbf{e}'_2 \\ \dot{\mathbf{e}}'_2 &= \boldsymbol{\omega} \times \mathbf{e}'_2 = \omega_1 \mathbf{e}'_3 - \omega_3 \mathbf{e}'_1 \\ \dot{\mathbf{e}}'_3 &= \boldsymbol{\omega} \times \mathbf{e}'_3 = -\omega_1 \mathbf{e}'_2 + \omega_2 \mathbf{e}'_1 \end{aligned} \qquad [2.5.17]$$

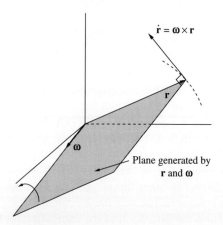

Figure 2.18

Note that the magnitude of the time derivative of a unit vector, $|\dot{\mathbf{e}}'_i|$ ($i = 1, 2, 3$), is not equal to unity.

The preceding definition of angular velocity is not the only way angular velocity can be defined. In the following, we present a more abstract definition. Consider a moving reference frame which is rotating with respect to a fixed reference frame. The angular velocity of the rotating frame is defined as the vector $\boldsymbol{\omega}$, which, when crossed into any vector fixed in the moving frame, gives the rate of change of that vector viewed from the inertial frame. The angular velocity of the relative frame $\boldsymbol{\omega}$ is the quantity that makes the relationship [2.5.16] hold.

Using the unit vectors of the moving reference frame, which in this section we have taken as $\mathbf{e}'_1$, $\mathbf{e}'_2$, and $\mathbf{e}'_3$, and their rates of change, one can define the angular velocity vector as

$$\boldsymbol{\omega} = (\dot{\mathbf{e}}'_2 \cdot \mathbf{e}'_3)\mathbf{e}'_1 + (\dot{\mathbf{e}}'_3 \cdot \mathbf{e}'_1)\mathbf{e}'_2 + (\dot{\mathbf{e}}'_1 \cdot \mathbf{e}'_2)\mathbf{e}'_3 \qquad [2.5.18]$$

This definition can be verified by analyzing the expressions for the rates of change of the unit vectors from Eqs. [2.5.17]. While this definition is more abstract than the way we arrived at Eq. [2.5.12], it is mathematically more sound, and it can be substituted more easily in mathematical operations that involve angular velocity. In Chapter 7 we will see yet another definition of angular velocity.

Note that in this section so far, we have defined angular velocity in a number of ways, discussed what it is physically, and derived expressions for rotating reference frames. What we have not done is to come up with a general way to quantify angular velocity as a function of rotational parameters. We will analyze the quantification issue for the general case in Chapter 7, within the context of rigid bodies.

Now let us discuss a special case of angular velocity. Previously, we defined angular velocity as a vector with certain properties and stated that it is not the derivative of any quantity but rather it is a defined one. There is an exception to this. When angular velocity is along a fixed direction, then angular velocity is called *simple angular velocity,* and it becomes the time derivative of the rotation angle about the fixed direction.

If we denote the unit vector along this fixed direction by, say, $\mathbf{J}$, we can express the angular velocity by $\boldsymbol{\omega} = \omega\mathbf{J}$, and can write ω as an exact differential in the form

$$\omega = \frac{d\theta}{dt} \qquad [2.5.19]$$

Here, θ is the angular displacement about the fixed axis. The commonly studied special cases of plane motion and rotation about a fixed axis involve simple angular velocity.

Let us next consider more than one relative reference frame. We begin with a fixed frame XYZ and rotate it by an angle θ_1 about the X axis to obtain the $x'y'z'$ frame. The angular velocity of the $x'y'z'$ frame with respect to the inertial frame is recognized as simple angular velocity. Denoting it by $\boldsymbol{\omega}_1$, we can write

$$\boldsymbol{\omega}_1 = \dot{\theta}_1\mathbf{I} = \dot{\theta}_1\mathbf{i}' \qquad [2.5.20]$$

We then rotate the $x'y'z'$ frame about the z' axis by θ_2 and obtain the xyz frame. The angular velocity of the xyz frame with respect to the $x'y'z'$ frame, which we will denote by $\boldsymbol{\omega}_2$, is also "simple" when this second rotation is considered by itself, thus

$$\omega_2 = \dot{\theta}_2 \mathbf{k}' = \dot{\theta}_2 \mathbf{k} \qquad [\mathbf{2.5.21}]$$

The angular velocity of the rotated frame xyz can be written by adding the angular velocity terms associated with the two rotations as

$$\omega = \omega_1 + \omega_2 = \dot{\theta}_1 \mathbf{I} + \dot{\theta}_2 \mathbf{k} \qquad [\mathbf{2.5.22}]$$

Let us express the angular velocity in terms of the different reference frames. First, consider the final relative frame. We find $\mathbf{i}'$ by reading the first column of Eq. [2.4.19] as $\mathbf{I} = \mathbf{i}' = \cos\theta_2 \mathbf{i} - \sin\theta_2 \mathbf{j}$. We introduce this into Eq. [2.5.22] and obtain

$$\omega = \omega_1 + \omega_2 = \dot{\theta}_1 \cos\theta_2 \mathbf{i} - \dot{\theta}_1 \sin\theta_2 \mathbf{j} + \dot{\theta}_2 \mathbf{k} \qquad [\mathbf{2.5.23}]$$

We observe that ω cannot be expressed as the derivative of another vector. Hence, it cannot be classified as simple angular velocity, although both ω_1 and ω_2 are simple angular velocities when considered individually. This can be explained by noting that while ω_1 is about a set of fixed axes, ω_2 is actually with respect to a set of rotating axes. The situation does not change when we express the angular velocity vector using the unit vectors of the inertial frame. Indeed, if we use, from Eq. [2.4.17], the relation $\mathbf{k} = \mathbf{k}' = -\sin\theta_1 \mathbf{J} + \cos\theta_1 \mathbf{K}$ and substitute it into Eq. (2.5.22), we obtain

$$\omega = \omega_1 + \omega_2 = \dot{\theta}_1 \mathbf{I} - \dot{\theta}_2 \sin\theta_1 \mathbf{J} + \dot{\theta}_2 \cos\theta_1 \mathbf{K} \qquad [\mathbf{2.5.24}]$$

We hence conclude that for a sequence of rotations about nonparallel axes, the combined angular velocity will not be "simple."

Example 2.6

A momentum wheel is a useful classroom tool to demonstrate angular momentum conservation. It is basically like a bicycle wheel with a thickened rim and handles along the spin axis. The angular momentum principles are illustrated by asking a student to hold the wheel and spin it, and then to move the wheel around or stand on a platform that is free to rotate, as shown in Fig. 2.19.

At a given instant, the momentum wheel is spinning counterclockwise (viewed from the right) with angular velocity of 3 rad/s, and the student holding the wheel is leaning left with

Figure 2.19

2.5 INFINITESIMAL ROTATIONS, ANGULAR VELOCITY

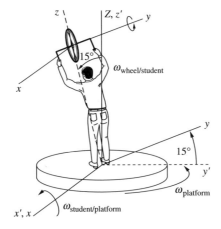

Figure 2.20

an angular velocity of 0.2 rad/s and making an angle of 15° with the vertical. At the same time, the platform is rotating with a clockwise angular velocity of 0.5 rad/s. Find the total angular velocity of the wheel.

Solution

Figure 2.20 illustrates the reference frames involved. We attach the frame $x'y'z'$ to the platform. The inertial Z axis is in the vertical, and it is aligned with the z' axis. The orientation of the wheel can be obtained by rotating the $x'y'z'$ axes by an angle of 15° counterclockwise about the x' axis. Referring to this coordinate system by xyz, the momentum wheel's spin is in the y direction. We write the angular velocity as

$$\boldsymbol{\omega} = \boldsymbol{\omega}_{\text{platform}} + \boldsymbol{\omega}_{\text{student/platform}} + \boldsymbol{\omega}_{\text{wheel/student}} = -0.5\mathbf{k}' + 0.2\mathbf{i}' + 3\mathbf{j} \text{ rad/s} \quad \textbf{[a]}$$

At the instant shown, the unit vectors in the xyz and $x'y'z'$ coordinate frames are related by

$$\begin{bmatrix} \mathbf{i} \\ \mathbf{j} \\ \mathbf{k} \end{bmatrix} = \begin{bmatrix} 1 & 0 & 0 \\ 0 & \cos 15° & \sin 15° \\ 0 & -\sin 15° & \cos 15° \end{bmatrix} \begin{bmatrix} \mathbf{i}' \\ \mathbf{j}' \\ \mathbf{k}' \end{bmatrix} \quad \textbf{[b]}$$

Using the inverse of Eq. [b], we have $\mathbf{k}' = \sin 15°\mathbf{j} + \cos 15°\mathbf{k} = 0.2588\mathbf{j} + 0.9659\mathbf{k}$ and $\mathbf{i}' = \mathbf{i}$, so we can express the angular velocity in terms of a set of coordinates attached to the momentum wheel as

$$\boldsymbol{\omega} = -0.5(0.2588\mathbf{j} + 0.9659\mathbf{k}) + 0.2\mathbf{i} + 3\mathbf{j} = 0.2\mathbf{i} + 2.8706\mathbf{j} - 0.4830\mathbf{k} \text{ rad/s} \quad \textbf{[c]}$$

Note that to express $\boldsymbol{\omega}$ in terms of the unit vectors associated with an inertial reference frame XYZ requires that the exact relationship between the coordinates xyz and XYZ be known. In this problem, we conducted an instantaneous analysis and did not specify the inertial XYZ axes, except for the vertical direction. Expressing $\boldsymbol{\omega}$ in terms of a moving coordinate system is more meaningful.

Example 2.7

The robot arm makes an angle of 40° with the rotating shaft, which oscillates about the y' axis with the relation $\theta(t) = \frac{\pi}{20} \cos 2t$ rad, as shown in Figs. 2.21 through 2.23. The shaft has an angular velocity of $\omega_1 = 0.5$ rad/s. At the tip of the arm, there is another shaft. A disk is

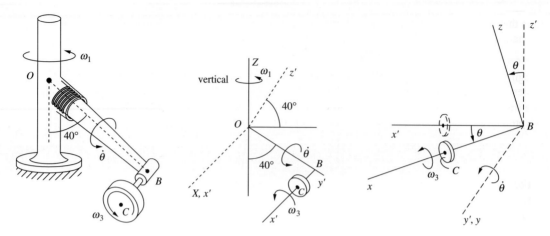

Figure 2.21   **Figure 2.22**   **Figure 2.23**

spinning counterclockwise with $\omega_3 = 7$ rad/s about this shaft. Find the angular velocity of the disk at $t = 3$ s.

Solution

As shown in Fig. 2.22, the Z axis is the vertical, and the $X'Y'Z$ axes are attached to the shaft. The $x'y'z'$ axes are obtained by rotating the $X'Y'Z'$ axes about the X' axis. The y' axis is along the robot arm. We use a rotation about the y' axis by θ to go from the $x'y'z'$ axes to the xyz axes of the second shaft, about which the disk turns. We have

$$\mathbf{i} = \cos\theta \mathbf{i}' - \sin\theta \mathbf{k}' \qquad \mathbf{j} = \mathbf{j}' \qquad \mathbf{k} = \sin\theta \mathbf{i}' + \cos\theta \mathbf{k}' \qquad [\mathbf{a}]$$

The angular velocity of the disk can be written as

$$\boldsymbol{\omega} = \boldsymbol{\omega}_1 + \boldsymbol{\omega}_2 + \boldsymbol{\omega}_3 \qquad [\mathbf{b}]$$

in which

$$\boldsymbol{\omega}_1 = 0.5\mathbf{K} = 0.5(\sin 40°\mathbf{k}' - \cos 40°\mathbf{j}') = -0.3830\mathbf{j}' + 0.3214\mathbf{k}' \text{ rad/s}$$

$$\boldsymbol{\omega}_2 = \dot{\theta}(3)\mathbf{j}' = -\frac{\pi}{10}\sin 6\mathbf{j} = 0.08778\mathbf{j} \text{ rad/s} \qquad \boldsymbol{\omega}_3 = 7\mathbf{i} \text{ rad/s} \qquad [\mathbf{c}]$$

At $t = 3$ s, $\theta(3) = 0.1508$ rad, so that

$$\mathbf{k}' = -\sin\theta \mathbf{i} + \cos\theta \mathbf{k} = -0.1502\mathbf{i} + 0.9887\mathbf{k} \qquad [\mathbf{d}]$$

and we can express the total angular velocity in terms of the xyz coordinates as

$$\boldsymbol{\omega} = 0.3214(-0.1502\mathbf{i} + 0.9887\mathbf{k}) - 0.3830\mathbf{j} + 0.08778\mathbf{j} + 7\mathbf{i}$$

$$= 6.952\mathbf{i} - 0.2952\mathbf{j} + 0.3178\mathbf{k} \text{ rad/s} \qquad [\mathbf{e}]$$

Note that in this example, as well as in the previous one, we did not express the angular velocity in terms of a set of coordinates attached to the rotating body. Rather, we used the xyz coordinates attached to the spin axis of the disk. This is commonly done when analyzing axisymmetric bodies, as we will see in Chapters 7 and 8.

Example 2.8

Obtain the derivatives of the unit vectors in the cylindrical, spherical, and normal-tangential coordinates using Eq. [2.5.16] and compare with the results in Chapter 1.

Solution

We begin by examining the angular parameters associated with the coordinate systems. First consider cylindrical coordinates, as shown in Fig. 1.9. The z direction is fixed and the coordinate frame rotates in the xy plane with angular speed $\dot{\theta}$, so that $\boldsymbol{\omega} = \dot{\theta}\mathbf{k}$. The unit vectors are expressed as the mutually orthogonal triad $\mathbf{e}_r$, $\mathbf{e}_\theta$, and $\mathbf{k}$. For the time derivatives we then have

$$\dot{\mathbf{e}}_r = \dot{\theta}\mathbf{k} \times \mathbf{e}_r = \dot{\theta}\mathbf{e}_\theta \qquad \dot{\mathbf{e}}_\theta = \dot{\theta}\mathbf{k} \times \mathbf{e}_\theta = -\dot{\theta}\mathbf{e}_r \qquad \dot{\mathbf{k}} = \dot{\theta}\mathbf{k} \times \mathbf{k} = 0 \qquad \text{[a]}$$

For spherical coordinates (Fig. 1.12), the mutually orthogonal unit vectors are $\mathbf{e}_R$, $\mathbf{e}_\phi$, and $\mathbf{e}_\theta$. Denoting the unit vectors in the inertial frame by $\mathbf{i}, \mathbf{j}, \mathbf{k}$ and those in the rotated frame by $\mathbf{i}', \mathbf{j}', \mathbf{k}'$, from Fig. 1.12 we can write

$$\mathbf{i}' = \mathbf{e}_\phi \qquad \mathbf{j}' = \mathbf{e}_\theta \qquad \mathbf{k}' = \mathbf{e}_R \qquad \text{[b]}$$

From Fig. 1.12, there are two angular components, θ about the z axis and ϕ about the polar axis. The combined body-fixed rotation matrix $[R]$ is

$$[R] = \begin{bmatrix} c\phi & 0 & -s\phi \\ 0 & 1 & 0 \\ s\phi & 0 & c\phi \end{bmatrix} \begin{bmatrix} c\theta & s\theta & 0 \\ -s\theta & c\theta & 0 \\ 0 & 0 & 1 \end{bmatrix} = \begin{bmatrix} c\phi c\theta & c\phi s\theta & -s\phi \\ -s\theta & c\theta & 0 \\ s\phi c\theta & s\phi s\theta & c\phi \end{bmatrix} \qquad \text{[c]}$$

Using Eq. [2.4.9], we obtain for the unit vectors

$$\mathbf{e}_\phi = \cos\phi\cos\theta\,\mathbf{i} + \cos\phi\sin\theta\,\mathbf{j} - \sin\phi\,\mathbf{k}$$
$$\mathbf{e}_\theta = -\sin\theta\,\mathbf{i} + \cos\theta\,\mathbf{j}$$
$$\mathbf{e}_R = \sin\phi\cos\theta\,\mathbf{i} + \sin\phi\sin\theta\,\mathbf{i} + \cos\phi\,\mathbf{k} \qquad \text{[d]}$$

which are the same as Eqs. [1.3.54] and [1.3.56]. The angular velocity vector is a superposition of the two angular velocities, so that

$$\boldsymbol{\omega} = \dot{\theta}\mathbf{k} + \dot{\phi}\mathbf{e}_\theta \qquad \text{[e]}$$

Expressing the unit vector in the z direction as $\mathbf{k} = \cos\phi\,\mathbf{e}_R - \sin\phi\,\mathbf{e}_\phi$, we arrive at the angular velocity of the coordinate frame in terms of the unit vectors in spherical coordinates as

$$\boldsymbol{\omega} = \dot{\theta}\cos\phi\,\mathbf{e}_R - \dot{\theta}\sin\phi\,\mathbf{e}_\phi + \dot{\phi}\mathbf{e}_\theta \qquad \text{[f]}$$

leading to the derivative expressions

$$\dot{\mathbf{e}}_R = (\dot{\theta}\cos\phi\,\mathbf{e}_R - \dot{\theta}\sin\phi\,\mathbf{e}_\phi + \dot{\phi}\mathbf{e}_\theta)\times\mathbf{e}_R = \dot{\theta}\sin\phi\,\mathbf{e}_\theta + \dot{\phi}\mathbf{e}_\phi$$
$$\dot{\mathbf{e}}_\phi = (\dot{\theta}\cos\phi\,\mathbf{e}_R - \dot{\theta}\sin\phi\,\mathbf{e}_\phi + \dot{\phi}\mathbf{e}_\theta)\times\mathbf{e}_\phi = \dot{\theta}\cos\phi\,\mathbf{e}_\theta - \dot{\phi}\mathbf{e}_R$$
$$\dot{\mathbf{e}}_\theta = (\dot{\theta}\cos\phi\,\mathbf{e}_R - \dot{\theta}\sin\phi\,\mathbf{e}_\phi + \dot{\phi}\mathbf{e}_\theta)\times\mathbf{e}_\theta = -\dot{\theta}\cos\phi\,\mathbf{e}_\phi - \dot{\theta}\sin\phi\,\mathbf{e}_R \qquad \text{[g]}$$

These are the same as Eqs. [1.3.59] through [1.3.61].

For the normal and tangential coordinates, the unit vectors are $\mathbf{e}_t$, $\mathbf{e}_n$, and $\mathbf{e}_b = \mathbf{e}_t \times \mathbf{e}_n$. We showed in Chapter 1 that

$$\dot{\mathbf{e}}_t = \frac{v}{\rho}\mathbf{e}_n \qquad \text{[h]}$$

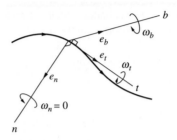

Figure 2.24

where ρ is the radius of curvature. We can write the angular velocity vector in its general form as

$$\boldsymbol{\omega} = \omega_t \mathbf{e}_t + \omega_n \mathbf{e}_n + \omega_b \mathbf{e}_b \qquad [\mathbf{i}]$$

where we have yet to determine ω_t, ω_n, and ω_b, the components of the angular velocity in the tangential, normal, and binormal directions. Fig. 2.24 shows the angular velocities. We obtain the time derivative of $\mathbf{e}_t$ using Eq. [2.5.16] as

$$\dot{\mathbf{e}}_t = (\omega_t \mathbf{e}_t + \omega_n \mathbf{e}_n + \omega_b \mathbf{e}_b) \times \mathbf{e}_t = -\omega_n \mathbf{e}_b + \omega_b \mathbf{e}_n \qquad [\mathbf{j}]$$

Comparing Eqs. [h] and [j], we conclude that

$$\omega_b = \frac{v}{\rho} \qquad \omega_n = 0 \qquad [\mathbf{k}]$$

The above relations can be explained physically and by inspecting Fig. 2.24 more closely. Because the binormal direction is perpendicular to the osculating plane, the component of the angular velocity in the binormal direction is the speed divided by the radius of curvature, or the rate at which the path bends. That $\omega_n = 0$ can be deduced from the same argument. Because at any point the motion of the particle can be considered as going along a circular path whose center is the center of curvature, there is no rotation in the normal direction.

We next consider the time derivatives of the normal and binormal unit vectors, and write

$$\dot{\mathbf{e}}_n = (\omega_t \mathbf{e}_t + \omega_b \mathbf{e}_b) \times \mathbf{e}_n = \omega_t \mathbf{e}_b - \omega_b \mathbf{e}_t \qquad [\mathbf{l}]$$

$$\dot{\mathbf{e}}_b = (\omega_t \mathbf{e}_t + \omega_b \mathbf{e}_b) \times \mathbf{e}_b = -\omega_t \mathbf{e}_n \qquad [\mathbf{m}]$$

Recalling the definition of the torsion of the curve as $|d\mathbf{e}_b| = ds/\tau$, and how the torsion is linked to the twisting of the osculating plane, we obtain the component of the angular velocity in the tangential direction as

$$\omega_t = \frac{v}{\tau} \qquad \dot{\mathbf{e}}_b = -\frac{v}{\tau} \mathbf{e}_n \qquad [\mathbf{n}]$$

As the torsion τ gets larger, the plane of the curve twists less. Equation [l] indicates that the rate of change of the unit vector in the normal direction depends on the way the curve bends as well as on the amount by which it twists, an expected result. The angular velocity of the reference frame can thus be written as

$$\boldsymbol{\omega} = \omega_t \mathbf{e}_t + \omega_b \mathbf{e}_b = v\left(\frac{1}{\tau}\mathbf{e}_t + \frac{1}{\rho}\mathbf{e}_b\right) \qquad [\mathbf{o}]$$

2.6 RATE OF CHANGE OF A VECTOR, ANGULAR ACCELERATION

Consider a vector **u** observed from a moving coordinate system xyz. The coordinate system is rotating with angular velocity $\boldsymbol{\omega}$. The vector **u** is expressed as

$$\mathbf{u} = u_x\mathbf{i} + u_y\mathbf{j} + u_z\mathbf{k} \qquad [2.6.1]$$

The time derivative of **u** can be found by differentiating Eq. [2.6.1] as

$$\dot{\mathbf{u}} = \dot{u}_x\mathbf{i} + \dot{u}_y\mathbf{j} + \dot{u}_z\mathbf{k} + u_x\dot{\mathbf{i}} + u_y\dot{\mathbf{j}} + u_z\dot{\mathbf{k}} \qquad [2.6.2]$$

The first three terms on the right side of this equation denote the change in **u** as viewed by an observer on the moving frame. Hence, the differentiation is carried out in the moving frame. We denote this local derivative term by

$$\left(\frac{d\mathbf{u}}{dt}\right)_{\text{rel}} = \dot{u}_x\mathbf{i} + \dot{u}_y\mathbf{j} + \dot{u}_z\mathbf{k} \qquad [2.6.3]$$

The next three terms on the right side of Eq. [2.6.2] denote the change in **u** due to the rotation of the coordinate system. Considering Eq. [2.5.16], we can express them as

$$u_x\dot{\mathbf{i}} + u_y\dot{\mathbf{j}} + u_z\dot{\mathbf{k}} = u_x\boldsymbol{\omega}\times\mathbf{i} + u_y\boldsymbol{\omega}\times\mathbf{j} + u_z\boldsymbol{\omega}\times\mathbf{k} = \boldsymbol{\omega}\times\mathbf{u} \qquad [2.6.4]$$

leading to the relation

$$\dot{\mathbf{u}} = \frac{d\mathbf{u}}{dt} = \left(\frac{d\mathbf{u}}{dt}\right)_{\text{rel}} + \boldsymbol{\omega}\times\mathbf{u} \qquad [2.6.5]$$

This relation is known as the *transport theorem*. In column vector format we can write it as

$$\frac{d}{dt}\{u\} = \{\dot{u}\}_{\text{rel}} + [\tilde{\omega}]\{u\} \qquad [2.6.6]$$

In operator notation the transport theorem is written as

$$\frac{d}{dt}(\ ) = \frac{d}{dt}(\ )_{\text{rel}} + \boldsymbol{\omega}\times(\ ) \qquad [2.6.7]$$

The physical interpretation of the transport theorem is that the rate of change of a vector is a different quantity when viewed from different reference frames. When dealing with moving reference frames, one must be careful that the differentiation operation is carried out in the proper reference frame.

A natural application of the transport theorem is the calculation of the derivative of the angular velocity, known as the *angular acceleration*. The angular acceleration of a coordinate frame, denoted by $\boldsymbol{\alpha}$, is defined as

$$\boldsymbol{\alpha} = \frac{d}{dt}\boldsymbol{\omega} \qquad [2.6.8]$$

Note that the time derivative is being taken here in the inertial reference frame. We write the angular velocity and acceleration in terms of the unit vectors of the relative frame as

$$\boldsymbol{\omega} = \omega_x \mathbf{i} + \omega_y \mathbf{j} + \omega_z \mathbf{k} \qquad \boldsymbol{\alpha} = \alpha_x \mathbf{i} + \alpha_y \mathbf{j} + \alpha_z \mathbf{k} \qquad [\textbf{2.6.9}]$$

Differentiating the angular velocity, we obtain

$$\boldsymbol{\alpha} = \dot{\omega}_x \mathbf{i} + \dot{\omega}_y \mathbf{j} + \dot{\omega}_z \mathbf{k} + \boldsymbol{\omega} \times \boldsymbol{\omega} = \dot{\omega}_x \mathbf{i} + \dot{\omega}_y \mathbf{j} + \dot{\omega}_z \mathbf{k} \qquad [\textbf{2.6.10}]$$

If we write the angular velocity in terms of the unit vectors of the inertial frame as $\boldsymbol{\omega} = \omega_X \mathbf{I} + \omega_Y \mathbf{J} + \omega_Z \mathbf{K}$, the angular acceleration has the form

$$\boldsymbol{\alpha} = \frac{d}{dt}\boldsymbol{\omega} = \dot{\omega}_X \mathbf{I} + \dot{\omega}_Y \mathbf{J} + \dot{\omega}_Z \mathbf{K} \qquad [\textbf{2.6.11}]$$

In both Eqs. [2.6.10] and [2.6.11], the components of the angular acceleration are the rates of change of the angular velocity, $\alpha_i = \dot{\omega}_i$ ($i = x, y, z$, or $i = X, Y, Z$). We draw the following important conclusion:

If the angular velocity components of a moving coordinate frame are expressed in terms of inertial coordinates or in terms of the coordinates of the moving frame, the components of the angular acceleration can be obtained by a simple differentiation of the angular velocity components.

When the angular velocity of the reference frame is expressed in terms of the unit vectors of another frame that is not attached to the relative frame and is rotating with angular velocity, say, $\boldsymbol{\Omega}$, where $\boldsymbol{\Omega} \neq \boldsymbol{\omega}$, then the expression for the angular acceleration has the form

$$\boldsymbol{\alpha} = \dot{\boldsymbol{\omega}}_{\text{rel}} + \boldsymbol{\Omega} \times \boldsymbol{\omega} \qquad [\textbf{2.6.12}]$$

Note that in this case, $\alpha_i \neq \dot{\omega}_i$.

An interesting application of the transport theorem is in systems involving more than one reference frame. Consider the disk in Fig. 2.25. As the disk spins, the axis about which it rotates also turns. We attach a moving reference frame to the axis and find the angular acceleration of the wheel.

Denote the angular velocity of the disk with respect to its axis by $\boldsymbol{\omega}_2$ and the angular velocity of the axis as $\boldsymbol{\omega}_1$. An observer sitting on the axis of the disk sees it rotating with angular velocity $\boldsymbol{\omega}_2$. The total angular velocity of the wheel is $\boldsymbol{\omega} = \boldsymbol{\omega}_1 + \boldsymbol{\omega}_2$.

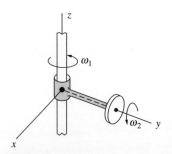

Figure 2.25

2.6 RATE OF CHANGE OF A VECTOR, ANGULAR ACCELERATION

To find the angular acceleration of the wheel, we differentiate the angular velocity expression

$$\boldsymbol{\alpha} = \frac{d}{dt}(\boldsymbol{\omega}_1 + \boldsymbol{\omega}_2) = \frac{d}{dt}\boldsymbol{\omega}_1 + \frac{d}{dt}\boldsymbol{\omega}_2 \qquad [2.6.13]$$

The first term on the right side of Eq. [2.6.13] is the angular acceleration of the axis about which the wheel is rotating. This term can be obtained by straightforward differentiation.

To obtain the angular velocity of the second frame, we invoke the transport theorem. Using Eq. [2.6.5], we obtain

$$\frac{d}{dt}\boldsymbol{\omega}_2 = \left(\frac{d\boldsymbol{\omega}_2}{dt}\right)_{\text{rel}} + \boldsymbol{\omega} \times \boldsymbol{\omega}_2 = \dot{\boldsymbol{\omega}}_{2\text{rel}} + \boldsymbol{\omega}_1 \times \boldsymbol{\omega}_2 \qquad [2.6.14]$$

We get the total angular acceleration by adding Eqs. [2.6.13] and [2.6.14]; thus

$$\boldsymbol{\alpha} = \dot{\boldsymbol{\omega}}_1 + \dot{\boldsymbol{\omega}}_{2\text{rel}} + \boldsymbol{\omega}_1 \times \boldsymbol{\omega}_2 \qquad [2.6.15]$$

Let us contrast the difference between Eqs. [2.6.15] and [2.6.10]. In Eq. [2.6.10] we have a straightforward form for the angular acceleration, because the angular velocity of the frame (the disk) was expressed in terms of the reference frame attached to the disk only. The derivations that led to Eq. [2.6.15] are based on an intermediate frame attached to the disk axis, rotating with $\boldsymbol{\omega}_1$.

To illustrate the point further, consider the coordinate frame transformation in the previous section with an inertial frame XYZ, an intermediate frame $x'y'z'$ obtained by a rotation θ_1 about X, and the final relative frame xyz obtained by a rotation θ_2 about z'. From Eq. [2.5.22], the angular velocities of the two frames are

$$\boldsymbol{\omega}_1 = \dot{\theta}_1 \mathbf{I} \qquad \boldsymbol{\omega}_2 = \dot{\theta}_2 \mathbf{k}' \qquad [2.6.16]$$

in which $\mathbf{I} = \mathbf{i}' = \cos\theta_2 \mathbf{i} - \sin\theta_2 \mathbf{j}$ so that using Eq. [2.6.15], the angular acceleration has the form

$$\boldsymbol{\alpha} = \dot{\boldsymbol{\omega}}_1 + \dot{\boldsymbol{\omega}}_{2\text{rel}} + \boldsymbol{\omega}_1 \times \boldsymbol{\omega}_2 = \ddot{\theta}_1 \mathbf{i}' + \ddot{\theta}_2 \mathbf{k} + \dot{\theta}_1 \mathbf{i}' \times \dot{\theta}_2 \mathbf{k}$$
$$= \ddot{\theta}_1(\cos\theta_2 \mathbf{i} - \sin\theta_2 \mathbf{j}) + \ddot{\theta}_2 \mathbf{k} + \dot{\theta}_1\dot{\theta}_2(\cos\theta_2 \mathbf{i} - \sin\theta_2 \mathbf{j}) \times \mathbf{k}$$
$$= (\ddot{\theta}_1 \cos\theta_2 - \dot{\theta}_1\dot{\theta}_2 \sin\theta_2)\mathbf{i} - (\ddot{\theta}_1 \sin\theta_2 + \dot{\theta}_1\dot{\theta}_2 \cos\theta_2)\mathbf{j} + \ddot{\theta}_2 \mathbf{k} \qquad [2.6.17]$$

Next, let us obtain the angular acceleration by direct differentiation of Eq. [2.5.23], and write

$$\boldsymbol{\alpha} = (\ddot{\theta}_1 \cos\theta_2 - \dot{\theta}_1\dot{\theta}_2 \sin\theta_2)\mathbf{i} - (\ddot{\theta}_1 \sin\theta_2 + \dot{\theta}_1\dot{\theta}_2 \cos\theta_2)\mathbf{j} + \ddot{\theta}_2 \mathbf{k} \qquad [2.6.18]$$

which, of course, is the same answer as Eq. [2.6.17].

The transport theorem is most often the preferred approach for obtaining derivatives, especially for complex problems, and it is more adaptable to implementation by digital computers.

When analyzing the relative motion of bodies and especially when studying three-dimensional rotation problems, one may need to transform velocities and

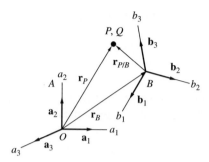

Figure 2.26

accelerations from one reference frame to another several times. The resulting expressions can become quite complicated. To avoid confusion, we are going to adopt the following notation for dealing with multiple reference frames.

We will denote reference frames by capital letters, such as A, B, C, and so on. In general, the A frame will be inertial and the B frame will be relative, and when rigid bodies are involved, this frame will be attached to the body. The coordinate axes in these frames will be defined with lowercase letters corresponding to the frames. For example, for the A frame, the axes will be denoted by a_1, a_2, and a_3, and the corresponding unit vectors by $\mathbf{a}_1$, $\mathbf{a}_2$, and $\mathbf{a}_3$. The origins of the relative frames will usually be denoted by the same capital letter used to denote the reference frame. The origin of the inertial frame will usually be denoted by O. The frames A and B are illustrated in Fig. 2.26.

The angular velocity and angular acceleration of one frame with respect to another, say of frame B with respect to frame A, will be denoted by $^A\boldsymbol{\omega}^B$ and $^A\boldsymbol{\alpha}^B$, respectively. When describing translational velocities and accelerations, as well as the differentiation operation, the frame in which the differentiation is performed will be denoted by a superscript on the left side of these vectors. The differentiation operation can be written as $^A\frac{d}{dt}\mathbf{u}$, $^B\frac{d}{dt}\mathbf{u}$, where $\mathbf{u}$ is a vector. Expressions for the velocity of point P, where the position vector is $\mathbf{r} = \mathbf{r}_P$, have the form

$$^A\mathbf{v}_P = {^A\frac{d}{dt}}\mathbf{r} \qquad {^B\mathbf{v}_P} = {^B\frac{d}{dt}}\mathbf{r} \qquad [2.6.19]$$

Consider the transport theorem, Eq. [2.6.5]. Using the notation introduced above and using the A and B frames as the inertial and relative frames, we can write Eq. [2.6.5] as

$$^A\frac{d}{dt}\mathbf{u} = {^B\frac{d}{dt}}\mathbf{u} + {^A\boldsymbol{\omega}^B} \times \mathbf{u} \qquad [2.6.20]$$

It is clear that the transport theorem can be used to relate derivatives in any two reference frames. We will denote the expression on the left side of Eq. [2.6.20] as the *rate of change of the vector in frame A*, the first expression on the right as the *rate of change of the vector in frame B*, and the second term on the right side as the

2.6 RATE OF CHANGE OF A VECTOR, ANGULAR ACCELERATION

transport term. In operator form we can write the transport theorem as

$$^A\frac{d}{dt}(\ ) = {^B}\frac{d}{dt}(\ ) + {^A}\omega^B \times (\ ) \qquad [2.6.21]$$

Equations [2.6.20] and [2.6.21] describe the transport theorem between any two reference frames with no need to have or to identify an inertial frame. Considering how we defined angular velocity previously, the notation introduced here is more general. For example, the transport theorem between the B and A frames becomes

$$^B\frac{d}{dt}(\ ) = {^A}\frac{d}{dt}(\ ) + {^B}\omega^A \times (\ ) \qquad [2.6.22]$$

which, when compared with Eq. [2.6.21], leads to the expected conclusion that $^B\omega^A = -{^A}\omega^B$.

When there are a number of intermediate frames of reference between the A and B frames, say $A_2, A_3, \ldots, A_N$, one can express the angular velocity of frame B with respect to the A frame as

$$^A\omega^B = {^A}\omega^{A_2} + {^{A_2}}\omega^{A_3} + \ldots + {^{A_N}}\omega^B \qquad [2.6.23]$$

The second derivative of a vector $\mathbf{u}$ in the A and B frames using this notation is $^A\frac{d}{dt}\left({^A}\frac{d}{dt}\mathbf{u}\right)$, $^B\frac{d}{dt}\left({^B}\frac{d}{dt}\mathbf{u}\right)$. Note that in both expressions, all the derivative terms are consistently in the same frame. Following this argument one can write the acceleration of point P as

$$^A\mathbf{a}_P = {^A}\frac{d}{dt}\left({^A}\frac{d}{dt}\mathbf{r}\right) \qquad {^B}\mathbf{a}_P = {^B}\frac{d}{dt}\left({^B}\frac{d}{dt}\mathbf{r}\right) \qquad [2.6.24]$$

In dynamics, one frequently encounters the need to take the derivative of an expression in one frame that has been derived by differentiation in another frame. In such cases we again invoke the transport theorem. For example, given a vector obtained by taking the derivative of a vector $\mathbf{u}$ in the B frame, we use the transport theorem and obtain

$$^A\frac{d}{dt}\left({^B}\frac{d}{dt}\mathbf{u}\right) = {^B}\frac{d}{dt}\left({^B}\frac{d}{dt}\mathbf{u}\right) + {^A}\omega^B \times \left({^B}\frac{d}{dt}\mathbf{u}\right) \qquad [2.6.25]$$

One can show that

$$^A\frac{d}{dt}\left({^B}\frac{d}{dt}\mathbf{u}\right) \neq {^B}\frac{d}{dt}\left({^A}\frac{d}{dt}\mathbf{u}\right) \qquad [2.6.26]$$

When more than one derivative is taken in different reference frames, changing the order of the differentiation gives different results. This applies not only to differentiation with respect to time, but to differentiation with respect to other variables as well. Consider, for example, a vector $\mathbf{u}$ that is a function of the variables $q_1, q_2, \ldots, q_n$. We denote the time derivative of $\mathbf{u}$ in frame A by

$$^A\dot{\mathbf{u}} = \sum_{k=1}^{n} {^A}\frac{\partial \mathbf{u}}{\partial q_k}\dot{q}_k + {^A}\frac{\partial \mathbf{u}}{\partial t} \qquad [2.6.27]$$

When considering the second derivatives of **u**, one can show that

$$\frac{^A\partial \mathbf{u}}{\partial q_i} \neq \frac{^B\partial \mathbf{u}}{\partial q_i} \qquad \frac{^A\partial}{\partial q_i}\left(\frac{^B\partial \mathbf{u}}{\partial q_j}\right) \neq \frac{^B\partial}{\partial q_j}\left(\frac{^A\partial \mathbf{u}}{\partial q_i}\right) \qquad i, j = 1, 2, \ldots, n \quad [\mathbf{2.6.28}]$$

Also, be aware that a vector **u** may be a function of a variable q_j in one reference frame and not in another.

Consider next the derivative of Eq. [2.6.23]. We differentiate each term individually to find the angular acceleration. The first term is measured from the inertial frame, thus its derivative is straightforward. To obtain the derivatives of the subsequent terms we use the above relation. For example, for two intermediate frames, we have

$$^A\boldsymbol{\alpha}^B = {}^A\frac{d}{dt}\left({}^A\boldsymbol{\omega}^B\right) = {}^A\frac{d}{dt}\left({}^A\boldsymbol{\omega}^{A_2} + {}^{A_2}\boldsymbol{\omega}^B\right)$$

$$= {}^A\frac{d}{dt}\left({}^A\boldsymbol{\omega}^{A_2}\right) + {}^{A_2}\frac{d}{dt}\left({}^{A_2}\boldsymbol{\omega}^B\right) + {}^A\boldsymbol{\omega}^{A_2} \times {}^{A_2}\boldsymbol{\omega}^B$$

$$= {}^A\boldsymbol{\alpha}^{A_2} + {}^{A_2}\boldsymbol{\alpha}^B + {}^A\boldsymbol{\omega}^{A_2} \times {}^{A_2}\boldsymbol{\omega}^B \qquad [\mathbf{2.6.29}]$$

The expression for the angular acceleration for the general case of several frames is left as an exercise.

We end this section with an important note. It is crucial that one be able to distinguish between the reference frame in which a derivative is taken and the coordinates of the reference frame in which the differentiated vector is resolved. Usually, one expresses a vector to be differentiated in a particular reference frame in terms of the unit vectors of the frame. However, exceptions to this general procedure do exist.

Example 2.9

Consider the robot arm in Example 2.7 and find the angular acceleration of the disk, given that ω_1 and ω_3 are both constant.

Solution

From Example 2.7, the angular velocity of the disk is written as

$$\boldsymbol{\omega} = \boldsymbol{\omega}_1 + \boldsymbol{\omega}_2 + \boldsymbol{\omega}_3 \qquad [\mathbf{a}]$$

in which

$$\boldsymbol{\omega}_1 = 0.5\mathbf{K} = -0.04827\mathbf{i} - 0.3830\mathbf{j} + 0.3178\mathbf{k} \text{ rad/s}$$
$$\boldsymbol{\omega}_2 = 0.08778\mathbf{j} \text{ rad/s} \qquad \boldsymbol{\omega}_3 = 7\mathbf{i} \text{ rad/s} \qquad [\mathbf{b}]$$

We use the transport theorem to get the angular acceleration, which gives

$$\boldsymbol{\alpha}_1 = \dot{\boldsymbol{\omega}}_1 = \mathbf{0}$$
$$\boldsymbol{\alpha}_2 = \dot{\boldsymbol{\omega}}_{2\text{rel}} + \boldsymbol{\omega}_1 \times \boldsymbol{\omega}_2$$
$$\boldsymbol{\alpha}_3 = \dot{\boldsymbol{\omega}}_{3\text{rel}} + (\boldsymbol{\omega}_1 + \boldsymbol{\omega}_2) \times \boldsymbol{\omega}_3 = (\boldsymbol{\omega}_1 + \boldsymbol{\omega}_2) \times \boldsymbol{\omega}_3 \qquad [\mathbf{c}]$$

so that the angular acceleration becomes

$$\boldsymbol{\alpha} = \boldsymbol{\alpha}_1 + \boldsymbol{\alpha}_2 + \boldsymbol{\alpha}_3 = \dot{\boldsymbol{\omega}}_{2\text{rel}} + \boldsymbol{\omega}_1 \times (\boldsymbol{\omega}_2 + \boldsymbol{\omega}_3) + \boldsymbol{\omega}_2 \times \boldsymbol{\omega}_3 \qquad [\mathbf{d}]$$

We now evaluate the individual terms in the above equation. Given that $\theta(t) = \frac{\pi}{20}\cos 2t$ rad, we have at $t = 3$ s

$$\dot{\omega}_{2\text{rel}} = -\frac{\pi}{5}\cos 2t\mathbf{j} \text{ rad/s}^2 = -\frac{\pi}{5}\cos 6\mathbf{j} \text{ rad/s}^2 = -0.6033\mathbf{j} \text{ rad/s}^2$$

$$\boldsymbol{\omega}_1 \times (\boldsymbol{\omega}_2 + \boldsymbol{\omega}_3) = (-0.04827\mathbf{i} - 0.3830\mathbf{j} + 0.3178\mathbf{k}) \times (7\mathbf{i} + 0.08778\mathbf{j})$$
$$= -0.02789\mathbf{i} + 2.225\mathbf{j} + 2.677\mathbf{k} \text{ rad/s}^2$$

$$\boldsymbol{\omega}_2 \times \boldsymbol{\omega}_3 = 0.08778\mathbf{j} \times 7\mathbf{i} = -0.6145\mathbf{k} \text{ rad/s}^2 \qquad [e]$$

Adding the individual terms, we obtain the total acceleration as

$$\boldsymbol{\alpha} = -0.02789\mathbf{i} + 1.622\mathbf{j} + 2.063\mathbf{k} \text{ rad/s}^2 \qquad [f]$$

Example 2.10

Find the angular acceleration of the disk shown in Fig. 2.27, which is spinning at the constant rate of $60/\pi$ rpm. The disk is attached to a collar, which is rotating at the rate of $3/\pi$ rpm, with the rotation rate increasing by $0.6/\pi$ rpm/min. A rod connects the disk to the collar and it is pinned to the collar. It makes an angle of 30° with the vertical, which is increasing at the constant rate of $18/\pi$°/sec. Express the angular acceleration in terms of a reference frame attached to the collar.

Solution

We attach an $x'y'z'$ coordinate system to the collar, with the $Z = z'$ direction denoting the fixed vertical. The y coordinate attached to the arm is obtained by rotating the $x'y'z'$ axes about the x' axis counterclockwise by an angle of 60°, so that $\mathbf{j} = \sin 30°\mathbf{j}' - \cos 30°\mathbf{k}'$. We write the total angular velocity of the disk as

$$\boldsymbol{\omega}_{\text{disk}} = \boldsymbol{\omega}_{\text{collar}} + \boldsymbol{\omega}_{\text{rod/collar}} + \boldsymbol{\omega}_{\text{disk/rod}} \qquad [a]$$

where

$$\boldsymbol{\omega}_{\text{collar}} = \frac{3}{\pi}\mathbf{K} \text{ rpm} = \frac{3}{\pi}\left(\frac{2\pi}{60}\right)\mathbf{k}' \text{ rad/s} = 0.1\mathbf{k}' \text{ rad/s}$$

$$\boldsymbol{\omega}_{\text{rod/collar}} = \frac{18}{\pi}\mathbf{i}'°/s = \frac{18}{\pi}\left(\frac{\pi}{180}\right)\mathbf{i}' \text{ rad/s} = 0.1\mathbf{i}' \text{ rad/s}$$

$$\boldsymbol{\omega}_{\text{disk/rod}} = \frac{60}{\pi}\mathbf{j} \text{ rpm} = \frac{60}{\pi}(\sin 30°\mathbf{j}' - \cos 30°\mathbf{k}')\frac{2\pi}{60} \text{ rad/s} = \mathbf{j}' - \sqrt{3}\mathbf{k}' \text{ rad/s} \qquad [b]$$

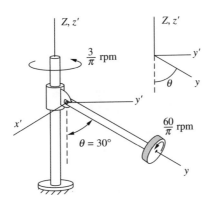

Figure 2.27

To find the angular accelerations, we differentiate each of the angular velocity terms separately. Since the angular velocity of the collar is measured from a fixed frame, its derivative is obtained through straightforward differentiation as

$$\boldsymbol{\alpha}_{\text{collar}} = \dot{\boldsymbol{\omega}}_{\text{collar}} = \left(\frac{0.6}{\pi}\text{ rpm/min}\right)\mathbf{K} = \frac{0.6}{\pi}\left(\frac{2\pi}{60}\right)\left(\frac{1}{60}\right) = 3.333(10^{-4})\mathbf{k}'\text{ rad/s}^2 \qquad \textbf{[c]}$$

The angular velocity of the rod is measured from a frame that is rotating with the angular velocity of the collar, so we can express the angular acceleration as

$$\boldsymbol{\alpha}_{\text{rod/collar}} = \dot{\boldsymbol{\omega}}_{\text{rod/collar rel}} + \boldsymbol{\omega}_{\text{collar}} \times \boldsymbol{\omega}_{\text{rod/collar}} = 0 + 0.1\mathbf{k}' \times 0.1\mathbf{i}' = 0.01\mathbf{j}'\text{ rad/s}^2 \qquad \textbf{[d]}$$

The angular velocity of the disk is relative to the rod; thus we can write its angular acceleration as

$$\boldsymbol{\alpha}_{\text{disk/rod}} = \dot{\boldsymbol{\omega}}_{\text{disk/rod rel}} + (\boldsymbol{\omega}_{\text{collar}} + \boldsymbol{\omega}_{\text{rod/collar}}) \times \boldsymbol{\omega}_{\text{disk/rod}}$$
$$= 0 + (0.1\mathbf{k}' + 0.1\mathbf{i}') \times (\mathbf{j}' - \sqrt{3}\mathbf{k}')$$
$$= -0.1\mathbf{i}' + \frac{\sqrt{3}}{10}\mathbf{j}' + 0.1\mathbf{k}'\text{ rad/s}^2 \qquad \textbf{[e]}$$

Adding Eqs. [c]–[e], we obtain the total angular acceleration of the disk as

$$\boldsymbol{\alpha}_{\text{disk}} = -0.1\mathbf{i}' + 0.1832\mathbf{j}' + 0.1003\mathbf{k}'\text{ rad/s}^2 \qquad \textbf{[f]}$$

Note that, as discussed before, this selection of the coordinate axes makes it much easier to visualize the motion than would a reference frame attached to the disk.

2.7 Relative Velocity and Acceleration

Consider the two reference frames shown in Fig. 2.26. The position of point P is expressed as

$$\mathbf{r}_P = \mathbf{r}_B + \mathbf{r}_{P/B} \qquad \textbf{[2.7.1]}$$

The vectors $\mathbf{r}_P$ and $\mathbf{r}_B$ are measured from the inertial frame, and $\mathbf{r}_{P/B}$ is measured from the relative frame. To obtain the velocity, we differentiate Eq. [2.7.1] for

$$\mathbf{v}_P = \mathbf{v}_B + \mathbf{v}_{P/B} \qquad \textbf{[2.7.2]}$$

We find the expression for $\mathbf{v}_{P/B}$ by means of the transport theorem as

$$\mathbf{v}_{P/B} = \dot{\mathbf{r}}_{P/B} = \mathbf{v}_{P/B_{\text{rel}}} + \boldsymbol{\omega} \times \mathbf{r}_{P/B} \qquad \textbf{[2.7.3]}$$

The difference in derivatives is because $\mathbf{r}_P$ and $\mathbf{r}_B$ are measured from the inertial frame, while $\mathbf{r}_{P/B}$ is measured from the rotating frame. Introducing Eq. [2.7.3] into Eq. [2.7.2], we obtain the relative velocity expression, written

$$\mathbf{v}_P = \mathbf{v}_B + \mathbf{v}_{P/B} = \mathbf{v}_B + \mathbf{v}_{P/B_{\text{rel}}} + \boldsymbol{\omega} \times \mathbf{r}_{P/B} \qquad \textbf{[2.7.4]}$$

The first term on the right side of this equation, $\mathbf{v}_B$, is known as the *base velocity*; it denotes the absolute velocity of the origin of the moving frame. The second term, $\mathbf{v}_{P/B_{\text{rel}}}$, is known as the *relative velocity,* as it denotes the velocity of point P as viewed by an observer attached to the relative frame. The third term, $\boldsymbol{\omega} \times \mathbf{r}_{P/B}$, is called the *transport velocity*; it describes the change in the position vector $\mathbf{r}_{P/B}$ as the relative frame rotates.

Equation [2.7.4] can also be written in terms of a point Q, which is coincident with point P but is not moving with respect to the rotating frame. We write

$$\mathbf{v}_P = \mathbf{v}_Q + \mathbf{v}_{P/Q} \qquad [2.7.5]$$

where $\mathbf{v}_Q$ is the absolute velocity of point Q, and $\mathbf{v}_{P/Q}$ is the relative velocity of P with respect to Q, having the forms

$$\mathbf{v}_Q = \mathbf{v}_B + \boldsymbol{\omega} \times \mathbf{r}_{P/B} \qquad \mathbf{v}_{P/Q} = \mathbf{v}_{P/B_{\text{rel}}} \qquad [2.7.6a,b]$$

When analyzing the relative motion of two points both fixed on the same reference frame, the relative velocity term $\mathbf{v}_{P/Q}$ vanishes, $P = Q$ represents the same point, and we use Eq. [2.7.6a] to relate the velocities.

To find the acceleration of point P, we differentiate Eq. [2.7.4] once more, with the result

$$\mathbf{a}_P = \mathbf{a}_B + \frac{d}{dt}\mathbf{v}_{P/B_{\text{rel}}} + \frac{d}{dt}(\boldsymbol{\omega} \times \mathbf{r}_{P/B}) \qquad [2.7.7]$$

Differentiation of the left side of Eq. [2.7.7] and the first term on the right side is straightforward:

$$\mathbf{a}_P = \frac{d}{dt}\mathbf{v}_P = \frac{d^2}{dt^2}\mathbf{r}_P \qquad \mathbf{a}_B = \frac{d}{dt}\mathbf{v}_B = \frac{d^2}{dt^2}\mathbf{r}_B \qquad [2.7.8]$$

Differentiation of the second and third terms requires that we invoke the transport theorem for each of these terms, with the result

$$\frac{d}{dt}\mathbf{v}_{P/B_{\text{rel}}} = \mathbf{a}_{P/B_{\text{rel}}} + \boldsymbol{\omega} \times \mathbf{v}_{P/B_{\text{rel}}}$$

$$\frac{d}{dt}(\boldsymbol{\omega} \times \mathbf{r}_{P/B}) = \boldsymbol{\alpha} \times \mathbf{r}_{P/B} + \boldsymbol{\omega} \times \mathbf{v}_{P/B_{\text{rel}}} + \boldsymbol{\omega} \times (\boldsymbol{\omega} \times \mathbf{r}_{P/B}) \qquad [2.7.9]$$

Introducing Eqs. [2.7.8]–[2.7.9] into Eq. [2.7.7] and combining terms, we obtain

$$\mathbf{a}_P = \mathbf{a}_B + \boldsymbol{\alpha} \times \mathbf{r}_{P/B} + \boldsymbol{\omega} \times (\boldsymbol{\omega} \times \mathbf{r}_{P/B}) + \mathbf{a}_{P/B_{\text{rel}}} + 2\boldsymbol{\omega} \times \mathbf{v}_{P/B_{\text{rel}}} \qquad [2.7.10]$$

The term $\boldsymbol{\alpha} \times \mathbf{r}_{P/B}$ is due to the angular acceleration of the rotating frame, while $\boldsymbol{\omega} \times (\boldsymbol{\omega} \times \mathbf{r}_{P/B})$ is the centripetal acceleration of point P. For the general case of three-dimensional motion $\boldsymbol{\omega} \times (\boldsymbol{\omega} \times \mathbf{r}_{P/B})$ lies on the plane generated by the angular velocity $\boldsymbol{\omega}$ and $\mathbf{r}_{P/B}$. For the special case of plane motion, the centripetal acceleration takes the form

$$\boldsymbol{\omega} \times (\boldsymbol{\omega} \times \mathbf{r}_{P/B}) = -\omega^2 \mathbf{r}_{P/B} \qquad [2.7.11]$$

The fourth term $\mathbf{a}_{P/B_{\text{rel}}} = (\ddot{\mathbf{r}}_{P/B})_{\text{rel}}$ is the acceleration of point P as measured by an observer located on the moving frame. The fifth term, $2\boldsymbol{\omega} \times \mathbf{v}_{P/B_{\text{rel}}}$, is known as the *Coriolis acceleration*. It is due to two effects: a directional change in $\mathbf{v}_{P/B_{\text{rel}}}$, and in $\boldsymbol{\omega} \times \mathbf{r}_{P/B}$, a change in magnitude of $\mathbf{r}_{P/B}$. Both terms contributing to the Coriolis effect arise because there is translational motion with respect to a relative frame.

The direction of the Coriolis acceleration is perpendicular to the plane generated by $\boldsymbol{\omega}$ and $(\mathbf{v}_{P/B})_{\text{rel}}$, so that it always results in a change in direction from $(\mathbf{v}_{P/B})_{\text{rel}}$, as shown in Fig. 2.28. Even in cases when the magnitude of this acceleration is small, because it always causes a change in direction the Coriolis acceleration must be considered in the analysis of several systems.

Equation [2.7.10] can also be expressed in terms of a point Q coincident with point P but not moving with respect to the coordinate frame as

$$\mathbf{a}_P = \mathbf{a}_Q + \mathbf{a}_{P/Q} \qquad [2.7.12]$$

in which

$$\mathbf{a}_Q = \mathbf{a}_B + \boldsymbol{\alpha} \times \mathbf{r}_{P/B} + \boldsymbol{\omega} \times (\boldsymbol{\omega} \times \mathbf{r}_{P/B}) \qquad \mathbf{a}_{P/Q} = \mathbf{a}_{P/B_{\text{rel}}} + 2\boldsymbol{\omega} \times \mathbf{v}_{P/B_{\text{rel}}}$$
$$[2.7.13a,b]$$

The term $\mathbf{a}_Q$ is the absolute acceleration of point Q. The term $\mathbf{a}_{P/Q}$ is the acceleration of point P due to its motion with respect to the reference frame. Unlike $\mathbf{v}_{P/Q}$, it contains two terms. The difference is the Coriolis acceleration.

When there is no motion with respect to the moving frame, such as with the motion of two points fixed on a rigid body, the relative motion equations reduce to Eq. [2.7.13a], and one replaces Q with P

$$\mathbf{a}_P = \ddot{\mathbf{r}}_B + \boldsymbol{\alpha} \times \mathbf{r}_{P/B} + \boldsymbol{\omega} \times (\boldsymbol{\omega} \times \mathbf{r}_{P/B}) \qquad [2.7.14]$$

Now let us write the relative velocity and acceleration expressions using the notation introduced in the previous section. For the relative velocity expression from Eq. [2.7.4] we write

$$^A\mathbf{v}_P = {}^A\mathbf{v}_B + {}^B\mathbf{v}_P + {}^A\boldsymbol{\omega}^B \times {}^B\mathbf{r}_P \qquad [2.7.15]$$

and, for the relative acceleration from Eq. [2.7.10], we write

$$^A\mathbf{a}_P = {}^A\mathbf{a}_B + {}^A\boldsymbol{\alpha}^B \times \mathbf{r}_P + {}^A\boldsymbol{\omega}^B \times ({}^A\boldsymbol{\omega}^B \times \mathbf{r}_P) + {}^B\mathbf{a}_P + 2{}^A\boldsymbol{\omega}^B \times {}^B\mathbf{v}_P \qquad [2.7.16]$$

with

$$^A\mathbf{a}_P = {}^A\frac{d}{dt}({}^A\mathbf{v}_P) = {}^A\frac{d}{dt}\left({}^A\frac{d}{dt}\mathbf{r}_P\right) \qquad {}^A\mathbf{a}_B = {}^A\frac{d}{dt}({}^A\mathbf{v}_B) = {}^A\frac{d}{dt}\left({}^A\frac{d}{dt}\mathbf{r}_B\right)$$
$$[2.7.17]$$

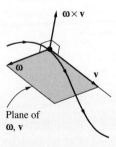

Figure 2.28

The most effective way of dealing with problems where more than one reference frame is involved is to systematically break the problem into several parts and to calculate the terms for each part individually.

We next discuss two very important issues associated with the kinematics of relative motion: how to select the origin of the relative frame(s), and how to select the orientation of the relative frame(s). There is no clear-cut answer to these questions. One guideline is to select, if possible, the origin and orientation of the relative frame so that it minimizes the number of expressions in the relative motion equations. Considering Fig. 2.26, if B is selected so that it coincides with P ($B = Q$), then $\mathbf{r}_{P/B} = \mathbf{0}$. If B is selected such that it coincides with the origin of the coordinate system O, then $\mathbf{v}_B = \mathbf{0}$ and $\mathbf{a}_B = \mathbf{0}$. Another guideline is to select the relative frames so that the number of relative frames is minimized and the angles that have to be calculated are simple. The way to learn how to select reference frames is by gaining experience and solving problems.

When there is more than one reference system involved, the relative motion expressions can become lengthy and complicated. One way to avoid confusion is to select a tabulation approach when obtaining the components of the relative motion expression.

Example 2.11

The platform in Fig. 2.29 is rotating with a constant angular velocity of $\omega = 0.2$ rad/s. Pivoted on the platform is a tube oscillating according to the relationship $\theta(t) = \frac{\pi}{6} \sin 2t$ rad. A particle of mass m slides without friction inside the tube. The particle is attached to the ends of the tube by a spring of constant k and dashpot of constant c. Find the velocity of the particle at $t = 3.6$ s, at which point it is given that $y = 40$ cm and $\dot{y} = -30$ cm/s. Also find a general expression for its acceleration.

Solution

Consider an $X'Y'Z'$ coordinate system moving with the platform and an xyz coordinate system attached to the tube, as shown in Fig. 2.30. The $Z' = Z$ axis is the vertical. When $\theta = 0$, the y and Y' axes coincide. The coordinate axes are related to each other by

$$x = X' \qquad y = Y' \cos\theta + Z' \sin\theta \qquad z = -Y' \sin\theta + Z' \cos\theta \qquad \textbf{[a]}$$

We have

$$\boldsymbol{\omega} = \omega\mathbf{K} + \dot{\theta}\mathbf{I}' = \dot{\theta}\mathbf{i} + \omega\sin\theta\mathbf{j} + \omega\cos\theta\mathbf{k} \qquad \mathbf{r} = y\mathbf{j} \qquad \mathbf{v}_{\text{rel}} = \dot{y}\mathbf{j} \qquad \textbf{[b]}$$

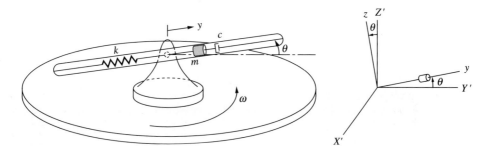

Figure 2.29   **Figure 2.30**

Because the origins of the coordinate frames coincide, the relative velocity expression becomes

$$\mathbf{v} = \mathbf{v}_{rel} + \boldsymbol{\omega} \times \mathbf{r} = \dot{y}\mathbf{j} + (\dot{\theta}\mathbf{i} + \omega \sin\theta \mathbf{j} + \omega \cos\theta \mathbf{k}) \times y\mathbf{j}$$
$$= -y\omega \cos\theta \mathbf{i} + \dot{y}\mathbf{j} + y\dot{\theta}\mathbf{k} \qquad \text{[c]}$$

At $t = 3.6$ s, $\theta(3.6) = 0.4156$ rad, $\sin\theta = 0.4037$, $\cos\theta = 0.9149$, $\dot{\theta}(3.6) = \frac{\pi}{3}\cos 7.2 = 0.6371$ rad/s. Substituting in these values, we obtain for the velocity at $t = 3.6$ s

$$\mathbf{v} = -0.07319\mathbf{i} - 0.3\mathbf{j} + 0.2548\mathbf{k} \text{ m/s} \qquad \text{[d]}$$

The relative acceleration expression for this problem is

$$\mathbf{a} = \mathbf{a}_{rel} + \boldsymbol{\alpha} \times \mathbf{r} + \boldsymbol{\omega} \times (\boldsymbol{\omega} \times \mathbf{r}) + 2\boldsymbol{\omega} \times \mathbf{v}_{rel} \qquad \text{[e]}$$

in which

$$\mathbf{a}_{rel} = \ddot{y}\mathbf{j} \qquad \boldsymbol{\alpha} = \ddot{\theta}\mathbf{i} + \omega\dot{\theta}\cos\theta \mathbf{j} - \omega\dot{\theta}\sin\theta \mathbf{k} \qquad \text{[f]}$$

Note that we obtained the angular acceleration expression by direct differentiation of the angular velocity expression in Eq. [b], rather than using the transport theorem. It was possible to do this because in Eq. [b] the components of $\boldsymbol{\omega}$ were expressed in terms of the coordinates of the relative frame.

We perform the cross products, writing

$$\boldsymbol{\alpha} \times \mathbf{r} = (\ddot{\theta}\mathbf{i} + \omega\dot{\theta}c\theta\mathbf{j} - \omega\dot{\theta}s\theta\mathbf{k}) \times y\mathbf{j} = \omega\dot{\theta}ys\theta\mathbf{i} + \ddot{\theta}y\mathbf{k}$$

$$\boldsymbol{\omega} \times (\boldsymbol{\omega} \times \mathbf{r}) = (\dot{\theta}\mathbf{i} + \omega s\theta\mathbf{j} + \omega c\theta\mathbf{k}) \times (-\omega yc\theta\mathbf{i} + \dot{\theta}y\mathbf{k})$$
$$= \omega\dot{\theta}ys\theta\mathbf{i} + (-\dot{\theta}^2 y - \omega^2 yc^2\theta)\mathbf{j} + \omega^2 ys\theta c\theta\mathbf{k}$$

$$2\boldsymbol{\omega} \times \mathbf{v}_{rel} = 2(\dot{\theta}\mathbf{i} + \omega s\theta\mathbf{j} + \omega c\theta\mathbf{k}) \times \dot{y}\mathbf{j} = -2\dot{y}\omega c\theta\mathbf{i} + 2\dot{y}\dot{\theta}\mathbf{k} \qquad \text{[g]}$$

Introducing these expressions into the relative acceleration, we obtain

$$\ddot{\mathbf{r}} = (2\omega\dot{\theta}ys\theta - 2\dot{y}\omega c\theta)\mathbf{i} + (\ddot{y} - \dot{\theta}^2 y - \omega^2 yc^2\theta)\mathbf{j} + (2\dot{y}\dot{\theta} + \omega^2 ys\theta c\theta + \ddot{\theta}y)\mathbf{k} \qquad \text{[h]}$$

Example 2.12

An airplane, shown in Fig. 2.31, is moving with a speed of 420 mph. A flight attendant who weighs 120 lb is standing 15 ft from the center of mass. To avoid a turbulent region, the pilot initiates an emergency maneuver. The aircraft begins to pitch upward at the constant rate of 0.1 rad/sec, and it begins to pursue a curved trajectory toward the left of the pilot with a radius of curvature of 30,000 ft. The speed of the center of mass of the airplane does not change with these maneuvers. Find the forces exerted on the flight attendant's feet if the attendant wishes to move forward with a speed of 2 ft/sec.

Solution

We consider two reference frames. The first frame is associated with the curved trajectory and has an angular velocity of $420(88/60)/(30,000) = 0.02053$ rad/s. The second frame is attached to the airplane, and its angular velocity is the pitch rate of 0.1 rad/s.[3] At the instant considered, the two reference frames coincide. We can write the angular velocity and angular

---

[3] In order to simplify this problem, we do not consider any roll of the aircraft. In general, when an airplane makes a turn the pilot rolls the aircraft so that the resultant acceleration vector due to the turn and due to gravity lies as much as possible along the local vertical direction, through the spinal cords of the passengers. This way, passengers experience less discomfort.

2.7 Relative Velocity and Acceleration

acceleration as

$$\boldsymbol{\omega} = 0.1\mathbf{j} - 0.02053\mathbf{k} \text{ rad/sec} \qquad \boldsymbol{\alpha} = -0.02053\mathbf{k} \times 0.1\mathbf{j} = 0.002053\mathbf{i} \text{ rad/sec}^2 \quad \text{[a]}$$

We write the relative acceleration relation as

$$\mathbf{a}_P = \mathbf{a}_B + \boldsymbol{\alpha} \times \mathbf{r}_{P/B} + \boldsymbol{\omega} \times \boldsymbol{\omega} \times \mathbf{r}_{P/B} + (\mathbf{a}_{P/B})_{\text{rel}} + 2\boldsymbol{\omega} \times (\mathbf{v}_{P/B})_{\text{rel}} \quad \text{[b]}$$

and identify each term. The acceleration of the center of mass is due to the change in curvature and can be written as (60 mph = 88 ft/sec)

$$\mathbf{a}_B = \mathbf{a}_n = -\frac{v^2}{\rho}\mathbf{j} = -\frac{420^2(88/60)^2}{30,000}\mathbf{j} = -12.65\mathbf{j} \text{ ft/sec}^2 \quad \text{[c]}$$

or, in aeronautics terminology, $12.65/32.17 = 0.3932g$. The angular acceleration term is very small and can be ignored. The centripetal acceleration term becomes

$$\boldsymbol{\omega} \times \boldsymbol{\omega} \times \mathbf{r}_{P/B} = (0.1\mathbf{j} - 0.02053\mathbf{k}) \times (0.1\mathbf{j} - 0.02053\mathbf{k}) \times 15\mathbf{i} = -0.1563\mathbf{i} \text{ ft/sec}^2 \quad \text{[d]}$$

The next term, $(\mathbf{a}_{P/B})_{\text{rel}}$, is zero, because we are assuming a constant speed for the flight attendant inside the aircraft. The last term is the Coriolis term, which has the form

$$2\boldsymbol{\omega} \times (\mathbf{v}_{P/B})_{\text{rel}} = 2(-0.02053\mathbf{k} + 0.1\mathbf{j}) \times 2\mathbf{i} = -0.08212\mathbf{j} - 0.4\mathbf{k} \text{ ft/sec}^2 \quad \text{[e]}$$

As can be seen from Eqs. [b] through [e], the dominant term in the acceleration is the normal acceleration due to the change in curvature. The Coriolis term has the smallest magnitude, and it is also in the direction of the gravitational attraction. The total acceleration is

$$\mathbf{a}_P = -0.1563\mathbf{i} - 12.73\mathbf{j} - 0.4\mathbf{k} \text{ ft/sec}^2 \quad \text{[f]}$$

Let us now draw a free-body diagram of the flight attendant (Fig. 2.32), treating the attendant as a point mass. The forces that the airplane exerts on the flight attendant are transmitted by the normal force N and the friction forces F_x and F_y. Using Newton's second law, we have

$$m\mathbf{a}_P = m(-0.1563\mathbf{i} - 12.73\mathbf{j} - 0.4\mathbf{k}) = F_x\mathbf{i} + F_y\mathbf{j} + (mg - N)\mathbf{k} \text{ lb} \quad \text{[g]}$$

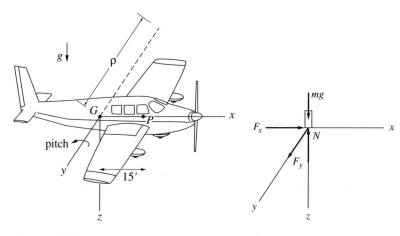

Figure 2.31   **Figure 2.32**

Solving for the unknowns

$$N = mg + 0.4m = 120 + 0.4\frac{120}{32.17} = 121.5 \text{ lb}$$

$$F_x = -0.1563m = -0.5830 \text{ lb} \qquad F_y = -12.73m = -47.49 \text{ lb} \qquad \text{[h]}$$

Note that the coefficient of friction between the attendant's shoes and the airplane floor must be large enough to permit the resultant of F_x and F_y to be less than the normal force times the coefficient of friction. Even then, because of the large forces acting on the attendant, it is very difficult for the attendant to keep standing or to walk. The Coriolis acceleration makes the flight attendant feel heavier, and for someone further away from the center of mass of the airplane the centrifugal force becomes much larger. Even for passengers who are sitting down, any change in the curvature of the path of the airplane causes a substantial amount of discomfort. It is for all these reasons that pilots navigate aircraft such that the path followed by the center of mass of the airplane is as close to a straight path as possible and any angular velocity is very small.

Example 2.13

Consider the robot in Fig. 2.33 mounted on a rotating shaft. The robot arm is attached to the shaft with a pin joint (in robotics terminology, a *revolute joint*). With a motion similar to that of an automobile antenna, a second arm can extend from the outer end of the first (in robotics terminology, a prismatic joint). Given that the shaft angle $\theta(t)$ and first arm angle $\phi(t)$ vary with the relationships $\theta(t) = 0.2t$ rad, $\phi(t) = \pi/4 (1 + \sin \pi t)$ rad and that the second arm is extending with the relation $r = 3t$ cm, find the angular velocity and angular acceleration of the robot arm as well as the velocity and acceleration of the tip.

Solution

We solve this problem using two approaches. In the first approach we use two relative frames, one attached to the shaft and rotating with $\dot\theta$, the other attached to the robot arm and rotating with $\dot\phi$ with respect to the shaft. In the second approach we use a single relative frame attached to the robot arm.

First Approach The two relative coordinate frames are denoted by H and B, as depicted in Fig. 2.34. The inertial frame is denoted by A. The H frame is attached to the rotating shaft,

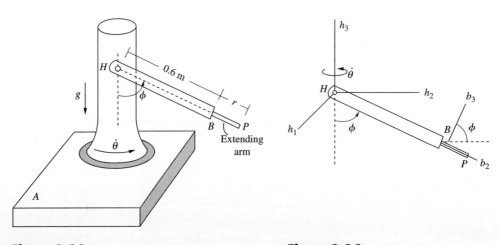

Figure 2.33   **Figure 2.34**

2.7 RELATIVE VELOCITY AND ACCELERATION

and the B frame is taken such that the b_2 axis is along the extending arm and attached to the tip of the first arm. The angular velocities of the reference frames can be written as

$$^A\boldsymbol{\omega}^H = \dot{\theta}\mathbf{h}_3 = 0.2\mathbf{h}_3 \text{ rad/s} \qquad ^H\boldsymbol{\omega}^B = \dot{\phi}\mathbf{h}_1 = \frac{\pi^2}{4}\cos\pi t\,\mathbf{h}_1 \text{ rad/s} \qquad [\mathbf{a}]$$

Considering first the B frame, we can express the position and relative velocity of P as

$$\mathbf{r}_{P/B} = 0.03t\mathbf{b}_2 \text{ m} \qquad ^B\mathbf{v}_P = 0.03\mathbf{b}_2 \text{ m/s} \qquad [\mathbf{b}]$$

where the unit vectors of the B and H frames are related by

$$\mathbf{b}_1 = \mathbf{h}_1 \qquad \mathbf{b}_2 = \sin\phi\mathbf{h}_2 - \cos\phi\mathbf{h}_3 \qquad \mathbf{b}_3 = \cos\phi\mathbf{h}_2 + \sin\phi\mathbf{h}_3 \qquad [\mathbf{c}]$$

The relative velocity expression for the H and B frames is

$$^H\mathbf{v}_P = {}^H\mathbf{v}_B + {}^H\boldsymbol{\omega}^B \times \mathbf{r}_{P/B} + {}^B\mathbf{v}_P \qquad [\mathbf{d}]$$

in which

$$^H\mathbf{v}_B = 0.6\dot{\phi}\mathbf{b}_3 \text{ rad/s} \qquad ^B\mathbf{v}_P = 0.03\mathbf{b}_2 \text{ m/s}$$

$$^H\boldsymbol{\omega}^B \times \mathbf{r}_{P/B} = \dot{\phi}\mathbf{b}_1 \times 0.03t\mathbf{b}_2 = 0.03t\dot{\phi}\mathbf{b}_3 \text{ m/s} \qquad [\mathbf{e}]$$

so that

$$^H\mathbf{v}_P = (0.6 + 0.03t)\dot{\phi}\mathbf{b}_3 + 0.03\mathbf{b}_2 \text{ m/s} \qquad [\mathbf{f}]$$

Now transfer the velocity of point P to the inertial frame. The relative velocity equation between the A and H frames is

$$^A\mathbf{v}_P = {}^A\mathbf{v}_H + {}^A\boldsymbol{\omega}^H \times \mathbf{r}_{P/H} + {}^H\mathbf{v}_P \qquad [\mathbf{g}]$$

in which

$$^A\mathbf{v}_H = \mathbf{0} \qquad ^A\boldsymbol{\omega}^H \times \mathbf{r}_{P/H} = \dot{\theta}\mathbf{a}_3 \times (0.6 + 0.03t)\mathbf{b}_2 \qquad [\mathbf{h}]$$

Noting that $\mathbf{a}_3 = -\cos\phi\mathbf{b}_2 + \sin\phi\mathbf{b}_3$, we obtain for the velocity of point P

$$^A\mathbf{v}_P = -\dot{\theta}\sin\phi(0.6 + 0.03t)\mathbf{b}_1 + 0.03\mathbf{b}_2 + (0.6 + 0.03t)\dot{\phi}\mathbf{b}_3 \text{ m/s} \qquad [\mathbf{i}]$$

Second Approach The angular velocity of the single frame is

$$^A\boldsymbol{\omega}^B = {}^A\boldsymbol{\omega}^H + {}^H\boldsymbol{\omega}^B = \dot{\theta}(t)\mathbf{h}_3 + \dot{\phi}(t)\mathbf{b}_1 = \dot{\phi}\mathbf{b}_1 - \dot{\theta}\cos\phi\mathbf{b}_2 + \dot{\theta}\sin\phi\mathbf{b}_3 \qquad [\mathbf{j}]$$

and the relative velocity expression is

$$^A\mathbf{v}_P = {}^A\mathbf{v}_B + {}^A\boldsymbol{\omega}^B \times \mathbf{r}_{P/B} + {}^B\mathbf{v}_P \qquad [\mathbf{k}]$$

We find

$$^A\mathbf{v}_B = {}^A\boldsymbol{\omega}^B \times \mathbf{r}_{B/H} = (\dot{\phi}\mathbf{b}_1 - \dot{\theta}\cos\phi\mathbf{b}_2 + \dot{\theta}\sin\phi\mathbf{b}_3) \times 0.6\mathbf{b}_2 = -0.6\dot{\theta}\sin\phi\mathbf{b}_1 + 0.6\dot{\phi}\mathbf{b}_3$$

$$^A\boldsymbol{\omega}^B \times \mathbf{r}_{P/B} = (\dot{\phi}\mathbf{b}_1 - \dot{\theta}\cos\phi\mathbf{b}_2 + \dot{\theta}\sin\phi\mathbf{b}_3) \times 0.03t\mathbf{b}_2 = -0.03t\dot{\theta}\sin\phi\mathbf{b}_1 + 0.03t\dot{\phi}\mathbf{b}_3$$

$$^B\mathbf{v}_P = 0.03\mathbf{b}_2 \text{ m/s} \qquad [\mathbf{l}]$$

which, when added up, yields Eq. [i]. Note that if we attach the relative frame to point H, $^A\mathbf{v}_B = \mathbf{0}$. Or, if the relative frame is attached to the tip of the protruding arm, then $\mathbf{r}_{P/B} = \mathbf{0}$.

Acceleration To find the acceleration of point P, let us use the single coordinate frame approach. The angular acceleration between the A and B frames has the form

$$^A\alpha^B = {}^A\alpha^H + {}^H\alpha^B + {}^A\omega^H \times {}^H\omega^B \qquad [\text{m}]$$

in which

$$^A\alpha^H = 0 \qquad {}^H\alpha^B = -\frac{\pi^3}{4}\sin \pi t \mathbf{h}_1$$

$$^A\omega^H \times {}^H\omega^B = \dot{\theta}\mathbf{h}_3 \times \dot{\phi}\mathbf{h}_1 = \dot{\theta}\dot{\phi}\mathbf{h}_2 = 0.05\pi^2 \cos \pi t \mathbf{h}_2 \qquad [\text{n}]$$

so that we find

$$^A\alpha^B = -\frac{\pi^3}{4}\sin \pi t \mathbf{h}_1 + 0.05\pi^2 \cos \pi t \mathbf{h}_2 \qquad [\text{o}]$$

The expression for the acceleration then becomes

$$^A\mathbf{a}_P = {}^A\mathbf{a}_B + {}^A\alpha^B \times \mathbf{r}_{P/B} + {}^A\omega^B \times {}^A\omega^B \times \mathbf{r}_{P/B} + 2{}^A\omega^B \times {}^B\mathbf{v}_P + {}^B\mathbf{a}_P \qquad [\text{p}]$$

The term $^B\mathbf{a}_P = 0$, as $^B\mathbf{v}_P$ is constant in the B frame. We find the absolute acceleration of point B using

$$^A\mathbf{a}_B = {}^A\alpha^B \times \mathbf{r}_{B/H} + {}^A\omega^B \times {}^A\omega^B \times \mathbf{r}_{B/H} \qquad [\text{q}]$$

so that the first three terms on the right in Eq. [p] can be expressed as

$$^A\mathbf{a}_B + {}^A\alpha^B \times \mathbf{r}_{P/B} + {}^A\omega^B \times {}^A\omega^B \times \mathbf{r}_{P/B} = {}^A\alpha^B \times \mathbf{r}_{P/H} + {}^A\omega^B \times {}^A\omega^B \times \mathbf{r}_{P/H} \qquad [\text{r}]$$

Evaluating the individual terms, we have

$$^A\alpha^B \times \mathbf{r}_{P/H} = \left[-\frac{\pi^3}{4}\sin \pi t \mathbf{b}_1 + 0.05\pi^2(\sin \phi \mathbf{b}_2 + \cos \phi \mathbf{b}_3)\right] \times (0.6 + 0.03t)\mathbf{b}_2$$

$$= (0.6 + 0.03t)\left[-0.05\pi^2 \cos \phi \mathbf{b}_1 - \frac{\pi^3}{4}\sin \pi t \mathbf{b}_3\right] \text{m/s}^2$$

$$^A\omega^B \times {}^A\omega^B \times \mathbf{r}_{P/H} = (\dot{\phi}\mathbf{b}_1 - \dot{\theta}\cos \phi \mathbf{b}_2 + \dot{\theta}\sin \phi \mathbf{b}_3) \times (\dot{\phi}\mathbf{b}_1 - \dot{\theta}\cos \phi \mathbf{b}_2 + \dot{\theta}\sin \phi \mathbf{b}_3)$$

$$\times (0.6 + 0.03t)\mathbf{b}_2$$

$$= (0.6 + 0.03t)[-\dot{\theta}\dot{\phi}\cos \phi \mathbf{b}_1 - (\dot{\phi}^2 + \dot{\theta}^2 \sin^2 \phi)\mathbf{b}_2 - \dot{\theta}^2 \cos \phi \sin \phi \mathbf{b}_3]$$

$$2{}^A\omega^B \times {}^B\mathbf{v}_P = 2(\dot{\phi}\mathbf{b}_1 - \dot{\theta}\cos \phi \mathbf{b}_2 + \dot{\theta}\sin \phi \mathbf{b}_3) \times 0.03\mathbf{b}_2 = 0.06\dot{\phi}\mathbf{b}_3 - 0.06\dot{\theta}\sin \phi \mathbf{b}_1 \text{m/s}^2 \qquad [\text{s}]$$

so that the acceleration of the tip of the robot becomes

$$^A\mathbf{a}_P = -[(0.6 + 0.03t)(0.05\pi^2 \cos \phi + \dot{\theta}\dot{\phi}\cos \phi) + 0.06\dot{\theta}\sin \phi]\mathbf{b}_1$$

$$- [(0.6 + 0.03t)(\dot{\phi}^2 + \dot{\theta}^2 \sin^2 \phi)]\mathbf{b}_2$$

$$+ \left[-(0.6 + 0.03t)\left(\frac{\pi^3}{4}\sin \pi t + \dot{\theta}^2 \cos \phi \sin \phi\right) + 0.06\dot{\phi}\right]\mathbf{b}_3 \text{m/s}^2 \qquad [\text{t}]$$

2.8 Observations from a Moving Frame

In many cases, either by choice or by necessity, one observes motion from a moving reference frame and manipulates the equations of motion within that reference frame. Typical examples are motion with respect to the earth and measurement taken from a moving platform. When one cannot measure absolute velocities and accelerations, one has to take relative measurements and to use the relative velocity and acceleration equations to find the actual accelerations. Similarly, when analyzing or integrating the equations of motion, one has to use the variables associated with the relative motion.

We first apply the relative motion equations to Newton's second law, which is valid for inertial frames. Given the inertial and relative coordinates from the preceding section, we write the equations of motion for P as

$$\mathbf{F} = m\mathbf{a}_P = m(\mathbf{a}_B + \mathbf{a}_{P/B_{rel}} + \boldsymbol{\alpha} \times \mathbf{r}_{P/B} + \boldsymbol{\omega} \times \boldsymbol{\omega} \times \mathbf{r}_{P/B} + 2\boldsymbol{\omega} \times \mathbf{v}_{P/B_{rel}}) \quad [\mathbf{2.8.1}]$$

where $\mathbf{F}$ is the sum of all external forces acting on the particle. We can rewrite the above equation as

$$m\mathbf{a}_{P/B_{rel}} = \mathbf{F} + \mathbf{F}^* \quad [\mathbf{2.8.2}]$$

where $\mathbf{F}^* = -m(\mathbf{a}_B + \boldsymbol{\alpha} \times \mathbf{r}_{P/B} + \boldsymbol{\omega} \times \boldsymbol{\omega} \times \mathbf{r}_{P/B} + 2\boldsymbol{\omega} \times \mathbf{v}_{P/B_{rel}})$, which is the resultant of all forces that need to be considered due to the motion of the moving reference frame. In general, it is preferable to write the equations of motion by placing on the left side every term that includes variables associated with the relative frame. Doing so, we obtain for Newton's second law

$$m(\mathbf{a}_{P/B_{rel}} + \boldsymbol{\alpha} \times \mathbf{r}_{P/B} + \boldsymbol{\omega} \times \boldsymbol{\omega} \times \mathbf{r}_{P/B} + 2\boldsymbol{\omega} \times \mathbf{v}_{P/B_{rel}}) = \mathbf{F} - m\mathbf{a}_B \quad [\mathbf{2.8.3}]$$

A word of caution is in order. Often, we analyze motion with respect to a moving frame by assuming that the motion characteristics of the moving frame are known and that *the motion of the body with respect to the relative frame does not affect the motion characteristics of that frame.* For example, for a car traveling on the earth, we can safely assume that the motion of the car does not affect the rotation of the earth. While this assumption is valid where the mass of the body to which the relative frame is attached is much larger than the mass of the body whose motion is analyzed, the assumption begins to lose its validity as the bodies involved become comparable in mass. Be cautious when assuming that the motion observed from a relative frame does not change the characteristics of that frame.

One of the most common applications of analyzing motion from a moving frame is motion with respect to the rotating earth. As stated earlier, motion over short distances or with small velocities and involving short time periods can be analyzed relatively accurately without considering the motion of the earth. Otherwise, the earth's rotation needs to be included in calculations.

Consider a particle near the surface of the earth, as shown in Fig. 2.35, and attach the moving frame B to the surface of the earth using an xyz coordinate system. The z direction is the vertical, the x direction is toward the north, and the y direction is

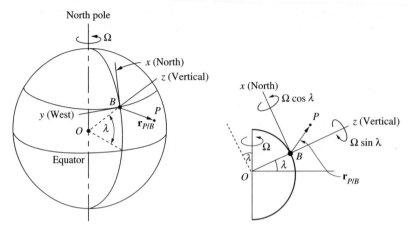

Figure 2.35   **Figure 2.36**

toward the west. We assume that the earth is rotating about its own axis with constant angular velocity Ω. Fig. 2.36 shows the coordinate system from the side view. To calculate the spin rate of the earth, we note that it takes the earth about 365.25 days to orbit the sun, and that the earth rotates about its own axis at the rate of one revolution per day.[4] Both rotations are counterclockwise, which leads to

$$\Omega = \left(\frac{2\pi}{24(60)60}\right)\left(1 + \frac{1}{365.25}\right) = 7.2921 \times 10^{-5} \text{ rad/s} \qquad [\mathbf{2.8.4}]$$

so that, considering Fig. 2.36, one can describe the angular velocity of the earth in vector form as

$$\mathbf{\Omega} = \Omega(\sin \lambda \mathbf{k} + \cos \lambda \mathbf{i}) \qquad [\mathbf{2.8.5}]$$

where λ is the latitude. We ignore the angular acceleration of the earth and set $\boldsymbol{\alpha} = \mathbf{0}$. This assumption and the assumption that the rotation rate of the earth are constant are not exactly true. The earth's rotation about itself is not along a fixed axis. The axis about which the earth rotates exhibits a small wobbling motion with a period of 433 days, primarily because the earth is not totally rigid and not totally spherical. The rate of the earth's rotation is not constant; it is slowing down at an extremely low rate. In addition, we ignore the inclination between the equatorial plane (the plane generated by the equator) and the ecliptic plane (the plane generated by the orbit of the earth around the sun). We also ignore any subsequent relative motion of the sun with respect to the fixed stars.

---

[4]The measured orbital period of the earth is 365.256 360 5 days. The difference in the third decimal place from the commonly used value of 365.25 days leads to a difference of one day about every 150 years. The Gregorian calendar was adopted to compensate for this difference. In this system, in a 400-year period, three years that normally should be leap years are not considered as leap years. These years are selected at the beginning of centuries. For example, the year 2000 C.E. is a leap year, while the years 2100, 2200, 2300 will not be considered as leap years. The year 2400 will be a leap year.

Using these assumptions, we calculate acceleration of a point on the surface of the earth (origin of the relative frame) as

$$\mathbf{a}_B = \mathbf{\Omega} \times (\mathbf{\Omega} \times \mathbf{r}_B) \qquad [2.8.6]$$

where $\mathbf{r}_B = r_e \mathbf{k}$, denoting the vector from the center of the earth to the surface of the earth. Using Newton's second law, the absolute acceleration of point P is written as

$$m\mathbf{a}_P = \mathbf{F} + \mathbf{F}'_g \qquad [2.8.7]$$

where $\mathbf{F}'_g$ is the gravitational force and $\mathbf{F}$ denotes the sum of all other external forces acting on the particle. Introducing Eqs. [2.8.6] and [2.8.7] into Eq. [2.8.3], we obtain the equations of motion in terms of the relative coordinates as

$$m\mathbf{a}_{P/B_{\text{rel}}} + 2m\mathbf{\Omega} \times \mathbf{v}_{P/B_{\text{rel}}} + m\mathbf{\Omega} \times (\mathbf{\Omega} \times \mathbf{r}_{P/B}) = \mathbf{F} + \mathbf{F}'_g - m\mathbf{\Omega} \times (\mathbf{\Omega} \times \mathbf{r}_B)$$
$$[2.8.8]$$

As discussed in Chapter 1, we include the centrifugal force $-m\mathbf{\Omega} \times (\mathbf{\Omega} \times \mathbf{r}_B)$ in the gravitational force and define the term $\mathbf{F}_g$ as the *augmented gravitational force* and approximate it as

$$\mathbf{F}_g = \mathbf{F}'_g - m\mathbf{\Omega} \times (\mathbf{\Omega} \times \mathbf{r}_B) \approx -mg\mathbf{k} \qquad [2.8.9]$$

This simplification is possible because the radius of the earth is almost a constant, making the term $\mathbf{\Omega} \times (\mathbf{\Omega} \times \mathbf{r}_B) = \Omega^2 r_e(-\cos^2 \lambda \mathbf{k} + \sin \lambda \cos \lambda \mathbf{i})$ nearly constant in magnitude. The component along the vertical (z direction) is used to augment the gravitational force. The component in the x direction (North) is usually ignored. The maximum value of the centripetal acceleration is $\Omega^2 r_e = 3.39$ cm/s^2, which is achieved at the equator.

The effect of the centripetal acceleration is to make the earth more squat. However, because the earth is not made up of fluid only, the flattening occurs mostly around the equator. Around the equator, the combined effects of flattening and centripetal acceleration make the value of g about 0.53% less than its value at the poles.

One can then write the equations of motion of a particle in the vicinity of the earth as

$$m\mathbf{a}_{P/B_{\text{rel}}} + 2m\mathbf{\Omega} \times \mathbf{v}_{P/B_{\text{rel}}} + m\mathbf{\Omega} \times (\mathbf{\Omega} \times \mathbf{r}_{P/B}) = \mathbf{F} - mg\mathbf{k} \quad [2.8.10]$$

Writing the equations of motion in terms of relative quantities is mathematically equivalent to applying a different force than the external forces. The equations of motion can be written as

$$m\mathbf{a}_{P/B_{\text{rel}}} = \mathbf{F}_{\text{eff}} + \mathbf{F}_g \qquad [2.8.11]$$

where the effective force $\mathbf{F}_{\text{eff}}$ denotes the force felt on the surface of the earth. It has the form

$$\mathbf{F}_{\text{eff}} = \mathbf{F} - m\boldsymbol{\omega} \times (\boldsymbol{\omega} \times \mathbf{r}_{P/B}) - 2m\boldsymbol{\omega} \times \mathbf{v}_{P/B_{\text{rel}}} \qquad [2.8.12]$$

After all the simplifications have been made with regard to the nature of the angular velocity of the earth and with regard to gravity, the difference between the actual and effective forces consists of a *centrifugal force*, $-m\boldsymbol{\omega} \times (\boldsymbol{\omega} \times \mathbf{r}_{P/B})$, and the *Coriolis force*, $-2m\boldsymbol{\Omega} \times \mathbf{v}_{P/B_{rel}}$. The centrifugal force is very small: it is a function of the square of the earth's rotational rate, and $\mathbf{r}_{P/B}$ is usually very small compared to the radius of the earth.

The Coriolis force has a distinctively more pronounced effect. The magnitude of the Coriolis force is dependent on both the velocity of the particle and on the latitude. Considering Eq. [2.8.5], for a particle moving in the east-west direction, maximum values of the Coriolis force are observed at the North and South Poles. The maximum value of the Coriolis force per unit mass is $2\Omega v \approx 1.5 \times 10^{-4} v$, where v is the speed with respect to the earth.

While the magnitude of the Coriolis force is very small, its direction is *always* perpendicular to the velocity; thus, this force *causes a change in direction*. If there are no significant forces acting perpendicular to the velocity, the effect of the Coriolis force builds over time. The Coriolis force for the Northern Hemisphere is depicted in Fig. 2.37. It causes a particle to veer to the right. Note that this analysis ignores the vertical component of the Coriolis force, as this component is much less than the effect of gravity.

The Coriolis effect is used to account for several kinds of natural and physical phenomena. Pertinent to weather analysis, the motion of air masses in the atmosphere is affected by the Coriolis force. For example, in the Northern Hemisphere, the spin of the air masses in cyclones and hurricanes is counterclockwise as shown in Fig. 2.38. A hurricane occurs when a low pressure center attracts air particles inward with large speeds. In the Southern Hemisphere, the spin of a cyclone is clockwise. Rossby waves, which include the Coriolis effect, are widely used to predict wave motion in the oceans as well as for weather analysis. The motion of projectiles such as missiles is affected substantially by Coriolis forces. And the Coriolis effect influences the whirl of water as it goes down the sink.

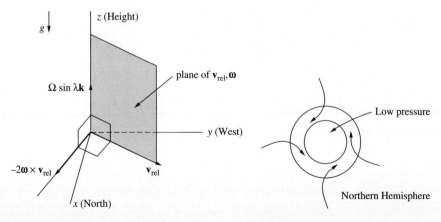

Figure 2.37   Coriolis force

Figure 2.38   Hurricane formation

We next consider the motion of a particle of mass m acted upon by an external force $\mathbf{F} = F_x \mathbf{i} + F_y \mathbf{j} + F_z \mathbf{k}$. Writing the position vector as $\mathbf{r}_{P/B} = x\mathbf{i} + y\mathbf{j} + z\mathbf{k}$ and introducing these terms into Eq. [2.8.10], and separating into components in the x, y, and z directions, we obtain

$$\ddot{x} - 2\Omega \dot{y} \sin \lambda - \Omega^2 (x \sin^2 \lambda - z \sin \lambda \cos \lambda) = \frac{F_x}{m} \quad [\mathbf{2.8.13a}]$$

$$\ddot{y} + 2\Omega (\dot{x} \sin \lambda - \dot{z} \cos \lambda) - \Omega^2 y = \frac{F_y}{m} \quad [\mathbf{2.8.13b}]$$

$$\ddot{z} + 2\Omega \dot{y} \cos \lambda + \Omega^2 (-z \cos^2 \lambda + x \sin \lambda \cos \lambda) = -g + \frac{F_z}{m} \quad [\mathbf{2.8.13c}]$$

We can perform a qualitative analysis to investigate the magnitudes of the components of the motion with respect to the rotating earth. Consider, for example, the case of free motion, $F_x = F_y = F_z = 0$ and a particle thrown upward, with initial conditions in the z direction only. Because Ω is a small quantity, terms of order Ω, namely the Coriolis terms, will dominate terms of order Ω^2. We thus ignore the centrifugal terms. In Eq. [2.8.13c], the gravitational acceleration g will dominate the response. We conclude that the component of the motion in the z direction is of order 1.

From Eq. [2.8.13b], the response in the y direction will be influenced by the motion in the x and z directions. That is,

$$O(y) = O(\Omega O(z) + \Omega O(x)) = \Omega + \Omega O(x) \quad [\mathbf{2.8.14}]$$

Investigation of Eq. [2.8.13a] shows that the component of the motion in the x direction is influenced by the motion in the y direction and can be considered as being of order $O(x) = O(\Omega O(y))$. This implies that motion in the x direction will be much smaller than the motion in the y direction. Hence, from Eq. [2.8.14], motion in the y direction will be dominated by the motion in the z direction and it will be of order Ω. It follows that the motion in the x direction will be of order Ω^2. Because the y axis denotes the west and the x direction denotes the north, for a freely moving particle the Coriolis effect will be much more significant in the east-west direction, and much less along the north-south direction.

The results of the above dimensional analysis will be different when there are external forces in the x, y, or z directions and for initial conditions.

Example 2.14

Find the equation of motion of the mass sliding in the tube in Example 2.11.

Solution

We will make use of Eq. [2.8.3]. We introduce the acceleration expression from Eq. [h] of Example 2.11 into Eq. [2.8.3], which yields

$$[m(2\omega\dot{\theta} y s\theta - 2\omega\dot{\theta} y c\theta)\mathbf{i} + m(\ddot{y} - \dot{\theta}^2 y - \omega^2 y c^2\theta)\mathbf{j} + m(2\dot{y}\dot{\theta} + \omega^2 y s\theta c\theta + \ddot{\theta} y)\mathbf{k}] = \mathbf{F} \quad [\mathbf{a}]$$

To find the forces acting on the mass, we draw a free-body diagram (Fig. 2.39). The forces acting on the particle can be classified into three groups:

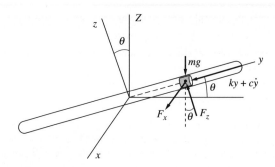

Figure 2.39

1. Gravity: $-mg\mathbf{K} = -mg\sin\theta\mathbf{j} - mg\cos\theta\mathbf{k}$
2. Normal reaction forces that the tube exerts on the particle: $F_x\mathbf{i} + F_z\mathbf{k}$
3. Spring and dashpot: $-(ky + c\dot{y})\mathbf{j}$

The total force thus becomes

$$\mathbf{F} = F_x\mathbf{i} + (-mg\sin\theta - ky - c\dot{y})\mathbf{j} + (F_z - mg\cos\theta)\mathbf{k} \quad [\mathbf{b}]$$

The reaction forces arise from the fact that the particle is constrained to move inside the tube. Introducing Eq. [b] into Eq. [a] and separating into the components along the x, y, and z directions, we find the force balances to be

In x direction: $\quad m(2\omega\dot{\theta}y\sin\theta - 2\omega\dot{\theta}y\cos\theta) = F_x \quad [\mathbf{c}]$

In y direction: $\quad m(\ddot{y} - \dot{\theta}^2 y - \omega^2 y\cos^2\theta) = -mg\sin\theta - ky - c\dot{y} \quad [\mathbf{d}]$

In z direction: $\quad m(2\dot{y}\dot{\theta} + \omega^2 y\sin\theta\cos\theta + \ddot{\theta}y) = F_z - mg\cos\theta \quad [\mathbf{e}]$

Of the three force balances, Eq. [d] represents the equation of motion and Eqs. [c] and [e] give expressions for the reaction forces F_x and F_z, that is, the constraint forces. Note that we have a single degree of freedom, because the motion of the platform and tube are defined as known quantities. The only variable is the motion of the particle inside the tube, described by y. In essence, Eqs. [c]–[e] represent an equation of motion and two constraint equations.

As stated earlier, it is customary to write the equation of motion by placing all of the dependent variables on the left side of the equation. Doing so, we rewrite Eq. [d] as

$$m\ddot{y} + c\dot{y} + (k - m\dot{\theta}^2 - m\omega^2\cos^2\theta)y = -mg\sin\theta \quad [\mathbf{f}]$$

Example 2.15

MOTION ANALYSIS USING PERTURBATION THEORY Equations [2.8.13] are a set of nonlinear equations, which cannot be integrated analytically. If quadratic terms in the angular velocity Ω are neglected, these equations become linear. Their qualitative analysis requires solution of an eigenvalue problem, and their integration requires simultaneous integration of three equations. The integrals of the motion for this set of equations do not give much additional insight. One can conduct a numerical integration of the equations of motion, but doing so gives answers for the particular set of forcing and initial conditions. It turns out that there is yet another analytical approach to analyzing the equations of motion. Because Ω is a very small quantity, Eqs. [2.8.13] can be analyzed by means of a perturbation approach by treating Ω as the perturbation parameter.

Here, we carry out what is known as a *straightforward expansion* and demonstrate perturbation analysis.[5] We assume that the solutions in x, y, and z can be expressed in a perturbation series in terms of the perturbation parameter Ω as

$$x(t) = x_0 + \Omega x_1 + \Omega^2 x_2 + \ldots \qquad y(t) = y_0 + \Omega y_1 + \Omega^2 y_2 + \ldots$$
$$z(t) = z_0 + \Omega z_1 + \Omega^2 z_2 + \ldots \qquad \textbf{[a]}$$

where the subscripts indicate the order of the solution (zeroth, first, second, ...). Given the initial conditions $x(0)$, $\dot{x}(0)$, $y(0)$, $\dot{y}(0)$, $z(0)$, $\dot{z}(0)$, we consider them as the initial conditions associated with the zeroth-order problem. All initial conditions associated with the higher-order solutions are zero. Substituting Eqs. [a] into Eqs. [2.8.13] and neglecting terms of order higher than Ω^2, we obtain

$$\ddot{x}_0 + \Omega \ddot{x}_1 + \Omega^2 \ddot{x}_2 + \ldots - 2\Omega \sin\lambda (\dot{y}_0 + \Omega \dot{y}_1 + \ldots)$$
$$- \Omega^2 (x_0 \sin^2 \lambda - z_0 \sin\lambda \cos\lambda + \ldots) = \frac{F_x}{m}$$

$$\ddot{y}_0 + \Omega \ddot{y}_1 + \Omega^2 \ddot{y}_2 + \ldots + 2\Omega (\dot{x}_0 \sin\lambda + \Omega \dot{x}_1 \sin\lambda$$
$$- \dot{z}_0 \cos\lambda - \Omega \dot{z}_1 \cos\lambda + \ldots) - \Omega^2 (y_0 + \ldots) = \frac{F_y}{m}$$

$$\ddot{z}_0 + \Omega \ddot{z}_1 + \Omega^2 \ddot{z}_2 + \ldots + 2\Omega \cos\lambda (\dot{y}_0 + \Omega \dot{y}_1 + \ldots)$$
$$+ \Omega^2 (-z_0 \cos^2 \lambda + x_0 \sin\lambda \cos\lambda + \ldots) = -g + \frac{F_z}{m} \qquad \textbf{[b]}$$

Collecting terms of like orders in Ω, we obtain a set of linear differential equations in the form

Order Ω^0

$$\ddot{x}_0 = \frac{F_x}{m} \qquad x_0(0) = x(0) \qquad \dot{x}_0(0) = \dot{x}(0)$$
$$\ddot{y}_0 = \frac{F_y}{m} \qquad y_0(0) = y(0) \qquad \dot{y}_0(0) = \dot{y}(0)$$
$$\ddot{z}_0 = -g + \frac{F_z}{m} \qquad z_0(0) = z(0) \qquad \dot{z}_0(0) = \dot{z}(0) \qquad \textbf{[c]}$$

Order Ω

$$\ddot{x}_1 = 2\sin\lambda \dot{y}_0$$
$$\ddot{y}_1 = -2(\dot{x}_0 \sin\lambda - \dot{z}_0 \cos\lambda)$$
$$\ddot{z}_1 = -2\cos\lambda \dot{y}_0 \qquad \textbf{[d]}$$

Order Ω^2

$$\ddot{x}_2 = 2\sin\lambda \dot{y}_1 + x_0 \sin^2\lambda - z_0 \sin\lambda \cos\lambda$$
$$\ddot{y}_2 = -2(\dot{x}_1 \sin\lambda - \dot{z}_1 \cos\lambda) + y_0$$
$$\ddot{z}_2 = -2\cos\lambda \dot{y}_1 + z_0 \cos^2\lambda - x_0 \sin\lambda \cos\lambda \qquad \textbf{[e]}$$

---

[5]The field of perturbation analysis is very broad and several perturbation techniques exist. An in-depth treatment of the subject is beyond the scope of this text. The straightforward expansion we use here is the simplest form of perturbation expansions.

We observe that x_0 and y_0 depend on the external force, and z_0 depends on both the external force and gravity. The first- and second-order equations indicate which terms dominate the motion. The order Ω equations are due to the Coriolis effect, and the order Ω^2 equations are due to both Coriolis and centrifugal effects.

Let us select the case of projectile motion, and consider a projectile launched in the Northern Hemisphere with speed v toward the west and at an angle of θ with the vertical. We have the following initial conditions:

$$x(0) = y(0) = z(0) = 0 \qquad \dot{x}(0) = 0, \qquad \dot{y}(0) = v \sin \theta \qquad \dot{z}(0) = v \cos \theta \qquad \text{[f]}$$

We assume that the external force consists of gravity only, neglecting the effects of wind resistance. The zeroth-order problem yields

$$\begin{aligned}
\ddot{x}_0 &= 0 & x_0(0) &= 0 & \dot{x}_0(0) &= 0 & &\Rightarrow x_0(t) = 0 \\
\ddot{y}_0 &= 0 & y_0(0) &= 0 & \dot{y}_0(0) &= v \sin \theta & &\Rightarrow y_0(t) = v \sin(\theta) t \\
\ddot{z}_0 &= -g & z_0(0) &= 0 & \dot{z}_0(0) &= v \cos \theta & &\Rightarrow z_0(t) = v \cos(\theta) t - gt^2/2
\end{aligned} \qquad \text{[g]}$$

Recalling that the higher-order solutions all have zero initial conditions, substitution of the zeroth-order solution into the first-order equations yields

$$\begin{aligned}
\ddot{x}_1 &= 2 \sin \lambda \, \dot{y}_0 = 2v \sin \lambda \sin \theta & &\Rightarrow x_1(t) = v \sin \lambda \sin(\theta) t^2 \\
\ddot{y}_1 &= -2(\dot{x}_0 \sin \lambda - \dot{z}_0 \cos \lambda) = 2 \cos \lambda (v \cos \theta - gt) & &\Rightarrow y_1(t) = \cos \lambda (v \cos(\theta) t^2 - gt^3/3) \\
\ddot{z}_1 &= -2 \cos \lambda \, \dot{y}_0 = -2v \cos \lambda \sin \theta & &\Rightarrow z_1(t) = -v \cos \lambda \sin(\theta) t^2
\end{aligned} \qquad \text{[h]}$$

Combining Eqs. [g] and [h], we obtain the first-order approximation to the solution as

$$x(t) = \Omega v \sin \lambda \sin(\theta) t^2$$

$$y(t) = v \sin(\theta) t + \Omega \cos \lambda \left(v \cos(\theta) t^2 - \frac{gt^3}{3} \right)$$

$$z(t) = v \cos(\theta) t - \frac{gt^2}{2} - \Omega v \cos \lambda \sin(\theta) t^2 \qquad \text{[i]}$$

This above solution is valid for $0 \leq t \leq t_f$, with t_f denoting the time at which the projectile falls to the ground. The first-order approximation to t_f can be found by setting $z(t) = 0$ in the above equation, which gives

$$0 = z(t_f) = v \cos(\theta) t_f - \frac{gt_f^2}{2} - \Omega v \cos \lambda \sin(\theta) t_f^2 \Rightarrow v \cos \theta - \left(\frac{g}{2} + \Omega v \cos \lambda \sin \theta \right) t_f = 0 \qquad \text{[j]}$$

which yields the final time as

$$t_f = \frac{v \cos \theta}{g/2 + \Omega v \cos \lambda \sin \theta} \qquad \text{[k]}$$

Note that in the absence of the Coriolis effect, the final time becomes the well-known result of $t_f = 2v \cos \theta / g$. The Coriolis effect, for this particular set of initial conditions, reduces the time the projectile stays in the air.

The Coriolis deflection in the x direction can be found by introducing the expression for the final time to the response, with the result

$$x(t_f) = \Omega v \sin \lambda \sin \theta \, t_f^2 = \Omega v \sin \lambda \sin \theta \left(\frac{v \cos \theta}{g/2 + \Omega v \cos \lambda \sin \theta} \right)^2 \qquad \text{[l]}$$

As expected, the magnitude of $x(t_f)$ is dictated by the latitude angle λ. In the Northern Hemisphere, where λ is greater than zero, $x(t_f)$ is positive, indicating a Coriolis deflection to the right. By contrast, in the Southern Hemisphere, where λ is negative, $x(t_f)$ is less than zero, indicating a deflection to the left. These results agree with the discussion on the Coriolis deflection in the previous section.

FOUCAULT'S PENDULUM   Another interesting application of motion with respect to the rotating earth is the Foucault pendulum (Fig. 2.40). This pendulum consists of a large concentrated mass, usually in the form of a sphere, suspended from a very high ceiling by a thin wire.[5] As the pendulum swings, its swing plane rotates slowly in the clockwise direction. This rotating motion may seem hard to explain, because the motion of a pendulum is expected to be along a fixed plane or a fixed elliptical path. (The initial conditions dictate whether the swing motion is along a plane or an ellipse.)

Example 2.16

We see the swing plane of the pendulum rotate because of the rotation of the earth and the resulting Coriolis acceleration. To an observer in an inertial frame, the pendulum executes swing motion along a fixed plane or a fixed elliptical path.

From the free-body diagram in Fig. 2.41, the external forces F_x, F_y, and F_z are due to the tension in the wire. Denoting the tension in the wire by T, and its components by T_x, T_y, and T_z, we have

$$F_x = -T_x = -\frac{x}{L}T \qquad F_y = -T_y = -\frac{y}{L}T \qquad F_z = T_z = \frac{L-z}{L}T \qquad \text{[a]}$$

where x, y, and z denote the coordinates of the pendulum. We next make use of the great length of the pendulum and assume that $(L-z)/L \approx 1$. We hence treat the amplitude of the motion in the z direction as negligible, and write $z/L \approx 0$, $\dot{z} \approx 0$, $\ddot{z} \approx 0$. In essence, we are assuming that the pendulum is moving on the xy plane only. In addition, we assume that

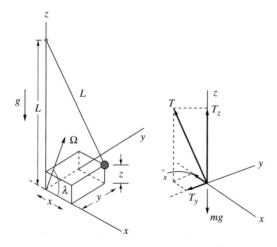

Figure 2.40   **Figure 2.41**

---

[5]In the United States, there is a Foucault pendulum in the United Nations building in New York City, at the American Museum of Natural History in Washington, D.C., at the Franklin Institute in Philadelphia, and at the Museum of Science and Industry in Chicago.

because the wire is very thin, its mass is negligible compared to the mass of the sphere, and that air resistance is negligible. Using these assumptions, introducing Eqs. [a] into Eqs. [2.8.13], and neglecting terms quadratic in Ω, we obtain

$$\ddot{x} - 2\Omega \sin \lambda \dot{y} = -\frac{x}{mL}T \qquad \ddot{y} + 2\Omega \sin \lambda \dot{x} = -\frac{y}{mL}T \qquad \text{[b]}$$

$$2\Omega \dot{y} \cos \lambda + g = \frac{T}{m} \qquad \text{[c]}$$

Rewriting Eq. [c] as $T = m(g + 2\Omega \dot{y} \cos \lambda)$, introducing it into Eq. [b] and ignoring terms quadratic in x and y, we obtain the linearized equations of motion as

$$\ddot{x} - 2\Omega \sin \lambda \dot{y} + \frac{g}{L}x = 0 \qquad \ddot{y} + 2\Omega \sin \lambda \dot{x} + \frac{g}{L}y = 0 \qquad \text{[d]}$$

Equations [d] represent a gyroscopic system where the rate of rotation is governed by $\Omega \sin \lambda$. We note that $\Omega \sin \lambda$ is the component of the angular velocity of the earth in the vertical direction. To analyze the nature of the response, we assume a solution in the form

$$x(t) = Xe^{i\omega t} \qquad y(t) = Ye^{i\omega t} \qquad \text{[e]}$$

where X and Y are amplitudes. Introducing Eqs. [e] into Eqs. [d] and collecting powers of $e^{i\omega t}$, we obtain

$$\begin{bmatrix} -\omega^2 + \frac{g}{L} & -i2\omega\Omega \sin \lambda \\ i2\Omega\omega \sin \lambda & -\omega^2 + \frac{g}{L} \end{bmatrix} \begin{bmatrix} X \\ Y \end{bmatrix} = \begin{bmatrix} 0 \\ 0 \end{bmatrix} \qquad \text{[f]}$$

In order for Eq. [f] to hold, the determinant of the coefficient matrix must vanish. Setting the determinant equal to zero leads to the characteristic equation

$$\left(-\omega^2 + \frac{g}{L}\right)^2 - (2\omega\Omega \sin \lambda)^2 = 0 \qquad \text{[g]}$$

Solving the characteristic equation for ω, we obtain four values as

$$\omega = \pm \Omega \sin \lambda \pm \sqrt{\frac{g}{L} + \Omega^2 \sin^2 \lambda} \qquad \text{[h]}$$

The response $x(t)$ and $y(t)$ is harmonic with frequency ω. The quantity inside the radical in Eq. [h] is always positive. It follows that all values of ω are real and very close to each other. After carrying out the algebra, the response can be expressed in the general form

$$x(t) = C_1 \cos(\omega_1 t + \phi_1) + C_2 \cos(\omega_2 t + \phi_2)$$
$$y(t) = -C_1 \sin(\omega_1 t + \phi_1) + C_2 \sin(\omega_2 t + \phi_2) \qquad \text{[i]}$$

where C_i and ϕ_i ($i = 1, 2$) depend on the initial conditions, and

$$\omega_1 = \sqrt{\frac{g}{L} + \Omega^2 \sin^2 \lambda} + \Omega \sin \lambda \qquad \omega_2 = \sqrt{\frac{g}{L} + \Omega^2 \sin^2 \lambda} - \Omega \sin \lambda \qquad \text{[j]}$$

are the frequencies. Because the angular velocity of the earth is much smaller compared to g/L, g/L dominates the roots of the characteristic equation. Indeed, approximating ω_1 and ω_2 using a Taylor series expansion, we get

$$\omega_1 = \sqrt{\frac{g}{L} + \Omega^2 \sin^2 \lambda} + \Omega \sin \lambda \approx \sqrt{\frac{g}{L}} + \Omega \sin \lambda + \tfrac{1}{2}\Omega^2 \sin^2 \lambda \sqrt{\frac{L}{g}}$$

$$\omega_2 = \sqrt{\frac{g}{L} + \Omega^2 \sin^2 \lambda} - \Omega \sin \lambda \approx \sqrt{\frac{g}{L}} - \Omega \sin \lambda + \tfrac{1}{2}\Omega^2 \sin^2 \lambda \sqrt{\frac{L}{g}} \qquad \text{[k]}$$

If we ignore the rotation of the earth and set $\Omega = 0$, then $\omega_1 = \omega_2$, and it becomes clear from Eqs. [i] that the motion of the pendulum is an ellipse or a straight line.

We next examine the effects of the rotating earth. We concluded earlier that the two roots ω_1 and ω_2 are very close to each other. The type of motion of a system when two of its frequencies are very close to each other is the classical case of the *beat phenomenon*. We introduce the expressions

$$\omega_a = \tfrac{1}{2}(\omega_1 + \omega_2) \approx \sqrt{\frac{g}{L}}\left(1 + \frac{L\Omega^2 \sin^2 \lambda}{2g}\right) \qquad \omega_b = \frac{1}{2}(\omega_1 - \omega_2) = \Omega \sin \lambda \qquad [\text{l}]$$

where ω_a is the *average frequency* and ω_b the *beat frequency*. The average frequency is the frequency of the pendulum as observed from an inertial reference frame (of order 1) plus a very small change (of order Ω^2) due to the rotation of the earth. The beat frequency is of order Ω, the same as the coefficient of the middle terms in Eqs. [d] and the component of the angular velocity of the earth in the z direction. One can express ω_1 and ω_2 in terms of the average and beat frequencies as

$$\omega_1 = \omega_a + \omega_b \qquad \omega_2 = \omega_a - \omega_b \qquad [\text{m}]$$

Without loss of generality, we consider the case where the local motion of the pendulum is on a swing plane. For this, we specify the initial conditions on the velocity as

$$\dot{x} = 0 \quad \dot{y} = 0 \quad \text{at} \quad t = 0 \qquad [\text{n}]$$

Furthermore, at $t = 0$ only one of x or y has to be nonzero. We select the swing plane as the xz plane, so that the remaining initial conditions are

$$x = C, \quad y = 0 \qquad [\text{o}]$$

Introducing these initial conditions into Eqs. [i] and solving, we obtain

$$\phi_1 = \phi_2 = 0$$

$$C_1 = \frac{C\omega_2}{(\omega_1 + \omega_2)} = \frac{C(\omega_a - \omega_b)}{2\omega_a} \qquad C_2 = \frac{C\omega_1}{(\omega_1 + \omega_2)} = \frac{C(\omega_a + \omega_b)}{2\omega_a} \qquad [\text{p}]$$

Using Eqs. [l] and [m], the constants [p], as well as the trigonometric identities $\cos(a + b) = \cos a \cos b - \sin a \sin b$, $\sin(a + b) = \sin a \cos b + \cos a \sin b$, we can express Eqs. [i] as

$$x(t) = \frac{C(\omega_a - \omega_b)}{2\omega_a}\cos\omega_1 t + \frac{C(\omega_a + \omega_b)}{2\omega_a}\cos\omega_2 t =$$

$$C\left(\cos\omega_a t \cos\omega_b t + \frac{\omega_b}{\omega_a}\sin\omega_a t \sin\omega_b t\right)$$

$$y(t) = -\frac{C(\omega_a - \omega_b)}{2\omega_a}\sin\omega_1 t + \frac{C(\omega_a + \omega_b)}{2\omega_a}\sin\omega_2 t =$$

$$-C\left(\cos\omega_a t \sin\omega_b t - \frac{\omega_b}{\omega_a}\sin\omega_a t \cos\omega_b t\right) \qquad [\text{q}]$$

To get a feel for the motion, we note that ω_b is an order of magnitude smaller than ω_a, and we ignore the terms with the coefficient ω_b/ω_a from the right side of the above equations. This simplification leads to an interesting explanation of the motion to a first-order approximation: The motion of the Foucault pendulum can be explained as an amplitude modulated swing motion for both x and y where the amplitude of the swing varies with the rate $\omega_b = \Omega \sin \lambda$.

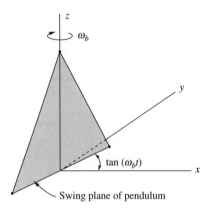

Figure 2.42 Rotation of Foucault's pendulum

Furthermore, if we take the ratio of $y(t)$ over $x(t)$, we obtain

$$\frac{y(t)}{x(t)} \approx -\tan(\omega_b t) = -\tan(\Omega \sin \lambda t) \qquad [\mathbf{r}]$$

which indicates that the plane of the pendulum rotates at the angular rate $\omega_b = \Omega \sin \lambda$, as depicted in Fig. 2.42. Because of the negative sign, the direction of rotation in the Northern Hemisphere is clockwise and opposite to the rotation of the earth, a conclusion we can visualize easily. One intuitively expects the pendulum to rotate in the opposite direction of the rotation of the earth. While the terms that were ignored in Eq. [r] change this interpretation slightly, the main result is the same.

At the North Pole, as λ becomes 90°, the Coriolis effect is the most pronounced. At the equator, the Coriolis effect disappears.

References

Ginsberg, J. H. *Advanced Engineering Dynamics*. New York: Harper & Row, 1988.
Goldstein, H. *Classical Mechanics*. 2nd ed. Reading, MA: Addison-Wesley, 1980.
Greenwood, D. T. *Principles of Dynamics*. 2nd ed. Englewood Cliffs, NJ: Prentice-Hall, 1988.
Kane, T., and D. Levinson. *Dynamics, Theory and Applications*. New York: McGraw-Hill, 1985.
Meirovitch, L. *Methods of Analytical Dynamics*. New York: McGraw-Hill, 1970.
———. *Elements of Vibration Analysis*. 2nd ed. New York: McGraw-Hill, 1986.

Homework Exercises

Section 2.2

1. A coordinate system XYZ is transformed into a coordinate system xyz by the following series of transformations: First, a counterclockwise rotation of 45° about the Y axis, resulting in the $x'y'z'$ coordinate system, and then a clockwise

rotation of 30° about the x' axis, resulting in the xyz coordinate system. Find the coordinates in the inertial frame of a vector which originally is $\mathbf{r} = 2\mathbf{I} + 3\mathbf{J}$ and moves with the transformed coordinate system. Then, find the components of a vector $\mathbf{p} = -3\mathbf{i} + 4\mathbf{k}$ in terms of the inertial system and the components of the vector $\mathbf{v} = -4\mathbf{j} + 1.2\mathbf{K}$ in both the XYZ and xyz coordinate systems.

2. A coordinate system XYZ is transformed into an xyz coordinate system, first by a rotation of θ_1 about the Z axis, which gives the $x'y'z'$ coordinates, and then by a rotation θ_2 about the y' axis. Express the unit vectors $\mathbf{i}$, $\mathbf{j}$, and $\mathbf{k}$ and their derivatives in terms of the unit vectors of the XYZ coordinates.

SECTION 2.3

3. Given the column vector $\{q\} = [q_1 \; q_2 \; q_3 \; q_4]^T$, the matrix $[D]$ as

$$[D] = \begin{bmatrix} 5 & 1 & 0 & -1 \\ 1 & 4 & 1 & 0 \\ 0 & 1 & 6 & -1 \\ -1 & 0 & -1 & 3 \end{bmatrix}$$

and the quadratic form $S = \{q\}^T[D]\{q\}$, evaluate S. Then, calculate $dS/d\{q\}$ and show that the value you obtain is identical to what would be obtained using Eq. [2.3.20].

4. The motion of the double pendulum in Fig. 2.43 is described by the angles θ_1 and θ_2. Find expressions for the position vector $\mathbf{r}$ associated with the motion of the tip of the pendulum, as well as the velocity $\mathbf{v}$ and acceleration $\mathbf{a}$ of the tip, using Eqs. [2.3.23]–[2.3.28], and verify Eq. [2.3.29].

5. Given the column vector $\{q\} = [q_1 \; q_2]^T$, the matrix $[D]$ is

$$[D] = \begin{bmatrix} 4q_2 & -2q_1 \\ -2q_1 & 6 \end{bmatrix}$$

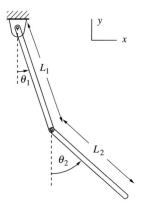

Figure 2.43

and the quadratic form $S = \{q\}^T[D]\{q\}$, evaluate S. Then calculate $dS/d\{q\}$ and show that the value you obtain is identical to what would be obtained using Eq. [2.3.21].

SECTION 2.4

6. Show (independently of the arguments in this chapter) that the determinant of the direction cosine matrix $[c]$ between two coordinate systems is always equal to 1.

7. Two coordinate systems XYZ and xyz are related to each other as shown in Fig. 2.44. Find the direction cosine matrix between the two coordinate systems.

8. The direction cosine matrix resulting from a 3-1-3 body fixed transformation (θ_1 about x_3, θ_2 about y_1, and θ_3 about z_3) is given below. Find the values of the rotation angles θ_1, θ_2, and θ_3.

$$[c] = \begin{bmatrix} -0.1768 & -0.9186 & -0.3536 \\ 0.8839 & -0.3062 & 0.3536 \\ -0.4330 & -0.2500 & 0.8660 \end{bmatrix}$$

9. The rectangular box in Fig. 2.14 is first rotated clockwise by 15° about the line OA and then by 45° counterclockwise about line OB. Find the coordinates of point C after this rotation sequence.

10. The rectangular box in Fig. 2.14 is rotated counterclockwise by 45° about a line passing through points A and B (viewed from B). Find the coordinates of point C after this rotation sequence.

11. The rectangular box shown in Fig. 2.45 is rotated counterclockwise by an angle of 30° about the axis passing through points O and B. Find the coordinates of point A after this rotation in terms of the inertial coordinates.

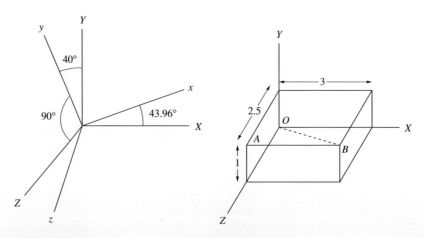

Figure 2.44   **Figure 2.45**

12. Consider the double-link robot mounted on a rotating base shown in Fig. 2.46, with the angles θ_1 and θ_2 measured from the vertical. Find the position of the tip of the robot arm in terms of inertial coordinates when $\phi = 30°$, $\theta_1 = 60°$, and $\theta_2 = -15°$. The XYZ coordinates are inertial.

SECTIONS 2.5 AND 2.6

13. A robotic sander shown in Fig. 2.47 has a sanding disk that spins at the constant rate of 1500 rpm. The arms AD and DB, which are used to position the sander, make angles of ϕ and ψ with the vertical. At the instant shown, their values are $\phi = 90°$, $\psi = 60°$, and they are moving with the constant angular speeds of $\dot{\phi} = 0.2$ rad/s and $\dot{\psi} = -0.3$ rad/s. Find the angular velocity and angular acceleration of the sanding disk in terms of inertial coordinates.

14. Consider the airplane in Fig. 2.31. The airplane is moving with speed of 600 mph in a curved trajectory $\rho = 3000$ ft. At the same time, the airplane is pitching upwards at the rate of 0.1 rad/s (constant). The propeller is spinning with the constant counterclockwise angular rate of 4000 rpm. Find the total angular acceleration of the propeller.

15. The disk shown in Fig. 2.48 rotates with angular velocity $\dot{\theta}_2 = \omega_2 = 15$ rad/s and angular acceleration $\dot{\omega}_2 = 1.2$ rad/s^2 about a rotating shaft. The shaft is bent, and it rotates with angular velocity $\omega_1 = 4$ rad/s and angular acceleration $\dot{\omega}_1 = -3$ rad/s^2. At this instant, the center of the disk coincides with its undeformed position. Find the angular acceleration of the disk.

16. Find a general expression for the angular acceleration $^A\alpha^B$, given the relation $^A\omega^B = {}^A\omega^{A_2} + {}^{A_2}\omega^{A_3} + \ldots + {}^{A_N}\omega^B$.

17. A single gimbal gyroscope (inner gimbal not moving), such as the one shown in Fig. 2.49, is used to measure the angular motion of vehicles. The spin rate of the

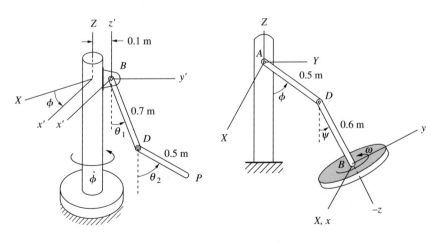

Figure 2.46 **Figure 2.47**

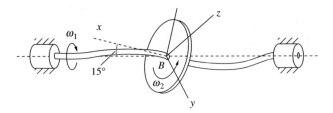

Figure 2.48

rotor is denoted by $\dot{\psi}$ while the gimbal makes an angle of θ with the platform. The angular velocities of the platform are ω_X, ω_Y, and ω_Z. Find expressions for the angular velocity and angular acceleration of the rotor using a set of relative coordinates attached to the outer gimbal.

18. Find the angular acceleration of the momentum wheel in Example 2.6, given that all angular velocities are constant.

19. Consider Problem 2 and solve for the angular velocity and angular acceleration of the moving frame using the transport theorem.

20. The flywheel of the gyroscope shown in Fig. 2.50 has a constant angular velocity of $\omega_3 = 5000$ rpm about its axis. The outer gimbal has an angular velocity of $\omega_1 = 3$ rad/s, which is decreasing at the rate of 1.8 rad/s². The inner gimbal is at a position such that the angle between the outer gimbal axis and flywheel axis is $\theta = 75°$, with $\dot{\theta} = 0, \ddot{\theta} = 3$ rad/s². Find the angular acceleration of the flywheel.

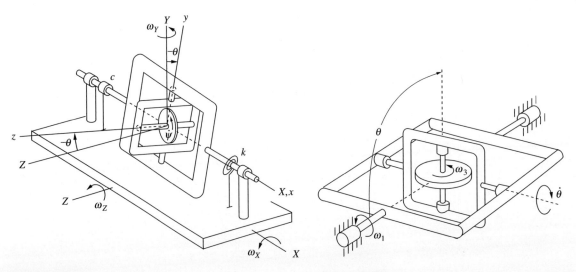

Figure 2.49   **Figure 2.50**

Section 2.7

21. A bead of mass m is free to slide on a hoop of radius R (Fig. 2.51). The hoop spins with a constant angular velocity Ω about the vertical axis. Find the acceleration of the bead.

22. Consider Example 2.1, and using the relative velocity and acceleration relations find the velocity and acceleration of a water particle as it exits the rotating sprinkler.

23. Consider Example 2.3. This time, the ant has started moving from point C along a circular path with point D as the center of the circle with constant speed v. Find the acceleration of the ant.

24. A spring pendulum is attached to a rotating shaft by an arm of distance d, as shown in Fig. 2.52. At the instant shown, the shaft is rotating with the constant angular velocity $\Omega = 0.4$ rad/s, $\theta = 30°$, $\dot{\theta} = 0.3$ rad/s, $\ddot{\theta} = 2$ rad/s². The length of the pendulum is $L = 1.3$ m and it is getting shorter at the constant rate of 0.1 m/s. Given also is that $d = 0.8$ m. Find the acceleration of the tip of the pendulum.

25. Consider the bent shaft in problem 17. Now, the shaft is bent such that its deformation can be expressed by the relation $d(x) = 0.18\sin(\pi x/2)$ m, as shown in Fig. 2.53. The disk is located at $x = 0.9$ m. Find the velocity and acceleration of point B. Then derive an expression for the acceleration of an arbitrary point on the edge of the disk. Assume the angular velocity of ω_1 is constant and the disk is of radius R.

26. A hunter shown in Fig. 2.54 is aiming at a moving target with her rifle. The hunter is moving the rifle upwards with a constant angular velocity of $\dot{\theta} = 10°$/sec and pulls the trigger when $\theta = 105°$. The bullet, which weighs 1/8 lb, leaves the barrel with a constant speed of 900 mph relative to the rifle. Find the total velocity and acceleration of the bullet.

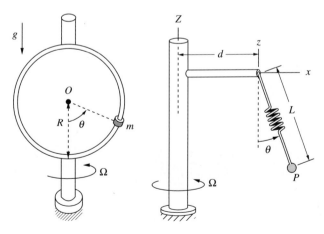

Figure 2.51   **Figure 2.52**

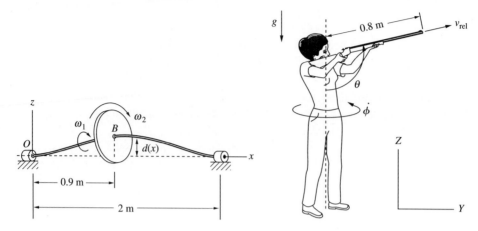

Figure 2.53 **Figure 2.54**

27. Consider the previous problem and find the average force exerted by the rifle on the hunter as the hunter pulls the trigger, assuming that it takes the bullet 0.01 seconds to reach its speed of 900 mph.

28. Consider the hunter in Problem 26 and find the velocity and acceleration of the bullet if the hunter is also turning at the constant angular speed of $\dot{\phi} = 15°/s$ at the instant the bullet is about to leave the rifle.

29. Consider the two-link robot mounted on a rotating base in Fig. 2.46. Given that the base and arm angles vary with the relationships $\phi(t) = 0.1 \sin t$ rad, $\theta_1(t) = 0.3t$ rad, $\theta_2(t) = -0.1t^2$ rad, find the angular velocity and angular acceleration of the robot arms and the velocity of the tip at $t = 2$ seconds.

30. A disk of radius R shown in Fig. 2.55 spins at the constant rate of $\omega_2 = \dot{\theta}$ about an axle held by a fork-ended horizontal rod that rotates itself at the rate $\omega_1 = \dot{\phi}$. An ant is walking toward the center of the disk with constant speed v with respect to the disk. Find the acceleration of the ant as a function of the angle θ and when the ant is at the edge of the disk (at point P).

31. An airplane, shown in Fig. 2.56, is flying at a speed of 500 km/h. The airplane has a constant pitch rate (ω_y) of 0.05 rad/s and a constant roll rate (ω_x) of 0.01 rad/s with no yaw, $\omega_z = 0$. The trailing edge flaps are being extended to give the airplane more lift at the constant rate of 0.04 m/s. Find the velocity and acceleration of point C on the flap.

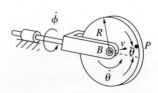

Figure 2.55

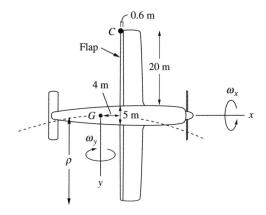

Figure 2.56

32. An airplane, shown in Fig. 2.56, is executing a circular turn with a radius of 4000 m, while flying at a speed of 600 km/h. The airplane has a constant pitch rate of $\omega_y = 0.04$ rad/s and a constant roll rate of $\omega_x = 0.01$ rad/s. The trailing edge flaps are being extended to give the airplane more lift at the constant rate of 0.05 m/s. Find the velocity and acceleration of point C on the flap.

Section 2.8

33. Find the equation of motion of the bead in Problem 21 using Newton's second law.
34. Figure 2.57 shows a bead of mass m sliding without friction over a thin wire shaped in the form of a parabola governed by the equation $z = x^2/2$. The thin wire is rotating about the z axis with the constant angular velocity Ω. There is a spring acting on the bead of constant k that deforms only in the vertical direction. Derive the equation of motion of the bead using Newton's second law.

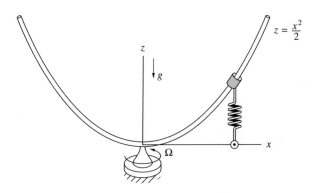

Figure 2.57

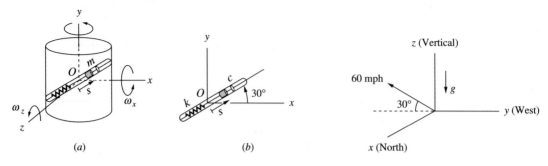

Figure 2.58

Figure 2.59

35. A satellite in space has angular velocities ω_x, ω_y, and ω_z about the x, y, and z axes, which are attached to the satellite at point O, as shown in Fig 2.58. On the xy plane, making an angle of 30° with the x axis and going through O, is a tube, inside which a particle slides without friction. The particle is attached by a spring and a dashpot to each end of the tube. The spring is unstretched when the particle is at point O. Find the equation of motion of the particle, using Newton's second law.

36. Consider a 90-ft-long bowling alley in Rio de Janeiro, Brazil. A bowler releases the ball with a constant velocity of 20 ft/sec, aimed directly at the pocket. What is the Coriolis deflection, and in which direction is it?

37. Integrate the perturbation expressions to find the deflection of a stone thrown in the air vertically with speed v as it falls to the ground. Then compare this result with the exact solution.

chapter 3

DYNAMICS OF A SYSTEM OF PARTICLES

3.1 INTRODUCTION

This chapter serves three purposes. First, it extends the developments of Chapter 1 to systems with more than one particle. Second, it prepares the reader for the analysis of rigid bodies, addressing the most basic form of rigid body motion, that is, plane kinetics of a rigid body. Third, it introduces basic concepts in orbital mechanics.

Analysis of systems of particles and rigid bodies is greatly simplified when the concept of center of mass is introduced. Newton's laws of translational motion and Euler's law of rotational motion for a single particle can be expressed in the same form for a system of particles as well as for rigid bodies in terms of the center of mass.

This chapter is written such that one can skip it and go directly into analytical mechanics or into rigid body dynamics. This chapter actually belongs with Chapter 1, as it is an extension of the basic concepts studied in that chapter. For pedagogical considerations the developments here are presented separately.

The sections in this chapter on plane kinetics are intended to serve as a review of this special case of rigid body motion. This review is essential, especially for those who have not studied the plane kinetics of rigid bodies for a long time. Plane kinetics is relevant to Chapter 4, where we carry out the developments in analytical mechanics for particle motion or rigid body motion on a plane.

3.2 EQUATIONS OF MOTION

In this section, we extend the developments in Newtonian particle mechanics to systems consisting of several particles. Consider a system of N particles, as shown in

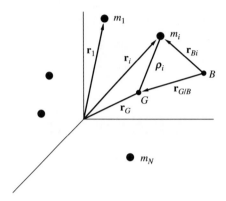

Figure 3.1

Fig. 3.1. Each particle is of mass m_i ($i = 1, 2, \ldots, N$) and is located by the displacement vector $\mathbf{r}_i$. The motion of the individual particles is not necessarily independent of each other, as there may be forces that relate the behavior of some of the particles to each other.

We next introduce the concept of *center of mass*. Denoting by m the total mass of the system

$$m = \sum_{i=1}^{N} m_i \qquad [3.2.1]$$

Denoting the center of mass by the point G and locating it by the vector $\mathbf{r}_G$, it is defined as

$$\mathbf{r}_G = \frac{1}{m} \sum_{i=1}^{N} m_i \mathbf{r}_i \qquad [3.2.2]$$

We can express the position of the ith particle relative to the center of mass as

$$\mathbf{r}_i = \mathbf{r}_G + \boldsymbol{\rho}_i \qquad [3.2.3]$$

in which $\boldsymbol{\rho}_i$ is the vector connecting the center of mass with the ith particle. Introducing Eq. [3.2.3] into Eq. [3.2.2], we obtain

$$\mathbf{r}_G = \frac{1}{m} \sum_{i=1}^{N} m_i (\mathbf{r}_G + \boldsymbol{\rho}_i) = \mathbf{r}_G + \frac{1}{m} \sum_{i=1}^{N} m_i \boldsymbol{\rho}_i \qquad [3.2.4]$$

which leads to the conclusion

$$\sum_{i=1}^{N} m_i \boldsymbol{\rho}_i = \mathbf{0} \qquad [3.2.5]$$

One can differentiate the above equations to find expressions for the velocity and acceleration of the center of mass, which we will denote by $\mathbf{v}_G$ and $\mathbf{a}_G$, respectively.

Differentiating Eq. [3.2.2] with respect to time, we obtain

$$\mathbf{v}_G = \dot{\mathbf{r}}_G = \frac{1}{m}\sum_{i=1}^{N} m_i \mathbf{v}_i \qquad [\mathbf{3.2.6}]$$

which leads to the relation for the relative velocity

$$\sum_{i=1}^{N} m_i \boldsymbol{\rho}_i = 0 \qquad [\mathbf{3.2.7}]$$

From this we find the linear momentum $\mathbf{p}$ of a system of particles to have the form

$$\mathbf{p} = \sum_{i=1}^{N} \mathbf{p}_i = \sum_{i=1}^{N} m_i \mathbf{v}_i = m\mathbf{v}_G \qquad [\mathbf{3.2.8}]$$

where $\mathbf{p}_i = m_i \mathbf{v}_i$ ($i = 1, 2, \ldots, N$) is the linear momentum of the ith particle.

In a similar fashion, we find the acceleration of the center of mass as

$$\mathbf{a}_G = \ddot{\mathbf{r}}_G = \frac{1}{m}\sum_{i=1}^{N} m_i \mathbf{a}_i \qquad [\mathbf{3.2.9}]$$

leading to the relation for the relative acceleration terms

$$\sum_{i=1}^{N} m_i \ddot{\boldsymbol{\rho}}_i = 0 \qquad [\mathbf{3.2.10}]$$

Next, we apply Newton's second law to a system of particles. We separate the total force acting on the ith particle into two parts:[1] (1) forces acting on the ith particle from outside the system of particles, referred to as the *external* or *impressed* forces and denoted by $\mathbf{F}_i$ ($i = 1, 2, \ldots, N$), and (2) forces exerted on m_i by the other particles within the system, referred to as *internal* or *constraint* forces, and denoted by $\mathbf{F}'_i$. Newton's second law for each particle is

$$m_i \mathbf{a}_i = \mathbf{F}_i + \mathbf{F}'_i \qquad i = 1, 2, \ldots, N \qquad [\mathbf{3.2.11}]$$

The N equations above are usually not independent of each other, so that it is not possible to analyze the motion of each particle individually. The number of degrees of freedom, denoted by n, is in general smaller than the number of coordinates (for the case here $3N$). The reduction is due to the action of one particle on the other and from the restrictions in the motion of the particles that they cause. These actions constrain the motion of the particles within the system. The internal forces $\mathbf{F}'_i$ ($i = 1, 2, \ldots, N$) are the forces associated with the constraints.

Considering the system as a whole, we sum the force balances for all of the particles, thus

$$\sum_{i=1}^{N} m_i \mathbf{a}_i = \sum_{i=1}^{N} (\mathbf{F}_i + \mathbf{F}'_i) \qquad [\mathbf{3.2.12}]$$

---

[1] We will see this separation also in Chapter 4, when we study the principle of virtual work.

156 CHAPTER 3 • DYNAMICS OF A SYSTEM OF PARTICLES

Substitution of Eq. [3.2.3] to the left side of this equation yields

$$\sum_{i=1}^{N} m_i \mathbf{a}_i = \sum_{i=1}^{N} m_i(\mathbf{a}_G + \ddot{\boldsymbol{\rho}}_i) = \sum_{i=1}^{N} m_i \mathbf{a}_G = m\mathbf{a}_G = \frac{d}{dt}\mathbf{p} \quad [\textbf{3.2.13}]$$

Now evaluating the right side of Eq. [3.2.12], because $\mathbf{F}'_i$ ($i = 1, 2, \ldots, N$) represents the forces that one particle exerts on another, their sum over all particles must be zero, by virtue of Newton's third law. Defining by $\mathbf{F}$ the sum of all external forces over all particles,

$$\mathbf{F} = \sum_{i=1}^{N}(\mathbf{F}_i + \mathbf{F}'_i) = \sum_{i=1}^{N} \mathbf{F}_i \quad [\textbf{3.2.14}]$$

we obtain Newton's second law for a system of particles as

$$m\mathbf{a}_G = \frac{d}{dt}\mathbf{p} = \mathbf{F} \quad [\textbf{3.2.15}]$$

Another way of writing Newton's second law for a system of particles is to combine Eqs. [3.2.13] and [3.2.15] to yield

$$\sum_{i=1}^{N} m_i \mathbf{a}_i = \mathbf{F} \quad [\textbf{3.2.16}]$$

Depending on the need and on the problem, one can use Eqs. [3.2.11], [3.2.15], or [3.2.16] to describe the force balance of a system of particles.

Example 3.1

The two masses m_1 and m_2 are connected by a massless rod, and they are acted upon by a force P, as shown in Fig. 3.2. Write Newton's second law using Eqs. [3.2.11], [3.2.15], and [3.2.16], and evaluate if these equations qualify as equations of motion.

Solution

We first separate the two masses and draw free-body diagrams, shown in Fig. 3.3. We denote the displacements of the masses by x_i and y_i ($i = 1, 2$). F_x and F_y are internal reaction forces. Using Eqs. [3.2.11] we obtain the four equations

$$m_1 \ddot{x}_1 = F_x \qquad m_2 \ddot{x}_2 = -F_x \qquad m_1 \ddot{y}_1 = F_y + P \qquad m_2 \ddot{y}_2 = -F_y \quad [\textbf{a}]$$

in terms of the reaction forces.

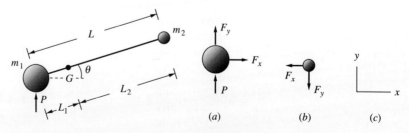

Figure 3.2

Figure 3.3 Free body diagrams

To use Eq. [3.2.15], we first need to locate the center of mass G. It is easy to show that the distance from m_1 to the center of mass is $L_1 = m_2 L/(m_1 + m_2)$. Considering the system as a whole, the only external force is P. Denoting the displacements of the center of mass by x_G and y_G, Newton's second law becomes

$$(m_1 + m_2)\ddot{x}_G = 0 \qquad (m_1 + m_2)\ddot{y}_G = P \qquad \textbf{[b]}$$

Finally, we add the first two of Eqs. [a] and the last two, to obtain Newton's second law in the form of Eq. [3.2.16] as

$$m_1 \ddot{x}_1 + m_2 \ddot{x}_2 = 0 \qquad m_1 \ddot{y}_1 + m_2 \ddot{y}_2 = P \qquad \textbf{[c]}$$

Let us now investigate the nature of these equations. Eqs. [a] have four equations and each of Eqs. [b] and [c] have two. To find if any of these equations qualify as equations of motion we calculate the number of degrees of freedom, which can be shown to be 3. Hence, we need three differential equations, void of internal forces, to describe the system. We conclude that Eqs. [b] represent two of the equations of motion, in terms of x_G and y_G, associated with the translation of the center of mass. Similarly, Eqs. [c] represent the same, but in terms of x_i and y_i ($i = 1, 2$). We still need a third equation.

To find the equations of motion, we need to find expressions for F_x and F_y in Eq. [a] and write one of x_1, x_2, y_1, or y_2 in terms of the other three variables. The procedure is tedious at best, especially if one realizes that this problem is ideally suited for treatment with angular coordinates and angular momentum balances. We discuss this approach in the next section.

3.3 LINEAR AND ANGULAR MOMENTUM

The linear momentum of a system of particles is defined in Eq. [3.2.8] and, as discussed in Chapter 1, it is an absolute quantity. We can integrate the equations of motion for a system of particles over time to obtain impulse-momentum relationships. Indeed, following the approach in Section 1.6, we can integrate Eqs. [3.2.11], [3.2.15], and [3.2.16] over a time period (t_1, t_2) to obtain the linear impulse-momentum relationships, writing

$$m_i \mathbf{v}_i(t_1) + \int_{t_1}^{t_2} (\mathbf{F}_i(t) + \mathbf{F}'_i(t))dt = m_i \mathbf{v}_i(t_2) \qquad i = 1, 2, \ldots, N$$

$$\sum_{i=1}^{N} m_i \mathbf{v}_i(t_1) + \int_{t_1}^{t_2} \mathbf{F}(t)dt = \sum_{i=1}^{N} m_i \mathbf{v}_i(t_2)$$

$$m\mathbf{v}_G(t_1) + \int_{t_1}^{t_2} \mathbf{F}(t)dt = m\mathbf{v}_G(t_2) \qquad \textbf{[3.3.1a, b, c]}$$

In order to solve for the individual velocities of each particle, one has to integrate all of Eqs. [3.3.1b]. One may not be able to eliminate the $\mathbf{F}'_i(t)$ terms directly; hence, a number of constraint forces may have to be solved for.

When the sum of all external forces acting on a system of particles is zero or its integral over a time period is zero, the linear momentum of the system remains unchanged, which is the statement of the *principle of conservation of linear momentum for a system of particles*. The principle can be written as

$$\mathbf{p}(t_1) = \mathbf{p}(t_2) \qquad [3.3.2]$$

or

$$m\mathbf{v}_G(t_1) = m\mathbf{v}_G(t_2) \qquad \sum_{i=1}^{N} m_i \mathbf{v}_i(t_1) = \sum_{i=1}^{N} m_i \mathbf{v}_i(t_2) \qquad [3.3.3]$$

These equations are vector relationships. If the sum of forces or their integral over a time period is zero along a certain direction, the linear momentum is conserved only along that particular direction. For example, if linear momentum is conserved along a certain direction, and we express the unit vector in that direction by $\mathbf{e}$, the conservation of linear momentum equations become

$$m\mathbf{v}_G(t_1) \cdot \mathbf{e} = m\mathbf{v}_G(t_2) \cdot \mathbf{e} \qquad \sum_{i=1}^{N} m_i \mathbf{v}_i(t_1) \cdot \mathbf{e} = \sum_{i=1}^{N} m_i \mathbf{v}_i(t_2) \cdot \mathbf{e} \qquad [3.3.4]$$

Impulsive forces are treated the same way as in Chapter 1.

We next define the angular momentum, or moment of the linear momentum, of a particle m_i about a point B by

$$\mathbf{H}_{Bi} = \mathbf{r}_{Bi} \times m_i \mathbf{v}_i = \mathbf{r}_{Bi} \times \mathbf{p}_i \qquad i = 1, 2, \ldots, N \qquad [3.3.5]$$

where $\mathbf{r}_{Bi}$ is the vector connecting point B and the ith particle (Fig. 3.1). Unlike linear momentum, which is an absolute quantity, angular momentum is relative: its value depends on the point about which it is calculated. The total angular momentum of a system of particles about point B is denoted by $\mathbf{H}_B$ and is expressed as

$$\mathbf{H}_B = \sum_{i=1}^{N} \mathbf{H}_{Bi} = \sum_{i=1}^{N} \mathbf{r}_{Bi} \times m_i \mathbf{v}_i \qquad [3.3.6]$$

We next relate the angular momentum of a system of particles to the center of mass motion. From Fig. 3.1, we write $\mathbf{r}_{Bi}$ in terms of the center of mass as $\mathbf{r}_{Bi} = \mathbf{r}_{G/B} + \boldsymbol{\rho}_i$ ($i = 1, 2, \ldots, N$). For the ith particle, we write the angular momentum expressions as

$$\mathbf{H}_{Bi} = (\mathbf{r}_{G/B} + \boldsymbol{\rho}_i) \times m_i (\mathbf{v}_G + \dot{\boldsymbol{\rho}}_i) \qquad [3.3.7]$$

We sum the individual angular momenta about point B and obtain the angular momentum of the system of particles about point B as

$$\mathbf{H}_B = \sum_{i=1}^{N} \mathbf{H}_{Bi} = \sum_{i=1}^{N} (\mathbf{r}_{G/B} + \boldsymbol{\rho}_i) \times m_i (\mathbf{v}_G + \dot{\boldsymbol{\rho}}_i) \qquad [3.3.8]$$

Considering the definition of the center of mass, this equation reduces to

$$\mathbf{H}_B = \mathbf{r}_{G/B} \times m\mathbf{v}_G + \sum_{i=1}^{N} \boldsymbol{\rho}_i \times m_i \dot{\boldsymbol{\rho}}_i \qquad [3.3.9]$$

where the first term on the right is associated with the motion of the center of mass, and the second is due to motion with respect to the center of mass. The second term is also referred to as the *apparent angular momentum*. Differentiation of Eq. [3.3.9] yields

$$\dot{\mathbf{H}}_B = \mathbf{r}_{G/B} \times m\mathbf{a}_G + \mathbf{v}_{G/B} \times m\mathbf{v}_G + \sum_{i=1}^{N}[\boldsymbol{\rho}_i \times m_i\ddot{\boldsymbol{\rho}}_i + \dot{\boldsymbol{\rho}}_i \times m_i\dot{\boldsymbol{\rho}}_i]$$

$$= \mathbf{r}_{G/B} \times m\mathbf{a}_G + \mathbf{v}_{G/B} \times m\mathbf{v}_G + \sum_{i=1}^{N}\boldsymbol{\rho}_i \times m_i\ddot{\boldsymbol{\rho}}_i \quad [\mathbf{3.3.10}]$$

The moment about point B of all the forces acting on the ith particle is defined by

$$\mathbf{M}_{Bi} = \mathbf{r}_{Bi} \times \mathbf{F}_i = r_{Bi} \times m_i\mathbf{a}_i \qquad i = 1, 2, \ldots, N \quad [\mathbf{3.3.11}]$$

The total of all of the moments acting on the system of particles can be obtained by summing the individual moments, which yields

$$\mathbf{M}_B = \sum_{i=1}^{N}\mathbf{r}_{Bi} \times \mathbf{F}_i = \sum_{i=1}^{N}\mathbf{r}_{Bi} \times m_i\mathbf{a}_i = \sum_{i=1}^{N}(\mathbf{r}_{G/B} + \boldsymbol{\rho}_i) \times m_i(\mathbf{a}_G + \ddot{\boldsymbol{\rho}}_i) \quad [\mathbf{3.3.12}]$$

Invoking the definition of the center of mass, Eq. [3.3.12] reduces to

$$\mathbf{M}_B = \mathbf{r}_{G/B} \times m\mathbf{a}_G + \sum_{i=1}^{N}\boldsymbol{\rho}_i \times m_i\ddot{\boldsymbol{\rho}}_i \quad [\mathbf{3.3.13}]$$

The second term on the right side of Eq. [3.3.10] can be written as

$$\mathbf{v}_{G/B} \times m\mathbf{v}_G = (\mathbf{v}_G - \mathbf{v}_B) \times m\mathbf{v}_G = m\mathbf{v}_G \times \mathbf{v}_B \quad [\mathbf{3.3.14}]$$

so that introducing Eqs. [3.3.13] and [3.3.14] into Eq. [3.3.10], we obtain the *angular momentum balance* for a system of particles as

$$\dot{\mathbf{H}}_B = \mathbf{M}_B + m\mathbf{v}_G \times \mathbf{v}_B \quad [\mathbf{3.3.15}]$$

Equation [3.3.15] is a general relation describing the angular momentum balance about a point B, whether B is fixed or moving. Under certain circumstances, and depending on the choice of point B, the equation can be further simplified:

1. When the point B is selected as the center of mass, $B = G$, then $\mathbf{v}_{G/B}$ vanishes and we have

$$\dot{\mathbf{H}}_G = \mathbf{M}_G \quad [\mathbf{3.3.16}]$$

2. When the point B is fixed in an inertial coordinate frame, then $\mathbf{v}_B = \mathbf{0}$, the second term in Eq. [3.3.15] vanishes, and we get

$$\dot{\mathbf{H}}_B = \mathbf{M}_B \quad [\mathbf{3.3.17}]$$

3. When $\mathbf{v}_{G/B}$ is parallel to $\mathbf{v}_G$ (for any reason), the cross product in Eq. [3.3.15] vanishes. This mathematical possibility is not of any physical significance.

Integration of the moment balance over time yields the *angular impulse-momentum relationships*. For each particle m_i and considering a fixed point B or about the center of mass (denoting such a point by D),

$$\mathbf{H}_{Di}(t_1) + \int_{t_1}^{t_2}\mathbf{M}_{Di}(t)\,dt = \mathbf{H}_{Di}(t_2) \qquad i = 1, 2, \ldots, N \quad [\mathbf{3.3.18}]$$

Evaluation of this equation for each mass m_i ($i = 1, 2, \ldots, N$) requires that the internal forces acting on the particles be calculated, reducing its usefulness. When we consider the entire system and either a fixed point B or the center of mass G, summation of Eq. [3.3.18] over all particles yields

$$\mathbf{H}_D(t_1) + \int_{t_1}^{t_2} \mathbf{M}_D(t)\,dt = \mathbf{H}_D(t_2) \qquad [3.3.19]$$

If the applied moment about the fixed point B or center of mass is zero, or the integral of $\mathbf{M}_D(t)$ over the interval (t_1, t_2) vanishes, we have

$$\mathbf{H}_D(t_1) = \mathbf{H}_D(t_2) \qquad [3.3.20]$$

which is the *principle of conservation of angular momentum* for a system of particles.

Example 3.2

Consider Example 3.1. The system is initially at rest with $\theta = 30°$ when it is hit by the impulsive force $\hat{P}$. Find the velocities of the two masses immediately after the impulsive force acts.

Solution

It is more convenient to use center of mass coordinates x_G and y_G. Hence, Eqs. [b] in Example 3.1 become two of the equations of motion. To find the third equation, we make use of the angle θ. We write the displacements of the two masses in terms of the coordinates of the center of mass as

$$x_1 = x_G - L_1 \cos\theta \qquad y_1 = y_G - L_1 \sin\theta$$
$$x_2 = x_G + L_2 \cos\theta \qquad y_2 = y_G + L_2 \sin\theta \qquad [a]$$

where

$$L_1 = \frac{m_2 L}{m_1 + m_2} \qquad L_2 = \frac{m_1 L}{m_1 + m_2} \qquad [b]$$

so that the velocities of the masses can be written as

$$\mathbf{v}_1 = (\dot{x}_G + L_1\dot{\theta}\sin\theta)\mathbf{i} + (\dot{y}_G - L_1\dot{\theta}\cos\theta)\mathbf{j}$$
$$\mathbf{v}_2 = (\dot{x}_G - L_2\dot{\theta}\sin\theta)\mathbf{i} + (\dot{y}_G + L_2\dot{\theta}\cos\theta)\mathbf{j} \qquad [c]$$

The position vectors from the center of mass to the individual masses are

$$\mathbf{r}_1 = -L_1\cos\theta\,\mathbf{i} - L_1\sin\theta\,\mathbf{j} \qquad \mathbf{r}_2 = L_2\cos\theta\,\mathbf{i} + L_2\sin\theta\,\mathbf{j} \qquad [d]$$

so that the angular momentum about G becomes (after a few manipulations)

$$\mathbf{H}_G = \mathbf{r}_1 \times m_1\mathbf{v}_1 + \mathbf{r}_2 \times m_2\mathbf{v}_2 = (m_1 L_1^2 + m_2 L_2^2)\dot{\theta}\,\mathbf{k} \qquad [e]$$

The moment generated about the center of mass is simply $\mathbf{M}_G = -PL_1\cos\theta\,\mathbf{k}$, so we can say that the impulsive moment due to the impulsive force is $\hat{\mathbf{M}}_G = -\hat{P}L_1\cos\theta$.

From the linear impulse-momentum theorem applied to Eqs. [b] of Example 3.1, we obtain immediately after the impulse

$$\dot{x}_G = 0 \qquad \dot{y}_G = \frac{\hat{P}}{m_1 + m_2} \qquad [f]$$

The angular impulse-momentum problem yields the third equation

$$\dot{\theta} = -\frac{\hat{P}L_1 \cos\theta}{m_1 L_1^2 + m_2 L_2^2} = -\frac{\sqrt{3}\hat{P}L_1}{2(m_1 L_1^2 + m_2 L_2^2)} \quad \text{[g]}$$

Substituting Eqs. [f] and [g] into Eqs. [c] yields the velocities of the individual masses right after the impulse, thus

$$\mathbf{v}_1 = -\frac{\sqrt{3}\hat{P}L_1^2}{4(m_1 L_1^2 + m_2 L_2^2)}\mathbf{i} + \left(\frac{\hat{P}}{m_1 + m_2} + \frac{3\hat{P}L_1^2}{4(m_1 L_1^2 + m_2 L_2^2)}\right)\mathbf{j}$$

$$\mathbf{v}_2 = \frac{\sqrt{3}\hat{P}L_1 L_2}{4(m_1 L_1^2 + m_2 L_2^2)}\mathbf{i} + \left(\frac{\hat{P}}{m_1 + m_2} - \frac{3\hat{P}L_1 L_2}{4(m_1 L_1^2 + m_2 L_2^2)}\right)\mathbf{j} \quad \text{[h]}$$

This example illustrates the considerable advantage of using the center of mass. If we wanted to solve this problem using Eqs. [a] in Example 3.1, we would have to first solve for the impulsive reaction forces and substitute into the impulse-momentum relations, which is a tedious procedure.

3.4 WORK AND ENERGY

From Chapter 1, the incremental work done by all forces acting on the ith particle is denoted by dW_i ($i = 1, 2, \ldots, N$) and has the form

$$dW_i = (\mathbf{F}_i + \mathbf{F}'_i) \cdot d\mathbf{r}_i \qquad i = 1, 2, \ldots, N \quad \text{[3.4.1]}$$

From the work-energy theorem, we write for the ith particle

$$dW_i = dT_i \quad \text{[3.4.2]}$$

where T_i is the kinetic energy associated with the ith particle

$$T_i = \frac{1}{2} m_i \mathbf{v}_i \cdot \mathbf{v}_i \quad \text{[3.4.3]}$$

To express the work done by all forces on the system of particles, write $d\mathbf{r}_i$ in terms of the center of mass as

$$d\mathbf{r}_i = d\mathbf{r}_G + d\boldsymbol{\rho}_i \quad \text{[3.4.4]}$$

Introducing this expression into Eq. [3.4.1] and separating the total force into its internal and external components, we obtain

$$dW_i = (\mathbf{F}_i + \mathbf{F}'_i) \cdot (d\mathbf{r}_G + d\boldsymbol{\rho}_i) \quad \text{[3.4.5]}$$

We find the total work done by all forces acting on the system by summing the individual incremental work for each particle and integrating over the displacement of each particle. Denoting by $\mathbf{r}_{i1}$ and $\mathbf{r}_{i2}$ the initial and final locations of the ith particle, the total work is

$$W_{1\to 2} = \int \sum_{i=1}^{N} dW_i = \sum_{i=1}^{N} \int_{\mathbf{r}_{i1}}^{\mathbf{r}_{i2}} (\mathbf{F}_i + \mathbf{F}'_i) \cdot (d\mathbf{r}_G + d\boldsymbol{\rho}_i) \quad \text{[3.4.6]}$$

Using Eq. [3.2.14], we can simplify the expression for work to

$$W_{1\to 2} = \int_{\mathbf{r}_{G1}}^{\mathbf{r}_{G2}} \mathbf{F} \cdot d\mathbf{r}_G + \sum_{i=1}^{N} \int_{\mathbf{r}_{i1}}^{\mathbf{r}_{i2}} (\mathbf{F}_i + \mathbf{F}_i') \cdot d\boldsymbol{\rho}_i \qquad [3.4.7]$$

The first term on the right in this equation describes the work done due to the motion of the center of mass. The second term includes the contribution due to the motion with respect to the center of mass. Note that this second term contains $\mathbf{F}_i'$ ($i = 1, 2, \ldots, N$), indicating that the constraint forces between the particles may contribute to the work done. Although these internal forces cancel each other when we are summing forces to get the equations of motion, they do not necessarily vanish when the work is calculated.

As we did for a single particle, we can write the expression for work as an integral over time. Multiplying and dividing each term in Eq. [3.4.7] by dt, we obtain

$$W = \int_{t_1}^{t_2} \mathbf{F} \cdot \mathbf{v}_G \, dt + \sum_{i=1}^{N} \int_{t_1}^{t_2} (\mathbf{F}_i + \mathbf{F}_i') \cdot \dot{\boldsymbol{\rho}}_i \, dt \qquad [3.4.8]$$

where t_1 and t_2 denote the times at which the particles are at positions $\mathbf{r}_{i1}$ and $\mathbf{r}_{i2}$, respectively.

We next evaluate the expression for kinetic energy by summing the individual kinetic energies

$$T = \sum_{i=1}^{N} T_i \qquad [3.4.9]$$

in which T_i is defined in Eq. [3.4.3]. Substituting the expression $\mathbf{v}_i = \mathbf{v}_G + \dot{\boldsymbol{\rho}}_i$ in the above equation, we obtain

$$T = \frac{1}{2} \sum_{i=1}^{N} m_i (\mathbf{v}_G + \dot{\boldsymbol{\rho}}_i) \cdot (\mathbf{v}_G + \dot{\boldsymbol{\rho}}_i) = \frac{1}{2} \sum_{i=1}^{N} m_i (\mathbf{v}_G \cdot \mathbf{v}_G + 2\dot{\boldsymbol{\rho}}_i \cdot \mathbf{v}_G + \dot{\boldsymbol{\rho}}_i \cdot \dot{\boldsymbol{\rho}}_i)$$

$$= \frac{1}{2}(m\mathbf{v}_G \cdot \mathbf{v}_G) + \frac{1}{2} \sum_{i=1}^{N} m_i \dot{\boldsymbol{\rho}}_i \cdot \dot{\boldsymbol{\rho}}_i = T_{\text{tran}} + T_{\text{rot}} \qquad [3.4.10]$$

The first term, $T_{\text{tran}} = m\mathbf{v}_G \cdot \mathbf{v}_G/2$, is due to the translation of the center of mass, and the second term, T_{rot}, is due to the motion of the individual particles with respect to the center of mass.

The potential energy of a system of particles does not lend itself to a special formulation, and we use the expressions in Chapter 1. In general, the potential energy is due to elastic forces between particles, such as interconnecting springs, and due to gravity. The gravitational potential energy for a system of particles can be expressed in terms of the potential energy of each particle,

$$V_i = m_i g h_i \qquad i = 1, 2, \ldots, N \qquad [3.4.11]$$

in which h_i is the height of each particle from a common datum. One can show that the potential energy of a system of particles can also be written in terms of the height

h_G of its center of mass as

$$V = \sum_{i=1}^{N} V_i = mgh_G \qquad [3.4.12]$$

The work-energy theorem for a system of particles can be expressed in terms of each particle as

$$T_{i1} + V_{i1} + W_{inc_{1\to 2}} = T_{i2} + V_{i2} \qquad i = 1, 2, \ldots, N \qquad [3.4.13]$$

or in terms of the system of particles as

$$T_1 + V_1 + W_{nc_{1\to 2}} = T_2 + V_2 \qquad [3.4.14]$$

where $W_{nc_{1\to 2}}$ denotes the work done by all nonconservative forces and the subscripts denote the initial and final stages of the motion. When all the forces acting on a system of particles are conservative, the total energy of a system of particles is conserved, and we write

$$E = T + V = \text{constant} \qquad [3.4.15]$$

A bullet of mass 0.05 kg is shot with speed 400 m/s at a target on wheels of mass 3 kg, shown in Fig. 3.4, which has a plate in it that is attached to a spring of constant $k = 80{,}000$ N/m. The target is initially at rest and it can slide without friction. Find the maximum compression of the spring, after the bullet hits the target.

| **Example** |
| 3.3 |

Solution

The motion takes place as follows. The bullet hits the target and begins to compress the spring. The maximum compression of the spring is reached when the target and bullet are moving with the same speed. Because there are no forces external to the bullet-target system, linear momentum is conserved along the horizontal. At the point of maximum compression

$$m_b v_b = (m_b + m_t) v \qquad [a]$$

where the subscripts b and t correspond to the bullet and target, respectively, and v is the common velocity at the instant of maximum deflection of the spring. Substituting in the values, we obtain

$$v = \frac{m_b v_b}{m_b + m_t} = \frac{0.05 \times 400}{3.05} = 6.557 \text{ m/s} \qquad [b]$$

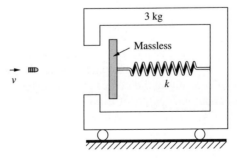

Figure 3.4

There is no loss of energy, as the energy transfer between the bullet and the target involves the spring and the bullet does not get lodged in the target. We thus invoke the conservation of energy between the initial state and the maximum deformation of the spring as

$$T_1 = T_2 + V_2 \qquad [c]$$

in which

$$T_1 = \frac{1}{2}m_b v_b^2 \qquad T_2 = \frac{1}{2}(m_b + m_t)v^2 \qquad V_2 = \frac{1}{2}k\Delta^2 \qquad [d]$$

so that the deflection of the spring is found as

$$\Delta = \sqrt{\frac{m_b v_b^2 - (m_b + m_t)v^2}{k}} \qquad [e]$$

Substituting in the values, we obtain

$$\Delta = \sqrt{\frac{(0.05 \times 400^2) - (3.05 \times 6.557^2)}{80{,}000}} = 0.3136 \text{ m} \qquad [f]$$

3.5 Impact of Particles

There are many ways two bodies come in contact with each other. The contact may take place at an isolated point, over a line, or over a surface. The contact between two bodies generates a constraint or reaction force on both bodies. From Newton's third law, the constraint force on each body is equal in magnitude and opposite in direction. In this section, we consider the special case of collision of two particles, that of *impact*. Chapter 8 conducts a general study of impact of rigid bodies.

In order to have a better understanding of impact, we initially assume that each particle is a solid sphere. We consider two spheres of masses m_1 and m_2 that have velocities (of their centers of mass) of $\mathbf{v}_1$ and $\mathbf{v}_2$ before impact and $\mathbf{u}_1$ and $\mathbf{u}_2$ after impact, as shown in Fig. 3.5. The changes in velocity, as they occur in a very short period of time, must be caused by impulsive forces, from which we conclude that impact generates an impulsive reaction (or impulsive constraint force) $\hat{\mathbf{F}}$ on each of the two spheres. This impulsive constraint lies along the line connecting the center points of the spheres and the point of contact; it is illustrated in Fig. 3.6. The line joining the two centers is commonly referred to as the *line of impact* and its direction the *normal direction*. The normal direction here should not be confused with the normal direction associated with normal and tangential coordinates. Because the impulsive constraint force is along the center of each sphere, there is no change in the linear momentum about the centers of mass of both spheres.

The impulsive force $\hat{\mathbf{F}}$ can be expressed as

$$\hat{\mathbf{F}} = \hat{F}\mathbf{n} \qquad [3.5.1]$$

where $\mathbf{n}$ is the unit vector along the line of impact.

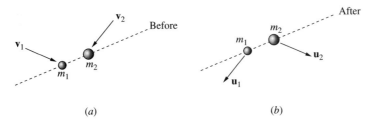

Figure 3.5 Colliding particles

The linear impulse-momentum relation for the two masses can be written as

$$m_1\mathbf{v}_1 - \hat{\mathbf{F}} = m_1\mathbf{u}_1 \qquad m_2\mathbf{v}_2 + \hat{\mathbf{F}} = m_2\mathbf{u}_2 \qquad [3.5.2]$$

We add the two equations above to obtain the conservation of linear momentum relationship for the system of the colliding particles as

$$m_1\mathbf{v}_1 + m_2\mathbf{v}_2 = m_1\mathbf{u}_1 + m_2\mathbf{u}_2 \qquad [3.5.3]$$

Equation [3.5.3] can also be written directly by using the impulse-momentum relationship for a system of particles, Eq. [3.3.1b]. Noting that during impact there is no impulse external to the system of the two spheres, the linear momentum of the system of two particles is conserved. Equation [3.5.3] indicates that the center of mass of the system of two spheres does not change position during the impact. Because the impulse is only in the normal direction, the components of the velocities of both masses in the plane perpendicular to the normal direction do not change, either. We separate the components of the velocities along the line of impact as $\mathbf{v}_i = v_i\mathbf{n} + \mathbf{v}_{ip}$, $\mathbf{u}_i = u_i\mathbf{n} + \mathbf{u}_{ip}$ ($i = 1, 2$), in which v_i and u_i denote the components of $\mathbf{v}$ and $\mathbf{u}$ along the line of impact. From Eqs. [3.5.2] and [3.5.3], we write for the velocities perpendicular to the line of impact

$$\mathbf{v}_{1p} = \mathbf{u}_{1p} \qquad \mathbf{v}_{2p} = \mathbf{u}_{2p} \qquad [3.5.4]$$

and along the line of impact, the linear momentum expression becomes

$$m_1 v_1 + m_2 v_2 = m_1 u_1 + m_2 u_2 \qquad [3.5.5]$$

Equation [3.5.5] is a relation in terms of two unknowns, u_1 and u_2. To solve for the two unknowns, we need another relation. This relation is derived from *Poisson's hypothesis*. Poisson's hypothesis is based on the assumption that the contacting bodies are not exactly rigid, and it states that the impact takes place in two stages. In the first stage, called the period of *compression*, the bodies compress each other until the relative velocity between the colliding particles becomes zero along the

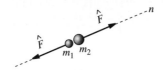

Figure 3.6

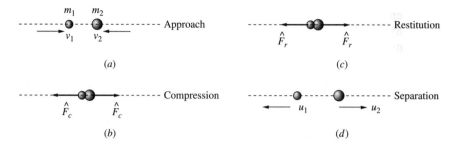

Figure 3.7 The stages of impact

line of impact. In the next stage, called the period of *restitution*, the bodies regain their original shapes, as shown in Fig. 3.7. The ratio of the strength of the two impulses is denoted by the *coefficient of restitution e*. We separate the impulsive force $\hat{\mathbf{F}}$ into two parts associated with the compression and restitution periods as

$$\hat{\mathbf{F}}_c \quad \text{during the compression stage}$$
$$\mathbf{F}_r \quad \text{during the restitution stage} \quad [3.5.6]$$

Because the entire impact is in one direction, the absolute values of the strengths can be expressed as $\hat{F}$, $\hat{F}_c$, and $\hat{F}_r$.

The coefficient of restitution is defined as

$$e = \frac{\hat{F}_r}{\hat{F}_c} \quad [3.5.7]$$

which leads to the relations

$$\hat{F}_c = \frac{\hat{F}}{1+e} \qquad \hat{F}_r = \frac{\hat{F}e}{1+e} \quad [3.5.8]$$

in which $\hat{F} = \hat{F}_c + \hat{F}_r$ is the total strength of the impact. To find the velocities of the colliding bodies along the line of impact, we integrate the equations of motion for the two stages of impact. Denoting the velocity at the end of the compression period by v_c, the linear momentum balances along the line of impact become

Mass 1   Compression: $m_1 v_1 - \hat{F}_c = m_1 v_c$ Restitution: $m_1 v_c - \hat{F}_r = m_1 u_1$

Mass 2   Compression: $m_2 v_2 + \hat{F}_c = m_2 v_c$ Restitution: $m_2 v_c + \hat{F}_r = m_2 u_2$

[3.5.9]

Introducing Eqs. [3.5.8] to these equations and eliminating the total force $\hat{F}$, we obtain

$$u_1 - u_2 = e(v_2 - v_1) \quad [3.5.10]$$

which is the commonly seen relation. The interpretation of this equation is that the coefficient of restitution represents the ratio of the relative velocity after impact to the relative velocity before impact. The relation [3.5.10] was first observed by Newton.

Poisson generalized Newton's result to bodies of any shape. We will make use of Eqs. [3.5.8] in Chapter 8, when dealing with the impact of rigid bodies.

Solving Eqs. [3.5.5] and [3.5.10] simultaneously yields the results

$$u_1 = \frac{1}{m}[(m_1 - em_2)v_1 + (1 + e)m_2v_2]$$

$$u_2 = \frac{1}{m}[(m_2 - em_1)v_2 + (1 + e)m_1v_1] \qquad \textbf{[3.5.11]}$$

where $m = m_1 + m_2$ is the total mass of the colliding particles.

The coefficient of restitution e is a quantity in the range $0 \leq e \leq 1$ that is dependent on the material properties of the colliding particles, as well as on the relative speed of the colliding masses. The coefficient of restitution decreases in value as the relative speed of impact gets higher. The special case of $e = 1$ is known as *perfectly elastic impact*. In this case, the strength of the impact is the same in the compression and restitution stages, and there is no energy loss. The case of $e = 0$ is referred to as *plastic impact*. Setting $e = 0$ in Eq. [3.5.11], we obtain

$$u_1 = \frac{1}{m}(m_1v_1 + m_2v_2) \qquad u_2 = \frac{1}{m}(m_2v_2 + m_1v_1) \qquad \textbf{[3.5.12]}$$

leading to the conclusion that in this case $u_1 = u_2$. After impact, the colliding particles have the same velocity along the line of impact.

We next examine the energy loss associated with impact. The energy loss occurs because the strength of the impact diminishes in the restitution phase. The lost energy gets transferred to the colliding bodies through internal vibrations as well as a temperature increase.

One can show that the energy loss is due to the change in the relative velocities of the colliding particles. For perfectly elastic impact there is no energy loss. On the other hand, when there is plastic impact, $e = 0$, all of the kinetic energy associated with the relative motion of the colliding masses is lost.

Be aware that the above derivation of impact relations represents a gross simplification of what actually happens when two bodies collide. First, we assume that the collision takes place in a very short period of time. This is possible if the speeds associated with the impact are large. Then, we implicitly assume that there is no material damage due to impact. For this to hold, the speeds involved should not be large. In realistic impact situations, the impact takes place over a finite time period, albeit small. As we discussed in Chapter 1, one should always check the validity of the assumptions used during impact. The coefficient of restitution itself is an approximation, as it is determined experimentally. As discussed earlier, its value depends on a variety of factors.

Another assumption whose validity comes into question is regarding the component of the motion orthogonal to the line of impact, especially when frictional forces are involved. For example, consider a ball thrown from a certain height with a horizontal velocity, as shown in Fig. 3.8. As the ball collides with the ground, it has both a horizontal as well as a vertical velocity. The line of impact is perpendicular to the ground. The impact results in an impulsive normal force. The impulsive normal force leads to an impulsive friction force. In this section, we assume that the impulsive friction force is zero. This assumption may not be valid in all cases.

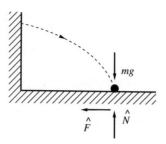

Figure 3.8

Example 3.4

A sphere of mass m and radius R is dropped from a height h onto another sphere of mass $2m$ and radius $1.5R$, which is at the bottom of an inextensible cord, as shown in Fig. 3.9. Find the velocities of the spheres immediately after impact, given that the coefficient of restitution is $e = 0.9$.

Solution

Figure 3.10 shows the geometry of the impact and the free body diagram. Denoting the spheres by A and B, the angle ϕ with which impact takes place is

$$\phi = \sin^{-1}\left(\frac{R}{R + \frac{3R}{2}}\right) = \sin^{-1}(0.4) = 23.58° \quad \textbf{[a]}$$

The line of impact joins the centers of the spheres, and the components of velocities before and after impact are

$$v_{At} = -v\sin\phi = -0.4v \quad v_{Bt} = 0$$

$$v_{An} = v\cos\phi = 0.9165v \quad v_{Bn} = 0 \quad \textbf{[b]}$$

where the subscripts n and t denote the components along the line of impact and perpendicular to it. After impact, the velocity of B is in the horizontal direction.

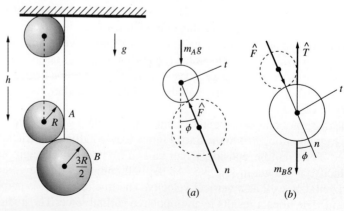

Figure 3.9 **Figure 3.10**

Consider the free-body diagrams of spheres A and B. Of the two forces acting on A, only one force is impulsive, $\hat{F}$, and of the three forces acting on B, two are impulsive, $\hat{F}$ and the impulsive tension in the string, denoted by $\hat{T}$. Hence, there is no conservation of linear momentum for the two spheres. We need to write the impulse-momentum relationships individually for each sphere.

We denote the components of the velocity after impact by u_A and u_B. Because u_B is horizontal, it can be expressed in terms of its components as

$$u_{Bt} = u_B \cos\phi = 0.9165 u_B \quad u_{Bn} = u_B \sin\phi = 0.4 u_B \qquad [c]$$

We can write the total impulsive force as

$$\hat{F}\mathbf{n} - \hat{T}\cos\phi\mathbf{n} + \hat{T}\sin\phi\mathbf{t} \qquad [d]$$

Considering the linear momentum balances in the n and t directions, we have

Sphere A, n direction $\qquad mv_{An} - \hat{F} = mu_{An}$ [e]

Sphere A, t direction $\qquad mv_{At} = mu_{At}$ [f]

Sphere B, n direction $\qquad 2mv_{Bn} + \hat{F} - \hat{T}\cos\phi = 2mu_{Bn} = 2mu_B \sin\phi$ [g]

Sphere B, t direction $\qquad 2mv_{Bt} + \hat{T}\sin\phi = 2mu_{Bt} = 2mu_B \cos\phi$ [h]

The components of the velocities along the line of impact are related by

$$u_{Bn} - u_{An} = e(v_{An} - v_{Bn}) = ev_{An} \qquad [i]$$

We then solve equations [e]–[i] for the unknowns, u_B, u_{An}, u_{At}, $\hat{F}$, and $\hat{T}$. Equation [f] gives u_{At} directly. Multiplying Eq. [g] by $\sin\phi$ and Eq. [h] by $\cos\phi$ and adding the two equations, we obtain

$$\hat{F} = \frac{2mu_B}{\sin\phi} \qquad [j]$$

which, when introduced into Eq. [e] yields

$$mv_{An} - \frac{2mu_B}{\sin\phi} = mu_{An} \qquad [k]$$

Equations [k] and [i] can be solved together for u_B as

$$u_B = \frac{v_{An}(1+e)}{D} = \frac{v\cos\phi(1+e)}{D} \qquad [l]$$

where $D = 2/\sin\phi + \sin\phi = 5.4$, so that the velocity of sphere B immediately after impact is

$$u_B = v\, 0.9165 \left(\frac{1.9}{5.4}\right) = 0.3225 v \qquad [m]$$

The value of v can be obtained using the work-energy relationship of

$$v = \sqrt{2gh} \qquad [n]$$

To find u_{An} we make use of Eq. [i]

$$u_{An} = u_{Bn} - ev_{An} = u_B \sin\phi - ev\cos\phi = 0.3225(0.4)v - 0.9(0.9165)v = -0.6959 v \qquad [o]$$

Hence, the velocity of sphere A immediately after the impact is

$$u_A = \sqrt{u_{At}^2 + u_{An}^2} = v\sqrt{0.4^2 + 0.6959^2} = 0.8027v \qquad \text{[p]}$$

3.6 VARIABLE MASS AND MASS FLOW SYSTEMS

Two interesting applications of impulse-momentum principles are variable mass and mass flow systems. In such systems, either the mass of the bodies involved changes or mass flows in and it flows out of a body at a certain rate. A typical application of variable mass systems is the rocket problem, depicted in Fig. 3.11. A rocket gains speed not only because of the thrust generated but also because the thrust results in loss of mass. In mass flow systems, air enters the system in a certain direction with a certain speed and comes out in a different direction with a different speed.

To analyze variable mass systems, we use the linear impulse-momentum relationship in its general form

$$\mathbf{p}(t_1) + \int_{t_1}^{t_2} \mathbf{F}(t)\,dt = \mathbf{p}(t_2) \qquad \text{[3.6.1]}$$

Consider a single particle of mass $m(t)$, acted upon by an external force $\mathbf{F}(t)$, as shown in Fig. 3.11. The initial mass of the particle at $t = t_1$ is $m(t_1)$, and its initial velocity is $\mathbf{v}(t_1)$. The initial linear momentum is $\mathbf{p}(t_1) = m(t_1)\mathbf{v}(t_1)$. The velocity of the particle at time t_2 is denoted by $\mathbf{v}(t_2)$ and its mass by $m(t_2)$. The relationship between the mass at t_1 and the mass at t_2 can be written as $m(t_1) + \Delta m = m(t_2)$. Note from the figure that Δm is defined as a negative quantity and that we are treating the body and the lost mass as one system. The linear momentum of the system at $t = t_2$ is

$$\mathbf{p}(t_2) = m(t_2)\mathbf{v}(t_2) - \Delta m \mathbf{v}_e(t_2) \qquad \text{[3.6.2]}$$

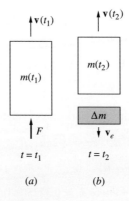

Figure 3.11

where $\mathbf{v}_e$ is the exit velocity of the mass that has left the particle. We can write the linear momentum balance as

$$m(t_1)\mathbf{v}(t_1) + \int_{t_1}^{t_2} \mathbf{F}(t)dt = m(t_2)\mathbf{v}(t_2) - \Delta m \mathbf{v}_e(t_2)$$
$$= m(t_1)\mathbf{v}(t_2) - \Delta m[\mathbf{v}_e(t_2) - \mathbf{v}(t_2)]$$
$$= m(t_1)\mathbf{v}(t_2) + \Delta m \mathbf{v}_{\text{rel}}(t_2) \quad [\mathbf{3.6.3}]$$

with $\mathbf{v}_{\text{rel}}(t) = \mathbf{v}(t) - \mathbf{v}_e(t)$ denoting the relative velocity of the exiting mass.

To solve the above equation, we need to know how the change in mass takes place and the exit velocity of the mass leaving the system. This requirement of knowledge about the characteristics of the separation is analogous to the need to know the coefficient of restitution of colliding bodies. We need to have information on how the separation takes place. Now, assume that the time interval is small and introduce the notation $\Delta t = t_2 - t_1$, and approximate the integral in the above equation by $\mathbf{F}(t_1)\Delta t$. This approximation is valid as long as $\mathbf{F}$ is not impulsive. Introducing next the notation $\Delta \mathbf{v} = \mathbf{v}(t_2) - \mathbf{v}(t_1)$ into Eq. [3.6.3] and dropping the subscripts 1 and 2, we obtain

$$\mathbf{F}(t)\Delta t = m(t)\Delta \mathbf{v} + \Delta m \mathbf{v}_{\text{rel}} \quad [\mathbf{3.6.4}]$$

Dividing both sides of this equation by Δt and taking the limit as Δt approaches zero gives

$$m(t)\dot{\mathbf{v}}(t) + \dot{m}(t)\mathbf{v}_{\text{rel}}(t) = \mathbf{F}(t) \quad [\mathbf{3.6.5}]$$

This is the general equation of motion for a variable mass system. Note that this equation is different than $\mathbf{F}(t) = d[m(t)\mathbf{v}(t)]/dt$. This is because the body which is losing mass and the lost mass are considered together as one system.

Next, consider mass flow systems. We restrict our analysis to steady mass flow systems, such that the rate of mass flowing into the system is the same as the rate of mass flowing out. Consider, for example, the air-blowing machine shown in Fig. 3.12. Air enters the container through a duct of cross-sectional area A_1 with speed v_1 and density ρ_1. It leaves through a duct of cross-sectional area A_2 with speed v_2 and density ρ_2. We denote the mass flow rate by m', recognizing that m'

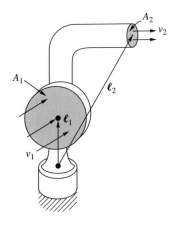

Figure 3.12

is not the rate of change of mass, but it is the rate of mass flow in and out of the system. The mass flow rate can be expressed as

$$m' = A_1 \rho_1 v_1 = A_2 \rho_2 v_2 \qquad [3.6.6]$$

The external forces acting on the system are the reaction and support forces that hold the container in its place. These forces counteract the change in linear and angular momentum due to the flow of mass. We denote the resultant of all forces by $\mathbf{F}$ and the resultant of all moments through a fixed point O by $\mathbf{M}_O$ (or about the center of mass). The change in linear momentum is due to the change in speed and direction of the mass flow. Denote the amount of mass that flows in and out of the system in a time period Δt by Δm, such that

$$m' = \lim_{\Delta t \to 0} \frac{\Delta m}{\Delta t} \qquad [3.6.7]$$

The change in linear momentum during Δt can be expressed as

$$\Delta \mathbf{p} = \Delta m \mathbf{v}_2 - \Delta m \mathbf{v}_1 \qquad [3.6.8]$$

Dividing the above equation by Δt, taking the limit as Δt approaches zero, and equating the change in linear momentum to the resultant force yields

$$\mathbf{F} = \dot{\mathbf{p}} = m'(\mathbf{v}_2 - \mathbf{v}_1) \qquad [3.6.9]$$

We relate the resultant moment about a fixed point O (or center of mass) to the change in angular momentum in a similar fashion. Denoting by $\boldsymbol{\ell}_1$ and $\boldsymbol{\ell}_2$ the vectors from O to the centers of the entry and exit ducts, the sum of moments about O becomes

$$\mathbf{M}_O = \dot{\mathbf{H}}_O = m'(\boldsymbol{\ell}_2 \times \mathbf{v}_2 - \boldsymbol{\ell}_1 \times \mathbf{v}_1) \qquad [3.6.10]$$

Example 3.5

Find an expression for the speed of the rocket fired vertically shown in Fig. 3.13.

Solution

We assume that the motion takes place along a straight line and drop the vector notation. We recognize the term $\dot{m}(t)\mathbf{v}_{\text{rel}}$ as the thrust. The external force F acting on the rocket is mainly due to three sources: the pressure differential between the exit nozzle and air, friction due to air resistance, and gravity. The force due to the pressure differential can be expressed as

$$F = p_e A \qquad [\mathbf{a}]$$

where p_e is the pressure difference at the exit nozzle and A is the cross-sectional area of the exit nozzle. It is customary to assume that the rate of change of mass is constant and that therefore the mass of the rocket can be written as

$$m(t) = m_0 - bt \qquad [\mathbf{b}]$$

where m_0 is the initial mass and b is the mass loss rate. It follows that $\dot{m}v_{\text{rel}} = -bv_{\text{rel}}$. The force of gravity is simply $-m(t)g$. Introducing this and Eqs. [a] and [b] into Eq. [3.6.5], we obtain, for a rocket moving vertically,

$$m(t)\dot{v}(t) + m(t)g = F_s = p_e A + bv_{\text{rel}} \qquad [\mathbf{c}]$$

The term F_s is referred to as the *static thrust*.

Figure 3.13

We next calculate the velocity of the rocket when all the fuel is expended (*burnout*). Denoting the mass of the propellant by m_p, and using the constant mass loss rate $\dot{m} = b$, it takes the rocket a time of $t_b = m_p/b$ to use up all its fuel. We write Eq. [c] in terms of $\dot{m}(t)$ and divide by $m(t)$, which yields

$$\frac{dv(t)}{dt} + \frac{dm}{dt}\frac{v_{\text{rel}}}{m(t)} + g = \frac{p_e A}{m(t)} \qquad [\text{d}]$$

Multiplying both sides by dt and noting from Eq. [b] that $dt = -dm/b$ we obtain

$$dv = -\left(v_{\text{rel}} + \frac{p_e A}{b}\right)\frac{dm}{m} + \frac{g\,dm}{b} \qquad [\text{e}]$$

Both sides in the above equation are expressed in differential form, so that Eq. [e] can be integrated to yield

$$v = v_0 - \left(v_{\text{rel}} + \frac{p_e A}{b}\right)\ln\frac{m}{m_0} + g\frac{m - m_0}{b} \qquad [\text{f}]$$

where v_0 is the initial velocity. Substituting the expression for $m(t)$ in the above equation gives

$$v = v_0 - \left(v_{\text{rel}} + \frac{p_e A}{b}\right)\ln\frac{m_0 - bt}{m_0} - gt \qquad [\text{g}]$$

and the final velocity is found by substituting $t_b = m_p/b$ into this equation.

Equation [g] may look misleading at first, because of the two negative terms. However, m/m_0 is always less than unity and its natural logarithm is a negative quantity, thus giving the rocket its upward velocity. It should also be noted that Eqs. [f] and [g] are derived by assuming that the gravitational constant remains the same during ascent of the rocket. A more accurate expression for the velocity should include the change in the gravitational attraction as a result of increased altitude, as well as the air resistance.

3.7 CONCEPTS FROM ORBITAL MECHANICS: THE TWO BODY PROBLEM

A key problem in mechanics is that of the motion of two celestial bodies moving under the gravitational attraction of each other. The class of problems is referred to as the *two body problem*. The two bodies are depicted in Fig. 3.14. Typical examples include the earth–moon, the earth–satellite, and the sun–planet pairs. For our purposes we assume that the only force acting on each body is the gravitational attraction due to the other. We ignore the gravitational force exerted by other bodies. For example, if we are considering the earth–satellite problem, we ignore the effects of the moon and the sun. This assumption is valid only as long as the distance between the satellite and earth is much smaller than the distance between the earth and moon, or any other planet, or the sun.

The only force acting on the particles is the gravitational attraction in the form

$$\mathbf{F} = \frac{Gm_1 m_2}{r^3}\mathbf{r} \qquad [3.7.1]$$

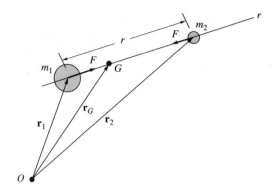

Figure 3.14

where $\mathbf{r} = \mathbf{r}_2 - \mathbf{r}_1$ is the vector connecting the two masses. Newton's second law for each particle is

$$m_1\ddot{\mathbf{r}}_1 = \mathbf{F} = \frac{Gm_1m_2}{r^3}\mathbf{r} \qquad m_2\ddot{\mathbf{r}}_2 = -\mathbf{F} = -\frac{Gm_1m_2}{r^3}\mathbf{r} \qquad [\text{3.7.2a,b}]$$

To manipulate the equations of motion, consider the center of mass of the two particle system. From Eq. [3.2.2] we have that

$$m_1\mathbf{r}_1 + m_2\mathbf{r}_2 = (m_1 + m_2)\mathbf{r}_G \qquad [\text{3.7.3}]$$

Adding Eqs. [3.7.2] and differentiating Eq. [3.7.3] with respect to time, we conclude that

$$m_1\ddot{\mathbf{r}}_1 + m_2\ddot{\mathbf{r}}_2 = (m_1 + m_2)\ddot{\mathbf{r}}_G = \mathbf{0} \qquad [\text{3.7.4}]$$

We can see that the center of mass of the system of two particles does not have an acceleration, an expected result because we originally assumed that there are no forces acting externally on the system of two particles.

Equation [3.7.4] hints that one may simplify the governing equations, by considering the motion of the two masses relative to each other. Actually, we are more interested in this relative motion of the masses than the absolute motion of each body. Let us express the equations of motion in terms of the vector $\mathbf{r}$ connecting the two bodies.

Dividing Eqs. [3.7.2a] and [3.7.2b] by m_1 and m_2, respectively, and subtracting Eq. [3.7.2a] from [3.7.2b], we obtain

$$\ddot{\mathbf{r}}_2 - \ddot{\mathbf{r}}_1 = \ddot{\mathbf{r}} = -\frac{Gm_1}{r^3}\mathbf{r} - \frac{Gm_2}{r^3}\mathbf{r} = -\frac{G(m_1 + m_2)}{r^3}\mathbf{r} \qquad [\text{3.7.5}]$$

If we introduce the gravitational parameter $\mu = G(m_1 + m_2)$, we can express the relative motion as the differential equation

$$\ddot{\mathbf{r}} + \frac{\mu}{r^3}\mathbf{r} = \mathbf{0} \qquad [\text{3.7.6}]$$

3.7 CONCEPTS FROM ORBITAL MECHANICS: THE TWO BODY PROBLEM

The question may be asked as to what information was given up when we reduced the two equations of motion into a single equation. The answer is: the location and the motion of the center of mass of the two body system. Using Eq. [3.7.6] alone, one cannot analyze the motion of the center of mass. In many orbital mechanics problems, one is not interested in the location of the center of mass but in the relative motions of the two masses. Also, in several celestial mechanics applications, the mass of one body is much smaller than the mass of the other, and one can safely ignore the mass of the smaller body when calculating the center of mass. The center of the larger body is assumed to be the center of mass of the two body system. Kepler made this assumption when he stated his laws of planetary motion.

Equation [3.7.6] represents a central force problem, as discussed in Chapter 1. It is conveniently analyzed by polar coordinates. Attaching a set of polar coordinates to the center of mass and separating the motion into the radial and transverse components, we write the radial and transverse components of the equation of motion as

$$\ddot{r} - r\dot{\theta}^2 = -\frac{\mu}{r^2} \qquad r\ddot{\theta} + 2\dot{r}\dot{\theta} = 0 \qquad \textbf{[3.7.7a,b]}$$

Example 1.12 showed that the angular momentum associated with a central force is conserved and that $r^2\dot{\theta} = h$ is constant. That is, the angular momentum about the center of mass is conserved. As we will see later, this is the mathematical statement of Kepler's second law. Note that the angular momentum considered here is actually the apparent angular momentum. Because the acting force is conservative, the energy of this system is also conserved. Indeed, introducing the expression $\dot{\theta} = h/r^2$ into Eq. [3.7.7a] we obtain

$$\ddot{r} = \frac{h^2}{r^3} - \frac{\mu}{r^2} \qquad \textbf{[3.7.8]}$$

which can be integrated to yield the energy integral

$$\frac{1}{2}\dot{r}^2 + \frac{1}{2}\frac{h^2}{r^2} - \frac{\mu}{r} = E = \text{constant} \quad \text{or} \quad \frac{1}{2}v^2 - \frac{\mu}{r} = E = \text{constant} \qquad \textbf{[3.7.9]}$$

where $v^2 = \dot{r}^2 + h^2/r^2$ is recognized as the square of the total velocity. This expression of the energy integral is for the relative motion with respect to the center of mass. However, because the center of mass executes motion with constant velocity, the energy associated with the center of mass motion can be absorbed in the constant on the right side of Eq. [3.7.9]. (This is entirely analogous to Eq. [3.4.10], which gives the kinetic energy for a system of particles.) It should also be noted that the energy and momentum expressions are per unit mass.

We can calculate the energy integral from the expressions for the kinetic and potential energies. The kinetic energy has the form

$$T = \frac{1}{2}\frac{m_1 m_2}{m_1 + m_2}\mathbf{v} \cdot \mathbf{v} = \frac{1}{2}\frac{m_1 m_2}{m_1 + m_2}(\dot{r}^2 + r^2\dot{\theta}^2) = \frac{1}{2}\frac{m_1 m_2}{m_1 + m_2}(\dot{r}^2 + \frac{h^2}{r^2})$$

$$\textbf{[3.7.10]}$$

The potential energy is obtained by integrating Eq. [3.7.1]. The gravitational force is in the radial direction and $\mathbf{r} = r\mathbf{e}_r$, so that $d\mathbf{r} = dr\mathbf{e}_r + r\,d\theta\mathbf{e}_\theta$. We select the datum position for the potential energy as the distant stars, $\mathbf{r}_0 = \infty$, so that

$$V = -\int \mathbf{F} \cdot d\mathbf{r} = -\int_{\mathbf{r}_0=\infty}^{\mathbf{r}} -\frac{Gm_1m_2}{r^3}\mathbf{r}\cdot d\mathbf{r} = \int_{r_0=\infty}^{r} \frac{Gm_1m_2}{r^2}dr = -\frac{Gm_1m_2}{r} \quad [3.7.11]$$

Adding Eqs. [3.7.10] and [3.7.11], and dividing by $m_1m_2/(m_1 + m_2)$ yields Eq. [3.7.9].

As Chapter 1 mentions, many celestial mechanics problems require higher accuracy than most engineering problems. Many times one manipulates very large numbers and obtains a very small number as a result. For this reason, it may be necessary to use more than four significant figures. Here are pertinent celestial data:

Universal gravitational constant: $6.668462(10^{-11})$ m^3/kg s^2

Mass of earth: $5.977414(10^{24})$ kg, Average radius of earth: 6,378.1 km

For earth, $\mu = 3.986(10^{14})$ m^3/s^2, Mass of sun: $1.987323(10^{30})$ kg

Example 3.6

A particle is thrown from the surface of the earth with a very large initial velocity, as shown in Fig. 3.15. Calculate the maximum height reached by the particle.

Solution

Because the initial velocity is very high, we will dispense with standard projectile motion equations, and we will use concepts from celestial mechanics. We denote the initial components of the velocity by v_x and v_z. Actually, at the onset of motion the x and z axes are basically the radial and transverse directions. When the particle reaches its maximum height, the component of the velocity in the vertical direction is zero. The velocity now is purely in the transverse direction. We denote its vertical velocity by w.

We have conservation of energy as well as conservation of angular momentum about the center of the earth. Denoting the maximum altitude reached by L, we describe the conservation of angular momentum by

$$mRv_x = m(R + L)w \quad [a]$$

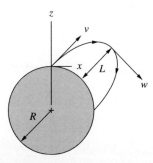

Figure 3.15

where R is the radius of the earth. This yields an expression for the velocity at maximum altitude by

$$w = \frac{Rv_x}{R+L} \qquad [b]$$

The components of the kinetic and potential energy are

$$T_1 = \frac{1}{2}m(v_x^2 + v_z^2) \qquad V_1 = -\frac{Gm_e m}{R}$$

$$T_2 = \frac{1}{2}mw^2 = \frac{1}{2}m\left(\frac{R}{R+L}\right)^2 v_x^2 \qquad V_2 = -\frac{Gm_e m}{R+L} \qquad [c]$$

The work-energy theorem gives

$$\frac{1}{2}m(v_x^2 + v_z^2) - \frac{Gm_e m}{R} = \frac{1}{2}m\left(\frac{R}{R+L}\right)^2 v_x^2 - \frac{Gm_e m}{R+L} \qquad [d]$$

which represents a nonlinear equation for L. To obtain an approximate, but still meaningful, solution, we note that L is much less than R. Defining by $e = L/R$, we linearize the terms involving L as

$$\left(\frac{R}{R+L}\right)^2 = \left(\frac{1}{1+e}\right)^2 \approx (1 - 2e) \qquad \frac{1}{R+L} \approx \frac{1}{R}(1-e) \qquad [e]$$

Introducing Eqs. [e] into Eq. [d] and solving for e we obtain

$$e = \frac{v_z^2}{2\left(\dfrac{Gm_e}{R} - v_x^2\right)} \qquad [f]$$

so the maximum height reached is

$$L = \frac{v_z^2 R}{2\left(\dfrac{Gm_e}{R} - v_x^2\right)} \qquad [g]$$

It is interesting to note that if v_x is set to zero, that is, if the particle is launched vertically, the height it reaches becomes

$$L = \frac{v_z^2 R^2}{2Gm_e} = \frac{v_z^2}{2g} \qquad [h]$$

which is the same equation obtained from projectile motion analysis. The value for the height L in Eq. [g] is higher than the value for L in Eq. [h], an expected result.

When we consider the total motion of the body, we see that it has some sort of orbital motion, although the particle cannot complete an orbit but falls to the earth. This situation is sometimes encountered in the launching of spacecraft, when the booster rockets malfunction and fail to place the spacecraft into a sufficiently high orbit.

3.8 THE NATURE OF THE ORBIT

We now wish to determine what kind of motion the masses execute with respect to each other. We will accomplish this by solving the equations of motion, Eqs. [3.7.7], and by making use of the energy and momentum integrals. Consider the equation of motion in the radial direction, Eq. [3.7.7a]. This equation is nonlinear. It turns out that Eq. [3.7.7] can be put in a simpler form if we introduce the transformation

$$u = \frac{1}{r} \qquad [3.8.1]$$

and replace differentiation with respect to time with differentiation with respect to the transverse variable θ. Noting that the rate of change of the transverse angle can be written as $\dot{\theta} = h/r^2$, we express the time derivative of r as

$$\dot{r} = \frac{dr}{d\theta}\frac{d\theta}{dt} = \frac{dr}{d\theta}\frac{h}{r^2} \qquad [3.8.2]$$

Taking the derivative of u with respect to θ

$$\frac{du}{d\theta} = \frac{d}{d\theta}\left(\frac{1}{r}\right) = -\frac{1}{r^2}\frac{dr}{d\theta} \qquad [3.8.3]$$

so that

$$\frac{dr}{dt} = \frac{h}{r^2}\frac{dr}{d\theta} = \frac{h}{r^2}\left(-r^2\frac{du}{d\theta}\right) = -h\frac{du}{d\theta} \qquad [3.8.4]$$

In a similar fashion, we write the second derivative of r with respect to time as

$$\ddot{r} = -h\frac{d}{dt}\left(\frac{du}{d\theta}\right) = -h\frac{d^2u}{d\theta^2}\frac{d\theta}{dt} = -h\frac{d^2u}{d\theta^2}\frac{h}{r^2} = -h^2u^2\frac{d^2u}{d\theta^2} \qquad [3.8.5]$$

The remaining terms in Eq. [3.7.7a] can be expressed in terms of u as

$$\dot{\theta} = \frac{h}{r^2} = hu^2 \qquad \frac{\mu}{r^2} = \mu u^2 \qquad [3.8.6]$$

which, when introduced together with Eq. [3.8.5] in Eq. [3.7.7a], yield

$$\frac{d^2u}{d\theta^2} + u = \frac{\mu}{h^2} \qquad [3.8.7]$$

This relationship is in the form of a second-order differential equation with constant coefficients. It is considerably simpler to solve than Eq. [3.7.7a]. The solution of Eq. [3.8.7] can be written as

$$u(\theta) = \frac{\mu}{h^2} + \alpha \cos(\theta - \theta_i) \qquad [3.8.8]$$

where α and θ_i depend on the initial conditions. This solution is used to find $r(\theta)$. We are interested in the value of the magnitude of the radial distance r as a function of the transverse angle θ, as well as r as a function of time.

The energy integral in Eq. [3.7.9] can be expressed in terms of u as

$$\frac{1}{2}h^2\left(\frac{du}{d\theta}\right)^2 + \frac{1}{2}h^2u^2 - \mu u = E \qquad [3.8.9]$$

3.8 THE NATURE OF THE ORBIT

This expression can also be obtained by integrating Eq. [3.8.7] over θ. Recall that the energy expressions here have units of energy per unit mass. We explore the relationship between the energy E and the amplitude α. Introducing Eq. [3.8.8] into Eq. [3.8.9] and carrying out the algebra, we obtain

$$\frac{1}{2}h^2[\alpha \sin(\theta - \theta_i)]^2 + \frac{1}{2}h^2\left[\frac{\mu}{h^2} + \alpha \cos(\theta - \theta_i)\right]^2 - \mu\left[\frac{\mu}{h^2} + \alpha \cos(\theta - \theta_i)\right]$$
$$= \frac{1}{2}\left[\alpha^2 h^2 - \frac{\mu^2}{h^2}\right] = E \qquad [3.8.10]$$

which we can rewrite as

$$\alpha = \frac{\mu}{h^2}\sqrt{1 + \frac{2Eh^2}{\mu^2}} \qquad [3.8.11]$$

Introducing the variable $\varepsilon = \sqrt{1 + \frac{2Eh^2}{\mu^2}}$, the above equation can be expressed as

$$\alpha = \frac{\mu}{h^2}\varepsilon \qquad [3.8.12]$$

and substituting Eq. [3.8.12] into Eq. [3.8.8], we write the solution for u,

$$u(\theta) = \frac{\mu}{h^2}[1 + \varepsilon \cos(\theta - \theta_i)] \qquad [3.8.13]$$

It follows that the solution for r is

$$r(\theta) = \frac{1}{u(\theta)} = \frac{h^2}{\mu}\left[\frac{1}{1 + \varepsilon \cos(\theta - \theta_i)}\right] \qquad [3.8.14]$$

Equation [3.8.14] represents the equation of a conic section, with the value of ε, referred to as the *eccentricity* of the section, determining its shape, with the values of $\varepsilon = 0$ and $\varepsilon = 1$ being the critical values. If $\varepsilon < 1$, the conic section is a closed one, in the form of an ellipse, as the values for $r(\theta)$ remain finite. The conic section becomes a circle when $\varepsilon = 0$. When $\varepsilon \geq 1$, $r(\theta)$ assumes infinite values, as in a parabola or hyperbola. The term h^2/μ is called the *semilatus rectum;* it defines the size of the conic section. The semilatus rectum is a function of the angular momentum. The initial condition θ_i is usually selected as zero.

Using Eq. [3.8.12], we can write the energy as

$$E = \frac{(\varepsilon^2 - 1)\mu^2}{2h^2} \qquad [3.8.15]$$

or, we can make use of the nondimensional ratio $E^* = 2Eh^2/\mu^2 = \varepsilon^2 - 1$.

We next consider the different types of conic sections. When $\varepsilon < 1$, the trajectory is an ellipse, with the center of mass as a focus of the ellipse, as shown in Fig. 3.16, and the motion is periodic with a closed orbit. An ellipse is defined by its two foci, its *semimajor axis a* and *semiminor axis b*. The point on the ellipse closest to a focus is called *pericenter,* and the point farthest away, *apocenter.* The pericenter and apocenter are known as the apsides (plural of apsis) of the ellipse. An *apsis* is defined as the point where $dr/d\theta = 0$. This definition of apsis is valid for any conic

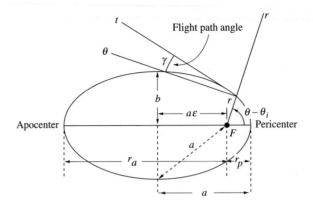

Figure 3.16 Elliptic orbit

section and is not limited to ellipses. Considering the motion of the body on the orbit, the angle γ between the transverse and tangential directions is referred to as the *flight path angle*. This angle is easier to measure than other orbital quantities and, hence, is useful in determining the nature of an orbit.

When dealing with two masses of comparable magnitude, the focus of the ellipse is the center of mass of the two body system; the ellipse describes the trajectory of the distance between the two masses. For sun–planet or earth–satellite type problems, because of the very large difference in the mass ratios, one can assume that the larger mass is located on a focus. The situation is different when the earth–moon pair is considered. The mass ratio between the earth and moon is 81.30, and the mean distance between the two bodies is 384,400 km. The center of mass of the earth–moon system lies 4670 km from the center of the earth, roughly 2/3 the mean radius of the earth.

It has become customary to use different names for the apsides when motion about the earth and sun are considered. Table 3.1 gives the names used for the apsides and the names of commonly used coordinate systems when the earth and sun are assumed to be on a focus.

Denoting the distances from the focus to the pericenter and apocenter by r_p and r_a, respectively, and without loss of generality selecting the initial condition as $\theta - \theta_i = 0$, so that θ is measured from the pericenter, we obtain

$$r_p = \frac{h^2}{\mu}\left[\frac{1}{1+\varepsilon}\right] \quad r_a = \frac{h^2}{\mu}\left[\frac{1}{1-\varepsilon}\right] \qquad [\text{3.8.16a,b}]$$

Table 3.1 Nomenclature for celestial motion problems

Types of Orbit	Apsis Near	Apsis Far Away	Coordinate System
Generic ellipse	Pericenter	Apocenter	
Earth as focus	Perigee	Apogee	Geocentric-equatorial or perifocal
Sun as focus	Perihelion	Aphelion	Heliocentric-ecliptic

From these equations, and noting from simple geometry that $a = (r_p + r_a)/2$, we obtain for the semimajor axis

$$a = \frac{1}{2}\frac{h^2}{\mu}\left[\frac{1}{1+\varepsilon}\right] + \frac{1}{2}\frac{h^2}{\mu}\left[\frac{1}{1-\varepsilon}\right] = \frac{h^2}{\mu}\left[\frac{1}{1-\varepsilon^2}\right] \qquad [3.8.17]$$

If we consider Eqs. [3.7.9] and [3.8.15], we can express the energy as

$$E = \frac{1}{2}v^2 - \frac{\mu}{r} = -\frac{\mu}{2a} \qquad [3.8.18]$$

so that the semimajor axis is a measure of the energy in the system. In a similar fashion, one can show that the semiminor axis is of length $b = a\sqrt{1-\varepsilon^2}$. Also, using Eqs. [3.8.16] and [3.8.17], we obtain

$$r_p = a(1-\varepsilon) \quad r_a = a(1+\varepsilon) \qquad [3.8.19]$$

We can then write the equation for the ellipse as

$$r = \frac{a(1-\varepsilon^2)}{1+\varepsilon\cos\theta} \qquad [3.8.20]$$

At this point, we consider Kepler's laws of planetary motion. The astronomer Johannes Kepler, after studying the observations of Galileo and Tyco Brahe and the motion of planets, developed the following laws for the motion of planets (his calculations were based on the motion of Mars) in the solar system.

Law 1: Each planet revolves in an elliptic orbit about the sun, with the sun at one focus of the ellipse.

Law 2: Equal areas are swept per unit time by the line joining the sun to the planet.

Law 3: The squares of the periods of the planets are proportional to the cubes of the semimajor axes of the ellipses.

The first two laws of Kepler were published in 1609 and the third in 1619. These laws predate Newton's laws, (first published in 1687) by over 70 years. Kepler did not know about Newton's laws of motion when he formulated his laws. Newton's three laws of motion and his law of gravitation are based in part on Kepler's laws. It is interesting that we traditionally study Newton's laws first and then Kepler's laws, in more natural order but in reverse historical order.

Kepler's first law becomes correct when the mass of the smaller second body (e.g., planet in sun–planet, moon in earth–moon, satellite in earth–satellite) is neglected from μ.

To demonstrate Kepler's second law, consider Fig. 3.17 and write an expression for the area swept by the position vector **r** as

$$\text{Area } A = \int dA = \int \frac{1}{2}rr\,d\theta = \frac{1}{2}\int r^2 d\theta = \frac{1}{2}\int r^2\dot\theta\,dt \qquad [3.8.21]$$

The *areal rate*—the area swept per unit time—is found by differentiating the area with respect to time. Noting from earlier that $r^2\dot\theta = h =$ constant, we obtain the

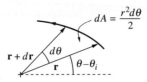

Figure 3.17

areal rate

$$\dot{A} = \frac{h}{2} = \text{constant} \qquad [3.8.22]$$

which is proof of Kepler's second law.

The period of the orbit can be obtained from Kepler's second law by noting that the period τ is equal to the area of the elliptic orbit divided by the areal rate,

$$\tau = \frac{A}{\dot{A}} = \frac{\pi ab}{h/2} \qquad [3.8.23]$$

where πab is the area of the ellipse. Introducing $b = a\sqrt{1-\varepsilon^2}$ into the above equation and noting from Eq. [3.8.17] that

$$h = \sqrt{\mu a(1-\varepsilon^2)} \qquad [3.8.24]$$

we obtain

$$\tau = \frac{2\pi a^2 \sqrt{1-\varepsilon^2}}{\sqrt{\mu a(1-\varepsilon^2)}} = 2\pi\sqrt{\frac{a^3}{\mu}} \qquad [3.8.25]$$

This is the mathematical statement of Kepler's third law.

The special case of $\varepsilon = 0$ corresponds to a circular orbit. Circular orbits became very important in the second half of the 20th century, as satellites were placed into circular orbits around the earth. A circular orbit in which the relative position of the satellite with respect to the earth does not change is called *geosynchronous*. As of this writing there were several communications satellites in geosynchronous orbit about the equator.

In a circular orbit, the body has constant velocity, which can easily be obtained from the energy expression. From Eq. [3.8.15] for a circular orbit, $E = -\mu^2/2h^2$. We verify this by writing the energy expression as

$$E = \frac{v^2}{2} - \frac{\mu}{r} = \frac{v^2}{2} - \frac{\mu^2}{h^2} = -\frac{\mu^2}{2h^2} \qquad [3.8.26]$$

so that we obtain the velocity in circular orbit, v_c, as

$$v_c = \sqrt{\frac{\mu}{r}} \qquad [3.8.27]$$

We next consider the case when the eccentricity is greater than or equal to 1, resulting in an open orbit. When $\varepsilon = 1$ the orbit is parabolic, and when $\varepsilon > 1$ the orbit is hyperbolic. Parabolic orbits rarely exist, as they describe a limiting case between an open and closed orbit. Fig. 3.18 depicts a parabolic orbit. From Eqs. [3.8.16], as ε approaches 1, the semimajor axis becomes longer. When $\varepsilon = 1$, the orbit becomes an open one. The orbit equation, setting $\varepsilon = 1$ in Eq. [3.8.14], becomes

$$r = \frac{h^2/\mu}{1 + \cos\theta} \qquad [3.8.28]$$

The expression for energy for a parabolic orbit gives insight into the nature of the orbit. Setting $\varepsilon = 1$ in Eq. [3.8.15] yields, for the energy, $E = 0$, so that we can write

$$E = \frac{v^2}{2} - \frac{\mu}{r} = 0 \qquad [3.8.29]$$

The velocity required to achieve a parabolic orbit and escape the gravitational attraction of the large mass is called the *escape velocity*. From the above equation, for a spacecraft to transfer from an elliptic to a parabolic orbit it must have the escape velocity, denoted by v_e, as

$$v_e = \sqrt{2\frac{\mu}{r}} \qquad [3.8.30]$$

The minimum value of the escape velocity is at the perigee. Many orbital maneuvers that involve a change of orbits are carried out at the perigee. It is interesting to note what happens as the satellite keeps moving in a parabolic orbit. As r approaches infinity, from Eq. [3.8.30] the velocity approaches zero. This implies that if a body can achieve a true parabolic orbit and there are no other disturbances acting on it, that body would eventually come to a rest. Physically this never happens.

When the eccentricity is greater than 1, the resulting orbit is hyperbolic. Spacecraft launched on interplanetary missions are given hyperbolic orbits. Fig. 3.19 shows a hyperbolic orbit. From Eq. [3.8.15], the energy for a hyperbolic orbit is

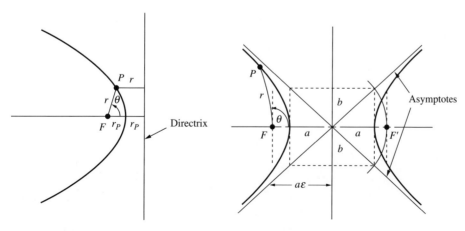

Figure 3.18 Parabolic orbit   **Figure 3.19** Hyperbolic orbit

Table 3.2 Types of orbits and associated eccentricities

Type of Orbit	Eccentricity ϵ	Energy Ratio E^*
Circular	0	-1
Elliptic	$0 < \epsilon < 1$	$-1 < E^* < 0$
Parabolic	1	0
Hyperbolic	$\epsilon > 1$	$E^* > 0$

positive, and the energy expression can be written as

$$E = \frac{v^2}{2} - \frac{\mu}{r} > 0 \qquad [3.8.31]$$

implying that as r approaches infinity, the speed of the body is not zero, but approaches a finite value. This value is called the *hyperbolic excess speed*. Table 3.2 summarizes the relation between orbits and eccentricity.

Two essential problems in space mechanics are

1. To place a satellite in a desired orbit and to change the path of the satellite from one orbit to another.
2. To determine the orbit of a satellite or a planet from measurements of its location and motion.

Analysis of both subjects is very lengthy and beyond the scope of this text. For more details on these subjects, the reader is referred to texts on orbital mechanics. In the following examples we illustrate two simple approaches.

Example 3.7

At the burnout of a rocket, the following data are given: r = 6500 km, v = 9750 m/s, flight path angle γ = 16.4°. Has the rocket achieved an earth orbit or will it crash into the earth?

Solution

The orbit of the rocket is shown in Fig. 3.20. We need to find the properties of the orbit and determine if the distance to the perigee is less than the radius of the earth. The angular momentum and energy per unit mass are

$$h = rv \cos \gamma = (6.5 \times 10^6)(9.75 \times 10^3)(\cos 16.4°) = 6.0797 \times 10^{10} \text{ m}^2/\text{s} \qquad [a]$$

$$E = \frac{v^2}{2} - \frac{\mu}{r} = \frac{(9.75 \times 10^3)^2}{2} - \frac{3.986 \times 10^{14}}{6.5 \times 10^6} = -1.3792 \times 10^7 \text{ m}^2/\text{s}^2 \qquad [b]$$

We next calculate the eccentricity. We have

$$\frac{2Eh^2}{\mu^2} = \frac{2(-1.3792 \times 10^7)(6.0797 \times 10^{10})^2}{(3.986 \times 10^{14})^2} = -0.64172 \qquad [c]$$

so that from Eq. [3.8.15]

$$\varepsilon = \sqrt{1 - 0.64172} = 0.59857 \qquad [d]$$

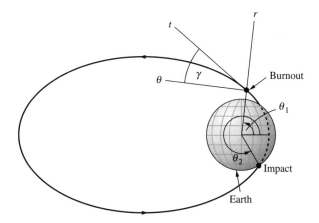

Figure 3.20

From Eq. [3.8.18], the semimajor axis is

$$a = -\frac{\mu}{2E} = \frac{-3.986 \times 10^{14}}{2(-1.3972 \times 10^6)} = 14.450 \times 10^6 \text{m} \quad \text{[e]}$$

and from Eq. [3.8.19b], the distance to perigee is

$$r_P = a(1 - \varepsilon) = 14.45 \times 10^6(1 - 0.59857) = 5.8008 \times 10^6 \text{m} \quad \text{[f]}$$

which, unfortunately, is smaller than the radius of the earth. The rocket crashes into the earth on its way back.

Example 3.8

HOHMANN TRANSFER A *Hohmann transfer* is a common way of moving a spacecraft in a circular orbit to another circular orbit that lies on the same plane. The transfer involves applying two impulses to the spacecraft. Denote the radii of the first and second orbits by r_1 and r_2, as shown in Fig 3.21. The first impulse changes the initial orbit to an elliptic orbit, called the *transfer orbit*, whose semimajor axis is of length $(r_1 + r_2)/2$.

Consider the case when $r_2 > r_1$. The location of the first impulse becomes the perigee of the transfer orbit (apogee when $r_2 < r_1$). The second impulse, applied at the apogee (perigee when $r_2 < r_1$) of the transfer orbit, moves the spacecraft to the desired circular orbit.

The velocity in the first circular orbit, denoted by v_1, is found noting that $a = r_1$, with the result

$$E_1 = \frac{v_1^2}{2} - \frac{\mu}{r_1} = -\frac{\mu}{2r_1} \rightarrow v_1 = \sqrt{\frac{\mu}{r_1}} \quad \text{[a]}$$

Immediately after the impulse is applied to move the spacecraft into the transfer orbit, $r = r_p = r_1$ and $2a = r_1 + r_2$, so that denoting the velocity at perigee by v_p, the energy integral becomes

$$E_{\text{trans}} = \frac{v_p^2}{2} - \frac{\mu}{r_1} = -\frac{\mu}{r_1 + r_2} \quad \text{[b]}$$

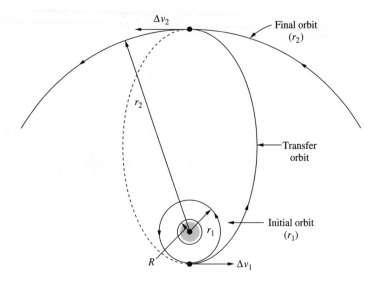

Figure 3.21   Hohmann transfer

which can be solved for v_p as

$$v_p = \sqrt{\frac{2\mu}{r_1} - \frac{2\mu}{r_1 + r_2}} = \sqrt{\frac{\mu}{r_1}} \sqrt{\frac{2r_2}{r_1 + r_2}} \qquad [\text{c}]$$

so that the impulse needed at the first step is

$$m\Delta v_1 = m(v_p - v_1) = m\sqrt{\frac{\mu}{r_1}} \left(\sqrt{\frac{2r_2}{r_1 + r_2}} - 1 \right) \qquad [\text{d}]$$

The second impulse is applied at the apogee of the transfer orbit. The velocity at apogee before the impulse can be found from the angular momentum conservation

$$r_1 v_p = r_2 v_a \qquad [\text{e}]$$

where v_a denotes the velocity at apogee and has the form

$$v_a = \frac{r_1}{r_2} v_p = \frac{r_1}{r_2} \sqrt{\frac{\mu}{r_1}} \sqrt{\frac{2r_2}{r_1 + r_2}} = \sqrt{\frac{\mu}{r_2}} \sqrt{\frac{2r_1}{r_1 + r_2}} \qquad [\text{f}]$$

Because the desired final orbit is circular, the velocity in orbit should be

$$v_2 = \sqrt{\frac{\mu}{r_2}} \qquad [\text{g}]$$

so that the second impulse is of magnitude

$$m\Delta v_2 = m(v_2 - v_a) = m\sqrt{\frac{\mu}{r_2}} \left(1 - \sqrt{\frac{2r_1}{r_1 + r_2}} \right) \qquad [\text{h}]$$

Consider the issue of what happens when the impulses are not applied at the proper places in the orbit. For the first impulse, the location of the impulse defines the perigee of the transfer orbit. This location is not critical. The critical impulse is the second. If the second impulse is not applied at apogee, the resulting orbit will not be circular.

3.9 ORBITAL ELEMENTS

In the previous section, we showed that the general solution of the orbital problem is a conic section. Based on the two integrals of the motion, angular momentum and energy, we developed two quantities, semimajor axis a and eccentricity ε. Given initial conditions, one can determine the nature of the orbit by calculating a and ε. The orbital problem as given in Eq. [3.7.6] requires six initial conditions or constants of integration, as it is a second-order differential equation in three dimensions.

Given a set of initial conditions, one can solve Eq. [3.7.6] and obtain a solution. This solution, while being a mathematical answer, does not describe the properties of the orbit. It is desirable to express the position and velocity of the body in orbit in terms of parameters that lend themselves to a physical explanation. It turns out that we can find four parameters in addition to a and ε that orient the orbit in space and the body on the orbit. Of these four parameters, two describe the orientation of the orbital plane, one describes the orientation of the orbit on the orbital plane, and the last one locates the body on the orbit and introduces time. The six parameters are referred to as the *orbital parameters* or *orbital elements*.

We begin by describing the position of the body on the orbit. Our primary interest is in closed orbits. The angle θ, referred to as the *true anomaly*, is measured from one of the foci of the ellipse. It is the angle between the line joining the focus to the body and the semimajor axis, relative to the pericenter. A more convenient way of measuring the orientation of the particle is to introduce the variable *eccentric anomaly*, denoted by $\mathscr{E}$ and shown in Fig. 3.22. We draw a circle of radius a with the same center as the ellipse. The eccentric anomaly is measured positive counterclockwise from the semimajor axis. It is the angle between the semimajor axis and the line that connects point P' and the center of the ellipse, in which point P' is the intersection between the circle and the vertical line perpendicular to the semimajor axis. This line goes through point P, with P denoting the position of the mass.

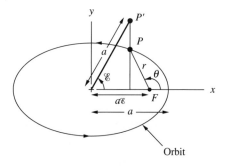

Figure 3.22

Denoting the coordinates of P by x and y, from Fig. 3.22 we can write

$$y = r\sin\theta \quad x = a\cos\mathcal{E} \quad a\cos\mathcal{E} = a\varepsilon + r\cos\theta \quad [3.9.1a,b,c]$$

Recalling the equation of the ellipse $x^2/a^2 + y^2/b^2 = 1$, and that $b = a\sqrt{1-\varepsilon^2}$, and introducing Eq. [3.9.1a] into the ellipse equation, we obtain

$$r\sin\theta = a\sqrt{1-\varepsilon^2}\sin\mathcal{E} \qquad [3.9.2]$$

Combining Eqs. [3.9.1c] and [3.9.2] gives

$$r = \sqrt{r^2\sin^2\theta + r^2\cos^2\theta} = a(1 - \varepsilon\cos\mathcal{E}) \qquad [3.9.3]$$

which is a simpler equation to deal with than Eq. [3.8.20]. In a similar fashion, we relate the true and eccentric anomalies. Introducing Eq. [3.9.3] into Eq. [3.9.1c], we obtain

$$\cos\theta = \frac{a(\cos\mathcal{E} - \varepsilon)}{r} = \frac{(\cos\mathcal{E} - \varepsilon)}{(1 - \varepsilon\cos\mathcal{E})} \qquad [3.9.4]$$

Using Eq. [3.9.4] with the trigonometric identity $\tan^2(\theta/2) = (1-\cos\theta)/(1+\cos\theta)$, we write

$$\tan^2\left(\frac{\theta}{2}\right) = \frac{1+\varepsilon}{1-\varepsilon}\tan^2\left(\frac{\mathcal{E}}{2}\right) \qquad [3.9.5]$$

from which we conclude that

$$\tan\left(\frac{\theta}{2}\right) = \sqrt{\frac{1+\varepsilon}{1-\varepsilon}}\tan\left(\frac{\mathcal{E}}{2}\right) \qquad [3.9.6]$$

We now introduce time into the formulation. Differentiating the two expressions for r in Eqs. [3.9.3] and [3.8.20] and using the expression for the momentum integral $h = r^2\dot\theta$, we obtain

$$\dot{r} = a\varepsilon\sin\mathcal{E}\,\dot{\mathcal{E}}$$

$$\dot{r} = \frac{a\varepsilon(1-\varepsilon^2)\sin\theta\,\dot\theta}{(1+\varepsilon\cos\theta)^2} = \frac{r^2\dot\theta\varepsilon\sin\theta}{a(1-\varepsilon^2)} = \frac{h\varepsilon\sin\theta}{a(1-\varepsilon^2)} \qquad [3.9.7a,b]$$

We rewrite Eq. [3.9.2] as

$$\sin\theta = \frac{a\sqrt{1-\varepsilon^2}\sin\mathcal{E}}{r} = \frac{\sqrt{1-\varepsilon^2}\sin\mathcal{E}}{(1-\varepsilon\cos\mathcal{E})} \qquad [3.9.8]$$

introduce it into Eq. [3.9.7b], and equate the result with Eq. [3.9.7a]. We express $\dot{\mathcal{E}}$ as $d\mathcal{E}/dt$ and collect all the terms involving $d\mathcal{E}$ on one side, with the result

$$(1 - \varepsilon\cos\mathcal{E})\,d\mathcal{E} = \frac{h}{a^2}\frac{dt}{\sqrt{1-\varepsilon^2}} \qquad [3.9.9]$$

Define by $n = 2\pi/\tau$ the *mean angular velocity,* in which τ is the period of the orbit given by Eq. [3.8.25], $\tau = 2\pi a^{3/2}/\mu^{1/2}$. Recalling that $h = \sqrt{\mu a(1-\varepsilon^2)}$, we can express the right side of Eq. [3.9.9] as

$$\frac{h}{a^2}\frac{dt}{\sqrt{1-\varepsilon^2}} = n\,dt \qquad [3.9.10]$$

It remains to integrate Eq. [3.9.9]. To this end, we select the initial conditions to coincide with the position of the body at the pericenter. At that location $\mathcal{E} = 0$, and the time is selected as $\mathcal{T}$, defined as the *time of pericenter passage*. We then have

$$\int_0^{\mathcal{E}} (1 - \varepsilon \cos \mathcal{E})\,d\mathcal{E} = \int_{\mathcal{T}}^{t} n\,dt \qquad [3.9.11]$$

with the result

$$n(t - \mathcal{T}) = \mathcal{E} - \varepsilon \sin \mathcal{E} \qquad [3.9.12]$$

This is known as *Kepler's equation*. We define the *mean anomaly* $\mathcal{M}$ by

$$\mathcal{M} = n(t - \mathcal{T}) \qquad [3.9.13]$$

so that Kepler's equation becomes

$$\mathcal{M} = \mathcal{E} - \varepsilon \sin \mathcal{E} \qquad [3.9.14]$$

The mean anomaly basically describes the angle that would have been described by the position of the particle if the motion was uniform, as in a circular orbit with mean angular velocity n. Using Kepler's equation, one can calculate the mean anomaly $\mathcal{M}$ if the eccentric anomaly $\mathcal{E}$ and the eccentricity ε are given. Often, one knows ε and can measure $\mathcal{M}$ more easily than the other parameters, so one needs to calculate $\mathcal{E}$ and θ. This requires the solution of Eq. [3.9.14], which usually is accomplished numerically.

We have defined a constant of integration that permits us to find the exact location of a mass on the ellipse. We have yet to locate the ellipse on the orbital plane. Denoting the plane of the orbit as the $x'y'$ plane, the xy axes, which are along the semimajor and semiminor axes of the ellipse, are obtained by a counterclockwise rotation about the z' axis. The angle of rotation is denoted by ω (not to be confused with angular velocity) and is called *argument of the pericenter*.

Next, we orient the orbital plane with respect to an inertial frame. We need to define both the origin and orientation of the inertial frame. There are several choices, depending on the bodies being analyzed:

1. For planetary motion, the sun is selected as the origin and the reference frame is called *heliocentric-ecliptic*. The coordinate axes, denoted by XYZ, are selected such that the XY plane coincides with the plane of the orbit of the earth around the sun (the *ecliptic plane*), shown in Fig. 3.23. The XY plane is also referred to as the *fundamental plane,* and the positive Z axis is referred to as the *north polar axis*. The X axis is selected as being toward the *vernal equinox,* also called the *first point of Aries*. On the first day of autumn, the line joining the centers of the sun and earth points toward the vernal equinox.

190 CHAPTER 3 • DYNAMICS OF A SYSTEM OF PARTICLES

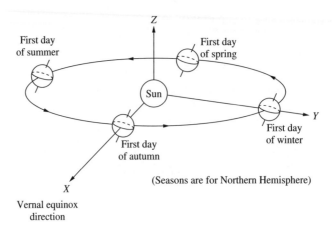

Figure 3.23 Heliocentric-ecliptic coordinate system

Because the XYZ plane is assumed to be inertial, the vernal equinox is assumed to be fixed as well, even though it is actually precessing very slowly. The causes of the precession are the disturbing effects of the moon, as well as the fact that the earth is not a perfect sphere, both of which add terms to the equations of motion describing the motion of the earth with respect to the sun. The period of precession can be shown to be, and has been observed as, about 26,000 years. When the vernal equinox was first defined, it was pointing toward the constellation Aries. At the writing of this text, it was pointing toward Pisces.

2. For satellite motion, a *geocentric-equatorial* or a *perifocal* coordinate system is often used. We describe here the geocentric-equatorial system. The origin of the coordinate system is selected as the center of the earth. The Z axis is selected toward the North Pole. The XY plane is the equatorial plane, and the X direction is selected toward the vernal equinox.

We are ready to orient the orbit. It turns out that this orientation is accomplished by a 3-1-3 coordinate transformation. Consider an inertial frame as shown in Fig. 3.24 and a sphere. The first rotation is about the Z axis. The rotation angle is called the *longitude of the ascending node* and is denoted by Ω. The ascending node is defined as the point at which the orbital plane crosses the fundamental plane with a northerly velocity, that is, moving in the positive Z direction. The *descending node* is defined as the point at which the orbital plane crosses the fundamental plane, moving in the negative Z direction. The line joining the ascending and descending nodes is called the *line of nodes*. Denoting the rotated coordinates by $X'Y'Z'$, the line of nodes is the X' axis, and it describes the intersection of the orbital plane and the fundamental plane. Historically, the term *line of nodes* originated from this coordinate transformation sequence.

The second rotation is about the X' axis. The rotation angle, denoted by i, is called the *orbital inclination*. The resulting coordinate axes are denoted by $x'y'z'$,

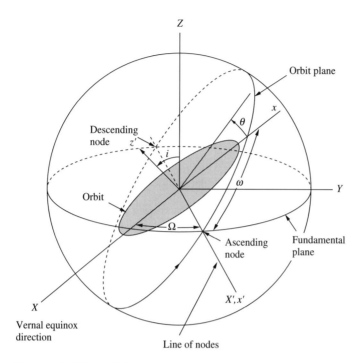

Figure 3.24 Orbital parameters

with the $x'y'$ plane describing the orbital plane. The angle i hence describes the inclination between the fundamental plane and the orbital plane (or the angle between the Z and z' axes). The angular momentum is in the z' direction. Finally, the $x'y'z'$ axes are rotated about the z' axis by the *argument of the pericenter,* denoted by ω to orient the orbit on the orbital plane.

The remaining three orbital parameters are the semimajor axis a, eccentricity ε, and perigee passage $\mathcal{T}$. In many cases, and especially in problems involving orbits about the sun, the mean anomaly $\mathcal{M}$ is used instead of $\mathcal{T}$.

The transformation from the three displacement and three velocity coordinates of the body on the orbit to the orbital parameters is a unique transformation. That is, given all three displacement and velocity components at a certain instant, one can determine the orbital parameters, and vice versa. For the ideal two body problem, where there are no disturbing effects of other bodies, the orbital plane as well as the size and orientation of the orbit are all constants. However, when there are disturbing functions, such as gravitational forces of other bodies and atmospheric drag, the right side of Eq. [3.7.6] is no longer zero. In addition, the equations derived in this section assume that the bodies involved are perfect and homogeneous spheres. Any deviation in geometry from a homogeneous sphere results in additional gravitational terms. Furthermore, the rotational motion of the bodies involved also must be considered. The analysis of these effects is beyond the scope of this text.

Example 3.9

Consider Example 3.7 and calculate the time elapsed after burnout until the rocket crashes.

Solution

We will make use of the eccentric anomaly to solve this problem. First, we find the transverse angle θ the rocket makes at burnout. Denoting this angle by θ_1, we use Eq. [3.8.20] to find it as

$$\cos\theta_1 = \frac{a(1-\varepsilon^2) - r}{r\varepsilon} = \frac{14450(1 - 0.59857^2) - 6500}{6500(0.59857)} = 0.71266 \quad \text{[a]}$$

from which we get

$$\theta_1 = \cos^{-1}(0.71266) = 0.77751 \text{ rad} = 44.548° \quad \text{[b]}$$

From Eq. [3.9.6] we have the eccentric anomaly at burnout as

$$\tan\left(\frac{\mathscr{E}_1}{2}\right) = \sqrt{\frac{1-\varepsilon}{1+\varepsilon}} \tan\left(\frac{\theta_1}{2}\right) = \sqrt{\frac{0.40143}{1.59857}} \tan(22.274°) = 0.50112(0.40962) = 0.20527 \quad \text{[c]}$$

so that

$$\mathscr{E}_1 = 2\tan^{-1}(0.20527) = 0.40492 \text{ rad} = 23.200° \quad \text{[d]}$$

We now make use of Kepler's equation, Eq. [3.9.12]. Setting the initial time as $\mathscr{T} = 0$, and using Eq. [3.8.24], we can solve for time as

$$t = \frac{\mathscr{E} - \varepsilon\sin\mathscr{E}}{n} = \sqrt{\frac{a^3}{\mu}}(\mathscr{E} - \varepsilon\sin\mathscr{E}) \quad \text{[e]}$$

and $\sqrt{\dfrac{a^3}{\mu}} = \sqrt{\dfrac{(1.445 \times 10^7)^3}{3.986 \times 10^{14}}} = \sqrt{7.5695 \times 10^6} = 2.7513 \times 10^3$ s. Denoting by t_1 the time at burnout and using Eq. [e], we obtain

$$t_1 = \sqrt{\frac{a^3}{\mu}}(\mathscr{E}_1 - \varepsilon\sin\mathscr{E}_1)$$

$$= 2.7513 \times 10^3 (0.40492 - 0.59857\sin(0.40492)) = 465.3 \text{ seconds} \quad \text{[f]}$$

We next find the above parameters at impact, and denote them with the subscript 2. To find the angle θ_2 we make use of Eq. [a], with r replaced by $R = 6378.1$ km

$$\cos\theta_2 = \frac{a(1-\varepsilon^2) - R}{R\varepsilon} = \frac{14450(1 - 0.59857^2) - 6378.1}{6378.1(0.59857)} = 0.75821 \quad \text{[g]}$$

Noting that the rocket crashes in the fourth quadrant, we can find the angle θ_2 from

$$\theta_2 = \cos^{-1}(0.75821) + \frac{3\pi}{2} = 5.4226 \text{ rad} = 310.69° \quad \text{[h]}$$

The corresponding eccentric anomaly is found from

$$\mathscr{E}_2 = 2\tan^{-1}\left(\sqrt{\frac{1-\varepsilon}{1+\varepsilon}} \tan\left(\frac{\theta_2}{2}\right)\right)$$

$$= 2\tan^{-1}((0.50112)(-0.45900)) = -0.45216 \text{ rad} = -25.910° \quad \text{[i]}$$

or, by adding 2π to the value we obtain

$$\mathcal{E}_2 = -0.45216 + 2\pi = 5.8310 \text{ rad} = -25.910° + 360° = 334.09° \qquad [\text{j}]$$

Introducing the mean anomaly to Kepler's equation, we obtain for the time

$$t_2 = \sqrt{\frac{a^3}{\mu}}(\mathcal{E}_2 - \varepsilon \sin \mathcal{E}_2) = 2.7513 \times 10^3 (5.8310 - 0.59857 \sin 5.8310) \qquad [\text{k}]$$

$$= 16,762 \text{ seconds}$$

Subtracting t_1 from t_2, we obtain the time the rocket stays in orbit before it crashes as

$$t_2 - t_1 = 16,762 - 465 \text{ s} = 16,297 \text{ s} = 4.527 \text{ hr} \qquad [\text{l}]$$

3.10 PLANE KINETICS OF RIGID BODIES

In this section we analyze the kinetics of rigid bodies that move on a plane. In essence, this section and the ones following it constitute a review of sophomore-level dynamics. This review is primarily aimed at refamiliarizing the reader with some basic concepts, as the developments in the following chapter are described in terms of particles or plane motion of rigid bodies. A detailed analysis of the kinematics and kinetics of three-dimensional rigid body motion is carried out in Chapters 7 and 8.

We denote the inertial frame by XYZ and consider the XY plane as the plane of motion. The moving reference frame xyz, obtained by rotating the XYZ frame about the Z axis by θ, is attached to the body. The angular velocity and angular acceleration have one component each:

$$\boldsymbol{\omega} = \omega \mathbf{K} = \dot{\theta} \mathbf{K} \qquad \boldsymbol{\alpha} = \alpha \mathbf{K} = \ddot{\theta} \mathbf{K} \qquad [\mathbf{3.10.1}]$$

We begin with defining the center of mass. Consider a system of N particles, as shown in Fig. 3.1. The center of mass is denoted by the point G and defined by Eq. [3.2.2]. A rigid body can be considered as a collection of particles in which the number of particles approaches infinity and in which the distance between the individual masses remains the same. As N approaches infinity, each particle is treated as a differential mass element, $m_i \to dm$, and the summation in Eq. [3.2.2] is replaced by integration over the body. We then define the center of mass G as

$$\mathbf{r}_G = \frac{1}{m} \int_{\text{body}} \mathbf{r} \, dm \qquad [\mathbf{3.10.2}]$$

where $\mathbf{r}$ is the vector from the origin O to differential element dm and

$$m = \int_{\text{body}} dm \qquad [\mathbf{3.10.3}]$$

is the mass of the body. Considering Figs. 3.1 and 3.25, for a system of particles we write $\mathbf{r}_i = \mathbf{r}_G + \boldsymbol{\rho}_i$, and for a rigid body, $\mathbf{r} = \mathbf{r}_G + \boldsymbol{\rho}$. Introducing this term in

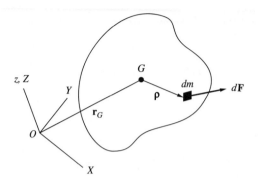

Figure 3.25

Eq. [3.10.2], we obtain

$$\mathbf{r}_G = \frac{1}{m}\int_{\text{body}} \mathbf{r}\,dm = \frac{1}{m}\int_{\text{body}} (\mathbf{r}_G + \boldsymbol{\rho})\,dm = \mathbf{r}_G + \frac{1}{m}\int_{\text{body}} \boldsymbol{\rho}\,dm \qquad [3.10.4]$$

leading to the conclusion that

$$\int_{\text{body}} \boldsymbol{\rho}\,dm = 0 \qquad [3.10.5]$$

This equation indicates that the weighted average of the displacement vector about the center of mass is zero.

The center of mass is a very important quantity, as its use simplifies the analysis of rigid bodies considerably. To see this, let us write the rigid body equivalent of the combined equations of motion. Considering the differential element and its equation of motion

$$\mathbf{a}\,dm = d\mathbf{F} \qquad [3.10.6]$$

where $d\mathbf{F}$ is the total force acting on the differential element. We write for the entire body,

$$\int_{\text{body}} \mathbf{a}\,dm = \int_{\text{body}} d\mathbf{F} = \mathbf{F} \qquad [3.10.7]$$

in which $\mathbf{F}$ is the resultant of all forces. This resultant contains contributions only from the external forces, as the internal forces cancel each other. Introducing Eq. [3.10.4] into Eq. [3.10.7] gives the translational equations of motion of a rigid body, that is, Newton's second law for a rigid body, as

$$m\mathbf{a}_G = \mathbf{F} \qquad [3.10.8]$$

The derivation of the translational equations of motion above is valid for the general motion of a rigid body. For plane motion, we define the moment of all forces acting on the body about the center of mass as

$$\mathbf{M}_G = \sum_{i=1}^{N} \boldsymbol{\rho}_i \times \mathbf{F}_i = \int_{\text{body}} \boldsymbol{\rho} \times d\mathbf{F} = M_G \mathbf{K} \qquad [3.10.9]$$

where M_G is the resultant moment about the center of mass. Next, introduce Eq. [3.10.6] to the above equation. Before doing that, we express the velocity and acceleration of the differential element as

$$\mathbf{v} = \mathbf{v}_G + \boldsymbol{\omega} \times \boldsymbol{\rho} \qquad \mathbf{a} = \mathbf{a}_G + \boldsymbol{\alpha} \times \boldsymbol{\rho} - \omega^2 \boldsymbol{\rho} \qquad [3.10.10]$$

and proceed, so that

$$\mathbf{M}_G = \int_{\text{body}} \boldsymbol{\rho} \times d\mathbf{F} = \int_{\text{body}} \boldsymbol{\rho} \times \mathbf{a} \, dm = \int_{\text{body}} \boldsymbol{\rho} \times (\mathbf{a}_G + \boldsymbol{\alpha} \times \boldsymbol{\rho} - \omega^2 \boldsymbol{\rho}) \, dm \qquad [3.10.11]$$

The first term inside the brackets vanishes because of the definition of the center of mass and the third term vanishes because of the cross product. To evaluate the middle term, consider that the cross product $\boldsymbol{\alpha} \times \boldsymbol{\rho}$ is on the plane of motion, perpendicular to $\boldsymbol{\rho}$, and it has the magnitude $|\alpha \rho|$. Hence, the cross product between $\boldsymbol{\rho}$ and $\boldsymbol{\alpha} \times \boldsymbol{\rho}$ is perpendicular to the plane of motion, parallel to $\boldsymbol{\alpha}$, and with magnitude $\alpha \rho^2$. Thus, Eq. [3.10.11] can be written as

$$\mathbf{M}_G = \int_{\text{body}} \boldsymbol{\rho} \times (\boldsymbol{\alpha} \times \boldsymbol{\rho}) \, dm = \int_{\text{body}} \alpha \rho^2 \, dm \, \mathbf{K} = \alpha \int_{\text{body}} \rho^2 \, dm \, \mathbf{K} \qquad [3.10.12]$$

The integral on the right side of this equation is not dependent on time or any displacement variable, so that it can be evaluated independently of the angular acceleration. We define it as the *mass moment of inertia* of the body about the center of mass, and denote it by I_G, so that

$$I_G = \int_{\text{body}} \rho^2 \, dm \qquad [3.10.13]$$

The mass moment of inertia is a property of the body itself; it is a measure of how the mass of the body is distributed about an axis passing through the center of mass and perpendicular to the plane of motion. We discuss ways of calculating the mass moment of inertia in Chapter 6. Figure 3.26 gives the centroidal mass moment of inertia for two common shapes. Appendix C gives a more comprehensive list.

Combining Eqs. [3.10.12] and [3.10.13], we obtain the rotational equation of motion

$$I_G \alpha = M_G \qquad [3.10.14]$$

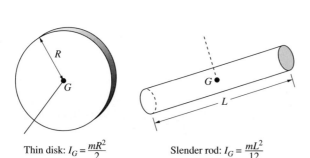

Thin disk: $I_G = \frac{mR^2}{2}$ Slender rod: $I_G = \frac{mL^2}{12}$

Figure 3.26

This equation, together with the two equations in Eq. [3.10.8], gives the three equations of motion for the plane motion of a rigid body. The rotational equation of motion can also be represented in terms of the angular momentum. Indeed, we define the angular momentum about the center of mass as

$$\mathbf{H}_G = \int_{\text{body}} \boldsymbol{\rho} \times \mathbf{v}\, dm = \int_{\text{body}} \boldsymbol{\rho} \times (\mathbf{v}_G + \boldsymbol{\omega} \times \boldsymbol{\rho})\, dm = \int_{\text{body}} \boldsymbol{\rho} \times (\boldsymbol{\omega} \times \boldsymbol{\rho})\, dm \qquad [3.10.15]$$

For plane motion, considering the integral in Eq. [3.10.12] and the definition of the mass moment of inertia, we can write

$$\mathbf{H}_G = \int_{\text{body}} \omega \rho^2 \, dm\, \mathbf{K} = I_G \omega \mathbf{K} \qquad [3.10.16]$$

and the rotational equation of motion can be written as

$$\dot{H}_G = M_G \quad \text{or} \quad \dot{\mathbf{H}}_G = \mathbf{M}_G \qquad [3.10.17]$$

just as the translational equations of motion can be written as

$$\dot{\mathbf{p}} = \mathbf{F} \qquad [3.10.18]$$

in which $\mathbf{p} = m\mathbf{v}_G$ is the linear momentum of the body. It should be noted that while we derived Eq. [3.10.17] for plane motion, this relationship in vector form holds for the general three-dimensional rotational motion of a body. As discussed in Chapter 1, there is debate about whether Eq. [3.10.17] is a derived relationship, or a stated law of motion.

The equations of motion can be illustrated by means of equivalent free-body and resultant force diagrams, as shown in Fig. 3.27. The sum of the external forces is equal to the resultant, which is equal to the rates of change of the linear and angular momentum.

Up to now, we considered the angular momentum and sum of moments about the center of mass. It turns out that under certain circumstances, it becomes more convenient to write the rotational equations of motion about another point. For example, such a case arises when motion about a fixed point is considered. To analyze the moment equation about an arbitrary point, consider the third part of Fig. 3.27 and sum moments about an arbitrary point D. We have

$$M_D = I_G \alpha + m a_G d \quad \text{or} \quad \mathbf{M}_D = I_G \alpha \mathbf{K} + \mathbf{r}_{G/D} \times m\mathbf{a}_G \qquad [3.10.19]$$

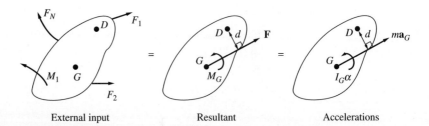

Figure 3.27

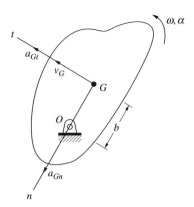

Figure 3.28 Rotation about a fixed point

in which d is the perpendicular distance from D to the acceleration vector for the center of mass G.

A special application of the above equation is for rotation about a fixed point. Consider Fig. 3.28, where the rigid body rotates about point O. It follows that the acceleration of the center of mass can be expressed in terms of the distance between points O and G, which we denote by b, as

$$\mathbf{a}_G = \mathbf{a}_{Gn} + \mathbf{a}_{Gt} = b\omega^2 \mathbf{e}_n + b\alpha \mathbf{e}_t \qquad [3.10.20]$$

The component of $\mathbf{a}_G$ in the normal direction does not affect the moment about O. Hence, we can write the sum of moments about O as

$$M_O = I_G \alpha + mb^2 \alpha = (I_G + mb^2)\alpha = I_O \alpha \qquad [3.10.21]$$

I_O is the mass moment of inertia about point O, and we have written the *parallel axis theorem,* described in detail in Chapter 6. This theorem essentially relates the mass moments of inertia of a body about the center of mass G and another point D by

$$I_D = I_G + md^2 \qquad [3.10.22]$$

where d is the distance between the two points. Another way of describing the inertia properties of a body is by means of the *radius of gyration,* denoted by κ, such that $I = m\kappa^2$. The radius of gyration is often used when dealing with complex bodies, and it is usually listed as a property of a body.

When solving plane kinetics problems, one should select the moment center such that the number of reactions to be solved and the total number of calculations become a minimum. The procedure in solving plane kinetics problems is the same as the procedure outlined in Chapter 1 for particle motion. The only difference is that one must invoke the rotational equations of motion.

Example 3.10

Rod OA is of length 0.4 m and has a mass of 3 kg, while rod AB is of mass 12 kg and length 1 m. The two rods are released from rest from the configuration shown in Fig. 3.29. Find the angular accelerations of both rods at this instant. There is no friction at point B.

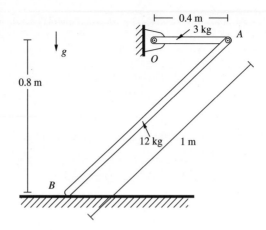

Figure 3.29

Solution

We first isolate the two bodies and draw free-body diagrams, shown in Fig. 3.30. Note that triangle *DBA* is a 3-4-5 triangle. The next task is to write the force and moment balances. For rod *OA* it is convenient to write the sum of moments about point *O* as (counterclockwise positive)

$$\sum M_O = I_O \alpha_1 = -0.2(3g) + 0.4 A_y \qquad \text{[a]}$$

The force balance equations merely give relations for the reactions for O_x and O_y, so they are not of much use, unless one wishes to calculate those reactions. Introducing the expressions $I_O = I_G + m(\frac{L}{2})^2 = \frac{1}{12}mL^2 + \frac{1}{4}mL^2 = mL^2/3 = 3(0.4^2)/3 = 0.16$ kg·m² into Eq. [a]

$$0.16\alpha_1 = 0.4 A_y - 0.6g \qquad \text{[b]}$$

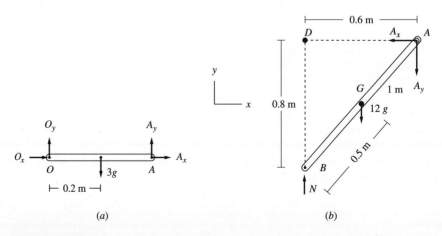

Figure 3.30

For rod AB, there are three unknown reactions, A_x, A_y, and N. Considering that the sum of forces in the horizontal and vertical directions and the sum of moments about the center of mass involve all three unknowns, we prefer writing the moment balance about point D, a point about which N and A_x do not exert a moment. To do this, we need an expression for the acceleration of point G. We will obtain the acceleration of point G by first writing a relative acceleration relation between points A and B and then a relation between B and G.

Consider points A and B first. Because motion is just being initiated, the angular velocities of OA and AB are both zero. Also, point B moves in the horizontal plane. We hence write the accelerations of points A and B as

$$\mathbf{a}_A = a_A \mathbf{j} = 0.4\alpha_1 \mathbf{j} \quad \mathbf{a}_B = a_B \mathbf{i} \qquad [\mathbf{c}]$$

and the relative acceleration relation becomes

$$\mathbf{a}_A = \mathbf{a}_B + \boldsymbol{\alpha}_2 \times \mathbf{r}_{A/B} \rightarrow a_A \mathbf{j} = a_B \mathbf{i} + \alpha_2 \mathbf{k} \times (0.6\mathbf{i} + 0.8\mathbf{j}) = (a_B - 0.8\alpha_2)\mathbf{i} + 0.6\alpha_2 \mathbf{j} \qquad [\mathbf{d}]$$

from which we conclude that

$$a_A = 0.4\alpha_1 = 0.6\alpha_2 \quad a_B = 0.8\alpha_2 \qquad [\mathbf{e}]$$

Next, write the relative acceleration relation between points B and G

$$\mathbf{a}_G = \mathbf{a}_B + \boldsymbol{\alpha}_2 \times \mathbf{r}_{G/B} = 0.8\alpha_2 \mathbf{i} + \alpha_2 \mathbf{k} \times (0.3\mathbf{i} + 0.4\mathbf{j}) = 0.4\alpha_2 \mathbf{i} + 0.3\alpha_2 \mathbf{j} \qquad [\mathbf{f}]$$

We are now in a position to write the moment equation about point D. Because $\mathbf{a}_D$ has components in both x and y directions, we write Eq. [3.10.21] in vector form as

$$\sum M_D \mathbf{k} = I_G \alpha_2 \mathbf{k} + \mathbf{r}_{G/D} \times m\mathbf{a}_G = (-0.3 \times 12\,g - 0.6\,A_y)\mathbf{k} \qquad [\mathbf{g}]$$

in which I_G is the mass moment of inertia of rod AB, having the value of $12 \times 1^2/12 = 1$ kg·m^2 and $\mathbf{r}_{G/D} = 0.3\mathbf{i} - 0.4\mathbf{j}$ m. Substituting in these values to the above equation,

$$1\alpha_2 \mathbf{k} + (0.3\mathbf{i} - 0.4\mathbf{j}) \times 12(0.4\alpha_2 \mathbf{i} + 0.3\alpha_2 \mathbf{j}) = (-3.6g - 0.6A_y)\mathbf{k} \qquad [\mathbf{h}]$$

which reduces to

$$(1 + 1.08 + 1.92)\alpha_2 = 4\alpha_2 = -3.6g - 0.6A_y \qquad [\mathbf{i}]$$

Equations [b] and [i] are two equations with the unknowns α_1, α_2, and A_y. From the first part of Eq. [e] we have

$$\alpha_1 = 1.5\alpha_2 \qquad [\mathbf{j}]$$

so that Eq. [b] can be written as

$$0.24\alpha_2 = 0.4A_y - 0.6g \qquad [\mathbf{k}]$$

Multiplying Eq. [k] by 1.5 and adding it to Eq. [i], we obtain for α_2

$$\alpha_2 = \frac{-4.5g}{4.36} = \frac{4.5 \times 9.807}{4.36} = 10.12 \text{ rad/s} \qquad [\mathbf{l}]$$

Using Eq. [j] we obtain α_1 as $\alpha_1 = 1.5\alpha_2 = 15.18$ rad/s.

3.11 Instant Centers and Rolling

An important concept associated with the plane motion of rigid bodies is that of an instant center. At any instant of the motion, there exists an axis perpendicular to the plane of motion, called the *instantaneous axis of zero velocity*, such that the body can be viewed as rotating about that axis at that instant. The intersection of this axis and the plane of motion is called the *instantaneous center of zero velocity*, or *instant center*. In general, the instant center of a rigid body is located by visual inspection. To establish the location of the instant center, one needs to know the velocities of two points on the body. If the velocities are not in the same direction, one draws two lines, beginning at the points at which the velocities are known and perpendicular to the velocities. Their intersection is the instant center. Figure 3.31 illustrates. If the velocities of the two points are in the same direction, one again draws two lines: one joining the points at which the velocities are known and the other joining the tips of the velocity vectors (drawn to scale). Their intersection gives the instant center, as shown in Fig. 3.32. In Chapter 7, we prove the existence of the instant center for plane motion.

It should be noted that while the instant center has zero velocity, its location at every time instant is different, and its acceleration is not zero. Hence, Eq. [3.10.21], the moment equation about a fixed point, cannot be written about an instant center.

An interesting case of motion is that of a body *rolling* over another body or over a fixed surface. For rolling to take place between two bodies, a continuous sequence of points on one of the bodies must be in continuous contact with a continuous sequence of points on the other body. For continuous contact to take place, the contacting bodies must have smooth contours, with no jumps.

Rolling can occur in a variety of ways, as Chapter 7 will show in detail. In this section we consider roll of a circular body over a fixed surface. The surface can be planar or curved. Consider Fig. 3.33. We denote by $\boldsymbol{\omega} = -\omega \mathbf{k}$ the angular velocity. The contact point is denoted by C, with velocity of $\mathbf{v}_C$. Let $\mathbf{n}$ denote the unit vector perpendicular to the plane of contact. The kinematic relation describing rolling is

$$\mathbf{v}_C \cdot \mathbf{n} = 0 \qquad [3.11.1]$$

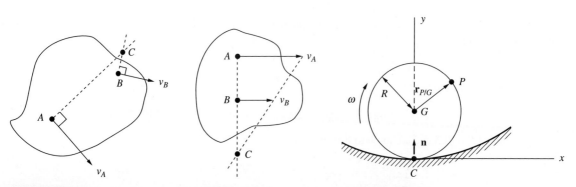

Figure 3.31 **Figure 3.32** **Figure 3.33**

3.11 INSTANT CENTERS AND ROLLING

If the contact point has a finite velocity along the plane of contact, the motion is referred to as *roll with slip*. If the contact point has zero velocity, the motion is referred to as *roll without slip*. Whether slipping exists or not depends on the forces acting on the rolling bodies, as well as the friction between the rolling bodies. Mathematically, the no slip condition is represented as

$$\mathbf{v}_C = \mathbf{0} \qquad [3.11.2]$$

Because the velocity of the contact point is zero, the contact point can be treated as an instant center. However, the acceleration of the contact point is not zero, whether there is slipping or not. The contact point approaches the plane of contact, it has contact, and then it moves away. The constraint associated with the contact is applied to the contacting points only during the instant of contact.

Consider rolling without slipping and a point P on the body. The velocity of a point P on the body can be expressed as

$$\mathbf{v}_P = \mathbf{v}_G + \boldsymbol{\omega} \times \mathbf{r}_{P/G} \qquad [3.11.3]$$

We write the above equation for the point of contact C, with the result

$$\mathbf{v}_C = \mathbf{0} = \mathbf{v}_G + \boldsymbol{\omega} \times \mathbf{r}_{C/G} \qquad [3.11.4]$$

which we can use to express the velocity of the center of mass as

$$\mathbf{v}_G = \boldsymbol{\omega} \times \mathbf{r}_{G/C} = -\omega \mathbf{k} \times R\mathbf{j} = R\omega\mathbf{i} = R\dot{\theta}\mathbf{i} \qquad [3.11.5]$$

This equation can be physically explained by noting that the point of contact C is the instant center.

Example 3.11

A sphere of mass m and radius r rolls without slip inside a circular curved surface with radius R, as shown in Fig. 3.34. Obtain the equation of motion as a function of θ.

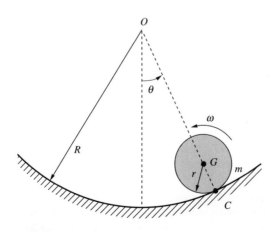

Figure 3.34

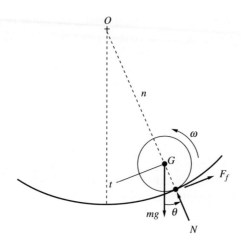

Figure 3.35

Solution

The free-body diagram of the sphere is given in Fig. 3.35. We write the force balance using normal and tangential coordinates, thus

$$ma_{Gt} = mg\sin\theta - F_f \quad \text{[a]}$$

$$ma_{Gn} = N - mg\cos\theta = \frac{mv_G^2}{\rho} \quad \text{[b]}$$

where ρ is the radius of curvature for point G and F_f is the friction force. The center of curvature is the center O of the curved surface and the radius of curvature is constant and has the value $\rho = R - r$. The moment balance about the center of mass is (counterclockwise positive)

$$I_G\alpha = F_f r \quad \text{[c]}$$

Because we assume rolling without slipping, the speed and tangential acceleration of the center of the sphere are

$$v_G = r\omega \qquad a_{Gt} = r\alpha \quad \text{[d]}$$

where ω is the angular speed of the sphere and $\alpha = \dot{\omega}$. Solving for F_f in Eq. [a] and using the expression for a_{Gt} in Eq. [d], we obtain

$$F_f = mg\sin\theta - mr\alpha \quad \text{[e]}$$

which, when introduced into Eq. [c], yields

$$(I_G + mr^2)\alpha = mgr\sin\theta \quad \text{[f]}$$

To obtain the equation of motion, we need to express the angular acceleration of the sphere in terms of θ. We make use of the property that the center of mass of the sphere can also be considered as moving about the center of curvature of the surface, so that we can write

$$v_G = r\omega = -(R - r)\dot{\theta} \quad \text{[g]}$$

from which we obtain

$$\alpha = -\frac{R-r}{r}\ddot{\theta} \qquad \text{[h]}$$

And introducing Eq. [h] into Eq. [f], we write the equation of motion as

$$(I_G + mr^2)(R - r)\ddot{\theta} + mgr^2 \sin\theta = 0 \qquad \text{[i]}$$

Note that whenever one assumes roll without slip, one must check the validity of the assumption. This can be done by calculating the magnitude of the friction force and comparing it with the maximum value of the friction force. If this maximum is exceeded, one must redefine the problem as a roll with slip problem.

3.12 ENERGY AND MOMENTUM

The kinetic energy of a rigid body is defined as

$$T = \frac{1}{2}\int_{body} \mathbf{v}\cdot\mathbf{v}\, dm \qquad \text{[3.12.1]}$$

If we introduce the expression of the velocity in terms of the center of mass to this equation, we obtain

$$T = \frac{1}{2}\int_{body} (\mathbf{v}_G + \boldsymbol{\omega}\times\boldsymbol{\rho})\cdot(\mathbf{v}_G + \boldsymbol{\omega}\times\boldsymbol{\rho})\, dm \qquad \text{[3.12.2]}$$

$$= \frac{1}{2}\int_{body} \left(\mathbf{v}_G\cdot\mathbf{v}_G + 2\mathbf{v}_G\cdot(\boldsymbol{\omega}\times\boldsymbol{\rho}) + (\boldsymbol{\omega}\times\boldsymbol{\rho})\cdot(\boldsymbol{\omega}\times\boldsymbol{\rho})\right) dm$$

The first term on the right side of this equation gives $m\mathbf{v}_G\cdot\mathbf{v}_G/2$. The second term vanishes due to the definition of the center of mass. To evaluate the third term, we note that the cross product of $\boldsymbol{\omega}$ and $\boldsymbol{\rho}$ is a vector of magnitude $\rho\omega$, so that

$$\int_{body} (\boldsymbol{\omega}\times\boldsymbol{\rho})\cdot(\boldsymbol{\omega}\times\boldsymbol{\rho})\, dm = \int_{body} \rho^2\omega^2\, dm = I_G\omega^2 \qquad \text{[3.12.3]}$$

and the kinetic energy of a rigid body undergoing plane motion can be written as

$$T = \frac{1}{2}m\mathbf{v}_G\cdot\mathbf{v}_G + \frac{1}{2}I_G\omega^2 \qquad \text{[3.12.4]}$$

When the rigid body is rotating about a fixed point C, its kinetic energy becomes

$$T = \frac{1}{2}I_C\omega^2 \qquad \text{[3.12.5]}$$

The kinetic energy can be written in this form about the instant center, as the instant center has zero velocity. When the body is rotating about a point C that is moving, Eq. [3.12.5] is not valid.

The gravitational potential energy for a rigid body is

$$V = mgh_G \qquad \text{[3.12.6]}$$

in which h_G is the perpendicular distance between the center of mass and the datum line. All the other potential energy expressions are the same as we have seen them before for particles. The work-energy theorem discussed in Chapter 1 is valid for all types of bodies.

An interesting property of bodies undergoing rolling without slipping is that the friction force does no work. This can be easily shown using the definition of work as a time integral. Because the friction force is always applied to a point with zero velocity, the power of the force $P = \mathbf{F}_f \cdot \mathbf{v}_C$ is zero. Hence, the work done, which is the integral of power over time, is zero.

We obtain the impulse-momentum relations the same way we did for particles, by integrating the equations of motion over time. Doing so between two time points t_1 and t_2 gives

$$m\mathbf{v}_G(t_1) + \int_{t_1}^{t_2} \mathbf{F}\, dt = m\mathbf{v}_G(t_2) \qquad \text{[3.12.7]}$$

$$I_G \omega(t_1) + \int_{t_1}^{t_2} M_G\, dt = I_G \omega(t_2) \qquad \text{[3.12.8]}$$

The integrals in Eqs. [3.12.7] and [3.12.8] are known as the *impulse* and the *angular impulse*, respectively. A very interesting application of the impulse-momentum relationships is in cases where the integral of the sum of moments or the sum of forces vanishes. In such cases, we have *conservation of linear momentum* or *conservation of angular momentum*.

Example 3.12

A solid uniform sphere of mass m and radius R is placed on top of a fixed sphere of the same radius, and it is slightly tipped (Fig. 3.36). Find the value of the angle θ at which sliding begins as a function of the coefficient of friction μ.

Solution

The displaced position of the sphere and its free-body diagram are depicted in Fig. 3.37. We denote by θ the angle made by the line joining the centers of the spheres and the vertical. By ϕ we denote the angular displacement of the top sphere. The friction force F is less than μN for no slipping, where N is the normal force. When there is slipping, $F = \mu N$. The speed of the center of the sphere is given by

$$v_G = 2R\dot\theta \qquad \text{[a]}$$

and for no slip

$$v_G = R\dot\phi \qquad \text{[b]}$$

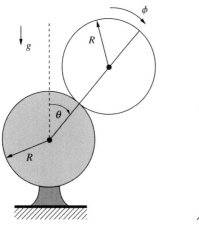
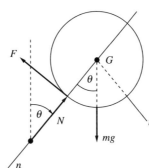

Figure 3.36 **Figure 3.37**

so that for the no-slip case the two angles are related to each other by

$$\dot{\phi} = 2\dot{\theta} \quad \text{[c]}$$

Continuing to consider no slip, and noting from Appendix C that the centroidal mass moment of inertia of a sphere is $I_G = 2mR^2/5$, we can write the kinetic energy about the point of rolling contact

$$T = \frac{1}{2}I_B\dot{\phi}^2 = \frac{7}{10}mR^2\dot{\phi}^2 = \frac{14}{5}mR^2\dot{\theta}^2 \quad \text{[d]}$$

where $I_B = I_G + mR^2$ is the mass moment of inertia about the point of contact. Considering the center of the lower sphere as the datum, the potential energy has the form

$$V = 2mgR\cos\theta \quad \text{[e]}$$

The sum of forces in the normal direction is

$$\sum F_n = ma_n \rightarrow mg\cos\theta - N = \frac{mv_G^2}{\rho} = \frac{m(2R\dot{\theta})^2}{2R} = 2mR\dot{\theta}^2 \quad \text{[f]}$$

The sum of the kinetic and potential energies is constant. Therefore,

$$\frac{14}{5}mR^2\dot{\theta}^2 + 2mgR\cos\theta = E \quad \text{[g]}$$

and noting that when $\dot{\theta} = 0$, θ is also equal to zero, we can evaluate E as

$$E = 2mgR \quad \text{[h]}$$

Now using Eqs. [g] and [h], the relation between $\dot{\theta}$ and θ can be expressed as

$$\dot{\theta}^2 = \frac{5g}{7R}(1 - \cos\theta) \quad \text{[i]}$$

Substituting Eq. [i] into Eq. [f] we obtain an expression for the normal force N in terms of θ as

$$N = mg\cos\theta - 2mR\dot{\theta}^2 = mg\cos\theta - 2mR\frac{5g}{7R}(1 - \cos\theta) = \frac{mg}{7}(17\cos\theta - 10) \quad \text{[j]}$$

Summing moments about the center of mass of the sphere we obtain

$$\sum M_G = I_G \ddot{\phi} \rightarrow FR = \frac{2}{5} mR^2 \ddot{\phi} \qquad [k]$$

which relates the friction force F to $\ddot{\phi}$. We sum forces in the tangential direction, thus

$$ma_t = -F + mg \sin \theta \qquad [l]$$

Noting that $a_t = 2R\ddot{\theta}$ and introducing Eq. [k] into Eq. [l] we obtain

$$2mR\ddot{\theta} = -\frac{4}{5} mR\ddot{\theta} + mg \sin \theta \rightarrow \ddot{\theta} = \frac{5g}{14R} \sin \theta \qquad [m]$$

Combining Eqs. [k] and [m] we write an expression for the friction force in terms of θ as

$$F = \frac{2}{5} mR\ddot{\phi} = \frac{4}{5} mR\ddot{\theta} = \frac{4}{5} mR \frac{5g}{14R} \sin \theta = \frac{2}{7} mg \sin \theta \qquad [n]$$

The instant slipping begins, $F = \mu N$. Hence, considering Eqs. [j] and [n], we obtain a relation for θ at the instant slip begins,

$$\frac{2}{7} mg \sin \theta = \mu \frac{mg}{7} (17 \cos \theta - 10) \qquad [o]$$

This can be solved for θ given a value of μ.

It should be noted that the solution is independent of the mass and radius of the sphere. For very small values of μ, we can solve Eq. [o] using a small angles assumption of $\sin \theta \approx \theta$, $\cos \theta \approx 1$, which yields the result $\theta = 7\mu/2$. For larger values of the coefficient of friction, one can use a numerical approach. For example, when $\mu = 0.5$, θ can be found to be $\theta = 41.5°$. The case of having a rough contact between the spheres, that is, when the coefficient of friction is $\mu = \infty$, is of interest. Dividing both sides of Eq. [o] by μ and setting $\mu = \infty$, we obtain

$$\frac{mg}{7} (17 \cos \theta - 10) = 0 \qquad [p]$$

which has the answer $\cos \theta = 10/17$, or $\theta = 53.97°$. We observe from Eqs. [j] and [o] that Eq. [o] basically implies that the normal force is zero. Therefore, at $\theta = 53.97°$, the top sphere loses contact with the lower sphere and goes into a free-fall mode.

REFERENCES

Bate, R.B., D.D. Mueller, and J.E. White, *Fundamentals of Astrodynamics.* New York: Dover Publications, 1971.

Greenwood, D.T. *Principles of Dynamics.* 2nd ed. Englewood Cliffs, NJ: Prentice-Hall, 1988.

Kaplan, M.H. *Modern Spacecraft Dynamics and Control.* New York: John Wiley, 1976.

Meirovitch, L. *Introduction to Dynamics and Control.* New York: John Wiley, 1985.

——*Methods of Analytical Dynamics.* New York: McGraw-Hill, 1970.

Homework Exercises

Section 3.2

1. The blocks shown in Fig. 3.38 are released from rest. Find the acceleration of each block. The pulleys are massless.
2. A chain of length 0.7 m is at rest over two slopes, as shown in Fig. 3.39. First, assume there is no friction between the chain and slopes. If the chain is released from rest, which end of it will rise, and what will its speed be when it reaches the top? Then, assume that there is friction and calculate the minimum amount of friction necessary to prevent the chain from moving.
3. The pulley system in Fig. 3.40 is released from rest with the spring unstretched. Find the acceleration of each block. The pulleys and the cord are massless and frictionless.

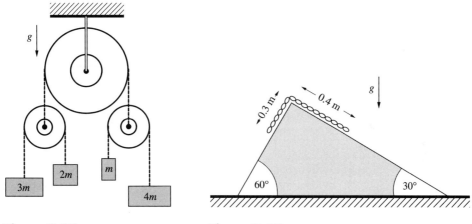

Figure 3.38 **Figure 3.39**

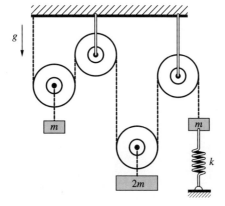

Figure 3.40

Section 3.3

4. The three masses in Fig. 3.41 are connected by links of negligible mass. A force of magnitude F is applied as shown. Find the acceleration of the center of mass, as well as the acceleration of the individual masses.

Section 3.4

5. Find the direction of sliding in Problem 2 using energy methods for the (case of no friction).

Section 3.5

6. A ball is released from a height h from a plane. The coefficient of restitution between the ball and the plane is e. Show that the time it takes for the ball to stop bouncing is $t = \sqrt{2h/g}(1+e)/(1-e)$.

7. Two spheres, made of the same material, the lower with radius $2b$ and upper of radius b, are dropped from a height h, as shown in Fig. 3.42. Assuming the centers of the spheres lie on a vertical line and all collisions between the spheres and the ground are elastic, find the maximum height the upper sphere reaches after impact. Assume that the lower sphere collides with the ground first and it then collides with the second sphere.

8. Four small spheres of equal size and weight are aligned in a straight line and are spaced equally (L), as shown in Fig. 3.43. Sphere A is given an initial velocity v along the line. The coefficient of restitution $e = 0.8$ is the same for all the spheres. Find the velocities of the spheres after 5 collisions have taken place, and the position of sphere A.

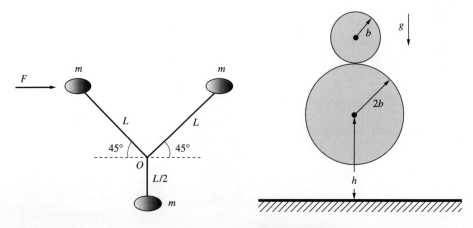

Figure 3.41

Figure 3.42

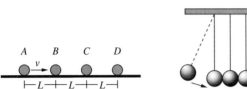
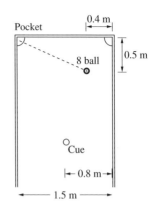

Figure 3.43   **Figure 3.44**   **Figure 3.45**

9. Figure 3.44 shows five pendulums of equal mass and length arranged so that they touch each other when they are at rest. If someone takes the pendulum at the very left, swings it, and lets it go, only one pendulum at the very right will swing out. If one takes N pendulums and swings them, N pendulums swing out. Explain why. Also, explain why, if two of the pendulums are glued together and then released, the above phenomenon is not observed.

10. In a pool game, a player must hit the cue ball and bounce it from a wall ($e = 0.9$), so that the cue ball hits the 8 ball head on and the 8 ball falls into the pocket, as illustrated in Fig. 3.45. Find the direction in which the player must hit the cue ball and location of cue ball on the table.

11. A mass m is attached to a massless rod. The mass is at rest in the position shown in Fig. 3.46 when it is hit by another body of mass $2m$ and velocity v ($e = 0.8$). Find the angular velocity of the rod immediately after impact.

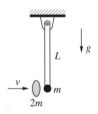

Figure 3.46

Section 3.6

12. Using the results of Example 3.5, calculate the altitude reached by a vertically fired rocket at burnout. Half of the rocket's weight is its propellant. Assume that $p_e A/b = 0.1$ $v_{\text{rel}} = $ constant.

13. A rocket is fired vertically such that dm/dt and v_{rel} are both constant. Derive expressions for the velocity and position as functions of time given that initial value of the acceleration is zero.

14. A bucket of mass 1.5 kg is filled with 3 kg of water and is being pulled up a well. The tension T in the rope is kept constant at $T = 30$ N. The bucket has a hole in it which lets the water leak out at a steady rate, and the bucket becomes empty in 38 seconds. Find the speed of the bucket at this instant.

15. A spacecraft of mass m_0 and cross-sectional area A is moving with speed v_0, as it encounters a dust cloud, as shown in Fig. 3.47. The dust cloud has density of ρ. As the spacecraft moves inside the cloud, dust begins to stick on its cross section. Derive an expression for the speed of the spacecraft as a function of time. Assume the dust offers no resistance to the spacecraft.

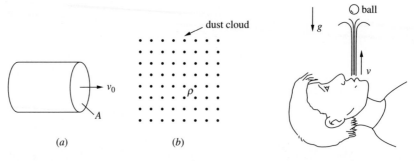

Figure 3.47   Figure 3.48

16. A performer wants to keep a ping pong ball afloat by blowing on it, as shown in Fig. 3.48. Given is that when he blows on the ball, the opening of his mouth is of area A, the speed is v, and the density of the air is ρ. Find the largest mass of the ball that he can keep afloat.

17. The cart in Fig. 3.49 is of mass 1250 kg and is moved by the action of a water jet. The water jet is of radius 0.08 m, and it squirts water with the speed of 200 m/s. Find the speed of the cart 3 seconds after the water jet is turned on. The cart has a coefficient of friction of $\mu = 0.08$. *Hint:* First develop an expression for the force acting on the cart as a function of the flow rate.

Section 3.7

18. Calculate the gravitational potential energy of a small mass inside a uniform sphere of mass m and radius R (Fig. 3.50). The distance of the mass from the center of the sphere is h, with $h < R$. *Hint:* Consider the sphere as consisting of an infinite number of thin spherical shells and calculate the potential energy of each shell.

19. A tube of length 500 km is dug inside the earth from one city to another, as shown in Fig. 3.51 (not drawn to scale). The inside of the tube is frictionless. A mass is released from rest from one end of the tube. Find the maximum speed the mass attains in the tube.

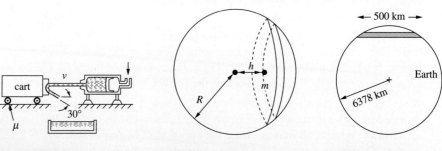

Figure 3.49   Figure 3.50   Figure 3.51

Section 3.8

20. Calculate the period of the earth's rotation around the sun using Kepler's third law and Eq. [3.8.25] and compare with the value given in Chapter 2.

21. Calculate the altitude of a satellite in geosynchronous orbit (geo) about the equator. Then, calculate how much time it takes for an electrical signal to travel between the equator and the satellite. Determine the minimum number of satellites in geo required to make it possible for any two people on earth to have a telephone conversation via satellite.

22. A satellite is launched into earth orbit. Burnout is at the perigee at an altitude of $2.2(10^5)$ m, at which point the satellite is parallel to the earth's surface and has a speed of $v = 9000$ m/s. Find the properties of the orbit.

23. A satellite of mass 800 kg is to be placed at circular orbit around the earth of altitude 400 km. Find the energy required, as well as the period of the orbit.

24. A spacecraft is in elliptic orbit with $\varepsilon = 0.5$ and $r_p = 2R$, where R is the radius of the earth. It is desired to change this orbit into a circular one with radius r_a. Find the impulse required to accomplish this maneuver. Then, calculate the properties of the resulting orbit if the impulse is applied incorrectly at $\theta = 185°$.

25. Calculate the speed and flight path angle a rocket must have at burnout at altitude 200 km so that it can achieve an orbit with $r_p = 5400$ km and $\varepsilon = 0.7$.

26. A meteorite is at a circular orbit around the earth with an altitude of 3500 km. The meteorite collides with another meteorite, and in doing so loses 3 percent of its kinetic energy. Find the properties of the orbit after this collision.

Section 3.9

27. A satellite is in orbit and its position and velocity at a certain instant are measured to be $\mathbf{r} = 3000\mathbf{I} + 2000\mathbf{J} + 6500\mathbf{K}$ km and $\mathbf{v} = 5222\mathbf{I} - 4000\mathbf{J} + 6000\mathbf{K}$ m/s where the origin is at a focus. Find the orbital parameters associated with this orbit.

28. A satellite is in an elliptic earth orbit with $\varepsilon = 0.6$ and $r_a = 8R$, where R is the radius of the earth. Calculate how long it takes for the satellite to go from $\theta = 15°$ to $\theta = 135°$. Compare your answer with the period of the orbit.

Section 3.10

29. Consider the mechanism shown in Fig. 3.52, consisting of a rod of mass m and length L and disk of mass m and radius $R = L/4$. The mechanism is held in place with the pin joint at O and the string. Suddenly, the string breaks. Find the angular acceleration of the rod at that instant if
    a. The disk is welded to the rod.
    b. The disk is attached to the rod with a pin joint.

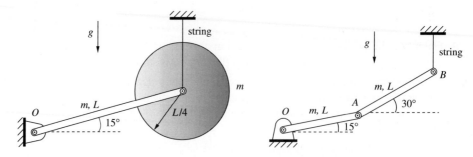

Figure 3.52 **Figure 3.53**

30. The two bars of equal length and mass are connected by pin joints and they are supported by a string at point B, as shown in Fig. 3.53. Find the angular acceleration of each bar when the string breaks.

31. Figure 3.54 shows the schematic of a car. Explain why the body of the car rotates counterclockwise when the car is accelerated.

32. The disk in Fig. 3.55 rotates with constant angular velocity $\omega = 5$ rad/s. The pin attached to it moves in the slotted bar OA. Bar OA has a mass of 10 kg and centroidal mass moment of inertia of 1.4 kg m^2. Determine magnitude of the force exerted by the pin on rod OA when rod OA makes an angle of 10° with the horizontal. The coefficient of friction between the pin and the slot is $\mu = 0.15$.

Section 3.11

33. A bar of mass m and length L (Fig. 3.56) is connected to a disk of mass $2m$ and radius $R = L/2$. The assembly is released from rest with $\theta = 30°$. Given that friction between the disk and the surface is sufficient to prevent slipping, find the angular acceleration of the disk at this instant.

34. Consider Fig. 3.53. Let the string be broken and the following parameters apply: $L = 0.6$ m, $v_B = 0.1$ m/s downwards. Find the instant center and angular velocity of the link AB.

35. The arm AOC rotates with angular velocity 3 rad/s ccw, with point O stationary, as shown in Fig. 3.57. Gears A, B, and C are of the same radius R. Use instant centers to find the angular velocity of gear B if (a) gear D is fixed, and (b) gear D is not fixed but is rotating clockwise with $\omega_D = 2.5$ rad/s.

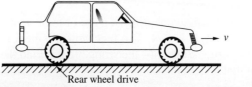

Figure 3.54

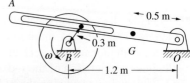

Figure 3.55

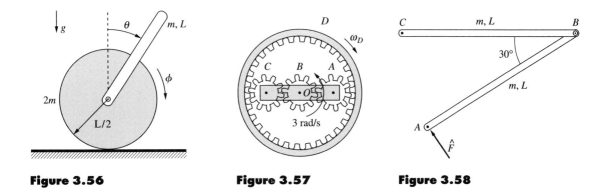

Figure 3.56 **Figure 3.57** **Figure 3.58**

36. Consider Example 3.11. Find the minimum value of the coefficient of friction that will prevent slipping as a function of θ.

SECTION 3.12

37. Consider the bar attached to a disk in Problem 35, which is released from rest at $\theta = 30°$. Find the velocity of the center of the disk when the rod becomes horizontal.

38. Two rods of equal length and mass are connected by a pin joint and they are at rest, as shown in Fig. 3.58. An impulsive force $\hat{F}$ is applied at point A perpendicular to the line AB. Find the angular velocities of the rods immediately after the impulse.

39. Consider Example 3.12, and assume that friction is sufficient to prevent slip at all times.
    a. If the spheres involved were of different diameter, would the angle at which the spheres lose contact be different than in Example 3.12?
    b. Consider now that the cylinder is rolling over a cylinder. How would the final results change if (*a*) the sphere and cylinder have the same radius and (*b*) the sphere and cylinder have different radii?

chapter 4

ANALYTICAL MECHANICS: BASIC CONCEPTS

4.1 INTRODUCTION

This chapter and Chapter 5 introduce analytical techniques for describing the motion of dynamical systems. The dynamical system is considered as a whole and scalar quantities such as energy and work are used. Constraint forces and moments are treated differently than in Newtonian mechanics. Constraint forces that do no work do not appear in the formulation, and they are accounted for by appropriately selecting the variables used to describe the motion. Sometimes, one may need to find out the magnitudes of the constraint forces. This can be accomplished by calculating the magnitudes of the constraint forces after the problem is solved, or by leaving the constraints in the system formulation by means of Lagrange multipliers. The approaches described in this chapter are analytical approaches and they are based on the principles of variational calculus. Appendix B provides a more detailed look at variational principles. Generalized coordinates, which do not necessarily have to be physical coordinates, are used as motion variables. This makes the analytical approach more flexible than the Newtonian, as the Newtonian approach is implemented using physical coordinates.

We derive the analytical equations of motion in this chapter for particles and for plane motion of rigid bodies, though these equations are valid for three-dimensional rigid body motion and deformable bodies as well. Chapter 8 will deal with D'Alembert's principle and Lagrange's equations for the general three-dimensional motion of rigid bodies.

One question often asked is whether it is more convenient to use a Newtonian technique or an analytical one when obtaining the equations of motion. There is no set answer to this question, with the possible exception of dynamical systems consisting of several interconnected components. When the number of coordinates needed to describe the system is much less than the number of components, it is usually preferable to use analytical techniques. When amplitudes of reaction forces are sought, it is usually better to use a Newtonian analysis. Looking at the problem from *both* the Newtonian and analytical points of view gives one more insight and a better understanding.

Analytical techniques use scalar functions like work and energy in the formulation, rather than vector quantities. While this approach makes a lot of sense, the experiences of dynamicists in recent years have shown that vector approaches combined with analytical techniques are more desirable when modeling complex systems. One advantage of a vector approach is that it can be implemented on a digital computer more readily.

4.2 GENERALIZED COORDINATES

A system of N particles requires $3N$ physical coordinates to specify the system's position. Consider an inertial coordinate system and let the vector $\mathbf{r}_i = \mathbf{r}_i(x_i, y_i, z_i)$ be the mapping of the ith particle in this coordinate system.[1] We express $\mathbf{r}_i$ as (Fig. 4.1)

$$\mathbf{r}_i = x_i\mathbf{i} + y_i\mathbf{j} + z_i\mathbf{k} \qquad i = 1, 2, \ldots, N \qquad \text{[4.2.1]}$$

The $3N$ coordinates required to represent the system span a $3N$-dimensional space, which is called the *configuration space* of dimension $n = 3N$. In many cases, as we will soon see, it is more advantageous to use a different set of variables than the physical coordinates to describe the motion. This approach is analogous to that of using different coordinate systems that we saw in Chapter 1. We introduce a set of variables $q_1, q_2, \ldots, q_n$, related to the physical coordinates by

$$x_1 = x_1(q_1, q_2, \ldots, q_n)$$
$$y_1 = y_1(q_1, q_2, \ldots, q_n)$$
$$z_1 = z_1(q_1, q_2, \ldots, q_n)$$
$$x_2 = x_2(q_1, q_2, \ldots, q_n)$$
$$\vdots$$
$$z_n = z_n(q_1, q_2, \ldots, q_n) \qquad \text{[4.2.2]}$$

We will refer to a set of variables that can completely describe the position of a dynamical system as *generalized coordinates*. The space spanned by the generalized coordinates is the *configuration space*. As an illustration, consider the spherical pendulum in Fig. 4.2, whose length can change. The motion of the pendulum can be described by the Cartesian coordinates x, y, and z, or by q_1, q_2, and q_3, where $q_1 = L$ describes the length of the pendulum, and $q_2 = \theta$ and $q_3 = \phi$ describe the angular

---

[1] If a noninertial coordinate system is used, one has to include the variables describing the motion of the reference frame in the set of coordinates that describe the motion, unless the characteristics of the reference frame are treated as known.

4.2 GENERALIZED COORDINATES

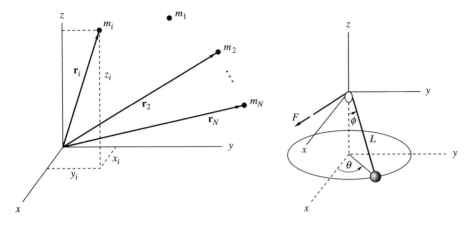

Figure 4.1   A system of N particles

Figure 4.2   A spherical pendulum whose length changes

displacement. The choice of L, θ, and ϕ as generalized coordinates is equivalent to using spherical coordinates. The two sets of coordinates are related by

$$x = q_1 \cos q_2 \sin q_3 = L \cos \theta \sin \phi \qquad y = q_1 \sin q_2 \sin q_3 = L \sin \theta \sin \phi$$

$$z = -q_1 \cos q_3 = -L \cos \phi \qquad [4.2.3]$$

If the length of the pendulum is constant, $q_1 = L = $ constant, we do not need to use it as a variable; $q_2 = \theta$ and $q_3 = \phi$ are sufficient. If we use the coordinates x, y, and z to describe the motion, we have to relate them employing the *constraint* relation

$$x^2 + y^2 + z^2 = L^2 = \text{constant} \qquad [4.2.4]$$

Constraint relations, such as the one in this equation, indicate that the generalized coordinates are related to each other, and that the system can be analyzed by a smaller number of coordinates. The double link in Fig. 4.3, where the lengths of the rods are constant, requires at least two generalized coordinates to describe the configuration of the two rods. One can conveniently select them as the angles θ_1 and θ_2.

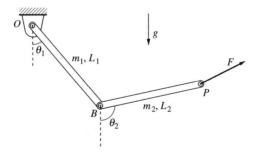

Figure 4.3   A double link

We hence need to distinguish between sets of generalized coordinates where each coordinate is independent of the others and where these variables are not independent.[2] In general, if a system of N particles has m constraint equations acting on it, we can describe the system uniquely by p independent generalized coordinates q_k, $(k = 1, 2, \ldots, p)$, where

$$p = 3N - m = n - m \qquad [4.2.5]$$

in which p is called the number of degrees of freedom of the system. The term *degree of freedom* can be defined as the minimum number of *independent* coordinates necessary to describe a system *uniquely*. Sets of generalized coordinates where each coordinate is not independent of the others are called *constrained generalized coordinates* or *dependent generalized coordinates*. The number of degrees of freedom is a characteristic of the dynamical system and is independent of the coordinates used to describe the motion. While one can select the number and types of generalized coordinates and associated constraints in more than one way, $p = n - m$ is invariant.

The rate of change of a generalized coordinate with respect to time is called the *generalized velocity* and is denoted by $\dot{q}_k(t)$ $(k = 1, 2, \ldots, n)$. The $2n$-dimensional space spanned by the generalized coordinates and generalized velocities is called the *state space*.

For the pendulum in Fig. 4.2 we generated two sets of generalized coordinates. We could select other sets of generalized coordinates as well. For example, we could select the generalized coordinates as L, ϕ, and x. However, this would introduce some ambiguity into the description of the pendulum, as x has the same value when the angle θ is positive or negative. Such coordinates are known as *ambiguous generalized coordinates*. Another example of ambiguous generalized coordinates would be to use the coordinates x_P and y_P of the endpoint P of the double link in Fig. 4.3. One can easily show that a given coordinate of the endpoint can be reached by two different configurations of the links, the two being mirror images about the line joining points O and P, as shown in Fig. 4.4.

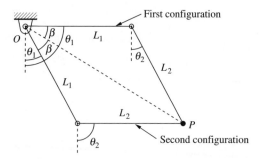

Figure 4.4

---

[2]In this regard, the definition of *generalized coordinate* here is slightly different than the traditional definition in older texts, which often restrict the term's meaning to only an independent set.

We draw two conclusions from the above. First, the generalized coordinates, whether they are independent or not, do not constitute a unique set. This actually is a tremendous advantage, as it gives a lot of flexibility. Second, one must exercise care when selecting generalized coordinates, especially independent generalized coordinates, to avoid redundancies and ambiguities. A poor choice of generalized coordinates can make the problem formulation and solution unnecessarily difficult.

The discussion here with regards to generalized coordinates is similar to the analysis of coordinate systems in Chapter 1. When we go from Cartesian to cylindrical or spherical coordinates, all we are doing is going from one set of generalized coordinates to another. We choose the coordinate system so that it simplifies the formulation.

4.3 CONSTRAINTS

In this section we analyze constraints that act on dynamical systems. We describe the constraints in terms of physical as well as generalized coordinates. The interest is primarily in equality constraints.

In dynamical systems, constraints are usually encountered as a result of contact between two (or more) bodies. Constraints restrict the motion of the bodies on which they act. Associated with a constraint are a *constraint equation* and a *constraint force*. The constraint equation describes the geometry and/or kinematics of the contact. The constraint force is the contact force, also called the *reaction*. (Constraint equations can also be written when the motion is viewed from a moving reference frame and there is no contact. The relative motion equation becomes the constraint equation.)

Consider Fig. 4.5 and a particle moving on a smooth surface whose shape is described by

$$f(x, y, z, t) = 0 \qquad [4.3.1]$$

where f has continuous second derivatives in all its variables. The motion of the particle over the surface can be viewed as the motion of an otherwise free particle subjected to the constraint of moving on that particular surface. Hence, $f(x, y, z, t) = 0$ represents a constraint equation. The constraint equation [4.3.1] is referred to as a *configuration constraint*. For a system described in terms of n generalized

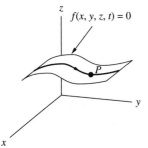

Figure 4.5   A particle moving on a smooth surface

coordinates, we can express a configuration constraint as

$$f(q_1, q_2, \ldots, q_n, t) = 0 \qquad [4.3.2]$$

The differential of the constraint f (in terms of physical and generalized coordinates) is

$$df = \frac{\partial f}{\partial x}dx + \frac{\partial f}{\partial y}dy + \frac{\partial f}{\partial z}dz + \frac{\partial f}{\partial t}dt = 0 \qquad [4.3.3a]$$

$$df = \frac{\partial f}{\partial q_1}dq_1 + \frac{\partial f}{\partial q_2}dq_2 + \cdots + \frac{\partial f}{\partial q_n}dq_n + \frac{\partial f}{\partial t}dt = 0 \qquad [4.3.3b]$$

The expressions [4.3.3] are said to be *constraint relations in Pfaffian form*. (A constraint in Pfaffian form is one that is expressed in the form of differentials.) Dividing these equations by dt, we write the *constraint equations in velocity form* (also called *velocity constraints* or *motion constraints*) as

$$\frac{df}{dt} = \frac{\partial f}{\partial x}\dot{x} + \frac{\partial f}{\partial y}\dot{y} + \frac{\partial f}{\partial z}\dot{z} + \frac{\partial f}{\partial t} = 0 \qquad [4.3.4a]$$

$$\frac{df}{dt} = \frac{\partial f}{\partial q_1}\dot{q}_1 + \frac{\partial f}{\partial q_2}\dot{q}_2 + \ldots + \frac{\partial f}{\partial q_n}\dot{q}_n + \frac{\partial f}{\partial t} = 0 \qquad [4.3.4b]$$

The general form of a velocity constraint can be written in terms of physical coordinates as

$$a_x \dot{x} + a_y \dot{y} + a_z \dot{z} + a_0 = 0 \qquad [4.3.5]$$

and, in terms of a system with n generalized coordinates subjected to m constraints,

$$\sum_{k=1}^{n} a_{jk}\dot{q}_k + a_{j0} = 0 \qquad j = 1, 2, \ldots, m \qquad [4.3.6]$$

where a_x, a_y, a_z, a_0, and a_{jk} and a_{j0} ($j = 1, 2, \ldots, m; k = 1, 2, \ldots, n$) are functions of the generalized coordinates and time, for example, $a_{jk} = a_{jk}(q_1, q_2, \ldots, q_n, t)$. Note that once the constraints are imposed to a set of independent generalized coordinates, these coordinates are no longer independent.

A constraint that can be expressed as both a configuration constraint as well as in velocity form is called *holonomic*. Constraints that do not have this property are called *nonholonomic*. In other words, nonholonomic constraints cannot be expressed as configuration constraints.

4.3.1 HOLONOMIC CONSTRAINTS

An unconstrained dynamical system or one subjected to a holonomic constraint that is not an explicit function of time, for example, $f_j(q_1, q_2, \ldots, q_n) = 0$, is called a *scleronomic* system. If the holonomic constraint is an explicit function of time, the system is called *rheonomic*. Throughout this text we will deal mostly with scleronomic systems, as they constitute the majority of situations encountered in engineering applications.

4.3 CONSTRAINTS

Consider the single particle discussed above and the case when the holonomic constraint f is not an explicit function of time. That is, the plane defined by the constraint is fixed. Elimination of the $\partial f/\partial t$ term from Eq. [4.3.4a] yields

$$\frac{df}{dt} = \frac{\partial f}{\partial x}\dot{x} + \frac{\partial f}{\partial y}\dot{y} + \frac{\partial f}{\partial z}\dot{z} = 0 \qquad [4.3.7]$$

Denote the position and velocity of the particle by $\mathbf{r}(t) = x(t)\mathbf{i} + y(t)\mathbf{j} + z(t)\mathbf{k}$ and $\mathbf{v}(t) = \dot{\mathbf{r}}(t) = \dot{x}(t)\mathbf{i} + \dot{y}(t)\mathbf{j} + \dot{z}(t)\mathbf{k}$. The gradient of the constraint is

$$\nabla f = \frac{\partial f}{\partial x}\mathbf{i} + \frac{\partial f}{\partial y}\mathbf{j} + \frac{\partial f}{\partial z}\mathbf{k} \qquad [4.3.8]$$

Taking the dot product between the gradient of the constraint and velocity $\mathbf{v}(t)$ gives

$$\nabla f \cdot \mathbf{v} = \frac{\partial f}{\partial x}\dot{x} + \frac{\partial f}{\partial y}\dot{y} + \frac{\partial f}{\partial z}\dot{z} \qquad [4.3.9]$$

which, when compared with Eq. [4.3.4a], yields

$$\nabla f \cdot \mathbf{v} = \frac{df}{dt} = 0 \qquad [4.3.10]$$

with the expected result that the particle velocity is always tangent to the surface.[3] The same relation can be derived for generalized coordinates.

Given the holonomic constraint of a particle moving on a surface, the question then arises as to what keeps the particle on the surface. The answer is a constraint force normal to the surface, as shown in Fig. 4.6. To every constraint relation corresponds a constraint force. Considering a single particle and denoting the constraint force by $\mathbf{F}'$, one can express it as

$$\mathbf{F}' = F'\mathbf{n} \qquad [4.3.11]$$

where $\mathbf{n}$ is a unit vector representing the direction perpendicular to the surface, usually referred to as the normal direction. (This direction is similar to the normal direction in normal-tangential coordinates, but here it can be taken as in either direction perpendicular to the surface.) Since $\mathbf{F}'$ is perpendicular to the surface, it must be perpendicular to the velocity. It follows from Eq. [4.3.9] that the unit vector $\mathbf{n}$, which is

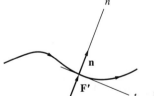

Figure 4.6   Constraint force for a holonomic constraint

---

[3] Recall the derivation in Chapter 1 when analyzing path variables that the particle velocity is always tangent to the path.

normal to the surface, should be parallel to ∇f. One can define **n** as

$$\mathbf{n} = \frac{\pm \nabla f}{|\nabla f|} = \pm \frac{\frac{\partial f}{\partial x}\mathbf{i} + \frac{\partial f}{\partial y}\mathbf{j} + \frac{\partial f}{\partial z}\mathbf{k}}{\left[\left(\frac{\partial f}{\partial x}\right)^2 + \left(\frac{\partial f}{\partial y}\right)^2 + \left(\frac{\partial f}{\partial z}\right)^2\right]^{1/2}} \quad [4.3.12]$$

Since the constraint force is expressed as

$$\mathbf{F}' = F'_x \mathbf{i} + F'_y \mathbf{j} + F'_z \mathbf{k} \quad [4.3.13]$$

when we compare Eqs. [4.3.12] and [4.3.13] we conclude that the components of the constraint force must be proportional to the partial derivatives of the constraint, or

$$\frac{F'_x}{\left(\frac{\partial f}{\partial x}\right)} = \frac{F'_y}{\left(\frac{\partial f}{\partial y}\right)} = \frac{F'_z}{\left(\frac{\partial f}{\partial z}\right)} \quad [4.3.14]$$

Now, consider the work done by the constraint force as the particle moves from position $\mathbf{r}$ to $\mathbf{r} + d\mathbf{r}$. Denoting this incremental work by dW and considering Eqs. [4.3.11] and [4.3.12], we obtain

$$dW = \mathbf{F}' \cdot d\mathbf{r} = F'\mathbf{n} \cdot d\mathbf{r} = \frac{F'}{|\nabla f|} \nabla f \cdot d\mathbf{r} = 0 \quad [4.3.15]$$

This relation indicates that *the work done by a holonomic constraint force which is independent of time in any possible displacement is zero.* Such constraints are referred to as *workless constraints.* This result is to be expected, because the constraint force is always perpendicular to the velocity.

Note that, while the total work done by the constraint forces that are independent of time is zero, the individual constraint forces are doing work themselves. This work is in the form of transferring energy from one component of the system to the other. The sum of the transferred energies is zero. To visualize this, consider the double link in Fig. 4.3, whose free-body diagram is given in Fig. 4.7. If the first link is given an initial motion, the second link will begin moving, and vice versa. The motion of the second link is initiated by the constraint forces acting at point B.

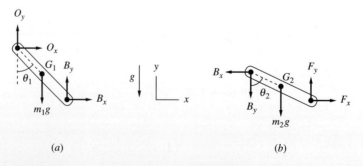

Figure 4.7   Free-body diagram of double link

Considering Fig. 4.7, reaction forces, such as the forces at the pin at O and at point B, are holonomic constraint forces. Normal forces are also holonomic constraint forces. However, friction forces are not constraint forces, even though their magnitude is directly dependent on a constraint force. Nevertheless, for static problems one can treat friction as a reaction force, because in such cases friction prevents motion.

Next consider a holonomic constraint that is an explicit function of time. For the particle considered earlier, this implies that the surface is moving and the constraint is in the form $f = f(x, y, z, t)$. Using Eqs. [4.3.4a] and [4.3.9] we obtain

$$\nabla f \cdot \mathbf{v} = \frac{\partial f}{\partial x}\dot{x} + \frac{\partial f}{\partial y}\dot{y} + \frac{\partial f}{\partial z}\dot{z} = -\frac{\partial f}{\partial t} \qquad [4.3.16]$$

which implies that $\nabla f \cdot d\mathbf{r} \neq 0$. It follows that the incremental work dW, which now is *not* a perfect differential as time is explicitly involved, is not zero. The incremental work has the form

$$dW = \mathbf{F}' \cdot d\mathbf{r} = F'\mathbf{n} \cdot d\mathbf{r} = \frac{F'}{|\nabla f|} \nabla f \cdot d\mathbf{r} \neq 0 \qquad [4.3.17]$$

When the holonomic constraint is time dependent, the work performed by the corresponding constraint force is not zero. The path followed by the particle can no longer be described by the path variables associated with the surface. The vector $\mathbf{n}$ describes the normal to the surface, but it is not the normal to the path followed by the particle.

4.3.2 NONHOLONOMIC CONSTRAINTS

When the constraint is nonholonomic, it can only be expressed in the form of Eqs. [4.3.5] or [4.3.6], as an integrating factor does not exist to permit expression in the form of Eqs. [4.3.1] or [4.3.2]. Consequently, none of the preceding results we obtained regarding the work done by the constraint force are valid for nonholonomic constraints. The constraint force associated with a nonholonomic constraint cannot be expressed as a force normal to a surface, as the nonholonomic constraint does not define a surface. One can go into the space spanned by $q_i(t)$ and $\dot{q}_i(t)$ ($i = 1, 2, \ldots, n$) and define a surface there, but this does not give any physical insight or significant results. Hence, there is no general expression for the constraint force when the constraint is nonholonomic.

A common example of a nonholonomic constraint is the rolling without slipping of a body with no sharp corners or edges, such as a disk or a sphere.

In general, constraint equations in terms of relative velocities and especially those involving angular velocities that are not "simple" turn out to be nonholonomic. Recall the discussion of angular velocity in Chapter 2. When a reference frame is described by successive rotations about nonparallel axes, the resulting angular velocity cannot be described as the derivative of a vector.

Other examples of nonholonomic systems are from vehicle dynamics. Included in this category are the motions of ships, missiles, airplanes, automobiles, wheelbarrows, shopping carts, and sleds. Figure 4.8 is a simplified illustration of such a vehicle undergoing plane motion, such as a sled. Vehicles usually have a plane

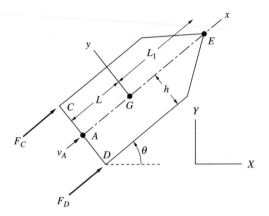

Figure 4.8 Generic model of a vehicle

of symmetry, and they are propelled in a way that the guiding forces act primarily along the symmetry plane, with a very small component of the force used to change direction. A steering mechanism usually accomplishes the change in direction.

One then makes the assumption that there is a point along the plane of symmetry, denoted by A, such that the velocity of point A is always along the plane of symmetry. The location of this point depends on the vehicle and the types of forces that prevent point A from having a velocity component perpendicular to the plane of symmetry. In a tricycle or automobile, the point A is in the middle between the rear wheels. In a boat, the hydrodynamic forces determine the location of A.

Consider the vehicle in Fig. 4.8. The configuration of this system can be described by the coordinates of point A, X_A and Y_A, and by the angle the body makes with the inertial X axis, denoted by θ. The nonholonomic constraint is associated with the translational velocity of point A. Denoting this velocity by $\mathbf{v}_A$, we write it as

$$\mathbf{v}_A = \dot{X}_A \mathbf{I} + \dot{Y}_A \mathbf{J} \qquad [4.3.18]$$

The constraint is written as

$$\mathbf{v}_A \cdot \mathbf{j} = 0 \qquad [4.3.19]$$

where $\mathbf{j} = \cos\theta \mathbf{J} - \sin\theta \mathbf{I}$. Introducing Eq. [4.3.19] into Eq. [4.3.18], we obtain

$$\mathbf{v}_A \cdot \mathbf{j} = (\dot{X}_A \mathbf{I} + \dot{Y}_A \mathbf{J}) \cdot (-\sin\theta \mathbf{I} + \cos\theta \mathbf{J})$$
$$= -\dot{X}_A \sin\theta + \dot{Y}_A \cos\theta = 0 \qquad [4.3.20]$$

This equation can conveniently be expressed as

$$\dot{X}_A - \frac{\dot{Y}_A}{\tan\theta} = 0 \qquad [4.3.21]$$

It is clear that this constraint is nonholonomic. The associated constraint force is basically the resistance of point A to have any motion perpendicular to the line of motion. In an automobile, for example, this force would be the friction force between the rear tires and the road surface in the direction perpendicular to the velocity of the tires. A very strong wind in the lateral direction, collision with another vehicle, or taking a turn with high speed would violate this constraint.

In general, the constraint force associated with a nonholonomic constraint performs work. A special case when this is not valid is rolling without slipping, where the friction force is applied to a point with zero velocity. For roll without slip, friction becomes a constraint force, as it reduces the number of degrees of freedom.

We next look into determining whether a constraint is holonomic or not. In general, whether it is or is not can be ascertained by visual inspection. Mathematically, in order for a constraint in Pfaffian or velocity form to be integrable to configuration form, the constraint relation must satisfy differentiability conditions. The constraint must represent an exact differential. Consider Eq. [4.3.6]. If the jth constraint equation is holonomic, one should be able to write it as $f_j(q_1, q_2, \ldots, q_n, t) = 0$. Taking the differential of f_j and, for the most general case, dividing it by an integrating factor $g_j(q_1, q_2, \ldots, q_n)$ we obtain Eq. [4.3.3b]. Comparing Eq. [4.3.4b] with Eq. [4.3.6], we obtain for the general case of a holonomic constraint

$$\frac{\partial f_j}{\partial q_k} = g_j a_{jk} \qquad \frac{\partial f_j}{\partial t} = g_j a_{j0} \qquad k = 1, 2, \ldots, n \qquad [4.3.22]$$

For a constraint given by Eq. [4.3.6] to be holonomic, there must be a function f_j and an integrating factor $g_j(q_1, q_2, \ldots, q_n)$ ($j = 1, 2, \ldots, m$) where the partial derivatives of f_j ($j = 1, 2, \ldots, m$) satisfy Eq. [4.3.22]. To check this, we evaluate the second derivatives of f_j. Indeed, considering an index r, we obtain

$$\frac{\partial^2 f_j}{\partial q_k \partial q_r} = \frac{\partial}{\partial q_r}(g_j a_{jk}) \quad \text{and} \quad \frac{\partial^2 f_j}{\partial q_k \partial q_r} = \frac{\partial}{\partial q_k}(g_j a_{jr}) \qquad [4.3.23]$$

$$\frac{\partial^2 f_j}{\partial q_r \partial t} = \frac{\partial}{\partial q_r}(g_j a_{j0}) \quad \text{and} \quad \frac{\partial^2 f_j}{\partial q_r \partial t} = \frac{\partial}{\partial t}(g_j a_{jr})$$

$$k, r = 1, \ldots, n; \, j = 1, 2, \ldots, m \qquad [4.3.24]$$

From Eqs. [4.3.23] and [4.3.24] if an integrating factor g_j exists such that a_{jk} and a_{j0} satisfy the relations

$$\frac{\partial}{\partial q_r}(g_j a_{jk}) = \frac{\partial}{\partial q_k}(g_j a_{jr}) \qquad \frac{\partial}{\partial q_r}(g_j a_{j0}) = \frac{\partial}{\partial t}(g_j a_{jr})$$

$$r = 1, \ldots, n; \, j = 1, 2, \ldots, m \qquad [4.3.25]$$

then the constraint is holonomic. The problem with using the above procedure is that it may not be easy to find the integrating factor, especially for systems having more than three degrees of freedom.

A constraint of the form $f(q_1, q_2, \ldots, q_n, t) \geq 0$, or $\sum a_k \dot{q}_k + a_0 \geq 0$, that is, an inequality constraint, is nonholonomic because it cannot be reduced to a form $f(q_1, q_2, \ldots, q_n, t) = 0$. Such constraints require a different treatment than equality constraints. We also encounter constraints that are valid in some positions of the body or during certain intervals of the motion. Such constraints can also be classified as inequality constraints. They can be found in problems involving contact.

Consider now a system that originally has n degrees of freedom and is subjected to m holonomic constraints. Introduction of m constraints reduces the degrees of freedom by m to $p = n - m$, resulting in a set of m excess, or surplus, coordinates.

It is possible, at least mathematically, to eliminate the surplus coordinates from the formulation, which results in an unconstrained system of order $n - m$. Because of this, unconstrained systems are referred to as holonomic.

By contrast, a nonholonomic constraint constrains only the generalized velocities, without affecting the generalized coordinates. In such systems there are n independent generalized coordinates and $n - m$ independent generalized velocities.

Example 4.1

A bead is sliding in a tube, whose shape is given by the equation $y = 1 - x^2$, as shown in Fig. 4.9. Find the direction of the normal to the tube.

Solution

One can solve this problem in a number of ways. We first consider the problem from a physical standpoint. Because the bead is sliding in the tube, the equation defining the shape of the tube becomes the constraint equation, and it has the form

$$f(x, y) = y - 1 + x^2 = 0 \quad \text{[a]}$$

Taking the partial derivatives of f, we obtain

$$\frac{\partial f}{\partial x} = 2x \qquad \frac{\partial f}{\partial y} = 1 \quad \text{[b]}$$

so that, using Eq. [4.3.8], the gradient of f has the form $\nabla f = 2x\mathbf{i} + \mathbf{j}$. From Eq. [4.3.12], the unit vector in the normal direction (chosen, for convenience, positive outward) becomes

$$\mathbf{n} = \frac{2x\mathbf{i} + \mathbf{j}}{\sqrt{1 + 4x^2}} \quad \text{[c]}$$

As expected, because the constraint is not an explicit function of time, neither is the direction of the constraint force. The constraint is, of course, holonomic.

To solve this problem geometrically, we define the angle θ between the horizontal and the tangent to the curve. The tangent of $\theta(x)$ describes the slope of the tube, and

$$\tan \theta = \frac{dy}{dx} = -2x \quad \text{[d]}$$

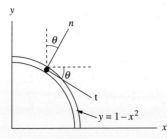

Figure 4.9 Bead sliding inside a tube

We find the sine and cosine of θ by

$$\sin\theta = \frac{2x}{\sqrt{1+4x^2}} \qquad \cos\theta = -\frac{1}{\sqrt{1+4x^2}} \qquad \text{[e]}$$

Now we can express the unit vector describing the normal as

$$\mathbf{n} = \sin\theta\mathbf{i} - \cos\theta\mathbf{j} = \frac{2x\mathbf{i} + \mathbf{j}}{\sqrt{1+4x^2}} \qquad \text{[f]}$$

We can also determine the normal direction directly from the geometry, using the approach in Chapter 1, without going into any constraint equations. Denoting the path variable by x, we write the position vector as

$$\mathbf{r} = x\mathbf{i} + (1 - x^2)\mathbf{j} \qquad \text{[g]}$$

and the expressions for the slope and the unit vector in the tangential direction become

$$\mathbf{r}' = \frac{d\mathbf{r}}{dx} = \mathbf{i} - 2x\mathbf{j} \qquad \mathbf{e}_t = \frac{\mathbf{i} - 2x\mathbf{j}}{\sqrt{1+4x^2}} \qquad s' = \sqrt{1+4x^2} \qquad \text{[h]}$$

Use of Eq. [1.3.36] yields $\mathbf{n}$. When the path parameters associated with the motion of a body are specified, in essence a constraint has been imposed on an otherwise free body.

Example 4.2

A block of mass m is attached to a cord of original length L and is rotating about a thin hub, as shown in Fig. 4.10. Friction is negligible. Find the constraint force if (*a*) the cord is not wrapping around the hub, and (*b*) the cord is wrapping around the hub.

Solution

a. When the cord is not wrapping around the hub, the constraint is holonomic and independent of time. The constraint equation basically describes that the length of the cord is constant, and it has the form

$$f(x, y) = x^2 + y^2 - L^2 = 0 \qquad \text{[a]}$$

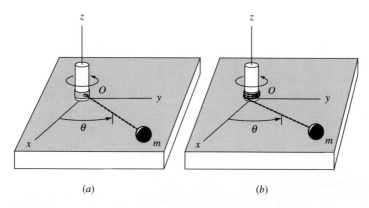

Figure 4.10   Mass rotating around a thin hub (a) Cord is not wrapping around hub (b) Cord is wrapping around hub

Once motion is initiated, the mass keeps rotating with the same speed and the energy of the particle does not change. The constraint force is the tension in the rope, and it does no work.

b. The situation is quite different when the rope wraps around the hub. Assuming that the hub radius is very small, the tension in the rope is directed toward point O. Summing moments about O, we obtain

$$\sum M_O = 0 \qquad [\mathbf{b}]$$

so that the angular momentum about O is conserved. In essence, we have a central force problem. Let us use polar coordinates r and θ. Consider that the length of the rope, denoted by r, reduces continuously by the relation

$$r = L - r_o \theta \qquad [\mathbf{c}]$$

where r_o is the radius of the hub. In one revolution of the mass, the rope shortens by $2\pi r_o$.

The angular momentum about O is given by $H_O = mr^2 \dot\theta$. Because the angular momentum is conserved,

$$r^2 \dot\theta = \text{constant} = h \qquad [\mathbf{d}]$$

where we note that the constant h is always greater than zero, $h > 0$, and that h is a function of the initial condition. Differentiating the relation between r and θ, we write

$$\dot r = -r_o \dot\theta \rightarrow \dot\theta = -\frac{\dot r}{r_o} \qquad [\mathbf{e}]$$

and substituting the above relation into Eq. [d], we obtain

$$r^2 \dot\theta = \frac{-r^2 \dot r}{r_o} = h \qquad [\mathbf{f}]$$

or

$$r^2 \dot r = -r_o r^2 \dot\theta = -r_o h = C \qquad C < 0 \qquad [\mathbf{g}]$$

where C is constant. Now, let us find the response $r(t)$. We can rewrite Eq. [g] as

$$r^2 \, dr = C \, dt \qquad [\mathbf{h}]$$

which, when integrated, gives

$$\frac{r^3}{3} = Ct + D \qquad [\mathbf{i}]$$

where D is a constant of integration, determined from the initial conditions. We note that at $t = 0$, $r = L$, and from Eq. [i] $r^3/3 = D$, so that $D = L^3/3$. Considering that the length of the rope is related to x and y by $r^2 = x^2 + y^2$, we can write Eq. [i] as

$$f(x, y, t) = \frac{(x^2 + y^2)^{3/2}}{3} - \frac{L^3}{3} - Ct = 0 \qquad [\mathbf{j}]$$

The constraint is a time-dependent holonomic constraint, that is, a rheonomic constraint. The constraint force, which is the tension in the rope, does perform work. To show that the constraint force does indeed perform work, we consider the configuration vector $\mathbf{r}$ and its derivative

$$\mathbf{r} = r \mathbf{e}_r \qquad \dot{\mathbf{r}} = \dot r \mathbf{e}_r + r \dot\theta \mathbf{e}_\theta \qquad [\mathbf{k}]$$

The constraint force (the tension in the rope) can be expressed as $\mathbf{F}' = -F\mathbf{e_r}$, so that the dot product between the constraint force and the particle velocity becomes

$$\mathbf{F}' \cdot \dot{\mathbf{r}} = -F\dot{r} \qquad \text{[I]}$$

which is not zero. Note that for the case when the length of the rope is not changing, $\dot{r} = 0$, and the work done by the constraint is zero. Also note that in order to find an explicit expression for $r(t)$, the initial angular velocity must be specified.

Example 4.3

Given a system with generalized coordinates q_1 and q_2 and the constraint equation

$$\left(3q_1 \sin q_2 + \frac{q_2^2}{q_1} + 2\right)dq_1 + (q_1^2 \cos q_2 + 2q_2)\,dq_2 = 0$$

determine whether the constraint is holonomic or not.

Solution

The constraint equation is holonomic if there exists an integrating factor $g(q_1, q_2)$, such that Eq. [4.3.22] holds, or

$$\frac{\partial f}{\partial q_1} = 3gq_1 \sin q_2 + \frac{gq_2^2}{q_1} + 2g \qquad \frac{\partial f}{\partial q_2} = gq_1^2 \cos q_2 + 2gq_2 \qquad \text{[a]}$$

We observe that if $g(q_1, q_2) = q_1$, then

$$\frac{\partial f}{\partial q_1} = 3q_1^2 \sin q_2 + q_2^2 + 2q_1 \qquad \frac{\partial f}{\partial q_2} = q_1^3 \cos q_2 + 2q_1 q_2 \qquad \text{[b]}$$

Integrating the two expressions, we obtain

$$f = q_1^3 \sin q_2 + q_1 q_2^2 + q_1^2 + h_1(q_2) + C_1 \qquad f = q_1^3 \sin q_2 + q_1 q_2^2 + h_2(q_1) + C_2 \quad \text{[c]}$$

where h_1 and h_2 are functions that appear as a result of the integration over q_1 and q_2, respectively, and C_1 and C_2 are constants. Comparing the two integrated terms, we conclude that $h_2(q_1) = q_1^2$ and $h_1(q_2) = 0$ and that the constants are related by $C_1 = C_2$. The constraint, therefore, is holonomic and has the form

$$f(q_1, q_2) = q_1^3 \sin q_2 + q_1 q_2^2 + q_1^2 + C = 0 \qquad \text{[d]}$$

where C is a constant. For this problem the integrating factor was found by visual inspection. In general, there are no set guidelines for finding the integrating factor.

Example 4.4

The tip of the double-link mechanism in Fig. 4.11 is constrained to lie on the inclined plane. Derive the constraint equation and express it in velocity form.

Solution

This is a single degree of freedom system. We use θ_1 and θ_2 as generalized coordinates. Hence, we need one constraint equation. We can simplify the formulation by expressing the position of the tip along the incline by the variable s. To derive the constraint equation we write the position vector of the tip in two ways: using the links and using the incline. Using the links, the position vector has the form

$$\mathbf{r}_P = (L_1 \cos \theta_1 + L_2 \cos \theta_2)\mathbf{i} + (L_1 \sin \theta_1 + L_2 \sin \theta_2)\mathbf{j} \qquad \text{[a]}$$

230 CHAPTER 4 • ANALYTICAL MECHANICS: BASIC CONCEPTS

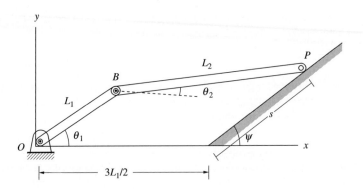

Figure 4.11

and using the incline, it has the form

$$\mathbf{r}_P = \frac{3L_1}{2}\mathbf{i} + s\cos\psi\,\mathbf{i} + s\sin\psi\,\mathbf{j} \qquad \text{[b]}$$

We equate the above two expressions and separate components in the x and y directions, thus

$$L_1 \cos\theta_1 + L_2 \cos\theta_2 = \frac{3L_1}{2} + s\cos\psi \qquad \text{[c]}$$

$$L_1 \sin\theta_1 + L_2 \sin\theta_2 = s\sin\psi \qquad \text{[d]}$$

To obtain the constraint equation, we eliminate s by multiplying Eq. [c] by $\sin\psi$ and Eq. [d] by $-\cos\psi$ and adding the two equations. Dividing the result by $L_1 \sin\psi$, we obtain

$$\cos\theta_1 + \frac{L_2}{L_1}\cos\theta_2 - \frac{1}{\tan\psi}\sin\theta_1 - \frac{L_2}{L_1 \tan\psi}\sin\theta_2 = \frac{3}{2} \qquad \text{[e]}$$

which is recognized as the holonomic constraint equation. To express this constraint in velocity form, we differentiate Eq. [e] with respect to time, with the result

$$\left(\sin\theta_1 + \frac{\cos\theta_1}{\tan\psi}\right)\dot\theta_1 + \frac{L_2}{L_1}\left(\sin\theta_2 + \frac{\cos\theta_2}{\tan\psi}\right)\dot\theta_2 = 0 \qquad \text{[f]}$$

4.4 VIRTUAL DISPLACEMENTS AND VIRTUAL WORK

At this point, we introduce the variational notation. The variational notation is ideally suited for dynamics problems because it makes the formulation concise, and it has a meaningful physical interpretation. When applied to dynamical systems, the variations of displacements are known as *virtual displacements,* denoted by δx, δy, δz, etc. In terms of generalized coordinates, the virtual displacements have the form $\delta q_1, \delta q_2, \ldots, \delta q_n$. The variations of the velocities are denoted by $\delta \dot x$, $\delta \dot y$, $\delta \dot z$ for physical coordinates and $\delta \dot q_k$ ($k = 1, 2, \ldots, n$) for generalized velocities.

Virtual displacements have the following properties:

- They are infinitesimal displacements.
- They are consistent with the system constraints, but are arbitrary otherwise.
- The variation of displacements (or velocities, etc.) is obtained by holding time fixed; therefore, virtual displacements can be considered as occurring instantaneously, and time is not involved in their applications.

Dealing with virtual displacements is like imagining the system in a different position that is physically realizable, while freezing time. It is as if a different set of forces were applied and, as a result, the system moved to another location by one of the admissible paths it can follow.

The rules for calculating virtual displacements are intimately related to the rules of differentiation. For the position vector $\mathbf{r} = x(t)\mathbf{i} + y(t)\mathbf{j} + z(t)\mathbf{k}$, or $\mathbf{r} = \mathbf{r}(q_1, q_2, \ldots, q_n, t)$, the variation of $\mathbf{r}$ becomes

$$\delta \mathbf{r} = \delta x \mathbf{i} + \delta y \mathbf{j} + \delta z \mathbf{k} \quad \text{or} \quad \delta \mathbf{r} = \frac{\partial \mathbf{r}}{\partial q_1} \delta q_1 + \frac{\partial \mathbf{r}}{\partial q_2} \delta q_2 + \ldots + \frac{\partial \mathbf{r}}{\partial q_n} \delta q_n$$

[4.4.1a,b]

Figure 4.12 depicts the concept of a variation (for the coordinate y). When expressing the motion $\mathbf{r} = x\mathbf{i} + y\mathbf{j} + z\mathbf{k}$ in which x, y, and z are all functions of the generalized coordinates, the variation of $\mathbf{r}$ has the form

$$\delta \mathbf{r} = \sum_{k=1}^{n} \left(\frac{\partial x}{\partial q_k} \mathbf{i} + \frac{\partial y}{\partial q_k} \mathbf{j} + \frac{\partial z}{\partial q_k} \mathbf{k} \right) \delta q_k \qquad [4.4.2]$$

As discussed in Appendix B, we distinguish between dependent and independent variables. For dynamical systems, time is the independent variable. The coordinates x, y, z, as well $q_1, q_2, \ldots, q_n$ are functions of time and are referred to as the dependent variables. The term *dependent* is used here to denote explicit dependence of the generalized coordinates on time, rather than on each other. It follows that one can interchange the time differentiation and the variation operators. That is, $\delta \dot{q}_k = \delta(dq_k/dt) = d(\delta q_k)/dt$ ($k = 1, 2, \ldots, n$).

The variation of a position vector can be obtained in two different ways. One way is by obtaining an analytical expression for the position vector and taking its variation

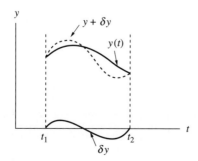

Figure 4.12   Variation of y

by differentiating with respect to the generalized coordinates. Basically this is the use of Eq. [4.4.1b]; it is known as the *analytical approach*. This approach may lead to lengthy expressions for certain complex problems. When **r** is expressed in terms of the coordinates of a moving reference frame, one must also take the variation of the unit vectors of the moving reference frame. The exception to this is when the motion of the relative frame is prespecified as a known quantity and is not treated as a motion variable.

In the second way, known as the *kinematical approach*, one explores similarities between velocities and virtual displacements. When taking the variation of an expression, the independent variable is not varied. We use this property, as time is the independent variable. The time derivative of **r** is

$$\dot{\mathbf{r}} = \dot{x}\mathbf{i} + \dot{y}\mathbf{j} + \dot{z}\mathbf{k} + \frac{\partial \mathbf{r}}{\partial t} \qquad \dot{\mathbf{r}} = \frac{\partial \mathbf{r}}{\partial q_1}\dot{q}_1 + \frac{\partial \mathbf{r}}{\partial q_2}\dot{q}_2 + \cdots + \frac{\partial \mathbf{r}}{\partial q_n}\dot{q}_n + \frac{\partial \mathbf{r}}{\partial t}$$

[4.4.3a,b]

Elimination of the partial derivative of **r** with respect to time, elimination of all expressions explicit in time, and replacement of $\dot{x}$ by δx, $\dot{y}$ by δy, $\dot{z}$ by δz in Eq. [4.4.3a] and of $\dot{q}_k$ ($k = 1, 2, \ldots, n$) with δq_k in Eq. [4.4.3b] yields the variation of **r**. This implies that if the expression for the velocity is known, the associated virtual displacement can be obtained directly from it. This approach of calculating virtual displacements from velocities is especially useful when the velocity of a point can be found using an instant center or a relative velocity expression, such as

$$\mathbf{v}_B = \mathbf{v}_A + \boldsymbol{\omega} \times \mathbf{r}_{B/A} + \mathbf{v}_{Brel} \qquad [4.4.4]$$

The variation of the displacement of point B is

$$\delta \mathbf{r}_B = \delta \mathbf{r}_A + \boldsymbol{\delta\theta} \times \mathbf{r}_{B/A} + \delta \mathbf{r}_{Brel} \qquad [4.4.5]$$

where we note that the $\mathbf{r}_{B/A}$ term is left intact and that $\boldsymbol{\delta\theta}$ represents the variation of an infinitesimal rotation. Also, keeping in line with the developments in Chapter 2, we extend the boldface to the entire term $\boldsymbol{\delta\theta}$ to denote that $\boldsymbol{\delta\theta}$ is a variation of a rotation and that it is not obtained by differentiating a vector.

Consider Eq. [4.4.3b] and the derivative of $\dot{\mathbf{r}}$ with respect to $\dot{q}_k$. Of all the terms in Eq. [4.4.3b] only one survives and we obtain the important relationship

$$\frac{\partial \dot{\mathbf{r}}}{\partial \dot{q}_k} = \frac{\partial \mathbf{r}}{\partial q_k} \qquad k = 1, 2, \ldots, n \qquad [4.4.6]$$

so that the variation of **r** can be expressed as

$$\delta \mathbf{r} = \sum_{k=1}^{n} \frac{\partial \dot{\mathbf{r}}}{\partial \dot{q}_k} \delta q_k \qquad [4.4.7]$$

Note that Eq. [4.4.7] is in essence the mathematical representation of the kinematical method of calculating virtual displacements. Next, consider the holonomic constraint $f(x, y, z, t) = 0$ and obtain its variation, which has the form

$$\delta f = \frac{\partial f}{\partial x}\delta x + \frac{\partial f}{\partial y}\delta y + \frac{\partial f}{\partial z}\delta z = 0 \qquad [4.4.8]$$

Because time is held fixed while f is varied, δf has the same form whether the constraint is time dependent or not. When a constraint is given in velocity form by Eqs. [4.3.5] and [4.3.6], in terms of physical coordinates the virtual displacements satisfy

$$a_x \delta x + a_y \delta y + a_z \delta z = 0 \qquad [4.4.9]$$

and, in terms of generalized coordinates and the jth constraint, they satisfy

$$\delta f_j = a_{j1} \delta q_1 + a_{j2} \delta q_2 + \cdots + a_{jn} \delta q_n \qquad j = 1, 2, \ldots, m \quad [4.4.10]$$

Let us next consider the work done by a force over a virtual displacement. Consider a body acted upon by a force $\mathbf{F}$ and the virtual displacement associated with the point at which the force $\mathbf{F}$ is applied. We define the work done by the force over the virtual displacement $\delta \mathbf{r}$ as the *virtual work* or *variation of work* and denote it by δW. Hence

$$\delta W = \mathbf{F} \cdot \delta \mathbf{r} \qquad [4.4.11]$$

We will examine the virtual work associated with a general force in the next section. For now, let us consider the holonomic constraint $f(x, y, z, t) = 0$ and the associated virtual work. Recall that whether the constraint is time dependent or not is immaterial. From Eqs. [4.3.11]–[4.3.14], the constraint force $\mathbf{F}'$ has the form

$$\mathbf{F}' = F'_x \mathbf{i} + F'_y \mathbf{j} + F'_z \mathbf{k} = \frac{F'}{|\nabla f|} \left(\frac{\partial f}{\partial x} \mathbf{i} + \frac{\partial f}{\partial y} \mathbf{j} + \frac{\partial f}{\partial z} \mathbf{k} \right) = \frac{F'}{|\nabla f|} \nabla f \quad [4.4.12]$$

We define by $\delta W'$ the work performed by the constraint force in any virtual displacement *virtual work due to constraint forces*, as

$$\delta W' = \mathbf{F}' \cdot \delta \mathbf{r} = F'_x \delta x + F'_y \delta y + F'_z \delta z \qquad [4.4.13]$$

Using Eqs. [4.4.1] and [4.4.8] we conclude that

$$\delta W' = \mathbf{F}' \cdot \delta \mathbf{r} = \frac{F'}{|\nabla f|} \nabla f \cdot \delta \mathbf{r} = \frac{F'}{|\nabla f|} \delta f = 0 \qquad [4.4.14]$$

so that *the work performed by a holonomic constraint force in any virtual displacement is zero.*

Example 4.5

A disk of radius R rolls without slipping on a rod of length L pivoted at one end, as shown in Fig. 4.13. Denoting the pivot angle by θ and the angular displacement of the disk by ϕ, find the virtual displacement of the center of the disk.

Solution

We will solve this problem using both a kinematical and an analytical approach. We begin with the kinematical approach. We select an inertial frame XYZ and a relative frame xyz, such that the xyz axes are obtained by rotating the XYZ frame by an angle θ counterclockwise about the Z axis.

The velocity of point G can be written as

$$\mathbf{v}_G = \mathbf{v}_B + \boldsymbol{\omega} \times \mathbf{r}_{G/B} + \mathbf{v}_{\text{rel}} \qquad [a]$$

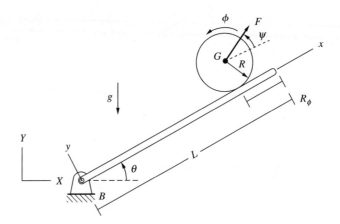

Figure 4.13 Disk rolling over bar

in which $\mathbf{v}_B = 0$, $\boldsymbol{\omega} = \dot{\theta}\mathbf{k}$, and

$$\mathbf{r}_{G/B} = \mathbf{r}_G = (L - R\phi)\mathbf{i} + R\mathbf{j} \qquad \mathbf{v}_{rel} = -R\dot{\phi}\mathbf{i} \qquad [\mathbf{b, c}]$$

Substituting the above values into Eq. [a] we obtain

$$\mathbf{v}_G = \dot{\theta}\mathbf{k} \times [(L - R\phi)\mathbf{i} + R\mathbf{j}] - R\dot{\phi}\mathbf{i} = -R(\dot{\phi} + \dot{\theta})\mathbf{i} + (L\dot{\theta} - R\phi\dot{\theta})\mathbf{j} \qquad [\mathbf{d}]$$

Thus, we write the variation of $\mathbf{r}_G$ as

$$\delta\mathbf{r}_G = -R(\delta\phi + \delta\theta)\mathbf{i} + (L\,\delta\theta - R\phi\,\delta\theta)\mathbf{j} \qquad [\mathbf{e}]$$

Now we will find the variation of $\mathbf{r}_G$ analytically. The position vector $\mathbf{r}_G$ is given in Eq. [b]. There are two ways to obtain its variation. In the first, we express $\mathbf{r}_G$ in terms of the inertial coordinate frame and then differentiate. In the second, we take the variation of Eq. [b] directly, which requires the variation of the unit vectors $\mathbf{i}$ and $\mathbf{j}$ of the moving frame. The relation between the unit vectors of the inertial and relative frames is

$$\mathbf{k} = \mathbf{K} \qquad \begin{bmatrix} \mathbf{i} \\ \mathbf{j} \end{bmatrix} = \begin{bmatrix} \cos\theta & \sin\theta \\ -\sin\theta & \cos\theta \end{bmatrix} \begin{bmatrix} \mathbf{I} \\ \mathbf{J} \end{bmatrix} \qquad [\mathbf{f}]$$

Introducing this into Eq. [b], we obtain

$$\mathbf{r}_G = [(L - R\phi)\cos\theta - R\sin\theta]\mathbf{I} + [(L - R\phi)\sin\theta + R\cos\theta]\mathbf{J} \qquad [\mathbf{g}]$$

The virtual displacement then becomes

$$\delta\mathbf{r}_G = [-(L - R\phi)\sin\theta\,\delta\theta - R\cos\theta\,\delta\phi - R\cos\theta\,\delta\theta]\mathbf{I}$$
$$+ [(L - R\phi)\cos\theta\,\delta\theta - R\sin\theta\,\delta\phi - R\sin\theta\,\delta\theta]\mathbf{J} \qquad [\mathbf{h}]$$

To convert the virtual displacement in terms of the relative frame, we introduce the relationships

$$\mathbf{I} = \cos\theta\mathbf{i} - \sin\theta\mathbf{j} \qquad \mathbf{J} = \sin\theta\mathbf{i} + \cos\theta\mathbf{j} \qquad [\mathbf{i}]$$

into Eq. [h], which gives Eq. [e].

Next, take the variation of Eq. [b] directly, with the result

$$\delta \mathbf{r}_G = -R\,\delta\phi\,\mathbf{i} + (L - R\phi)\,\delta\mathbf{i} + R\,\delta\mathbf{j} \qquad [\mathbf{j}]$$

From Eq. [f], the variations of the unit vectors have the form

$$\delta\mathbf{i} = -\sin\theta\,\delta\theta\,\mathbf{I} + \cos\theta\,\delta\theta\,\mathbf{J} = \delta\theta\mathbf{j} \qquad \delta\mathbf{j} = -\cos\theta\,\delta\theta\,\mathbf{I} - \sin\theta\,\delta\theta\,\mathbf{J} = -\delta\theta\mathbf{i} \qquad [\mathbf{k}]$$

Equations [k] can also be obtained directly from the rates of change of the unit vectors. Indeed, recalling that the angular velocity is $\boldsymbol{\omega} = \dot{\theta}\mathbf{k}$, the derivatives of the unit vectors are

$$\frac{d\mathbf{i}}{dt} = \boldsymbol{\omega} \times \mathbf{i} = \dot{\theta}\mathbf{j} \qquad \frac{d\mathbf{j}}{dt} = \boldsymbol{\omega} \times \mathbf{j} = -\dot{\theta}\mathbf{i} \qquad [\mathbf{l}]$$

from which the variations can be calculated easily. Introducing Eqs. [k] into Eq. [j], we obtain Eq. [e].

We have thus obtained the variation of $\mathbf{r}_G$ three different ways. It is clear that the number of manipulations is the least when we obtain the variation of $\mathbf{r}_G$ from the velocity expressions.

Consider the two-link mechanism in Fig. 4.3. A force $\mathbf{F}$ is acting at point P. Find the virtual work expression for each link and demonstrate that Eq. [4.4.14] holds.

Example 4.6

Solution

The free-body diagrams of the link are shown in Fig. 4.7. For the first link, the forces that contribute to the virtual work are the reactions at point B and the force of gravity at the mass center G_1. The forces can be expressed in vector form as

$$\mathbf{F}_B = B_x\mathbf{i} + B_y\mathbf{j} \qquad \mathbf{F}_{G1} = -m_1 g\mathbf{j} \qquad [\mathbf{a}]$$

The associated displacement vectors are

$$\mathbf{r}_B = L_1 s\theta_1 \mathbf{i} - L_1 c\theta_1 \mathbf{j} \qquad \mathbf{r}_{G_1} = \frac{L_1}{2} s\theta_1 \mathbf{i} - \frac{L_1}{2} c\theta_1 \mathbf{j} \qquad [\mathbf{b}]$$

so that the virtual displacements become

$$\delta\mathbf{r}_B = L_1 c\theta_1\,\delta\theta_1\mathbf{i} + L_1 s\theta_1\,\delta\theta_1\mathbf{j} \qquad \delta\mathbf{r}_{G_1} = \frac{L_1}{2} c\theta_1\,\delta\theta_1\mathbf{i} + \frac{L_1}{2} s\theta_1\,\delta\theta_1\mathbf{j} \qquad [\mathbf{c}]$$

We thus find the virtual work for the first link as

$$\delta W_{\text{link1}} = \mathbf{F}_B \cdot \delta\mathbf{r}_B - m_1 g\mathbf{j} \cdot \delta\mathbf{r}_{G_1} = B_x L_1 c\theta_1\,\delta\theta + B_y L_1 s\theta_1\,\delta\theta_1 - \frac{m_1 g L_1}{2} s\theta_1\,\delta\theta_1 \qquad [\mathbf{d}]$$

For the second link, the virtual work is due to the reactions at point B, gravity acting through the center of mass of the link G_2, and the external force $\mathbf{F}$ at the tip P. The forces at B are equal and opposite of $\mathbf{F}_B$. The other forces can be expressed by

$$\mathbf{F} = F_x\mathbf{i} + F_y\mathbf{j} \qquad \mathbf{F}_{G_2} = -m_2 g\mathbf{j} \qquad [\mathbf{e}]$$

with associated displacements

$$\mathbf{r}_P = (L_1 s\theta_1 + L_2 s\theta_2)\mathbf{i} - (L_1 c\theta_1 + L_2 c\theta_2)\mathbf{j}$$

$$\mathbf{r}_{G_2} = \left(L_1 s\theta_1 + \frac{L_2}{2} s\theta_2\right)\mathbf{i} - \left(L_1 c\theta_1 + \frac{L_2}{2} c\theta_2\right)\mathbf{j} \qquad [\mathbf{f}]$$

whose variations are

$$\delta \mathbf{r}_P = (L_1 c\theta_1 \delta\theta_1 + L_2 c\theta_2 \delta\theta_2)\mathbf{i} + (L_1 s\theta_1 \delta\theta_1 + L_2 s\theta_2 \delta\theta_2)\mathbf{j}$$

$$\delta \mathbf{r}_{G_2} = \left(L_1 c\theta_1 \delta\theta_1 + \frac{L_2}{2} c\theta_2 \delta\theta_2\right)\mathbf{i} + \left(L_1 s\theta_1 \delta\theta_1 + \frac{L_2}{2} s\theta_2 \delta\theta_2\right)\mathbf{j} \quad [\text{g}]$$

We now find the virtual work for the second link as

$$\delta W_{\text{link2}} = -\mathbf{F}_B \cdot \delta \mathbf{r}_B + \mathbf{F} \cdot \delta \mathbf{r}_P - m_2 g \mathbf{j} \cdot \delta \mathbf{r}_{G_2} \quad [\text{h}]$$

The virtual work of the entire system is found by adding Eqs. [d] and [h], with the result

$$\delta W = \delta W_{\text{link1}} + \delta W_{\text{link2}} = \mathbf{F} \cdot \delta \mathbf{r}_P - m_2 g \mathbf{j} \cdot \delta \mathbf{r}_{G_2} - m_1 g \mathbf{j} \cdot \delta \mathbf{r}_{G_1} \quad [\text{i}]$$

The only terms that contribute to the virtual work are those associated with the external forces. The contribution of the holonomic constraint forces (in this case the reaction at point B) to the virtual work is zero.

Taking the dot products, we write Eq. [i] in terms of the generalized coordinates as

$$\delta W = F_x(L_1 c\theta_1 \delta\theta_1 + L_2 c\theta_2 \delta\theta_2) + F_y(L_1 s\theta_1 \delta\theta_1 + L_2 s\theta_2 \delta\theta_2)$$

$$- m_2 g\left(L_1 s\theta_1 \delta\theta_1 + \frac{1}{2}L_2 s\theta_2 \delta\theta_2\right) - \frac{1}{2}m_1 g L_1 s\theta_1 \delta\theta_1$$

$$= \left(F_x L_1 c\theta_1 + F_y L_1 s\theta_1 - m_2 g L_1 s\theta_1 - \frac{1}{2}m_1 g L_1 s\theta_1\right)\delta\theta_1$$

$$+ \left(F_x L_2 c\theta_2 + F_y L_2 s\theta_2 - \frac{1}{2}m_2 g L_2 s\theta_2\right)\delta\theta_2 \quad [\text{j}]$$

Consider next the problem of having not a pinned joint at point B, but a joint that permits sliding motion, such as the collar shown in Fig. 4.14. Such a joint, as we will see in more detail in Chapter 7, is called a *prismatic joint*. The free-body diagram is illustrated in Fig. 4.15. The friction force at the sliding joint must be considered, and the forces that the two bodies exert on each other are split into two parts: a normal force N and a friction force F_f. Introducing the unit vectors $\mathbf{e}_1$ and $\mathbf{e}_2$ along and perpendicular to the link, we express the normal and

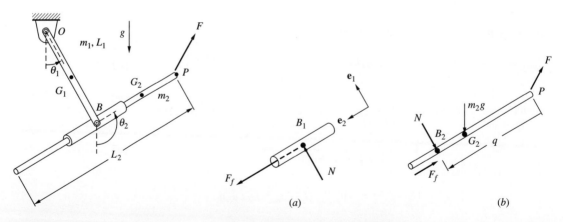

Figure 4.14   Prismatic joint   **Figure 4.15**   Free-body diagrams

friction forces on the two rods as

$$\mathbf{F}_N = N\mathbf{e}_1 \qquad \mathbf{F}_f = F_f \mathbf{e}_2 \qquad [\mathbf{k}]$$

Note also that because of sliding, the points B on the first link and on the collar do not have the same velocity. Denoting these points by B_1 and B_2 and introducing a generalized coordinate q to describe the sliding of link 2, the position vectors for B_1 and B_2 become

$$\mathbf{r}_{B_1} = \mathbf{r}_B \qquad \mathbf{r}_{B_2} = \mathbf{r}_B + q\mathbf{e}_2 \qquad [\mathbf{l}]$$

The virtual work expression has the form

$$\delta W_{\text{link1}} = (\mathbf{F}_N + \mathbf{F}_f) \cdot \delta \mathbf{r}_{B_1} - m_1 g \mathbf{j} \cdot \delta \mathbf{r}_{G_1}$$

$$\delta W_{\text{link2}} = -(\mathbf{F}_N + \mathbf{F}_f) \cdot \delta \mathbf{r}_{B_2} + \mathbf{F} \cdot \delta \mathbf{r}_P - m_2 g \mathbf{j} \cdot \delta \mathbf{r}_{G_2} \qquad [\mathbf{m}]$$

so that the virtual work for the entire system is

$$\delta W = \delta W_{\text{link1}} + \delta W_{\text{link2}} = -F_f \delta q + \mathbf{F} \cdot \delta \mathbf{r}_P - m_2 g \mathbf{j} \cdot \delta \mathbf{r}_{G_2} - m_1 g \mathbf{j} \cdot \delta \mathbf{r}_{G_1} \qquad [\mathbf{n}]$$

The contribution of the friction force to the virtual work is clear. Note that in order to determine the magnitude of the friction force, we need to have the normal force, which is absent from the above expression. This, basically, is the typical problem encountered when formulating problems involving friction. Also note that the position vectors for the center of mass and for the tip of the second rod change when the sliding joint is introduced to the problem.

4.5 GENERALIZED FORCES

Consider the system of particles in Fig. 4.16. The jth particle exerts a force of $\mathbf{F}_{ij}$ on the ith particle ($i, j = 1, 2, \ldots, N$). The resultant of all forces acting on the ith particle is denoted by $\mathbf{R}_i$ and has the form

$$\mathbf{R}_i = \mathbf{F}_i + \mathbf{F}'_i = \mathbf{F}_i + \sum_{j=1}^{N} \mathbf{F}_{ij} \qquad i = 1, 2, \ldots, N \qquad [\mathbf{4.5.1}]$$

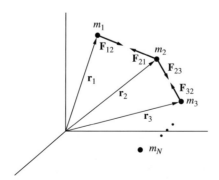

Figure 4.16

where $\mathbf{F}_i$ denotes the sum of all external (impressed, applied) forces exerted on the ith particle and $\mathbf{F}'_i$ is the sum of all internal forces (constraint or reaction forces that one particle exerts on the other).

The virtual work for each particle is defined as

$$\delta W_i = \mathbf{R}_i \cdot \delta \mathbf{r}_i \qquad [4.5.2]$$

One obtains the virtual work for the entire system by summing over the individual particles

$$\delta W = \sum_{i=1}^{N} \delta W_i = \sum_{i=1}^{N} \mathbf{R}_i \cdot \delta \mathbf{r}_i \qquad [4.5.3]$$

Substituting Eq. [4.5.1] into Eq. [4.5.3],

$$\delta W = \sum_{i=1}^{N} \mathbf{F}_i \cdot \delta \mathbf{r}_i + \sum_{i=1}^{N} \mathbf{F}'_i \cdot \delta \mathbf{r}_i \qquad [4.5.4]$$

We showed in Eq. [4.4.14] that the total work performed by the constraint forces in any virtual displacement is zero. It follows that the second term on the right side of the above equation vanishes because

$$\sum_{i=1}^{N} \mathbf{F}'_i \cdot \delta \mathbf{r}_i = 0 \qquad [4.5.5]$$

and the expression for the virtual work becomes

$$\delta W = \sum_{i=1}^{N} \mathbf{F}_i \cdot \delta \mathbf{r}_i \qquad [4.5.6]$$

It is of interest to examine the virtual work in terms of generalized coordinates. We express the displacement of each particle in terms of a set of n generalized coordinates q_k ($k = 1, 2, \ldots, n$) as $\mathbf{r}_i = \mathbf{r}_i(q_1, q_2, \ldots, q_n, t)$, ($i = 1, 2, \ldots, N$). The variation of $\mathbf{r}_i$ is

$$\delta \mathbf{r}_i = \sum_{i=1}^{N} \frac{\partial \mathbf{r}_i}{\partial q_k} \delta q_k \qquad [4.5.7]$$

Substitution of Eq. [4.5.7] into the expression for virtual work yields

$$\delta W = \sum_{i=1}^{N} \mathbf{F}_i \cdot \delta \mathbf{r}_i = \sum_{i=1}^{N} \mathbf{F}_i \cdot \sum_{k=1}^{n} \frac{\partial \mathbf{r}_i}{\partial q_k} \delta q_k = \sum_{k=1}^{n} \left(\sum_{i=1}^{N} \mathbf{F}_i \cdot \frac{\partial \mathbf{r}_i}{\partial q_k} \right) \delta q_k \qquad [4.5.8]$$

We define the term inside the fences in the above equation as *generalized forces* and write

$$Q_k = \sum_{i=1}^{N} \mathbf{F}_i \cdot \frac{\partial \mathbf{r}_i}{\partial q_k} \qquad k = 1, 2, \ldots, n \qquad [4.5.9]$$

where Q_k is the *generalized force associated with the kth generalized coordinate.* We can then express the virtual work as

$$\delta W = \sum_{k=1}^{n} Q_k \delta q_k \quad [4.5.10]$$

The relation between a generalized coordinate and a generalized force is analogous to the relation between a physical coordinate and the force applied in the direction of that coordinate. Also, the dimensional relation between generalized coordinates and generalized forces is worth noting. The product of Q_k and δq_k has the same units as the variation of energy. For example, if the generalized coordinate describes a displacement, the generalized force has the units of force. If the generalized coordinate describes a rotation, the generalized force becomes a moment.

Recalling from the previous section that the variations are often calculated with more ease by velocity relations, we make use of Eq. [4.4.6]

$$\frac{\partial \mathbf{r}}{\partial q_k} = \frac{\partial \dot{\mathbf{r}}}{\partial \dot{q}_k} \quad k = 1, 2, \ldots, n \quad [4.5.11]$$

to express the generalized forces as

$$Q_k = \sum_{i=1}^{N} \mathbf{F}_i \cdot \frac{\partial \mathbf{r}_i}{\partial q_k} = \sum_{i=1}^{N} \mathbf{F}_i \cdot \frac{\partial \dot{\mathbf{r}}_i}{\partial \dot{q}_k} \quad [4.5.12]$$

Another way of calculating generalized forces is based on the nature of the applied forces. For a conservative system, because dW is a perfect differential, the virtual work can be written as the variation of the negative of the potential energy, or

$$\delta W = \sum_{i=1}^{N} \mathbf{F}_i \cdot \delta \mathbf{r}_i = -\delta V \quad [4.5.13]$$

in which V is the *potential function,* or the *potential energy.* The variation of the potential energy in terms of physical coordinates is

$$\delta V = \sum_{i=1}^{N} \left(\frac{\partial V}{\partial x_i} \delta x_i + \frac{\partial V}{\partial y_i} \delta y_i + \frac{\partial V}{\partial z_i} \delta z_i \right) \quad [4.5.14]$$

When there are no constraints acting on the system, x_i, y_i, and z_i ($i = 1, 2, \ldots, N$) are independent. It follows that δx_i, δy_i and δz_i are arbitrary, and using Eqs. [4.5.13] and [4.5.14], we obtain

$$\frac{\partial V}{\partial x_i} = -F_{x_i} \qquad \frac{\partial V}{\partial y_i} = -F_{y_i} \qquad \frac{\partial V}{\partial z_i} = -F_{z_i} \quad [4.5.15]$$

In terms of independent generalized coordinates, and when all the applied forces are conservative, the virtual work expression can be written as

$$\delta V = -\delta W = \sum_{k=1}^{n} \frac{\partial V}{\partial q_k} \delta q_k \quad [4.5.16]$$

Comparing Eqs. [4.5.16] and [4.5.10], and considering the independence of the variations of the generalized coordinates, we conclude that the generalized forces are related to the potential energy by

$$Q_k = -\frac{\partial V}{\partial q_k} \qquad [4.5.17]$$

In the presence of both conservative and nonconservative forces, the virtual work and generalized forces can be written as

$$\delta W = -\delta V + \delta W_{nc} \qquad [4.5.18]$$

$$Q_k = Q_{kc} + Q_{knc} = -\frac{\partial V}{\partial q_k} + \sum_{i=1}^{N} \mathbf{F}_{inc} \cdot \frac{\partial \mathbf{r}_i}{\partial q_k} \qquad [4.5.19]$$

where the notation is obvious. When they are constant, nonconservative forces can be treated as conservative.

In summary, one can use a number of ways to calculate generalized forces:

1. Write Eq. [4.5.6] and after the virtual work is calculated collect coefficients of δq_k ($k = 1, 2, \ldots, n$).
2. Calculate $\partial \mathbf{r}_i/\partial q_k$ (or $\partial \dot{\mathbf{r}}_i/\partial \dot{q}_k$) and use Eq. [4.5.12].
3. Take advantage of the potential energy and use Eq. [4.5.17] for the conservative forces.

The reader is encouraged to use and compare all three approaches.

Example 4.7

Find the generalized forces for the mechanism in Fig. 4.3 (Example 4.6).

Solution

The generalized coordinates are θ_1 and θ_2. We will calculate the generalized forces in a number of ways. First, we take the expression for virtual work from Eq. [j] in Example 4.6, thus

$$\delta W = \left(F_x L_1 c\theta_1 + F_y L_1 s\theta_1 - m_2 g L_1 s\theta_1 - m_1 g \frac{L_1}{2} s\theta_1 \right) \delta\theta_1$$

$$+ \left(F_x L_2 c\theta_2 + F_y L_2 s\theta_2 - m_2 g \frac{L_2}{2} s\theta_2 \right) \delta\theta_2 \qquad [\mathbf{a}]$$

so that we can identify the generalized forces as

$$Q_1 = F_x L_1 c\theta_1 + F_y L_1 s\theta_1 - m_2 g L_1 s\theta_1 - \frac{1}{2} m_1 g L_1 s\theta_1$$

$$Q_2 = F_x L_2 c\theta_2 + F_y L_2 s\theta_2 - \frac{1}{2} m_2 g L_2 s\theta_2 \qquad [\mathbf{b}]$$

Next, consider each force and $\partial \mathbf{r}_i/\partial q_k$. For the first link we have one external force, gravity, and

$$\mathbf{F}_{G_1} = -m_1 g \mathbf{j} \qquad \mathbf{r}_{G_1} = \frac{L_1}{2} s\theta_1 \mathbf{i} - \frac{L_1}{2} c\theta_1 \mathbf{j} \qquad [\mathbf{c}]$$

so that

$$\frac{\partial \mathbf{r}_{G_1}}{\partial \theta_1} = \frac{L_1}{2} c\theta_1 \mathbf{i} + \frac{L_1}{2} s\theta_1 \mathbf{j} \qquad \frac{\partial \mathbf{r}_{G_1}}{\partial \theta_2} = 0 \qquad \text{[d]}$$

There are two external forces acting on the second link, written

$$\mathbf{F} = F_x \mathbf{i} + F_y \mathbf{j} \qquad \mathbf{r}_P = (L_1 s\theta_1 + L_2 s\theta_2)\mathbf{i} - (L_1 c\theta_1 + L_2 c\theta_2)\mathbf{j}$$

$$\mathbf{F}_{G_2} = -m_2 g \mathbf{j} \qquad \mathbf{r}_{G_2} = \left(L_1 s\theta_1 + \frac{L_2}{2} s\theta_2\right)\mathbf{i} - \left(L_1 c\theta_1 + \frac{L_2}{2} c\theta_2\right)\mathbf{j} \qquad \text{[e]}$$

so that

$$\frac{\partial \mathbf{r}_P}{\partial \theta_1} = L_1 c\theta_1 \mathbf{i} + L_1 s\theta_1 \mathbf{j} \qquad \frac{\partial \mathbf{r}_P}{\partial \theta_2} = L_2 c\theta_2 \mathbf{i} + L_2 s\theta_2 \mathbf{j}$$

$$\frac{\partial \mathbf{r}_{G_2}}{\partial \theta_1} = L_1 c\theta_1 \mathbf{i} + L_1 s\theta_1 \mathbf{j} \qquad \frac{\partial \mathbf{r}_{G_2}}{\partial \theta_2} = \frac{1}{2} L_2 c\theta_2 \mathbf{i} + \frac{1}{2} L_2 s\theta_2 \mathbf{j} \qquad \text{[f]}$$

Applying Eq. [4.5.12] we obtain

$$Q_1 = \mathbf{F}_{G_1} \cdot \frac{\partial \mathbf{r}_{G_1}}{\partial \theta_1} + \mathbf{F} \cdot \frac{\partial \mathbf{r}_P}{\partial \theta_1} + \mathbf{F}_{G_2} \cdot \frac{\partial \mathbf{r}_{G_2}}{\partial \theta_1}$$

$$= -\frac{1}{2} m_1 g L_1 s\theta_1 - m_2 g L_1 s\theta_1 + F_x L_1 c\theta_1 + F_y L_1 s\theta_1$$

$$Q_2 = \mathbf{F}_{G_1} \cdot \frac{\partial \mathbf{r}_{G_1}}{\partial \theta_2} + \mathbf{F} \cdot \frac{\partial \mathbf{r}_P}{\partial \theta_2} + \mathbf{F}_{G_2} \cdot \frac{\partial \mathbf{r}_{G_2}}{\partial \theta_2}$$

$$= -\frac{1}{2} m_2 g L_2 s\theta_2 + F_x L_2 c\theta_2 + F_y L_2 s\theta_2 \qquad \text{[g]}$$

which are the same as Eq. [b].

Finally, we make use of the potential energy to calculate the portion of the generalized forces associated with the gravitational forces. Taking point O as the datum, we write the potential energy as

$$V = -m_1 g \frac{L_1}{2} c\theta_1 - m_2 g L_1 c\theta_1 - m_2 g \frac{L_2}{2} c\theta_2 \qquad \text{[h]}$$

hence, the generalized forces due to the conservative forces become

$$Q_{1c} = -\frac{\partial V}{\partial \theta_1} = -\frac{1}{2} m_1 g L_1 s\theta_1 - m_2 g L_1 s\theta_1$$

$$Q_{2c} = -\frac{\partial V}{\partial \theta_2} = -\frac{1}{2} m_2 g L_2 s\theta_2 \qquad \text{[i]}$$

It is easy to see that the use of potential energy simplifies the calculation of the generalized forces.

4.6 PRINCIPLE OF VIRTUAL WORK FOR STATIC EQUILIBRIUM

Let us now consider *static equilibrium*. For a dynamical system, static equilibrium is described as the state where all components of the system are at rest, with zero velocity and zero acceleration. To find the equilibrium position, one can write the equilibrium equations using Newton's second law and solve these equations. The disadvantage of doing so is that if the motions of any two components are related to each other with a constraint relation, then the associated constraint forces must be calculated in the process. This may become tedious for systems with several interconnected components. Encouraged by the results of the previous section, we seek a different solution to the equilibrium problem that does not require one to solve for the constraint equations.

At equilibrium, the resultant force on each component of a system must be zero. Hence, we have $\mathbf{R}_i = \mathbf{0}\,(i = 1, 2, \ldots, N)$. It follows from Eq. [4.5.3] that since every resultant $\mathbf{R}_i = \mathbf{0}$, the virtual work must vanish as well and we must have $\delta W = 0$. Introducing this into Eq. [4.5.6] gives

$$\delta W = \sum_{i=1}^{N} \mathbf{F}_i \cdot \delta \mathbf{r}_i = 0 \qquad [4.6.1]$$

The above equation, first formulated by Johann Bernoulli, is known as the *principle of virtual work for static equilibrium.* It basically states that, at static equilibrium, the work performed by the external, impressed forces through virtual displacements compatible with the system constraints is zero. It can easily be extended to rigid bodies if we consider $\mathbf{r}_i$ to be the displacement of the point on the body to which the force $\mathbf{F}_i$ is applied.

Let us consider the principle of virtual work in terms of generalized forces. It follows from Eq. [4.5.10] that at equilibrium

$$\delta W = \sum_{k=1}^{n} Q_k \, \delta q_k = 0 \qquad [4.6.2]$$

When the system is represented in terms of independent generalized coordinates, because the generalized coordinates are independent of each other, their variations δq_k also are independent. Therefore, for Eq. [4.6.2] to hold, each of the coefficients of δq_k, that is, Q_k, must vanish individually. We write

$$Q_k = \sum_{i=1}^{N} \mathbf{F}_i \cdot \frac{\partial \mathbf{r}_i}{\partial q_k} = \sum_{i=1}^{N} \mathbf{F}_i \cdot \frac{\partial \dot{\mathbf{r}}_i}{\partial \dot{q}_k} = 0 \qquad k = 1, 2, \ldots, n \qquad [4.6.3]$$

In the presence of conservative forces we can take advantage of the potential energy and write

$$-\frac{\partial V}{\partial q_k} + Q_{knc} = 0 \qquad [4.6.4]$$

The above results can also be interpreted as follows: Because independent generalized coordinates represent the independent motion of each degree of freedom, their corresponding generalized forces must vanish at equilibrium.

As in the previous section, one can follow two approaches when solving static equilibrium problems using the principle of virtual work:

1. One can work with physical coordinates and use Eq. [4.6.1].
2. One can select a set of generalized coordinates, calculate the associated generalized forces, and use Eq. [4.6.3] or [4.6.4].

In the second approach, Eq. [4.6.4] is usually recommended over Eq. [4.6.3] in the presence of conservative forces, as it makes use of the potential energy. On the other hand, computation of $\partial \mathbf{r}_i/\partial q_k$ or $\partial \dot{\mathbf{r}}_i/\partial \dot{q}_k$ ($i = 1, 2, \ldots, N$; $k = 1, 2, \ldots, n$) in Eq. [4.6.3] can be done in a systematic fashion and tabulated, thereby mechanizing the derivation of the equilibrium equations.

Next, consider the principle of virtual work in terms of constrained generalized coordinates. To this end, write the constraint equations in Pfaffian form as

$$\sum_{k=1}^{n} a_{jk} dq_k + a_{j0} dt = 0 \qquad j = 1, 2, \ldots, m \qquad [4.6.5]$$

Now write the variation of the generalized coordinates as

$$\sum_{k=1}^{n} a_{jk} \delta q_k = 0 \qquad [4.6.6]$$

We add this relation to the principle of virtual work via the Lagrange multipliers λ_j ($j = 1, 2, \ldots, m$), resulting in the expression for the augmented virtual work as

$$\delta \hat{W} = \delta W - \sum_{j=1}^{m} \lambda_j \left(\sum_{k=1}^{n} a_{jk} \delta q_k \right) = \sum_{k=1}^{n} Q_k \delta q_k - \sum_{j=1}^{m} \lambda_j \left(\sum_{k=1}^{n} a_{jk} \delta q_k \right) = 0 \qquad [4.6.7]$$

Rearranging this equation as

$$\delta \hat{W} = \sum_{k=1}^{n} \left(Q_k - \sum_{j=1}^{m} \lambda_j a_{jk} \right) \delta q_k = 0 \qquad [4.6.8]$$

and by selecting the Lagrange multipliers such that the coefficients of δq_k vanish individually, we write the equilibrium equations as

$$Q_k = \sum_{j=1}^{m} \lambda_j a_{jk} \qquad k = 1, 2, \ldots, n \qquad [4.6.9]$$

In the presence of conservative forces, we introduce Eq. [4.6.4] to this equation, which leads to

$$Q_{knc} = \frac{\partial V}{\partial q_k} + \sum_{j=1}^{m} \lambda_j a_{jk} \qquad [4.6.10]$$

Example 4.8

Find the equilibrium position of the two links in Fig. 4.17. The springs are unstretched when both rods are horizontal. Both springs deflect only vertically.

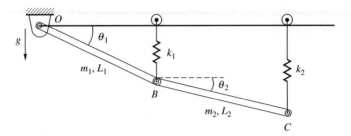

Figure 4.17

Solution

Because this problem involves two interconnected bodies and it is conservative, it is preferable to use potential energy to find the equilibrium position. Noting that the spring deflections are $L_1 \sin\theta_1$ and $L_1 \sin\theta_1 + L_2 \sin\theta_2$, and taking as the datum position the horizontal position of both links, the potential energy is

$$V = -m_1 g \frac{L_1}{2} s\theta_1 + \frac{1}{2}k_1(L_1 s\theta_1)^2 - m_2 g\left(L_1 s\theta_1 + \frac{L_2}{2} s\theta_2\right) + \frac{1}{2}k_2(L_1 s\theta_1 + L_2 s\theta_2)^2 \quad \text{[a]}$$

The equilibrium positions are found from

$$\frac{\partial V}{\partial \theta_1} = 0 \qquad \frac{\partial V}{\partial \theta_2} = 0 \quad \text{[b]}$$

and, taking the partial derivatives of V, we obtain

$$\frac{\partial V}{\partial \theta_1} = -\frac{1}{2}m_1 g L_1 c\theta_1 + k_1 L_1^2 s\theta_1 c\theta_1 - m_2 g L_1 c\theta_1 + k_2(L_1 s\theta_1 + L_2 s\theta_2)L_1 c\theta_1 \quad \text{[c]}$$

$$\frac{\partial V}{\partial \theta_2} = -\frac{1}{2}m_2 g L_2 c\theta_2 + k_2(L_1 s\theta_1 + L_2 s\theta_2)L_2 c\theta_2$$

We introduce Eqs. [c] into Eqs. [b]. Because $\cos\theta_1$ and $\cos\theta_2$ are common to the first and second of Eqs. [c], respectively, we eliminate them from Eqs. [c] and obtain

$$(k_1 + k_2)L_1^2 s\theta_1 + k_2 L_1 L_2 s\theta_2 = \frac{1}{2}m_1 g L_1 + m_2 g L_1 \quad \text{[d]}$$

$$k_2 L_1 L_2 s\theta_1 + k_2 L_2^2 s\theta_2 = \frac{1}{2}m_2 g L_2$$

Note that by eliminating $\cos\theta_1$ and $\cos\theta_2$ from the formulation, we are concluding that $\cos\theta_1 = 0$ and $\cos\theta_2 = 0$ represent equilibrium positions themselves. This basically is the vertical position of the links. At equilibrium either both links can be vertical, or one can. If link 1 is vertical, then the equilibrium position for link 2 is found by solving the second of Eqs. [d], and vice versa. To find the equilibrium positions where neither link is vertical, we solve Eqs. [d] simultaneously. To this end, we express Eqs. [d] in matrix form by

$$\begin{bmatrix} (k_1+k_2)L_1^2 & k_2 L_1 L_2 \\ k_2 L_1 L_2 & k_2 L_2^2 \end{bmatrix} \begin{bmatrix} \sin\theta_1 \\ \sin\theta_2 \end{bmatrix} = \begin{bmatrix} \frac{1}{2}m_1 g L_1 + m_2 g L_1 \\ \frac{1}{2}m_2 g L_2 \end{bmatrix} \quad \text{[e]}$$

which can be written as $[K]\{q\} = \{Q\}$, and whose solution is $\{q\} = [K]^{-1}\{Q\}$. The solution can be shown to be

$$\sin\theta_1 = \frac{g}{2k_1 L_1}(m_1 + m_2) \qquad [f]$$

$$\sin\theta_2 = -\frac{g}{2k_1 L_2}(m_1 + m_2) + \frac{g}{2k_2 L_2}m_2$$

An interesting case arises when k_1 is set to zero, or when there is no spring attached to the middle link. In this case, $\det[K] = 0$, which implies that one cannot solve for the equilibrium position by inverting Eq. [e]. The double link can assume an infinite number of equilibrium positions.

4.7 D'ALEMBERT'S PRINCIPLE

D'Alembert's principle extends the principle of virtual work from the static to the dynamic case. Consider the system of N particles discussed in the previous sections. If the system is not at rest, we can write Newton's second law for the ith particle as

$$\mathbf{R}_i = m_i \mathbf{a}_i = \frac{d}{dt}\mathbf{p}_i \qquad i = 1, 2, \ldots, N \qquad [4.7.1]$$

where $\mathbf{p}_i = m\mathbf{v}_i$ is the linear momentum of the ith particle and $\mathbf{R}_i$ is the resultant of all forces acting on the ith particle. As in the static case, we split the resultant $\mathbf{R}_i$ into the sum of the externally applied and constraint forces as

$$\mathbf{R}_i = \mathbf{F}_i + \mathbf{F}'_i \qquad [4.7.2]$$

Introducing Eq. [4.7.2] into Eq. [4.7.1], we obtain

$$\mathbf{F}_i + \mathbf{F}'_i - \dot{\mathbf{p}}_i = \mathbf{0} \qquad [4.7.3]$$

This equation is known as the *dynamic equilibrium* relation, where the negative of the rate of change of linear momentum, $-\dot{\mathbf{p}}_i = -m_i \mathbf{a}_i$, is treated as a force, referred to as the *inertia force,* that provides equilibrium. We can now treat the dynamic system as if it is a static system and invoke the principle of virtual work. Equation [4.7.3] is sometimes referred to as D'Alembert's principle. We proceed with the dot product of Eq. [4.7.3] and the variation in the displacement, and write

$$(\mathbf{F}_i + \mathbf{F}'_i - m_i \mathbf{a}_i) \cdot \delta \mathbf{r}_i = 0 \qquad [4.7.4]$$

Summing over all the particles gives

$$\sum_{i=1}^{N}(\mathbf{F}_i + \mathbf{F}'_i - m_i \mathbf{a}_i) \cdot \delta \mathbf{r}_i = 0 \qquad [4.7.5]$$

Recalling from Section 4.4 that work done by the constraint forces over virtual displacements is zero, or

$$\sum_{i=1}^{N} \mathbf{F}'_i \cdot \delta \mathbf{r}_i = 0 \qquad [\mathbf{4.7.6}]$$

and subtracting Eq. [4.7.6] from Eq. [4.7.5], we arrive at

$$\sum_{i=1}^{N} (\mathbf{F}_i - m_i \mathbf{a}_i) \cdot \delta \mathbf{r}_i = 0 \qquad [\mathbf{4.7.7}]$$

This we call the *generalized principle of D'Alembert,* or *D'Alembert's principle.* We observe immediately that the principle of virtual work, given in Eq. [4.6.1], becomes a special case of D'Alembert's principle.

D'Alembert's principle is a fundamental principle that provides a complete formulation of all of the problems of mechanics. Hamilton's principle and Lagrange's equations are all derived from D'Alembert's principle, as will be shown in the next sections. The advantage of using D'Alembert's principle over a Newtonian approach is that constraint forces and interacting forces between particles are eliminated from the formulation. This advantage becomes more pronounced for systems with several degrees of freedom.

We next extend D'Alembert's principle to rigid bodies. We consider here plane motion only (the general three-dimensional case will be derived in Chapter 8). We treat a rigid body as a collection of particles, so that in Eq. [4.7.7], N approaches infinity. Define the angular velocity of the rigid body as $\omega = \dot{\theta}$. Also, we express the position, velocity, and acceleration in terms of the center of mass motion as

$$\mathbf{r}_i = \mathbf{r}_G + \boldsymbol{\rho}_i \qquad \mathbf{v}_i = \mathbf{v}_G + \boldsymbol{\omega} \times \boldsymbol{\rho}_i$$
$$\mathbf{a}_i = \mathbf{a}_G + \boldsymbol{\alpha} \times \boldsymbol{\rho}_i - \omega^2 \boldsymbol{\rho}_i \qquad i = 1, 2, \ldots, N \qquad [\mathbf{4.7.8}]$$

where $\boldsymbol{\omega} = \dot{\theta}\mathbf{k}$, $\boldsymbol{\alpha} = \ddot{\theta}\mathbf{k}$, so that the variation of $\mathbf{r}_i$ can be written as

$$\delta \mathbf{r}_i = \delta \mathbf{r}_G + \delta \theta \mathbf{k} \times \boldsymbol{\rho}_i \qquad [\mathbf{4.7.9}]$$

and we recognize that $\delta \boldsymbol{\theta} = \delta \theta\, \mathbf{k}$. Introducing Eqs. [4.7.8] and [4.7.9] into D'Alembert's principle, we obtain

$$\sum_{i=1}^{N} (\mathbf{F}_i - m_i \mathbf{a}_G - m_i \boldsymbol{\alpha} \times \boldsymbol{\rho}_i + m_i \omega^2 \boldsymbol{\rho}_i) \cdot (\delta \mathbf{r}_G + \delta \theta \, \mathbf{k} \times \boldsymbol{\rho}_i) = 0 \qquad [\mathbf{4.7.10}]$$

Now, consider that the number of particles approaches infinity. The summation is replaced by integration, and m_i, $\boldsymbol{\rho}_i$, and $\mathbf{F}_i$ are replaced by dm, $\boldsymbol{\rho}$, and $d\mathbf{F}$, respectively. Evaluating the individual terms and using the definitions of center of mass and mass moment of inertia, we obtain

$$\int d\mathbf{F} \cdot \delta \mathbf{r}_G = \mathbf{F} \cdot \delta \mathbf{r}_G \qquad \int d\mathbf{F} \cdot (\delta \theta \mathbf{k} \times \boldsymbol{\rho}) = M_G \delta \theta \qquad \int dm \mathbf{a}_G \cdot \delta \mathbf{r}_G = m \mathbf{a}_G \cdot \delta \mathbf{r}_G$$

$$\int dm(\ddot{\theta}\mathbf{k} \times \boldsymbol{\rho}) \cdot (\delta \theta \mathbf{k} \times \boldsymbol{\rho}) = \ddot{\theta}\, \delta \theta \int \rho^2\, dm = I_G \ddot{\theta}\, \delta \theta \qquad [\mathbf{4.7.11}]$$

where m is the total mass, $\mathbf{F}$ is the resultant of all forces, I_G is the centroidal mass moment of inertia, and M_G is the sum of moments about the center of mass. All other remaining terms in Eq. [4.7.10] are zero. It follows that D'Alembert's principle for a rigid body in plane motion is

$$(\mathbf{F} - m\mathbf{a}_G) \cdot \delta \mathbf{r}_G + (M_G - I_G \ddot{\theta}) \delta \theta = 0 \quad \text{[4.7.12]}$$

For a system of N rigid bodies in plane motion, D'Alembert's principle becomes

$$\sum_{i=1}^{N} [(\mathbf{F}_i - m_i \mathbf{a}_{G_i}) \cdot \delta \mathbf{r}_{G_i} + (M_{G_i} - I_{G_i} \ddot{\theta}_i) \delta \theta_i] = 0 \quad \text{[4.7.13]}$$

where the subscript i now denotes the ith rigid body.

Up until the second half of the 20th century, the property of D'Alembert's principle being a vector relationship was usually viewed as a disadvantage, and D'Alembert's principle was primarily considered as a tool to obtain Hamilton's principle and Lagrange's equations. Equations [4.7.7] or [4.7.13] were rarely used in the form given here. The need to deal with complex multibody problems and the availability of digital computers has led scientists and engineers to take another look at D'Alembert's principle as a primary method of solution. For example, if we introduce Eq. [4.4.1b] into Eq. [4.7.7], we obtain

$$\sum_{i=1}^{N} (\mathbf{F}_i - m_i \mathbf{a}_i) \cdot \left(\sum_{k=1}^{n} \frac{\partial \mathbf{r}_i}{\partial q_k} \delta q_k \right) = \sum_{k=1}^{n} \left[\sum_{i=1}^{N} (\mathbf{F}_i - m_i \mathbf{a}_i) \cdot \frac{\partial \mathbf{r}_i}{\partial q_k} \right] \delta q_k = 0 \quad \text{[4.7.14]}$$

When we have a set of independent generalized coordinates, the coefficients of δq_k must vanish independently, with the result

$$\sum_{i=1}^{N} (\mathbf{F}_i - m_i \mathbf{a}_i) \cdot \frac{\partial \mathbf{r}_i}{\partial q_k} = 0 \qquad k = 1, 2, \ldots, n \quad \text{[4.7.15]}$$

Extending this to the case of N rigid bodies in plane motion, we obtain

$$\sum_{i=1}^{N} \left[(\mathbf{F}_i - m_i \mathbf{a}_{G_i}) \cdot \frac{\partial \mathbf{r}_{G_i}}{\partial q_k} + (M_{G_i} - I_{G_i} \ddot{\theta}_i) \frac{\partial \theta_i}{\partial q_k} \right] = 0 \quad \text{[4.7.16]}$$

Equations [4.7.15] and [4.7.16] represent direct use of D'Alembert's principle to derive equations of motion.

Example 4.9

Consider a bead of mass m free to slide on a ring (hoop) of radius R, as shown in Fig. 4.18. The ring is rotating with the constant angular velocity Ω. Find the equation of motion using D'Alembert's principle.

Solution

Because we are dealing with a single particle, we drop the subscript in Eq. [4.7.7] and write it as

$$(\mathbf{F} - m\mathbf{a}) \cdot \delta \mathbf{r} = 0 \quad \text{[a]}$$

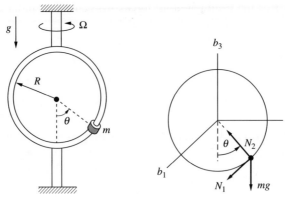

Figure 4.18 Bead on a rotating ring

Figure 4.19 Free-body diagram

The free-body diagram is given in Fig. 4.19. The $\mathbf{b}_1\mathbf{b}_2\mathbf{b}_3$ axes are attached to the hoop. The generalized coordinate is selected as θ. We first derive an expression for the acceleration. The moving frame is attached to the ring. The position vector is

$$\mathbf{r} = R\sin\theta\,\mathbf{b}_2 - R\cos\theta\,\mathbf{b}_3 \qquad [\mathbf{b}]$$

so its variation is

$$\delta\mathbf{r} = R\cos\theta\,\delta\theta\,\mathbf{b}_2 + R\sin\theta\,\delta\theta\,\mathbf{b}_3 \qquad [\mathbf{c}]$$

Because the motion of the relative frame, that is, of the hoop, is treated as a known quantity, its variation is zero. Hence, it is possible to calculate the variation of $\mathbf{r}$ in the relative frame. To see this better, write the velocity of the bead as

$$\mathbf{v} = \mathbf{v}_{\text{rel}} + \boldsymbol{\omega}\times\mathbf{r} = R\dot\theta\cos\theta\,\mathbf{b}_2 + R\dot\theta\sin\theta\,\mathbf{b}_3 + \Omega\mathbf{b}_3\times(R\sin\theta\,\mathbf{b}_2 - R\cos\theta\,\mathbf{b}_3) \qquad [\mathbf{d}]$$
$$= R\dot\theta\cos\theta\,\mathbf{b}_2 + R\dot\theta\sin\theta\,\mathbf{b}_3 - R\Omega\sin\theta\,\mathbf{b}_1$$

Since Ω is a constant and it is not the derivative of a motion variable, it cannot be expressed in terms of a variation. Consequently, the third term on the right side of Eq. [d] does not contribute to the virtual displacement.

Because the angular velocity is constant, the expression for the acceleration has the form

$$\mathbf{a} = \mathbf{a}_{\text{rel}} + \boldsymbol{\omega}\times\boldsymbol{\omega}\times\mathbf{r} + 2\boldsymbol{\omega}\times\mathbf{v}_{\text{rel}} = R\ddot\theta\cos\theta\,\mathbf{b}_2 + R\ddot\theta\sin\theta\,\mathbf{b}_3 - R\dot\theta^2\sin\theta\,\mathbf{b}_2 + R\dot\theta^2\cos\theta\,\mathbf{b}_3$$
$$+ \Omega\mathbf{b}_3\times\Omega\mathbf{b}_3\times(R\sin\theta\,\mathbf{b}_2 - R\cos\theta\,\mathbf{b}_3) + 2\Omega\mathbf{b}_3\times(R\dot\theta\cos\theta\,\mathbf{b}_2 + R\dot\theta\sin\theta\,\mathbf{b}_3)$$
$$\mathbf{a} = -2R\dot\theta\Omega\cos\theta\,\mathbf{b}_1 + (-R\sin\theta(\dot\theta^2 + \Omega^2)) + R\ddot\theta\cos\theta)\mathbf{b}_2 + (R\dot\theta^2\cos\theta + R\ddot\theta\sin\theta)\mathbf{b}_3 \qquad [\mathbf{e}]$$

The only force acting on the system which is not a constraint force is gravity, and it has the form $\mathbf{F} = -mg\mathbf{b}_3$.

Substituting Eqs. [c] and [e] into the generalized principle of D'Alembert yields

$$(\mathbf{F} - m\mathbf{a})\cdot\delta\mathbf{r} = [-mg\mathbf{b}_3 + m(R\sin\theta(\dot\theta^2 + \Omega^2) - R\ddot\theta\cos\theta)\mathbf{b}_2$$
$$- m(R\dot\theta^2\cos\theta + R\ddot\theta\sin\theta)\mathbf{b}_3 - 2mR\dot\theta\Omega\cos\theta\,\mathbf{b}_1]$$
$$\cdot(R\cos\theta\,\delta\theta\,\mathbf{b}_2 + R\sin\theta\,\delta\theta\,\mathbf{b}_3) = 0 \qquad [\mathbf{f}]$$

After evaluating the dot product and setting the coefficient of $\delta\theta$ equal to zero, we obtain the equation of motion as

$$\ddot{\theta} + \sin\theta\left(\frac{g}{R} - \Omega^2\cos\theta\right) = 0 \qquad [\text{g}]$$

Let us compare the procedure we used in this example with a Newtonian approach. From the free-body diagram, there are two normal (reaction) forces, N_1 and N_2. After applying Newton's second law, we get three equations and we need to eliminate the reactions. It is obvious that using D'Alembert's principle is simpler. The difference becomes more pronounced where there are several degrees of freedom.

4.8 HAMILTON'S PRINCIPLES

From D'Alembert's principle we develop the scalar variational principles that provide a complete formulation of the problems of mechanics. These principles were stated for the most general case of motion by Sir William Rowan Hamilton.

Consider a system of N particles and D'Alembert's principle

$$\sum_{i=1}^{N}(m_i\ddot{\mathbf{r}}_i - \mathbf{F}_i)\cdot\delta\mathbf{r}_i = 0 \qquad [4.8.1]$$

We denote by $\delta W = \sum \mathbf{F}_i\cdot\delta\mathbf{r}_i$ the virtual work of all the impressed forces. To manipulate the first term in the above equation, consider the expression

$$\frac{d}{dt}(\dot{\mathbf{r}}_i\cdot\delta\mathbf{r}_i) = \ddot{\mathbf{r}}_i\cdot\delta\mathbf{r}_i + \dot{\mathbf{r}}_i\cdot\delta\dot{\mathbf{r}}_i \qquad i = 1, 2, \ldots, N \qquad [4.8.2]$$

The second term on the right in Eq. [4.8.2] can be recognized as

$$\dot{\mathbf{r}}_i\cdot\delta\dot{\mathbf{r}}_i = \frac{\delta(\dot{\mathbf{r}}_i\cdot\dot{\mathbf{r}}_i)}{2} \qquad [4.8.3]$$

The kinetic energy of the ith particle is

$$T_i = \frac{1}{2}m_i\dot{\mathbf{r}}_i\cdot\dot{\mathbf{r}}_i \qquad [4.8.4]$$

so that the variation of the kinetic energy of the ith particle becomes

$$\delta T_i = \frac{1}{2}m_i\,\delta(\dot{\mathbf{r}}_i\cdot\dot{\mathbf{r}}_i) = m_i\dot{\mathbf{r}}_i\cdot\delta\dot{\mathbf{r}}_i \qquad [4.8.5]$$

and we can express Eq. [4.8.2] as

$$m_i\frac{d}{dt}(\dot{\mathbf{r}}_i\cdot\delta\mathbf{r}_i) = m_i\ddot{\mathbf{r}}_i\cdot\delta\mathbf{r}_i + m_i\dot{\mathbf{r}}_i\cdot\delta\dot{\mathbf{r}}_i = m_i\ddot{\mathbf{r}}_i\cdot\delta\mathbf{r}_i + \delta T_i \qquad [4.8.6]$$

The variation in the total kinetic energy of the system is

$$\delta T = \sum_{i=1}^{N} \delta T_i = \sum_{i=1}^{N} \frac{1}{2} m_i \delta(\dot{\mathbf{r}}_i \cdot \dot{\mathbf{r}}_i) \qquad [4.8.7]$$

Using Eq. [4.8.6], we express D'Alembert's principle as

$$\sum_{i=1}^{N} (m_i \ddot{\mathbf{r}}_i - \mathbf{F}_i) \cdot \delta \mathbf{r}_i = 0 = -\delta T + \sum_{i=1}^{N} m_i \frac{d}{dt}(\dot{\mathbf{r}}_i \cdot \delta \mathbf{r}_i) - \delta W \qquad [4.8.8]$$

so that we have an expression for the variation of the kinetic and potential energies

$$\delta T + \delta W = \sum_{i=1}^{N} m_i \frac{d}{dt}(\dot{\mathbf{r}}_i \cdot \delta \mathbf{r}_i) \qquad [4.8.9]$$

Next, we integrate the right side of Eq. [4.8.9] over two points in time, say, t_1 and t_2, thus

$$\int_{t_1}^{t_2} (\delta T + \delta W) \, dt = \int_{t_1}^{t_2} \sum_{i=1}^{N} m_i \frac{d}{dt}(\dot{\mathbf{r}}_i \cdot \delta \mathbf{r}_i) \, dt$$

$$= \int \sum_{i=1}^{N} m_i \, d(\dot{\mathbf{r}}_i \cdot \delta \mathbf{r}_i) = \sum_{i=1}^{N} m_i \dot{\mathbf{r}}_i \cdot \delta \mathbf{r}_i \Big|_{t_1}^{t_2} \qquad [4.8.10]$$

The term $m_i \dot{\mathbf{r}}_i$ is recognized as the partial derivative of T_i with respect to $\dot{\mathbf{r}}_i$, so that we may write

$$\sum_{i=1}^{N} m_i \dot{\mathbf{r}}_i \cdot \delta \mathbf{r}_i \Big|_{t_1}^{t_2} = \sum_{i=1}^{N} \frac{\partial T_i}{\partial \dot{\mathbf{r}}_i} \cdot \delta \mathbf{r}_i \Big|_{t_1}^{t_2} \qquad [4.8.11]$$

which, when introduced back into Eq. [4.8.10], yields

$$\int_{t_1}^{t_2} (\delta T + \delta W) \, dt - \sum_{i=1}^{N} \frac{\partial T_i}{\partial \dot{\mathbf{r}}_i} \cdot \delta \mathbf{r}_i \Big|_{t_1}^{t_2} = 0 \qquad [4.8.12]$$

This equation is known as *Hamilton's principle (or law) of varying action*. One can put this principle into more general form, by expressing it in terms of generalized coordinates alone. Introducing Eq. [4.4.7] into Eq. [4.8.11], we obtain

$$\sum_{i=1}^{N} \frac{\partial T_i}{\partial \dot{\mathbf{r}}_i} \cdot \delta \mathbf{r}_i \Big|_{t_1}^{t_2} = \sum_{i=1}^{N} \frac{\partial T_i}{\partial \dot{\mathbf{r}}_i} \cdot \sum_{k=1}^{n} \frac{\partial \dot{\mathbf{r}}_i}{\partial \dot{q}_k} \delta q_k \Big|_{t_1}^{t_2} = \sum_{k=1}^{n} \frac{\partial T}{\partial \dot{q}_k} \delta q_k \Big|_{t_1}^{t_2} \qquad [4.8.13]$$

Introducing Eq. [4.8.13] into Eq. [4.8.12] we write Hamilton's principle of varying action as

$$\int_{t_1}^{t_2} (\delta T + \delta W) \, dt - \sum_{k=1}^{n} \frac{\partial T}{\partial \dot{q}_k} \delta q_k \Big|_{t_1}^{t_2} = 0 \qquad [4.8.14]$$

Note that the derivation above does not put any restrictions on the time instances t_1 and t_2.

A special case of Hamilton's principle of varying action is obtained when we consider the variation of $\mathbf{r}_i$ as time is held fixed. We reexamine Fig. 4.12, which is analogous to Fig. B1 in Appendix B. The varied path can take any value within the set of admissible displacements of the system, and it coincides with the true path at the end points. It follows that the variation of the displacement $\delta\mathbf{r}_i$ and of the generalized coordinates have values of zero at $t = t_1$ and $t = t_2$, provided $\mathbf{r}$ is specified at t_1 and t_2. Of interest is the case when $\mathbf{r}_i$ ($i = 1, 2, \ldots, N$) are specified, which eliminates the integrated term in Hamilton's principle of varying action, resulting in

$$\int_{t_1}^{t_2} (\delta T + \delta W)\, dt = 0 \qquad [4.8.15]$$

This equation is known as the *extended Hamilton's principle*. Writing the virtual work as $\delta W = \delta W_{nc} - \delta V$, one can express the extended Hamilton's principle also as

$$\int_{t_1}^{t_2} (\delta T - \delta V + \delta W_{nc})\, dt = 0 \qquad [4.8.16]$$

Even though we derived it here for a system of particles, the extended Hamilton's principle is valid both for particles and for rigid or elastic bodies. It is, again, a fundamental principle of mechanics from which the motion of *all* bodies can be described. In this sense, the extended Hamilton's principle is not exactly a derived principle. Rather, it is more like a law of nature, in the same way that Newton's second law is a law of nature. Further, only scalar quantities like work and energy are needed. No acceleration terms need to be calculated to invoke this principle.

Introduce the Lagrangian L such that $L = T - V$. For conservative systems, $\delta W = -\delta V$, and we can write

$$\int_{t_1}^{t_2} \delta L\, dt = 0 \qquad [4.8.17]$$

and Eq. [4.8.17] is referred to as *Hamilton's principle*. This principle was first stated by Lagrange and originally called *Principle of least action*. When the system is holonomic, one can interchange the integration and variation operations, which yields

$$\delta \int_{t_1}^{t_2} L\, dt = 0 \qquad [4.8.18]$$

Hamilton's principle for a holonomic system basically states that among all the paths that a system can take, the actual path followed renders the definite integral $I = \int_{t_1}^{t_2} L\, dt$ stationary. This integral is also known as the *action integral*.

The implementation of the extended Hamilton's principle for finding the equations of motion requires the evaluation of the variations of the kinetic and potential energies. The procedure can become tedious, primarily because of the large number of integrations by parts that one must perform to relate the variations of generalized velocities to the variations of the generalized coordinates. A simpler and more general procedure for deriving the equations of motion for systems with a finite number of degrees of freedom is by means of Lagrange's equations, as we will see in the next section.

The direct use of the extended Hamilton's principle is effective when deriving the equations of motion of deformable bodies, such as for the vibrations of beams, plates, and shells. In such problems, the extended Hamilton's principle yields the equations of motion in the form of partial differential equations with accompanying boundary conditions. We will investigate the dynamics of deformable bodies in Chapter 11. Hamilton's principle is also used in transformation theory and in optimal control theory.

One may wonder why we list two major principles in this section that encompass nonconservative forces when the first, Hamilton's law of varying action, is lengthier and has the appearance of being redundant when compared with the extended principle. The difference between the two principles is in how they treat the time instances t_1 and t_2.

If we view t_1 and t_2 as arbitrary time instances, we obtain the extended Hamilton's principle from Hamilton's law of varying action and the two principles become the same. But if we view t_1 as a point at which we know the values of the generalized coordinates, then we can make use of Eq. [4.8.14] to find the values of the generalized coordinates at time t_2. To do this we do *not* need to derive any equations of motion, just the variation of the Lagrangian and the virtual work. This approach comes in handy in numerical integration, as t_2 can be taken as $t_1 + \Delta$, in which Δ is a small time increment.

Example 4.10

Obtain the equation of motion of the bead problem in Example 4.9 using the extended Hamilton's principle.

Solution

To find the kinetic energy, we need the velocity of the bead. From Example 4.9 we have

$$\mathbf{r} = R \sin \theta \mathbf{b}_2 - R \cos \theta \mathbf{b}_3 \qquad [\mathbf{a}]$$

$$\mathbf{v} = -R\Omega \sin \theta \mathbf{b}_1 + R\dot{\theta} \cos \theta \mathbf{b}_2 + R\dot{\theta} \sin \theta \mathbf{b}_3 \qquad [\mathbf{b}]$$

The kinetic energy is

$$T = \frac{1}{2} m \mathbf{v} \cdot \mathbf{v} = \frac{1}{2} m[(\Omega R \sin \theta)^2 + (R\dot{\theta} \cos \theta)^2 + (R\dot{\theta} \sin \theta)^2] = \frac{mR^2}{2} \Omega^2 \sin^2 \theta + \frac{mR^2}{2} \dot{\theta}^2 \qquad [\mathbf{c}]$$

Using the position of the bead at the bottom of the ring ($\theta = 0$) as the datum, the potential energy becomes

$$V = mgR(1 - \cos \theta) \qquad [\mathbf{d}]$$

so that the Lagrangian has the form

$$L = T - V = \frac{mR^2}{2} \Omega^2 \sin^2 \theta + \frac{mR^2}{2} \dot{\theta}^2 - mgR(1 - \cos \theta) \qquad [\mathbf{e}]$$

The variation of the Lagrangian is

$$\delta L = \frac{\partial L}{\partial \theta}\delta\theta + \frac{\partial L}{\partial \dot\theta}\delta\dot\theta = mR^2\left[\Omega^2 \sin\theta\cos\theta - \frac{g}{R}\sin\theta\right]\delta\theta + mR^2\dot\theta\,\delta\dot\theta \qquad \text{[f]}$$

The second term in this equation is in terms of $\delta\dot\theta$. To invoke the extended Hamilton principle, we have to express all the terms in terms of $\delta\theta$. To accomplish this, we integrate this second term by parts and write

$$\int_{t_1}^{t_2} \dot\theta\,\delta\dot\theta\,dt = \int_{t_1}^{t_2} \dot\theta\,\frac{d}{dt}(\delta\theta)\,dt = \dot\theta\,\delta\theta\Big|_{t_1}^{t_2} - \int_{t_1}^{t_2} \ddot\theta\,\delta\theta\,dt \qquad \text{[g]}$$

The integrated term on the right side of Eq. [g] vanishes by virtue of the definition of the variation operation. (The values of the variation at the beginning and end of the path are zero.) The second term, when used with Eq. [f] and the Extended Hamilton's Principle, yields

$$\int_{t_1}^{t_2}\left[-mR^2\ddot\theta + mR^2\left(\Omega^2\sin\theta\cos\theta - \frac{g}{R}\sin\theta\right)\right]\delta\theta\,dt = 0 \qquad \text{[h]}$$

In order for the equality to hold, the integrand must vanish at all times. Because $\delta\theta$ is arbitrary, for the integrand to be zero the coefficient of $\delta\theta$ must be identically zero. Thus we recognize as the equation of motion

$$\ddot\theta + \sin\theta\left(\frac{g}{R} - \Omega^2\cos\theta\right) = 0 \qquad \text{[i]}$$

Let us review the operations we carried out. After obtaining the kinetic and potential energies and taking the partial derivatives, we performed an integration by parts on the term $\dot\theta\,\delta\dot\theta$. We could have done the integration by parts on the general expression $\frac{\partial L}{\partial\dot\theta}\delta\dot\theta$ rather than the corresponding specific term in this problem, $\dot\theta\,\delta\dot\theta$. The question then arises as to whether, manipulating the extended Hamilton's principle, one can perform the integrations by part in advance and develop a general form for the equations of motion. This is the question we will explore in the next section.

4.9 LAGRANGE'S EQUATIONS

From Hamilton's principle, we derive Lagrange's equations, which present themselves as a convenient way of deriving the equations of motion. The extended Hamilton's principle can be expressed as

$$\int_{t_1}^{t_2}(\delta T - \delta V + \delta W_{nc})\,dt = \int_{t_1}^{t_2}\delta L\,dt + \int_{t_1}^{t_2}\delta W_{nc}\,dt = 0 \qquad \text{[4.9.1]}$$

The Lagrangian L can be written in terms of generalized coordinates q_k and generalized velocities $\dot q_k$ ($k = 1, 2, \ldots, n$) as $L = L(q_1, q_2, \ldots, q_n, \dot q_1, \dot q_2, \ldots, \dot q_n, t)$.

The variation of L is

$$\delta L = \sum_{k=1}^{n} \left(\frac{\partial L}{\partial q_k} \delta q_k + \frac{\partial L}{\partial \dot{q}_k} \delta \dot{q}_k \right) \qquad [4.9.2]$$

and, using Eqs. [4.5.10] and [4.5.19], the variation of the nonconservative work is written in terms of the generalized forces as

$$\delta W_{nc} = \sum_{k=1}^{n} Q_{knc} \delta q_k \qquad [4.9.3]$$

Making use of the property that the variation and differentiation (with regard to time) operations can be interchanged, we integrate by parts the second term in Eq. [4.9.2] and obtain

$$\int_{t_1}^{t_2} \frac{\partial L}{\partial \dot{q}_k} \delta \dot{q}_k \, dt = \int_{t_1}^{t_2} \frac{\partial L}{\partial \dot{q}_k} \frac{d}{dt} (\delta q_k) \, dt = \left. \frac{\partial L}{\partial \dot{q}_k} \delta q_k \right|_{\delta q_k(t_1)}^{\delta q_k(t_2)} - \int_{t_1}^{t_2} \frac{d}{dt} \left(\frac{\partial L}{\partial \dot{q}_k} \right) \delta q_k \, dt \qquad [4.9.4]$$

The integrated term requires evaluation of δq_k ($k = 1, 2, \ldots, n$) at the beginning and the end of the time intervals. By the definition of the variation, the varied path vanishes at the end points, thus $\delta q_k(t_1) = \delta q_k(t_2) = 0$ for all values of k. Considering this, and introducing Eqs. [4.9.2]–[4.9.4] into the extended Hamilton's principle, we obtain

$$\int_{t_1}^{t_2} (\delta T - \delta V + \delta W_{nc}) \, dt = \int_{t_1}^{t_2} \sum_{k=1}^{n} \left[-\frac{d}{dt} \left(\frac{\partial L}{\partial \dot{q}_k} \right) + \frac{\partial L}{\partial q_k} + Q_{knc} \right] \delta q_k \, dt = 0 \qquad [4.9.5]$$

For the integral over time to vanish at all times, the integrand must be identically equal to zero, which can be expressed as

$$\sum_{k=1}^{n} \left[-\frac{d}{dt} \left(\frac{\partial L}{\partial \dot{q}_k} \right) + \frac{\partial L}{\partial q_k} + Q_{knc} \right] \delta q_k = 0 \qquad [4.9.6]$$

It should be noted that this equation can be directly obtained from D'Alembert's principle, without using Hamilton's principle. Because of this, Eq. [4.9.6] is sometimes referred to as *Lagrange's form of D'Alembert's principle*.

Consider now a set of independent generalized coordinates. It follows that the only way Eq. [4.9.6] can be equal to zero is if the coefficients of δq_k vanish individually for all values of the index k. Setting the coefficients equal to zero, we obtain *Lagrange's equations* of motion

$$\frac{d}{dt} \left(\frac{\partial L}{\partial \dot{q}_k} \right) - \frac{\partial L}{\partial q_k} = Q_{knc} \qquad k = 1, 2, \ldots, n \qquad [4.9.7]$$

Equation [4.9.7] is the most general form of Lagrange's equations. They can also be expressed in terms of the kinetic and potential energies. Noting that the potential energy is not a function of the generalized velocities (except for electromagnetic systems), we write Eq. [4.9.7] as

$$\frac{d}{dt}\left(\frac{\partial T}{\partial \dot{q}_k}\right) - \frac{\partial T}{\partial q_k} + \frac{\partial V}{\partial q_k} = Q_{knc} \qquad k = 1, 2, \ldots, n \qquad \textbf{[4.9.8]}$$

This form of Lagrange's equations is preferred by many, as it reduces the possibility of making a sign error when evaluating the partial derivatives. It also is similar to the format Lagrange first presented these equations in 1788. Under certain circumstances it is more convenient to write Lagrange's equations in terms of the kinetic energy alone, in the form of

$$\frac{d}{dt}\left(\frac{\partial L}{\partial \dot{q}_k}\right) - \frac{\partial T}{\partial q_k} = Q_k \qquad \textbf{[4.9.9]}$$

where the values of Q_k contain contributions from the conservative as well as nonconservative forces. The principle of virtual work given by Eq. [4.6.4], is a special case of Lagrange's equations. In the static case, the first two terms in Eq. [4.9.8] vanish.

For a holonomic conservative system, one can use Eq. [4.8.15] directly in conjunction with the Euler-Lagrange equation in Appendix B to derive Lagrange's equations. The order of variation and integration can be exchanged, and one seeks the stationary values of the integral $I = \int_{t_1}^{t_2} L \, dt$, leading to

$$\frac{d}{dt}\left(\frac{\partial L}{\partial \dot{q}_k}\right) - \frac{\partial L}{\partial q_k} = 0 \qquad \textbf{[4.9.10]}$$

Lagrange's equations can conveniently be expressed in column vector format. Introducing the n-dimensional generalized coordinate and generalized force vectors

$$\{q\} = [q_1 \quad q_2 \quad \ldots \quad q_n]^T \qquad \{Q_{nc}\} = [Q_{1nc} \quad Q_{2nc} \quad \ldots \quad Q_{nnc}]^T \qquad \textbf{[4.9.11]}$$

we can write Lagrange's equations as

$$\frac{d}{dt}\left(\frac{\partial L}{\partial \{\dot{q}\}}\right) - \frac{\partial L}{\partial \{q\}} = \{Q_{nc}\}^T \qquad \textbf{[4.9.12]}$$

Let us now compare the steps involved in obtaining the equations of motion using Lagrange's equations and using the Newtonian approach. When using Newton's second law, we

1. Isolate the different bodies involved.
2. Select a coordinate system and draw free-body diagrams.
3. Relate the sum of forces and sum of moments to the translational and angular accelerations.
4. Use kinematics to express the accelerations in terms of translational and angular parameters.
5. Eliminate the constraint and reaction forces and derive the equations of motion.

When using the Lagrangian approach, we

1. Determine the number of degrees of freedom and select a set of independent generalized coordinates. The free-body diagram is a useful tool for this.
2. Use the kinematical relations to find the velocities and virtual displacements involved.
3. Identify the forces that are conservative and those that are not.
4. Write the kinetic and potential energies, as well as the virtual work.
5. Apply Lagrange's equations.

There are two distinct differences between the two approaches. The first difference is in the order of the steps involved: In the Newtonian approach, one first writes the force and moment balances for all bodies separately and then uses kinematical relations and the constraint forces to reduce the number of equations. In the Lagrangian approach, one considers the constraints and kinematics of the problem first. Then, the equations of motion are written, one for each degree of freedom. The bulk of the work involved in Lagrangian mechanics is to find a proper set of generalized coordinates and to express the kinematics. Once this is done, the rest is straightforward.

The second difference is that the Lagrangian approach uses velocities and scalar quantities, whereas the Newtonian approach uses accelerations and vector quantities. Dealing with velocities involves considerably less algebra than dealing with accelerations.

It may appear, from the above discussion, that Lagrange's equations should be preferable to the Newtonian approach at all times; but this is not so. By eliminating the constraint forces from the formulation, the Lagrangian approach does not calculate the amplitudes of these forces. While this may be acceptable for classroom examples, it certainly is not in many real-life applications, where one must know the amplitudes of the reaction and other contact forces acting on a body. Furthermore, for certain geometries a Newtonian approach is more suitable. The best way to determine which approach is most suited to one's needs is by gaining experience in solving mechanics problems. In many cases, looking at a problem from *both* a Lagrangian and Newtonian point of view increases the physical insight and makes it easier to understand the characteristics of the system.

We should add here that the historical development of analytical mechanics did not follow the sequence in which it is presented in this chapter. Lagrange's equations were derived before the extended Hamilton's principle, and they were derived for conservative systems only. It was Hamilton, born after Lagrange, who put together the developments in variational mechanics and Lagrange's equations to develop a general scalar principle from which all the equations of motion can be derived.

Example 4.11

For the system in Fig. 4.20, find the equations of motion using Lagrange's equations. Assume that the spring and dashpot deflect only horizontally and that the force F is always applied horizontally.

4.9 LAGRANGE'S EQUATIONS

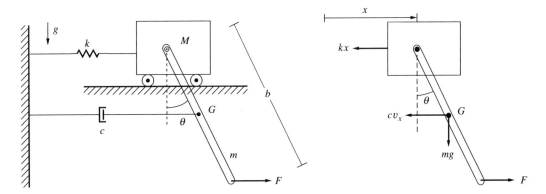

Figure 4.20   **Figure 4.21**   Free body diagram

Solution

This is a two degree of freedom system, and we select the generalized coordinates as the displacement of the mass x and rotation of the bar θ. The free-body diagram of the entire system is shown in Fig. 4.21. The kinetic energy of the cart is $T_{\text{cart}} = \frac{1}{2} M \dot{x}^2$. The kinetic energy of the bar is due to the translation and rotation and can be expressed as

$$T_{\text{bar}} = \frac{1}{2} I_G \dot{\theta}^2 + \frac{1}{2} m (v_x^2 + v_y^2) \quad \text{[a]}$$

where I_G is the mass moment of inertia about the center of mass, $I_G = mb^2/12$, and v_x and v_y are the velocities of the center of mass of the bar, found as

$$v_x = \frac{d}{dt} x_G = \frac{d}{dt}\left(x + \frac{b}{2}\sin\theta\right) = \dot{x} + \frac{b}{2}\dot{\theta}\cos\theta$$

$$v_y = \frac{d}{dt} y_G = \frac{d}{dt}\left(-\frac{b}{2}\cos\theta\right) = \frac{b}{2}\dot{\theta}\sin\theta \quad \text{[b]}$$

The total kinetic energy is

$$T = \frac{1}{2} M \dot{x}^2 + \frac{1}{24} mb^2 \dot{\theta}^2 + \frac{1}{2} m \left[\left(\dot{x} + \frac{b}{2}\dot{\theta}\cos\theta\right)^2 + \left(\frac{b}{2}\dot{\theta}\sin\theta\right)^2\right]$$

$$= \frac{1}{2}(M + m)\dot{x}^2 + \frac{1}{2} mb\dot{\theta}\dot{x}\cos\theta + \frac{1}{6} mb^2 \dot{\theta}^2 \quad \text{[c]}$$

The potential energy is due to the deflection of the spring and the vertical movement of the center of mass of the bar, written

$$V = \frac{1}{2} kx^2 - mg\frac{b}{2}\cos\theta \quad \text{[d]}$$

The virtual work of the nonconservative forces is due to the external force F and the dashpot, so

$$\delta W_{nc} = F\,\delta r - cv_x\,\delta x_G = F\,\delta(x + b\sin\theta) - c\left[\dot{x} + \frac{b}{2}\dot{\theta}\cos\theta\right]\left(\delta x + \frac{b}{2}\cos\theta\,\delta\theta\right)$$

$$= F\,\delta x + Fb\cos\theta\,\delta\theta - c\dot{x}\,\delta x - \frac{1}{2} cb\dot{\theta}\cos\theta\,\delta x - \frac{1}{2} cb\dot{x}\cos\theta\,\delta\theta - \frac{1}{4} cb^2\dot{\theta}\cos^2\theta\,\delta\theta$$

$$\text{[e]}$$

from which we recognize the generalized forces as

$$Q_x = F - c\dot{x} - \frac{1}{2}cb\cos\theta \qquad Q_\theta = Fb\cos\theta - \frac{1}{4}cb^2\dot{\theta}\cos^2\theta - \frac{1}{2}cb\dot{x}\cos\theta \qquad \text{[f]}$$

Taking the appropriate derivatives, we obtain

$$\frac{\partial T}{\partial \dot{x}} = (M+m)\dot{x} + \frac{1}{2}mb\dot{\theta}\cos\theta \qquad \frac{\partial T}{\partial \dot{\theta}} = \frac{1}{2}mb\dot{x}\cos\theta + \frac{1}{3}mb^2\dot{\theta}$$

$$-\frac{\partial T}{\partial x} = 0 \qquad \frac{\partial V}{\partial x} = kx \qquad -\frac{\partial T}{\partial \theta} = \frac{1}{2}mb\dot{\theta}\dot{x}\sin\theta \qquad \frac{\partial V}{\partial \theta} = \frac{1}{2}mgb\sin\theta \qquad \text{[g]}$$

Substituting the above values into Lagrange's equations we obtain the equations of motion as

$$(M+m)\ddot{x} + \frac{1}{2}mb\ddot{\theta}\cos\theta - \frac{1}{2}mb\dot{\theta}^2\sin\theta + c\dot{x} + \frac{1}{2}cb\dot{\theta}\cos\theta + kx = F$$

$$\frac{1}{3}mb^2\ddot{\theta} + \frac{1}{2}mb\ddot{x}\cos\theta + \frac{1}{2}cb\dot{x}\cos\theta + \frac{1}{4}cb^2\dot{\theta}\cos^2\theta + \frac{1}{2}mgb\sin\theta = Fb\cos\theta \qquad \text{[h]}$$

Example 4.12

Figure 4.22 shows a collar of mass m sliding outside a long, slender rod of mass M and length L. The coefficient of friction between the rod and collar is μ. There is a force F acting at the tip of the rod. Find the equations of motion.

Solution

We will solve this problem as a two degree of freedom unconstrained system. Polar coordinates are suitable as generalized coordinates. The free-body diagrams are given in Fig. 4.23. There are four external forces: two gravity forces, which we will account for in the potential energy, the friction force, and the force at the tip.

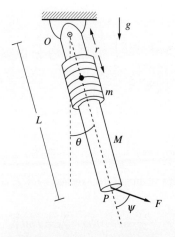
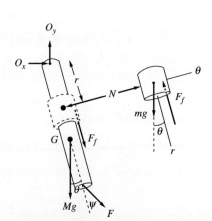

Figure 4.22 Collar sliding on a rod

Figure 4.23 Free-body diagram

The position and velocity of the collar are

$$\mathbf{r} = r\mathbf{e}_r \qquad \mathbf{v} = \dot{r}\mathbf{e}_r + r\dot{\theta}\mathbf{e}_\theta \qquad \text{[a]}$$

The virtual work associated with the two external forces can be written as

$$\delta W = \mathbf{F} \cdot \delta \mathbf{r}_P + \mathbf{F}_f \cdot \delta \mathbf{r} \qquad \text{[b]}$$

in which

$$\mathbf{F} = F\cos\psi\,\mathbf{e}_r + F\sin\psi\,\mathbf{e}_\theta \quad \mathbf{F}_f = -F_f\,\text{sign}(\dot{r})\mathbf{e}_r \quad \delta\mathbf{r} = \delta r\,\mathbf{e}_r + r\delta\theta\,\mathbf{e}_\theta \quad \delta\mathbf{r}_P = L\delta\theta\,\mathbf{e}_\theta \qquad \text{[c]}$$

so that the virtual work becomes

$$\delta W = -F_f\,\text{sign}(\dot{r})\,\delta r + FL\sin\psi\,\delta\theta = Q_r\,\delta r + Q_\theta\,\delta\theta \qquad \text{[d]}$$

with $Q_r = -F_f\,\text{sign}(\dot{r})$ and $Q_\theta = FL\sin\psi$ as the generalized forces due to the nonconservative forces.

The kinetic energy is

$$T = T_{\text{rod}} + T_{\text{collar}} = \frac{1}{6}ML^2\dot{\theta}^2 + \frac{1}{2}m(\dot{r}^2 + r^2\dot{\theta}^2) \qquad \text{[e]}$$

and the potential energy is

$$V = -mgr\cos\theta - Mg\frac{L}{2}\cos\theta \qquad \text{[f]}$$

Application of Lagrange's equations yields the equations of motion as

$$m\ddot{r} - mr\dot{\theta}^2 - mg\cos\theta = -F_f\,\text{sign}(\dot{r}) \qquad \text{[g]}$$

$$\left(\frac{1}{3}ML^2 + mr^2\right)\ddot{\theta} + 2mr\dot{r}\dot{\theta} + \left(mr + \frac{1}{2}ML\right)g\sin\theta = FL\sin\psi \qquad \text{[h]}$$

The friction force is related to the normal force N between the collar and rod by $F_f = \mu N$. However, at this point we do not know what the normal force is. To find the normal force, we need to go to a Newtonian analysis. Reconsidering the free-body diagram and summing forces along the transverse direction, we obtain

$$\sum F_\theta = m(r\ddot{\theta} + 2\dot{r}\dot{\theta}) = N - mg\sin\theta \qquad \text{[i]}$$

from which we obtain the magnitude of the normal force as

$$N = m(r\ddot{\theta} + 2\dot{r}\dot{\theta} + g\sin\theta) \qquad \text{[j]}$$

We can eliminate the normal force from the equations of motion by introducing Eq. [j] into Eq. [g]. Note that the friction force is always a positive quantity, as it is proportional to the magnitude of the normal force. The expression involving N in Eq. [j] can lead to both positive and negative values. Therefore, we express the friction force as

$$F_f = \mu|N| = \mu m|r\ddot{\theta} + 2\dot{r}\dot{\theta} + g\sin\theta| \qquad \text{[k]}$$

and use Eq. [k] in the equations of motion.

The preceding example illustrates the problems that one encounters when dealing with problems involving friction. As stated earlier, friction is not a constraint force, but its magnitude depends on a constraint force. If we select a set of unconstrained generalized coordinates to describe the motion, as we did in this example, we cannot obtain the magnitudes of the friction force without an additional Newtonian analysis. In the next section, we will see an analytical approach that calculates magnitudes of constraint forces.

4.10 Lagrange's Equations for Constrained Systems

The formulation of Lagrange's equations in the previous section was for unconstrained systems and for constrained systems where the generalized coordinates are selected such that all constraints are accounted for and the surplus coordinates eliminated. This approach is not feasible under a number of circumstances:

1. When the constraints are nonholonomic. Because nonholonomic constraints involve velocity expressions that cannot be integrated to displacement expressions, one cannot find a set of unconstrained generalized coordinates.

2. When the constraints are holonomic and one cannot eliminate the surplus coordinates easily, for one of the following reasons:
   a. The constraint equation is complicated.
   b. Finding the transformations that lead to unconstrained equations makes the equations of motion very complicated.
   c. Some of the forces acting on the system are functions of constraint forces.

3. When the constraints are holonomic but one does not want to eliminate the surplus coordinates from the formulation, usually because of the need to know the amplitudes of the reaction forces.

Consider a system originally of n degrees of freedom, to which m constraints are applied. For the most general case, we express the constraints in velocity form as

$$\sum_{k=1}^{n} a_{jk} \dot{q}_k + a_{j0} = 0 \qquad j = 1, 2, \ldots, m \qquad \textbf{[4.10.1]}$$

whose variation is

$$\sum_{k=1}^{n} a_{jk} \delta q_k = 0 \qquad \textbf{[4.10.2]}$$

Multiplying Eq. [4.10.2] by the Lagrange multipliers λ_j ($j = 1, 2, \ldots, m$) and introducing these constraints to the extended Hamilton's principle, we obtain

$$\int_{t_1}^{t_2} \delta L \, dt + \int_{t_1}^{t_2} \delta W_{nc} \, dt - \int_{t_1}^{t_2} \sum_{j=1}^{m} \sum_{k=1}^{n} \lambda_j a_{jk} \delta q_k \, dt = 0 \qquad \textbf{[4.10.3]}$$

When the constraints are holonomic, the coordinates $q_1, q_2, \ldots, q_n$ no longer constitute a set of independent generalized coordinates. They are now *constrained* generalized coordinates. When the constraints are nonholonomic, only the generalized velocities are constrained, while the generalized coordinates are still independent. In both cases, the variations of the generalized coordinates are constrained. Following the same procedure as when deriving Lagrange's equations for the unconstrained case, we take the appropriate partial derivatives and perform the integrations by parts to obtain

$$\sum_{k=1}^{n} \int_{t_1}^{t_2} \left[-\frac{d}{dt}\left(\frac{\partial L}{\partial \dot{q}_k}\right) + \frac{\partial L}{\partial q_k} + Q_{knc} - \sum_{j=1}^{m} \lambda_j a_{jk} \right] \delta q_k \, dt = 0 \quad \textbf{[4.10.4]}$$

As in the static case, we select the Lagrange multipliers λ_j such that the coefficients of δq_k ($k = 1, 2, \ldots, n$) vanish, which leads to a modified form of Lagrange's equations, written

$$\frac{d}{dt}\left(\frac{\partial L}{\partial \dot{q}_k}\right) - \frac{\partial L}{\partial q_k} + \sum_{j=1}^{m} \lambda_j a_{jk} = Q_{knc} \qquad k = 1, 2, \ldots, n \quad \textbf{[4.10.5]}$$

where $\lambda_j a_{jk}$ are the *generalized constraint forces*. They have the same units as the generalized forces (which do not necessarily have the units of force). In column vector notation, Eq. [4.10.5] is expressed as

$$\frac{d}{dt}\left(\frac{\partial L}{\partial \{\dot{q}\}}\right) - \frac{\partial L}{\partial \{q\}} + \{\lambda\}^T [a] = \{Q_{nc}\}^T \quad \textbf{[4.10.6]}$$

in which $[a]$ is a matrix of order $m \times n$ whose entries are a_{jk} and $\{\lambda\}$ is a column vector of order m that contains the Lagrange multipliers.

After obtaining the equations of motion, one has two courses of action for finding a solution. The first is to eliminate the Lagrange multipliers from the equations of motion and obtain a set of $n - m$ unconstrained equations. One accomplishes this by algebraic manipulation of the equations of motion. Many times, such an approach results in complicated expressions.

The second course of action is to take the n equations of motion in Eq. [4.10.5] and the m constraint relations in Eq. [4.10.1] and then to solve them together for the $n + m = p + 2m$ unknowns $q_1, q_2, \ldots, q_n, \lambda_1, \lambda_2, \ldots, \lambda_m$. The resulting $n + m$ equations are not a set of differential equations, as there is no derivative of the Lagrange multipliers involved. Such equations are known as *differential-algebraic* equations. Their analysis requires a different treatment than that for differential equations.

When the constraints are holonomic and expressed in the configuration form [4.3.2], one can add them to the extended Hamilton's principle by

$$\int_{t_1}^{t_2} \delta L \, dt + \int_{t_1}^{t_2} \delta W_{nc} \, dt - \int_{t_1}^{t_2} \sum_{j=1}^{m} \lambda_j \delta c_j(q_1, q_2, \ldots, q_n) = 0 \quad \textbf{[4.10.7]}$$

and obtain the contribution of the constraint by replacing a_{jk} in Eq. [4.10.5] with $\partial c_j/\partial q_k$. Or, we can add them directly to the Lagrangian as

$$\hat{L} = T - V - \sum_{j=1}^{m} \lambda_j c_j(q_1, q_2, \ldots, q_n) \quad \textbf{[4.10.8]}$$

262 CHAPTER 4 • ANALYTICAL MECHANICS: BASIC CONCEPTS

When the objective is to obtain the amplitude of a constraint force, an analytical approach that can be used is the *constraint relaxation* method. This method is mathematically equivalent to the Lagrange multiplier approach. However, it is more intuitive and it is particularly useful when dealing with holonomic constraints expressed in configuration form. Following is a description of the method.

We relax the constraint from the formulation and represent the effects of the constraint by a constraint force. Then, we write the Lagrangian and virtual work. The constraint force enters the formulation via the virtual work. We invoke Lagrange's equations and obtain the equations of motion. We next impose the constraint, which enables us to calculate the magnitude of the constraint force.

Example 4.13

Consider Example 4.12, in which a collar of mass m is sliding on a rod of mass M and length L. The coefficient of friction between the rod and collar is μ. Obtain the equations of motion using constrained generalized coordinates and find the value of the normal force N.

Solution

To describe this system in terms of constrained generalized coordinates, consider the rod and the collar separately. We express the motion of the collar using polar coordinates, r and θ, as in Example 4.12. To express the motion of the rod, we introduce another angle, ϕ. The constraint equation is

$$\theta - \phi = 0 \qquad [a]$$

The kinetic and potential energy has the same form as in Example 4.12. We write them here in terms of the constrained generalized coordinates as

$$T = \frac{1}{6}ML^2\dot{\phi}^2 + \frac{1}{2}m(\dot{r}^2 + r^2\dot{\theta}^2) \qquad V = -mgr\cos\theta - Mg\frac{L}{2}\cos\phi \qquad [b]$$

The normal force N acts in the transverse direction and it contributes to the virtual work. Considering that the velocity of the collar in the transverse direction is $v_\theta = r\dot{\theta}\mathbf{e}_\theta$, we write the virtual work expression as

$$\delta W = -F_f \text{sign}(\dot{r})\,\delta r + FL\sin\psi\,\delta\theta + Nr(\delta\theta - \delta\phi) \qquad [c]$$

We obtain the Lagrange's equations as

$$\text{For } r \rightarrow m\ddot{r} - mr\dot{\theta}^2 - mg\cos\theta = -F_f \text{sign}(\dot{r}) \qquad [d]$$

$$\text{For } \theta \rightarrow mr^2\ddot{\theta} + 2mr\dot{r}\dot{\theta} + mgr\sin\theta = Nr \qquad [e]$$

$$\text{For } \phi \rightarrow \frac{1}{3}ML^2\ddot{\phi} + \frac{1}{2}MgL\sin\phi = -Nr + FL\sin\psi \qquad [f]$$

These equations have to be solved together with Eq. [a]. Equation [d] is the same as Eq. [g] in Example 4.12, and if we add Eqs. [e] and [f] and use the constraint equation [a] we eliminate the normal force and obtain Eq. [h] in Example 4.12. From Eq. [e], we find the normal force as

$$N = mr\ddot{\theta} + 2m\dot{r}\dot{\theta} + mg\cos\theta \qquad [g]$$

which is the same value obtained in Example 4.12. The friction force is given in Eq. [k] of Example 4.12.

4.10 LAGRANGE'S EQUATIONS FOR CONSTRAINED SYSTEMS

The difference between the approach here and the approach in Example 4.12 is that here we calculated the normal force directly from the Lagrange's equations, while in Example 4.12 we conducted a force balance in addition to the Lagrange's equations.

Example 4.14

Consider the vehicle in Fig. 4.8. Given that the velocity of point A is along the line of symmetry of the vehicle, derive the equations of motion. Gravity acts perpendicular to the plane of motion.

Solution

We denote the coordinates of the center of mass by X and Y and select the generalized coordinates as X, Y, and θ. The kinetic energy of the vehicle is

$$T = \frac{1}{2}m(\dot{X}^2 + \dot{Y}^2) + \frac{1}{2}I_G \dot{\theta}^2 \qquad [a]$$

where m is the mass and I_G is the centroidal mass moment of inertia. There is no potential energy, and the virtual work expression involves the two forces F_C and F_D. We can find the virtual work conveniently by calculating the velocities of points C and D. Defining a coordinate system xy attached to the vehicle, we write the velocities of points G and A as

$$\mathbf{v}_G = \dot{X}\mathbf{I} + \dot{Y}\mathbf{J} = (\dot{X}\cos\theta + \dot{Y}\sin\theta)\mathbf{i} + (-\dot{X}\sin\theta + \dot{Y}\cos\theta)\mathbf{j} \qquad [b]$$

$$\mathbf{v}_A = \mathbf{v}_G + \dot{\theta}\mathbf{k} \times -L\mathbf{i} = (\dot{X}\cos\theta + \dot{Y}\sin\theta)\mathbf{i} + (-\dot{X}\sin\theta + \dot{Y}\cos\theta - L\dot{\theta})\mathbf{j} \qquad [c]$$

The constraint is defined as

$$f = \mathbf{v}_A \cdot \mathbf{j} = -\dot{X}\sin\theta + \dot{Y}\cos\theta - L\dot{\theta} = 0 \qquad [d]$$

thus the velocity of A reduces to

$$\mathbf{v}_A = (\dot{X}\cos\theta + \dot{Y}\sin\theta)\mathbf{i} \qquad [e]$$

and the variation of the constraint becomes

$$\delta f = \sin\theta \, \delta X - \cos\theta \, \delta Y + L \, \delta\theta = 0 \qquad [f]$$

Hence, the velocities of C and D become

$$\mathbf{v}_C = \mathbf{v}_A + \dot{\theta}\mathbf{k} \times h\mathbf{j} = (\dot{X}\cos\theta + \dot{Y}\sin\theta - h\dot{\theta})\mathbf{i} \qquad [g]$$

$$\mathbf{v}_D = \mathbf{v}_A + \dot{\theta}\mathbf{k} \times (-h\mathbf{j}) = (\dot{X}\cos\theta + \sin\theta + h\dot{\theta})\mathbf{i} \qquad [h]$$

The external forces are $\mathbf{F}_C = F_C\mathbf{i}$, $\mathbf{F}_D = F_D\mathbf{i}$, so the virtual work expression becomes

$$\delta \hat{W} = \mathbf{F}_C \cdot \delta \mathbf{r}_C + \mathbf{F}_D \cdot \delta \mathbf{r}_D + \lambda \, \delta f$$
$$= (F_C + F_D)\cos\theta \, \delta X + (F_C + F_D)\sin\theta \, \delta Y + (F_D - F_C)h \, \delta\theta$$
$$+ \lambda(\delta X \sin\theta - \delta Y \cos\theta + L \, \delta\theta) \qquad [i]$$

The physical interpretation of the Lagrange multiplier is that it is the resultant of all forces that keep $\mathbf{v}_A$ along the x axis, and it acts in a direction perpendicular to this axis. Introducing Eqs. [a] and [i] into Lagrange's equations, we obtain the equations of motion as

$$mÿ\ddot{X} = (F_C + F_D)\cos\theta + \lambda\sin\theta \qquad \text{[j]}$$

$$m\ddot{Y} = (F_C + F_D)\sin\theta - \lambda\cos\theta \qquad \text{[k]}$$

$$I_G\ddot{\theta} = (F_D - F_C)h + L\lambda \qquad \text{[l]}$$

Note that while deriving the equations of motion we did not introduce Eq. [d] directly into the expression for the kinetic energy, thus eliminating one of the generalized coordinates from the outset. Had we done so, we would have eliminated the contribution due to the variation of that coordinate and ended up with an incorrect representation. This procedure is crucial to the treatment of nonholonomic constraints.

Even if we eliminated one of the generalized velocities and the Lagrange multiplier, writing the equations of motion in terms of Y and θ (or X and θ) would not give the most meaningful description of the motion. A quantity critical to the understanding of the motion is the speed of point A. If the equations of motion can be expressed in terms of that speed, one gets a clearer picture of the nature of the motion. One can introduce v_A to Eqs. [j]–[l] with a substitution.

Indeed, if we multiply Eq. [j] by $\cos\theta$ and Eq. [k] by $\sin\theta$ and add the two we obtain

$$m(\ddot{X}\cos\theta + \ddot{Y}\sin\theta) = F_C + F_D \qquad \text{[m]}$$

Recalling from Eq. [e] that $v_A = (\dot{X}\cos\theta + \dot{Y}\sin\theta)$, differentiating this expression we obtain

$$\dot{v}_A = \ddot{X}\cos\theta + \ddot{Y}\sin\theta + \dot{\theta}(-\dot{X}\sin\theta + \dot{Y}\cos\theta) \qquad \text{[n]}$$

Introducing Eq. [d] to Eq. [n] we can write

$$\ddot{X}\cos\theta + \ddot{Y}\sin\theta = \dot{v}_A - L\dot{\theta}^2 \qquad \text{[o]}$$

so that Eq. [m] can be written as

$$m(\dot{v}_A - L\dot{\theta}^2) = F_C + F_D \qquad \text{[p]}$$

which is recognized as the force balance along the x direction.

We next find an expression for the Lagrange multiplier λ and introduce it to Eq. [e]. To this end, we multiply Eq. [j] with $\sin\theta$ and Eq. [k] by $-\cos\theta$ and add the two, with the result

$$m(\ddot{X}\sin\theta - \ddot{Y}\cos\theta) = \lambda \qquad \text{[q]}$$

We can introduce this relationship to Eq. [l], but a more meaningful expression can be generated if we consider Eq. [d] and differentiate it

$$\ddot{X}\sin\theta - \ddot{Y}\cos\theta + \dot{\theta}(\dot{X}\cos\theta + \dot{Y}\sin\theta) = -L\ddot{\theta} \qquad \text{[r]}$$

Considering Eq. [q], we express the Lagrange multiplier as

$$\lambda = -mv_A\dot{\theta} - mL\ddot{\theta} \qquad \text{[s]}$$

Introducing this equation into Eq. [l] we obtain

$$(I_G + mL^2)\ddot{\theta} + mLv_A\dot{\theta} = (F_D - F_C)h \qquad \text{[t]}$$

which we recognize as the moment balance about point A. Equations [p] and [t] are the two independent equations of motion of the vehicle.

REFERENCES

Brenan, K. E., S. L. Campbell, and L. Petzold. *Numerical Solution of Initial-Value Problems in Differential-Algebraic Equations.* New York: Elsevier North Holland, 1989.

Ginsberg, J. H. *Advanced Engineering Dynamics.* New York: Harper & Row, 1988.

Goldstein, H. *Classical Mechanics.* 2nd ed. Reading, MA: Addison-Wesley, 1980.

Greenwood, D. T. *Principles of Dynamics.* 2nd ed. Englewood Cliffs, NJ: Prentice-Hall, 1988.

Haug, E. J. *Intermediate Dynamics.* Englewood Cliffs, NJ: Prentice-Hall, 1992.

Lanczos, C. *The Variational Principles of Mechanics.* 4th ed. New York: Dover Publications, 1970.

Meirovitch, L. *Methods of Analytical Dynamics.* New York: McGraw-Hill, 1970.

HOMEWORK EXERCISES

SECTION 4.3

1. The four-bar linkage in Fig. 4.24 is a single degree of freedom system. Show that this is so by separating the mechanism into its three components and by writing the constraint equations that relate the configurations of the links.

2. A bead slides up a spiral of constant radius R and height h, as shown in Fig. 4.25. It takes the bead six full turns to reach the top. Express the characteristics of the path of the bead as a constraint relation.

3. A particle slides inside a smooth paraboloid of revolution described by $z = r^2/b$, as shown in Fig. 4.26. Using cylindrical coordinates, find an expression for the constraint force on the particle.

4. The radar tracking of a moving vehicle by another moving vehicle is a common problem. Consider the two vehicles A and B in Fig. 4.27. The orientation of vehicle A must always be toward vehicle B. Express the constraint relation between the velocities and distance between the two vehicles and determine whether this is a holonomic constraint or not.

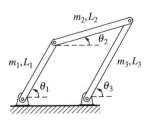

Figure 4.24 Four-bar linkage

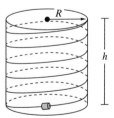

Figure 4.25

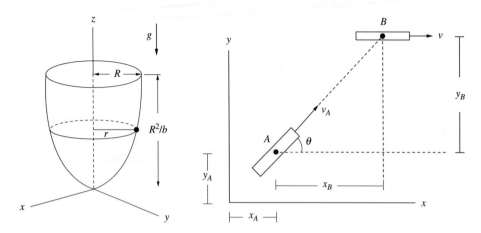

Figure 4.26   **Figure 4.27**

5. Consider the double pendulum in Fig. 4.3. It is desired to have the velocity of the tip of the pendulum point toward the pinned end O. Express this condition as a constraint and determine whether the constraint is holonomic or not.

Section 4.4

6. Consider a particle moving along a path and the description of motion by path variables. Express the force keeping the particle moving along the path in terms of its components in the tangential, normal, and binormal directions and evaluate the virtual work expression. Identify which of these forces are constraint forces and verify Eq. [4.4.14].

7. Find the virtual displacement of point P in Fig. 4.28. The mass is suspended from an arm which is attached to a rotating column. The pendulum swings in the plane generated by the column and arm.

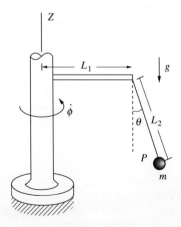

Figure 4.28

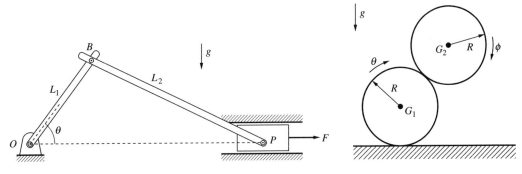

Figure 4.29 Slider-crank mechanism

Figure 4.30

8. Express the virtual displacement of the slider in the slider-crank mechanism shown in Fig. 4.29 using (a) the relative velocity relations, and (b) the analytical expressions.
9. A uniform solid cylinder of radius R rolls without slip on a horizontal plane and an identical cylinder rolls without slip on it (Fig. 4.30). Find the virtual displacements of the centers of the cylinders.
10. Consider Fig. 4.2 and the case when the cord is getting pulled down by an external force, such that the length of the cord varies by $L(t) = L_0 e^{-0.2t}$. Find the virtual displacement of the mass.

Section 4.5

11. Find the generalized force associated with the system in Fig. 4.29.
12. The spherical pendulum of mass m shown in Fig. 4.2 has its length being reduced by a force F, according to the relationship $L(t) = L_0 - bt$, where L_0 is the initial length and b is a constant. Calculate the generalized forces using spherical coordinates as generalized coordinates.
13. Consider Fig. 4.13, and calculate the associated generalized forces. The disk is of mass m and the rod is of mass $2m$. There is a moment M acting on the rod at the pin joint.

Section 4.6

14. For the two links attached to a spring as shown in Fig. 4.31, find the equilibrium position. The spring is not stretched when the rods are horizontal.
15. Find the equilibrium position of the rod of mass m and length L sliding in the guide bars shown in Fig. 4.32. The spring is not stretched when the rod is vertical. The sliders are massless and the contact between the horizontal slider and the guide bar involves friction with coefficient μ.
16. Find the equilibrium position for the system shown in Fig. 4.33, with the middle mass equal to zero. Assume that the displacements are small, and that the springs

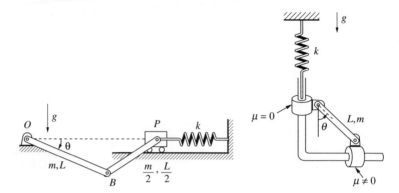

Figure 4.31

Figure 4.32

deflect only in the vertical direction. Use as generalized coordinates the translation of the center of the rod and the rotation of the rod. Then, use the deflections of the springs at A and B as generalized coordinates and obtain the equilibrium configuration. Compare the results.

17. Consider the two systems in Figs. 4.3 and 4.14 and set up the equations to find the magnitude and direction of the force **F** necessary to keep the systems at equilibrium at $\theta_1 = 30°, \theta_2 = 15°$.
18. Find the equilibrium position of the system in Fig. 4.34.
19. Find the equilibrium position of the pulley system in Fig. 1.77 (Problem 1.36). Use the constrained coordinate approach.

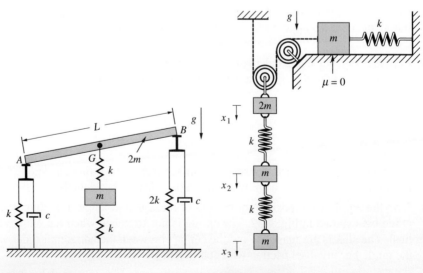

Figure 4.33

Figure 4.34

SECTION 4.7

20. Find the equation of motion of the rod in Fig. 4.32 using D'Alembert's principle.
21. Find the equations of motion of the pulley system in Fig. 3.40 using D'Alembert's principle. The pulleys are massless.

SECTION 4.8

22. Find the equation of motion of the system mechanism in Fig. 4.34 by Hamilton's principle.
23. Find the equation of motion of the rod in Fig. 4.32 by Hamilton's principle.

SECTION 4.9

24. Find the equations of motion of the system in Fig. 4.35 using Lagrange's equations.
25. Use Lagrange's equations to derive the equations of motion for the Foucault's pendulum in Chapter 2.
26. Figure 4.36 depicts a simplified illustration of a spacecraft to which a robot arm is attached at the center of mass. The robot arm moves by a moment T exerted to it at the pin joint by a motor on the spacecraft. Considering only plane motion for both the spacecraft and the robot arm, derive the equations of motion by
    a. Separating the two masses, writing force and moment balances, and eliminating constraints.
    b. Using Lagrange's equations. Compare the complexity in both cases.
27. A block of mass m and length L is positioned over a semicircular block (Fig. 4.37). It is given that friction is sufficient to prevent slipping. Derive the equation of the rod as it rocks over the semicircular block.

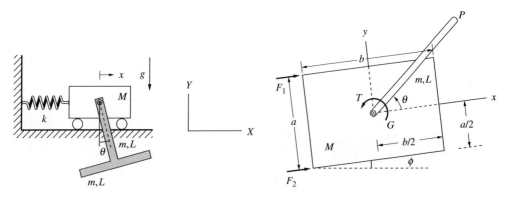

Figure 4.35   **Figure 4.36**

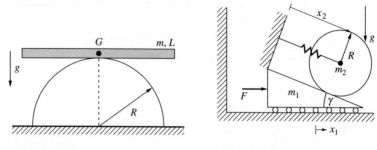

Figure 4.37   **Figure 4.38**

28. A cylinder of mass m_2 and radius R rolls without slipping on a wedge of mass m_1 (Fig. 4.38). The wedge is moving under the influence of the force F with no friction. Obtain the equations of motion.
29. Find the equation of motion of the rod in Fig. 4.32 using Lagrange's equations.
30. Find the equations of motion of the pulley system in Fig. 3.40 using Lagrange's equations.
31. Find the equation of motion of the bead in Fig. 4.39 sliding without friction in the parabolic tube described by $z = x^2/4$, while the tube is rotating about the z axis with constant angular velocity Ω.
32. Find the equations of motion of the system in Fig. 4.33, using Lagrange's equations. Assume small motions and that the springs and dashpots deflect only vertically.
33. A particle of mass m is constrained to slide without friction down a channel attached to a cone spinning with constant angular velocity Ω, as shown in Fig. 4.40. Derive the following:
    a. A differential equation of motion describing the motion of P using Newton's second law.
    b. The equation of motion using Lagrange's equations.

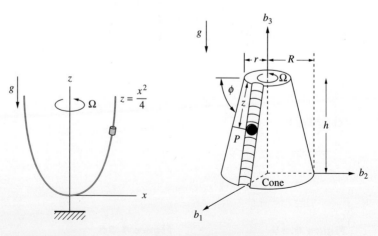

Figure 4.39   **Figure 4.40**

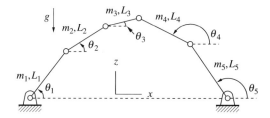

Figure 4.41

34. Consider the pendulum in Fig. 4.28 of mass m, $L_1 = L$, $L_2 = 2L$. The pendulum swings in the plane generated by the column and arm. Derive the equations of motion, using the pendulum angle θ and rotation angle ϕ of the shaft as generalized coordinates. The combined mass moment of inertia of the column and arm about the Z axis is I.

35. Consider the particle in Problem 3. Find the equations of motion.

SECTION 4.10

36. Consider the four-bar linkage mechanism in Fig. 4.24 and derive the equations of motion using constrained generalized coordinates. Consider each link separately. A moment M acts on the first link.

37. Consider the platform robot in Fig. 4.41. Derive the equations of motion using constrained coordinates for the following cases: *(a)* each link is considered separately; and *(b)* links 1 and 2 and links 4 and 5 are each considered as one system each.

38. Given the two-link system in Fig. 4.3, derive the equations of motion using the constrained coordinate formulation and using as generalized coordinates θ_1, θ_2, x_P, and y_P. Calculate the kinetic energy of the second link using the motion of its center of mass, expressed in terms of θ_2, x_P, and y_P.

39. Consider the vehicle in Fig. 4.8. We are given the constraint that the velocity of the tip of the vehicle, that is, point E, is along the x direction. Derive the equations of motion using constrained generalized coordinates.

40. Consider the pendulum in Fig. 4.28. The angular velocity of the column is kept constant at $\dot\phi = \Omega$ by a motor that generates a moment about the Z axis. Find the equation of motion of the pendulum and using the constraint relaxation method find the moment necessary to maintain the constant angular velocity of the column.

41. Find the equation of motion of the rod in Fig. 4.32 using the constraint relaxation method.

chapter 5

ANALYTICAL MECHANICS: ADDITIONAL TOPICS

5.1 INTRODUCTION

This chapter continues with the analytical techniques discussed in the previous chapter and introduces additional concepts. We first discuss natural and nonnatural systems, and revisit the concept of equilibrium. We consider small motions around equilibrium and derive the linearized equations of motion about equilibrium. A new way of describing damped systems is introduced. We analyze the response of linearized systems, which in essence is multidegree of freedom vibrations. We look at cyclic coordinates and generalized momenta and discuss integrals of the motion. We develop Routh's method to eliminate cyclic coordinates. Then we revisit the idea of impulsive motion from a Lagrangian perspective.

We continue with Hamilton's canonical equations. These constitute an alternative to Lagrange's equations and have the advantage of being first-order equations. The chapter ends with a look at computational issues involving the derivation and integration of the equations of motion and with the introduction of additional differential variational principles, namely Jourdain's and Gauss's. These additional variational principles make it easier to analyze nonholonomic systems.

The concepts in this chapter enhance the understanding of the principles studied in Chapter 4, as well as providing additional insight and perspective. The reader is always encouraged to analyze a problem using more than one perspective.

5.2 NATURAL AND NONNATURAL SYSTEMS, EQUILIBRIUM

Consider the kinetic energy of a system of N particles that has n degrees of freedom

$$T = \frac{1}{2} \sum_{i=1}^{N} m_i \mathbf{v}_i \cdot \mathbf{v}_i \qquad [5.2.1]$$

For the most general case, the position vector of particle i can be expressed as

$$\mathbf{r}_i = \mathbf{r}_i(q_1, q_2, \ldots, q_n, t) \quad [\mathbf{5.2.2}]$$

We have not excluded the possibility that displacements $\mathbf{r}_i$ ($i = 1, 2, \ldots, N$) have an explicit dependence on time. This may seem strange at first, as the generalized coordinates should be sufficient to describe the evolution of the system completely. Viewing $\mathbf{r}_i$ also as explicit functions of time implies that the system has a prescribed motion or that part of the motion is treated as known. From Newton's third law, the motion of each body is related to the motion of every other body that it interacts with; thus, in absolute reality such prescribed motion does not exist. However, in several cases reasonable approximations can be made. For example, when we analyze the motion of a body on the surface of the earth, the motion of that body is influenced by the motion of the earth. The motion of the earth is not affected by the motion of the body, and it is treated as known.

The time derivative of $\mathbf{r}_i$ is

$$\mathbf{v}_i = \frac{d\mathbf{r}_i}{dt} = \frac{\partial \mathbf{r}_i}{\partial q_1}\dot{q}_1 + \frac{\partial \mathbf{r}_i}{\partial q_2}\dot{q}_2 + \ldots + \frac{\partial \mathbf{r}_i}{\partial q_n}\dot{q}_n + \frac{\partial \mathbf{r}_i}{\partial t} \qquad i = 1, 2, \ldots, N \quad [\mathbf{5.2.3}]$$

or

$$\mathbf{v}_i = \sum_{k=1}^{n} \frac{\partial \mathbf{r}_i}{\partial q_k}\dot{q}_k + \frac{\partial \mathbf{r}_i}{\partial t} \quad [\mathbf{5.2.4}]$$

Introducing Eq. [5.2.4] into Eq. [5.2.1], we obtain

$$T = \frac{1}{2}\sum_{i=1}^{N} m_i \left[\left(\sum_{k=1}^{n} \frac{\partial \mathbf{r}_i}{\partial q_k}\dot{q}_k + \frac{\partial \mathbf{r}_i}{\partial t} \right) \cdot \left(\sum_{s=1}^{n} \frac{\partial \mathbf{r}_i}{\partial q_s}\dot{q}_s + \frac{\partial \mathbf{r}_i}{\partial t} \right) \right]$$

$$= \frac{1}{2}\sum_{i=1}^{N} m_i \left(\sum_{k=1}^{n}\sum_{s=1}^{n} \frac{\partial \mathbf{r}_i}{\partial q_k} \cdot \frac{\partial \mathbf{r}_i}{\partial q_s}\dot{q}_k\dot{q}_s + 2\frac{\partial \mathbf{r}_i}{\partial t} \cdot \sum_{k=1}^{n}\frac{\partial \mathbf{r}_i}{\partial q_k}\dot{q}_k + \frac{\partial \mathbf{r}_i}{\partial t} \cdot \frac{\partial \mathbf{r}_i}{\partial t} \right) \quad [\mathbf{5.2.5}]$$

Denoting by

$$\alpha_{ks} = \sum_{i=1}^{N} m_i \frac{\partial \mathbf{r}_i}{\partial q_k} \cdot \frac{\partial \mathbf{r}_i}{\partial q_s} \qquad k, s = 1, 2, \ldots, n$$

$$\beta_k = \sum_{i=1}^{N} m_i \frac{\partial \mathbf{r}_i}{\partial t} \cdot \frac{\partial \mathbf{r}_i}{\partial q_k} \qquad \tau = \frac{1}{2}\sum_{i=1}^{N} m_i \frac{\partial \mathbf{r}_i}{\partial t} \cdot \frac{\partial \mathbf{r}_i}{\partial t} \quad [\mathbf{5.2.6}]$$

where α_{ks}, β_k, and τ are functions of the generalized coordinates and time, one can express the kinetic energy as

$$T = T_2 + T_1 + T_0 \quad [\mathbf{5.2.7}]$$

in which

$$T_2 = \frac{1}{2}\sum_{k=1}^{n}\sum_{s=1}^{n}\alpha_{ks}\dot{q}_k\dot{q}_s \qquad T_1 = \sum_{k=1}^{n}\beta_k\dot{q}_k \qquad T_0 = \tau \quad [\mathbf{5.2.8}]$$

Examining the nature of the terms that contribute to the kinetic energy, we notice that T_2 is quadratic in the generalized velocities. T_1 is linear in the generalized velocities and it is usually related to Coriolis effects. T_0 has no generalized velocity terms and is usually related to centrifugal effects. Both T_1 and T_0 exist because of the explicit time dependence of $\mathbf{r}_i$ due to a component of the motion being treated as known.

The kinetic energy can be expressed in matrix form as

$$T = \frac{1}{2}\{\dot{q}\}^T [M]\{\dot{q}\} + \{\beta\}^T \{\dot{q}\} + \tau \qquad [5.2.9]$$

in which $\{\dot{q}\} = [\dot{q}_1 \dot{q}_2 \ldots \dot{q}_n]^T$ is the generalized velocity vector, $\{\beta\} = [\beta_1 \beta_2 \ldots \beta_n]^T$ is referred to as the *gyroscopic vector,* and the matrix

$$[M] = \begin{bmatrix} \alpha_{11} & \alpha_{12} & \ldots & \alpha_{1n} \\ \alpha_{21} & \alpha_{22} & \ldots & \alpha_{2n} \\ \vdots & \vdots & & \vdots \\ \alpha_{n1} & \alpha_{n2} & \ldots & \alpha_{nn} \end{bmatrix} \qquad [5.2.10]$$

is referred to as the *mass matrix* or the *inertia matrix*. One can show that $[M]$ is symmetric and positive definite.

When $T_1 = T_0 = 0$, the system is called a *natural system,* and when T_1 or T_0 are not zero the system is called *nonnatural*. The name *nonnatural* is associated with the fact that in a nonnatural system, a component of the motion is treated and/or assumed as known; hence, the system being observed is artificial and not a natural one. One should always keep in mind the procedures used and the assumptions made when viewing a system as nonnatural and evaluate whether the assumptions are realistic or not.

The Foucault's pendulum studied in Chapter 2 represents a nonnatural system, as we treat the motion of the earth as known and unaffected by the motion of the pendulum. So is the bead on a hoop example considered in Chapter 4, where we assume that the motion of the bead does not affect the rotation of the hoop.

As an illustration of what constitutes a nonnatural system, consider a particle of mass m sliding along an incline σ with speed $\dot{q}$, as shown in Fig. 5.1. The incline (of mass M) is moving in the horizontal direction with speed $\dot{w}$. The total

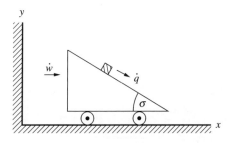

Figure 5.1   Mass sliding on a moving wedge

velocity of the particle is

$$\mathbf{v} = \dot{w}\mathbf{i} + \dot{q}\cos\sigma\mathbf{i} - \dot{q}\sin\sigma\mathbf{j} \qquad [5.2.11]$$

so the kinetic energy becomes

$$T = \frac{1}{2}m\mathbf{v}\cdot\mathbf{v} + \frac{1}{2}M\dot{w}^2 = \frac{1}{2}m[(\dot{w} + \dot{q}\cos\sigma)^2 + (\dot{q}\sin\sigma)^2] + \frac{1}{2}M\dot{w}^2$$

$$= \frac{1}{2}m[\dot{q}^2 + 2\dot{q}\dot{w}\cos\sigma + \dot{w}^2] + \frac{1}{2}M\dot{w}^2 \qquad [5.2.12]$$

Whether this system is natural or not depends on how the velocity of the inclined surface is treated. If we consider the velocity $\dot{w}$ to be known *a priori*, then there is only one generalized coordinate, q, and the system is nonnatural. The term $2\dot{q}\dot{w}\cos\theta$ becomes linear in the generalized velocities. In addition, $\dot{w}^2$ has no generalized velocity terms. On the other hand, if we treat w as a variable, the system has two degrees of freedom and every term in the kinetic energy is quadratic in the generalized velocities. The system becomes a natural system.

Nonnatural systems can be categorized into two distinct groups: (1) when both T_1 and T_0 are nonzero, and (2) when only T_0 is nonzero. The second case can basically be treated as an otherwise natural system with an equivalent kinetic energy T_2 and equivalent potential energy $U = V - T_0$, referred to as *modified potential energy* or *dynamic potential*. The Lagrangian for such a system can be written as

$$L = T - V = T_2 + T_0 - V = T_2 - U \qquad [5.2.13]$$

Recall that T_2 is quadratic in generalized velocities. The case when $T_1 \neq 0$ usually describes more complex problems; its response characteristics are quite different than when $T_1 = 0$.

We now reexamine the equilibrium of a dynamical system. We considered static equilibrium in Chapters 1 and 4 and defined it as the state at which velocities and accelerations of all parts of the system are zero. All of the components of the system are at rest. Of course, because all physical velocities and accelerations are zero, so are the generalized velocities and accelerations.

For a nonnatural system *equilibrium* is defined as *the state where all generalized velocities and accelerations are zero*. The velocities and accelerations of certain parts of the system are not zero at equilibrium. For example, for a nonnatural system where there is motion with respect to a moving frame, at equilibrium, only the motion viewed from the moving frame is at rest.

Consider an n degree of freedom system with generalized coordinates $q_1, q_2, \ldots, q_n$. At equilibrium, all generalized velocities are zero, and all forces and moments acting on the system are not functions of time, that is, $Q_{knc} \neq f(\text{time})$. It follows that Lagrange's equations become

$$-\frac{\partial L}{\partial q_k} = Q_{knc} \qquad k = 1, 2, \ldots, n \qquad [5.2.14]$$

For a nonnatural system, because T_2 and T_1 are functions of the generalized velocities, they vanish at equilibrium, which results in the equilibrium condition

5.2 NATURAL AND NONNATURAL SYSTEMS, EQUILIBRIUM

$$-\frac{\partial(T_0 - V)}{\partial q_k} = \frac{\partial U}{\partial q_k} = Q_{knc} \qquad [5.2.15]$$

For a natural system $T = T_2$, $T_1 = T_0 = 0$, so Eq. [5.2.15] reduces to Eq. [4.6.4], and the equilibrium position is given by

$$\frac{\partial V}{\partial q_k} = Q_{knc} \qquad [5.2.16]$$

When all forces acting on the system are derivable from a potential, the equilibrium position for a nonnatural system corresponds to the point where the dynamic potential has a stationary value, that is,

$$\frac{\partial U}{\partial q_k} = 0 \qquad [5.2.17]$$

For natural systems this relation reduces to $\partial V/\partial q_k = 0$, which is the relation obtained in Chapter 4 when deriving the principle of virtual work for static equilibrium.

Example 5.1

We return to the bead problem considered in Chapter 4. However, here we treat the rotation of the hoop as another degree of freedom. Let us write the rotation rate Ω as $\dot{\phi}$. The Lagrangian, which we derived in Example 4.10, can now be written as

$$L = T - V = \frac{1}{2}I\dot{\phi}^2 + \frac{1}{2}mR^2\dot{\phi}^2 \sin^2\theta + \frac{1}{2}mR^2\dot{\theta}^2 - mgR(1 - \cos\theta) \qquad [a]$$

where I is the mass moment of inertia of the hoop about the axis of rotation. With the term $\frac{1}{2}I\dot{\phi}^2$ we have included the kinetic energy associated with the rotation of the hoop.

The hoop-bead system can now be treated as a natural system. Observe that ϕ is absent from the Lagrangian. This puts the problem at hand into a special category, which we will analyze later on in this chapter. We evaluate the partial derivatives of the Lagrangian, writing

$$\frac{\partial L}{\partial \theta} = mR^2\dot{\phi}^2 \sin\theta \cos\theta - mgR\sin\theta \qquad \frac{\partial L}{\partial \dot{\theta}} = mR^2\dot{\theta} \qquad [b]$$

$$\frac{\partial L}{\partial \phi} = 0 \qquad \frac{\partial L}{\partial \dot{\phi}} = I\dot{\phi} + mR^2\dot{\phi}\sin^2\theta = (I + mR^2\sin^2\theta)\dot{\phi} \qquad [c]$$

Introducing Eqs. [b] into Lagrange's equations we obtain the equation of motion for θ as

$$\ddot{\theta} + \sin\theta\left(\frac{g}{R} - \dot{\phi}^2\cos\theta\right) = 0 \qquad [d]$$

To evaluate the equation of motion for ϕ, we note that $\partial L/\partial \phi = 0$, thus

$$\frac{d}{dt}\frac{\partial L}{\partial \dot{\phi}} = \frac{d}{dt}(I\dot{\phi} + mR^2\dot{\phi}\sin^2\theta) = 0 \qquad [e]$$

We then have an integral of the motion as

$$\frac{\partial L}{\partial \dot{\phi}} = I\dot{\phi} + mR^2\dot{\phi}\sin^2\theta = \text{constant} \qquad [f]$$

Equations [d] and [f] need to be integrated simultaneously to obtain the response. However, we can simplify this problem if we take advantage of the fact that $\partial L/\partial \dot{\phi} =$ constant. Denoting this constant by C, where C depends on the initial conditions, we can express $\dot{\phi}$ as

$$\dot{\phi} = \frac{C}{I + mR^2 \sin^2 \theta} \qquad \text{[g]}$$

The constant C is an integral of the motion. Introducing Eq. [g] into Eq. [d], we rewrite the equation of motion for θ as

$$\ddot{\theta} + \sin \theta \left[\frac{g}{R} - \left(\frac{C}{I + mR^2 \sin^2 \theta} \right)^2 \cos \theta \right] = 0 \qquad \text{[h]}$$

Equation [h] can be solved by itself, and the value of $\dot{\phi}$ at any point in time can be ascertained by substituting the value of θ into Eq. [g].

We can treat the angular velocity of the hoop as a constant in two ways. The first is in the presence of a motor that keeps the angular velocity $\dot{\phi}$ constant. In this case, we have a nonnatural system. In the second way, we assume that $\dot{\phi}$ remains constant on its own. Let us analyze the accuracy of this assumption. The mass moment of inertia of the hoop is $I = MR^2/2$, where M is the mass of the hoop. Introducing the mass ratio $\mu = m/M$, Eq. [g] can be rewritten as

$$\dot{\phi} = \frac{2}{MR^2} \frac{C}{1 + 2\mu \sin^2 \theta} \qquad \text{[i]}$$

When μ is small, one can expand Eq. [h] in a Taylor series

$$\dot{\phi} \approx \frac{2}{MR^2} C(1 - 2\mu \sin^2 \theta) \qquad \text{[j]}$$

It is clear from Eq. [j] that how much of a variable $\dot{\phi}$ is depends on the mass ratio. When μ is very small, one can assume that $\dot{\phi}$ is constant and treat the problem as nonnatural with a single degree of freedom. The initial conditions on θ also affect the accuracy of this approximation.

Let us next analyze the equilibrium positions for this problem. When we consider the rotation of the hoop as a variable, the equilibrium positions are

$$\dot{\phi} = \text{constant} \qquad \sin \theta = 0 \qquad \text{[k]}$$

That is, the hoop is not rotating, and $\theta_e = 0$ or π, that is, the bead is either in the top or at the bottom of the hoop.

When we consider the motion of the hoop as a known quantity, say Ω, the dynamic potential U is

$$U = V - T_0 = mgR(1 - \cos \theta) - \frac{1}{2} mR^2 \Omega^2 \sin^2 \theta \qquad \text{[l]}$$

Taking the variation of U and setting it to zero, we obtain the position describing equilibrium as

$$\delta U = \frac{\partial U}{\partial \theta} \delta \theta = (mgR \sin \theta - mR^2 \Omega^2 \sin \theta \cos \theta) \delta \theta = 0 \qquad \text{[m]}$$

leading to the equilibrium equation

$$mgR\sin\theta\left(1 - \frac{R}{g}\Omega^2\cos\theta\right) = 0 \qquad [\text{n}]$$

Note that for this problem, setting the partial derivative of V with respect to θ equal to zero to find the equilibrium position would give incorrect results. Solving for the equilibrium positions, we obtain

$$\sin\theta = 0 \qquad \cos\theta = \frac{g}{R\Omega^2} \qquad [\text{o}]$$

leading to the equilibrium angles

$$\theta_e = 0 \qquad \theta_e = \pi \qquad \theta_e = \cos^{-1}\left(\frac{g}{R\Omega^2}\right) \qquad [\text{p}]$$

Let us next analyze the moment that is necessary to maintain a constant angular velocity $\dot{\phi}$. To this end, we make use of the constraint relaxation method described in Section 4.10 and consider that a moment T is acting on the hoop. The associated virtual work expression is $\delta W = T\,\delta\phi$, so that considering Eq. [f] the equation of motion of the hoop becomes

$$\frac{d}{dt}(I\dot{\phi} + MR^2\dot{\phi}\sin^2\theta) = T \qquad [\text{q}]$$

Imposing the constraint that $\dot{\phi} = \Omega = $ constant, the moment required to maintain the constant angular velocity becomes

$$T = 2mR^2\Omega\dot{\theta}\sin\theta\cos\theta \qquad [\text{r}]$$

5.3 SMALL MOTIONS ABOUT EQUILIBRIUM

As discussed in Chapter 1, the behavior of a system in the neighborhood of equilibrium is of utmost interest. Here, we extend the developments of Chapter 1 to multi-degree of freedom systems. We derive the equations of motion, find the equilibrium positions, denoted by q_{re} and $\dot{q}_{re} = 0$ ($r = 1, 2, \ldots, n$), and linearize by a Taylor series expansion.[1] As we saw in Chapter 1, if the linearized equations exhibit significant behavior, then the nature of the motion in the neighborhood of equilibrium is governed by this significant behavior. Otherwise, one must perform higher-level stability analyses.

Another physically intuitive way of analyzing motion in the neighborhood of equilibrium is to expand the kinetic and potential energies in the neighborhood of equilibrium. Let us consider a system with no nonconservative forces and express the potential energy V (or modified potential energy U) in a Taylor series expansion about the equilibrium position. Noting that $V = V(q_1, q_2, \ldots, q_n)$, and denoting the equilibrium positions by $q_{1e}, q_{2e}, \ldots, q_{ne}$, the Taylor series expansion of V has the

---

[1] Note that we are switching to the index r for the generalized coordinates, to avoid confusion with the symbol k that denotes stiffness.

form

$$V(q_1, q_2, \ldots, q_n) = V_e + \sum_{r=1}^{n} \left(\frac{\partial V}{\partial q_r}\right)_e (q_r - q_{re})$$
$$+ \frac{1}{2} \sum_{r=1}^{n} \sum_{s=1}^{n} \left(\frac{\partial^2 V}{\partial q_r \partial q_s}\right)_e (q_r - q_{re})(q_s - q_{se}) + \text{h.o.t.} \quad [\mathbf{5.3.1}]$$

where $V_e = V(q_{1e}, q_{2e}, \ldots, q_{ne})$ is the value of the potential energy at equilibrium. Without loss of generality, we select the datum position for the potential energy as being zero at equilibrium, so that $V_e = 0$. We measure the generalized coordinates from equilibrium, so that $q_{re} = 0$ ($r = 1, 2, \ldots, n$). Recalling that for a conservative system at static equilibrium all the first derivatives of V vanish,

$$\frac{\partial V}{\partial q_r} = 0 \quad [\mathbf{5.3.2}]$$

so that Eq. [5.3.1] reduces to

$$V(q_1, q_2, \ldots, q_n) = \frac{1}{2} \sum_{r=1}^{n} \sum_{s=1}^{n} k_{rs} q_r q_s + \text{h.o.t.} \quad [\mathbf{5.3.3}]$$

in which

$$k_{rs} = \left(\frac{\partial^2 V}{\partial q_r \partial q_s}\right)_e \quad r, s = 1, 2, \ldots, n \quad [\mathbf{5.3.4}]$$

are referred to as *stiffness coefficients*. This name is used in analogy with the potential energy of a spring. For nonnatural systems, at equilibrium $\partial U/\partial q_r$ vanishes and the stiffness coefficients have the form

$$k_{rs} = \left(\frac{\partial^2 U}{\partial q_r \partial q_s}\right)_e \quad [\mathbf{5.3.5}]$$

Making use of the generalized coordinate vector $\{q\} = [q_1 \ q_2 \ \ldots \ q_n]^T$, one can write the quadratic approximation to the potential energy in matrix form as

$$V \approx \frac{1}{2} \{q\}^T [K] \{q\} \quad [\mathbf{5.3.6}]$$

in which $[K]$ is known as the *stiffness matrix*. The stiffness matrix is symmetric. Furthermore, if for a conservative system $[K]$ is positive definite, the potential energy has a minimum at the equilibrium configuration. This can be concluded by comparing the potential energy V with the developments in Appendix B regarding the minimization of a function. The Hessian matrix in Appendix B becomes the stiffness matrix when the function whose stationary values are sought is the potential energy. A positive semidefinite stiffness matrix is usually an indication that the system possesses rigid body motion. It should be reiterated that Eq. [5.3.6] is valid when the generalized coordinates assume small values, and that it is for small motions about equilibrium only.

A theorem from stability theory states that for a natural conservative system, if the potential energy has a local minimum at equilibrium, then the equilibrium position is stable, implying that if the system is disturbed from its equilibrium position it either returns to the equilibrium position or oscillates around it. The instability theorem states that if the potential energy does not have a local minimum at equilibrium, the equilibrium position is unstable. We conclude that the equilibrium position of a natural conservative system is stable if the stiffness matrix associated with that equilibrium position is positive definite.

For nonnatural conservative systems, the corresponding stability theorem is different. When $T_1 = 0$, the system is treated as an otherwise natural system with kinetic energy T_2 and potential energy U. However, the situation changes when $T_1 \neq 0$. The theorem states that when $U = V - T_0$ is a local minimum at equilibrium, the equilibrium position is stable. But when U does not have a local minimum at equilibrium, the equilibrium position is not necessarily unstable. This concept can be explained physically by noting that gyroscopic effects usually increase the stability properties. Table 5.1 summarizes the stability theorems.

We next investigate linearization of the kinetic energy. For natural systems, the kinetic energy is

$$T = T_2 = \frac{1}{2}\{\dot{q}\}^T [M]\{\dot{q}\} \qquad [5.3.7]$$

where we note that it is already in quadratic form. It follows that, for small motions about equilibrium, the kinetic energy can be expressed as

$$T = T_2 \approx \frac{1}{2}\{\dot{q}\}^T [M_e]\{\dot{q}\} \qquad [5.3.8]$$

where the subscript e denotes that the inertia matrix is evaluated at the equilibrium position.

We conclude that, for small motions of a natural system about equilibrium, the Lagrangian can be written as

$$L = \frac{1}{2}\{\dot{q}\}^T [M_e]\{\dot{q}\} - \frac{1}{2}\{q\}^T [K]\{q\} \qquad [5.3.9]$$

Table 5.1 Summary of stability theorems for conservative dynamical systems

	Is Dynamic Potential U Minimum? Or, Is $[K]$ Positive Definite?	
	Yes	No
Natural systems	Stable	Unstable
Nonnatural systems		
$T_1 = 0$	Stable	Unstable
$T_1 \neq 0$	Stable	No conclusion

Lagrange's equations in column vector format are given in Eq. [4.9.12]. Using the properties in Chapter 2 of the derivative of a scalar with respect to a column vector, we have

$$\frac{d}{dt}\left(\frac{\partial L}{\partial \{\dot{q}\}}\right) = \{\ddot{q}\}^T[M_e] \qquad \frac{\partial L}{\partial \{q\}} = -\{q\}^T[K] \qquad [5.3.10]$$

Introducing Eqs. [5.3.10] into Eq. [4.9.12] we obtain the linearized equations of motion in matrix form as

$$[M_e]\{\ddot{q}\} + [K]\{q\} = \{Q_{nc}\} \qquad [5.3.11]$$

so that after calculating the kinetic and potential energies about equilibrium, one can use them directly to derive the equations of motion.

Next, consider nonnatural systems. The kinetic energy is $T = T_2 + T_1 + T_0$. The T_0 term is absorbed into the modified potential energy U, and it enters the stiffness matrix via Eq. [5.3.5]. The T_2 term is treated the same way the entire kinetic energy is treated in a natural system. It follows that, for the case where $T_1 = 0$, Eq. [5.3.11] is still the linearized equation of motion, with the entries of $[K]$ obtained by using Eq. [5.3.5] and $[M_e]$ obtained from T_2.

When $T_1 = \{\beta\}^T\{\dot{q}\} = \sum \beta_r \dot{q}_r$ is not zero, the equations of motion have an added term. Noting that the gyroscopic vector $\{\beta\} = [\beta_1 \beta_2 \ldots \beta_n]^T$ is a function of the generalized coordinates, we write the Taylor series expansion of T_1 as

$$T_1 = T_1(\{\dot{q}_e\}) + \sum_{r=1}^{n}\left(\frac{\partial T_1}{\partial q_r}\right)_e (q_r - q_{re}) + \sum_{r=1}^{n}\left(\frac{\partial T_1}{\partial \dot{q}_r}\right)_e \dot{q}_r$$

$$+ \sum_{r=1}^{n}\sum_{s=1}^{n}\frac{1}{2}\left(\frac{\partial^2 T_1}{\partial q_r \partial q_s}\right)_e (q_r - q_{re})(q_s - q_{se}) + \sum_{r=1}^{n}\sum_{s=1}^{n}\frac{1}{2}\left(\frac{\partial^2 T_1}{\partial \dot{q}_r \partial \dot{q}_s}\right)_e \dot{q}_r \dot{q}_s$$

$$+ \sum_{r=1}^{n}\sum_{s=1}^{n}\left(\frac{\partial^2 T_1}{\partial q_r \partial \dot{q}_s}\right)_e (q_r - q_{re})\dot{q}_s + \text{h.o.t.}$$

$$[5.3.12]$$

Of all the terms on the right side of this equation, only two survive: the third term and the last. The first, second, and fourth terms vanish because $\{\dot{q}_e\} = \{0\}$ at equilibrium, and the fifth term vanishes because T_1 is not quadratic in the generalized velocities. Introducing the notation β_{re} as the value of β_r at equilibrium and $B_{sr} = \partial \beta_s / \partial q_r$ evaluated at equilibrium, and noting that

$$\frac{\partial^2 T_1}{\partial q_r \partial \dot{q}_s} = \frac{\partial \beta_s}{\partial q_r} = B_{sr} \qquad r, s = 1, 2, \ldots, n \qquad [5.3.13]$$

we obtain the approximation to T_1 in the neighborhood of equilibrium as

$$T_1 \approx \sum_{r=1}^{n} \beta_{re} \dot{q}_r + \sum_{s=1}^{n}\sum_{r=1}^{n} B_{sr} \dot{q}_s q_r \qquad [5.3.14]$$

or, in column vector format,

$$T_1 \approx \{\beta_e\}^T\{\dot{q}\} + \{\dot{q}\}^T[B]\{q\} \qquad [5.3.15]$$

where the entries of $[B]$ are B_{rs}. We now introduce T_1 into the Lagrangian and write

$$L = \frac{1}{2}\{\dot{q}\}^T[M_e]\{\dot{q}\} + \{\beta_e\}^T\{\dot{q}\} + \{\dot{q}\}^T[B]\{q\} - \frac{1}{2}\{q\}^T[K]\{q\} \quad [\mathbf{5.3.16}]$$

Taking the appropriate derivatives, we obtain

$$\left(\frac{\partial L}{\partial \{\dot{q}\}}\right)^T = [M_e]\{\dot{q}\} + \{\beta_e\} + [B]\{q\} \qquad \left(\frac{\partial L}{\partial \{q\}}\right)^T = -[K]\{q\} + [B]^T\{\dot{q}\}$$

$$[\mathbf{5.3.17}]$$

which leads to the equations of motion as

$$[M_e]\{\ddot{q}\} + ([B] - [B]^T)\{\dot{q}\} + [K]\{q\} = \{Q_{nc}\} \quad [\mathbf{5.3.18}]$$

It is of interest to note the coefficient of the generalized velocity vector $\{\dot{q}\}$. A matrix subtracted from its transpose results in a skew-symmetric matrix (a null matrix if the original matrix is symmetric). Denoting $[B] - [B]^T$ by $[G]$, where $[G]$ is called the *gyroscopic matrix,* we write the linearized equations of motion for a nonnatural system as

$$[M_e]\{\ddot{q}\} + [G]\{\dot{q}\} + [K]\{q\} = \{Q_{nc}\} \quad [\mathbf{5.3.19}]$$

For relative motion problems when the motion of the relative frame is treated as known, T_1 leads to the Coriolis effect. For linear or linearized systems, the Coriolis effect is manifested in a skew-symmetric matrix. Note that in Example 2.16, the Foucault's pendulum, the linearized equations of motion in column vector format have a skew-symmetric matrix as the coefficient of the velocity vector. The same can be said about the projectile motion example, Example 2.15.

The first step associated with understanding the nature of motion in the neighborhood of equilibrium is to linearize the equations of motion about equilibrium and see if the linearized equations imply significant behavior. If there is no significant behavior, then higher-order analyses need to be conducted, including the stability theorems summarized in Table 5.1.

Example 5.2

Examine the stability of the equilibrium positions of the bead problem by analyzing the motion in the neighborhood of equilibrium.

Solution

For this problem, $T_1 = 0$ and the second derivative of U is

$$\frac{\partial^2 U}{\partial \theta^2} = mgR\left(\cos\theta - \frac{R}{g}\Omega^2\cos^2\theta + \frac{R}{g}\Omega^2\sin^2\theta\right) \quad [\mathbf{a}]$$

Noting that $T_2 = mR^2\dot{\theta}^2/2$, and $\varepsilon = \theta - \theta_e$, the linearized equation of motion about equilibrium has the form

$$mR^2\ddot{\varepsilon} + k\varepsilon = 0 \quad [\mathbf{b}]$$

Substituting the values of θ at equilibrium, we obtain

$$\text{For } \theta_e = 0 \qquad k = \frac{\partial^2 U}{\partial \theta^2} = mgR\left(1 - \frac{R}{g}\Omega^2\right) \quad [\mathbf{c}]$$

For $\theta_e = \pi$
$$k = \frac{\partial^2 U}{\partial \theta^2} = -mgR\left(1 + \frac{R}{g}\Omega^2\right) \quad \text{[d]}$$

For $\theta_e = \cos^{-1}\left(\frac{g}{R\Omega^2}\right)$
$$k = \frac{\partial^2 U}{\partial \theta^2}$$
$$= mgR\left[\frac{g}{R\Omega^2} - \frac{R}{g}\Omega^2\left(\frac{g}{R\Omega^2}\right)^2 + \frac{R}{g}\Omega^2\left(1 - \left(\frac{g}{R\Omega^2}\right)^2\right)\right]$$
$$= mR^2\Omega^2\left[1 - \left(\frac{g}{R\Omega^2}\right)^2\right] \quad \text{[e]}$$

For the equilibrium point to be stable, k must be positive. For the position $\theta_e = 0$, which corresponds to the bead lying on the bottom of the hoop, to be stable $1 - R\Omega^2/g$ must be positive, or

$$\Omega^2 < \frac{g}{R} \quad \text{[f]}$$

The second equilibrium position, $\theta_e = \pi$, which corresponds to the bead being at the top, is unstable, as the value of $\partial^2 U/\partial \theta^2$ is always negative. This result is easily verified from a physical point of view, as the bead will always fall from the highest point on the hoop.

From Eq. [e], the third equilibrium position is stable when

$$\Omega^2 > \frac{g}{R} \quad \text{[g]}$$

The motion of the bead can be explained as follows. When the angular velocity of the hoop is less than $\sqrt{g/R}$, the only stable equilibrium position is the bottom of the hoop, $\theta_e = 0$. The bead oscillates around that equilibrium point. As the angular velocity of the hoop becomes greater than $\sqrt{g/R}$, the stable equilibrium position becomes $\theta_e = \cos^{-1}(g/R\Omega^2)$ and the bead begins to oscillate about θ_e. The equilibrium point $\theta_e = 0$ is no longer a stable equilibrium point. The angular velocity $\Omega = \sqrt{g/R}$ is the critical angular velocity that dictates which equilibrium point is stable. Note that until the critical angular velocity is reached the equilibrium point is still $\theta_e = 0$. As the angular velocity increases over $\sqrt{g/R}$, the location of the stable equilibrium point moves up. The highest stable equilibrium position approaches $\theta_e = \pi/2$ for very high values of Ω.

Example 5.3

Consider the three mass-spring-damper system shown in Fig. 5.2, and obtain the equations of motion directly from the energy expressions.

Solution

We will solve for the equations of motion of this linear system by expressing the kinetic and potential energies in matrix form. The generalized coordinates q_1, q_2, and q_3 represent the

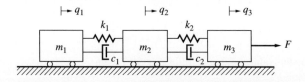

Figure 5.2 Free-body diagrams

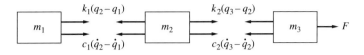

Figure 5.3 Free-body diagrams

displacements of the masses. From Fig. 5.3 the kinetic and potential energies have the form

$$T = \frac{1}{2}\{m_1\dot{q}_1^2 + m_2\dot{q}_2^2 + m_3\dot{q}_3^2\} \qquad V = \frac{1}{2}\{k_1(q_2 - q_1)^2 + k_2(q_3 - q_2)^2\} \qquad \textbf{[a]}$$

Introducing the column vector $\{q\} = [q_1 \; q_2 \; q_3]^T$, we can write the kinetic and potential energies in matrix form as

$$T = \frac{1}{2}\{\dot{q}\}^T[M]\{\dot{q}\} \qquad V = \frac{1}{2}\{q\}^T[K]\{q\} \qquad \textbf{[b]}$$

where the mass and stiffness matrices have the form

$$[M] = \begin{bmatrix} m_1 & 0 & 0 \\ 0 & m_2 & 0 \\ 0 & 0 & m_3 \end{bmatrix} \qquad [K] = \begin{bmatrix} k_1 & -k_1 & 0 \\ -k_1 & k_1 + k_2 & -k_2 \\ 0 & -k_2 & k_2 \end{bmatrix} \qquad \textbf{[c]}$$

Note that $[K]$ is positive semidefinite. The dashpots exert forces of magnitude $c_1(\dot{q}_2 - \dot{q}_1)$ and $c_2(\dot{q}_3 - \dot{q}_2)$ to the masses. The virtual work is given by

$$\delta W = -c_1(\dot{q}_2 - \dot{q}_1)\delta q_2 + c_1(\dot{q}_2 - \dot{q}_1)\delta q_1 - c_2(\dot{q}_3 - \dot{q}_2)\delta q_3 + c_2(\dot{q}_3 - \dot{q}_2)\delta q_2 + F\delta q_3 \qquad \textbf{[d]}$$

or

$$\delta W = Q_1 \delta q_1 + Q_2 \delta q_2 + Q_3 \delta q_3 \qquad \textbf{[e]}$$

in which

$$Q_1 = c_1(\dot{q}_2 - \dot{q}_1) \qquad Q_2 = -c_1(\dot{q}_2 - \dot{q}_1) + c_2(\dot{q}_3 - \dot{q}_2) \qquad Q_3 = -c_2(\dot{q}_3 - \dot{q}_2) + F \qquad \textbf{[f]}$$

so that the generalized force vector has the form

$$\{Q_{nc}\} = \begin{bmatrix} c_1(\dot{q}_2 - \dot{q}_1) \\ c_1(\dot{q}_1 - \dot{q}_2) + c_2(\dot{q}_3 - \dot{q}_2) \\ -c_2(\dot{q}_3 - \dot{q}_2) + F \end{bmatrix} \qquad \textbf{[g]}$$

Using Eq. [5.3.11], the equations of motion become

$$m_1\ddot{q}_1 + k_1 q_1 - k_1 q_2 + c_1\dot{q}_1 - c_1\dot{q}_2 = 0$$

$$m_2\ddot{q}_2 + (k_1 + k_2)q_2 - k_1 q_1 - k_2 q_3 + (c_1 + c_2)\dot{q}_2 - c_1\dot{q}_1 - c_2\dot{q}_3 = 0$$

$$m_3\ddot{q}_3 + k_2 q_3 - k_2 q_2 + c_2\dot{q}_3 - c_2\dot{q}_2 = F \qquad \textbf{[h]}$$

In the next section we will learn about a simpler way to handle the dissipative forces generated by a dashpot.

5.4 RAYLEIGH'S DISSIPATION FUNCTION

An important class of nonconservative forces is the class of forces that dissipate energy. A common way of approximating energy-dissipating forces is by friction. Two models of friction are widely used, even though both are crude simplifications of a complex phenomenon: The first is *dry* or *Coulomb friction*, where the friction force opposing the impending motion has the magnitude of the friction coefficient times the normal force, and it opposes the velocity. As we saw in Chapter 1,

$$\mathbf{F}_f = -\mu N \frac{\mathbf{v}}{|\mathbf{v}|} = -\mu N \mathbf{e}_t \qquad [5.4.1]$$

where $\mathbf{F}_f$ is the friction force, N is the normal force, $\mathbf{v}$ is the relative sliding velocity between the contacting points, and μ is the coefficient of friction. In general, μ is approximated by two values, the static coefficient of friction and the dynamic coefficient of friction. Equation [5.4.1] represents a nonlinear relationship.

The second approximation is *viscous damping*, where the damping force is modeled as opposing the velocity and proportional to it. The special case of linear proportionality is commonly used. In this case, the friction force is assumed to be in the form

$$\mathbf{F}_f = -c\mathbf{v} \qquad [5.4.2]$$

in which c is the viscous damping coefficient. This approximation is used especially when modeling light amounts of damping, because it is a linear approximation and easier to deal with mathematically. Its applications include modeling of shock absorbers (dashpots).

In analytical mechanics, a convenient way of treating viscous damping forces is by the use of *Rayleigh's dissipation function*. Consider a single particle moving in one direction, such as a mass-spring-dashpot system. In terms of the generalized coordinate x, the viscous damping force is given by $F_d = -c\dot{x}$. Rayleigh's dissipation function $\mathcal{F}$ is defined as

$$\mathcal{F} = \frac{1}{2}c\dot{x}^2 \qquad [5.4.3]$$

and the generalized force Q is obtained by

$$Q = -\frac{d\mathcal{F}}{d\dot{x}} \qquad [5.4.4]$$

Note the similarity between the expression for potential energy of springs ($kx^2/2$) and Rayleigh's dissipation function. We extend this definition to multiple dashpots by

$$\mathcal{F} = \sum_{i=1}^{N_d} \frac{1}{2}c_i(v_i - v_{i-1})^2 \qquad [5.4.5]$$

where $c_i (i = 1, 2, \ldots, N_d)$ are the viscous damping coefficients and v_i and v_{i-1} denote the components of the velocities of the points connected to the ith dashpot.

The Rayleigh's dissipation function can be written in terms of the generalized coordinates as

$$\mathcal{F} = \frac{1}{2} \sum_{k=1}^{n} \sum_{r=1}^{n} d_{kr} \dot{q}_k \dot{q}_r \qquad [5.4.6]$$

where the coefficients d_{kr} depend on the viscous damping coefficients. The contribution of viscous damping forces to Lagrange's equations can be obtained from Rayleigh's dissipation function by

$$\frac{\partial \mathcal{F}}{\partial \dot{q}_k} = \sum_{r=1}^{n} d_{kr} \dot{q}_r \qquad k = 1, 2, \ldots, n \qquad [5.4.7]$$

One can then express Lagrange's equations in the presence of viscous friction forces as

$$\frac{d}{dt}\left(\frac{\partial L}{\partial \dot{q}_k}\right) - \frac{\partial L}{\partial q_k} + \frac{\partial \mathcal{F}}{\partial \dot{q}_k} = Q_{knc} \qquad k = 1, 2, \ldots, n \qquad [5.4.8]$$

where Q_{knc} no longer include contributions from viscous damping forces. Note that, using the column vector $\{\dot{q}\} = [\dot{q}_1 \quad \dot{q}_2 \quad \ldots \quad \dot{q}_n]^T$, we can write Rayleigh's dissipation function as

$$\mathcal{F} = \frac{1}{2}\{\dot{q}\}^T [D]\{\dot{q}\} \qquad [5.4.9]$$

in which $[D]$ is the damping matrix with entries d_{kr}. It can be shown that the damping matrix is symmetric and positive semidefinite.

Another class of dissipative forces is the *circulatory forces*. Circulatory forces occur in power transmission devices, pipes, and as constraint damping in structures undergoing rotational motion. For systems subjected to circulatory forces, the dissipation function has the general form

$$\mathcal{F} = \frac{1}{2}\{\dot{q}\}^T [D]\{\dot{q}\} + \{\dot{q}\}^T [H]\{q\} \qquad [5.4.10]$$

where $[H]$ is the *circulatory matrix*.

To analyze the effect of viscous damping and circulatory forces for small motions in the neighborhood of equilibrium, we note that $\mathcal{F}$ is quadratic in the generalized coordinates and velocities. Considering the developments in Section 5.3, a Taylor series expansion of $\mathcal{F}$ in Eq. [5.4.10] up to quadratic terms has the form

$$\mathcal{F} \approx \frac{1}{2}\{\dot{q}\}^T [D_e]\{\dot{q}\} + \{\dot{q}\}^T [H_e]\{q\} \qquad [5.4.11]$$

where $[D_e]$ is the damping matrix and $[H_e]$ is the circulatory matrix, both evaluated at the equilibrium position. $[D_e]$ is symmetric and positive semidefinite and $[H_e]$ is skew symmetric. It follows that the equations of motion in the neighborhood of equilibrium can be expressed as

$$[M_e]\{\ddot{q}\} + ([D_e] + [G])\{\dot{q}\} + ([K] + [H_e])\{q\} = \{Q\} \qquad [5.4.12]$$

Finally, we examine the units of Rayleigh's dissipation function $\mathcal{F}$. Consider a single dashpot, with $\mathcal{F} = c\dot{x}^2/2$. The damping force is $F = -c\dot{x}$. It follows that the unit of Rayleigh's dissipation function is force $\times$ velocity, or power.

Example 5.4

Consider Example 5.3 and obtain the equations of motion using Rayleigh's dissipation function.

Solution

Rayleigh's dissipation function is given by

$$\mathcal{F} = \frac{1}{2}c_1(\dot{q}_2 - \dot{q}_1)^2 + \frac{1}{2}c_2(\dot{q}_3 - \dot{q}_2)^2 \qquad [\mathbf{a}]$$

which can be expressed in the matrix form

$$\mathcal{F} = \frac{1}{2}\{\dot{q}\}^T [D]\{\dot{q}\} \qquad [\mathbf{b}]$$

where the damping matrix is

$$[D] = \begin{bmatrix} c_1 & -c_1 & 0 \\ -c_1 & c_1 + c_2 & -c_2 \\ 0 & -c_2 & 0 \end{bmatrix} \qquad [\mathbf{c}]$$

It follows that the equations of motion are given by

$$[M]\{\ddot{q}\} + [D]\{\dot{q}\} + [K]\{q\} = \{Q_{nc}\} \qquad [\mathbf{d}]$$

where $[M]$ and $[K]$ are defined in Example 5.3. For this example, because all nonconservative forces are accounted for in the Rayleigh's dissipation function, the generalized forces vector is $\{Q_{nc}\} = [0\ 0\ F]^T$. A brief examination of Eq. [d] and Eq. [i] in Example 5.3 indicates that they are equivalent.

Example 5.5

Consider the system in Example 4.11. Obtain Rayleigh's dissipation function, linearize it around equilibrium, and obtain the $[D_e]$ matrix.

Solution

From Fig. 4.21 Rayleigh's dissipation function can be written as

$$\mathcal{F} = \frac{1}{2}cv_x^2 = \frac{1}{2}c\left(\dot{x} + \frac{b}{2}\dot{\theta}\cos\theta\right)^2 = \frac{1}{2}c\left(\dot{x}^2 + b\dot{x}\dot{\theta}\cos\theta + \frac{1}{4}b^2\dot{\theta}^2\cos^2\theta\right) \qquad [\mathbf{a}]$$

The equilibrium points can be obtained by visual inspection as $x_e = 0, \theta_e = 0, \pi$. We consider the position $\theta_e = 0$. In the neighborhood of equilibrium $\mathcal{F}$ becomes

$$\mathcal{F} \approx \frac{1}{2}c\left(\dot{x}^2 + b\dot{x}\dot{\theta} + \frac{1}{4}b^2\dot{\theta}^2\right) \qquad [\mathbf{b}]$$

and, considering that the generalized coordinate vector can be written as $\{q\} = [x\ \theta]^T$, the damping matrix becomes

$$[D_e] = c\begin{bmatrix} 1 & \dfrac{b}{2} \\ \dfrac{b}{2} & \dfrac{b^2}{4} \end{bmatrix} \qquad [\mathbf{c}]$$

5.5 EIGENVALUE PROBLEM FOR LINEARIZED SYSTEMS

In the previous sections we saw the linearized equations of motion for dynamical systems. Here, we obtain the solution of the linearized equations, analogous to the treatment of a single degree of freedom system. The developments in Sections 1.9 through 1.11 and the developments of this and the next two sections constitute the basis of linear vibration theory.

The equations of motion of a linearized undamped, nongyroscopic system are given in Eq. [5.3.11]. To obtain the response, we first consider the free vibration case, that is, $\{Q_{nc}\} = \{0\}$. Dropping subscript e from $[M_e]$, the equations of motion have the form

$$[M]\{\ddot{q}(t)\} + [K]\{q(t)\} = \{0\} \qquad [5.5.1]$$

where $[M]$ and $[K]$ are constant coefficient matrices of order $n \times n$. As in Chapter 1, we are interested in analyzing stable motion about equilibrium, so we consider cases when the stiffness matrix $[K]$ is positive definite or positive semidefinite. Such systems are basically vibratory systems. The case of a positive semidefinite $[K]$ is encountered in systems admitting rigid body motion.

The above equations represent n second-order coupled differential equations. To analyze the response, we make the *synchronous motion* assumption and express the solution in the form

$$\{q(t)\} = \{u\}e^{\lambda t} \qquad [5.5.2]$$

where $\{u\}$ is an n-dimensional vector and λ is a scalar. The synchronous motion assumption is based on the observation that the response of every component of the dynamical system has the same time variation. The amplitudes are different, but the nature of the motion is the same. Hence, λ describes the time variation and $\{u\}$ describes the different amplitudes.

We introduce Eq. [5.5.2] into [5.5.1] and collect terms, which gives

$$(\lambda^2[M] + [K])\{u\}e^{\lambda t} = \{0\} \qquad [5.5.3]$$

For a nontrivial solution, $\{u\}e^{\lambda t}$ cannot be zero. It follows that Eq. [5.5.3] vanishes only if

$$(\lambda^2[M] + [K])\{u\} = \{0\} \qquad [5.5.4]$$

Equation [5.5.4] converts the problem to a set of n homogenous algebraic equations for the parameters $\{u\}$ and λ^2. The task at hand is to find the values of λ^2 and corresponding values of $\{u\}$ for which Eq. [5.5.4] has a nontrivial solution. This problem is known as the *eigenvalue problem*.

For the set of n equations [5.5.4] to have a nontrivial solution, the determinant of the coefficient matrix must vanish, that is

$$\det(\lambda^2[M] + [K]) = |\lambda^2[M] + [K]| = 0 \qquad [5.5.5]$$

This is known as the *characteristic equation*, completely analogous to the developments and definitions in Sec. 1.9. It represents an nth order polynomial in λ^2, known as the *characteristic polynomial*. The characteristic polynomial has n roots, denoted by λ_r^2 ($r = 1, 2, \ldots, n$), and known as *eigenvalues* or *characteristic values*.[2]

It follows that for each root λ_r^2 of the characteristic equation, there is a vector $\{u_r\}$ such that

$$(\lambda_r^2[M] + [K])\{u_r\} = \{0\} \qquad r = 1, 2, \ldots, n \qquad [\textbf{5.5.6}]$$

is satisfied, in which $\{u_r\}$ is the solution of the eigenvalue problem corresponding to λ_r^2. The term $\{u_r\}$ is called the *eigenvector belonging to the rth eigenvalue*, or the *rth eigenvector*.

Let us examine the nature of the eigenvalues and eigenvectors. We know from the previous section that $[M]$ is positive definite, and we are considering cases where $[K]$ is positive definite or positive semidefinite. A theorem from linear algebra states that the eigenvalues and eigenvectors of a symmetric matrix are real. Further, if the matrix is positive definite, all eigenvalues > 0. Also, the eigenvalues of the general eigenvalue problem

$$\lambda[A]\{x\} = [B]\{x\} \qquad [\textbf{5.5.7}]$$

where $[A]$ is positive definite and $[B]$ is positive semidefinite, are real and nonnegative. It follows that the eigenvalues of Eq. [5.5.6], λ_r^2 ($r = 1, 2, \ldots, n$), are all real and nonpositive. Hence $\pm \lambda_r$ are all pure imaginary. This is to be expected, because we are dealing with a critically stable system. Introducing

$$\omega_r^2 = -\lambda_r^2 \qquad r = 1, 2, \ldots, n \qquad [\textbf{5.5.8}]$$

where ω_r is a positive quantity called the *rth natural frequency*, the response can be represented as a summation of the individual frequencies multiplied by the eigenvectors

$$\{q(t)\} = \sum_{r=1}^{n} \{u_r\}(X_r e^{i\omega_r t} + Y_r e^{-i\omega_r t}) \qquad [\textbf{5.5.9}]$$

where X_r and Y_r depend on the initial conditions. Because $\{q(t)\}$ and $\{u_r\}$ are real-valued quantities, following an argument similar to the one in Sec. 1.9 we conclude that X_r and Y_r are complex conjugates of each other, so that the free response can be written as

$$\{q(t)\} = \sum_{r=1}^{n} \{u_r\} A_r \cos(\omega_r t - \phi_r) \qquad [\textbf{5.5.10}]$$

with A_r and ϕ_r being the amplitude and the phase angle corresponding to the *r*th natural frequency, and their values depend on the initial conditions.

Before analyzing the amplitudes and phase angles, let us discuss the physical interpretation of the eigenvalue problem. The eigenvalue problem has as its solution n natural frequencies (for systems admitting rigid body motion, one or more of the natural frequencies is usually zero) and corresponding eigenvectors. The n natural frequencies describe the n unique ways (or *modes*) the system can vibrate. For a

---

[2]The word *eigenvalue* comes from the German word *eigen*, meaning characteristic.

single degree of freedom system we had one natural frequency, and for an n degree of freedom system there are n natural frequencies.

In each mode of the motion, the free response is harmonic (except for a rigid body mode with zero frequency). The eigenvector corresponding to a particular mode represents the amplitude ratios of the generalized coordinates. That is, it describes the shape of the motion when the system vibrates with that particular mode. The eigenvectors are also referred to as *modal vectors* or *natural modes*. Note that, from Eq. [5.5.6], the amplitudes of the modal vectors are not unique. A modal vector multiplied by a nonzero constant is still a modal vector. It is useful to adopt a procedure to specify the magnitudes of the modal vectors, and we will do exactly that in the next section, via a procedure called *normalization*.

Example 5.6

Obtain the natural frequencies and eigenvalues associated with the system in Example 5.3, for the values $m_1 = m_2 = m_3 = m$, $k_1 = k_2 = k$, $c_1 = c_2 = 0$.

Solution

From Eq. [5.5.4], the eigenvalue problem is

$$(\lambda^2 [M] + [K])\{u\} = \left(\lambda^2 m \begin{bmatrix} 1 & 0 & 0 \\ 0 & 1 & 0 \\ 0 & 0 & 1 \end{bmatrix} + k \begin{bmatrix} 1 & -1 & 0 \\ -1 & 2 & -1 \\ 0 & -1 & 1 \end{bmatrix} \right) \{u\} = \{0\} \quad \textbf{[a]}$$

Dividing the above equation by k, and introducing the quantity $\gamma = -\lambda^2 m/k = \omega^2 m/k$, the characteristic equation becomes

$$\det \left(\gamma \begin{bmatrix} 1 & 0 & 0 \\ 0 & 1 & 0 \\ 0 & 0 & 1 \end{bmatrix} - \begin{bmatrix} 1 & -1 & 0 \\ -1 & 2 & -1 \\ 0 & -1 & 1 \end{bmatrix} \right) = \det \begin{bmatrix} \gamma - 1 & 1 & 0 \\ 1 & \gamma - 2 & 1 \\ 0 & 1 & \gamma - 1 \end{bmatrix} = 0 \quad \textbf{[b]}$$

Evaluating the determinant gives a third-order polynomial in γ in the form

$$(\gamma - 1)(\gamma - 2)(\gamma - 1) - 2(\gamma - 1) = (\gamma - 1)(\gamma^2 - 3\gamma + 2 - 2) = (\gamma - 1)(\gamma^2 - 3\gamma) = 0 \quad \textbf{[c]}$$

whose roots are (listed for convenience in ascending order, which is the common way of ranking them)

$$\gamma_1 = 0 \qquad \gamma_2 = 1 \qquad \gamma_3 = 3 \quad \textbf{[d]}$$

It should be noted that the characteristic equation for most systems is much harder to solve than the example here. Recalling the definition of γ above, the natural frequencies can be written as $\omega_r = \sqrt{\gamma_r k/m}$ ($r = 1, 2, 3$), with the result

$$\omega_1 = 0 \qquad \omega_2 = \sqrt{\frac{k}{m}} \qquad \omega_3 = \sqrt{\frac{3k}{m}} \quad \textbf{[e]}$$

One of the eigenvalues is zero, indicating a rigid body mode. We can verify the presence of a rigid body mode visually, by noting that the two ends of the mass-spring system are free. Hence, if the system is given an initial velocity, it will keep translating, until opposing forces are applied or friction takes its toll.

Next, we calculate the eigenvectors. For each mode of the motion there exists a unique eigenvector. To solve for the eigenvectors, we make use of Eq. [5.5.6], which can be written as

$$\begin{bmatrix} \gamma_r - 1 & 1 & 0 \\ 1 & \gamma_r - 2 & 1 \\ 0 & 1 & \gamma_r - 1 \end{bmatrix} \{u_r\} = \{0\} \qquad r = 1, 2, 3 \qquad \text{[f]}$$

The coefficient matrix in the above equation is singular, so that the three equations obtained from Eq. [f] are not independent. Equation [f] gives only two independent equations, which raises the question of how to obtain the three elements of $\{u_r\}$. It turns out that only two of the elements of $\{u_r\}$ can be found, and the third is expressed as a ratio.

Writing $\{u_r\}$ as $\{u_r\} = [u_{1r} \quad u_{2r} \quad u_{3r}]^T$, we obtain for $r = 1, \gamma_1 = 0$

$$\begin{bmatrix} -1 & 1 & 0 \\ 1 & -2 & 1 \\ 0 & 1 & -1 \end{bmatrix} \begin{bmatrix} u_{11} \\ u_{21} \\ u_{31} \end{bmatrix} = \{0\} \qquad \text{[g]}$$

Taking the first and third equations as the independent ones, we obtain

$$-u_{11} + u_{21} = 0 \qquad u_{21} - u_{31} = 0 \qquad \text{[h]}$$

which leads to the conclusion

$$u_{11} = u_{21} = u_{31} \qquad \text{[i]}$$

This result can be confirmed by substituting in the second equation in Eq. [g]. Hence, the first eigenvector can be expressed as

$$\{u_1\} = \alpha_1 \begin{bmatrix} 1 \\ 1 \\ 1 \end{bmatrix} \qquad \text{[j]}$$

where α_1 is an amplitude ratio. Let us examine this first eigenvector. Recall that the elements of the eigenvectors denote the amplitudes of the generalized coordinates, the displacements of the three masses in this problem. For the rigid body mode, all of the elements of the eigenvector are the same, indicating that all three masses move with the same amplitude. It follows that the springs between the masses are not stretched, and the entire mass-spring system moves as one piece, as if it is a single rigid body with no moving parts.

For the second mode we have $\gamma_2 = 1$ and

$$\begin{bmatrix} 0 & 1 & 0 \\ 1 & -1 & 1 \\ 0 & 1 & 0 \end{bmatrix} \begin{bmatrix} u_{12} \\ u_{22} \\ u_{32} \end{bmatrix} = \{0\} \qquad \text{[k]}$$

and we take the first two equations from above

$$u_{22} = 0 \qquad u_{12} - u_{22} + u_{32} = 0 \qquad \text{[l]}$$

thus, the second eigenvector can be written as

$$\{u_2\} = \alpha_2 \begin{bmatrix} 1 \\ 0 \\ -1 \end{bmatrix} \qquad \text{[m]}$$

For the third mode we have $\gamma_3 = 3$ and

$$\begin{bmatrix} 2 & 1 & 0 \\ 1 & 1 & 1 \\ 0 & 1 & 2 \end{bmatrix} \begin{bmatrix} u_{13} \\ u_{23} \\ u_{33} \end{bmatrix} = \{0\} \qquad \text{[n]}$$

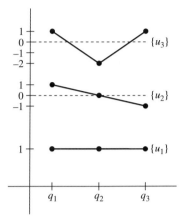

Figure 5.4 Plot of the eigenvectors

Using the first and second equations

$$2u_{13} + u_{23} = 0 \qquad u_{23} + 2u_{33} = 0 \qquad \textbf{[o]}$$

so that the third eigenvector becomes

$$\{u_3\} = \alpha_3 \begin{bmatrix} 1 \\ -2 \\ 1 \end{bmatrix} \qquad \textbf{[p]}$$

The three eigenvectors are plotted in Fig. 5.4. We see that all of the elements of the first eigenvector have the same sign, elements of the second eigenvector have one sign change, and in the third eigenvector there are two sign changes. This is a common feature of modal vectors, which can be generalized to a system of order n as: When the eigenvalues are arranged in ascending order, the rth eigenvector has $r - 1$ zero crossings.

5.6 ORTHOGONALITY AND NORMALIZATION

The natural modes possess an important property called *orthogonality*. Consider the eigenvalue problem in Eq. [5.5.5] and two of its solutions, say the rth and sth modes with $r \neq s$ and $\omega_r \neq \omega_s$. A theorem from linear algebra states that

$$\{u_r\}^T [M] \{u_s\} = 0 \qquad \{u_r\}^T [K] \{u_s\} = 0 \qquad r, s = 1, 2, \ldots, n, \; r \neq s, \omega_r \neq \omega_s$$
[5.6.1]

This relation is true for any two nonequal natural frequencies. In the case of repeated eigenvalues ($\omega_r = \omega_s, r \neq s$), the associated modal vectors are not unique. Any linear combination of two such modal vectors is a modal vector itself. However, it is possible to select the modal vectors associated with repeating eigenvalues such that Eq. [5.6.1] holds. In Chapter 6, we will see an interesting application of repeated eigenvalues.

When $r = s$, the product $\{u_r\}^T[M]\{u_r\} > 0$, and the magnitude of the product depends on the magnitude of the modal vector. Because these amplitudes are arbitrary, a common way to deal with them is to normalize them. One way of normalizing the eigenvectors is to set the above product to unity

$$\{u_r\}^T[M]\{u_r\} = 1 \qquad [5.6.2]$$

This equation is one way of normalizing the modal vectors. There are other approaches, such as $\{u_r\}^T\{u_r\} = 1$. In this text we use Eq. [5.6.2]. Using this equation, and Eq. [5.5.6], one can show that

$$\{u_r\}^T[K]\{u_r\} = \omega_r^2 \qquad [5.6.3]$$

Equations [5.6.1]–[5.6.3] can be combined to give the so-called *orthonormality* relations

$$\{u_r\}^T[M]\{u_s\} = \underline{\delta}_{rs} \qquad \{u_r\}^T[K]\{u_s\} = \omega_r^2 \underline{\delta}_{rs} \qquad [5.6.4]$$

where $\underline{\delta}_{rs}$ is the Kronecker delta, defined as

$$\underline{\delta}_{rs} = 1 \quad \text{when} \quad r = s$$
$$\underline{\delta}_{rs} = 0 \quad \text{when} \quad r \neq s \qquad [5.6.5]$$

A corollary of the result above is that the modal vectors constitute an independent set. This implies that any vector of order n can be expressed as a linear combination of the eigenvectors multiplied by appropriate coefficients. Hence, given an n-dimensional vector $\{z\}$, one can expand it as

$$\{z\} = \sum_{k=1}^{n} a_k\{u_k\} \qquad [5.6.6]$$

where a_k are the coefficients of the expansion. Note the similarity of Eqs. [5.6.6] and [5.5.10]. The question becomes that of given $\{z\}$, determining a_k ($k = 1, 2, \ldots, n$). For this, we make use of orthogonality, and left-multiply both sides of the above equation by $\{u_r\}^T[M]$ ($r = 1, 2, \ldots, n$) with the result

$$\{u_r\}^T[M]\{z\} = \sum_{k=1}^{n} a_k\{u_r\}^T[M]\{u_k\} = \sum_{k=1}^{n} a_k \delta_{rk} = a_r \qquad r = 1, 2, \ldots, n$$
$$[5.6.7]$$

We conclude that any vector $\{z\}$ of order n can be expressed in terms of the modal vectors

$$\{z\} = \sum_{r=1}^{n} a_r\{u_r\} \qquad a_r = \{u_r\}^T[M]\{z\} \qquad [5.6.8]$$

This equation is known as the *expansion theorem*. A very important application of the expansion theorem is in obtaining the response of a vibrating system, as we will discuss in the next section.

Example 5.7

Consider Examples 5.3 and 5.6, and demonstrate that the orthogonality relations hold. Normalize the eigenvectors. Then, take the vector $\{b\} = [1 \quad 2 \quad 5]^T$ and expand it in terms of the eigenvectors.

Solution

The mass matrix for this problem can be written as $[M] = m[1]$, where $[1]$ is the identity matrix. Hence, the orthogonality relations in Eq. [5.6.1] reduce to

$$\{u_r\}^T[M]\{u_s\} = m\{u_r\}^T\{u_s\} \qquad r,s = 1,2,3 \qquad \text{[a]}$$

Evaluating the possible combinations, we get

$$\{u_1\}^T[M]\{u_2\} = m\alpha_1\alpha_2(1+0-1) = 0 \qquad \{u_1\}^T[M]\{u_3\} = m\alpha_1\alpha_3(1-2+1) = 0$$

$$\{u_2\}^T[M]\{u_3\} = m\alpha_2\alpha_3(1-0+1) = 0 \qquad \text{[b]}$$

Next, we normalize the eigenvectors according to Eq. [5.6.2]. For the first mode we have

$$\{u_1\}^T[M]\{u_1\} = m\alpha_1^2(1+1+1) = 3m\alpha_1^2 = 1 \qquad \text{[c]}$$

which can be solved for α_1 as

$$\alpha_1 = \pm\sqrt{\frac{1}{3m}} \qquad \text{[d]}$$

The choice of the plus and minus sign is up to the analyst. Let us select the positive value.
For the second and third modes we have

$$\{u_2\}^T[M]\{u_2\} = 2m\alpha_1^2 = 1 \qquad \{u_3\}^T[M]\{u_3\} = 6m\alpha_1^2 = 1 \qquad \text{[e]}$$

from which we obtain

$$\alpha_2 = \pm\sqrt{\frac{1}{2m}} \qquad \alpha_3 = \pm\sqrt{\frac{1}{6m}} \qquad \text{[f]}$$

and we select the positive values for α_2 and α_3.
Let us now consider the vector $\{b\} = [1 \quad 2 \quad 5]^T$ and expand it as

$$\{b\} = a_1\{u_1\} + a_2\{u_2\} + a_3\{u_3\} \qquad \text{[g]}$$

where the coefficients a_i ($i = 1, 2, 3$) are found from Eq. [5.6.8]. We then have

$$a_1 = \{u_1\}^T[M]\{b\} = m\{u_1\}^T\{b\} = m\sqrt{\frac{1}{3m}}(1 \times 1 + 1 \times 2 + 1 \times 5) = 8\sqrt{\frac{m}{3}}$$

$$a_2 = \{u_2\}^T[M]\{b\} = m\{u_2\}^T\{b\} = m\sqrt{\frac{1}{2m}}(1 \times 1 + 0 \times 2 - 1 \times 5) = -4\sqrt{\frac{m}{2}}$$

$$a_3 = \{u_3\}^T[M]\{b\} = m\{u_3\}^T\{b\} = m\sqrt{\frac{1}{6m}}(1 \times 1 - 2 \times 2 + 1 \times 5) = 2\sqrt{\frac{m}{6}} \qquad \text{[h]}$$

so that $\{b\}$ is expanded as

$$\begin{bmatrix} 1 \\ 2 \\ 5 \end{bmatrix} = 8\sqrt{\frac{m}{3}}\left(\sqrt{\frac{1}{3m}}\right)\begin{bmatrix} 1 \\ 1 \\ 1 \end{bmatrix} - 4\sqrt{\frac{m}{2}}\left(\sqrt{\frac{1}{2m}}\right)\begin{bmatrix} 1 \\ 0 \\ -1 \end{bmatrix} + 2\sqrt{\frac{m}{6}}\left(\sqrt{\frac{1}{6m}}\right)\begin{bmatrix} 1 \\ -2 \\ 1 \end{bmatrix}$$
$$= \frac{8}{3}\begin{bmatrix} 1 \\ 1 \\ 1 \end{bmatrix} - 2\begin{bmatrix} 1 \\ 0 \\ -1 \end{bmatrix} + \frac{1}{3}\begin{bmatrix} 1 \\ -2 \\ 1 \end{bmatrix} \qquad \text{[i]}$$

5.7 MODAL EQUATIONS OF MOTION AND RESPONSE

Consider the undamped vibration problem, repeated here as

$$[M]\{\ddot{q}(t)\} + [K]\{q(t)\} = \{Q(t)\} \qquad \textbf{[5.7.1]}$$

subject to the initial conditions $\{q(0)\}$ and $\{\dot{q}(0)\}$. We will make use of the developments of the previous sections to find the response. We first solve the eigenvalue problem and find the natural frequencies and the modal vectors. We then apply the expansion theorem to the generalized coordinates and expand $\{q(t)\}$ as

$$\{q(t)\} = \sum_{k=1}^{n} \eta_k(t)\{u_k\} \qquad \textbf{[5.7.2]}$$

where $\eta_k(t)$ are known as *modal coordinates* or *principal coordinates*. Note that because $\{q(t)\}$ is a function of time, so are the coefficients of its expansion in terms of the modal vectors. In essence, the principal coordinates $\eta_1(t), \eta_2(t), \ldots \eta_n(t)$ constitute another set of generalized coordinates.

Next, we introduce Eq. [5.7.2] into Eq. [5.7.1]. Obviously, we wish to make use of the orthogonality properties, so we left-multiply both sides of Eq. [5.7.1] by $\{u_r\}^T (r = 1, 2, \ldots, n)$, with the result

$$\{u_r\}^T[M]\sum_{k=1}^{n}\ddot{\eta}_k(t)\{u_k\} + \{u_r\}^T[K]\sum_{k=1}^{n}\eta_k(t)\{u_k\} = \{u_r\}^T\{Q(t)\} \qquad \textbf{[5.7.3]}$$

which can be rewritten as

$$\sum_{k=1}^{n}\{u_r\}^T[M]\{u_k\}\ddot{\eta}_k(t) + \sum_{k=1}^{n}\{u_r\}^T[K]\{u_k\}\eta_k(t) = \{u_r\}^T\{Q(t)\} \qquad r = 1, 2, \ldots, n$$

$$\textbf{[5.7.4]}$$

Invoking the orthonormality relations in Eq. [5.6.4], we see that of all the terms on the left side of Eq. [5.7.4], two terms survive, those corresponding to $r = k$. We define the quantity *modal force* $\mathcal{N}_r(t)$, where

$$\mathcal{N}_r(t) = \{u_r\}^T\{Q(t)\} \qquad \textbf{[5.7.5]}$$

so that the describing equations become

$$\ddot{\eta}_r(t) + \omega_r^2\eta_r(t) = \mathcal{N}_r(t) \qquad \textbf{[5.7.6]}$$

We have converted the equations of motion from a set of n coupled second-order differential equations to n independent equations of motion. This conversion serves a number of purposes. It permits the user to view the equations of motion as a collection of single oscillators. It also makes it much simpler to obtain the solution. Indeed, rather than solving a set of n coupled differential equations, one solves n independent equations. For each modal coordinate, from Sec. 1.11, the response has the form of the convolution sum

$$\eta_r(t) = \eta_r(0)\cos\omega_r t + \frac{\dot{\eta}_r(0)}{\omega_r}\sin\omega_r t + \frac{1}{\omega_r}\int_0^t \mathcal{N}_r(t-\sigma)\sin\omega_r\sigma \, d\sigma \qquad \textbf{[5.7.7]}$$

where $\eta_r(0)$ and $\dot{\eta}_r(0)$ are the initial conditions. They can be obtained from the initial conditions on the generalized coordinates $\{q(0)\}$ and $\{\dot{q}(0)\}$ by means of the expansion theorem as

$$\eta_r(0) = \{u_r\}^T[M]\{q(0)\} \qquad \dot{\eta}_r(0) = \{u_r\}^T[M]\{\dot{q}(0)\} \qquad [\mathbf{5.7.8}]$$

Once the response of each modal coordinate is obtained, Eq. [5.7.2] is invoked and the response of the principal coordinates is found.

In the presence of a rigid body mode with zero frequency (denoted by R), the modal equation of motion is

$$\ddot{\eta}_R(t) = \mathcal{N}_R(t) \qquad [\mathbf{5.7.9}]$$

subject to the initial conditions $\eta_R(0) = \{u_R\}^T[M]\{q(0)\}$, $\dot{\eta}_R(0) = \{u_R\}^T[M]\{\dot{q}(0)\}$. The response can be shown to be

$$\eta_R(t) = \eta_R(0) + \dot{\eta}_R(0)t + \int_0^t \int_0^\tau \mathcal{N}_R(\sigma)\,d\sigma\,d\tau \qquad [\mathbf{5.7.10}]$$

For damped systems, the eigensolution associated with the undamped part of the system does not usually lead to a decoupled set of equations. Let us add a $[D]\{\dot{q}(t)\}$ term to the left side of Eq. [5.7.1], thus

$$[M]\{\ddot{q}(t)\} + [D]\{\dot{q}(t)\} + [K]\{q(t)\} = \{Q(t)\} \qquad [\mathbf{5.7.11}]$$

Introduce Eq. [5.7.2] into it and left-multiply by $\{u_r\}^T$. This yields

$$\{u_r\}^T[M]\sum_{k=1}^n \ddot{\eta}_k(t)\{u_k\} + \{u_r\}^T[D]\sum_{k=1}^n \dot{\eta}_k(t)\{u_k\} + \{u_r\}^T[K]\sum_{k=1}^n \eta_k(t)\{u_k\} = \{u_r\}^T\{Q(t)\}$$

$$[\mathbf{5.7.12}]$$

which can be reduced to

$$\ddot{\eta}_r(t) + \sum_{k=1}^n d_{kr}\dot{\eta}_k(t) + \omega_r^2\eta_r(t) = \mathcal{N}_r(t) \qquad [\mathbf{5.7.13}]$$

where

$$d_{kr} = \{u_k\}^T[D]\{u_r\} \qquad [\mathbf{5.7.14}]$$

As a result, the modal equations of motion are no longer independent but are coupled through the damping terms. There are cases when the damping matrix has a special form that decouples the modal equations. One such case is *proportional damping*, where the damping matrix is a linear combination of the mass and stiffness matrices in the form

$$[D] = \alpha_1[M] + \alpha_2[K] \qquad [\mathbf{5.7.15}]$$

so that the values of d_{kr} become

$$d_{kr} = \{u_k\}^T(\alpha_1[M] + \alpha_2[K])\{u_r\} = (\alpha_1 + \omega_r^2\alpha_2)\delta_{kr} \qquad [\mathbf{5.7.16}]$$

Actually, proportional damping is more of a mathematical convenience and not a very realistic model. When the amount of damping in a system is small, a simplifying assumption is to ignore the values of d_{kr} when $k \neq r$. This assumption decouples the modal equations. One can justify such an assumption by noting that the damping matrix $[D]$ is itself a gross approximation and that the damping is not known accurately.

The decoupled modal equations can then be written in the form

$$\ddot{\eta}_r(t) + 2\zeta_r\omega_r\dot{\eta}_r(t) + \omega_r^2\eta_r(t) = \mathcal{N}_r(t) \qquad [5.7.17]$$

where $2\zeta_r\omega_r = d_{rr}$ ($r = 1, 2, \ldots, n$). The response for each mode is obtained by Eq. [1.11.8].

Gyroscopic and circulatory systems require a different analysis. One has to express the equations of motion in state form and then solve the associated eigenvalue problem. This same analysis can be carried out for damped systems as well. It turns out that for undamped gyroscopic systems, there is a very elegant solution. The interested reader is referred to the texts by Meirovitch.

Example 5.8

Consider the undamped mass-spring system in the previous examples. A uniform force F is applied to the third mass. The initial conditions are $\{q(0)\} = L[1\ 0\ -1]^T$, $\{\dot{q}(0)\} = \{0\}$. Find the response.

Solution

The generalized force vector is written as $\{Q\} = [0\ 0\ F]^T$, and using Eq. [5.7.5], the modal forces become

$$\mathcal{N}_1 = \{u_1\}^T\{Q\} = \sqrt{\frac{1}{3m}}[1\ 1\ 1]\begin{bmatrix}0\\0\\F\end{bmatrix} = \sqrt{\frac{1}{3m}}F$$

$$\mathcal{N}_2 = \{u_2\}^T\{Q\} = \sqrt{\frac{1}{2m}}[1\ 0\ -1]\begin{bmatrix}0\\0\\F\end{bmatrix} = -\sqrt{\frac{1}{2m}}F$$

$$\mathcal{N}_3 = \{u_3\}^T\{Q\} = \sqrt{\frac{1}{6m}}[1\ -2\ 1]\begin{bmatrix}0\\0\\F\end{bmatrix} = \sqrt{\frac{1}{6m}}F \qquad [a]$$

The initial conditions are

$$\eta_1(0) = \{u_1\}^T[M]\{q(0)\} = \sqrt{\frac{1}{3m}}L[1\ 1\ 1]\begin{bmatrix}m&0&0\\0&m&0\\0&0&m\end{bmatrix}\begin{bmatrix}1\\0\\-1\end{bmatrix} = 0$$

$$\eta_2(0) = \{u_2\}^T[M]\{q(0)\} = \sqrt{\frac{1}{2m}}L[1\ 0\ -1]\begin{bmatrix}m&0&0\\0&m&0\\0&0&m\end{bmatrix}\begin{bmatrix}1\\0\\-1\end{bmatrix} = \sqrt{2mL}$$

$$\eta_3(0) = \{u_3\}^T[M]\{q(0)\} = \sqrt{\frac{1}{6m}}L[1\ -2\ 1]\begin{bmatrix}m&0&0\\0&m&0\\0&0&m\end{bmatrix}\begin{bmatrix}1\\0\\-1\end{bmatrix} = 0 \qquad [b]$$

$$\dot{\eta}_1(0) = \dot{\eta}_2(0) = \dot{\eta}_3(0) = 0 \qquad \text{[c]}$$

One can ascertain by simply looking at $\{q(0)\}$ that $\eta_1(0)$ and $\eta_3(0)$ should both be zero. This is because the initial condition is recognized as a constant times the second eigenvector, so that it should have no contribution to the first and third modes.

Hence, we have for the first mode,

$$\ddot{\eta}_1(t) = \sqrt{\frac{1}{3m}} F \qquad \eta_1(0) = 0, \dot{\eta}_1(0) = 0 \qquad \text{[d]}$$

whose response is

$$\eta_1(t) = \sqrt{\frac{1}{3m}} \left(\frac{Ft^2}{2}\right) \qquad \text{[e]}$$

For the second mode we have

$$\ddot{\eta}_2(t) + \omega_2^2 \eta_2(t) = -\sqrt{\frac{1}{2m}} F \qquad \eta_2(0) = \sqrt{2m} L \qquad \dot{\eta}_2(0) = 0 \qquad \text{[f]}$$

whose response can be shown to be

$$\eta_2(t) = \sqrt{2m} L \cos \omega_2 t - \sqrt{\frac{1}{2m}} \frac{F}{\omega_2} \int_0^t \sin \omega_2 \sigma \, d\sigma$$

$$= \sqrt{2m} L \cos \omega_2 t - \sqrt{\frac{1}{2m}} \frac{F}{\omega_2} \left(\frac{1}{\omega_2} - \frac{\cos \omega_2 t}{\omega_2}\right) \qquad \text{[g]}$$

which, upon introduction of the value of $\omega_2^2 = k/m$ can be written as

$$\eta_2(t) = \sqrt{2m} L \cos\left(\sqrt{\frac{k}{m}} t\right) - \frac{F}{k} \left(\sqrt{\frac{m}{2}}\right) \left(1 - \cos \sqrt{\frac{k}{m}} t\right) \qquad \text{[h]}$$

In a similar fashion, noting that $\omega_3^2 = 3k/m$, the response of the third mode becomes

$$\eta_3(t) = \frac{F}{3k} \left(\sqrt{\frac{m}{6}}\right) \left(1 - \cos \sqrt{\frac{3k}{m}} t\right) \qquad \text{[i]}$$

The total system response is then written as

$$\{q(t)\} = \{u_1\}\eta_1(t) + \{u_2\}\eta_2(t) + \{u_3\}\eta_3(t)$$

$$= \begin{bmatrix} 1 \\ 1 \\ 1 \end{bmatrix} \frac{Ft^2}{6m} + \begin{bmatrix} 1 \\ 0 \\ -1 \end{bmatrix} \left(L \cos \sqrt{\frac{k}{m}} t - \frac{F}{2k}\left(1 - \cos \sqrt{\frac{k}{m}} t\right)\right) + \begin{bmatrix} 1 \\ -2 \\ 1 \end{bmatrix} \left(\frac{F}{18k}\left(1 - \cos \sqrt{\frac{3k}{m}} t\right)\right)$$

[j]

Let us conduct a dimensional analysis of the terms in the response. The coefficient of the first eigenvector has units of force $\times$ time^2/mass = length. For the coefficients of the second eigenvector, the component due to the initial condition has units of length, and the component due to forcing has units of force/spring constant. But the spring constant has units of force/length, so that the component due to forcing has units of length, also. The same can be said about the contribution of the third mode, so that all of the terms in the above equation have the unit of length. As always, such dimensional analysis is a good way of spotting errors.

5.8 Generalized Momentum, First Integrals

As discussed in Chapter 1, first integrals are expressions that involve derivatives up to one order less than the highest derivative in the equations of motion. They come in handy when analyzing the behavior of a dynamical system qualitatively, instead of solving explicitly for the response. We investigate here first integrals associated with Lagrangian mechanics. We first define by π_k the *generalized momentum* associated with the kth generalized coordinate as

$$\pi_k = \frac{\partial L}{\partial \dot{q}_k} \qquad k = 1, 2, \ldots, n \qquad [\mathbf{5.8.1}]$$

The relationship between the generalized coordinates and the generalized momenta is very similar to the relationship between a translational coordinate and linear momentum or between a rotation angle and angular momentum. Because the potential energy does not contain any terms in the generalized velocities (except for in electromagnetic systems), one can express the generalized momentum as

$$\pi_k = \frac{\partial L}{\partial \dot{q}_k} = \frac{\partial T}{\partial \dot{q}_k} \qquad [\mathbf{5.8.2}]$$

Consider now a system where the lth generalized coordinate q_l does not appear in the Lagrangian. Such a coordinate is referred to as *cyclic* or *ignorable*. The name cyclic is due to the fact that such coordinates are encountered mostly in rotational motion. It follows that in the lth equation of motion, $\partial L/\partial q_l = 0$, and Lagrange's equations become

$$\frac{d}{dt}\left(\frac{\partial L}{\partial \dot{q}_l}\right) = \dot{\pi}_l = Q_{lnc} \qquad [\mathbf{5.8.3}]$$

When the generalized coordinate is cyclic, the rate of change of the generalized momentum is equal to the generalized force. One can integrate Eq. [5.8.3] over time to obtain the generalized momentum. In the special case when the generalized force associated with the ignorable coordinate is zero, we obtain from Eq. [5.8.3] that

$$\frac{d}{dt}\left(\frac{\partial L}{\partial \dot{q}_l}\right) = 0 = \frac{d}{dt}(\pi_l) \qquad \rightarrow \pi_l = \text{constant} \qquad [\mathbf{5.8.4}]$$

When a generalized coordinate is absent from the Lagrangian, and the external excitation is not a function of that generalized coordinate, the associated generalized momentum is conserved. This is an integral of the motion. One can then raise the question as to whether it is possible to take advantage of the cyclic coordinates and simplify the equations of motion. The answer to this is positive. If a system has n degrees of freedom and l cyclic coordinates, the n equations of motion can be reduced to $n - l$ equations of motion that can be solved independently of the cyclic coordinates, plus l first integrals associated with the cyclic coordinates. We actually carried out such a procedure in Example 5.1. One way to identify and separate ignorable coordinates is described in Section 5.9.

5.8 GENERALIZED MOMENTUM, FIRST INTEGRALS

We encounter an interesting integral of the motion when the Lagrangian is not an explicit function of time and when no nonconservative forces act on the system. In this case, considering that $L = L(q_1, q_2, \ldots, q_n, \dot{q}_1, \dot{q}_2, \ldots, \dot{q}_n)$, the time derivative of the Lagrangian becomes[3]

$$\frac{dL}{dt} = \sum_{k=1}^{n} \frac{\partial L}{\partial q_k} \dot{q}_k + \sum_{k=1}^{n} \frac{\partial L}{\partial \dot{q}_k} \frac{d}{dt}(\dot{q}_k) \qquad [5.8.5]$$

Next, we rewrite Lagrange's equations in the absence of nonconservative forces as

$$\frac{\partial L}{\partial q_k} = \frac{d}{dt}\left(\frac{\partial L}{\partial \dot{q}_k}\right) \qquad k = 1, 2, \ldots, n \qquad [5.8.6]$$

Introducing Eq. [5.8.6] into Eq. [5.8.5] we obtain

$$\frac{dL}{dt} = \sum_{k=1}^{n} \frac{d}{dt}\left(\frac{\partial L}{\partial \dot{q}_k}\right)\dot{q}_k + \sum_{k=1}^{n} \frac{\partial L}{\partial \dot{q}_k}\frac{d}{dt}(\dot{q}_k) = \sum_{k=1}^{n} \frac{d}{dt}\left(\frac{\partial L}{\partial \dot{q}_k}\dot{q}_k\right) \qquad [5.8.7]$$

or

$$\frac{d}{dt}\left(\sum_{k=1}^{n} \frac{\partial L}{\partial \dot{q}_k}\dot{q}_k - L\right) = 0 \qquad [5.8.8]$$

And, integrating this equation, we obtain the *Jacobi integral h*, defined as

$$h = \sum_{k=1}^{n} \frac{\partial L}{\partial \dot{q}_k}\dot{q}_k - L = \sum_{k=1}^{n} \pi_k \dot{q}_k - L = \text{constant} \qquad [5.8.9]$$

The Jacobi integral is yet another integral of the motion. But h can be expressed in a simpler form: We first write the generalized momenta in column vector format as $\{\pi\} = [\pi_1 \pi_2 \ldots \pi_n]^T$, and using Eqs. [5.8.2] and [5.2.9], we have

$$\{\pi\} = \left(\frac{\partial T}{\partial \{\dot{q}\}}\right)^T = [M]\{\dot{q}\} + \{\beta\} \qquad [5.8.10]$$

so that

$$\sum_{k=1}^{n} \pi_k \dot{q}_k = \{\dot{q}\}^T\{\pi\} = \{\dot{q}\}^T[M]\{\dot{q}\} + \{\dot{q}\}^T\{\beta\} = 2T_2 + T_1 \qquad [5.8.11]$$

Introducing this into Eq. [5.8.9], we obtain

$$h = \sum_{k=1}^{n} \pi_k \dot{q}_k - L = 2T_2 + T_1 - (T_2 + T_1 + T_0 - V) = T_2 - T_0 + V = T_2 + U$$

$$[5.8.12]$$

For a natural system $T_0 = 0$ and $T_2 = T$, so the Jacobi integral becomes

---

[3] Note the similarity between this approach and the procedure in Appendix B to find first integrals.

$$h = T_2 + V = T + V = \text{constant} \qquad [5.8.13]$$

Thus, for a natural system, the Jacobi integral is the system energy.

We next demonstrate that the generalized momenta are indeed a set of independent variables that are derivable from the generalized velocities, and vice versa. The relation between the generalized momenta and generalized velocities is given in Eq. [5.8.10]. Inverting it, we can express the generalized velocities in terms of the generalized momenta as

$$\{\dot{q}\} = [M]^{-1}(\{\pi\} - \{\beta\}) \qquad [5.8.14]$$

For Eq. [5.8.14] to hold, $[M]$ must be nonsingular. It was stated earlier that $[M]$ is positive definite, so it is guaranteed to be nonsingular. Equation [5.8.14] leads to the conclusion that the generalized momenta and generalized velocities are related to each other by a linear one-to-one relationship and that they can be used interchangeably in the problem formulation. We had earlier expressed the Lagrangian in terms of the generalized coordinates and generalized velocities as

$$L = L(q_1, q_2, \ldots, q_n, \dot{q}_1, \dot{q}_2, \ldots \dot{q}_n, t) \qquad [5.8.15]$$

Based on the discussions above, we can now also represent the Lagrangian as

$$L = L(q_1, q_2, \ldots, q_n, \pi_1, \pi_2, \ldots, \pi_n, t) \qquad [5.8.16]$$

It follows that one can use any combination of generalized velocities and generalized momenta. (We cannot use π_k and $\dot{q}_k (k = 1, 2, \ldots, n)$ together, though, as they would constitute a redundant set.)

Example 5.9

Consider the bead problem again. The Jacobi integral is given by Eq. [5.8.12] so that it has the form

$$h = T_2 + U = \frac{mR^2}{2}\dot{\theta}^2 + mgR(1 - \cos\theta) - \frac{mR^2}{2}\Omega^2 \sin^2\theta = \text{constant} \qquad [a]$$

which obviously is different than the sum of the kinetic and potential energies of the bead, signifying that for nonnatural systems the energy integral is not the sum of the kinetic and potential energies.

5.9 ROUTH'S METHOD FOR IGNORABLE COORDINATES

If a system has n degrees of freedom and l of those coordinates are cyclic, Routh's method permits one to reduce the n equations of motion to $n - l$ equations that can be solved separately from the cyclic coordinates. The reduced set of $n - l$ equations is usually easier to solve than the full set of n equations. The remaining l equations associated with the cyclic coordinates are expressed as first integrals.

Consider a scleronomic conservative system that has n degrees of freedom, with l of them cyclic. Order the generalized coordinates such that the first $n - l$ coordinates

5.9 ROUTH'S METHOD FOR IGNORABLE COORDINATES

q_k ($k = 1, 2, \ldots, n-l$) are not cyclic, and the next q_k ($k = n-l+1, n-l+2, \ldots, n$) are. The generalized momenta π_k ($k = n-l+1, n-l+2, \ldots, n$) associated with the cyclic coordinates are constant. We will write the Lagrangian in terms of the generalized coordinates and the generalized momenta. This is perfectly acceptable because, as demonstrated in the previous section, generalized momenta are independent variables.

For a natural system with no cyclic coordinates the Lagrangian can be expressed in terms of the generalized momenta as

$$L = L(q_1, q_2, \ldots, q_n, \pi_1, \pi_2, \ldots, \pi_n) \qquad [5.9.1]$$

In the presence of l cyclic coordinates, we write the Lagrangian in terms of the generalized velocities for the coordinates that are *not* cyclic and in terms of the generalized momenta for the coordinates that *are* cyclic. For the cyclic coordinates the generalized coordinates are absent from the Lagrangian

$$L = L(q_1, q_2, \ldots, q_{n-l}, \dot{q}_1, \dot{q}_2, \ldots, \dot{q}_{n-l}, \pi_{n-l+1}, \pi_{n-l+2}, \ldots, \pi_n) \qquad [5.9.2]$$

and

$$\pi_{n-l+k} = \text{constant} \qquad k = 1, 2, \ldots, l \qquad [5.9.3]$$

We next introduce the *Routhian*, denoted by $\mathcal{R}$, as

$$\mathcal{R} = L - \sum_{k=1}^{l} \pi_{n-l+k} \dot{q}_{n-l+k} \qquad [5.9.4]$$

Comparing the Lagrangian and Routhian, we observe for the coordinates that are not cyclic

$$\frac{\partial L}{\partial q_k} = \frac{\partial \mathcal{R}}{\partial q_k} \qquad \frac{\partial L}{\partial \dot{q}_k} = \frac{\partial \mathcal{R}}{\partial \dot{q}_k} \qquad k = 1, 2, \ldots, n-l \qquad [5.9.5]$$

and for the coordinates that are cyclic

$$\frac{\partial \mathcal{R}}{\partial \pi_{n-l+k}} = \frac{\partial L}{\partial \pi_{n-l+k}} - \dot{q}_{n-l+k} - \pi_{n-l+k} \frac{\partial \dot{q}_{n-l+k}}{\partial \pi_{n-l+k}} \qquad k = 1, 2, \ldots, l \qquad [5.9.6]$$

The last term on the right of this equation can be written using the chain law for differentiation as

$$\pi_{n-l+k} \frac{\partial \dot{q}_{n-l+k}}{\partial \pi_{n-l+k}} = \frac{\partial L}{\partial \dot{q}_{n-l+k}} \frac{\partial \dot{q}_{n-l+k}}{\partial \pi_{n-l+k}} = \frac{\partial L}{\partial \pi_{n-l+k}} \qquad [5.9.7]$$

which, when introduced into Eq. [5.9.6] yields

$$\frac{\partial \mathcal{R}}{\partial \pi_{n-l+k}} = -\dot{q}_{n-l+k} \qquad k = 1, 2, \ldots, l \qquad [5.9.8]$$

One can substitute Eqs. [5.9.5] into Lagrange's equations and write them as

$$\frac{d}{dt}\left(\frac{\partial \mathcal{R}}{\partial \dot{q}_k}\right) - \frac{\partial \mathcal{R}}{\partial q_k} = 0 \qquad k = 1, 2, \ldots, n-l \qquad [5.9.9]$$

Equations [5.9.9] and [5.9.8] represent two sets of equations, of orders $n - l$ and l, respectively, which can be solved separately from each other. One first solves Eqs. [5.9.9] for the generalized coordinates and velocities $q_1, q_2, \ldots, q_{n-l}, \dot{q}_1, \dot{q}_2, \ldots, \dot{q}_{n-l}$ associated with the noncyclic coordinates. The results are then substituted into Eqs. [5.9.8] to solve for the generalized velocities associated with the cyclic coordinates. The values for the constants $\pi_{n-l+1}, \pi_{n-l+2}, \ldots, \pi_n$ are obtained from the initial conditions.

The motivation behind defining the Routhian comes from Hamiltonian mechanics, as we will see in Sec. 5.11.

Example 5.10

Return again to the bead problem. As in Example 5.1, we treat the rotation of the hoop as another degree of freedom. Writing the rotation rate Ω as $\dot{\phi}$, the Lagrangian becomes

$$L = T - V = \frac{1}{2}I\dot{\phi}^2 + \frac{1}{2}mR^2\dot{\phi}^2 \sin^2\theta + \frac{1}{2}mR^2\dot{\theta}^2 - mgR(1 - \cos\theta) \quad \text{[a]}$$

where I is the mass moment of inertia of the hoop.

We observe that ϕ is absent from the Lagrangian, which makes it an ignorable coordinate. The generalized momentum associated with ϕ is constant. The partial derivatives of the Lagrangian are

$$\frac{\partial L}{\partial \theta} = mR^2\dot{\phi}^2 \sin\theta\cos\theta - mgR\sin\theta \qquad \frac{\partial L}{\partial \dot{\theta}} = mR^2\dot{\theta} \quad \text{[b]}$$

$$\frac{\partial L}{\partial \dot{\phi}} = I\dot{\phi} + mR^2\dot{\phi}\sin^2\theta = (I + mR^2\sin^2\theta)\dot{\phi} = \pi_\phi = \text{constant} \quad \text{[c]}$$

One can seek a physical explanation of the generalized momentum. Indeed, π_ϕ is the component of the angular momentum of the hoop-bead system along the axis of the hoop.

We derived the equations of motion in Example 5.1. To obtain the equation of motion using Routh's method, we calculate the Routhian as

$$\mathcal{R} = L - \dot{\phi}\pi_\phi = \frac{1}{2}I\dot{\phi}^2 + \frac{1}{2}mR^2\dot{\phi}^2 \sin^2\theta + \frac{1}{2}mR^2\dot{\theta}^2 - mgR(1 - \cos\theta) - \dot{\phi}\pi_\phi \quad \text{[d]}$$

Using the expression for π_ϕ the Routhian can be expressed without $\dot{\phi}$ by

$$\mathcal{R} = \frac{1}{2}mR^2\dot{\theta}^2 - mgR(1 - \cos\theta) - \frac{\dot{\phi}\pi_\phi}{2}$$

$$= \frac{1}{2}mR^2\dot{\theta}^2 - mgR(1 - \cos\theta) - \frac{\pi_\phi^2}{2(I + mR^2\sin^2\theta)} \quad \text{[e]}$$

5.10 IMPULSIVE MOTION

The linear impulse-momentum relationship for a particle of mass m acted upon by a force $\mathbf{F}(t)$ is obtained by integrating Newton's second law over time, for

$$m\mathbf{v}(t_1) + \int_{t_1}^{t_2} \mathbf{F}(t)\, dt = m\mathbf{v}(t_2) \qquad [5.10.1]$$

We saw in Chapter 1 that if a true impulsive force were applied, the particle velocity would change immediately after the impulse but the position would not change. Displacements require finite time to develop.

We now consider Lagrange's equations and impulsive external excitation. Similar to the approach in Chapter 1, we integrate the equations of motion. In general, integration of Lagrange's equations over time does not yield important results. But the situation is different when the applied forces are impulsive. We rewrite Lagrange's equations in the form of Eq. [4.9.8] as

$$\frac{d}{dt}\left[\frac{\partial T}{\partial \dot{q}_k}\right] - \frac{\partial T}{\partial q_k} + \frac{\partial V}{\partial q_k} = Q_{knc} \qquad k = 1, 2, \ldots, n \qquad [5.10.2]$$

Assume an impulsive force is applied at time $t = t_0$. Integrating the above equation from t_0 to $t_0 + \epsilon$ and taking the limit as ϵ approaches zero, we obtain for each term

$$\lim_{\epsilon \to 0} \int_{t_0}^{t_0+\epsilon} \frac{d}{dt}\left(\frac{\partial T}{\partial \dot{q}_k}\right) dt = \lim_{\epsilon \to 0} \int d\left(\frac{\partial T}{\partial \dot{q}_k}\right) \qquad [5.10.3]$$

$$= \lim_{\epsilon \to 0}(\pi_k)\big|_{t_0}^{t_0+\epsilon} = \lim_{\epsilon \to 0}[\pi_k(t_0+\epsilon) - \pi_k(t_0)] = \Delta\pi_k$$

$$\lim_{\epsilon \to 0} \int_{t_0}^{t_0+\epsilon} \frac{\partial T}{\partial q_k} dt = 0 \qquad \lim_{\epsilon \to 0} \int_{t_0}^{t_0+\epsilon} \frac{\partial V}{\partial q_k} dt = 0 \qquad [5.10.4]$$

$$\lim_{\epsilon \to 0} \int_{t_0}^{t_0+\epsilon} Q_{knc} = \hat{Q}_k \qquad k = 1, 2, \ldots, n \qquad [5.10.5]$$

Equations [5.10.4] are both equal to zero because $\partial T/\partial q_k$ and $\partial V/\partial q_k$ have finite magnitudes. We refer to $\hat{Q}_k$ as the *generalized impulse*. The only forces that contribute to the generalized impulse are impulsive forces, as the contribution of all other forces disappears as the duration of the impulse becomes infinitesimally small. Chapter 1 presented this same argument. It follows that

$$\Delta\pi_k = \hat{Q}_k \qquad k = 1, 2, \ldots, n \qquad [5.10.6]$$

so that the *change in generalized momentum is equal to the generalized impulse,* a result totally analogous to the impulse-momentum theorem for particles. The virtual work expression in terms of impulsive forces becomes

$$\delta \hat{W}_{nc} = \sum_{i=1}^{N} \hat{\mathbf{F}}_{inc} \cdot \delta \mathbf{r}_i = \sum_{i=1}^{N} \hat{\mathbf{F}}_i \cdot \sum_{k=1}^{n} \frac{\partial \mathbf{r}_i}{\partial q_k} \delta q_k$$

$$= \sum_{i=1}^{N}\left(\sum_{k=1}^{n} \hat{\mathbf{F}}_{inc} \cdot \frac{\partial \mathbf{r}_i}{\partial q_k} \delta q_k\right) = \sum_{k=1}^{n} \hat{Q}_k \delta q_k \qquad [5.10.7]$$

so that the generalized impulse has the form

$$\hat{Q}_k = \sum_{i=1}^{N} \hat{\mathbf{F}}_{inc} \cdot \frac{\partial \mathbf{r}_i}{\partial q_k} \qquad [5.10.8]$$

Equation [5.10.6] can be expressed in matrix form. Noting that $\{\pi\} = [M]\{\dot{q}\} + \{\beta\}$, and assuming that the inertia matrix and the $\{\beta\}$ vector do not change during the impulse, we can write Eq. [5.10.6] as

$$[M]\{\Delta\dot{q}\} = \{\hat{Q}\} \qquad [5.10.9]$$

in which $\{\hat{Q}\} = [\hat{Q}_1 \quad \hat{Q}_2 \quad \ldots \quad \hat{Q}_n]^T$ is the generalized impulse vector. Hence, given the impulsive forces, one can calculate the generalized impulses, and by inverting Eq. [5.10.9] one calculates the generalized velocities immediately after the impulse as

$$\{\Delta\dot{q}\} = [M]^{-1}\{\hat{Q}\} \qquad [5.10.10]$$

Note that if an impulsive force is applied to a body subject to constraints, the constraint forces, such as reaction forces, also become impulsive. One should exercise caution in determining those forces that are impulsive and those that are not. Also, as discussed in Chapter 1, always keep in mind the fact that Eq. [5.10.6] is an approximation, as it assumes the ideal situation of the impulse taking place in zero time. In reality, impulses take place over finite time, so that there is some change of position as well.

Example 5.11

The massless collar in Fig. 5.5 is free to slide over a guide bar. Attached to the collar with a pin joint is a rod of mass m and length L. A ball of mass M is attached to the tip of the rod. The system is at rest when an impulsive force $\hat{F}$ is applied to the mass at the tip of the rod in the horizontal direction. Find the velocity of the collar and angular velocity of the rod immediately after the impulsive force is applied.

Solution

This system has two degrees of freedom. We select the generalized coordinates as the translation of the collar and the rotation angle θ. The configuration of the system disturbed from equilibrium is shown in Fig. 5.6, together with the external forces. The kinetic energy is due to the kinetic energy of the ball and the kinetic energy of the rod

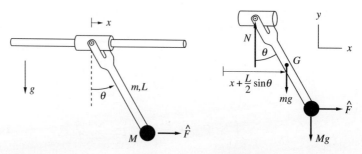

Figure 5.5   **Figure 5.6**

5.10 Impulsive Motion

$$T = \frac{1}{2}M(v_{Mx}^2 + v_{My}^2) + \frac{1}{2}m(v_x^2 + v_y^2) + \frac{1}{2}I_G\dot{\theta}^2 \qquad \textbf{[a]}$$

where v_{Mx} and v_{My} are the velocities of the ball in the x and y directions, v_x and v_y are the velocities of the center of mass of the rod, and I_G is the centroidal mass moment of inertia of the rod, $I_G = mL^2/12$. From Fig. 5.6, we can express the displacements of the center of mass of the rod and of the ball as

$$x_G = x + \frac{L}{2}\sin\theta \qquad y_G = -\frac{L}{2}\cos\theta \qquad x_M = x + L\sin\theta \qquad y_M = -L\cos\theta \qquad \textbf{[b]}$$

Differentiating the above terms, we obtain the velocities in the x and y directions as

$$v_x = \dot{x} + \frac{L}{2}\dot{\theta}\cos\theta \qquad v_y = \frac{L}{2}\dot{\theta}\sin\theta \qquad v_{Mx} = \dot{x} + L\dot{\theta}\cos\theta \qquad v_{My} = L\dot{\theta}\sin\theta \qquad \textbf{[c]}$$

Substituting Eqs. [c] into Eq. [a] we obtain

$$T = \frac{1}{2}M\left[\left(\dot{x} + L\dot{\theta}\cos\theta\right)^2 + \left(L\dot{\theta}\sin\theta\right)^2\right] + \frac{1}{2}m\left[\left(\dot{x} + \frac{L}{2}\dot{\theta}\cos\theta\right)^2 + \left(\frac{L}{2}\dot{\theta}\sin\theta\right)^2\right] + \frac{1}{2}\left(\frac{1}{12}\right)mL^2\dot{\theta}^2$$

$$= \frac{1}{2}M[\dot{x}^2 + L^2\dot{\theta}^2 + 2L\dot{\theta}\dot{x}\cos\theta] + \frac{1}{2}m\left[\dot{x}^2 + \frac{1}{3}L^2\dot{\theta}^2 + L\dot{\theta}\dot{x}\cos\theta\right] \qquad \textbf{[d]}$$

The change in potential energy during the impulse is zero, as we assume that the position of the system does not change during the impulse. Using Eq. [5.8.2], the generalized momenta have the form

$$\frac{\partial T}{\partial \dot{x}} = M\dot{x} + ML\dot{\theta}\cos\theta + m\dot{x} + \frac{1}{2}mL\dot{\theta}\cos\theta = \pi_x$$

$$\frac{\partial T}{\partial \dot{\theta}} = ML^2\dot{\theta} + ML\dot{x}\cos\theta + \frac{1}{3}mL^2\dot{\theta} + \frac{1}{2}mL\dot{x}\cos\theta = \pi_\theta \qquad \textbf{[e]}$$

The virtual work is

$$\delta W = F\,\delta x_M = F\,\delta(x + L\sin\theta) = F\,\delta x + FL\cos\theta\,\delta\theta = Q_x\,\delta x + Q_\theta\,\delta\theta \qquad \textbf{[f]}$$

so that the generalized impulses are

$$\hat{Q}_x = \hat{F} \qquad \hat{Q}_\theta = \hat{F}L\cos\theta \qquad \textbf{[g]}$$

At the point of application of the impulse, the rod is vertical and $\theta = 0$. Combining Eqs. [e] and [g], we obtain two equations for the two unknowns $\dot{x}$ and $\dot{\theta}$ as

$$(M + m)\dot{x} + \left(ML + \frac{1}{2}mL\right)\dot{\theta} = \hat{F}$$

$$\left(ML + \frac{1}{2}mL\right)\dot{x} + \left(ML^2 + \frac{1}{3}mL^2\right)\dot{\theta} = \hat{F}L \qquad \textbf{[h]}$$

Solving Eqs. [h] for the values of $\dot{x}$ and $\dot{\theta}$ immediately after the impulse, we obtain

$$\dot{x} = \frac{-2\hat{F}}{4M + m} \qquad \dot{\theta} = \frac{6\hat{F}}{(4M + m)L} \qquad \textbf{[g]}$$

Note that the velocity of the collar right after the impulse is in the opposite direction of the impulse.

5.11 HAMILTON'S EQUATIONS

Hamilton's equations have the property that they directly yield $2n$ first-order equations of motion that are in state form. They also find applications in the stability analysis of dynamical systems, in transformation theory, and in control systems.

Consider the generalized momenta, $\{\pi\} = [M]\{\dot{q}\} + \{\beta\}$. It follows that Lagrange's equations can be written in terms of the generalized momenta as

$$\{\dot{\pi}\} + \{g\} = \{Q_{nc}\} \qquad [5.11.1]$$

where $\{g\}^T = -\partial L/\partial\{q\}$. It follows that one can express Lagrange's equations in state form as

$$\{\dot{q}\} = [M]^{-1}(\{\pi\} - \{\beta\}) \qquad \{\dot{\pi}\} = -\{g\} + \{Q_{nc}\} \qquad [5.11.2a,b]$$

The equations in the above form can be derived from a scalar function called the *Hamiltonian,* defined in a way similar to the Jacobi integral and the Routhian as

$$\mathcal{H} = \sum_{k=1}^{n} \frac{\partial L}{\partial \dot{q}_k}\dot{q}_k - L = \sum_{k=1}^{n} \pi_k \dot{q}_k - L = \{\pi\}^T\{\dot{q}\} - L \qquad [5.11.3]$$

Considering Eq. [5.8.12] we can write

$$\mathcal{H} = T_2 + U = T_2 - T_0 + V \qquad [5.11.4]$$

As $\{\dot{q}\}$ can be expressed in terms of $\{\pi\}$, the variation of $\mathcal{H}$ can be written using $\{q\}$ and $\{\pi\}$ as the independent variables, thus

$$\delta\mathcal{H} = \frac{\partial \mathcal{H}}{\partial\{q\}}\{\delta q\} + \frac{\partial \mathcal{H}}{\partial\{\pi\}}\{\delta \pi\} \qquad [5.11.5]$$

If we take the variation of $\mathcal{H}$ using its definition in Eq. [5.11.3], we obtain

$$\delta\mathcal{H} = \{\delta\pi\}^T\{\dot{q}\} + \{\delta\dot{q}\}^T\{\pi\} - \frac{\partial L}{\partial\{\dot{q}\}}\{\delta\dot{q}\} - \frac{\partial L}{\partial\{q\}}\{\delta q\} \qquad [5.11.6]$$

Because the generalized momenta are defined as $\{\pi\}^T = \partial L/\partial\{\dot{q}\}$, the second and third terms on the right side of this expression cancel each other. Equating Eqs. [5.11.5] to the remaining terms in Eq. [5.11.6], we write

$$\{\dot{q}\}^T = \frac{\partial \mathcal{H}}{\partial\{\pi\}} \qquad -\frac{\partial L}{\partial\{q\}} = \frac{\partial \mathcal{H}}{\partial\{q\}} = -\{g\}^T \qquad [5.11.7a,b]$$

Now introducing Eq. [5.11.7b] into Eq. [5.11.2b], we obtain *Hamilton's canonical equations* written as

$$\{\dot{q}\}^T = \frac{\partial \mathcal{H}}{\partial\{\pi\}} \qquad \{\dot{\pi}\}^T = -\frac{\partial \mathcal{H}}{\partial\{q\}} + \{Q_{nc}\}^T \qquad [5.11.8]$$

or, in scalar form, as

$$\dot{q}_k = \frac{\partial \mathcal{H}}{\partial \pi_k} \qquad \dot{\pi}_k = -\frac{\partial \mathcal{H}}{\partial q_k} + Q_{knc} \qquad k = 1, 2, \ldots, n \qquad [5.11.9]$$

Equations [5.11.9] constitute $2n$ first-order equations with all time derivatives being on the left side. Hence, the equations of motion are in state form.

The rationale behind defining the Hamiltonian as in Eq. [5.11.3] is as follows. Lagrange's equations are a set of second-order differential equations, with the second-order derivative terms arising from the expression $\frac{d}{dt}\left(\frac{\partial L}{\partial \dot{q}_k}\right)(k = 1, 2, \ldots, n)$. One then ponders the possibility of having an augmented form of the Lagrangian, denoted by $\mathcal{H}$, such that $\partial \mathcal{H}/\partial \dot{q}_k = 0$. Consequently, second-order derivatives are eliminated. The Hamiltonian is defined in a similar way to the Routhian, but considering *all* the generalized coordinates. Another rationale comes from the definition of the Jacobi integral, which uses the expression $\pi_k \dot{q}_k$. Yet another motivation comes from Legendre's dual transformation, to which Lagrangian mechanics and generalized velocities and momenta are an application. It should be noted that Hamilton lived much before Routh and that it was Routh who was inspired from the Hamiltonian to develop the Routhian.

It can be shown easily that when all forces acting on the system can be derived from a potential, so that $Q_{knc} = 0$ $(k = 1, 2, \ldots, n)$, we have

$$\frac{\partial \mathcal{H}}{\partial t} = -\frac{\partial L}{\partial t} \qquad [5.11.10]$$

For a holonomic conservative system, if the Hamiltonian does not depend on time explicitly, its time derivative is zero, $d\mathcal{H}/dt = 0$, and the Hamiltonian becomes the Jacobi integral h, an integral of the motion. One advantage of Hamilton's equations is that they make it easier to identify integrals of the motion. Hamilton's equations are also useful in the stability analysis of dynamical systems. However, it may not be so simple to obtain Hamilton's equations, especially when a large number of degrees of freedom are involved. Also, from an algebraic standpoint, Hamilton's equations are usually lengthier than Lagrange's equations, limiting their usefulness to low-order systems. Hamilton's equations are viewed more as a tool to extract information from the system and to identify integrals of the motion.

Example 5.12

Figure 5.7 describes a spring pendulum, where the length of the pendulum varies because it is acted upon by two springs of constant $k/2$ each and a dashpot of constant c. Find the equations of motion using Hamilton's canonical equations.

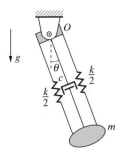

Figure 5.7   A swing pendulum

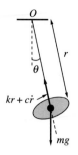

Figure 5.8
Free-body diagram

Solution

We use polar coordinates to describe the motion, as described in Fig. 5.8. We denote the unstretched length of the spring as a. The velocity of the mass is

$$\mathbf{v} = \dot{r}\mathbf{e}_r + r\dot{\theta}\mathbf{e}_\theta \qquad [\mathrm{a}]$$

leading to the kinetic energy expression

$$T = \frac{1}{2}m\mathbf{v} \cdot \mathbf{v} = \frac{1}{2}m(\dot{r}^2 + r^2\dot{\theta}^2) \qquad [\mathrm{b}]$$

The potential energy is due to the stretch of the spring and due to gravity, thus

$$V = \frac{1}{2}k(r-a)^2 - mgr\cos\theta \qquad [\mathrm{c}]$$

The dashpot gives rise to the Rayleigh's dissipation function term

$$\mathcal{F} = \frac{1}{2}c\dot{r}^2 \qquad [\mathrm{d}]$$

We next find the generalized momenta terms using Eq. [5.8.2], which yields

$$\pi_r = \frac{\partial L}{\partial \dot{r}} = \frac{\partial T}{\partial \dot{r}} = m\dot{r} \qquad \pi_\theta = \frac{\partial L}{\partial \dot{\theta}} = \frac{\partial T}{\partial \dot{\theta}} = mr^2\dot{\theta} \qquad [\mathrm{e}]$$

From Eqs. [e] we can express $\dot{r}$ and $\dot{\theta}$ as

$$\dot{r} = \frac{\pi_r}{m} \qquad \dot{\theta} = \frac{\pi_\theta}{mr^2} \qquad [\mathrm{f}]$$

and, using these expressions, we write the Hamiltonian in terms of the generalized coordinates and generalized momenta as

$$\mathcal{H} = \pi_r \dot{r} + \pi_\theta \dot{\theta} - L$$
$$= \frac{\pi_r^2}{m} + \frac{\pi_\theta^2}{mr^2} - \frac{1}{2}\frac{\pi_r^2}{m} - \frac{1}{2}\frac{\pi_\theta^2}{mr^2} + \frac{1}{2}k(r-a)^2 - mgr\cos\theta \qquad [\mathrm{g}]$$

The Rayleigh's dissipation function leads to the generalized forces associated with the nonconservative forces acting on the system as

$$Q_{rnc} = -\frac{\partial \mathcal{F}}{\partial \dot{r}} = -c\dot{r} = -\frac{c\pi_r}{m} \qquad Q_{\theta nc} = -\frac{\partial \mathcal{F}}{\partial \dot{\theta}} = 0 \qquad [\mathrm{h}]$$

Using the Hamilton's canonical equations in conjunction with Eqs. [g] and [h], we obtain the equations of motion

$$\dot{r} = \frac{\partial \mathcal{H}}{\partial \pi_r} = \frac{\pi_r}{m} \qquad \dot{\theta} = \frac{\partial \mathcal{H}}{\partial \pi_\theta} = \frac{\pi_\theta}{mr^2}$$

$$\dot{\pi}_r = -\frac{\partial \mathcal{H}}{\partial r} + Q_{rnc} = \frac{1}{mr^3}\pi_\theta^2 - k(r-a) + mg\cos\theta - \frac{c}{m}\pi_r$$

$$\dot{\pi}_\theta = -\frac{\partial \mathcal{H}}{\partial \theta} + Q_{\theta nc} = -mgr\sin\theta \qquad [\mathrm{i}]$$

5.12 COMPUTATIONAL CONSIDERATIONS: ALTERNATE DESCRIPTIONS OF THE MOTION EQUATIONS

With the advent of digital computers and powerful software, it has become possible to derive the equations of motion, as well as to solve them, using computational techniques. We discuss here issues associated with the derivation and integration of equations of motion from a computational standpoint. We begin with derivation of the equations of motion.

The classical derivation of the equations of motion in terms of Lagrange's equations is elegant, and it gives a lot of physical insight. As the number of degrees of freedom become larger, however, using Lagrange's equations directly to derive the equations of motion becomes increasingly difficult and time consuming. This is true for hand calculations as well as for derivations performed by a computer, using symbolic manipulation software. To make matters worse, an experienced user of Lagrange's equations will notice that some of the terms in the evaluation of $\partial T/\partial q_k$ ($k = 1, 2, \ldots, n$) cancel terms in $d(\partial T/\partial \dot{q}_k)/dt$, resulting in excess, wasted manipulations.

An alternative way to derive the equations of motion is to use D'Alembert's principle directly (Eq. [4.7.16]). The advantages of using D'Alembert's principle are that the kinetic energy does not need to be calculated, and, because the entire process involves vectors, the derivation process can be mechanized and accomplished more efficiently by digital computers. The disadvantage is that acceleration terms must be calculated, which requires more effort than the velocity terms used in Lagrange's equations.

Consider a system of N bodies having n degrees of freedom undergoing plane motion. We express the position of the center of mass of the ith body in column vector format $\{r_i\}$ ($i = 1, 2, \ldots, N$) where $\{r_i\} = \{r_i(q_1, q_2, \ldots, q_n, t)\}$ is a column vector of order 3. Similarly, the angles θ_i ($i = 1, 2, \ldots, N$) can be expressed in terms of the generalized coordinates as $\theta_i = \theta_i(q_1, q_2, \ldots, q_n, t)$. The time derivatives of $\{r_i\}$ and θ_i are

$$\{\dot{r}_i\} = \frac{\partial \{r_i\}}{\partial \{q\}}\{\dot{q}\} + \left\{\frac{\partial r_i}{\partial t}\right\} = [r_{iq}]\{\dot{q}\} + \left\{\frac{\partial r_i}{\partial t}\right\} \qquad \dot{\theta}_i = \{\theta_{iq}\}^T\{\dot{q}\} + \frac{\partial \theta_i}{\partial t} \qquad [5.12.1]$$

where $\{\theta_{iq}\}^T = [\partial\theta_i/\partial q_1 \quad \partial\theta_i/\partial q_2 \quad \ldots \quad \partial\theta_i/\partial q_n]$. The expression for $\{\dot{r}_i\}$ is the column vector representation of Eq. [5.2.4]. Note that $[r_{iq}]$ is a matrix of dimension $3 \times n$.

From the above equation, we write the variations of $\{r_i\}$ and θ_i as

$$\{\delta r_i\} = [r_{iq}]\{\delta q\} \qquad \delta\theta_i = \{\theta_{iq}\}^T\{\delta q\} \qquad [5.12.2]$$

In a similar fashion, we obtain the translational and angular acceleration of the ith body as

$$\{a_i\} = \frac{d^2}{dt^2}\{r_i\} = [r_{iq}]\{\ddot{q}\} + \frac{\partial}{\partial\{q\}}([r_{iq}]\{\dot{q}\})\{\dot{q}\} + \frac{\partial}{\partial t}[r_{iq}]\{\dot{q}\} + \frac{\partial}{\partial\{q\}}\left\{\frac{\partial r_i}{\partial t}\right\}\{\dot{q}\} + \left\{\frac{\partial^2 r_i}{\partial t^2}\right\}$$

$$= [r_{iq}]\{\ddot{q}\} + \frac{\partial}{\partial\{q\}}([r_{iq}]\{\dot{q}\})\{\dot{q}\} + 2\frac{\partial}{\partial t}[r_{iq}]\{\dot{q}\} + \left\{\frac{\partial^2 r_i}{\partial t^2}\right\} \qquad i = 1, 2, \ldots, N$$

$$\ddot{\theta}_i = \{\theta_{iq}\}^T\{\ddot{q}\} + \{\dot{q}\}^T[\theta_{iq}]\{\dot{q}\} + 2\frac{\partial}{\partial t}\{\theta_{iq}\}^T\{\dot{q}\} + \frac{\partial^2 \theta_i}{\partial t^2} \quad [5.12.3]$$

where $[\theta_{iq}]_{ks} = \partial^2 \theta_i / \partial q_k \, \partial q_s$ ($k, s = 1, 2, \ldots, n$). The generalized work expression can be written as

$$\delta W = \sum_{i=1}^{N}(\{F_i\}^T\{\delta r_i\} + M_{Gi}\,\delta\theta_i) = \sum_{i=1}^{N}\left(\{F_i\}^T[r_{iq}] + M_{Gi}\{\theta_{iq}\}^T\right)\{\delta q\}$$
$$= \{Q\}^T\{\delta q\} \quad [5.12.4]$$

We next combine all these relations. D'Alembert's principle in column vector format is

$$\sum_{i=1}^{N}\left((m_i\{a_i\} - \{F_i\})^T\{\delta r_i\} + (I_{Gi}\ddot{\theta}_i - M_{Gi})\,\delta\theta_i\right) = 0 \quad [5.12.5]$$

Introducing Eqs. [5.12.2]–[5.12.4] into this equation, and considering the general case of constrained systems, we obtain

$$\sum_{i=1}^{N}\left(m_i[r_{iq}]^T\left[[r_{iq}]\{\ddot{q}\} + \frac{\partial}{\partial\{q\}}([r_{iq}]\{\dot{q}\}) + 2\frac{\partial}{\partial t}[r_{iq}]\{\dot{q}\} + \left\{\frac{\partial^2 r_i}{\partial t^2}\right\}\right]\right.$$
$$\left. + I_{Gi}\{\theta_{iq}\}\left[\{\theta_{iq}\}^T\{\ddot{q}\} + \{\dot{q}\}^T[\theta_{iq}]\{\dot{q}\} + 2\frac{\partial}{\partial t}\{\theta_{iq}\}^T\{\dot{q}\} + \frac{\partial^2 \theta_i}{\partial t^2}\right]\right) + [a]^T\{\lambda\} = \{Q\}$$

$$[5.12.6]$$

in which $[a]$ is the constraint matrix of order $m \times n$. The above equation, of course, needs to be solved simultaneously with the constraint equation, Eq. [4.10.1]. It is clear that calculation of the $[r_{iq}]$ matrix and $\{\theta_{iq}\}$ vector is crucial to obtaining the equations of motion.

Considering the above derivation, one may question why we study Lagrange's equations in the first place, if the equations of motion can be more conveniently derived using Eq. [5.12.6]. There are several answers to this question. Lagrange's equations are based on D'Alembert's and Hamilton's principles, and one can derive the equations of motion using variational principles and from scalar functions. Doing so gives a very powerful result, putting in better perspective the relation between the equations of motion and energy. It points to an order in dynamical systems. It gives a better understanding of the nature of the motion, what the integrals of the motion are; and it makes it easier to establish these integrals.

Next, we consider numerical integration of the equations of motion. As discussed in Chapter 1, the equations of motion have to be cast in state form, which implies that they have to be first-order differential equations. In each equation there must be a single derivative term, on the left side. Neither Lagrange's equations nor the direct application of D'Alembert's principle yields equations of motion in state form. On the other hand, Hamilton's equations are in state form.

One way to convert Lagrange's equations into state form is to introduce the n variables $z_k = \dot{q}_k$ ($k = 1, 2, \ldots, n$). Then, realizing that the equations of motion can be written as

$$[M]\{\ddot{q}\} + \{g(\{q\},\{\dot{q}\})\} = \{Q\} \quad [5.12.7]$$

5.12 COMPUTATIONAL CONSIDERATIONS: ALTERNATE DESCRIPTIONS OF THE MOTION EQUATIONS

we write the equations as

$$[M]\{\dot{z}\} + \{g(\{q\},\{z\})\} = \{Q\} \qquad [5.12.8]$$

The $2n$ first-order equations are obtained as the inverse of Eq. [5.12.8], which provides n of the equations, together with the relationship between $\{z\}$ and $\{q\}$, which provides the other n equations. They can be expressed as

$$\{\dot{z}\} = -[M]^{-1}\{g\} + [M]^{-1}\{Q\} \qquad \{\dot{q}\} = \{z\} \qquad [5.12.9]$$

Recall that $[M]$ is a function of the generalized coordinates only, so its calculation and inversion do not involve velocity expressions.

Another way to convert the equations of motion into state form is to define $\{z\}$ as

$$\{z\} = [T]\{\dot{q}\} + \{y\} \qquad [5.12.10]$$

where $[T]$ is a nonsingular matrix of order $n \times n$ and $\{y\}$ is a vector of order n. We introduce this transformation into the equations of motion. When $[T] = [1]$ and $\{y\} = \{0\}$ we get Eqs. [5.12.9]. When $[T] = [M]$ and $\{y\} = \{\beta\}$, $\{z\}$ becomes the generalized momentum vector and we obtain Hamilton's equations. In general, the selection of $[T]$ depends on the nature of the problem.

We next discuss integration of the equations of motion of constrained systems. We repeat Eqs. [4.10.1] and [4.10.5] here as

$$\sum_{k=1}^{n} a_{jk}\dot{q}_k + a_{j0} = 0 \qquad j = 1, 2, \ldots, m$$

$$\frac{d}{dt}\left(\frac{\partial L}{\partial \dot{q}_k}\right) - \frac{\partial L}{\partial q_k} = Q_{knc} - \sum_{j=1}^{m} \lambda_j a_{jk} \qquad k = 1, 2, \ldots, n \qquad [5.12.11]$$

which can be written in matrix form as

$$[A]\{\dot{q}\} + \{b\} = \{0\} \qquad [M]\{\ddot{q}\} + [A]^T\{\lambda\} + \{g\} = \{Q_{nc}\} \qquad [5.12.12a,b]$$

in which $[A]$ is a matrix of order $m \times n$, whose elements are a_{jk}, as defined in Section 4.10. It is preferable to combine the two expressions in Eqs. [5.12.12]. The highest-order derivative in Eq. [5.12.12a] is of order one, while Eq. [5.12.12b] has time derivatives of order two. Differentiating Eq. [5.12.12a] with respect to time yields

$$[A]\{\ddot{q}\} = -[\dot{A}]\{\dot{q}\} - \{\dot{b}\} \qquad [5.12.13]$$

This can now be combined with Eq. [5.12.12b] to give one matrix equation of order $n + m$ as

$$\begin{bmatrix} [M] & [A]^T \\ [A] & [0] \end{bmatrix} \begin{bmatrix} \{\ddot{q}\} \\ \{\lambda\} \end{bmatrix} = \begin{bmatrix} \{Q_{nc}\} - \{g\} \\ -[\dot{A}]\{\dot{q}\} - \{\dot{b}\} \end{bmatrix} \qquad [5.12.14]$$

where all the terms not involving second derivatives of q_i ($i = 1, 2, \ldots, n$) are moved to the left side of Eq. [5.12.14]. The coefficient matrix on the left side of the above equation is symmetric, but it is no longer positive definite. However, it can be shown to be nonsingular. Equation [5.12.14] represents a set of hybrid equations, combining differential and algebraic equations. They are called *differential-algebraic equations*. Several new methods have been developed in recent years to solve these equations. The interested reader is referred to the texts by Brenan et al., and by Haug.

5.13 Additional Differential Variational Principles

Previous developments in this chapter and in Chapter 4 are based on variational principles, namely D'Alembert's and Hamilton's principles and the special case of static problems, the principle of virtual work. These principles are based on the variation of the displacement coordinates while keeping time fixed. In this section we analyze the variational principles that one can derive using variations of velocities and accelerations.

Consider D'Alembert's principle for a system of N particles

$$\sum_{i=1}^{N}(m_i\ddot{\mathbf{r}}_i - \mathbf{F}_i) \cdot \delta\mathbf{r}_i = 0 \qquad [5.13.1]$$

where we note that the variation is taken by keeping the dependent variable, which is the time t, constant. We now look at the position vector $\mathbf{r}_i(t)$ ($i = 1, 2, \ldots, N$) at an increment of time Δt later. Given the value of the position vector associated with the ith particle at time t, $\mathbf{r}_i(t)$, one can obtain its value at $t = t + \Delta t$, by means of a Taylor series approximation as

$$\mathbf{r}_i(t + \Delta t) = \mathbf{r}_i(t) + \dot{\mathbf{r}}_i(t)\Delta t + \ddot{\mathbf{r}}_i(t)\frac{(\Delta t)^2}{2} + \dddot{\mathbf{r}}_i(t)\frac{(\Delta t)^3}{3}! + \ldots \qquad [5.13.2]$$

The variation of $\mathbf{r}_i(t + \Delta t)$, using time fixed at t, becomes

$$\delta\mathbf{r}_i(t + \Delta t) = \delta\mathbf{r}_i(t) + \delta\dot{\mathbf{r}}_i(t)\Delta t + \delta\ddot{\mathbf{r}}_i(t)\frac{(\Delta t)^2}{2} + \delta\dddot{\mathbf{r}}_i(t)\frac{(\Delta t)^3}{3}! + \ldots \qquad [5.13.3]$$

Substitution of this equation into Eq. [5.13.1] yields

$$\sum_{i=1}^{N}(m_i\ddot{\mathbf{r}}_i - \mathbf{F}_i) \cdot \left(\delta\mathbf{r}_i(t) + \delta\dot{\mathbf{r}}_i(t)\Delta t + \delta\ddot{\mathbf{r}}_i(t)\frac{(\Delta t)^2}{2} + \delta\dddot{\mathbf{r}}_i(t)\frac{(\Delta t)^3}{3}! + \ldots\right) = 0$$

[5.13.4]

Obviously, taking the limit as $\Delta t \to 0$ gets one back to Eq. [5.13.1]. Now, consider a different variation: one obtained by setting $\delta\mathbf{r}_i(t) = \mathbf{0}$. We will denote this variation by the subscript 1, so that we write $\delta_1\mathbf{r}_i = \mathbf{0}$ ($i = 1, 2, \ldots, N$). Using this variation and keeping in mind that $\delta\mathbf{r}_i(t) = \mathbf{0}$, dividing Eq. [5.13.4] by Δt, and taking the limit as Δt approaches zero, we obtain

$$\sum_{i=1}^{N}(m_i\ddot{\mathbf{r}}_i - \mathbf{F}_i) \cdot \delta_1\dot{\mathbf{r}}_i = 0 \qquad (\delta_1\mathbf{r}_i = \mathbf{0}) \qquad [5.13.5]$$

where we note that the evaluation took place at $t = t + \Delta t$. Equation [5.13.5] is known as *Jourdain's variational principle*. Note that $\delta_1\dot{\mathbf{r}}_i \neq \delta\dot{\mathbf{r}}_i$ ($i = 1, 2, \ldots, N$), because $\delta\dot{\mathbf{r}}_i$, which are used in Hamilton's principle and Lagrange's equations, are obtained by taking the variation of $\dot{\mathbf{r}}_i$ holding only time fixed, with no restriction on $\mathbf{r}_i$.

Jourdain's variational principle can be used to deal with systems subjected to nonholonomic constraints. One can show that Eq. [5.13.5] leads to the relation

$$\sum_{k=1}^{n}\left[\frac{d}{dt}\left(\frac{\partial L}{\partial \dot{q}_k}\right) - \frac{\partial L}{\partial q_k} - Q_{knc}\right]\delta_1\dot{q}_k = 0 \qquad [5.13.6]$$

For a holonomic system that is unconstrained and q_k are chosen as independent, the variations of the generalized coordinates are also independent and the above equation yields Lagrange's equations. In the presence of nonholonomic constraints $\delta\dot{q}_k$ are no longer independent, but one can introduce the constraint into the formulation using Eq. [5.13.6]. Hence, Lagrange's equations can be considered as a special case of Eq. [5.13.6].

In a similar fashion, consider a second variation, denoted by the subscript 2, during which both $\mathbf{r}_i$ and $\dot{\mathbf{r}}_i$ are held fixed, such that $\delta_2\mathbf{r}_i = \mathbf{0}, \delta_2\dot{\mathbf{r}}_i = \mathbf{0}$. Divide Eq. [5.13.4] by $(\Delta t)^2$ and take the limit as Δt approaches zero to obtain

$$\sum_{i=1}^{N}(m_i\ddot{\mathbf{r}}_i - \mathbf{F}_i) \cdot \delta_2\ddot{\mathbf{r}}_i = 0 \qquad (\delta_2\mathbf{r}_i = \mathbf{0}, \delta_2\dot{\mathbf{r}}_i = \mathbf{0}) \qquad [5.13.7]$$

This is known as *Gauss's variational principle* or *Gauss's principle of least constraint*. This principle can be shown to lead to a least squares interpretation of the minimization of the quantity Z as

$$Z = \sum_{i=1}^{N}\frac{1}{2m_i}(m_i\ddot{\mathbf{r}}_i - \mathbf{F}_i) \cdot (m_i\ddot{\mathbf{r}}_i - \mathbf{F}_i) \qquad [5.13.8]$$

Indeed, taking the variation of Z we obtain Eq. [5.13.7]. The variations of $\mathbf{F}_i$ ($i = 1, 2, \ldots, N$) are zero as the forcing is a known quantity. We conclude that of all accelerations that are compatible with the constraint equations, the actual accelerations make the quantity Z a minimum and, hence, constitute the solution.

In practice, one uses the variational principle that yields the equations of motion with the most ease. For unconstrained holonomic systems, there is usually no need to use the Gauss or Jourdain variations. The advantage of using Jourdain's or Gauss's variational principle becomes more pronounced for systems that are nonholonomically constrained. Because nonholonomic constraints can only be written in velocity form, one can impose these constraints using the Gauss and Jourdain variations, without manipulating the generalized coordinates. We will discuss this issue further in Chapter 9.

It is interesting to note that D'Alembert's principle is dated to the year 1743, and Gauss's principle was stated in 1829, while Jourdain's principle—which establishes the link between the other two principles—was stated in 1909.

Example 5.13

Consider the vehicle in Example 4.14 and apply Jourdain's variational principle to it.

Solution

The vehicle is shown in Fig. 4.8. The nonholonomic constraint is that the velocity of point A can only be in the x direction. The constraint equation is written in terms of the generalized coordinates X, Y, and θ as

$$-\dot{X}\sin\theta + \dot{Y}\cos\theta - L\dot{\theta} = 0 \qquad [\mathbf{a}]$$

We will use Eq. [5.13.6]. To this end, we write the kinetic energy as

$$T = \frac{1}{2}m(\dot{X}^2 + \dot{Y}^2) + \frac{1}{2}I_G\dot{\theta}^2 \qquad [\mathbf{b}]$$

and the generalized forces as

$$Q_X = (F_C + F_D)\cos\theta \qquad Q_Y = (F_C + F_D)\sin\theta \qquad Q_\theta = (F_D - F_C)h \qquad [\mathbf{c}]$$

so that Eq. [5.13.6] has the form

$$(m\ddot{X} - (F_C + F_D)\cos\theta)\,\delta_1\dot{X}$$
$$+ (m\ddot{Y} - (F_C + F_D)\sin\theta)\,\delta_1\dot{Y} + (I_G\ddot{\theta} - (F_D - F_C)h)\,\delta_1\dot{\theta} = 0 \qquad [\mathbf{d}]$$

The Jourdain variation of the constraint has the form

$$-\sin\theta\,\delta_1\dot{X} + \cos\theta\,\delta_1\dot{Y} - L\,\delta_1\dot{\theta} = 0 \qquad [\mathbf{e}]$$

We can manipulate this expression and express the variation of one of the variables in terms of the other. Expressing the Jourdain variation of θ in terms of the Jourdain variations of X and Y is futile, because in this problem X and Y are not sufficient to describe the motion completely. Hence, we eliminate the variation of either X or Y. Let us select Y, so we can write

$$\cos\theta\,\delta_1\dot{Y} = \sin\theta\,\delta_1\dot{X} + L\,\delta_1\dot{\theta} \qquad [\mathbf{f}]$$

We multiply Eq. [d] by $\cos\theta$ and introduce Eq. [f] to it, which yields

$$(m\ddot{X} - (F_C + F_D)\cos\theta)\cos\theta\,\delta_1\dot{X} + (m\ddot{Y} - (F_C + F_D)\sin\theta)(\sin\theta\,\delta_1\dot{X} + L\,\delta_1\dot{\theta})$$
$$+ (I_G\ddot{\theta} - (F_D - F_C)h)\cos\theta\,\delta_1\dot{\theta} = 0 \qquad [\mathbf{g}]$$

Because the Jourdain variations of X and θ are independent, their coefficients must vanish independently, which yields

$$m\ddot{X}\cos\theta + m\ddot{Y}\sin\theta = F_C + F_D$$
$$I_G\ddot{\theta}\cos\theta + m\ddot{Y}L = (F_C + F_D)L\sin\theta + (F_D - F_C)h\cos\theta \qquad [\mathbf{h}]$$

These equations still contain derivatives of three variables. To simplify, consider incorporating the velocity of A as a motion variable. The velocity of point A can be written as

$$\mathbf{v}_A = v_A\mathbf{i} = (\dot{X}\cos\theta + \dot{Y}\sin\theta)\mathbf{i} \qquad [\mathbf{i}]$$

The acceleration of A then becomes

$$\mathbf{a}_A = \dot{v}_A\mathbf{i} + \boldsymbol{\omega}\times\mathbf{v}_A = \dot{v}_A\mathbf{i} + \dot{\theta}v_A\mathbf{j} \qquad [\mathbf{j}]$$

Differentiating v_A and using Eq. [a], we obtain

$$\dot{v}_A = \ddot{X}\cos\theta + \ddot{Y}\sin\theta - \dot{X}\dot{\theta}\sin\theta + \dot{Y}\dot{\theta}\cos\theta = \ddot{X}\cos\theta + \ddot{Y}\sin\theta + L\dot{\theta}^2 \qquad [\mathbf{k}]$$

which, when introduced into the first of Eqs. [h] yields

$$m(\dot{v}_A - L\dot{\theta}^2) = F_C + F_D \qquad [\mathbf{l}]$$

In a similar fashion, one can show that the second of Eqs. [h] can be written as

$$(I_G + mL^2)\ddot{\theta} + mL\dot{\theta}v_A = (F_D - F_C)h \qquad \textbf{[m]}$$

so that Eqs. [l] and [m] constitute the two equations of motion in terms of v_A and θ. Upon closer examination, one recognizes Eq. [l] as the force balance in the x direction, and Eq. [m] as the moment balance about point A. These equations are, of course, the same as the equations of motion obtained in Example 4.14.

References

Brenan, K. E., S. L. Campbell, and L. Petzold. *Numerical Solution of Initial-Value Problems in Differential-Algebraic Equations.* New York: Elsevier, 1989.
Ginsberg, J. H. *Advanced Engineering Dynamics.* New York: Harper & Row, 1988.
Goldstein, H. *Classical Mechanics.* 2nd ed. Reading, MA: Addison Wesley, 1980.
Greenwood, D. T. *Principles of Dynamics.* 2nd ed. Englewood Cliffs, NJ: Prentice-Hall, 1988.
Haug, E. J. *Intermediate Dynamics.* Englewood Cliffs, NJ: Prentice-Hall, 1992.
Lanczos, C. *The Variational Principles of Mechanics.* 4th ed. New York: Dover Publications, 1970.
Meirovitch, L. *Methods of Analytical Dynamics.* New York: McGraw-Hill, 1970.
——. *Principles and Techniques of Vibrations.* Englewood Cliffs, NJ: Prentice-Hall, 1997.

Homework Exercises

Section 5.2

1. A link of mass m and length L is pinned to the edge of a disk of mass M and radius R, as shown in Fig. 5.9. A servomotor keeps the angular velocity of the

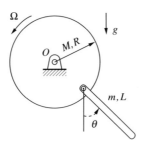

Figure 5.9

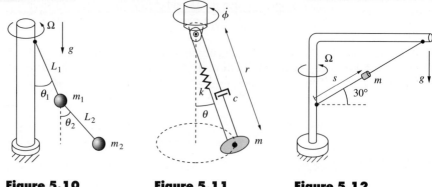

Figure 5.10   **Figure 5.11**   **Figure 5.12**

disk constant at a value of Ω. Find the equilibrium position for θ as a function of the angular velocity of the disk.

2. Find the equilibrium equations for the rotating double pendulum in Fig. 5.10. The angular velocity of the shaft Ω is constant.

3. Consider the rotating spring pendulum in Fig. 5.11 and obtain the equilibrium positions when $\dot{\phi} = \Omega$ is a constant and $c = 0$.

4. Find the equilibrium position of the bead shown in Fig. 5.12 for Ω = constant.

SECTION 5.3

5. Consider the equations of motion for the double pendulum in Fig. 4.3. Linearize them in the neighborhood of the equilibrium position(s).

6. Consider Example 4.11 and linearize the equations of motion about the equilibrium point $x_e = 0, \theta_e = 0$.

7. Consider Problem 3 and evaluate the stability of the equilibrium points.

8. Given the bead problem considered earlier, plot the dynamic potential U as a function of θ for the following conditions: (a) $\Omega^2 = 0.5g/R$, (b) $\Omega^2 = 1.5g/R$. Verify that the stability relations discussed in Sec. 5.3 hold.

9. A particle of mass m moves on a smooth surface described by the relation $z = x^2 + y^2 - xy$, with the z axis being the vertical. Derive the equations of motion and obtain the linearized equations about equilibrium.

SECTION 5.4

10. For the mass-spring system shown in Fig. 5.13, calculate the expressions for the kinetic and potential energies, as well as the Rayleigh's dissipation function. Then, write the equations of motion in matrix form directly from the energy terms, without using Lagrange's equations or Newton's second law.

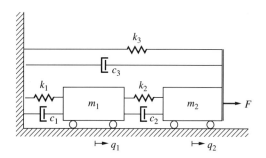
Figure 5.13

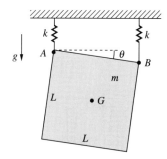
Figure 5.14

Section 5.5

11. A thin square metal plate, of mass m and sides L, hangs from its corners attached to two springs, each of constant k, as shown in Fig. 5.14. Find the equations of motion for small oscillations, as well as the natural frequencies. The horizontal motion of the center of mass is negligible.

12. Find the equations of motion, natural frequencies, and eigenvectors of the system shown in Fig. 5.13 for $k_1 = k_2 = k$, $k_3 = 2k$, $m_1 = m$, $m_2 = 2m$ and no damping.

13. Find the linearized equations of motion, natural frequencies, and eigenvectors of the system shown in Fig. 5.15. The system consists of two links of the same length and mass, attached by a spring. Assume small motions and consider only the horizontal deformation of the spring.

14. Consider Problem 5 and obtain the natural frequencies and modal vectors for $m_1 = m_2 = m$, $L_1 = L_2 = L$.

15. Consider Problem 6 and obtain the natural frequencies and modal vectors for $M = 2m$ and $k = mg/b$.

16. Four identical masses are connected to four identical springs and constrained to move in a circle, as shown in Fig. 5.16. The springs are unstretched when the masses are all equidistant. How many rigid body modes does this system have? Find the natural frequencies.

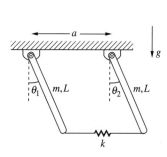
Figure 5.15

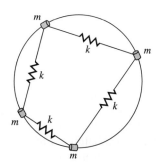
Figure 5.16

Section 5.6

17. Check for the orthogonality of the natural modes in Problem 12.
18. Check for the orthogonality of the natural modes in Problem 14.

Section 5.7

19. Consider Problem 13 and the case when the spring is weak, that is, the potential energy of the spring is much less than the potential energy due to gravity. Approximate the two natural frequencies and plot the response to zero initial conditions.
20. Obtain the response of the system in Problem 12 to a unit impulse on mass 1. Initial conditions are zero.
21. Consider the double pendulum in Fig 4.3. Set $m_1 = m_2 = m, L_1 = L_2 = L$ and find the response to the initial conditions $\theta_1(0) = 30°, \theta_2(0) = 0°$. Calculate the maximum value of the amplitude of $\theta_2(t)$.

Section 5.8

22. Identify the integrals of the motion for the pendulum in Problem 3 for (a) $\dot{\phi}$ is a variable, and (b) $\dot{\phi}$ is a constant $= \Omega$. Take $c = 0$ in both cases.
23. Identify the integrals of the motion for the Foucault's pendulum.
24. Find the integrals of the motion for the rotating double pendulum in Fig. 5.10.

Section 5.9

25. Consider the rotating spring pendulum in Problem 3 and obtain the Routhian.

Section 5.10

26. Obtain the velocity of the collar shown in Fig. 5.17 immediately after an impulsive force of $\hat{F}$ is applied to the bottom of the disk of radius $R = L/6$. At the instant the impulse is applied, $\theta = 0$, and the disk is not rotating.
27. Two rods of equal length and mass are connected by a pin joint, and they are at rest, as shown in Fig. 5.18. An impulsive force is applied at point B and perpendicular to the line AB. Find the angular velocities of the rods immediately after the impulse. *Hint:* Include the coordinates of point A in the generalized coordinates.
28. The double pendulum in Fig. 4.3 is at rest when an impulsive force $\hat{F}$ is applied in the horizontal direction to the bottom of the second rod. Find the ensuing angular velocities.

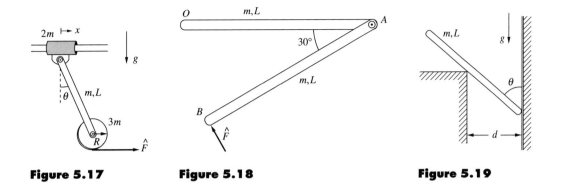

Figure 5.17 **Figure 5.18** **Figure 5.19**

Section 5.11

29. Derive the equation of motion of the bead on a hoop that we have considered earlier using Hamilton's equations. Consider the rotation of the hoop $\dot{\phi}$ as a variable.

Section 5.12

30. Consider the system in Example 4.11 and derive the equations of motion using the direct application of D'Alembert's principle by Eq. [5.12.6].
31. Solve Problem 30 using D'Alembert's principle, but without explicitly calculating the $[r_{iq}]$ matrices. Compare the ease with which the solution is obtained with the solution obtained by using Eq. [5.12.6] and by using Lagrange's equations.

Section 5.13

32. Solve Problem 4.41 using Jourdain's variational principle.
33. Find the equation of motion of the rod in Fig. 5.19 using Jourdain's variational principle. Friction is negligible and the rod maintains contact with the wall.

chapter 6

RIGID BODY GEOMETRY

6.1 INTRODUCTION

In this chapter we consider the geometrical properties of rigid bodies. The term *rigid* is in reality a mathematical idealization, because all bodies deform by a certain amount under the application of loads. If the deformation is small compared to the overall dimensions of the body, and energy dissipation due to elastic effects is negligible, the rigid body assumption can safely be used.

Let us distinguish between particles and rigid bodies. A particle was defined earlier as a body with no physical dimensions. This definition, of course, is also an idealization, as all bodies have physical dimensions. If the dimensions of the body are much smaller than the path followed by the body, it becomes possible to neglect the physical dimensions of the body. One does not consider any rotational motion, and three translational degrees of freedom are sufficient to describe the motion. Whether one can use this approximation depends on the conditions under which the motion takes place, the loading, as well as the material properties of the bodies involved.

A rigid body is defined as a body with physical dimensions where the distances between the particles that constitute the body remain unchanged. One needs to consider the rotational motion of a rigid body; thus, six degrees of freedom, three translational and three rotational, are required to completely describe its motion. In addition, one needs to develop quantities that give information regarding the distribution of mass along the body. Just as the mass of a body represents its resistance to translational motion, the distribution of the mass about a certain axis represents the body's resistance to rotational motion about that axis.

6.2 CENTER OF MASS

Consider a system of N particles viewed from a fixed reference frame with origin O, as shown in Fig. 6.1. The term m_i denotes the mass of the ith particle ($i = 1, 2, \ldots, N$) and $\mathbf{r}_i$ its distance from O. In Chapter 3, the center of mass of a system of

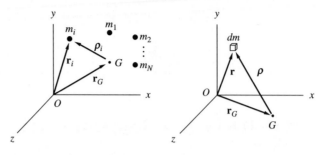

Figure 6.1 A system of particles

Figure 6.2 Differential element for a rigid body

particles was denoted by the point G, and its location was defined as

$$\mathbf{r}_G = \frac{1}{m} \sum_{i=1}^{N} m_i \mathbf{r}_i \qquad [6.2.1]$$

in which m denotes the total mass

$$m = \sum_{i=1}^{N} m_i \qquad [6.2.2]$$

A rigid body can be considered as a collection of particles in which the number of particles approaches infinity and in which the distances between the individual masses remain the same. As $N \to \infty$, each particle can be treated as a differential mass element, as shown in Fig. 6.2, $m_i \to dm$, and the summations in the above two equations are replaced by integrations over the body. We then define the location of the center of mass G as

$$\mathbf{r}_G = \frac{1}{m} \int_{\text{body}} \mathbf{r}\, dm \qquad [6.2.3]$$

where $\mathbf{r}$ is the vector from the origin to the differential element dm, and

$$m = \int_{\text{body}} dm \qquad [6.2.4]$$

is the mass of the rigid body. Considering Figs. 6.1 and 6.2, for a system of particles we can write $\mathbf{r}_i = \mathbf{r}_G + \boldsymbol{\rho}_i$, and for a rigid body $\mathbf{r} = \mathbf{r}_G + \boldsymbol{\rho}$. Introducing this term into Eq. [6.2.3], we obtain

$$\mathbf{r}_G = \frac{1}{m} \int_{\text{body}} \mathbf{r}\, dm = \frac{1}{m} \int_{\text{body}} (\mathbf{r}_G + \boldsymbol{\rho})\, dm = \mathbf{r}_G + \frac{1}{m} \int_{\text{body}} \boldsymbol{\rho}\, dm \qquad [6.2.5]$$

leading to the conclusion that

$$\int_{\text{body}} \boldsymbol{\rho}\, dm = 0 \qquad [6.2.6]$$

This equation indicates that the weighted average of the displacement vector about the center of mass is zero. Considering concepts from statistics, one can refer to the definition of the center of mass as the first moment of the mass distribution. Naturally, Eq. [6.2.6] is identical to its counterpart for a system of particles, Eq. [3.2.5].

The center of mass is a very important quantity, as its use simplifies the analysis of bodies considerably. One has to perform the integrations in Eqs. [6.2.3] and [6.2.4] in order to find the center of mass. These integrations are in general triple integrations. Many times, the body that one is dealing with consists of parts that are uniform in density and that have shapes whose masses and centers of mass are known in advance or found more easily. Under such circumstances, one can calculate the center of mass of the composite body by using the center of masses of the individual bodies. If the body is separated into N components, and the mass and center of mass of each component is denoted by m_i and $\mathbf{r}_{Gi}$ ($i = 1, 2, \ldots, N$) the center of mass can be found from

$$\mathbf{r}_G = \frac{1}{m} \sum_{i=1}^{N} m_i \mathbf{r}_{Gi} \qquad [6.2.7]$$

When dealing with several components, we will also use the notation of denoting the center of mass by an overbar. The velocity and acceleration of the center of mass can be obtained by differentiating the expression for position and integrating it over the body. Using the expression for $\mathbf{r}_G$ in Eq. [6.2.3] we obtain

$$\mathbf{v}_G = \frac{1}{m} \int_{body} \dot{\mathbf{r}} \, dm \qquad \mathbf{a}_G = \frac{1}{m} \int_{body} \ddot{\mathbf{r}} \, dm \qquad [6.2.8]$$

The velocity and acceleration of a point can be expressed in terms of the center of mass as

$$\mathbf{v} = \mathbf{v}_G + \dot{\boldsymbol{\rho}} \qquad \mathbf{a} = \mathbf{a}_G + \ddot{\boldsymbol{\rho}} \qquad [6.2.9]$$

Introducing these expressions to Eqs. [6.2.8] and performing the integrations gives

$$\int_{body} \dot{\boldsymbol{\rho}} \, dm = 0 \qquad \int_{body} \ddot{\boldsymbol{\rho}} \, dm = 0 \qquad [6.2.10]$$

Example 6.1

Figure 6.3 shows a decorative pencil holder, made of uniform material of thickness 0.4 cm. Find the center of mass.

Solution

We first select the coordinate frame. For convenience, we attach it to point O. The pencil holder can be analyzed as consisting of three parts: the base, the hollow cylinder that holds the pencils, and the circular hole. As the entire pencil holder is made of uniform material, we dispense with the density and base our calculations on the volume. For each component, we will determine the volume and location of center of mass. We thus have:

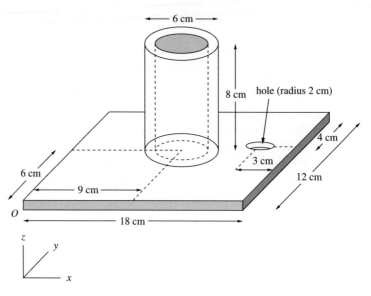

Figure 6.3

For the base: Volume $V_1 = 18(12)(0.4) = 86.4 \text{ cm}^3$ $\bar{r}_1 = 9\mathbf{i} + 6\mathbf{j} + 0.2\mathbf{k}$ cm

For the cylinder: Volume $V_2 = \pi(3^2 - 2.6^2)(8) = 56.30 \text{ cm}^3$ $\bar{r}_2 = 9\mathbf{i} + 6\mathbf{j} + 4.4\mathbf{k}$ cm

For the circular hole: Volume $V_3 = \pi(2^2)(0.4) = 5.027 \text{ cm}^3$ $\bar{r}_3 = 15\mathbf{i} + 8\mathbf{j} + 0.2\mathbf{k}$ cm

Combining all these values, we obtain

$$V = V_1 + V_2 - V_3 = 86.4 + 56.30 - 5.027 = 137.7 \text{ cm}^3 \quad \text{[a]}$$

so that the center of mass is located at

$$\mathbf{r}_G = \frac{86.4(9\mathbf{i} + 6\mathbf{j} + 0.2\mathbf{k}) + 56.30(9\mathbf{i} + 6\mathbf{j} + 4.4\mathbf{k}) - 5.027(15\mathbf{i} + 8\mathbf{j} + 0.2\mathbf{k})}{137.7}$$

$$= 8.780\mathbf{i} + 5.926\mathbf{j} + 1.917\mathbf{k} \text{ cm} \quad \text{[b]}$$

6.3 MASS MOMENTS OF INERTIA

While the center of mass provides valuable information and simplifies the analysis of translational motion, it gives no measure of the way the mass is distributed on the body. The mass of a body describes the amount of matter contained in the body and the resistance of the body to translational motion. We know of the need to develop a quantity that describes the resistance of a body to rotation. Such a quantity is dependent on how the mass is distributed. As the center of mass is located using the first moment of the mass distribution, we consider the second moment of the mass distribution.

6.3 MASS MOMENTS OF INERTIA

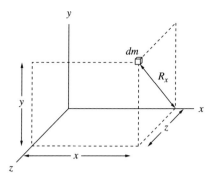

Figure 6.4

We select a coordinate system xyz fixed to a point on the body and describe the configuration of a differential mass element by the vector $\mathbf{r} = x\mathbf{i} + y\mathbf{j} + z\mathbf{k}$, also written as the column vector $\{r\} = [x \quad y \quad z]^T$. Such a selection of the coordinate frame is possible because the body is assumed to be rigid. We are primarily concerned with two types of quantities: the distribution of the mass with respect to a certain axis; and the distribution of the mass with respect to a certain plane. Consider the x axis first. From Fig. 6.4, the perpendicular distance of a differential element dm from the x axis is $R_x = \sqrt{y^2 + z^2}$. We define the *mass moment of inertia about the x axis* as

$$I_{xx} = \int_{\text{body}} R_x^2 \, dm = \int_{\text{body}} (y^2 + z^2) \, dm \qquad [6.3.1]$$

In a similar fashion, the mass moments of inertia about the y and z axes are defined as

$$I_{yy} = \int_{\text{body}} R_y^2 \, dm = \int_{\text{body}} (x^2 + z^2) \, dm$$

$$I_{zz} = \int_{\text{body}} R_z^2 \, dm = \int_{\text{body}} (x^2 + y^2) \, dm \qquad [6.3.2]$$

One quick observation is that the mass moment of inertia of a body about a certain axis becomes larger as the axis is selected further away from the body. This is an indication that mass moments of inertia will be useful in describing the rotational motion of a body.

Considering the distribution of the mass with respect to the xy, xz, and yz planes, we introduce the three *products of inertia*

$$I_{xy} = \int_{\text{body}} xy \, dm \qquad I_{yz} = \int_{\text{body}} yz \, dm \qquad I_{xz} = \int_{\text{body}} xz \, dm \qquad [6.3.3]$$

It is clear that $I_{xy} = I_{yx}$, and so forth. In general, the products of inertia do not contribute too much to the physical description of the mass distribution, unless there are certain symmetry properties with respect to the coordinate axes. If any two coordinate axes form a plane of symmetry for the body, then the products of inertia

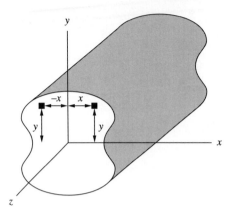

Figure 6.5 A body symmetric about the yz plane

associated with coordinates normal to that plane vanish. That is,

If there is symmetry with respect to the yz plane, then $I_{xy} = I_{xz} = 0$.
If there is symmetry with respect to the xz plane, then $I_{xy} = I_{yz} = 0$.
If there is symmetry with respect to the xy plane, then $I_{xz} = I_{yz} = 0$.

The concept is illustrated in Fig. 6.5. Assume there is symmetry with respect to the yz plane. Then, for each point defined by the coordinates (x, y, z) there corresponds the point $(-x, y, z)$. It follows that the products of inertia I_{xy} and I_{xz} vanish because both terms require integration over x, and the contribution of the differential mass on the left side of the yz plane is countered by the mass on the right side. If a body is almost symmetric about a certain plane, then the corresponding product of inertia will be a small quantity. However, the converse is not necessarily true. The sign of a product of inertia does not usually give much insight, either.

If a rigid body is symmetric about an axis then it must have symmetry about at least two planes. Thus for a body that has an axis of symmetry, all products of inertia vanish when one of the coordinate axes is along the symmetry axis. It should be noted that a body need not have planes or axes of symmetry for the products of inertia to vanish. A proper orientation of the xyz axes leads to the same result, as we will see later.

The moments and products of inertia form the so called *inertia matrix*, denoted by $[I]$ and defined as

$$[I] = \begin{bmatrix} I_{xx} & -I_{xy} & -I_{xz} \\ -I_{xy} & I_{yy} & -I_{yz} \\ -I_{xz} & -I_{yz} & I_{zz} \end{bmatrix} \quad [\mathbf{6.3.4}]$$

Note that the inertia matrix is symmetric. Furthermore, it is positive definite. To demonstrate the positive definiteness of $[I]$, one can use *Sylvester's criterion*, which states that for a symmetric matrix to be positive definite, all the diagonal elements and all principal minor determinants must be positive. The diagonal elements of $[I]$ are the principal moments of inertia and they are all positive quantities, each obtained by integration of a positive integrand. The first principal minor determinant is $I_{xx}I_{yy} - I_{xy}^2$. To show that it is greater than zero, consider the *Schwartz inequality*, which states that for any two functions $f(x)$ and $g(x)$ continuous over an interval $[a, b]$,

$$\left(\int_a^b f(x)g(x)\,dx\right)^2 \leq \int_a^b |f(x)|^2\,dx \int_a^b |g(x)|^2\,dx \qquad [6.3.5]$$

The equality holds only when $f(x) = \gamma g(x)$, where γ is a constant. Considering an integral over the rigid body, and the products of inertia, we can write

$$I_{xy}^2 = \left(\int_{\text{body}} xy\,dm\right)^2 \leq \int_{\text{body}} x^2\,dm \int_{\text{body}} y^2\,dm$$

$$< \int_{\text{body}} (x^2 + z^2)\,dm \int_{\text{body}} (y^2 + z^2)\,dm \qquad [6.3.6]$$

so that $I_{xy}^2 < I_{yy}I_{xx}$, which shows that the first principal minor determinant is greater than zero. In a similar fashion, one can show that the determinant of $[I] > 0$. In some cases, such as when dealing with slender rods or thin plates, the components of the inertia matrix about certain axes will be very small compared with the inertia components about other axes. Then, one can assume that the very small inertia components are negligible. For example, a slender rod is assumed to have zero moment of inertia about an axis along the rod.

The positive definiteness of the inertia matrix is to be expected, because it describes the distribution of the mass about a certain set of axes. The expression from statistics analogous to mass moments of inertia is *variance*. Two rigid bodies may have the same inertia matrix and not the same shape. Such bodies are called *equimomental*.

If the coordinate axes are selected such that the products of inertia vanish, the coordinate axes are referred to as *principal axes* and the corresponding mass moments of inertia are called *principal moments of inertia*. For an axisymmetric body the axis of symmetry is a principal axis.

The units of mass moments of inertia are mass × length squared. The moments of inertia of a body are usually written as the mass of the body multiplied by a length squared parameter and an appropriate constant. For example, the mass moment of inertia of a disk about the symmetry axis is written as $mR^2/2$, where m is the mass and R is the radius of the disk.

The mass, center of mass, and inertia matrix of a rigid body specify what are called the *internal properties* of the body completely. For an elastic body, one needs to know measures of the resistance of the body to deformation, in addition to the internal properties. We will study these in Chapter 11.

We next calculate the entries of the inertia matrix. This is dependent on the choice of the differential element to be integrated. We have two choices:

1. Select the differential element as a differential mass element and perform a triple integration. For Cartesian coordinates xyz, the differential element is $dm = \rho\,dx\,dy\,dz$, with ρ being the density. Polar coordinates come in handy for bodies with circular cross sections. The differential element is $dm = \rho r\,dr\,d\theta\,dz$.

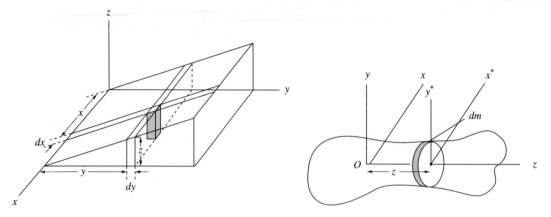

Figure 6.6 Differential element as a rod or a thin plate   **Figure 6.7** Differential element as a thin plate

2. One or two of the sides of the differential element has a finite size. Select the differential element as a thin rod or as a thin plate, as shown in Figs. 6.6 and 6.7. The differential element then has a finite mass moment of inertia about one or two axes. The expression for mass moment of inertia is written about such an axis as (say, about the z axis)

$$I_{zz} = \int dI_{zz} \qquad [6.3.7]$$

in which dI_{zz} is the mass moment of inertia of the differential element about the z axis.

The advantage of using such elements is that the number of integrations is reduced. When the differential element is a rod, one can perform a double integration, and when the differential element is a plate, one can evaluate the mass moment of inertia by a single integral.

Consider a thin flat plate of thickness t as shown in Fig. 6.8. We take a differential element in the form $dm = \rho t \, dx \, dy$. The mass moment of inertia about the z axis becomes

$$I_{zz} = \int_{\text{body}} (x^2 + y^2) \, dm = \rho t \int_{\text{area}} (x^2 + y^2) \, dx \, dy = \rho t J_z \qquad [6.3.8]$$

where J_z is recognized as the *area polar moment of inertia* about the z axis. Assuming that the thickness is small, the moment of inertia about the x axis can be approximated as

$$I_{xx} = \int_{\text{body}} (y^2 + z^2) \, dm = \rho t \int_{\text{area}} (y^2 + t^2) \, dx \, dy \approx \rho t \int_{\text{area}} y^2 \, dx \, dy = \rho t I_x$$
$$[6.3.9]$$

in which I_x is the *area moment of inertia* of the plate about the x axis. In a similar fashion, one can show that $I_{yy} = \rho t I_y$. From area moments of inertia, we know that $J_z = I_x + I_y$. Considering Eqs. [6.3.8] and [6.3.9], for a thin flat plate oriented as in Fig. 6.8, $I_{zz} = I_{xx} + I_{yy}$.

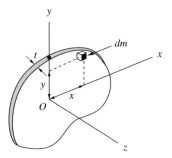

Figure 6.8 Differential element on a thin plate

Now, consider the differential element in Fig. 6.7, specifically a slice of the cross section with an infinitesimal thickness, say dz. Denoting by $x*y*z*$ the set of coordinate axes that are attached to its center of mass, the differential element has a moment of inertia of $dI_{zz} = dI_{x*x*} + dI_{y*y*}$. One then finds I_{zz} using

$$I_{zz} = \int dI_{zz} = \int (dI_{x*x*} + dI_{y*y*}) \qquad [6.3.10]$$

The mass moment of inertia I_{zz} is thus obtained by a single integral. If there is some sort of symmetry in the body, one can select the coordinate axes to exploit this symmetry and have $dI_{x*x*} = dI_{y*y*}$, such that $dI_{zz} = 2dI_{x*x*}$. This helps determine I_{xx} as well.

To calculate I_{xx} and I_{yy} we have to use the parallel axis theorem, which relates the moments of inertia of a body about two axes that are parallel to each other with one of the axes centroidal. The theorem was stated in Chapter 3, and its general form is derived in the Section 6.4. Here, we state it between the x and $x*$ axes, which are separated by a distance of z, as shown in Fig. 6.7. If the z axis is the symmetry axis, hence a centroidal axis, dI_{xx} can be expressed as

$$dI_{xx} = dI_{x*x*} + z^2 \, dm \qquad [6.3.11]$$

For a circular plate as the differential element, $dI_{x*x*} = dI_{zz}/2$ and

$$dI_{xx} = \frac{dI_{zz}}{2} + z^2 \, dm \qquad [6.3.12]$$

which, when integrated, gives the mass moment of inertia about the x axis (and y axis) as

$$I_{xx} = I_{yy} = \frac{I_{zz}}{2} + \int z^2 \, dm \qquad [6.3.13]$$

The above approach is most appealing when there is a symmetry axis, as the products of inertia vanish. For bodies where there is no symmetry axis, then $dI_{x*x*} \neq dI_{zz}/2$, and it must be evaluated independently. Further, as the products of inertia do not vanish in such cases, the above approach is of little use.

An interesting property of a mass moment of inertia is that it is an additive quantity. Given the task of finding the mass moment of inertia of a body (about an axis or a plane), one can accomplish it by taking advantage of the fact that an integral from one point to another can be expressed as the sum of two or more integrals, by appropriately separating the domain of the integration. The procedure is analogous to obtaining the center of mass of a composite body. The body is separated into more than one part. The moment of inertia of the entire body is calculated as the sum of the moments of inertia of the individual parts. To see this, take a body, separate it into N parts (body$_1$, body$_2$, ..., body$_N$), and consider any of the inertia expressions. We then have, say, for I_{xx}

$$I_{xx} = \int_{\text{body}} (y^2 + z^2) \, dm = \sum_{i=1}^{N} \int_{\text{body}_i} (y^2 + z^2) \, dm = \sum_{i=1}^{N} I_{ixx} \qquad [6.3.14]$$

in which I_{ixx} is the mass moment of inertia of the ith body about the x axis. For bodies that can be broken down into simple parts whose individual inertia properties are known or easily calculated, one can obtain the moments of inertia of each of the parts and add them together to find the moment of inertia of the entire body. This process is facilitated by use of the parallel axis theorem. Mass moments of inertia of bodies with common shapes are given in Appendix C.

A useful quantity when dealing with mass moments of inertia is the *radius of gyration*. The radius of gyration about the x, y, and z axes is defined as

$$\kappa_{xx} = \sqrt{\frac{I_{xx}}{m}} \qquad \kappa_{yy} = \sqrt{\frac{I_{yy}}{m}} \qquad \kappa_{zz} = \sqrt{\frac{I_{zz}}{m}} \qquad [6.3.15]$$

and it represents the distance that an equivalent particle of mass m would have to be placed from an axis so that it would have the same mass moment of inertia as the entire body. Hence, the mass moments of inertia are expressed as $I_{xx} = m\kappa_{xx}^2 \ldots$. The radius of gyration is frequently used in engineering practice when tabulating the inertia properties of commonly used components such as I-beams and brackets. One can define the radius of gyration for the products of inertia as well, thus

$$\kappa_{xy} = \sqrt{\frac{I_{xy}}{m}} \qquad \kappa_{yz} = \sqrt{\frac{I_{yz}}{m}} \qquad \kappa_{xz} = \sqrt{\frac{I_{xz}}{m}} \qquad [6.3.16]$$

with the understanding that $\kappa_{xx}\kappa_{yy} \neq \kappa_{xy}^2$.

Example 6.2

Find the mass moments of inertia of the paraboloid shown in Fig. 6.9 about the x and y axes. The paraboloid is obtained by rotating the parabola $x/L = (y/R)^2$ ($0 \leq y \leq R$) about the x axis.

Solution

The differential element for the paraboloid is taken as a thin disk of thickness dx. At a distance of x ($0 \leq x \leq L$), the radius of the disk is $r = R\sqrt{x/L}$. The mass and mass moment of inertia of the differential element are

$$dm = \rho A \, dx = \rho \pi r^2 \, dx = \frac{\rho \pi R^2}{L} x \, dx \qquad [a]$$

$$dI_{xx} = \frac{1}{2}(dm)r^2 = \frac{1}{2}(\rho \pi r^2 \, dx)r^2 = \frac{1}{2}\rho \pi r^4 \, dx = \frac{\rho \pi R^4}{2L^2} x^2 \, dx \qquad [b]$$

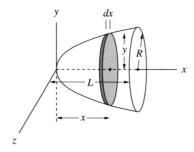

Figure 6.9

To find the mass moment of inertia I_{xx}, we integrate Eq. [b] over x, which yields

$$I_{xx} = \frac{\rho \pi R^4}{2L^2} \int_0^L x^2 \, dx = \frac{\rho \pi R^4 L}{6} \qquad \text{[c]}$$

The mass of the paraboloid is found by integrating Eq. [a] over x, thus

$$m = \int dm = \frac{\rho \pi R^2}{L} \int_0^L x \, dx = \frac{\rho \pi R^2 L}{2} \qquad \text{[d]}$$

Introduction of Eq. [d] into Eq. [c] yields the mass moment of inertia about the x axis as

$$I_{xx} = \frac{1}{3} m R^2 \qquad \text{[e]}$$

To find the moment of inertia about the y axis, attach an xy^*z^* axis to the center of the differential element. The y axis is separated by a distance x from the y^* axis, so that using Eq. [6.3.13] gives

$$I_{yy} = \frac{I_{xx}}{2} + \int x^2 \, dm \qquad \text{[f]}$$

Using Eq. [a], the second term on the right becomes

$$\int x^2 \, dm = \frac{\rho \pi R^2}{L} \int_0^L x^3 \, dx = \frac{\rho \pi R^2 L^3}{4} = \frac{1}{2} m L^2 \qquad \text{[g]}$$

and the mass moment of inertia about the y axis (and z axis) becomes

$$I_{yy} = I_{zz} = \frac{1}{6} m R^2 + \frac{1}{2} m L^2 \qquad \text{[h]}$$

Because of the axial symmetry, all products of inertia are zero.

Example 6.3

Find the mass moments of inertia and products of inertia of the right triangular prism of uniform density ρ shown in Fig. 6.10.

Solution

We first find the mass of the prism as $m = \rho V$, where the volume $V = abc/2$, so that the mass of the prism is

$$m = \frac{\rho abc}{2} \qquad \text{[a]}$$

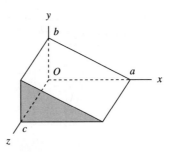

Figure 6.10

To find the moments of inertia, we need to calculate the expressions $\int x^2\,dm$, $\int y^2\,dm$, $\int z^2\,dm$. Using a rectangular differential element, we write

$$dm = \rho\,dz\,dy\,dx \qquad \text{[b]}$$

We now evaluate $\int x^2\,dm$, $\int y^2\,dm$, and $\int z^2\,dm$. Note that we perform the integrations in the order over the z, y, and x axes.

$$\int x^2\,dm = \int_0^a \int_0^{b-\frac{b}{a}x} \int_0^c \rho x^2 \,dz\,dy\,dx = \rho c \int_0^a x^2\left(b - \frac{b}{a}x\right)dx = \frac{\rho a^3 bc}{12} = \frac{ma^2}{6} \qquad \text{[c]}$$

$$\int y^2\,dm = \int_0^a \int_0^{b-\frac{b}{a}x} \int_0^c \rho y^2 \,dz\,dy\,dx = \frac{\rho c}{3} \int_0^a \left(b - \frac{b}{a}x\right)^3 dx = \frac{\rho ab^3 c}{12} = \frac{mb^2}{6} \qquad \text{[d]}$$

$$\int z^2\,dm = \int_0^a \int_0^{b-\frac{b}{a}x} \int_0^c \rho z^2 \,dz\,dy\,dx = \frac{\rho c^3}{3} \int_0^a \left(b - \frac{b}{a}x\right) dx = \frac{\rho abc^3}{6} = \frac{mc^2}{3} \qquad \text{[e]}$$

The mass moments of inertia then become

$$I_{xx} = m\left(\frac{b^2}{6} + \frac{c^2}{3}\right) \qquad I_{yy} = m\left(\frac{a^2}{6} + \frac{c^2}{3}\right) \qquad I_{zz} = m\left(\frac{a^2}{6} + \frac{b^2}{6}\right) \qquad \text{[f]}$$

Note the similarity between I_{xx} and I_{yy}. We next find the products of inertia

$$I_{xy} = \int_0^a \int_0^{b-\frac{b}{a}x} \int_0^c \rho xy\,dz\,dy\,dx = \frac{\rho c}{2}\int_0^a x\left(b - \frac{b}{a}x\right)^2 dx = \frac{\rho a^2 b^2 c}{24} = \frac{mab}{12} \qquad \text{[g]}$$

$$I_{xz} = \int_0^a \int_0^{b-\frac{b}{a}x} \int_0^c \rho xz\,dz\,dy\,dx = \frac{\rho c^2}{2}\int_0^a x\left(b - \frac{b}{a}x\right) dx = \frac{\rho a^2 b c^2}{12} = \frac{mac}{6} \qquad \text{[h]}$$

$$I_{yz} = \int_0^a \int_0^{b-\frac{b}{a}x} \int_0^c \rho yz\,dz\,dy\,dx = \frac{\rho c^2}{4}\int_0^a \left(b - \frac{b}{a}x\right)^2 dx = \frac{\rho ab^2 c^2}{12} = \frac{mbc}{6} \qquad \text{[i]}$$

Note again the similarity between I_{xz} and I_{yz}.

6.4 Transformation Properties

Given the inertia properties of a body about a point and a certain orientation of the set of axes, it is desirable to relate these properties to the inertia matrix obtained about a different point or a different orientation of the coordinate axes. This permits one to find the mass moments of inertia once and to then use the transformation equations to find the moments of inertia about other points and about other sets of axes, without having to perform the necessary integrations again. Also, moments of inertia read from tables are given in terms of a specific set of axes. We will first see how the inertia properties change as the coordinate system is translated and then as the coordinate system is rotated. Recall that in the previous section there were no restrictions on the location of the origin and on the orientation of the coordinate frame about which the inertia properties were calculated.

First, note that the inertia matrix can be expressed in column vector format and in terms of the position vector $\{r\}$ as

$$[I] = \int (\{r\}^T\{r\}[1] - \{r\}\{r\}^T)\,dm = \int (r^2[1] - \{r\}\{r\}^T)\,dm \qquad [\mathbf{6.4.1}]$$

in which $r^2 = \{r\}^T\{r\} = \mathbf{r}\cdot\mathbf{r}$. The inertia matrix can also be expressed as

$$[I] = \int [\tilde{r}]^T[\tilde{r}]\,dm \qquad [\mathbf{6.4.2}]$$

where $[\tilde{r}]$ is the skew-symmetric matrix formed from the elements of the column vector $\{r\} = [x \quad y \quad z]^T$. We will use the above definition of the inertia matrix when we discuss dynamics of rigid bodies.

Yet another way of expressing the elements of the inertia matrix is as follows. Denote the unit vectors associated with the coordinate frame by $\mathbf{e}_1$, $\mathbf{e}_2$, and $\mathbf{e}_3$, so that $\mathbf{r} = r_1\mathbf{e}_1 + r_2\mathbf{e}_2 + r_3\mathbf{e}_3$. The ijth element of the inertia matrix can be expressed as

$$I_{ij} = \int (\mathbf{r}\times\mathbf{e}_i)\cdot(\mathbf{r}\times\mathbf{e}_j)\,dm \qquad i,j = 1,2,3 \qquad [\mathbf{6.4.3}]$$

One can derive Eq. [6.4.1] from Eq. [6.4.3] by using the vector identity $(\mathbf{a}\times\mathbf{b})\cdot(\mathbf{c}\times\mathbf{d}) = (\mathbf{a}\cdot\mathbf{c})(\mathbf{b}\cdot\mathbf{d}) - (\mathbf{a}\cdot\mathbf{d})(\mathbf{b}\cdot\mathbf{c})$ and by setting $\mathbf{a} = \mathbf{c} = \mathbf{r}$, $\mathbf{b} = \mathbf{e}_i$, $\mathbf{d} = \mathbf{e}_j$.

6.4.1 TRANSLATION OF COORDINATES

Denote the origin of the coordinate system by O and consider a different coordinate system $x''y''z''$, which is parallel to the xyz system and has its origin at point B with coordinates (d_x, d_y, d_z), as shown in Fig. 6.11. That is, $x = x'' + d_x$, $y = y'' + d_y$, $z = z'' + d_z$, or $\mathbf{r}_{B/O} = d_x\mathbf{i} + d_y\mathbf{j} + d_z\mathbf{k}$. The moment of inertia about the x'' axis is found from

$$I_{Bx''x''} = \int (y''^2 + z''^2)\,dm = \int [(y - d_y)^2 + (z - d_z)^2]\,dm$$

$$= \int (y^2 + z^2)\,dm + \int (d_y^2 + d_z^2)\,dm - 2d_y\int z\,dm - 2d_z\int y\,dm \qquad [\mathbf{6.4.4}]$$

The first term on the right side of this expression is recognized as I_{xx}. The second term reduces to $m(d_y^2 + d_z^2)$. The third and fourth terms are basically the first moments of the mass distribution. If the axes xyz are selected as centroidal axes, $O = G$, these

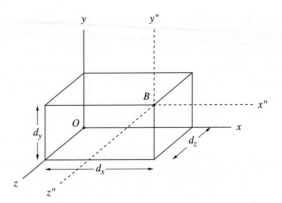

Figure 6.11 Translation of coordinates

terms vanish. We then obtain the *parallel axis theorem*, which states that

$$I_{Bx''x''} = I_{Gxx} + m(d_y^2 + d_y^2) \qquad [6.4.5]$$

where I_{Gxx} denotes that the moment of inertia is calculated with respect to the x axis passing through the center of mass. Equation [6.4.5] also indicates that the point about which the moments of inertia of a body are the smallest is the center of mass. Note that we have modified the notation for the mass moment of inertia by indicating the point about which it is calculated by the first subscript.

We calculate the remaining moments and products of inertia associated with the $x''y''z''$ axes in a similar fashion. For example, for the product of inertia term $I_{Bx''y''}$, we have

$$I_{Bx''y''} = \int x''y''\,dm = \int (x - d_x)(y - d_y)\,dm$$

$$= \int xy\,dm + d_x d_y \int dm - d_y \int x\,dm - d_x \int y\,dm$$

$$= I_{Gxy} + m d_x d_y \qquad [6.4.6]$$

We can express the parallel axis theorem in matrix form as

$$[I_B] = [I_G] + m \begin{bmatrix} d_y^2 + d_z^2 & -d_x d_y & -d_x d_z \\ -d_x d_y & d_x^2 + d_z^2 & -d_y d_z \\ -d_x d_z & -d_y d_z & d_x^2 + d_y^2 \end{bmatrix} \qquad [6.4.7]$$

Using column vector notation, denoting by $\{r''\}$ the position vector associated with a point with the primed axes, such that $\{r''\} = \{r\} - \{d\}$, in which $\{d\} = [d_x \quad d_y \quad d_z]^T$, we obtain

$$[I_B] = \int \left(r''^2 [1] - \{r''\}\{r''\}^T \right) dm$$

$$= \int \left[(r^2 + d^2 - 2\{r\}^T\{d\})[1] - (\{r\}\{r\}^T + \{d\}\{d\}^T - 2\{r\}\{d\}^T) \right] dm \qquad [6.4.8]$$

Considering the centroidal axes, $\int \{r\} \, dm$ vanishes, and we end up with

$$[I_B] = \int \left[(r^2 + d^2)[1] - \{r\}\{r\}^T - \{d\}\{d\}^T \right] dm$$

$$= [I_G] + \int (d^2[1] - \{d\}\{d\}^T) \, dm \qquad \text{[6.4.9]}$$

where $d = \{d\}^T\{d\}$.

The above relation is significant in that it permits a simple calculation of the moments of inertia about any set of axes parallel to the centroidal axes, as long as the inertia matrix about the centroidal axes is known. For bodies with complex shapes, one splits the body into smaller parts whose centroidal moments of inertia can be calculated easily or looked up in a handbook. The total moment of inertia of the body is found by using the parallel axis theorem and by adding up the individual moments of inertia. Appendix C gives the mass moments of inertia of bodies with commonly found shapes.

6.4.2 ROTATION OF COORDINATE AXES

Consider now a set of axes $x'y'z'$, obtained by applying a set of rotations to the original coordinate system xyz. The origins of the two coordinate systems coincide. The position vector of a point is $\{r'\}$. From Chapter 2, using the direction cosine matrix $[c]$ or transformation matrix $[R]$, one can relate the vectors $\{r\}$ and $\{r'\}$ by the relationship

$$\{r'\} = [c]^T\{r\} \qquad \text{or} \qquad \{r'\} = [R]\{r\} \qquad \text{[6.4.10]}$$

The distance of the differential element from the origin is the same in both coordinate frames, $r'^2 = \{r'\}^T\{r'\} = \{r\}^T[c][c]^T\{r\} = \{r\}^T\{r\} = r^2$. Consider now a differential element whose location is described by $\{r'\}$. The inertia matrix about the primed axes is defined as

$$[I'] = \int (r'^2[1] - \{r'\}\{r'\}^T) \, dm = \int (r^2[1] - \{r'\}\{r'\}^T) \, dm \qquad \text{[6.4.11]}$$

Substituting Eq. [6.4.10] into this expression, and noting that $r^2[1] = [c]^T r^2[1][c]$, we obtain

$$[I'] = \int ([c]^T r^2[1][c] - [c]^T\{r\}\{r\}^T[c]) \, dm = [c]^T \int (r^2[1] - \{r\}\{r\}^T) \, dm [c] \qquad \text{[6.4.12]}$$

Recognizing the term in the middle as $[I]$, we obtain the result

$$[I'] = [c]^T[I][c] \qquad \text{or} \qquad [I'] = [R][I][R]^T \qquad \text{[6.4.13a,b]}$$

Note that for a rotational transformation, it is not necessary to begin with a centroidal set of axes. Any set of axes will do. In general, when calculating mass moments of inertia, select the xyz axes such that the calculation of the moments of inertia is simpler (e.g., using symmetry axes or symmetry planes) and then use a coordinate transformation to obtain the moments of inertia about the desired axes.

A very important coordinate transformation is one that yields a diagonal inertia matrix. The question then arises as to whether and how a transformation matrix $[R]$ can be found such that the resulting inertia matrix is diagonal. We will show in Section 6.5 that the transformation matrix can be found by solving an eigenvalue problem.

When the coordinate axes must be both translated and rotated, the order of these operations does not affect the final result.

Example 6.4

Given the rod of mass m with the square cross section to which a concentrated mass $m/4$ is attached at point C, as shown in Fig. 6.12, find the mass moment of inertia about a set of axes $x''y''z''$ which are parallel to the sides of the rod and which pass through point A.

Solution

There are two approaches to solve this problem. The first is to find the center of mass of the rod combined with the concentrated mass and to then use the parallel axis theorem. The second is to treat the rod and the concentrated mass separately, find their mass moments of inertia about the $x''y''z''$ axes, and then add the two moments of inertia. We demonstrate the latter approach first as it is usually simpler, and is so for this example, because one needs to use the parallel axis theorem only once for each component.

Using the second approach, we find the inertia matrices associated with the bar and point mass separately and then use the parallel axis theorem. For the bar, we attach the xyz axes to the center of mass, as shown in Fig. 6.12. We denote the center of mass of the bar by

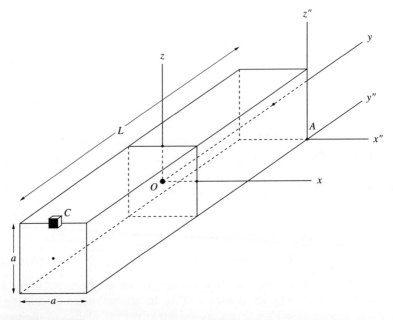

Figure 6.12

$O = G_{\text{bar}}$. From Appendix C, the inertia matrix is diagonal with its elements

$$[I_{G_{\text{bar}}}] = [I_O] = \text{diag}\left[\frac{m(a^2 + L^2)}{12} \quad \frac{ma^2}{6} \quad \frac{m(a^2 + L^2)}{12}\right] \quad \text{[a]}$$

Points A and O are related by

$$\mathbf{r}_{A/O} = \frac{a}{2}\mathbf{i} + \frac{L}{2}\mathbf{j} - \frac{a}{2}\mathbf{k} \quad \text{[b]}$$

so that the $x''y''z''$ axes are apart from the xyz axes by $d_x = a/2, d_y = L/2, d_z = -a/2$. Using Eq. [6.4.7] the combined inertia matrix becomes

$$[I_{A_{\text{bar}}}] = \frac{m}{12}\begin{bmatrix} a^2 + L^2 & 0 & 0 \\ 0 & 2a^2 & 0 \\ 0 & 0 & a^2 + L^2 \end{bmatrix} + \frac{m}{12}\begin{bmatrix} 3a^2 + 3L^2 & -3aL & 3a^2 \\ -3aL & 6a^2 & 3aL \\ 3a^2 & 3aL & 3a^2 + 3L^2 \end{bmatrix}$$

$$= \frac{m}{12}\begin{bmatrix} 4a^2 + 4L^2 & -3aL & 3a^2 \\ -3aL & 8a^2 & 3aL \\ 3a^2 & 3aL & 4a^2 + 4L^2 \end{bmatrix} \quad \text{[c]}$$

For the point mass, the inertia matrix about the center of mass is a null matrix. The coordinates of point C in the $x''y''z''$ axes are $\mathbf{r}_{C/A} = -a/2\mathbf{i} - L\mathbf{j} + a\mathbf{k}$, so that

$$d_x = \frac{-a}{2} \quad d_y = -L \quad d_z = a \quad \text{[d]}$$

Use of Eq. [6.4.7] yields the inertia matrix

$$[I_{A_{\text{mass}}}] = \frac{m}{4}\begin{bmatrix} a^2 + L^2 & \frac{-aL}{2} & \frac{a^2}{2} \\ \frac{-aL}{2} & \frac{5a^2}{4} & aL \\ \frac{a^2}{2} & aL & \frac{a^2}{4} + L^2 \end{bmatrix} \quad \text{[e]}$$

Addition of the two inertia matrices yields

$$[I_A] = [I_{A_{\text{bar}}}] + [I_{A_{\text{mass}}}] = \frac{m}{48}\begin{bmatrix} 28a^2 + 28L^2 & -18aL & 18a^2 \\ -18aL & 47a^2 & 24aL \\ 18a^2 & 24aL & 19a^2 + 28L^2 \end{bmatrix} \quad \text{[f]}$$

In the other approach, we first find the center of mass of the combined system, the associated inertia matrix, and then the inertia matrix about point A. To find the center of mass we select the xyz frame. The properties of the bar and point mass can now be written as

$$m_{\text{bar}} = m \quad \bar{\mathbf{r}}_{\text{bar}} = 0 \quad m_{\text{mass}} = \frac{m}{4} \quad \bar{\mathbf{r}}_{\text{mass}} = -\frac{L}{2}\mathbf{j} + \frac{a}{2}\mathbf{k} \quad \text{[g]}$$

where we have denoted the position of the centers of mass of the rod and point mass by an overbar. The center of mass of the combined system, denoted by G, has the coordinates

$$\mathbf{r}_G = \frac{m_{\text{bar}}\bar{\mathbf{r}}_{\text{bar}} + m_{\text{mass}}\bar{\mathbf{r}}_{\text{mass}}}{m_{\text{bar}} + m_{\text{mass}}} = \frac{-mL/8\mathbf{j} + ma/8\mathbf{k}}{5m/4} = -\frac{L}{10}\mathbf{j} + \frac{a}{10}\mathbf{k} \quad \text{[h]}$$

Next, we need to find the inertia matrix about the center of mass by using the parallel axis theorem on each component. The vectors from the centers of mass of the individual components to the center of mass are

For the rod: $\quad \mathbf{r}_{O/G} = \dfrac{L}{10}\mathbf{j} - \dfrac{a}{10}\mathbf{k}$

For the point mass: $\quad \mathbf{r}_{C/G} = -\dfrac{L}{2}\mathbf{j} + \dfrac{a}{2}\mathbf{k} - \left(-\dfrac{L}{10}\mathbf{j} + \dfrac{a}{10}\mathbf{k}\right) = -\dfrac{2L}{5}\mathbf{j} + \dfrac{2a}{5}\mathbf{k}$ [i]

so that for the bar we have $d_x = 0, d_y = L/10, d_z = -a/10$, and for the point mass $d_x = 0, d_y = -2L/5, d_z = 2a/5$. Introducing these values into the parallel axis theorem, we obtain

$$[I_G] = [I_{G_{\text{bar}}}] + [I_{G_{\text{mass}}}] \quad [j]$$

in which

$$[I_{G_{\text{bar}}}] = \dfrac{m}{12}\begin{bmatrix} a^2 + L^2 & 0 & 0 \\ 0 & 2a^2 & 0 \\ 0 & 0 & a^2 + L^2 \end{bmatrix} + \dfrac{m}{100}\begin{bmatrix} L^2 + a^2 & 0 & 0 \\ 0 & a^2 & aL \\ 0 & aL & L^2 \end{bmatrix} \quad [k]$$

and

$$[I_{G_{\text{mass}}}] = [0] + \dfrac{m}{4}\left(\dfrac{4}{25}\right)\begin{bmatrix} L^2 + a^2 & 0 & 0 \\ 0 & a^2 & aL \\ 0 & aL & L^2 \end{bmatrix} \quad [l]$$

To find the moment of inertia about point A, we calculate the distances between points G and A as

$$\mathbf{r}_{A/G} = \mathbf{r}_{A/O} - \mathbf{r}_{G/O} = \dfrac{a}{2}\mathbf{i} + \dfrac{L}{2}\mathbf{j} - \dfrac{a}{2}\mathbf{k} - \left(-\dfrac{L}{10}\mathbf{j} + \dfrac{a}{10}\mathbf{k}\right) = \dfrac{a}{2}\mathbf{i} + \dfrac{3L}{5}\mathbf{j} - \dfrac{3a}{5}\mathbf{k} \quad [m]$$

so that

$$d_x = \dfrac{a}{2} \qquad d_y = \dfrac{3L}{5} \qquad d_z = -\dfrac{3a}{5} \quad [n]$$

The inertia matrix about point A becomes

$$[I_A] = [I_G] + \dfrac{5m}{4}\begin{bmatrix} \dfrac{9(L^2 + a^2)}{25} & \dfrac{-3aL}{10} & \dfrac{3a^2}{10} \\ \dfrac{-3aL}{10} & \dfrac{61a^2}{100} & \dfrac{9aL}{25} \\ \dfrac{3a^2}{10} & \dfrac{9aL}{25} & \dfrac{a^2}{4} + \dfrac{9L^2}{25} \end{bmatrix} \quad [o]$$

Adding Eq. [o] to Eqs. [k] and [l] gives Eq. [f]. Note that this approach is much lengthier than when we dealt with the individual components directly. On the other hand, it gives us the location of the center of mass as well as the centroidal inertia matrix of the composite body.

Example 6.5

Given the mass moments of inertia of a body, find the resulting moments and products of inertia when the coordinate system is rotated about the z axis by an angle of θ.

Solution

We can approach this problem in two ways. The first is to find the direction cosine matrix between the two coordinate systems and to then use Eq. [6.4.13]. The second is to take advantage of the simplicity of the coordinate transformation and to obtain the moments and products of inertia by direct substitution. We will show the second approach first.

Consider the two sets of axes xyz and $x'y'z'$, with the primed axes obtained by rotating the unprimed axes by an angle θ about the z axis. The two sets of axes are related by

$$x' = x\cos\theta + y\sin\theta \qquad y' = -x\sin\theta + y\cos\theta \qquad z' = z \qquad \text{[a]}$$

Substitution of Eqs. [a] into the moment of inertia expressions about the primed axes gives

$$I_{x'x'} = \int (y'^2 + z'^2)\,dm = \int [(-x\sin\theta + y\cos\theta)^2 + z^2]\,dm$$

$$= \int (x^2\sin^2\theta + y^2\cos^2\theta - 2xy\sin\theta\cos\theta + z^2)\,dm \qquad \text{[b]}$$

We write z^2 as $z^2\sin^2\theta + z^2\cos^2\theta$, which, when substituted into Eq. [b], yields

$$I_{x'x'} = \sin^2\theta \int (x^2 + z^2)\,dm + \cos^2\theta \int (y^2 + z^2)\,dm - 2\sin\theta\cos\theta \int xy\,dm$$

$$= I_{xx}\sin^2\theta + I_{yy}\cos^2\theta - 2I_{xy}\sin\theta\cos\theta \qquad \text{[c]}$$

To simplify Eq. [c] further, we make use of the trigonometric identities

$$\sin^2\theta = \frac{1 - \cos 2\theta}{2} \qquad \cos^2\theta = \frac{1 + \cos 2\theta}{2} \qquad \text{[d]}$$

and substituting Eqs. [d] into Eq. [c] we get

$$I_{x'x'} = \frac{I_{xx} + I_{yy}}{2} + \frac{I_{yy} - I_{xx}}{2}\cos 2\theta - I_{xy}\sin 2\theta \qquad \text{[e]}$$

which are the familiar relations from the transformation of coordinates. In a similar way, we can obtain $I_{y'y'}$. Because the rotation of the axes is about the z axis, the mass moment of inertia about the z' axis is the same as the moment of inertia about the z axis.

The products of inertia are found the same way. For $I_{x'y'}$ we obtain

$$I_{x'y'} = \int x'y'\,dm = \int (x\cos\theta + y\sin\theta)(-x\sin\theta + y\cos\theta)\,dm$$

$$= \int [(y^2 - x^2)\sin\theta\cos\theta + xy(\cos^2\theta - \sin^2\theta)]\,dm$$

$$= I_{xy}\cos 2\theta + \frac{I_{xx} - I_{yy}}{2}\sin 2\theta \qquad \text{[f]}$$

The remaining products of inertia, $I_{x'z'}$ and $I_{y'z'}$, can be shown to become

$$I_{y'z'} = -I_{xz}\sin\theta + I_{yz}\cos\theta \qquad I_{x'z'} = I_{xz}\cos\theta + I_{yz}\sin\theta \qquad \text{[g]}$$

We can also solve this problem by generating the direction cosine matrix $[c]$ and by using Eq. [6.4.13a]. The direction cosine matrix $[c]$ has the form

$$[c] = [R]^T = \begin{bmatrix} \cos\theta & -\sin\theta & 0 \\ \sin\theta & \cos\theta & 0 \\ 0 & 0 & 1 \end{bmatrix} \qquad [\mathbf{h}]$$

which, when substituted into Eq. [6.4.13a], yield the inertia terms derived above.

Suppose now that we are told that I_{xz} and I_{yz} are both zero. Given the entries of the inertia matrix associated with the transformed system, we can calculate the rotation angle that will yield the principal axes by setting $I_{x'y'}$ equal to zero and solving for θ. Setting Eq. [f] equal to zero yields

$$\tan 2\theta = \frac{2I_{xy}}{I_{yy} - I_{xx}} \qquad [\mathbf{i}]$$

which is the same result obtained from a Mohr's circle analysis (of stress or area moments of inertia) for a two-dimensional system.

Example 6.6

Consider the right triangular prism in Example 6.3. Given that $m = 1, a = 2, b = 1, c = 3$, find the elements of the inertia matrix associated with a set of coordinates obtained by rotating the xyz axes about the z axis such that the x' axis is parallel with the incline.

Solution

The configuration is shown in Fig. 6.13. The rotation is clockwise with rotation angle $\theta = \tan^{-1}(b/a) = \tan^{-1}(1/2) = 26.57°$. From Example 6.3, the inertia matrix has the form

$$[I] = \frac{m}{6} \begin{bmatrix} b^2 + 2c^2 & -ab/2 & -ac \\ -ab/2 & a^2 + 2c^2 & -bc \\ -ac & -bc & a^2 + b^2 \end{bmatrix} = \frac{1}{6} \begin{bmatrix} 19 & -1 & -6 \\ -1 & 22 & -3 \\ -6 & -3 & 5 \end{bmatrix} \qquad [\mathbf{a}]$$

The rotation matrix between the xyz and $x'y'z'$ coordinates is

$$[R] = \begin{bmatrix} 0.8944 & -0.4472 & 0 \\ 0.4472 & 0.8944 & 0 \\ 0 & 0 & 1 \end{bmatrix} \qquad [\mathbf{b}]$$

From Eq. [6.4.13b], the inertia matrix associated with the transformed coordinates is

$$[I'] = [R][I][R]^T = \frac{1}{6} \begin{bmatrix} 20.40 & -1.800 & -4.025 \\ -1.800 & 20.60 & -5.367 \\ -4.025 & -5.367 & 5.000 \end{bmatrix} \qquad [\mathbf{c}]$$

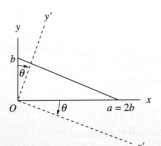

Figure 6.13   Side view of prism

As can be seen $I_{z'z'}$ is the same as I_{zz}, by virtue of the fact that the coordinate transformation is a rotation about the z axis. However, the products of inertia involving the z axis (I_{xz} and I_{yz}) do change, because of the change in the x and y coordinates.

6.5 PRINCIPAL MOMENTS OF INERTIA

As we will see in Chapter 8, it is desirable to work with an inertia matrix that is diagonal. The equations for rotational motion become substantially simpler. When the coordinate axes x, y, and z are chosen such that the products of inertia vanish, the x, y, and z axes are called the *principal axes*, and the moments of inertia I_{xx}, I_{yy}, and I_{zz} are called the *principal moments of inertia*.

There are two ways of finding the principal moments of inertia. The first is by visual inspection. For example, if the coordinate axes are selected such that they lie on a plane of symmetry, then at least two of the products of inertia vanish. If one of the coordinate axes is an axis of symmetry, all products of inertia vanish.

The second method of finding principal moments of inertia is by solving the eigenvalue problem associated with the inertia matrix. Since $[I]$ is symmetric and positive definite, all of its eigenvalues are real and positive. Its eigenvectors are real, as well.

Examining Eq. [6.4.13a] and considering the case where $[I']$ is a diagonal matrix, we observe that the transformation equation for the inertia matrix represents an orthogonal transformation. Because the direction cosine matrix is orthogonal, Eq. [6.4.13a] can be written as

$$[I'] = [c]^T[I][c] \qquad [c]^T[c] = [1] \qquad \textbf{[6.5.1]}$$

These equations are essentially the mathematical description of the eigenvalue problem associated with the matrix $[I]$. The diagonal matrix $[I']$ contains the eigenvalues in its diagonal elements, and $[c]$ is the normalized eigenvector matrix with the eigenvectors as the columns of $[c]$. To verify this, consider the eigenvalue problem associated with $[I]$,

$$[I]\{u\} = \lambda\{u\} \qquad \textbf{[6.5.2]}$$

where λ denotes the eigenvalues and $\{u\}$ the eigenvectors. Equation [6.5.2] can also be expressed as $([I] - \lambda[1])\{u\} = \{0\}$. A solution to the above equation exists if $([I] - \lambda[1])$ is singular, hence

$$\det([I] - \lambda[1]) = 0 \qquad \textbf{[6.5.3]}$$

This is called the *characteristic equation*, and it is a third-order polynomial in λ. Solution of the characteristic equation yields three eigenvalues λ_i ($i = 1, 2, 3$) and associated eigenvectors $\{u_i\}$. The eigenvectors are orthogonal to each other, such that

$$\{u_i\}^T\{u_j\} = 0 \qquad \{u_i\}^T[I]\{u_j\} = 0 \qquad i \neq j;\ i, j = 1, 2, 3 \qquad \textbf{[6.5.4]}$$

and, as we saw in Chapter 5, they can be normalized to yield $\{u_i\}^T\{u_i\} = 1$ ($i = 1, 2, 3$). It follows that $\{u_i\}^T[I]\{u_i\} = \lambda_i$. We then form the eigenvector matrix $[U] = [\{u_1\} \quad \{u_2\} \quad \{u_3\}]$. The eigenvector matrix is an orthogonal matrix, such that

$[U]^T[U] = [1]$. The orthogonality relations with respect to the inertia matrix can be written in compact form as

$$[U]^T[I][U] = [\lambda] = [I'] \qquad [U]^T[U] = [1] \qquad [\mathbf{6.5.5}]$$

in which $[\lambda]$ is a diagonal matrix containing the eigenvalues, which are the principal moments of inertia. Eq. [6.5.5] can be inverted to yield

$$[U][\lambda][U]^T = [I] \qquad [\mathbf{6.5.6}]$$

We conclude by comparing Eq. [6.5.5] and Eq. [6.5.1] that the direction cosine matrix $[c]$ that leads to the principal axes is the same as the eigenvector matrix $[U]$ obtained by solving the eigenvalue problem associated with the inertia matrix. The eigenvalues of $[I]$ are the principal moments of inertia. The eigenvectors $\{u_i\}$ ($i = 1, 2, 3$) are the direction cosines associated with the principal directions.

Care must be taken when relating the eigenvector matrix to the direction cosine matrix $[c]$ that leads to the principal axes. In Chapter 2, we learned that for $[c]$ to describe a proper rotation, its determinant must be equal to 1. The determinant of the normalized eigenvector matrix is ± 1, with the choice of plus or minus depending on the analyst. This is because if $\{u_i\}$ is an eigenvector associated with λ_i, so is $-\{u_i\}$. For $[U]$ to represent a proper transformation matrix, the eigenvectors should be normalized such that $\det [U] = 1$.

Another issue to consider is the ordering of the eigenvalues and eigenvectors. In general, in eigenvalue problems associated with real symmetric matrices, one writes the eigenvalues in ascending (or descending) order and constructs the eigenvector matrix accordingly. This is primarily done for convenience, as we did in the vibration problems in Chapter 5. In vibration problems, the lower frequencies are usually of more interest than the higher frequencies. When dealing with inertia matrices there is no such need. One can select the order of eigenvectors entirely arbitrarily. One guideline for selection is to look at the magnitudes of the diagonal elements of $[I]$ and then to order the eigenvalues such that they are similar to the order of the diagonal elements of $[I]$. When the products of inertia are small quantities, such an approach usually leads to the smallest rotation angles between the original coordinates and the principal axes.

An interesting special case in eigenvalue analysis is that of systems possessing repeated eigenvalues. In mathematics, a matrix that has repeated eigenvalues is referred to as *degenerate*. Inertia matrices having repeated eigenvalues have an elegant physical interpretation in dynamics.

Consider a symmetric matrix that has two repeated eigenvalues $\lambda_i = \lambda_j$, with corresponding eigenvectors $\{u_i\}$ and $\{u_j\}$. Because the eigenvalues are repeated, there is a certain amount of arbitrariness in $\{u_i\}$ and $\{u_j\}$. Any linear combination of these two eigenvectors is also an eigenvector belonging to λ_i, that is, $\beta_1\{u_i\} + \beta_2\{u_j\}$ is an eigenvector for any β_1 and β_2. There is no unique eigenvector $\{u_i\}$ or $\{u_j\}$. One usually selects these two eigenvectors so that $\{u_i\}^T\{u_j\} = 0$.

Consider next a rigid body that has two of its principal moments of inertia the same, say I_1 and I_2. Such a body is referred to as *inertially symmetric*. The most common case when such a situation is encountered is in *axisymmetric* bodies, such as disks, circular rods, or an American football. Mathematically, inertial symmetry can exist for bodies that have no physical symmetry properties. Using a set of axes $b_1 b_2 b_3$ attached to the body, the symmetry axis is the b_3 axis. The $b_1 b_2$ plane is referred to as the *principal plane*. If we take a rotation of the coordinate axes about the symmetry axis b_3, we see that the rotated coordinate axes also constitute a set of principal axes. There is no unique way of defining the principal axes, the same as with the eigenvectors associated with repeated eigenvalues.

We shall see in Chapter 8 that bodies possessing inertial symmetry have better stability properties than arbitrary bodies. Bodies that spin are usually designed to be axisymmetric, with the spin axis and the symmetry axis coinciding.

Example 6.7

Given a rigid body with the inertia matrix

$$[I] = \begin{bmatrix} 400 & 0 & -125 \\ 0 & 350 & 0 \\ -125 & 0 & 100 \end{bmatrix} \text{kg} \cdot \text{m}^2$$

find the principal moments of inertia and the transformation (or direction cosine) matrix that diagonalizes $[I]$.

Solution

Two of the products of inertia are zero, $I_{xy} = 0$ and $I_{yz} = 0$, in this problem, leading to the conclusion that we are probably dealing with a body which has symmetry about the xz plane, as depicted in Fig. 6.14. The eigenvalue problem is defined in Eq. [6.5.3] as

$$([I] - \lambda[1])\{u\} = \{0\} \qquad \text{[a]}$$

which requires that the determinant $\det([I] - \lambda[1]) = 0$, or

$$\det \begin{bmatrix} 400 - \lambda & 0 & -125 \\ 0 & 350 - \lambda & 0 \\ -125 & 0 & 100 - \lambda \end{bmatrix} = 0 \qquad \text{[b]}$$

leading to the characteristic equation

$$(400 - \lambda)(350 - \lambda)(100 - \lambda) - (350 - \lambda)125^2$$
$$= (350 - \lambda)[(400 - \lambda)(100 - \lambda) - 15{,}625] = 0 \qquad \text{[c]}$$

the solution of which yields the eigenvalues, which are the principal moments of inertia, as

$$\lambda_1 = I_1 = 445.26 \text{ kg} \cdot \text{m}^2 \qquad \lambda_2 = I_2 = 350 \text{ kg} \cdot \text{m}^2 \qquad \lambda_3 = I_3 = 54.74 \text{ kg} \cdot \text{m}^2 \qquad \text{[d]}$$

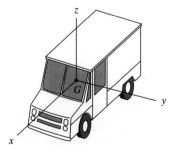

Figure 6.14 A vehicle that is symmetric about the xz axes

Note that we ordered the eigenvalues such that they follow the order of the diagonal elements of $[I]$, which in this case happens to be in descending order. To find the corresponding eigenvectors, we use Eq. [a] in conjunction with the eigenvalues. Using the notation $\{u_1\} = [u_{11} \ u_{21} \ u_{31}]^T$, we have for the first eigenvalue $\lambda_1 = 445.26 \text{ kg} \cdot \text{m}^2$

$$\begin{bmatrix} 400 - \lambda_1 & 0 & -125 \\ 0 & 350 - \lambda_1 & 0 \\ -125 & 0 & 100 - \lambda_1 \end{bmatrix} \begin{bmatrix} u_{11} \\ u_{21} \\ u_{31} \end{bmatrix} = \{0\} \qquad [\text{e}]$$

which leads to the simultaneous equations

$$-45.26 u_{11} - 125 u_{31} = 0 \qquad u_{21} = 0 \qquad -125 u_{11} - 345.26 u_{31} = 0 \qquad [\text{f}]$$

As discussed in Chapter 5, only two of the equations are independent, so that the eigenvectors can be determined to a multiplicative constant. Eqs. [f] yield

$$\{u_1\} = d_1 [1 \ 0 \ -0.3621]^T \qquad [\text{g}]$$

where d_1 is a normalization parameter. We normalize the eigenvectors such that $\{u_1\}^T \{u_1\} = 1$

$$\{u_1\}^T \{u_1\} = d_1^2 (1 + 0.3621^2) = 1.101 d_1^2 \qquad [\text{h}]$$

so that the value of $d_1 = \pm 0.9403$. When selecting the sign of $d_i (i = 1, 2, 3)$ we must take into consideration that the eigenvector matrix $[U] = [\{u_1\} \ \{u_2\} \ \{u_3\}]$ must have a determinant equal to 1.

The second eigenvalue is $350 \text{ kg} \cdot \text{m}^2$, unchanged from I_{22}. Actually, one can observe this from the symmetry about the xz plane. The second eigenvector is $\{u_2\} = \pm [0 \ 1 \ 0]^T$. The logical choice is $\{u_2\} = [0 \ 1 \ 0]^T$. This choice implies that the y axis is not changed after the rotation, whereas selecting $\{u_2\} = [0 \ -1 \ 0]^T$ would have resulted in a 180° rotation. The y axis is a principal axis. It follows that the coordinate transformation that leads to the principal axes is a rotation about the y axis, also known as a 2 rotation.

Using the same procedure as the one for the first eigenvalue, the third normalized eigenvector is found to be

$$\{u_3\} = \pm 0.9403 [0.3621 \ 0 \ 1]^T = \pm [0.3404 \ 0 \ 0.9403]^T \qquad [\text{i}]$$

We construct the eigenvector matrix keeping in mind that its determinant has to be equal to 1. Considering what a rotation about the y axis looks like, we write the eigenvector matrix as

$$[U] = [c] = [R]^T = \begin{bmatrix} 0.9403 & 0 & 0.3404 \\ 0 & 1 & 0 \\ -0.3404 & 0 & 0.9403 \end{bmatrix} \qquad [\text{j}]$$

Comparing $[U]$ with the transformation matrix for a 2 rotation, the cosine of the rotation angle is 0.9403 and its sine is 0.3404. The rotation is counterclockwise and the rotation angle turns out to be 19.90°. The rotation is shown in Fig. 6.15.

Had we selected $\{u_2\}$ as $[0 \ -1 \ 0]^T$, the sequence of rotations that leads to the principal coordinates would not be the single clockwise rotation about the y axis, but a more complicated one. Also, when the eigenvector matrix $[U]$ is fully populated there is no set way of finding the transformation to obtain the principal coordinates. This further demonstrates that there is no unique way of transforming one coordinate system into another.

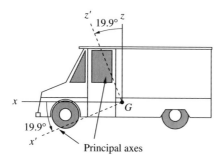

Figure 6.15 Principal axes of the vehicle

6.6 INERTIA ELLIPSOID

The eigenvalue problem associated with the inertia matrix is given in Eq. [6.5.2]. Left-multiplying it by $\{u\}^T$ gives

$$\{u\}^T[I]\{u\} = \lambda\{u\}^T\{u\} \qquad [6.6.1]$$

If we use a coordinate system xyz and the expansion $\{u\} = [x \ y \ z]^T$, Eq. [6.6.1] can be expressed as

$$I_{xx}x^2 + I_{yy}y^2 + I_{zz}z^2 - 2I_{xy}xy - 2I_{xz}xz - 2I_{yz}yz = \lambda(x^2 + y^2 + z^2) \qquad [6.6.2]$$

This equation represents the intersection of two closed quadratic surfaces, the left an ellipsoid and the right a sphere. This is because the expressions on either side of the equation are positive definite. The ellipsoid on the left side is called the *inertia ellipsoid*. The inertia ellipsoid is fixed to the body and is dependent on the point about which the inertia matrix is considered.

Let us next normalize the vector $\{u\}$ such that

$$\{u\}^T\{u\} = x^2 + y^2 + z^2 = 1 \qquad [6.6.3]$$

Now we have reduced our points of interest to the locus of all points that define a sphere of unit radius. As we vary the parameter λ, the shape of the ellipsoid does not change, but its size varies. For each value of λ, the ellipsoid and sphere intersect at different points. Without loss of generality, ordering the principal moments of inertia such that $I_1 \leq I_2 \leq I_3$, one can show that the values of λ are bounded by the principal moments of inertia

$$\lambda_{\min} = I_1 \qquad \lambda_{\max} = I_3 \qquad [6.6.4]$$

The smallest ellipsoid of inertia is obtained when $\lambda = I_1$. In this case, the longest axis of the ellipsoid has a length of unity and lies tangent to the sphere, and the entire ellipsoid lies inside the unit sphere. The largest ellipsoid is obtained when $\lambda = I_3$, when the shortest axis of the ellipsoid is of length unity and is tangent to the unit sphere, with all of the sphere lying inside the ellipsoid. When $\lambda = I_2$, the intermediate axis of the ellipsoid becomes tangent to the sphere, with the smallest axis of the ellipsoid lying entirely inside the sphere and the largest axis extending

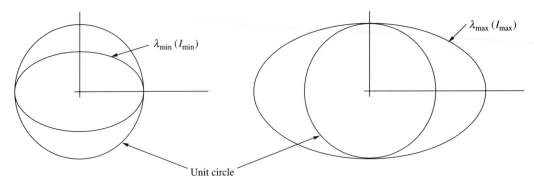

Figure 6.16 Inertia ellipsoid in two dimensions

outside the sphere. Fig. 6.16 illustrates this concept in two dimensions. We refer to the longest, shortest, and intermediate axes of the ellipsoid as the *major axes*.

The major axes of the ellipsoid correspond to the eigenvectors of the inertia matrix. One can ascertain this from the discussion above, as well as by writing Eq. [6.6.2] when the coordinate axes are selected as the principal axes. Denoting the coordinates associated with the principal axes as u_1, u_2 and u_3, the inertia ellipsoid equation reduces to

$$I_1 u_1^2 + I_2 u_2^2 + I_3 u_3^2 = \lambda \qquad [6.6.5]$$

The sphere equation is simply $u_1^2 + u_2^2 + u_3^2 = 1$. Now examine the relationship between the lengths of the axes of the ellipsoid, which are the principal axes, and the principal moments of inertia. At the tip of the ellipsoid along each principal direction, only one of u_1, u_2 or u_3 has a value of unity, and the rest are zero. This confirms that the axes of the ellipsoid are the principal axes.

The normalization in Eq. [6.6.3] and the associated inertia ellipsoid equation are not unique. The normalization we used gives a good physical explanation of the inertia properties, but it has the disadvantage of dealing with ellipsoids of different sizes. An alternate normalization is to use

$$\lambda \{u\}^T \{u\} = \lambda(x^2 + y^2 + z^2) = 1 \qquad [6.6.6]$$

which leads to the ellipsoid equation written

$$I_{xx} x^2 + I_{yy} y^2 + I_{zz} z^2 - 2I_{xy} xy - 2I_{xz} xz - 2I_{yz} yz = 1 \qquad [6.6.7]$$

or, in terms of principal coordinates

$$I_1 u_1^2 + I_2 u_2^2 + I_3 u_3^2 = 1 \qquad [6.6.8]$$

In this case, as shown in Fig. 6.17, the size of the ellipsoid is fixed but the sphere with which the ellipsoid intersects is no longer of unit radius, but of radius $\sqrt{1/\lambda}$. The length of the axes of this ellipsoid is equal to $\sqrt{1/I_i}$ ($i = 1, 2, 3$), so that the axis about which the moment of inertia is the largest will have the smallest axis in the inertia ellipsoid, and vice versa. Equation [6.6.7] is commonly used when relating the inertia ellipsoid and the kinetic energy of a rigid body.

6.6 INERTIA ELLIPSOID

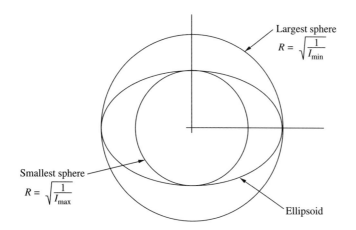

Figure 6.17   Inertia ellipsoid of fixed size

Construct the inertia ellipsoid for the rectangular prism shown in Fig. 6.18 about the center of mass, and about point A.

Example 6.8

Solution

Using the xyz axes attached to the center of mass, the inertia matrix is diagonal with

$$[I_G] = \frac{m}{12}\text{diag}(b^2 + c^2 \quad a^2 + c^2 \quad a^2 + b^2) \qquad \textbf{[a]}$$

Point A and the center of mass are separated by $d_x = a/2, d_y = 0, d_z = -c/2$, so that the inertia matrix about point A becomes

$$[I_A] = [I_G] + \frac{m}{4}\begin{bmatrix} c^2 & 0 & ac \\ 0 & a^2+c^2 & 0 \\ ac & 0 & a^2 \end{bmatrix} = \frac{m}{12}\begin{bmatrix} b^2+4c^2 & 0 & 3ac \\ 0 & 4a^2+4c^2 & 0 \\ 3ac & 0 & 4a^2+b^2 \end{bmatrix} \qquad \textbf{[b]}$$

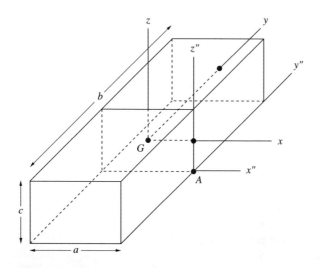

Figure 6.18

Using an $x''y''z''$ coordinate system attached to point A and parallel to the xyz coordinates, the inertia ellipsoid equation can be written as

$$I_{Ax''x''}x''^2 + I_{Ay''y''}y''^2 + I_{Az''z''}z''^2 - 2I_{Ax''z''}x''z'' = 1 \qquad [\text{c}]$$

The inertia ellipsoid about point G is relatively simple to draw. The ellipsoid has its major axes about the x, y, and z axes, as these axes are principal axes. The largest axis of the ellipsoid corresponds to the smallest moment of inertia.

The inertia ellipsoid associated with point A is harder to draw, as two of the axes of the ellipsoid no longer lie parallel to the x'', y'', or z'' axes. Only one of its axes is parallel to the y'' axis.

Looking at Eq. [b] and comparing it with the results of Example 6.7, we conclude that the principal axes are obtained by a rotation of the xyz axes about the y axis. The rotation angle can be found by solving the associated eigenvalue problem. Note also that because the inertia matrix is recognized as one obtained by rotating a set of principal axes about the y axis, the y component of the inertia ellipsoid is not going to yield any significant results. Because of this, and because visualization in two dimensions is simpler, we plot and compare the inertia ellipsoids associated with points G and A for the value $y = 0$. Doing so, and assigning the parameters $m = 1, a = 1, b = 2, c = 3$, we obtain the following ellipsoid equations:

For point G (with $y = 0$):

$$\frac{1}{12}(13x^2 + 5z^2) = 1 \qquad [\text{d}]$$

For point A (with $y'' = 0$):

$$\frac{1}{12}(40x''^2 + 8z''^2 - 18x''z'') = 1 \qquad [\text{e}]$$

Figures 6.19 and 6.20 show the ellipsoids associated with Eqs. [d] and [e]. In Fig. 6.20 the points in which the ellipsoid intersect the x and z axes are at $x = \pm\sqrt{12/13} = 0.9608$ and $z = \pm\sqrt{12/5} = 1.549$. To find where the ellipsoid associated with Fig. 6.20 intersects the principal axes, we can carry out a graphical analysis and visually find these points. Or, we solve the eigenvalue problem associated with the inertia matrix and determine the exact location of the principal axes. Solving the eigenvalue problem gives the eigenvalues as 3.550 and 1.283. Note again that we have ordered the eigenvectors similar to the diagonal elements

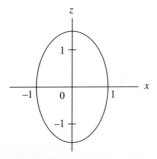

Figure 6.19   Inertia ellipsoid for $[I_G]$

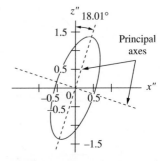

Figure 6.20   Inertia ellipsoid for $[I_A]$

of $[I_A]$. The normalized eigenvector matrix is written

$$[U] = [R]^T = \begin{bmatrix} 0.9510 & -0.3092 \\ 0.3092 & 0.9510 \end{bmatrix} \qquad \textbf{[f]}$$

Recalling that we are dealing with a rotation about the y axis, we conclude that the rotation from the $x'y'z'$ axes to the principal axes is a clockwise rotation of $\theta = \cos^{-1}(0.9510) = 18.01°$. One can verify the rotation angle and the lengths of the major axes of the ellipse from Fig. 6.20. A clockwise rotation of $71.99°$ also leads to a set of principal axes.

REFERENCES

Ginsberg, J. H. *Advanced Engineering Dynamics*. New York: Harper & Row, 1988.
Greenwood, D. T. *Principles of Dynamics*. 2nd ed. Englewood Cliffs, NJ: Prentice-Hall, 1988.
Kane, T. R., and D. A. Levinson. *Dynamics: Theory and Applications*. New York: McGraw-Hill, 1985.
Meirovitch, L. *Methods of Analytical Dynamics*. New York: McGraw-Hill, 1970.

HOMEWORK EXERCISES

SECTION 6.2

1. Find the center of mass of the body shown in Fig. 6.21.

SECTION 6.3

2. Find the elements of the inertia matrix associated with the figure shown in Fig. 6.22 about point O and using the xyz axes.

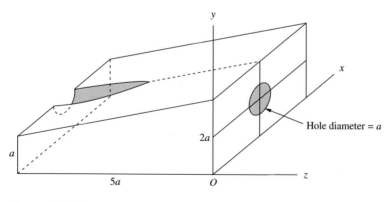

Figure 6.21

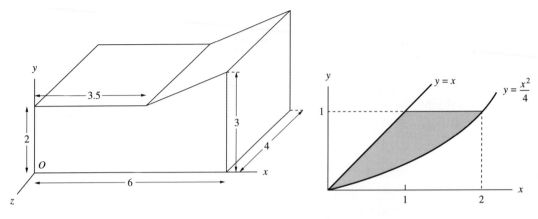

Figure 6.22 **Figure 6.23**

3. Calculate the mass moment of inertia about the x, y and z axes of the body of revolution generated by rotating about the y axis the area between the curves $y = x^2/4$ and $y = x$, where $0 \leq x \leq 2$, $0 \leq y \leq 1$, as shown in Fig. 6.23.

4. Find by direct integration the inertia matrix associated with the paraboloid in Example 6.2. Use a differential element in the form $dm = \rho r \, dr \, d\theta \, dx$.

5. Calculate the mass moment of inertia of a torus of mass m, mean radius R, and core radius of the inner cross section a about the generating axis (Fig. 6.24).

6. Find the mass moment of inertia of the body shown in Fig. 6.25 about the x, y, and z axes.

Section 6.4

7. Consider the composite shape in Fig. 6.21, without the hole. Determine the inertia matrix about point O.

8. Determine I_{zz} and I_{xy} of the square plate shown in Fig. 6.26 with three circular holes cut in it.

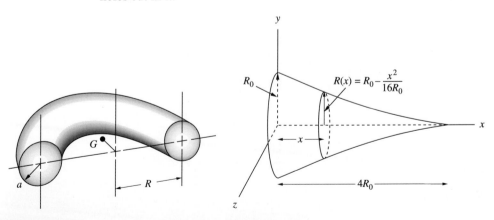

Figure 6.24 **Figure 6.25**

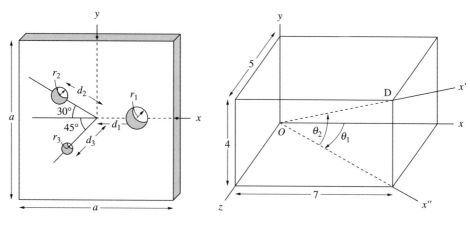

Figure 6.26 **Figure 6.27**

9. Consider the rectangular box in Fig. 6.27 and determine its inertia matrix about point O, using the $x'y'z'$ axes. The x' axis goes through points O and D. The $x'y'z'$ axes are obtained by rotating the xyz axes about the y axis by θ_1, and then rotating the resulting frame about the z'' axis by θ_2.

10. Consider the right circular cone in Fig. 6.28. Find the elements of the inertia matrix about point O and a coordinate system $x'y'z'$ obtained by rotating the xyz coordinates about the z axis, so that the x' axis goes through point B. Let $R/L = 0.2041$.

11. Calculate the inertia matrix of the body in Fig. 6.29 about point O, using the x, y, and z axes. The body is composed of thin slender rods.

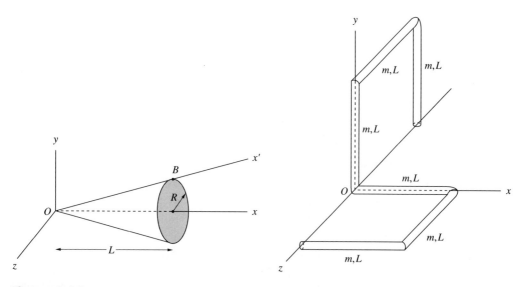

Figure 6.28 **Figure 6.29**

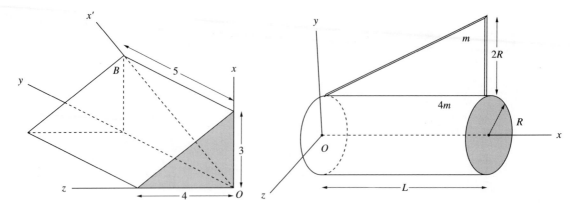

Figure 6.30 **Figure 6.31**

12. Using the $x'y'z'$ axes, find the mass moments of inertia of the triangular prism in Fig. 6.30 about O. The x' axis goes through points O and B and it is obtained by a counterclockwise rotation about the z axis.

13. Find the inertia matrix of the composite body shown in Figure 6.31 about point O and using the xyz axes. The body consists of a thin triangular plate attached to a solid shaft.

SECTIONS 6.5 AND 6.6

14. Find the principal moments of inertia of a body whose inertia matrix is given below for $a = 7$. Then, suggest a set of rotations that will transform the inertial axes into the principal axes. Next, find the principal moments of inertia and transformation matrix for $a = 0$ and compare the results.

$$[I] = \begin{bmatrix} 450 & 0 & 100 \\ 0 & 500 & a \\ 100 & a & 550 \end{bmatrix}$$

15. Find the principal moments of inertia of the box in Fig. 6.27 about point O. Sketch the principal axes.

16. Find the principal moments of inertia of the shape in Problem 7.

17. Find the principal moments of inertia of the composite body shown in Figure 6.31 about point O. Sketch the inertia ellipsoid.

18. Find the principal moments of inertia of the body in Fig. 6.30, about point O. Sketch the inertia ellipsoid.

chapter 7

RIGID BODY KINEMATICS

7.1 INTRODUCTION

In this chapter we consider kinematical relations that describe the motion of rigid bodies. The analysis follows the developments of reference frames and relative motion equations in Chapter 2. To extend these concepts to rigid bodies, we attach moving reference frames to the bodies and express the angular motion components using the moving frames. We also quantify the angular velocity by means of Euler angles, as well as Euler parameters, the latter being quantities that eliminate the singularities associated with the Euler angles. We then discuss commonly encountered constraints when rigid bodies are interconnected or when they move against each other. The chapter ends with a discussion of rolling.

The field of kinematics can be viewed as consisting of two major components: analysis and synthesis. The focus of this chapter is kinematic analysis. Kinematic synthesis is usually needed when designing interconnected bodies and mechanisms. It is mostly a specialty field and is beyond the scope of this text.

7.2 BASIC KINEMATICS OF RIGID BODIES

We can classify the general motion of a rigid body into three categories:

- Pure translation
- Pure rotation
- Combined translation and rotation

To describe the kinematics, we make use of the relative motion equations developed in Chapter 2. The velocity of a point P, whose motion is observed from a rotating coordinate system with origin at B, is

$$\mathbf{v}_P = \mathbf{v}_B + \mathbf{v}_{P/B} = \mathbf{v}_B + \boldsymbol{\omega} \times \mathbf{r}_{P/B} + \mathbf{v}_{\text{rel}} \qquad [7.2.1]$$

where $\mathbf{v}_B$ is the velocity of the origin of the reference frame, $\boldsymbol{\omega}$ is the angular velocity of the reference frame, and $\mathbf{v}_{\text{rel}}$ is the velocity of point P as observed from the moving reference frame. The expression for the acceleration of P is

$$\mathbf{a}_P = \mathbf{a}_B + \boldsymbol{\alpha} \times \mathbf{r}_{P/B} + \boldsymbol{\omega} \times (\boldsymbol{\omega} \times \mathbf{r}_{P/B}) + 2\boldsymbol{\omega} \times \mathbf{v}_{\text{rel}} + \mathbf{a}_{\text{rel}} \qquad [7.2.2]$$

with the terms having their obvious meaning.

These two equations have a significant application for rigid bodies. The moving reference frame can be attached to a point on the body and it moves with the body. In this configuration, the relative axes are referred to as the *body-fixed axes*, or the *body axes*. The origin of the reference frame is usually selected as the center of mass or, if it exists, the center of rotation.

The angular velocity and angular acceleration of the body are then the angular velocity and angular acceleration, respectively, of the reference frame and $\mathbf{v}_{\text{rel}}$ and $\mathbf{a}_{\text{rel}}$ become the velocity and acceleration of point P with respect to the body. If P is a point fixed on the rigid body, then $\mathbf{v}_{\text{rel}} = \mathbf{0}$ and $\mathbf{a}_{\text{rel}} = \mathbf{0}$.

In the majority of dynamics problems involving three-dimensional motion, one attaches the relative axes to the body. We will later see a special case of employing a relative frame not attached to the body. The study of axisymmetric bodies primarily makes use of this special case.

7.2.1 PURE TRANSLATION

In this case, the rigid body moves with no angular velocity and no angular acceleration, that is, $\boldsymbol{\omega} = \mathbf{0}, \boldsymbol{\alpha} = \mathbf{0}$. Every point on the body has the same translational velocity and acceleration, so that three translational parameters are sufficient to describe the motion. For the most general case we have three degrees of freedom. It should be noted that a rigid body is capable of moving along a curved trajectory without any rigid body rotation. An example of this is landing of an aircraft.

7.2.2 PURE ROTATION

Here, the motion of the rigid body is described using rotational parameters alone. The velocity and acceleration of any point on the body can be expressed in terms of the angular velocity, angular acceleration, and the distance of the point from the rotation center. We separate this type of motion into two categories:

- Rotation about a fixed axis.
- Rotation about a fixed point.

Rotation about a fixed axis is a special case of rotation about a fixed point. For two-dimensional motion, the above two categories coincide.

Rotation About a Fixed Axis Consider a body rotating about a fixed axis h. The direction of the angular velocity vector $\boldsymbol{\omega}$ is along the fixed axis, as shown in

Fig. 7.1. According to the definition in Chapter 2, the angular velocity is "simple." Denoting by $\mathbf{e}_h$ the unit vector along the fixed axis, we can write

$$\boldsymbol{\omega} = \omega \mathbf{e}_h \qquad \boldsymbol{\alpha} = \dot{\boldsymbol{\omega}} = \dot{\omega}\mathbf{e}_h = \alpha \mathbf{e}_h \qquad [7.2.3]$$

The unit vector $\mathbf{e}_h$ is similar to the binormal vector we considered in Chapter 1 in conjunction with normal and tangential coordinates, the difference here being that the direction of $\mathbf{e}_h$ is fixed. Consider a point P on the body and express its position in terms of its components along the h axis and the plane perpendicular to the h axis. We write $\mathbf{r}_P$ as

$$\mathbf{r}_P = \mathbf{h} + \mathbf{b} \qquad [7.2.4]$$

in which $\mathbf{h} = h\mathbf{e}_h$ and $\mathbf{b} = -b\mathbf{e}_n$, where b is the perpendicular distance from point P to the axis of rotation and $\mathbf{e}_n$ is the associated unit vector. We recognize that $\mathbf{e}_n$ is in the normal direction.

When P is fixed on the body, its velocity is

$$\mathbf{v}_P = \boldsymbol{\omega} \times \mathbf{r}_P = \omega \mathbf{e}_h \times (\mathbf{h}+\mathbf{b}) = \omega \mathbf{e}_h \times (h\mathbf{e}_h - b\mathbf{e}_n) = \omega b \mathbf{e}_t \qquad [7.2.5]$$

The magnitude of the velocity is $b\omega$, leading to the conclusion that the velocity of a point on the body is dependent only on the perpendicular distance between that point and the axis of rotation. Rotation about a fixed axis is a single degree of freedom problem.

We differentiate Eq. [7.2.5] and write the acceleration of point P as

$$\mathbf{a}_P = \frac{d}{dt}(\boldsymbol{\omega} \times \mathbf{r}_P) = \boldsymbol{\alpha} \times \mathbf{r}_P + \boldsymbol{\omega} \times (\boldsymbol{\omega} \times \mathbf{r}_P) \qquad [7.2.6]$$

which is recognized as the sum of the tangential plus normal components. We have

$$\mathbf{a}_P = \mathbf{a}_t + \mathbf{a}_n \qquad [7.2.7]$$

where

$$\mathbf{a}_t = \boldsymbol{\alpha} \times \mathbf{r}_P \qquad \mathbf{a}_n = \boldsymbol{\omega} \times (\boldsymbol{\omega} \times \mathbf{r}_P) \qquad [7.2.8]$$

The magnitude of the tangential component of the acceleration is $a_t = \alpha b$, and of the normal component it is $a_n = \omega^2 b$. The components of the acceleration can be

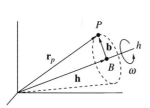

Figure 7.1 Rotation about a fixed axis

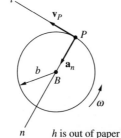

h is out of paper

Figure 7.2 Normal and tangential coordinates

expressed as

$$\mathbf{a}_t = b\alpha\mathbf{e}_t \qquad \mathbf{a}_n = -\omega^2\mathbf{b} = b\omega^2\mathbf{e}_n \qquad [7.2.9]$$

Rotation About a Fixed Point In this case, the angular velocity vector $\boldsymbol{\omega}$ does not lie on a fixed axis. The rate of change of the angular velocity depends on a change in direction as well as a change in magnitude. A body rotating about a fixed point has three degrees of freedom, and the angular velocity is usually a combination of two or more rotation components. Consider the cylinder in Fig. 7.3. The shaft is rotating about the fixed Z axis with angular velocity ω_1. The cylinder is spinning with angular velocity ω_3 about an axis h, which lies on the yz plane and makes an angle of β with the rod. The angular velocity of the cylinder can be expressed as

$$\boldsymbol{\omega} = \boldsymbol{\omega}_1 + \boldsymbol{\omega}_2 + \boldsymbol{\omega}_3 = \omega_1 \mathbf{K} + \dot{\beta}\mathbf{i} + \omega_3 \mathbf{h}$$
$$= \omega_1 \mathbf{K} + \dot{\beta}\mathbf{i} + \omega_3(\cos\beta\mathbf{k} - \sin\beta\mathbf{j}) \qquad [7.2.10]$$

In general, ω_3 is known as the *spin rate* and ω_1 as the *precession rate*. The angle β is referred to as the *nutation angle*. The rate of change of β is called the *nutation rate*. We will formally quantify these rotational parameters in the next two sections.

Even if the spin rate and precession rate are constant—that is, the components of the angular velocity are constant in magnitude—the angular acceleration of the body is not zero, because the direction of the angular velocity vector is changing. Indeed, applying the transport theorem we obtain

$$\boldsymbol{\alpha} = \dot{\boldsymbol{\omega}} = \dot{\boldsymbol{\omega}}_1 + \dot{\boldsymbol{\omega}}_2 + \dot{\boldsymbol{\omega}}_3 + \boldsymbol{\omega}_1 \times (\boldsymbol{\omega}_2 + \boldsymbol{\omega}_3) + \boldsymbol{\omega}_2 \times \boldsymbol{\omega}_3 \qquad [7.2.11]$$

The last two terms in this equation are also known as *gyroscopic effects*. Figure 7.4 shows this effect for $\boldsymbol{\omega}_2 = \mathbf{0}$. The terms $\dot{\boldsymbol{\omega}}_1 + \dot{\boldsymbol{\omega}}_3$ describe the change in the magnitude of the angular velocity vector, and $\boldsymbol{\omega}_1 \times \boldsymbol{\omega}_3$ describes the change in direction.

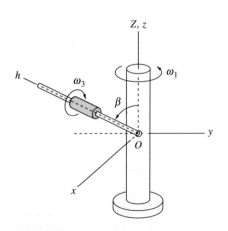

Figure 7.3 Rotation about a fixed point

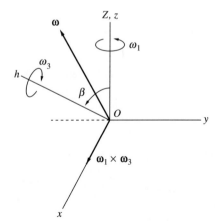

Figure 7.4 Angular velocities and gyroscopic effect

The line specifying the direction of the angular velocity vector $\boldsymbol{\omega}$ is known as the *instantaneous axis of acceleration,* or the *instantaneous axis of rotation.* The unit vector along this axis is defined as

$$\mathbf{n} = \frac{\boldsymbol{\omega}}{\omega} \qquad [7.2.12]$$

in which $\omega = |\boldsymbol{\omega}|$ is the magnitude of the angular velocity. The rigid body can be viewed as rotating about the axis defined by $\boldsymbol{\omega}(t)$ at a particular instant.

The trajectory of the instantaneous axis of rotation defines the *body and space cones.* Body and space cones are helpful in visualizing the motion of rigid bodies. When the trajectory of the axis of rotation is viewed from the rotating body (the body-fixed axes), the cone that is generated by the angular velocity vector is the *body cone.* When the trajectory of the axis of rotation is viewed from an inertial frame, the *space cone* is generated. The body and space cones are always in contact with each other. The line of contact is the angular velocity vector $\boldsymbol{\omega}$, which is also referred to as the *generatrix.* Figure 7.5 shows body and space cones for an arbitrary body. We will study body and space cones for axisymmetric bodies in Chapter 10.

7.2.3 COMBINED TRANSLATION AND ROTATION

A body undergoing combined translation and rotation requires both translational and angular parameters to describe its motion. The unrestricted three-dimensional motion of a rigid body is a six degree of freedom problem.

Combined translation and rotation of an arbitrary rigid body is too general to be described in broad terms. Chapter 3 presents the special case of plane motion. An interesting concept associated with plane motion is the existence of an instant center. Once the instant center is located, the velocity of a point P on the body can be found from the relation

$$\mathbf{v}_P = \boldsymbol{\omega} \times \mathbf{q} \qquad [7.2.13]$$

where $\mathbf{q} = \mathbf{r}_{P/C}$, and C is the instant center.

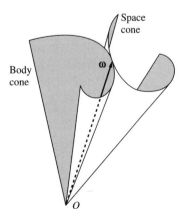

Figure 7.5 Body and space cones for an arbitrary body

The situation is quite different for three-dimensional motion. One can prove the existence of an instantaneous axis of rotation, as we will do in Section 7.3. However, barring a few exceptions, a center of instantaneous velocity does not exist for three-dimensional motion. In three dimensions, Eq. [7.2.13] has an infinite number of solutions. Using the column vector notation from Chapter 2, Eq. [7.2.13] can be written as

$$\{v_P\} = [\tilde{\omega}]\{q\} \qquad [7.2.14]$$

The matrix $[\tilde{\omega}]$ is a singular matrix of rank 2, another proof that Eq. [7.2.13] has an infinite number of solutions.

7.3 EULER'S AND CHASLES'S THEOREMS

A rigid body has three rotational degrees of freedom. From this, and from the developments of Chapter 2, we deduce that three rotations (about nonparallel axes) should be sufficient to describe the most general transformation from one coordinate system to another. According to Euler's theorem, one can view a transformation from one coordinate system into another as a single rotation about a certain axis. Specifically, Euler's theorem states that:

> The most general displacement of a rigid body with one point fixed can be described as a single rotation about some axis through that fixed point, called the *axis of rotation* or *principal line*.

The rotation angle is denoted by Φ. Proof of Euler's theorem can be carried out by an eigenvalue analysis. Consider a coordinate system A with coordinate axes $a_1 a_2 a_3$. We apply a sequence of rotations, resulting in the frame B with the axes $b_1 b_2 b_3$. For convenience, we use the same origin for both coordinate systems. We denote the unit vector along the principal line by **n**, and consider an arbitrary vector **r**. The coordinate systems and the principal line are sketched in Fig. 7.6. Note that

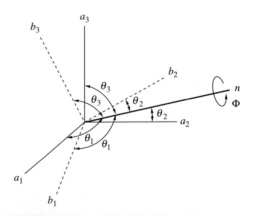

Figure 7.6 Rotation about principal line

according to Euler's theorem *the coordinates of the principal line remain unchanged when they are viewed from the A and B frames*. The angles θ_i ($i = 1, 2, 3$) that $\mathbf{n}$ makes with the a_i ($i = 1, 2, 3$) axes are the same as the angles $\mathbf{n}$ makes with the b_i axes.

We can express the vector $\mathbf{r}$ in terms of the components of the two coordinate systems as

$$\mathbf{r} = {}_A\mathbf{r} = {}_Ar_1\mathbf{a}_1 + {}_Ar_2\mathbf{a}_2 + {}_Ar_3\mathbf{a}_3 = {}_B\mathbf{r} = {}_Br_1\mathbf{b}_1 + {}_Br_2\mathbf{b}_2 + {}_Br_3\mathbf{b}_3 \quad [\mathbf{7.3.1}]$$

in which the subscript on the left of $\mathbf{r}$ identifies the coordinate axes in which the vector is resolved. The usual convention is that a vector is resolved along the coordinates associated with the reference frame from which it is being viewed. We will use the notation in Eq. [7.3.1] only when it is necessary to make a distinction between the reference frame from which a vector is viewed and the unit vectors used to resolve it.

In column vector format we have

$$\{_Br\} = [c]^T\{_Ar\} \quad [\mathbf{7.3.2}]$$

where $\{_Br\} = [{}_Br_1 \; {}_Br_2 \; {}_Br_3]^T$ and $\{_Ar\} = [{}_Ar_1 \; {}_Ar_2 \; {}_Ar_3]^T$. In a similar fashion, we can express the unit vector along the principal line as

$$\mathbf{n} = {}_A\mathbf{n} = {}_An_1\mathbf{a}_1 + {}_An_2\mathbf{a}_2 + {}_An_3\mathbf{a}_3 \qquad \mathbf{n} = {}_B\mathbf{n} = {}_Bn_1\mathbf{b}_1 + {}_Bn_2\mathbf{b}_2 + {}_Bn_3\mathbf{b}_3$$
$$[\mathbf{7.3.3}]$$

or, in column vector format, as

$$\{_Bn\} = [c]^T\{_An\} \quad [\mathbf{7.3.4}]$$

The expressions ${}_An_i$ and ${}_Bn_i$ ($i = 1, 2, 3$) are the direction cosines of the principal line, ${}_An_i = \mathbf{n} \cdot \mathbf{a}_i$, ${}_Bn_i = \mathbf{n} \cdot \mathbf{b}_i$, so they are equal to each other. Hence,

$$\{_Bn\} = \{_An\} = [c]^T\{_An\} = \{c\} \quad [\mathbf{7.3.5}]$$

where the vector $\{c\} = [c_1 \; c_2 \; c_3]^T$ is recognized as the direction cosine vector associated with the principal line. Substituting Eq. [7.3.5] into Eq. [7.3.4] we can write

$$\{c\} = [c]^T\{c\} \quad [\mathbf{7.3.6}]$$

Mathematically, Eq. [7.3.6] can be explained as $\{c\} = \{_Bn\} = \{_An\}$ being the eigenvector of $[c]^T$ associated with the eigenvalue $\lambda = 1$. To prove Euler's theorem we need to establish that $[c]$ has an eigenvalue equal to 1.

Consider the eigenvalue problem associated with $[c]^T = [R]$, where $[R]$ is the transformation matrix that relates the $b_1b_2b_3$ axes to the $a_1a_2a_3$ axes. We can write the left and right eigenvalue problems as

$$[R]\{x\} = \lambda\{x\} \qquad [R]^T\{y\} = \lambda\{y\} \quad [\mathbf{7.3.7}]$$

Solution of the eigenvalue theorem yields three eigenvalues $\lambda_1, \lambda_2, \lambda_3$; and corresponding right eigenvectors $\{x_1\}, \{x_2\}$, and $\{x_3\}$ and left eigenvectors $\{y_1\}, \{y_2\}$, and $\{y_3\}$. We know from Chapter 2 that the determinant of $[R] = [c]^T$ is equal to one. From a theorem in linear algebra the product of the eigenvalues of a matrix is equal to the determinant, so that $\lambda_1\lambda_2\lambda_3 = 1$.

The relations between the eigenvalues and right and left eigenvectors of $[R]$ can be written as

$$[R]\{x_i\} = \lambda_i\{x_i\} \qquad [R]^T\{y_i\} = \lambda_i\{y_i\} \qquad i = 1, 2, 3 \qquad \textbf{[7.3.8a,b]}$$

Now, consider the eigenvalue problem associated with the inverse of $[R]$. It is easy to show that the eigenvalues of the inverse of a matrix are the multiplicative inverses of the eigenvalues of the original matrix, while the eigenvectors of the original matrix and its inverse are the same. We can therefore write the right eigenvector formulation as

$$[R]^{-1}\{x_i\} = \frac{1}{\lambda_i}\{x_i\} \qquad \textbf{[7.3.9]}$$

One can arrive at Eq. [7.3.9] by left-multiplying Eq. [7.3.8a] by $[R]^{-1}$. But the inverse of $[R]$ is equal to its transpose, so that Eqs. [7.3.8b] and [7.3.9] define the same eigenvalue problem. Hence, for a unitary matrix, the left and right eigenvectors coincide, $\{x_i\} = \{y_i\}$ ($i = 1, 2, 3$). Furthermore, we have

$$\lambda_i = \frac{1}{\lambda_i} \qquad \textbf{[7.3.10]}$$

Because $\det[R] = 1$ and the product of the three eigenvalues is unity, there are two possible cases: (a) all eigenvalues are real with values 1, 1, 1 or 1, -1, -1, and (b) one eigenvalue is real and unity, and the other two are complex conjugates with moduli equal to 1. The first case when all eigenvalues are real does not make sense physically, as it would indicate at least two repeated roots and hence an infinity of axes about which a rotation can be performed. This is impossible, as one can always find an axis that will not yield the desired rotation. We are left with the second possibility. Hence, $[R]$ (or $[c]^T$) has one real eigenvalue equal to 1. The corresponding eigenvector gives the direction cosines of the principal line. Note that the eigenvector must be normalized using the relation $\{c\}^T\{c\} = 1$. Also, if $\{c\}$ is the eigenvector associated with the eigenvalue $\lambda_1 = 1$, so is $-\{c\}$. One can take the unit vector along the principal line along either of the two directions of the principal line. The difference will be the direction of the rotation angle Φ.

We next discuss determination of the rotation angle Φ. Consider an intermediate coordinate system $d_1d_2d_3$, where the d_1 axis is aligned with the principal line and the d_2 and d_3 axes are selected arbitrarily. We can write the relationship between the $d_1d_2d_3$ axes and the A frame as

$$\{a\} = [c^*]\{d\} \qquad \textbf{[7.3.11]}$$

where $[c^*]$ is the associated direction cosine matrix. Using the relationship between the A and B frames, we can write

$$\{b\} = [c]^T\{a\} = [c]^T[c^*]\{d\} \qquad \textbf{[7.3.12]}$$

Let us now perform the rotation by the angle Φ about the d_1 axis, which leads to the coordinate system $d_1d_2'd_3'$. It follows that the relation between the $a_1a_2a_3$ and $d_1d_2d_3$ axes has to be the same as the relationship between the $b_1b_2b_3$ and $d_1d_2'd_3'$

axes. Both coordinate axes were transformed in the same way. We can express this relationship as

$$\{b\} = [c^*]\{d'\} \qquad [7.3.13]$$

However, the relationship between $\{d\}$ and $\{d'\}$ is

$$\{d'\} = [R']\{d\} \qquad [7.3.14]$$

where $[R']$ is a rotation matrix in the form

$$[R'] = \begin{bmatrix} 1 & 0 & 0 \\ 0 & \cos\Phi & \sin\Phi \\ 0 & -\sin\Phi & \cos\Phi \end{bmatrix} \qquad [7.3.15]$$

Combining Eqs. [7.3.11]–[7.3.13], we can write

$$[c^*]^T[c]^T[c^*] = [R'] \qquad [7.3.16]$$

This equation represents a similarity transformation between $[c]^T$ and $[R']$; thus the traces (sum of the diagonal elements) of these two matrices are the same. We can now write

$$c_{11} + c_{22} + c_{33} = [R']_{11} + [R']_{22} + [R']_{33} = 1 + 2\cos\Phi \qquad [7.3.17]$$

which can be solved for Φ to yield the rotation angle.

It is interesting to note that in Chapter 2 we considered a three-parameter rotation to describe a general transformation from one set of coordinates to the other, whereas in this section we used Euler's theorem and accomplished the transformation using four parameters: the three direction cosines associated with the principal line, and the rotation angle Φ. It turns out that one can take advantage of the four-parameter description used here to simplify the kinematical equations, as we will see in Section 7.7. Before that, we will discuss the relationship between the angular velocity vector and the direction cosine matrix, and we will parametrize the angular velocity vector in terms of the transformation angles.

The above developments were for a body with one point fixed. If we consider an unrestrained rigid body, we can extend the results of Euler's theorem into what is known as *Chasles's theorem,* which states:

> The most general displacement of a rigid body is equivalent to a translation of some point in the body, plus a rotation about an axis through that point.

Depending on the point selected, the instantaneous axis of rotation will be different.

Example 7.1

The coordinate system $a_1 a_2 a_3$ is transformed into the $b_1 b_2 b_3$ system by first rotating by an angle $\phi = 30°$ about the a_3 axis, and then by rotating the resulting $a'_1 a'_2 a'_3$ frame about the a'_2 axis by $\theta = 45°$. Find the orientation of the principal line, as well as the rotation angle.

Solution

The rotation matrix associated with the transformation given is

$$[R] = [R_2][R_1] \qquad \text{[a]}$$

in which

$$[R_2] = \begin{bmatrix} 0.7071 & 0 & -0.7071 \\ 0 & 1 & 0 \\ 0.7071 & 0 & 0.7071 \end{bmatrix} \quad [R_1] = \begin{bmatrix} 0.8660 & 0.5 & 0 \\ -0.5 & 0.8660 & 0 \\ 0 & 0 & 1 \end{bmatrix} \qquad \text{[b]}$$

so that

$$[R] = \begin{bmatrix} 0.6124 & 0.3536 & -0.7071 \\ -0.5000 & 0.8660 & 0 \\ 0.6124 & 0.3536 & 0.7071 \end{bmatrix} \qquad \text{[c]}$$

The eigenvector associated with the eigenvalue $\lambda = 1$ can be found manually or numerically. This eigenvector can be shown to be

$$x = [-0.2195 \quad 0.8192 \quad 0.5299]^T \qquad \text{[d]}$$

The elements of the eigenvector are the direction cosines of the principal line. The angles the principal line makes with the inertial coordinates are then

$$\theta_1 = \cos^{-1}(-0.2195) = 102.7° \qquad \theta_2 = \cos^{-1}(0.8192) = 35.00°$$

$$\theta_3 = \cos^{-1}(0.5299) = 58.00° \qquad \text{[e]}$$

To find the rotation angle Φ, we make use of Eq. [7.3.17], thus

$$\Phi = \cos^{-1}\left(\frac{1}{2}(R_{11} + R_{22} + R_{33} - 1)\right) = \cos^{-1}(0.5928) = 53.65° \qquad \text{[f]}$$

The transformed coordinates, principal line, and rotation angle are shown in Figs. 7.7 and 7.8.

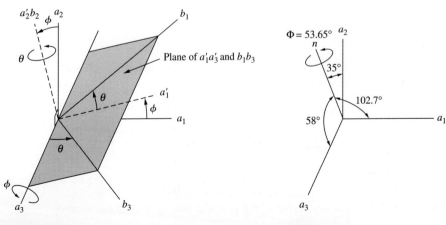

Figure 7.7   **Figure 7.8**

7.4 RELATION BETWEEN DIRECTION COSINES AND ANGULAR VELOCITY

In this section, we explore the relationship between the direction cosine matrix and the angular velocity vector. Consider two coordinate frames A and B and associated coordinate axes $a_1 a_2 a_3$ and $b_1 b_2 b_3$. The origin of the two frames coincide. The relative velocity expression for a point whose position is given by $\mathbf{r}$ is

$$^A\mathbf{v} = {}^B\mathbf{v} + {}^A\boldsymbol{\omega}^B \times \mathbf{r} \qquad [7.4.1]$$

Using the notation in Eq. [7.3.1] of denoting the reference frame in which a vector is resolved by a subscript on the left, we can relate the position and velocity vectors by

$$\{_B r\} = [c]^T \{_A r\} \qquad \{_A r\} = [c]\{_B r\} \qquad [7.4.2a,b]$$

$$\{^A_B v\} = [c]^T \{^A_A v\} \qquad \{^A_A v\} = [c]\{^A_B v\}$$

$$\{^B_B v\} = [c]^T \{^B_A v\} \qquad \{^B_A v\} = [c]\{^B_B v\} \qquad [7.4.3a,b,c,d]$$

where $[c]$ is the direction cosine matrix between the A and B frames.

We next find a relationship between $[c]$ and the angular velocity of the B frame with respect to the A frame, $^A\boldsymbol{\omega}^B$. To this end, we first express $^A\boldsymbol{\omega}^B$ along the components of the A frame as

$$^A_A\boldsymbol{\omega}^B = {}_A\omega_1 \mathbf{a}_1 + {}_A\omega_2 \mathbf{a}_2 + {}_A\omega_3 \mathbf{a}_3 \qquad [7.4.4]$$

and using the matrix

$$\left[{}^A_A\tilde{\omega}^B \right] = \begin{bmatrix} 0 & -{}_A\omega_3 & {}_A\omega_2 \\ {}_A\omega_3 & 0 & -{}_A\omega_1 \\ -{}_A\omega_2 & {}_A\omega_1 & 0 \end{bmatrix} \qquad [7.4.5]$$

we can write the transport theorem as

$$\{^A_A v\} = \{^B_A v\} + [{}^A_A\tilde{\omega}^B]\{_A r\} \qquad [7.4.6]$$

In a similar fashion, we write the velocity of a point in terms of the B frame as

$$\{^A_B v\} = \{^B_B v\} + [{}^A_B\tilde{\omega}^B]\{_B r\} \qquad [7.4.7]$$

Left-multiplying Eq. [7.4.6] by $[c]^T$ and using Eq. [7.4.7] relates the angular velocity matrices expressed in the A and B frames as

$$\{^A_B\tilde{\omega}^B\} = [c]^T[{}^A_A\tilde{\omega}^B][c] \qquad [7.4.8]$$

One should compare this equation, which represents a similarity transformation, with Eq. [6.4.13a]. We next obtain a relationship between the direction cosine matrix and the angular velocity vector. To this end, we consider Eq. [7.4.2b]. The left side of the equation is in terms of a vector resolved in the A frame, and the right side is in terms of a vector viewed in the B frame. Therefore, when we take the time derivative of Eq. [7.4.2b] we have to differentiate the left side in the A frame and the right side in the B frame. Doing so yields

$$\{^A_A v\} = [c]\{^B_B v\} + [\dot{c}]\{_B r\} \qquad [7.4.9]$$

366 CHAPTER 7 • RIGID BODY KINEMATICS

We convert Eq. [7.4.9] to one expressed in terms of the components of the A frame by introducing Eqs. [7.4.2b] and [7.4.3d] into it, which yields

$$\{_A^A v\} = \{_A^B v\} + [\dot{c}][c]^T \{_A r\} \qquad [7.4.10]$$

Comparing Eqs. [7.4.6] and [7.4.10], we conclude that

$$[_A^A \tilde{\omega}^B] = [\dot{c}][c]^T \qquad [7.4.11]$$

In terms of the rotation matrix $[R] = [c]^T$, we can express $[_A^A \tilde{\omega}^B]$ as

$$[_A^A \tilde{\omega}^B] = [\dot{R}]^T [R] \qquad [7.4.12]$$

Using Eqs. [7.4.8] and [7.4.12], we obtain the angular velocity matrix expressed in terms of its components in the B frame as

$$[_B^A \tilde{\omega}^B] = [c]^T [_A^A \tilde{\omega}^B][c] = [c]^T [\dot{c}][c]^T [c] = [c]^T [\dot{c}] = [R][\dot{R}]^T \qquad [7.4.13]$$

It is usually more convenient to write the components of the angular velocity in terms of the body coordinates, so that we will use Eq. [7.4.13] more extensively than Eq. [7.4.11].

Example 7.2

Consider two frames A and B, where the coordinates $b_1 b_2 b_3$ are obtained by rotating $a_1 a_2 a_3$ by an angle θ_1 about a_3 and then by an angle θ_2 about a'_2. Find $^A\omega^B$ using Eqs. [7.4.11] and [7.4.13].

Solution

Denoting the transformation matrices by $[R_1]$ and $[R_2]$, the combined transformation matrix is $[R] = [R_2][R_1]$. The direction cosine matrices are

$$[c_1] = \begin{bmatrix} c\theta_1 & -s\theta_1 & 0 \\ s\theta_1 & c\theta_1 & 0 \\ 0 & 0 & 1 \end{bmatrix} \qquad [c_2] = \begin{bmatrix} c\theta_2 & 0 & s\theta_2 \\ 0 & 1 & 0 \\ -s\theta_2 & 0 & c\theta_2 \end{bmatrix} \qquad [c] = [c_1][c_2] \qquad [\mathbf{a}]$$

Taking the derivative of $[c]$ with respect to time yields

$$[\dot{c}] = [c_1][\dot{c}_2] + [\dot{c}_1][c_2] \qquad [\mathbf{b}]$$

Let us first use Eq. [7.4.11] and evaluate $[\dot{c}][c]^T$. Right-multiplying Eq. [b] by $[c]^T = [c_2]^T [c_1]^T$, we obtain

$$[_A^A \tilde{\omega}^B] = [\dot{c}][c]^T = [c_1][\dot{c}_2][c_2]^T [c_1]^T + [\dot{c}_1][c_1]^T \qquad [\mathbf{c}]$$

The right side of Eq. [c] has an interesting interpretation. The second term is what one obtains for the angular velocity matrix when there is only one transformation. The first term contains the angular velocity expression for the second transformation, $[\dot{c}_2][c_2]^T$, with a proper coordinate transformation by the first rotation angle.

When the angular velocity is obtained in terms of the B frame we have

$$[_B^A \tilde{\omega}^B] = [c]^T [\dot{c}] = [c_2]^T [\dot{c}_2] + [c_2]^T [c_1]^T [\dot{c}_1][c_2] \qquad [\mathbf{d}]$$

which has a similar explanation as the one for Eq. [c].

It remains to calculate the derivative matrices and use them in Eqs. [c] and [d]. The derivatives of $[c_1]$ and $[c_2]$ are

$$[\dot{c}_1] = \begin{bmatrix} -\dot{\theta}_1 s\theta_1 & -\dot{\theta}_1 c\theta_1 & 0 \\ \dot{\theta}_1 c\theta_1 & -\dot{\theta}_1 s\theta_1 & 0 \\ 0 & 0 & 0 \end{bmatrix} \qquad [\dot{c}_2] = \begin{bmatrix} -\dot{\theta}_2 s\theta_2 & 0 & \dot{\theta}_2 c\theta_2 \\ 0 & 0 & 0 \\ -\dot{\theta}_2 c\theta_2 & 0 & -\dot{\theta}_2 s\theta_2 \end{bmatrix} \qquad \textbf{[e]}$$

Consider writing the angular velocity vector in terms of its components along the $a_1 a_2 a_3$ axes. We first evaluate $[c_1]^T[\dot{c}_1]$ and related terms. Using Eqs. [a] and [e], we have

$$[c_1]^T[\dot{c}_1] = \begin{bmatrix} c\theta_1 & s\theta_1 & 0 \\ -s\theta_1 & c\theta_1 & 0 \\ 0 & 0 & 1 \end{bmatrix} \begin{bmatrix} -\dot{\theta}_1 s\theta_1 & -\dot{\theta}_1 c\theta_1 & 0 \\ \dot{\theta}_1 c\theta_1 & -\dot{\theta}_1 s\theta_1 & 0 \\ 0 & 0 & 0 \end{bmatrix} = \begin{bmatrix} 0 & -\dot{\theta}_1 & 0 \\ \dot{\theta}_1 & 0 & 0 \\ 0 & 0 & 0 \end{bmatrix} \qquad \textbf{[f]}$$

which represents the angular velocity matrix associated with a 1 rotation. As expected, we find that $[\dot{c}_1][c_1]^T = [c_1]^T[\dot{c}_1]$. Performing the same matrix multiplications with θ_2, we obtain

$$[\dot{c}_2][c_2]^T = [c_2]^T[\dot{c}_2] = \begin{bmatrix} 0 & 0 & \dot{\theta}_2 \\ 0 & 0 & 0 \\ -\dot{\theta}_2 & 0 & 0 \end{bmatrix} \qquad \textbf{[g]}$$

which is the angular velocity matrix for a 2 rotation.

Carrying out the algebra we obtain in terms of the A frame

$$[{}^A_A\tilde{\omega}^B] = [\dot{c}][c]^T = [c_1][\dot{c}_2][c_2]^T[c_1]^T + [\dot{c}_1][c_1]^T = \begin{bmatrix} 0 & -\dot{\theta}_1 & \dot{\theta}_2 c\theta_1 \\ \dot{\theta}_1 & 0 & \dot{\theta}_2 s\theta_1 \\ -\dot{\theta}_2 c\theta_1 & -\dot{\theta}_2 s\theta_1 & 0 \end{bmatrix} \qquad \textbf{[h]}$$

We confirm this result using the formula for the angular velocity as

$$^A\omega^B = \dot{\theta}_1 \mathbf{a}_3 + \dot{\theta}_2 \mathbf{a}'_2 = \dot{\theta}_1 \mathbf{a}_3 + \dot{\theta}_2 (\cos\theta_1 \mathbf{a}_2 - \sin\theta_1 \mathbf{a}_1) \qquad \textbf{[i]}$$

so that

$$_A\omega_1 = -\dot{\theta}_2 \sin\theta_1 \qquad _A\omega_2 = \dot{\theta}_2 \cos\theta_1 \qquad _A\omega_3 = \dot{\theta}_1 \qquad \textbf{[j]}$$

These match the entries of the angular velocity matrix in Eq. [h]. In terms of the B frame we have

$$[{}^A_B\tilde{\omega}^B] = [c]^T[\dot{c}] = [c_2]^T[\dot{c}_2] + [c_2]^T[c_1]^T[\dot{c}_1][c_2] = \begin{bmatrix} 0 & -\dot{\theta}_1 c\theta_2 & \dot{\theta}_2 \\ \dot{\theta}_1 c\theta_2 & 0 & \dot{\theta}_1 s\theta_2 \\ -\dot{\theta}_2 & -\dot{\theta}_1 s\theta_2 & 0 \end{bmatrix} \qquad \textbf{[k]}$$

In a similar fashion, we write the angular velocity in terms of the coordinates of the B frame as

$$^A\omega^B = \dot{\theta}_1 \mathbf{a}_3 + \dot{\theta}_2 \mathbf{a}'_2 = \dot{\theta}_1(-\sin\theta_2 \mathbf{b}_1 + \cos\theta_2 \mathbf{b}_3) + \dot{\theta}_2 \mathbf{b}_2 \qquad \textbf{[l]}$$

so that

$$_B\omega_1 = -\dot{\theta}_1 \sin\theta_2 \qquad _B\omega_2 = \dot{\theta}_2 \qquad _B\omega_3 = \dot{\theta}_1 \cos\theta_2 \qquad \textbf{[m]}$$

which confirms our earlier results in terms of frame B, when we compare this with Eq. [k].

7.5 EULER ANGLES

We saw earlier that, at most, three successive rotations about nonparallel axes are sufficient to define a rotation transformation from one coordinate system to another. The transformation can be expressed in many ways and is not unique. In this section, we quantify the different choices for carrying out the transformation from one coordinate frame to the other.

7.5.1 EULER ANGLE SEQUENCES

Let us review what we have done in the previous chapters. In Chapter 1, we studied curvilinear coordinates and discussed the rate of change of the unit vectors. This discussion was in terms of specific coordinate systems, without a general formulation. In Chapter 2, we saw that three transformations about nonparallel axes are sufficient to describe the most general transformation from one coordinate system to another. We outlined two ways of accomplishing these rotations: body-fixed and space-fixed rotations. When the coordinate transformation angles are infinitesimal, the rotations can be viewed as vector operations, from which we defined the angular velocity vector. We obtained expressions for the rates of change of unit vectors. Most of the analysis in Chapter 2 was instantaneous.

In this section, we use body-fixed rotations and select the nonparallel axes about which the rotations are carried out as the coordinate axes of the rotated frames. The three angles used for transforming one coordinate set into another are commonly referred to as *Euler angles*.

In generating three sets of rotations to transform one set of coordinates into another, say, $a_1 a_2 a_3$ to $b_1 b_2 b_3$, there are twelve choices in which no two adjacent rotation indices are the same. We begin with the $a_1 a_2 a_3$ frame and rotate it about one of the axes to get the $a'_1 a'_2 a'_3$ frame. Here we have three choices. The next rotation is about one of the a'_1, a'_2, or a'_3 axes, excluding the axis that coincides with the previous rotation. That is, if the first rotation is about the a_2 axis, the second rotation should not be about the a'_2 axis. Hence, we have two possible rotations for each previous rotation. We follow the same procedure for the third transformation, resulting in two possible transformations for each previous rotation. As a result, we have $3 \times 2 \times 2 = 12$ possible combinations:

$$1\text{-}2\text{-}1, \ 1\text{-}2\text{-}3, \ 1\text{-}3\text{-}1, \ 1\text{-}3\text{-}2, \ 2\text{-}1\text{-}2, \ 2\text{-}1\text{-}3,$$
$$2\text{-}3\text{-}1, \ 2\text{-}3\text{-}2, \ 3\text{-}1\text{-}2, \ 3\text{-}1\text{-}3, \ 3\text{-}2\text{-}1, \ 3\text{-}2\text{-}3$$

The twelve sets are called *Euler angle sequences*. They can be classified into two main categories, each showing similar characteristics. The first category is when the first and third indices are the same (e.g., 3-2-3, 1-2-1) and the second category consists of rotations where the first and third indices are different (e.g., 1-2-3, 3-1-2). One selects the sequence depending on the application, such that the sequence used gives a better physical visualization and, as we will see later on in this section, leads to fewer singularities.

One of the most commonly used Euler angle sequences is the 3-1-3 sequence, often used to describe rotating rigid bodies. In recent years the 3-2-3 transformation has been used more widely for such problems. We learned in Chapter 3 that the 3-1-3 coordinate transformation has traditionally been used to locate the orbit of a body in space.

For purely historical reasons, we begin our discussion of Euler angles with a 3-1-3 transformation. (Readers are urged to familiarize themselves with other transformation sequences, especially the 3-2-3 and the 3-2-1. Properties of the 3-1-3, 3-2-3, and 3-2-1 sequences are summarized in Table 7.1 at the end of this chapter.)

In a 3-1-3 transformation, first, the axes $a_1 a_2 a_3$ are rotated about a_3 by an angle ϕ, to yield the $a_1' a_2' a_3'$ axes. Then, a rotation is performed about the a_1' axis by θ, yielding the $a_1'' a_2'' a_3''$ axes. The a_1' axis is also known as the *line of nodes* (see Chapter 3 for the source of this name). This axis describes the intersection of the $a_1 a_2$ and the $a_1'' a_2''$ planes. Finally, the $a_1'' a_2'' a_3''$ axes are rotated by ψ about the a_3'' axis, which results in the $b_1 b_2 b_3$ axes.

For a 3-1-3 transformation, the rotation angles ϕ, θ, and ψ are known as the *precession*, *nutation*, and *spin* angles, respectively. For a 3-2-3 transformation, the rotation angles are taken as ψ (precession), θ (nutation), and ϕ (spin).

To obtain the combined transformation from the A frame to the B frame, we use matrix notation and successively apply the transformations. For a 3-1-3 sequence the result is

$$\{a'\} = \begin{bmatrix} c\phi & s\phi & 0 \\ -s\phi & c\phi & 0 \\ 0 & 0 & 1 \end{bmatrix} \{a\} \qquad \{a''\} = \begin{bmatrix} 1 & 0 & 0 \\ 0 & c\theta & s\theta \\ 0 & -s\theta & c\theta \end{bmatrix} \{a'\}$$

$$\{b\} = \begin{bmatrix} c\psi & s\psi & 0 \\ -s\psi & c\psi & 0 \\ 0 & 0 & 1 \end{bmatrix} \{a''\} \qquad [7.5.1]$$

Combining the three transformations, as shown in Fig. 7.9, we obtain

$$\{b\} = [R_{3\text{-}1\text{-}3}]\{a\} \qquad [7.5.2]$$

where

$$[R_{3\text{-}1\text{-}3}] = \begin{bmatrix} c\phi c\psi - s\phi c\theta s\psi & s\phi c\psi + c\phi c\theta s\psi & s\theta s\psi \\ -c\phi s\psi - s\phi c\theta c\psi & -s\phi s\psi + c\phi c\theta c\psi & s\theta c\psi \\ s\phi s\theta & -c\phi s\theta & c\theta \end{bmatrix} \qquad [7.5.3]$$

7.5.2 Angular Velocity and Acceleration

We next obtain the angular velocity vector, which has a component due to each rotation as

$$^A\boldsymbol{\omega}^B = {}^A\boldsymbol{\omega}^{A'} + {}^{A'}\boldsymbol{\omega}^{A''} + {}^{A''}\boldsymbol{\omega}^B = \dot{\phi}\mathbf{a}_3 + \dot{\theta}\mathbf{a}_1' + \dot{\psi}\mathbf{a}_3'' \qquad [7.5.4]$$

We wish to express the angular velocity in terms of the coordinates of the transformed frame. From Eq. [7.5.3] we obtain

$$\mathbf{a}_3 = s\theta s\psi \mathbf{b}_1 + s\theta c\psi \mathbf{b}_2 + c\theta \mathbf{b}_3 \qquad \mathbf{a}_1' = c\psi \mathbf{b}_1 - s\psi \mathbf{b}_2 \qquad \mathbf{a}_3'' = \mathbf{b}_3 \qquad [7.5.5]$$

The inverse of $[R]$ is its transpose, $\{a\} = [R]^T\{b\}$, so an easy way to obtain $\mathbf{a}_3$ is to read down the third row of $[R]^T$ or the third column of $[R]$. To obtain $\mathbf{a}_1'$ we read the first row of $[R]^T$ (or the first column of $[R]$) while setting $\phi = 0$. Introducing Eqs. [7.5.5] into Eq. [7.5.4], we obtain

$$^A\boldsymbol{\omega}^B = \dot{\phi}(s\theta s\psi \mathbf{b}_1 + s\theta c\psi \mathbf{b}_2 + c\theta \mathbf{b}_3) + \dot{\theta}(c\psi \mathbf{b}_1 - s\psi \mathbf{b}_2) + \dot{\psi}\mathbf{b}_3$$

$$= (\dot{\phi}s\theta s\psi + \dot{\theta}c\psi)\mathbf{b}_1 + (\dot{\phi}s\theta c\psi - \dot{\theta}s\psi)\mathbf{b}_2 + (\dot{\phi}c\theta + \dot{\psi})\mathbf{b}_3 \qquad [7.5.6]$$

370   CHAPTER 7 • RIGID BODY KINEMATICS

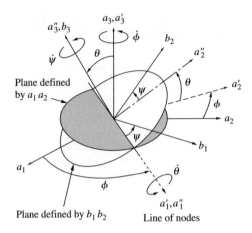

Figure 7.9   3-1-3 Euler angle sequence

which can be written in the matrix form as

$$\begin{bmatrix} {}^A\omega_1^B \\ {}^A\omega_2^B \\ {}^A\omega_3^B \end{bmatrix} = \begin{bmatrix} \dot\phi s\theta s\psi + \dot\theta c\psi \\ \dot\phi s\theta c\psi - \dot\theta s\psi \\ \dot\phi c\theta + \dot\psi \end{bmatrix} = \begin{bmatrix} s\theta s\psi & c\psi & 0 \\ s\theta c\psi & -s\psi & 0 \\ c\theta & 0 & 1 \end{bmatrix} \begin{bmatrix} \dot\phi \\ \dot\theta \\ \dot\psi \end{bmatrix} \quad [\textbf{7.5.7}]$$

in which ${}^A\omega_i^B$ is the ith component of ${}^A\boldsymbol\omega^B$ expressed in terms of the body-fixed axes. Equation [7.5.7] can also be written as $\{\omega\} = [B]\{\dot\theta\}$, where the notation is obvious. A few observations about the matrix $[B]$ that relates the angular velocities to the rates of the Euler angles are in order. First of all, $[B]$ is not orthogonal. This is because the rotations ϕ, θ, and ψ are performed about the a_3, a_1', and a_3'' axes, which do not form an orthogonal set, even though the rotation angles ϕ, θ, and ψ are independent of each other. We also observe that the sines or cosines of the precession angle ϕ are absent in Eq. [7.5.7].

Second, the matrix $[B]$ becomes singular when θ is equal to zero or to a multiple of π. This can be explained by noting that when $\sin\theta = 0$, the rotation reduces to a 3-3 sequence, and that the ϕ rotation cannot be distinguished from the ψ rotation. Because $[B]$ can become singular at times, the rates of change of the Euler angles cannot always be obtained from the components of the angular velocities, which causes problems when integrating the equations of motion. To visualize the singularity, we invert Eq. [7.5.7], to yield

$$\begin{bmatrix} \dot\phi \\ \dot\theta \\ \dot\psi \end{bmatrix} = \begin{bmatrix} \dfrac{s\psi}{s\theta} & \dfrac{c\psi}{s\theta} & 0 \\ c\psi & -s\psi & 0 \\ -\dfrac{s\psi}{t\theta} & -\dfrac{c\psi}{t\theta} & 1 \end{bmatrix} \begin{bmatrix} {}^A\omega_1^B \\ {}^A\omega_2^B \\ {}^A\omega_3^B \end{bmatrix} \quad [\textbf{7.5.8}]$$

in which $t\theta = \tan\theta$. Equation [7.5.8] is referred to as the *kinematic differential equations* relating the angular velocities to the Euler angles. In scalar form,

Eq. [7.5.8] is

$$\dot{\phi} = \frac{1}{\sin\theta}(\omega_1 \sin\psi + \omega_2 \cos\psi) \qquad \dot{\theta} = \omega_1 \cos\psi - \omega_2 \sin\psi$$

$$\dot{\psi} = \frac{1}{\sin\theta}(-\omega_1 \cos\theta \sin\psi - \omega_2 \cos\theta \cos\psi) + \omega_3 \qquad [\mathbf{7.5.9}]$$

where we have adopted the compact notation $\omega_i = {}^A\omega_i{}^B (i = 1, 2, 3)$. Existence of the singularity at $\theta = 0$ and at multiples of π is obvious. Moreover, Eqs. [7.5.9] are highly nonlinear, thus reducing their suitability for manipulation in rigid body dynamical equations, especially for numerical calculations. The singularity associated with a particular Euler angle sequence can be overcome by switching to another Euler angle sequence in the neighborhood of the singularity, but this makes the analysis cumbersome.

Each of the twelve Euler angle sequences have singularities at certain values of the second angle. One of the objectives when selecting an Euler angle sequence is to avoid singularities as much as possible. For example, in aircraft or other vehicle dynamics problems, usually a 3-2-1 sequence is used. The Euler angle sequences where no index is repeated (3-2-1, 1-3-2, etc.) all have a singularity when the second angle, θ, has the value $\theta = \pm\pi/2$. For the most common flight operations, all of these angles remain small quantities. The text by Junkins and Turner contains a tabulation of the transformation matrices and singular values for all twelve Euler angle sequences.

We next analyze the angular acceleration, extending the results in Chapter 2 to rigid bodies. If we express the components of the angular velocity in terms of an inertial frame or using a set of body-fixed coordinates, then the angular acceleration has a simple form. Mathematically,

$$^A\boldsymbol{\alpha}^B = {}^A\frac{d}{dt}({}^A\boldsymbol{\omega}^B) = {}^B\frac{d}{dt}({}^A\boldsymbol{\omega}^B) + {}^A\boldsymbol{\omega}^B \times {}^A\boldsymbol{\omega}^B = {}^B\frac{d}{dt}({}^A\boldsymbol{\omega}^B) \qquad [\mathbf{7.5.10}]$$

Given the angular velocity components along the body axes in Eq. [7.5.7], it follows that the components of the angular acceleration along the body axes are, for a 3-1-3 transformation,

$$\alpha_1 = \frac{d}{dt}(\dot{\phi} s\theta s\psi + \dot{\theta} c\psi) = \ddot{\phi} s\theta s\psi + \dot{\phi}\dot{\theta} c\theta s\psi + \dot{\phi}\dot{\psi} s\theta c\psi + \ddot{\theta} c\psi - \dot{\theta}\dot{\psi} s\psi$$

$$\alpha_2 = \frac{d}{dt}(\dot{\phi} s\theta c\psi - \dot{\theta} s\psi) = \ddot{\phi} s\theta c\psi + \dot{\phi}\dot{\theta} c\theta c\psi - \dot{\phi}\dot{\psi} s\theta s\psi - \ddot{\theta} s\psi - \dot{\theta}\dot{\psi} c\psi$$

$$\alpha_3 = \frac{d}{dt}(\dot{\phi} c\theta + \dot{\psi}) = \ddot{\phi} c\theta - \dot{\phi}\dot{\theta} s\theta + \ddot{\psi} \qquad [\mathbf{7.5.11}]$$

In general, evaluating the components of the angular acceleration along inertial axes yields complicated expressions and does not provide much insight. We hence continue to see that expressing the angular velocity by its components along body axes leads to a more compact form and a more meaningful understanding. This observation is totally in line with the results in Chapter 6, that most of the time, expressing the inertia properties of a body using a set of axes attached to the body is simpler and more meaningful.

7.5.3 AXISYMMETRIC BODIES

We next consider a case where it is more desirable to view the motion using a set of relative axes that do not coincide with the body axes. Such a case arises when dealing with axisymmetric bodies. A suitable choice for analyzing this type of motion is to use a reference frame that does not contain the spin angle. That is, if we are using, say, a 3-2-3 Euler angle transformation, the relative frame is obtained after the 3-2 transformation. The motivation behind selecting a reference frame different than the body frame stems from the fact that in many problems involving axial symmetry, the actual value of the spin angle is not of too much importance, while the spin rate is. One is more interested in the velocity and acceleration of the center of mass or of another point along the symmetry axis, as well as the angular velocity and angular acceleration, than one is in the motion of some specific point on the body. The precession and nutation rotations completely describe the orientation of the symmetry axis.

Consider a 3-1-3 transformation. Using the notation above of $a_1 a_2 a_3$ being the inertial axes, the coordinate axes associated with the new reference frame become the $a_1'' a_2'' a_3''$ axes, and we can find the angular velocity of the reference frame after the 3-1 rotation in Eq. [7.5.4] as

$$^A\boldsymbol{\omega}^{A''} = \dot{\phi}\mathbf{a}_3 + \dot{\theta}\mathbf{a}_1' = \dot{\phi}(\sin\theta \mathbf{a}_2'' + \cos\theta \mathbf{a}_3'') + \dot{\theta}\mathbf{a}_1'' \quad [7.5.12]$$

We will, from now on, refer to this relative reference frame as the *F frame*, denote the associated coordinate axes by $f_1 f_2 f_3$, and use the associated unit vectors $\mathbf{f}_1 \mathbf{f}_2 \mathbf{f}_3$. The concept is illustrated in Fig. 7.10 for a spinning top. For axisymmetric bodies, viewing the F frame is equivalent to viewing the general shape of a disk or top, without following the motion of a specific point on the body.

We can write the angular velocity of the body as

$$\boldsymbol{\omega} = {}^A\boldsymbol{\omega}^B = {}^A\boldsymbol{\omega}^F + {}^F\boldsymbol{\omega}^B \quad [7.5.13]$$

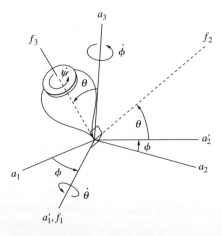

Figure 7.10   The F frame

where $^A\boldsymbol{\omega}^F = {}^A\boldsymbol{\omega}^{A''}$ and $^F\boldsymbol{\omega}^B$ is the spin, $^F\boldsymbol{\omega}^B = \dot{\psi}\mathbf{f}_3$. Where appropriate, the above notation will be shortened in $\boldsymbol{\omega}_b$, $\boldsymbol{\omega}_f$, and $\boldsymbol{\omega}_s$, where the subscripts b, f, and s stand for body, frame, and spin, respectively. We then write

$$\boldsymbol{\omega} = \boldsymbol{\omega}_b = \boldsymbol{\omega}_f + \boldsymbol{\omega}_s \quad [7.5.14]$$

For a 3-1-3 transformation, we can write the angular velocities of the body and of the F frame in terms of the components of the F frame as

$$\boldsymbol{\omega} = \boldsymbol{\omega}_b = \dot{\phi}(\sin\theta\mathbf{a}_2'' + \cos\theta\mathbf{a}_3'') + \dot{\theta}\mathbf{a}_1'' + \dot{\psi}\mathbf{a}_3''$$
$$= \dot{\theta}\mathbf{f}_1 + \dot{\phi}\sin\theta\mathbf{f}_2 + (\dot{\phi}\cos\theta + \dot{\psi})\mathbf{f}_3$$
$$\boldsymbol{\omega}_f = \dot{\phi}(\sin\theta\mathbf{a}_2'' + \cos\theta\mathbf{a}_3'') + \dot{\theta}\mathbf{a}_1'' = \dot{\theta}\mathbf{f}_1 + \dot{\phi}\sin\theta\mathbf{f}_2 + \dot{\phi}\cos\theta\mathbf{f}_3$$
$$\boldsymbol{\omega}_s = \dot{\psi}\mathbf{f}_3 \quad [7.5.15]$$

The expression for angular velocity above is noticeably simpler than its counterpart in Eq. [7.5.6].

Next, consider the expression for the angular acceleration. When using the F frame, the angular acceleration of the body can be found using the transport theorem. Noting the form of $\boldsymbol{\omega}_b$ from Eq. [7.5.14], we have

$$\boldsymbol{\alpha}_b = \frac{d}{dt}(\boldsymbol{\omega}_b)_{\text{rel}} + \boldsymbol{\omega}_f \times \boldsymbol{\omega}_b = \frac{d}{dt}(\boldsymbol{\omega}_b)_{\text{rel}} + \boldsymbol{\omega}_f \times \boldsymbol{\omega}_s$$

or

$$^A\boldsymbol{\alpha}^B = \frac{{}^Fd}{dt}{}^A\boldsymbol{\omega}^B + {}^A\boldsymbol{\omega}^F \times {}^A\boldsymbol{\omega}^B = \frac{{}^Fd}{dt}{}^A\boldsymbol{\omega}^B + {}^A\boldsymbol{\omega}^F \times {}^F\boldsymbol{\omega}^B \quad [7.5.16]$$

Expanding Eq. [7.5.16] in terms of the Euler angles using the 3-1-3 set, we obtain

$$\boldsymbol{\alpha}_b = \frac{d}{dt}(\boldsymbol{\omega}_b)_{\text{rel}} + \boldsymbol{\omega}_f \times \dot{\psi}\mathbf{f}_3$$
$$= \ddot{\theta}\mathbf{f}_1 + (\ddot{\phi}s\theta + \dot{\phi}\dot{\theta}c\theta)\mathbf{f}_2 + (\ddot{\phi}c\theta - \dot{\phi}\dot{\theta}s\theta + \ddot{\psi})\mathbf{f}_3$$
$$+ (\dot{\theta}\mathbf{f}_1 + \dot{\phi}s\theta\mathbf{f}_2 + \dot{\phi}c\theta\mathbf{f}_3) \times \dot{\psi}\mathbf{f}_3$$
$$= (\ddot{\theta} + \dot{\phi}\dot{\psi}s\theta)\mathbf{f}_1 + (\ddot{\phi}s\theta + \dot{\phi}\dot{\theta}c\theta - \dot{\theta}\dot{\psi})\mathbf{f}_2 + (\ddot{\phi}c\theta - \dot{\phi}\dot{\theta}s\theta + \ddot{\psi})\mathbf{f}_3$$

[7.5.17]

which is considerably simpler than its counterpart when the spin angle is included, Eq. [7.5.11]. The above result can be verified by setting $\psi = 0$ in Eq. [7.5.11], while retaining $\dot{\psi}$. Note that if one wishes to deal with the actual value of the spin angle ψ, one can still use Eq. [7.5.17]. When the manipulations are complete one can switch to the coordinate system that includes ψ.

Example 7.3

Derive the angular velocity expressions for aircraft dynamics problems, using a 3-2-1 transformation.

Solution

The coordinate system traditionally used with aircraft is shown in Fig. 7.11. We will denote the inertial coordinates by XYZ and the body-fixed ones by xyz. The z axis is the local vertical. The Euler angle transformations are:

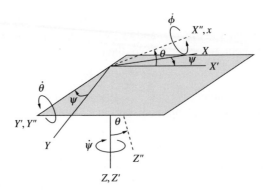

Figure 7.11   3-2-1 Euler angle sequence

a. A rotation about the Z axis, by angle ψ (the heading angle), leading to the $X'Y'Z'$ coordinate system.
b. A rotation about the Y' axis, by angle θ (the attitude angle), leading to the $X''Y''Z''$ coordinate system.
c. A rotation about the X'' axis, by angle ϕ (the bank angle), leading to the xyz coordinate system.

Using this configuration, the xz plane denotes the plane of symmetry of the aircraft, as shown in Fig. 7.12.

The rotation matrices are

$$[R_1] = \begin{bmatrix} c\psi & s\psi & 0 \\ -s\psi & c\psi & 0 \\ 0 & 0 & 1 \end{bmatrix} \quad [R_2] = \begin{bmatrix} c\theta & 0 & -s\theta \\ 0 & 1 & 0 \\ s\theta & 0 & c\theta \end{bmatrix} \quad [R_3] = \begin{bmatrix} 1 & 0 & 0 \\ 0 & c\phi & s\phi \\ 0 & -s\phi & c\phi \end{bmatrix} \quad [a]$$

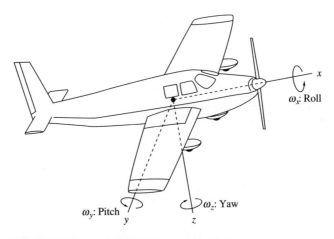

Figure 7.12   Body axes for an aircraft. The xz plane is the plane of symmetry.

so that the relation between the inertial and body-fixed coordinates is written as

$$\begin{bmatrix} x \\ y \\ z \end{bmatrix} = [R_3][R_2][R_1] \begin{bmatrix} X \\ Y \\ Z \end{bmatrix} = [R] \begin{bmatrix} X \\ Y \\ Z \end{bmatrix} \qquad [b]$$

and the combined transformation matrix $[R]$ is

$$[R] = \begin{bmatrix} c\psi c\theta & s\psi c\theta & -s\theta \\ -s\psi c\phi + c\psi s\theta s\phi & c\psi c\phi + s\psi s\theta s\phi & c\theta s\phi \\ s\psi s\phi + c\psi s\theta c\phi & -c\psi s\phi + s\psi s\theta c\phi & c\theta c\phi \end{bmatrix} \qquad [c]$$

The angular velocity vector is written as

$$\boldsymbol{\omega} = \dot{\psi}\mathbf{K} + \dot{\theta}\mathbf{J}' + \dot{\phi}\mathbf{i} \qquad [d]$$

Using the same approach as in the previous section, we obtain $\mathbf{K}$ and $\mathbf{J}'$ in terms of the unit vectors associated with the body axes by reading the third and second columns of $[R]$, respectively, and setting $\psi = 0$ when reading the second column, which yields

$$\boldsymbol{\omega} = (\omega_x\mathbf{i} + \omega_y\mathbf{j} + \omega_z\mathbf{k}) = \dot{\psi}(-s\theta\mathbf{i} + s\phi c\theta\mathbf{j} + c\phi c\theta\mathbf{k}) + \dot{\theta}(c\phi\mathbf{j} - s\phi\mathbf{k}) + \dot{\phi}\mathbf{i}$$
$$= (-\dot{\psi}s\theta + \dot{\phi})\mathbf{i} + (\dot{\psi}s\phi c\theta + \dot{\theta}c\phi)\mathbf{j} + (\dot{\psi}c\phi c\theta - \dot{\theta}s\phi)\mathbf{k} \qquad [e]$$

The body angular velocities ω_x, ω_y, and ω_z are commonly referred to as *roll*, *pitch*, and *yaw* rates, respectively, as shown in Fig. 7.12. We relate these angles to the heading, attitude, and bank angles ψ, θ, and ϕ using Eq. [e] as

$$\begin{bmatrix} \omega_x \\ \omega_y \\ \omega_z \end{bmatrix} = \begin{bmatrix} -s\theta & 0 & 1 \\ s\phi c\theta & c\phi & 0 \\ c\phi c\theta & -s\phi & 0 \end{bmatrix} \begin{bmatrix} \dot{\psi} \\ \dot{\theta} \\ \dot{\phi} \end{bmatrix} \qquad [f]$$

When $\theta = \pi/2$, the first and third columns of the coefficient matrix become similar, so that a singularity is reached. For civilian aircraft this is not a big problem, as θ rarely exceeds $\pi/2$ for most flight operations.

Example 7.4

The spinning symmetric top in Fig. 7.13 has the following rotational parameters: precession rate $\dot{\phi} = 0.3$ rad/s and increasing with the rate of 0.05 rad/s², nutation angle $\theta = 30°$, zero nutation rate $\dot{\theta} = 0$, $\ddot{\theta} = 0$, and constant spin rate of $\dot{\psi} = 5$ rad/s. Assuming that the bottom point of the top does not move, find the velocity and acceleration of the center of mass of the top, as well as its angular velocity and angular acceleration.

Solution

We use the 3-1-3 Euler angle transformation sequence and distinguish between the body frame and the relative frame, as we are interested in the velocity and acceleration of a point lying on the symmetry axis. We can write the angular velocity of the body frame and the relative frame in terms of the unit vectors of the F frame as

$$\boldsymbol{\omega}_b = \dot{\phi}(\sin\theta\mathbf{f}_2 + \cos\theta\mathbf{f}_3) + \dot{\theta}\mathbf{f}_1 + \dot{\psi}\mathbf{f}_3 = 0.15\mathbf{f}_2 + 5.2598\mathbf{f}_3 \text{ rad/s} \qquad [a]$$

$$\boldsymbol{\omega}_f = \dot{\phi}(\sin\theta\mathbf{f}_2 + \cos\theta\mathbf{f}_3) + \dot{\theta}\mathbf{f}_1 = 0.15\mathbf{f}_2 + 0.2598\mathbf{f}_3 \text{ rad/s} \qquad [b]$$

$$\boldsymbol{\omega}_s = \dot{\psi}\mathbf{f}_3 = 5\mathbf{f}_3 \text{ rad/s} \qquad [c]$$

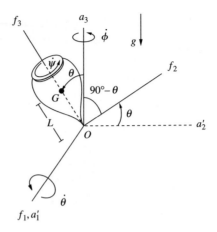

Figure 7.13

We can find the angular acceleration by using Eq. [7.5.17] as

$$\boldsymbol{\alpha}_b = \frac{d}{dt}(\boldsymbol{\omega}_b)_{rel} + \boldsymbol{\omega}_f \times \boldsymbol{\omega}_s \qquad [d]$$
$$= (\ddot{\theta} + \dot{\phi}\dot{\psi}s\,\theta)\mathbf{f}_1 + (\ddot{\phi}s\,\theta + \dot{\phi}\dot{\theta}c\,\theta - \dot{\theta}\dot{\psi})\mathbf{f}_2 + (\ddot{\phi}c\,\theta - \dot{\phi}\dot{\theta}s\,\theta + \ddot{\psi})\mathbf{f}_3$$

Substituting in the values, we obtain

$$\boldsymbol{\alpha}_b = 5\cdot 0.3\cdot 0.5\mathbf{f}_1 + 0.05\left(0.5\mathbf{f}_2 + \frac{\sqrt{3}}{2}\mathbf{f}_3\right) = 0.75\mathbf{f}_1 + 0.025\mathbf{f}_2 + 0.04330\mathbf{f}_3 \text{ rad/s}^2 \qquad [e]$$

We find the velocity of the center of mass as

$$\mathbf{v}_G = \boldsymbol{\omega}_b \times \mathbf{r}_{G/C} = [\dot{\phi}(\sin\theta\mathbf{f}_2 + \cos\theta\mathbf{f}_3) + \dot{\theta}\mathbf{f}_1 + \dot{\psi}\mathbf{f}_3] \times L\mathbf{f}_3$$
$$= L\dot{\phi}\sin\theta\mathbf{f}_1 - L\dot{\theta}\mathbf{f}_2 = 0.15L\mathbf{f}_1 \qquad [f]$$

To find the acceleration of the center of mass, one can use a number of approaches. One approach is to express the velocity in terms of the inertial coordinates and perform a straightforward differentiation. This is lengthy, and it requires that the results then be converted back to the body axes. We prefer to use the F frame and transport theorem, which gives

$$^A\mathbf{a}_G = \frac{^Fd}{dt}\mathbf{v}_G + {}^A\boldsymbol{\omega}^F \times \mathbf{v}_G$$
$$= L\ddot{\phi}\sin\theta\mathbf{f}_1 + L\dot{\phi}\dot{\theta}\cos\theta\mathbf{f}_1 - L\ddot{\theta}\mathbf{f}_2 + [\dot{\phi}(\sin\theta\mathbf{f}_2 + \cos\theta\mathbf{f}_3) + \dot{\theta}\mathbf{f}_1] \times (L\dot{\phi}\sin\theta\mathbf{f}_1 - L\dot{\theta}\mathbf{f}_2)$$
$$= (L\ddot{\phi}\sin\theta + L\dot{\phi}\dot{\theta}\cos\theta + L\dot{\phi}\dot{\theta}\cos\theta)\mathbf{f}_1$$
$$+ (-L\ddot{\theta} + L\dot{\phi}^2\sin\theta\cos\theta)\mathbf{f}_2 - (L\dot{\phi}^2\sin^2\theta + L\dot{\theta}^2)\mathbf{f}_3 \qquad [g]$$

Substituting in the appropriate values, we obtain

$$\mathbf{a}_G = 0.025L\mathbf{f}_1 + 0.03897L\mathbf{f}_2 - 0.0225L\mathbf{f}_3 \qquad [h]$$

A third approach to solve for the acceleration of the center of mass is to use the formula for the rotation about a fixed point, which is

$$\mathbf{a}_G = \boldsymbol{\alpha}_b \times \mathbf{r}_{G/O} + \boldsymbol{\omega}_b \times \boldsymbol{\omega}_b \times \mathbf{r}_{G/O} = \boldsymbol{\alpha}_b \times \mathbf{r}_{G/O} + \boldsymbol{\omega}_b \times \mathbf{v}_G \qquad [i]$$

Substituting in the above equation values for the angular velocity and acceleration parameters gives

$$\mathbf{a}_G = (0.75\mathbf{f}_1 + 0.025\mathbf{f}_2 + 0.4330\mathbf{f}_3) \times L\mathbf{f}_3 + (0.15\mathbf{f}_2 + 5.2598\mathbf{f}_3) \times 0.15L\mathbf{f}_1$$
$$= -0.75L\mathbf{f}_2 + 0.025L\mathbf{f}_1 - 0.0225L\mathbf{f}_3 + 0.7890L\mathbf{f}_2$$
$$= 0.025L\mathbf{f}_1 + 0.03897L\mathbf{f}_2 - 0.0225L\mathbf{f}_3 \qquad [\mathbf{i}]$$

which, of course, is the same as the result in Eq. [h].

Example 7.5

A gyropendulum, consisting of a disk of radius R attached to a shaft of length L, rotates with a spin rate of $\dot{\psi}$ about the shaft. The shaft is pivoted to another vertical shaft which itself rotates with the rate $\dot{\phi}$. The pivot angle is θ, as shown in Fig. 7.14. Find the inertia coefficients of the disk about point B, the angular velocity, and the angular acceleration.

Solution

Let us define the inertial frame XYZ with the Z-axis along the precessing shaft, a second frame $x'y'z'$ attached to the precessing shaft, rotating with $\dot{\phi}$, and a third frame xyz obtained by rotating the shaft about the y' axis by the nutation angle θ. In essence, we are using a 3-2 transformation. The resulting frame is the F frame.

The mass moments of inertia of the disk about its center of mass are

$$I_{Gzz} = \frac{1}{2}mR^2 \qquad I_{Gyy} = I_{Gxx} = \frac{1}{4}mR^2 \qquad [\mathbf{a}]$$

with all products of inertia zero. About the center of rotation B, the moments of inertia can be found using the parallel axis theorem, which yields

$$I_{Bzz} = \frac{1}{2}mR^2 \qquad I_{Byy} = I_{Bxx} = m\left(\frac{1}{4}R^2 + L^2\right) \qquad [\mathbf{b}]$$

Noting that the unit vector $\mathbf{K} = \mathbf{k'}$ can be expressed in the xyz frame as $\mathbf{k'} = -\sin\theta\mathbf{i} + \cos\theta\mathbf{k}$, the angular velocity of the disk and the frame can be written as

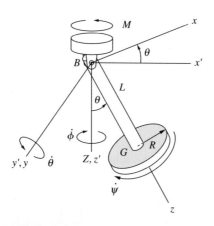

Figure 7.14

$$\boldsymbol{\omega}_b = \dot{\phi}\mathbf{K} + \dot{\theta}\mathbf{j}' + \dot{\psi}\mathbf{k} = \dot{\phi}\mathbf{k}' + \dot{\theta}\mathbf{j} + \dot{\psi}\mathbf{k} = -\dot{\phi}\sin\theta\mathbf{i} + \dot{\theta}\mathbf{j} + (\dot{\psi} + \dot{\phi}\cos\theta)\mathbf{k} \quad [\mathbf{c}]$$

$$\boldsymbol{\omega}_f = \dot{\phi}\mathbf{K} + \dot{\theta}\mathbf{j}' = \dot{\phi}\mathbf{k}' + \dot{\theta}\mathbf{j} = -\dot{\phi}\sin\theta\mathbf{i} + \dot{\theta}\mathbf{j} + \dot{\phi}\cos\theta\mathbf{k} \quad [\mathbf{d}]$$

$$\boldsymbol{\omega}_s = \dot{\psi}\mathbf{k} \quad [\mathbf{e}]$$

The angular acceleration of the disk is simply

$$\begin{aligned}
\boldsymbol{\alpha}_b &= \frac{d}{dt}(\boldsymbol{\omega}_b)_{\text{rel}} + \boldsymbol{\omega}_f \times \boldsymbol{\omega}_s = -(\ddot{\phi}\sin\theta + \dot{\phi}\dot{\theta}\cos\theta)\mathbf{i} + \ddot{\theta}\mathbf{j} + (\ddot{\psi} + \ddot{\phi}\cos\theta - \dot{\phi}\dot{\theta}\sin\theta)\mathbf{k} \\
&\quad + (-\dot{\phi}\sin\theta\mathbf{i} + \dot{\theta}\mathbf{j} + \dot{\phi}\cos\theta\mathbf{k}) \times (\dot{\psi}\mathbf{k}) \\
&= (-\ddot{\phi}\sin\theta - \dot{\phi}\dot{\theta}\cos\theta + \dot{\theta}\dot{\psi})\mathbf{i} + (\ddot{\theta} + \dot{\phi}\dot{\psi}\sin\theta)\mathbf{j} + (\ddot{\psi} + \ddot{\phi}\cos\theta - \dot{\phi}\dot{\theta}\sin\theta)\mathbf{k} \quad [\mathbf{f}]
\end{aligned}$$

7.6 ANGULAR VELOCITIES AS QUASI-VELOCITIES (GENERALIZED SPEEDS)

From Chapter 4, a generalized velocity is obtained by differentiating the corresponding generalized coordinate with respect to time. The three Euler angles that quantify the angular velocity are in essence generalized coordinates, and their time derivatives are the generalized velocities. However, the components of the angular velocity cannot be classified as generalized velocities, as there is no corresponding generalized coordinate. A similar statement is true for a nonholonomic constraint, which cannot be expressed as the derivative of a function.

The questions can then be posed as to how one can classify angular velocities and whether there are other cases in dynamics where one deals with such quantities. We introduce here a set of variables called *quasi-velocities* or *generalized speeds*. We define these as linear combinations of the generalized velocities, but they themselves are not necessarily derivatives of any coordinates and thus cannot always be integrated to generalized coordinates. Quasi-velocities are not perfect differentials, a fact from which their name is derived. We will use both terms, *generalized speeds* and *quasi-velocities,* interchangeably.

The introduction of generalized speeds increases the choice of parameters one can use to describe motion. Consider a holonomic system having n degrees of freedom, with generalized coordinates $q_1, q_2, \ldots, q_n$ and generalized velocities $\dot{q}_1, \dot{q}_2, \ldots, \dot{q}_n$. We define a set of quasi-velocities or generalized speeds $u_1, u_2, \ldots, u_n$ as

$$u_k = \sum_{j=1}^n Y_{kj}\dot{q}_j + Z_k \qquad k = 1, 2, \ldots, n \qquad [7.6.1]$$

where $Y_{kj} = Y_{kj}(q_1, q_2, \ldots, q_n, t)$, $Z_k = Z_k(q_1, q_2, \ldots, q_n, t)$ $(k, j = 1, 2, \ldots, n)$ are functions of the generalized coordinates and time. In column vector form we can write Eq. [7.6.1] as

$$\{u\} = [Y]\{\dot{q}\} + \{Z\} \qquad [7.6.2]$$

In order for the set of generalized speeds to completely describe a system, $[Y]$ must be nonsingular. We invert Eq. [7.6.2] to express the generalized velocities in

7.6 ANGULAR VELOCITIES AS QUASI-VELOCITIES (GENERALIZED SPEEDS)

terms of the generalized speeds as

$$\{\dot{q}\} = [Y]^{-1}\{u\} - [Y]^{-1}\{Z\} \qquad [7.6.3]$$

The angular velocities of a body can be classified as quasi-velocities. The relation between the body angular velocities and the Euler angles, Eq. [7.5.7], defines a valid set of generalized speeds.[1] By contrast, the angular velocities of the F frame do not constitute a set of quasi-velocities, as the relation between them and the Euler angles is described by a matrix that is singular at all times.

Representation of angular velocity as a set of generalized speeds is not the only application of these quantities. Generalized speeds are very useful when one deals with dynamical systems subjected to nonholonomic constraints and in cases when the use of generalized velocities makes the problem formulation cumbersome. We saw an example to this in Examples 4.14 and 5.13, where switching to the velocity of point A as a motion variable simplifies the equations.

Consider the position vector $\mathbf{r}$. In terms of the generalized coordinates, one can express $\mathbf{r}$ as $\mathbf{r} = \mathbf{r}(q_1, q_2, \ldots, q_n, t)$. Differentiating this expression with respect to time, we obtain

$$\dot{\mathbf{r}} = \frac{d\mathbf{r}}{dt} = \frac{\partial \mathbf{r}}{\partial q_1}\dot{q}_1 + \frac{\partial \mathbf{r}}{\partial q_2}\dot{q}_2 + \frac{\partial \mathbf{r}}{\partial q_n}\dot{q}_n + \frac{\partial \mathbf{r}}{\partial t} \qquad [7.6.4]$$

From Eq. [7.6.4], we can write the rate of change of $\mathbf{r}$ in terms of the quasi-velocities as

$$\mathbf{v} = \dot{\mathbf{r}} = \sum_{j=1}^{n} \frac{\partial \mathbf{r}}{\partial q_j} \sum_{k=1}^{n} ([Y]_{jk}^{-1} u_k - [Y]_{jk}^{-1} Z_k) + \frac{\partial \mathbf{r}}{\partial t} = \sum_{k=1}^{n} \mathbf{v}^k u_k + \mathbf{v}^t \qquad [7.6.5]$$

in which $\mathbf{v}^k$ ($k = 1, 2, \ldots, n$) and $\mathbf{v}^t$ are called *partial velocities* and have the form

$$\mathbf{v}^k = \sum_{j=1}^{n} \frac{\partial \mathbf{r}}{\partial q_j}[Y]_{jk}^{-1} \qquad \mathbf{v}^t = \sum_{j=1}^{n}\sum_{k=1}^{n} \frac{\partial \mathbf{r}}{\partial q_j}[Y]_{jk}^{-1} Z_k + \frac{\partial \mathbf{r}}{\partial t} \qquad [7.6.6]$$

Throughout this text, we will denote the index associated with partial velocities with a superscript. In a similar fashion, we can express the angular velocities associated with a body or systems of bodies in terms of the *partial angular velocities* $\boldsymbol{\omega}^k$ ($k = 1, 2, \ldots, n$) and $\boldsymbol{\omega}^t$ as

$$\boldsymbol{\omega} = \sum_{k=1}^{n} \boldsymbol{\omega}^k u_k + \boldsymbol{\omega}^t \qquad [7.6.7]$$

Describe the angular velocities associated with the F frame in terms of the quasi-coordinates associated with the body frame, and show that they do not constitute a set of generalized speeds. Then find the partial angular velocities associated with $\boldsymbol{\omega}$. | **Example 7.6**

---

[1] Eq. [7.5.7] has a singularity when the second Euler angle reaches a certain value. One can avoid this singularity by switching to a different set of Euler angles at that instant. Hence, the transformation between angular velocities and Euler angles is considered a valid one-to-one transformation between the generalized velocities and the generalized speeds.

Solution

We will use a 3-1-3 Euler angle set. From Eqs. [7.5.15], the angular velocities of the body and frame are

$$\boldsymbol{\omega} = \boldsymbol{\omega}_b = \dot{\theta}\mathbf{f}_1 + \dot{\phi}\sin\theta\mathbf{f}_2 + (\dot{\phi}\cos\theta + \dot{\psi})\mathbf{f}_3 \qquad \boldsymbol{\omega}_f = \dot{\theta}\mathbf{f}_1 + \dot{\phi}\sin\theta\mathbf{f}_2 + \dot{\phi}\cos\theta\mathbf{f}_3 \quad \textbf{[a]}$$

We define the generalized speeds as

$$u_1 = \omega_1 = \dot{\theta} \qquad u_2 = \omega_2 = \dot{\phi}\sin\theta \qquad u_3 = \omega_3 = \cos\theta + \dot{\psi} \quad \textbf{[b]}$$

The angular velocities associated with the F frame then can be related to the generalized speeds as

$$\omega_{f1} = \dot{\theta} = u_1 \qquad \omega_{f2} = \dot{\phi}\sin\theta = u_2 \qquad \omega_{f3} = \dot{\phi}\cos\theta = \frac{u_2}{\tan\theta} \quad \textbf{[c]}$$

As u_3 is not present in the description, the angular velocities of the F frame do not constitute a complete set of generalized speeds.

For the body angular velocity, using the generalized speeds defined in Eq. [b] we have

$$\boldsymbol{\omega} = \sum_{k=1}^{3} \boldsymbol{\omega}^k u_k + \boldsymbol{\omega}^t = \omega_1 \mathbf{f}_1 + \omega_2 \mathbf{f}_2 + \omega_3 \mathbf{f}_3 \quad \textbf{[d]}$$

so that the partial angular velocities are recognized as

$$\boldsymbol{\omega}^1 = \mathbf{f}_1 \qquad \boldsymbol{\omega}^2 = \mathbf{f}_2 \qquad \boldsymbol{\omega}^3 = \mathbf{f}_3 \qquad \boldsymbol{\omega}^t = 0 \quad \textbf{[e]}$$

Note that had we selected our generalized speeds as $u_1 = \dot{\theta}, u_2 = \dot{\phi}, u_3 = \dot{\psi}$, the partial angular velocities would have the form

$$\boldsymbol{\omega}^1 = \mathbf{f}_1 \qquad \boldsymbol{\omega}^2 = \sin\theta\mathbf{f}_2 + \cos\theta\mathbf{f}_3 \qquad \boldsymbol{\omega}^3 = \mathbf{f}_3 \qquad \boldsymbol{\omega}^t = 0 \quad \textbf{[f]}$$

7.7 EULER PARAMETERS

We studied Euler angles in the previous sections to relate the angular velocities to the rotation angles ϕ, θ, ψ and their rates. As we saw in Section 7.5, these equations are highly nonlinear and have singularities, which makes it difficult to integrate them and to do any analytical as well as numerical work with them.

To alleviate these difficulties it is preferable to work with another set, called *Euler parameters*. These parameters increase the number of variables one deals with from three to four, but they eliminate the nonlinearities and many of the numerical problems.

Because a set of three variables (ϕ, θ, ψ) is being expressed in terms of four variables, the use of Euler parameters introduces a redundancy. This implies that there is no unique way of expressing the Euler parameters, and one can easily come up with different sets of parameters. Other commonly used quantities include the Cayley-Klein parameters, Rodrigues parameters, and quaternions. All these quantities are, essentially, different forms of the Euler parameters. A comparison of these quantities can be found in the text by Junkins and Turner. The mathematics of the Euler parameters was first introduced by Hamilton in 1843. The vector formulation of Euler parameters, which is used with quaternions, was developed by Oliver Heaviside.

7.7 EULER PARAMETERS

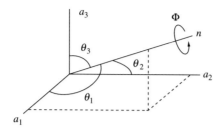

Figure 7.15 Principal line

Euler parameters are inspired from Euler's theorem, which states that any rotation of a rigid body about a point can be accomplished by a single rotation by an angle Φ (principal angle) about a line fixed in the body and passing through the center of rotation (principal line). Euler parameters are in essence a characterization of this rotation in terms of the principal angle and direction cosines of the principal line.

Let us describe the principal line by its direction cosines c_1, c_2, c_3, as shown in Fig 7.15. Denoting by $\theta_1, \theta_2, \theta_3$ the angles the principal line makes with the coordinate axes, we have

$$c_1 = \cos\theta_1 \qquad c_2 = \cos\theta_2 \qquad c_3 = \cos\theta_3 \qquad [7.7.1]$$

The Euler parameters are defined as the four parameters e_0, e_1, e_2, and e_3 such that

$$e_0 = \cos\left(\frac{\Phi}{2}\right) \quad e_1 = c_1 \sin\left(\frac{\Phi}{2}\right) \quad e_2 = c_2 \sin\left(\frac{\Phi}{2}\right) \quad e_3 = c_3 \sin\left(\frac{\Phi}{2}\right) \qquad [7.7.2]$$

We use the half angle so as to eliminate ambiguities associated with the rotation angle. We observe from Eq. [7.7.2] that

$$e_0^2 + e_1^2 + e_2^2 + e_3^2 = 1 \qquad [7.7.3]$$

We will relate the Euler parameters and their rates of change to the Euler angles, angular velocities, and direction cosines, and vice versa. That is, we will discuss the relationships

$$\begin{aligned} c_{ij} &= f(e_0, e_1, e_2, e_3) & i, j &= 1, 2, 3 \\ \omega_j &= f(e_0, e_1, e_2, e_3) & j &= 1, 2, 3 \\ \dot{e}_j &= f(e_0, e_1, e_2, e_3, \omega_1, \omega_2, \omega_3) & j &= 1, 2, 3, 4 \end{aligned} \qquad [7.7.4]$$

We denote the unit vector along the principal line by **n**. Section 7.3 discussed that the direction cosine matrix $[c]$ has an eigenvalue $\lambda = 1$ and that c_1, c_2, and c_3 are elements of the associated eigenvector. From the eigenvalue problem

$$[c]^T\{c\} = \lambda\{c\} \qquad [7.7.5]$$

we set $\lambda = 1$ and find $\{c\} = [c_1 \; c_2 \; c_3]^T$.

We now explore each of the relations in Eqs. [7.7.4] individually.

7.7.1 RELATING THE DIRECTION COSINES TO THE EULER PARAMETERS

Consider two coordinate frames $A\ (a_1 a_2 a_3)$ and $B\ (b_1 b_2 b_3)$, where $b_1 b_2 b_3$ is obtained by rotating $a_1 a_2 a_3$ by an angle of Φ about the unit vector $\mathbf{n}$. We use the notation in Section 7.3 and express the unit vector $\mathbf{n}$ in terms of its components in the two coordinate frames as

$$_A\mathbf{n} = {_A n_1}\mathbf{a}_1 + {_A n_2}\mathbf{a}_2 + {_A n_3}\mathbf{a}_3$$
$$_B\mathbf{n} = {_B n_1}\mathbf{b}_1 + {_B n_2}\mathbf{b}_2 + {_B n_3}\mathbf{b}_3 \qquad [\mathbf{7.7.6}]$$

Naturally, considering Fig. 7.15, $_A n_i = {_B n_i} = \cos\theta_i = c_i\ (i = 1, 2, 3)$. We can relate the unit vectors in the A and B frames and the components of $\mathbf{n}$ in the A and B frames as

$$\begin{bmatrix} \mathbf{b}_1 \\ \mathbf{b}_2 \\ \mathbf{b}_3 \end{bmatrix} = [c]^T \begin{bmatrix} \mathbf{a}_1 \\ \mathbf{a}_2 \\ \mathbf{a}_3 \end{bmatrix} \qquad \begin{bmatrix} _B n_1 \\ _B n_2 \\ _B n_3 \end{bmatrix} = [c]^T \begin{bmatrix} _A n_1 \\ _A n_2 \\ _A n_3 \end{bmatrix} \qquad [\mathbf{7.7.7}]$$

Let us consider the vectors $\mathbf{a}_1$ and $\mathbf{b}_1$ and think of an inverted cone with apex angle θ_1 and with the unit vector $\mathbf{n}$ as the axis of the cone, as shown in Fig. 7.16. Because frame B is obtained by rotating frame A around $\mathbf{n}$, this rotation defines a circle of radius $\sin\theta_1$ on a plane perpendicular to the vector $\mathbf{n}$. The angle between $\mathbf{a}_1$ and $\mathbf{n}$ is the same as the angle between $\mathbf{b}_1$ and $\mathbf{n}$, and the projections of $\mathbf{a}_1$ and $\mathbf{b}_1$ on $\mathbf{n}$ are identical. Mathematically,

$$\mathbf{n} \cdot \mathbf{a}_1 = \mathbf{n} \cdot \mathbf{b}_1 = \cos\theta_1 \qquad [\mathbf{7.7.8}]$$

We denote by Q the point at which the vector $\mathbf{n}$ intersects the circle defined by $\mathbf{a}_1$ and $\mathbf{b}_1$. The vector connecting points O and Q is defined as $\mathbf{x}_1$, and it can be expressed as $\mathbf{x}_1 = \cos\theta_1 \mathbf{n}$, as shown in Fig. 7.17. We introduce the vectors $\mathbf{y}_1$ and $\mathbf{z}_1$ such that $\mathbf{y}_1$ is along the line connecting point Q to the tip of $\mathbf{a}_1$, and $\mathbf{z}_1$ is perpendicular to $\mathbf{y}_1$ such that its tip coincides with the end of $\mathbf{b}_1$, as shown in Fig. 7.18. We can write

$$\mathbf{b}_1 = \mathbf{x}_1 + \mathbf{y}_1 + \mathbf{z}_1 \qquad [\mathbf{7.7.9}]$$

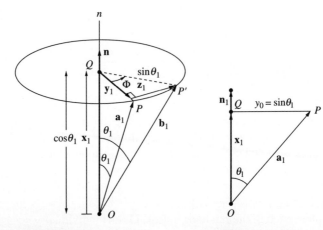

Figure 7.16   **Figure 7.17**   Side view

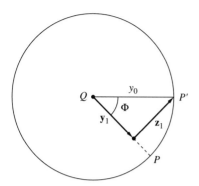

Figure 7.18 Top view

The objective is to express $\mathbf{b}_1$ in terms of the components of the A frame. To accomplish this, we consider Fig. 7.18 and the circular arc generated by rotating $\mathbf{a}_1$ into $\mathbf{b}_1$. Introducing the notation $y_0 = \sin\theta_1$, we can write

$$y_1 = |\mathbf{y}_1| = y_0 \cos\Phi \qquad z_1 = |\mathbf{z}_1| = y_0 \sin\Phi$$
$$y_0 = |\mathbf{a}_1 - \mathbf{x}_1| = |\mathbf{b}_1 - \mathbf{x}_1| \qquad [7.7.10]$$

from which we have

$$\mathbf{y}_1 = (\mathbf{a}_1 - \mathbf{x}_1)\cos\Phi = \cos\Phi\,\mathbf{a}_1 - \cos\theta_1 \cos\Phi\,\mathbf{n} \qquad [7.7.11]$$

The vector $\mathbf{z}_1$ is perpendicular to both $\mathbf{n}$ and $\mathbf{y}_1$. Denoting the unit vector along $\mathbf{z}_1$ by $\mathbf{e}_z$, we can express it as

$$\mathbf{e}_z = \frac{\mathbf{n}\times\mathbf{y}_1}{|\mathbf{n}\times\mathbf{y}_1|} = \frac{\mathbf{n}\times\mathbf{a}_1}{|\mathbf{n}\times\mathbf{a}_1|} \qquad [7.7.12]$$

However, since $|\mathbf{n}\times\mathbf{a}_1| = \sin\theta_1 = y_0$, and $|\mathbf{z}_1| = \sin\theta_1 \sin\Phi$, we express $\mathbf{z}_1$ as

$$\mathbf{z}_1 = |\mathbf{z}_1|\mathbf{e}_z = \sin\theta_1 \sin\Phi \frac{\mathbf{n}\times\mathbf{a}_1}{\sin\theta_1} = \sin\Phi\,\mathbf{n}\times\mathbf{a}_1 \qquad [7.7.13]$$

Now combining Eqs. [7.7.9], [7.7.11], and [7.7.13], we have

$$\mathbf{b}_1 = \mathbf{x}_1 + \mathbf{y}_1 + \mathbf{z}_1 = c_1\mathbf{n} + \cos\Phi\,\mathbf{a}_1 - c_1\cos\Phi\,\mathbf{n} + \sin\Phi\,\mathbf{n}\times\mathbf{a}_1 \qquad [7.7.14]$$

Using the half-angle formulas $\cos\Phi = 2\cos^2(\Phi/2) - 1$, $1 - \cos\Phi = 2\sin^2(\Phi/2)$, and $\sin\Phi = 2\sin(\Phi/2)\cos(\Phi/2)$, and Eq. [7.7.6], we can write the above equation as

$$\mathbf{b}_1 = \left(2c_1 \sin^2\frac{\Phi}{2}\right)(c_1\mathbf{a}_1 + c_2\mathbf{a}_2 + c_3\mathbf{a}_3) + \left(2\cos^2\frac{\Phi}{2} - 1\right)\mathbf{a}_1$$
$$+ \left(2\sin\frac{\Phi}{2}\cos\frac{\Phi}{2}\right)(c_1\mathbf{a}_1 + c_2\mathbf{a}_2 + c_3\mathbf{a}_3)\times\mathbf{a}_1 \qquad [7.7.15]$$

Introducing the definition of the Euler parameters given in Eq. [7.7.2] into Eq. [7.7.15], we obtain the relation between the vector $\mathbf{b}_1$ of the rotated axes and the unit vectors of the original coordinate frame in terms of the Euler parameters as

$$\mathbf{b}_1 = (2e_0^2 - 1 + 2e_1^2)\mathbf{a}_1 + (2e_1e_2 + 2e_0e_3)\mathbf{a}_2 + (2e_1e_3 - 2e_0e_2)\mathbf{a}_3 \qquad [7.7.16]$$

One can repeat the procedure used above to relate the vectors $\mathbf{b}_2$ and $\mathbf{b}_3$ to $\mathbf{a}_1$, $\mathbf{a}_2$, and $\mathbf{a}_3$ by the Euler parameters. The results can be expressed in matrix form as

$$\begin{bmatrix} \mathbf{b}_1 \\ \mathbf{b}_2 \\ \mathbf{b}_3 \end{bmatrix} = (2e_0^2 - 1)\begin{bmatrix} \mathbf{a}_1 \\ \mathbf{a}_2 \\ \mathbf{a}_3 \end{bmatrix} + 2\begin{bmatrix} e_1^2 & e_1 e_2 & e_1 e_3 \\ e_1 e_2 & e_2^2 & e_2 e_3 \\ e_1 e_3 & e_2 e_3 & e_3^2 \end{bmatrix}\begin{bmatrix} \mathbf{a}_1 \\ \mathbf{a}_2 \\ \mathbf{a}_3 \end{bmatrix} - 2e_0 \begin{bmatrix} 0 & -e_3 & e_2 \\ e_3 & 0 & -e_1 \\ -e_2 & e_1 & 0 \end{bmatrix}\begin{bmatrix} \mathbf{a}_1 \\ \mathbf{a}_2 \\ \mathbf{a}_3 \end{bmatrix}$$

[7.7.17]

Introducing the column vector $\{e\} = [e_1 \ e_2 \ e_3]^T$, we can write Eq. [7.7.17] as

$$\begin{bmatrix} \mathbf{b}_1 \\ \mathbf{b}_2 \\ \mathbf{b}_3 \end{bmatrix} = \left((2e_0^2 - 1)[1] + 2\{e\}\{e\}^T - 2e_0[\tilde{e}]\right)\begin{bmatrix} \mathbf{a}_1 \\ \mathbf{a}_2 \\ \mathbf{a}_3 \end{bmatrix} \qquad [\textbf{7.7.18}]$$

The first two terms on the right side of Eq. [7.7.18] are symmetric matrices, while the third term is a skew-symmetric matrix. A comparison of Eq. [7.7.18] and Eq. [7.7.7] indicates that the direction cosine matrix between the coordinate frames A and B, in terms of the Euler parameters, has the form

$$[c]^T = [R] = \left((2e_0^2 - 1)[1] + 2\{e\}\{e\}^T - 2e_0[\tilde{e}]\right)$$

$$= \begin{bmatrix} e_0^2 + e_1^2 - e_2^2 - e_3^2 & 2(e_1 e_2 + e_0 e_3) & 2(e_1 e_3 - e_0 e_2) \\ 2(e_1 e_2 - e_0 e_3) & e_0^2 - e_1^2 + e_2^2 - e_3^2 & 2(e_2 e_3 + e_0 e_1) \\ 2(e_1 e_3 + e_0 e_2) & 2(e_2 e_3 - e_0 e_1) & e_0^2 - e_1^2 - e_2^2 + e_3^2 \end{bmatrix} \qquad [\textbf{7.7.19}]$$

The elements of the direction cosine matrix are quadratic expressions in terms of the Euler parameters. When using Euler angles, the direction cosine matrix $[c]$, given in Eq. [7.5.3], contains trigonometric and other nonlinear terms, and it leads to singularities. There are no singularities in $[c]$ when it is expressed in terms of the Euler parameters. For any possible orientation of two coordinate frames there exists, and it is possible to find, associated Euler parameters. By contrast, every Euler angle transformation sequence has a nonlinearity at which point one cannot determine the values of the Euler angles.

7.7.2 Relating the Euler Parameters to Angular Velocities

In Section 7.4 we derived the relationship between the angular velocities of the coordinate frame and the direction cosines in Eq. [7.4.13], which we repeat here as

$$[{}^A_B\tilde{\omega}^B] = [c]^T[\dot{c}] \qquad [\textbf{7.7.20}]$$

Equation [7.7.19] gives the direction cosine matrix in terms of the Euler parameters. We use those results to obtain the angular velocities in terms of the Euler parameters. For example, let us calculate the component of the angular velocity in the b_1 direction, ω_1. From Eq. [7.4.13] ω_1 is the 3,2 element of $[{}^A_B\tilde{\omega}^B]$, and it can be obtained by multiplying the third row of $[c]^T$ with the second column of $[\dot{c}]$. Mathematically,

$$\omega_1 = \sum_{j=1}^{3} c_{j3}\dot{c}_{j2} \qquad [\textbf{7.7.21}]$$

Substituting the values for c_{ij} from Eq. [7.7.19] we obtain

$$\omega_1 = 4(e_1 e_3 + e_0 e_2)(e_1 \dot{e}_2 + e_2 \dot{e}_1 - e_0 \dot{e}_3 - e_3 \dot{e}_0)$$
$$+ 4(e_2 e_3 - e_0 e_1)(e_0 \dot{e}_0 - e_1 \dot{e}_1 + e_2 \dot{e}_2 - e_3 \dot{e}_3)$$
$$+ 2(e_0^2 - e_1^2 - e_2^2 + e_3^2)(e_2 \dot{e}_3 + e_3 \dot{e}_2 + e_0 \dot{e}_1 + e_1 \dot{e}_0) \quad \textbf{[7.7.22]}$$

Using the identity of Eq. [7.7.3] and its derivative, that is, $e_0^2 + e_1^2 + e_2^2 + e_3^2 = 1$, $e_0 \dot{e}_0 + e_1 \dot{e}_1 + e_2 \dot{e}_2 + e_3 \dot{e}_3 = 0$, Eq. [7.7.22] reduces to

$$\omega_1 = [{}^A_B \tilde{\omega}^B]_{23} = 2[-e_1 \dot{e}_0 + e_0 \dot{e}_1 + e_3 \dot{e}_2 - e_2 \dot{e}_3] \quad \textbf{[7.7.23]}$$

Similar expressions can be developed for ω_2 and ω_3. Introducing the vectors $\{\omega^*\} = [0 \ \omega_1 \ \omega_2 \ \omega_3]^T$ and $\{e\} = [e_0 \ e_1 \ e_2 \ e_3]^T$, we can express them in matrix form as

$$\begin{bmatrix} 0 \\ \omega_1 \\ \omega_2 \\ \omega_3 \end{bmatrix} = 2 \begin{bmatrix} e_0 & e_1 & e_2 & e_3 \\ -e_1 & e_0 & e_3 & -e_2 \\ -e_2 & -e_3 & e_0 & e_1 \\ -e_3 & e_2 & -e_1 & e_0 \end{bmatrix} \begin{bmatrix} \dot{e}_0 \\ \dot{e}_1 \\ \dot{e}_2 \\ \dot{e}_3 \end{bmatrix} \quad \textbf{[7.7.24]}$$

or, in compact form

$$\{\omega^*\} = 2[E^*]\{\dot{e}\} \quad \textbf{[7.7.25]}$$

where the notation is obvious. One can show that the coefficient matrix $[E^*]$ is orthonormal, that is, its inverse is equal to its transpose, $[E^*]^{-1} = [E^*]^T$. Using this property one can write

$$\{\dot{e}\} = 0.5[E^*]^{-1}\{\omega^*\} = 0.5[E^*]^T \{\omega^*\} \quad \textbf{[7.7.26]}$$

and can rearrange this as

$$\begin{bmatrix} \dot{e}_0 \\ \dot{e}_1 \\ \dot{e}_2 \\ \dot{e}_3 \end{bmatrix} = 0.5 \begin{bmatrix} 0 & -\omega_1 & -\omega_2 & -\omega_3 \\ \omega_1 & 0 & \omega_3 & -\omega_2 \\ \omega_2 & -\omega_3 & 0 & \omega_1 \\ \omega_3 & \omega_2 & -\omega_1 & 0 \end{bmatrix} \begin{bmatrix} e_0 \\ e_1 \\ e_2 \\ e_3 \end{bmatrix} \quad \textbf{[7.7.27]}$$

The equations relating the Euler parameters to their derivatives are linear, and they contain no singularities, making them easier to deal with both analytically and computationally. If Eq. [7.7.27] is compared with Eqs. [7.5.9], the derivatives of the Euler angles, the resulting simplification becomes clear. We also observe that the coefficient matrix in Eq. [7.7.27] is skew symmetric.

Noting that the first row of Eq. [7.7.24] is essentially $0 = 0$, we can generate a new matrix $[E]$ of order 3×4 by eliminating the first row of $[E^*]$, writing

$$[E] = \begin{bmatrix} -e_1 & e_0 & e_3 & -e_2 \\ -e_2 & -e_3 & e_0 & e_1 \\ -e_3 & e_2 & -e_1 & e_0 \end{bmatrix} \quad \textbf{[7.7.28]}$$

Thus, Eq. [7.7.25] can be expressed as

$$\{\omega\} = 2[E]\{\dot{e}\} \quad \textbf{[7.7.29]}$$

The matrix $[E]$ has a few interesting properties. One can show that

$$[E][E]^T = [1] \qquad [\tilde{\omega}] = 2[E][\dot{E}]^T \qquad [E]\{e\} = \{0\} \qquad \textbf{[7.7.30a,b,c]}$$

However, note that $[E]^T[E] \neq [1]$. Indeed, one can show that

$$[E]^T[E] = [1] - \{e\}\{e\}^T \qquad \textbf{[7.7.31]}$$

The relationships discussed so far have been for the Euler parameters that describe the transformation from one coordinate system to another. We next explore the relations among the Euler parameters when there is more than one transformation. We begin with a frame A and transform it to arrive at frame B. We denote the vector of Euler parameters that describe this transformation by $\{e\}$. Let us assume that the transformation from A to B is carried out in two steps: first a transformation from A to frame A' with associated Euler parameters $\{e'\}$, and then a second transformation from A' to frame B, with associated Euler parameters $\{e''\}$. We wish to relate $\{e\}$ to $\{e'\}$ and $\{e''\}$.

Denoting by $[E'^*]$ the counterpart of the $[E^*]$ matrix in Eq. [7.7.24] for the Euler parameters $\{e'\}$, we can show that the transformed Euler parameters obey the relationship

$$\{e\} = [E'^*]^T\{e''\} \qquad \textbf{[7.7.32]}$$

This relation can be derived by taking the coordinate transformation matrices between the A, A', and B frames.

7.7.3 Relating the Euler Parameters to the Direction Cosines

This process is in essence the reverse of the process in Sec. 7.7.1. As it is an inversion, the task can be accomplished in a number of ways. One way is to make use of the property of the direction cosine matrix and to note that the direction cosines of the principal line are elements of the eigenvector associated with the eigenvalue 1. The procedure to follow is this: Obtain the eigenvector matrix corresponding to the eigenvalue 1, and calculate the rotation angle Φ from Eq. [7.3.17].

Another and more straightforward way is to use Eqs. [7.7.2] and [7.7.19], which yield

$$c_{21} - c_{12} = 4e_0 e_3 = 4c_3 \cos\frac{\Phi}{2}\sin\frac{\Phi}{2} = 2c_3 \sin\Phi$$

$$c_{13} - c_{31} = 4e_0 e_2 = 4c_2 \cos\frac{\Phi}{2}\sin\frac{\Phi}{2} = 2c_2 \sin\Phi$$

$$c_{32} - c_{23} = 4e_0 e_1 = 4c_1 \cos\frac{\Phi}{2}\sin\frac{\Phi}{2} = 2c_1 \sin\Phi$$

$$c_{11} + c_{22} + c_{33} = 3e_0^2 - e_1^2 - e_2^2 - e_3^2 = 4\cos^2\frac{\Phi}{2} - 1 = 4e_0^2 - 1 \qquad \textbf{[7.7.33]}$$

These equations can be manipulated in a number of ways to yield expressions for $\cos\Phi$ and $\sin\Phi$. The goal in this algebraic exercise is to come up with relations that do not have singularity problems. For example, we can obtain an expression for $\sin\Phi$ directly from the first of Eq. [7.7.33] in the form

$$\sin\Phi = \frac{c_{21} - c_{12}}{2c_3} \quad [7.7.34]$$

which has a singularity when $c_3 = 0$. On the other hand, multiplying the first three of Eqs. [7.7.33] with c_3, c_2, and c_1, respectively, and adding them yields

$$[c_3(c_{21} - c_{12}) + c_2(c_{13} - c_{31}) + c_1(c_{32} - c_{23})] = 2\sin\Phi(c_3^2 + c_2^2 + c_1^2) = 2\sin\Phi \quad [7.7.35]$$

which exhibits no singularities. Another suitable choice is to use the half-angle formula $\cos\Phi = 2\cos^2(\Phi/2) - 1$ with the last of Eq. [7.7.33], which yields

$$c_{11} + c_{22} + c_{33} = 4\cos^2\frac{\Phi}{2} - 1 = 1 + 2\cos\Phi \quad [7.7.36]$$

This is identical to Eq. [7.3.17] and leads to

$$\cos\Phi = \frac{1}{2}(c_{11} + c_{22} + c_{33} - 1) \quad [7.7.37]$$

A commonly used set of expressions, provided e_0 is not equal to or close to zero, is derived by manipulating the diagonal elements of Eq. [7.7.19] as

$$4e_0^2 = c_{11} + c_{22} + c_{33} + 1 \qquad 4e_1^2 = c_{11} - c_{22} - c_{33} + 1$$
$$4e_2^2 = -c_{11} + c_{22} - c_{33} + 1 \qquad 4e_3^2 = -c_{11} - c_{22} + c_{33} + 1 \quad [7.7.38]$$

Usually, one first finds from Eqs. [7.7.38] the Euler parameter with the largest magnitude. This larger quantity is used together with the off-diagonal elements of the direction cosine matrix in Eq. [7.7.19] to solve for the Euler parameters. For example, if e_0 is the largest in magnitude, one uses the value of e_0 from the first of Eq. [7.7.38] and finds the remaining Euler parameters using Eqs. [7.7.33] as

$$e_0 = \frac{1}{2}\sqrt{c_{11} + c_{22} + c_{33} + 1} \qquad e_1 = \frac{c_{32} - c_{23}}{4e_0}$$
$$e_2 = \frac{c_{13} - c_{31}}{4e_0} \qquad e_3 = \frac{c_{21} - c_{12}}{4e_0} \quad [7.7.39]$$

7.7.4 RELATING THE EULER PARAMETERS TO THE EULER ANGLES AND VICE VERSA

This task can be accomplished by generating the direction cosine matrix for a specific Euler angle transformation, and using the relations given in Eqs. [7.7.39] for relating the Euler parameters to the direction cosines. The results, of course, vary from one set of transformation to another. We dispense with the algebra and give the results for a 3-1-3 Euler angle rotation sequence with the transformation angles ϕ, θ, ψ, which are

$$e_0 = \cos\frac{\phi+\psi}{2}\cos\frac{\theta}{2} \qquad e_1 = \cos\frac{\phi-\psi}{2}\sin\frac{\theta}{2}$$

$$e_2 = \sin\frac{\phi-\psi}{2}\sin\frac{\theta}{2} \qquad e_3 = \sin\frac{\phi+\psi}{2}\cos\frac{\theta}{2} \qquad [7.7.40]$$

Results for the 3-2-3 and 3-2-1 sequences are given in Table 7.1 at the end of the chapter. Once the Euler parameters are found, one can use Eq. [7.7.19] to calculate the direction cosine matrix. The reverse problem—determination of the Euler angles from the Euler parameters—is more complicated. It can be accomplished in a number of ways. In one way, we calculate the Euler parameters and the direction cosine matrix in terms of the Euler parameters. Then, we make use of the algebraic relations in the direction cosine matrix in terms of the Euler angles. This procedure is usually tedious.

Another approach for finding the Euler angles from the Euler parameters is by direct use of Eq. [7.7.40] or its counterpart for other sequences. For a 3-1-3 sequence we manipulate the Euler parameters as

$$\frac{e_3}{e_0} = \tan\frac{\phi+\psi}{2} \qquad \frac{e_2}{e_1} = \tan\frac{\phi-\psi}{2} \qquad [7.7.41]$$

After taking the inverse tangents of the above terms and manipulating, we obtain

$$\phi = \tan^{-1}\left(\frac{e_3}{e_0}\right) + \tan^{-1}\left(\frac{e_2}{e_1}\right) \qquad \psi = \tan^{-1}\left(\frac{e_3}{e_0}\right) - \tan^{-1}\left(\frac{e_2}{e_1}\right) \qquad [7.7.42]$$

To determine the exact values of these angles, that is, the quadrants they lie in, we can use Eqs. [7.7.40], or we can compare the results with the entries of the direction cosine matrix. The angle θ can be found by noting that

$$e_1^2 + e_2^2 = \sin^2\left(\frac{\theta}{2}\right) \qquad [7.7.43]$$

so that

$$\theta = 2\sin^{-1}\left(\sqrt{e_1^2 + e_2^2}\right) \qquad [7.7.44]$$

The same approach is used for other Euler angle sequences.

7.7.5 OTHER CONSIDERATIONS

Yet another advantage of using Euler parameters arises when conducting numerical computations. Because the Euler parameters are related to each other by Eq. [7.7.3], Eq. [7.7.3] can be used as a constant of the kinematic equations of motion, and it can be computed at each step of the numerical integration to check for accuracy and numerical stability. Other constants associated with this analysis are the diagonal elements of Eq. [7.7.20], which should be zero, by virtue of the skew symmetry of the angular velocity matrix. The values of the diagonal elements, $[{}_B^A\tilde{\omega}^B]_{ii}$ ($i = 1, 2, 3$), can be verified using

$$\sum_{j=1}^{3} c_{jk}\dot{c}_{jk} = 0 \qquad k = 1, 2, 3 \qquad [7.7.45]$$

These identities can also be checked to verify the accuracy of the numerical computation.

Disadvantages associated with using Euler parameters include the need to use four parameters and four differential equations instead of three for the case of Euler angles. However, the linearity of the resulting equations and lack of singularities far outweigh the complexities introduced by the additional differential equation. Another disadvantage is that Euler parameters do not lend themselves to a simple physical interpretation as Euler angles do. But here, too, one can always calculate the values of the Euler angles from the Euler parameters at every instant of the motion and seek a physical interpretation or visualization.

We conclude the discussion of Euler parameters with this important generalization and an additional definition of angular velocity. Assume we have a vector $\mathbf{r}$ and we are rotating $\mathbf{r}$ by an angle Φ about an axis that passes through the origin. Let $\mathbf{n}$ be the unit vector in the direction of the axis of rotation and $\mathbf{r}'$ be the rotated vector. We can write the relationship between $\mathbf{r}$ and $\mathbf{r}'$ by substituting $\mathbf{r}$ for $\mathbf{a}_1$ and $\mathbf{r}'$ for $\mathbf{b}_1$ in Eq. [7.7.14], which yields

$$\mathbf{r}' = (1 - \cos\Phi)(\mathbf{n}\cdot\mathbf{r})\mathbf{n} + \cos\Phi\,\mathbf{r} + \sin\Phi\,\mathbf{n}\times\mathbf{r} \qquad [7.7.46]$$

where we recognize that $c_1 = \cos\theta_1$ in Eq. [7.7.14] is equivalent to $\mathbf{n}\cdot\mathbf{r}/|\mathbf{r}|$. This equation is a very useful general relation for determining the trajectory of a vector that is being rotated. A special application of Eq. [7.7.46] is for small rotations over small time intervals for finding the rate at which $\mathbf{r}$ changes. Indeed, as Φ approaches zero, we have

$$\Phi \to d\Phi \qquad \cos\Phi \to 1 \qquad \sin\Phi \to d\Phi \qquad \mathbf{r}' - \mathbf{r} \to d\mathbf{r}$$

which, when substituted into Eq. [7.7.46], yields

$$d\mathbf{r} = d\Phi(\mathbf{n}\times\mathbf{r}) \qquad [7.7.47]$$

Because the infinitesimal rotation is taken about a single axis at that particular instant, we recognize that division of $d\Phi$ by dt yields the angular velocity expression in the form

$$\frac{d\Phi}{dt} = \omega \qquad \boldsymbol{\omega} = \omega\mathbf{n} = \frac{d\Phi}{dt}\mathbf{n} \qquad [7.7.48a,b]$$

Equation [7.7.48b] is yet another definition of the angular velocity vector. Dividing Eq. [7.7.47] by dt and using Eq. [7.7.48b] we obtain

$$\frac{d\mathbf{r}}{dt} = \boldsymbol{\omega}\times\mathbf{r} \qquad [7.7.49]$$

We have hence showed another way at arriving at the angular velocity of a reference frame and the relation describing how the angular velocity of a reference frame affects the rate of change of a vector fixed in that reference frame.

Example 7.7

Consider the transformation sequence in Example 7.1. Obtain the associated Euler parameters, and substitute them into Eq. [7.7.19] to calculate the values of $[R]$ to confirm your results.

Solution

From Example 7.1, the transformation matrix is

$$[R] = \begin{bmatrix} 0.6124 & 0.3536 & -0.7071 \\ -0.5000 & 0.8660 & 0 \\ 0.6124 & 0.3536 & 0.7071 \end{bmatrix} \quad \text{[a]}$$

The rotation angle was calculated as $\Phi = 53.64°$ and the direction cosines of the principal line were found to be

$$c_1 = -0.2195 \qquad c_2 = 0.8192 \qquad c_3 = 0.5299 \quad \text{[b]}$$

The sine and cosine of half the principal angle are

$$\sin\frac{\Phi}{2} = 0.4512 \qquad \cos\frac{\Phi}{2} = 0.8924 \quad \text{[c]}$$

The Euler parameters for this problem become

$$e_0 = \cos\frac{\Phi}{2} = 0.8924 \qquad e_1 = c_1 \sin\frac{\Phi}{2} = -0.2195(0.4512) = -0.09904$$

$$e_2 = c_2 \sin\frac{\Phi}{2} = 0.8192(0.4512) = 0.3696 \qquad e_3 = c_3 \sin\frac{\Phi}{2} = 0.5299(0.4512) = 0.2391 \quad \text{[d]}$$

We next calculate the entries of the transformation matrix. For example, for R_{11} and R_{23} we have, from Eq. [7.7.19],

$$R_{11} = 2(e_0^2 + e_1^2 - 0.5) = 2(0.8924^2 + 0.09904^2 - 0.5) = 0.6124$$

$$R_{23} = 2e_2e_3 + 2e_0e_1 = 2(0.3696)(0.2391) - 2(0.8924)(0.09904) = 0 \quad \text{[e]}$$

One can show that the rest of the entries of $[R]$ can be found using Eq. [7.7.19] as well. Now we will calculate the Euler parameters without using the eigensolution of $[R]$. To this end, it is preferable to find the largest of the Euler parameters and use calculations that involve division by that parameter. We can accomplish this by looking at the sines and cosines of half the principal angle. From Example 7.1 or using Eq. [7.7.35] we determine that $\Phi = 53.64°$. In Eq. [c] we determined the sine and cosine of the half angle. From that information, we conclude that $e_0 = \cos(\Phi/2) = 0.8924$ is the largest of the Euler parameters. We find the remaining Euler parameters using Eqs. [7.7.39] as

$$e_1 = \frac{c_{32} - c_{23}}{4e_0} = \frac{R_{23} - R_{32}}{4e_0} = \frac{(0 - 0.3536)}{4 \times 0.8924} = -0.09906$$

$$e_2 = \frac{c_{13} - c_{31}}{4e_0} = \frac{0.6124 + 0.7071}{4 \times 0.8924} = 0.3696$$

$$e_3 = \frac{c_{21} - c_{12}}{4e_0} = \frac{0.3536 + 0.5}{4 \times 0.8924} = 0.2391 \quad \text{[f]}$$

Note that there is a slight difference in the fourth significant figure of some of the results in Eqs. [d] and Eqs. [f], due to roundoff error.

7.8 CONSTRAINED MOTION OF RIGID BODIES, INTERCONNECTIONS

Constrained motion takes place because of contact between one or more points on one body with a point or a surface on another body. Such contact between two bodies can take place in a number ways:

Motion restriction or transmission through sliding and interconnections.
Impact.
Rolling.

In this section, we analyze constrained motion from a physical perspective and the interconnections that obtain such motion. Chapter 8 will analyze impact of bodies, as it cannot be treated using kinematics alone. And Section 7.9 will consider rolling.

The number of degrees of freedom of a system is equal to the number of coordinates minus the number of constraints. Constraints generated by contact of bodies reduce the number of degrees of freedom. To describe the general three-dimensional motion of a rigid body, one needs six degrees of freedom. Plane motion is a three degree of freedom problem.

A set of two or more bodies in contact with each other for the purpose of transmission of motion from one of the bodies to another is called a *mechanism*. Components of a mechanism are generally referred to as *links*.

7.8.1 ROTATIONAL CONTACT, JOINTS

A joint basically prevents the bodies it connects to translate with respect to each other, while permitting some type of rotational motion between the two bodies. The most common joint is the *pin joint,* also known as a *revolute* joint, shown in Fig. 7.19. The contact is over the external surface of the pin. The angular velocity of the second body, *as viewed from the first body,* is denoted by ω_2 and is along the axis of the pin. We express ω_2 as

$$\omega_2 = \omega_2 \mathbf{n} \qquad [7.8.1]$$

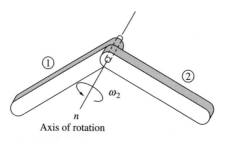

Figure 7.19   A revolute joint

where **n** is the unit vector along the axis of the pin. The constraint relation can be expressed as

$$\boldsymbol{\omega}_2 \times \mathbf{n} = \mathbf{0} \qquad [7.8.2]$$

which represents two rotational constraint equations.[2] Because the joint does not permit translational motion at all, the total number of constraints imposed is five. Hence, a rigid body connected to a system of rigid bodies by a revolute joint increases the degrees of freedom of the resulting system by one.

Denoting the angular velocity of the first body by $\boldsymbol{\omega}_1$, we can express the total angular velocity of the second body as the sum of the two angular velocities, thus

$$\boldsymbol{\omega} = \boldsymbol{\omega}_1 + \boldsymbol{\omega}_2 = \boldsymbol{\omega}_1 + \omega_2 \mathbf{n} \qquad [7.8.3]$$

Differentiation of Eq. [7.8.3] yields the angular acceleration expression as

$$\dot{\boldsymbol{\omega}} = \dot{\boldsymbol{\omega}}_1 + \dot{\omega}_2 \mathbf{n} + \boldsymbol{\omega}_1 \times \omega_2 \mathbf{n} \qquad [7.8.4]$$

For multilink bodies in plane motion, one can use *Gruebler's equation* to determine the number of degrees of freedom. The formula states that the number of degrees of freedom p is related to the number of active links and number of joints by

$$p = 3k - 2f$$

where k is the number of active links and f is the number of joints. For example, in the linkage shown in Fig. 4.41, there are five active links and six joints, leading to the conclusion that there are three degrees of freedom. Gruebler's equation can be extended to different types of joints as well. The interested reader is referred to texts on kinematics or mechanism theory. It should be noted that Gruebler's equation does not take into consideration the lengths of the links. One can find geometries where Gruebler's equation states that there are zero degrees of freedom and the mechanism can move.

Another commonly used joint is a *ball and socket joint,* also known as a *globular* or *spherical pair.* The joint is shown in Fig. 7.20. Ball and socket joints do not place a restriction on the angular velocity of the second body as viewed from the first. The angular velocity $\boldsymbol{\omega}_2$ can have components in any direction. Such joints impose three constraints that prevent translational motion at the joint.

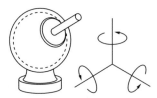

Figure 7.20   A ball and socket joint

---

[2] We deduce that Eq. [7.8.2] mathematically represents two constraint equations by considering Section 7.2. Equation [7.8.2] can be written in column vector notation as $[\tilde{\omega}]\{n\} = \{0\}$, which represents three scalar equations. Because $[\tilde{\omega}]$ is a singular matrix with rank 2, only two of the equations are independent.

7.8 CONSTRAINED MOTION OF RIGID BODIES, INTERCONNECTIONS

An important use of joints is in power transmission. In machine dynamics it is often necessary to transfer the rotational motion of a shaft onto another shaft rotating in a different direction. A common mechanism used to accomplish this is a *universal joint*. There are several kinds of universal joints available, such as the Cardan, Rzeppa, Weiss, and Devos joints. All these joints accomplish the task of power transmission in different ways; each is useful for specific engineering applications.

The Cardan joint shown in Fig. 7.21 is one of the common and low-cost universal joints. The joint consists of two shafts attached to forks, known as *yokes*, linked to each other by a cross. The two shafts are called the *input* and *output shafts*. The relative position of the cross with respect to the yokes can change, as the cross can rotate inside each of the yokes, making it possible to align the two shafts in almost any direction. This gives the universal joint its tremendous versatility and its name, because it can operate in transmission problems where the shafts connected to the yokes are not aligned, vibrate, or change orientation.

A universal joint is a single degree of freedom mechanism, because the angular velocity of one of the shafts determines the angular velocity of the other. To explore the relationship between the two angular velocities we split the universal joint into two parts, A and B, as shown in Fig. 7.22. We select the coordinate axes such that both shafts lie on the xy plane. The input shaft, shaft B, is aligned with the x direction, and its angular velocity is $-\dot{\theta}\mathbf{i}$, with θ measured from the z axis. We rotate the xy axes about the z direction by an angle of β to get the $x'y'$ axes and place the output shaft A along the negative x' axis. The angular velocity of the output shaft is $-\dot{\phi}\mathbf{i}'$, with ϕ measured from the z axis as well. We denote the unit vectors along the sides of the cross as $\mathbf{a}$ and $\mathbf{b}$, respectively, and express them as

$$\mathbf{b} = \sin\theta\mathbf{j} + \cos\theta\mathbf{k}$$

$$\mathbf{a} = \sin\phi\mathbf{j}' + \cos\phi\mathbf{k} = -\sin\phi\sin\beta\mathbf{i} + \sin\phi\cos\beta\mathbf{j} + \cos\phi\mathbf{k} \quad [\mathbf{7.8.5}]$$

It follows that the two unit vectors must be perpendicular to each other. The dot product between the two gives the constraint relation between the angles θ and ϕ.

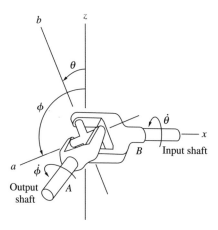

Figure 7.21 A Cardan joint

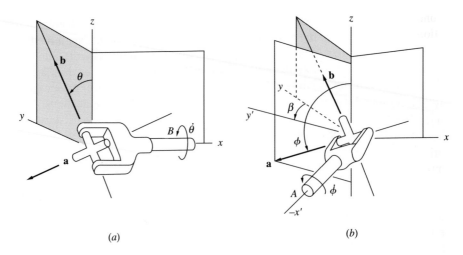

Figure 7.22

We have

$$\mathbf{a} \cdot \mathbf{b} = \sin\theta \sin\phi \cos\beta + \cos\theta \cos\phi = 0 \qquad [7.8.6]$$

It is of interest to compare the angular velocities of the input and output shafts. We express $\dot{\phi}$ in terms of $\dot{\theta}$ and θ. We rewrite Eq. [7.8.6] with the ϕ and θ terms on opposite sides of the equation as

$$\tan\phi = \frac{\sin\phi}{\cos\phi} = -\frac{\cos\theta}{\sin\theta \cos\beta} \qquad [7.8.7]$$

which, when differentiated, yields

$$\frac{\dot{\phi}}{\cos^2\phi} = \frac{\dot{\theta}}{\sin^2\theta \cos\beta} \qquad [7.8.8]$$

To eliminate ϕ from this equation we make use of the identity $\cos^2\phi = 1/(1 + \tan^2\phi)$. Substituting the value of $\tan\phi$ from Eq. [7.8.7] into this identity and using it with Eq. [7.8.8], we obtain the relation between the angular velocities as

$$\dot{\phi} = \frac{\cos\beta}{1 - \sin^2\theta \sin^2\beta}\dot{\theta} \qquad [7.8.9]$$

The angular velocities of the input and output shafts are not linearly related to each other, except when $\beta = 0$, in which case the joints are aligned. The Cardan joint is not what is called a *constant velocity joint,* making it impractical to use in applications such as the steering of front wheel drive automobiles, as it would lead to slipping tires and large stresses on power transmission components. In the extreme case, when $\beta = 90°$, the two shafts are perpendicular and no motion can be transmitted. This is known as *gimbal lock.* However, if two Cardan joints are used to join two parallel shafts, the angular velocities of the two parallel shafts will be the same. It should also be noted that there are several constant velocity joints available such as the ones discussed earlier. Many of these joints provide a near constant velocity relationship between the input and output shafts for a specific range of operation.

7.8.2 TRANSLATIONAL MOTION AND SLIDING CONTACT

Here one part of a body slides over another body. The simplest form of sliding is to have a point, a line, or a surface on a body sliding on a surface. This surface is the contact plane. The associated constraint is the same as the very first constraint relations we studied in Chapter 4. The velocity of the contact point does not have a component normal to the plane of contact, and the constraint can be written as

$$\mathbf{v}_P \cdot \mathbf{n} = 0 \qquad [7.8.10]$$

where $\mathbf{n}$ is the unit vector in the normal direction at the point of contact. If more than one line on the rigid body slides on a surface, then the angular velocity of the body can only have a component perpendicular to the plane, so that $\boldsymbol{\omega}$ is parallel to $\mathbf{n}$. The resulting two constraints can be described by

$$\boldsymbol{\omega} \times \mathbf{n} = \mathbf{0} \qquad [7.8.11]$$

A special case of sliding on a surface is sliding in a guide or in a slot (Fig. 7.23). The joint formed by the sliding components is referred to as a *prismatic* joint. In this case translational motion is possible in only one direction, and there are two constraints on the translational motion. Denoting the unit vector in the direction of the guide by $\mathbf{h}$, we can write the constraint as

$$\mathbf{v}_P \times \mathbf{h} = \mathbf{0} \qquad [7.8.12]$$

where we note that Eq. [7.8.12] represents two constraint relations. Further, no rotation is permitted, so a prismatic joint represents a total of five constraint equations. A prismatic joint for translational motion is basically the equivalent of a revolute joint for rotational motion.

In another form of sliding contact, a collar slides on a bar while turning around it. The simplest case here is when the collar is not attached to another body, as shown in Fig. 7.24. Such a connection is referred to as a *slider*. Denoting the unit vector along the rod by $\mathbf{h}$, we note that the collar can only have a velocity in the direction of $\mathbf{h}$, $\mathbf{v}_P = v_P \mathbf{h}$ so that the constraint associated with the translational motion is Eq. [7.8.12].

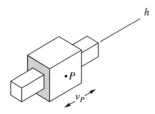

Figure 7.23   A prismatic joint

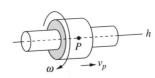

Figure 7.24   A slider

In addition, the collar can only have an angular velocity about the axis of the rod, $\boldsymbol{\omega}_{\text{collar}} = \omega_{\text{collar}}\mathbf{h}$, so that the constraint associated with the rotational motion is

$$\boldsymbol{\omega}_{\text{collar}} \times \mathbf{h} = \mathbf{0} \qquad [7.8.13]$$

There are a total of four constraints restricting the motion, so the slider has two degrees of freedom: one translational and one rotational.

7.8.3 Combined Sliding and Rotation

In this case, a body is attached to a slider by means of a pin, fork, or universal joint, as depicted in Fig. 7.25. These two elements in contact (rod and slider) are also referred to as a *cylindrical pair*. The collar is sliding on and turning around a guide bar. Consider the coordinate frame xyz where the x axis is aligned with the collar. The angular velocity of the collar becomes $\boldsymbol{\omega}_{\text{collar}} = \omega_{\text{collar}}\mathbf{i}$. The coordinate axes $xy'z'$ are obtained by rotating the xyz axes about the x axis by an angle ϕ. It follows that $\omega_{\text{collar}} = \dot{\phi}$. The xy' plane defines the plane of the collar to which the pin joint is attached. The body attached to the collar can have an angular velocity in the direction perpendicular to the plane of the collar, in the z' direction. We can then write

$$\boldsymbol{\omega}_{\text{rod}} = \dot{\phi}\mathbf{i} + \dot{\theta}\mathbf{k}' = \omega_{\text{collar}}\mathbf{i} + \omega_{\text{rod/collar}}\mathbf{k}' = \omega_{\text{collar}}\mathbf{i}' + \omega_{\text{rod/collar}}\mathbf{k}' \qquad [7.8.14]$$

Thus, the angular velocity of the rod cannot have a component in the y' direction. The constraint for the total angular velocity can then be written as

$$\boldsymbol{\omega}_{\text{rod}} \cdot \mathbf{j}' = 0 \qquad [7.8.15]$$

The total number of constraints on the rod becomes three, two for the translational motion and one for the rotational. The angular acceleration of the rod can be obtained by differentiating Eq. [7.8.14],

$$\boldsymbol{\alpha}_{\text{rod}} = \alpha_{\text{collar}}\mathbf{i}' + \alpha_{\text{rod/collar}}\mathbf{k}' + \omega_{\text{collar}}\mathbf{i}' \times \omega_{\text{rod/collar}}\mathbf{k}' = \ddot{\phi}\mathbf{i}' + \ddot{\theta}\mathbf{k}' - \dot{\phi}\dot{\theta}\mathbf{j}' \qquad [7.8.16]$$

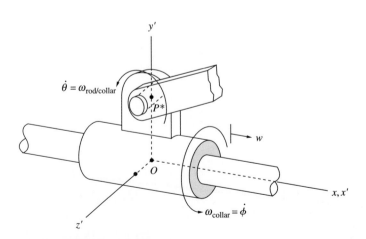

Figure 7.25 A cylindrical pair

7.8 CONSTRAINED MOTION OF RIGID BODIES, INTERCONNECTIONS

When the collar is attached to the rod by a ball and socket joint, there is no restriction on the angular motion of the rod with respect to the collar. The rod then has four degrees of freedom: one translational and three rotational.

As an illustration of the constraint relations for combined sliding and rotation, first consider plane motion and the rod AB where points A and B rest on collars that slide on bars perpendicular to each other, as shown in Fig. 7.26. Each collar introduces one constraint on the motion, resulting in a one degree of freedom system. The two constraint equations are

$$\mathbf{v}_A \cdot \mathbf{i} = 0 \qquad \mathbf{v}_B \cdot \mathbf{j} = 0 \qquad [7.8.17]$$

These constraint relations can be expressed in different forms as well. Also, the rotational constraint relations associated with the pin did not come into the picture, because the rotational motion of the rod is restricted to be perpendicular to the plane of motion.

We next extend the discussion to the three-dimensional case, where we consider a rod AB in which points A and B are attached to collars and the collars slide on perpendicular guide bars (Fig. 7.27). However, in this case the guide bars are not on the same plane.

Let us analyze the attachments of the points to the collars. If both points A and B are attached to the collars with pin joints, we end up with six constraint equations, three for each pin joint. It follows that the resulting system has zero degrees of freedom. No motion of the rod AB is possible.

When one of the joints is a ball and socket joint, say, the joint at A, the number of constraints is reduced to five, two translational constraints for joint A and three for joint B. The system has one degree of freedom. If the motion of any point on rod AB is specified, one can determine the characteristics of the motion of the rod, as well as the collars. In order to be able to do this we have to generate an expression for the angular velocity of the rod. It is preferable to work with collar B, as we will make use of the properties of the angular velocity of the rod with respect to the collar. From

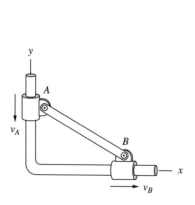

Figure 7.26 Guide bars on same plane

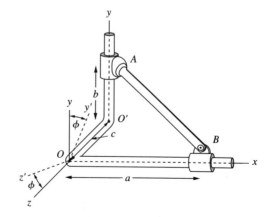

Figure 7.27 Guide bars not on same plane

Eq. [7.8.14] we can write

$$\omega_{rod} = \omega_{collar} + \omega_{rod/collar} \qquad [7.8.18]$$

The angular velocity of the collar B is

$$\omega_{collar} = \omega_{collar}\mathbf{i} \qquad [7.8.19]$$

The angular velocity of the rod as viewed from the pin joint is perpendicular to the plane defined by rod AB and the guide bar. The unit vector perpendicular to the plane of symmetry of collar B can be obtained by taking the cross product of any two vectors along rod AB and the guide bar. For example, taking the vectors as $\mathbf{r}_{B/A}$ and $\mathbf{r}_{B/O}$, we obtain the unit vector, denoted by $\mathbf{e}$, as

$$\mathbf{e} = \pm \frac{\mathbf{r}_{A/B} \times \mathbf{r}_{B/O}}{|\mathbf{r}_{A/B} \times \mathbf{r}_{B/O}|} \qquad [7.8.20]$$

The sign of the unit vector can be chosen by convenience. In the $xy'z'$ coordinate system (Fig. 7.27) the unit vector in the z' direction, $\mathbf{k}'$, can be found using the relation

$$\mathbf{k}' = \frac{\mathbf{i} \times \mathbf{r}_{A/B}}{|\mathbf{i} \times \mathbf{r}_{A/B}|} \qquad [7.8.21]$$

so that

$$\omega_{rod/collar} = \omega_{rod/collar}\mathbf{k}' \qquad [7.8.22]$$

Equation [7.8.18] can also be analyzed from an Euler angle point of view. Using a 1-3-1 coordinate transformation (similar to 3-1-3 but with x and z changing places, and we are starting with coordinates xyz and going into $x'y'z'$), the angular velocity of the collar becomes the precession with the angle ϕ and nutation with the angle θ, such that $\dot{\theta} = \omega_{rod/collar}$. Because of the constraints imposed on the rod, the precession and nutation rates are related to each other. There is no spin. When both the joints are ball and socket joints, the system has two degrees of freedom, with the spin of the rod about its own axis constituting the second degree of freedom.

Example 7.8

The rod AB in Fig. 7.27 is constrained to move between the horizontal and vertical guides. Given that the velocity of point A is 2 m/s downward, and $a = 6$ m, $b = 3$ m, $c = 4$ m, find the velocity of point B and the angular velocity of the rod.

Solution

We write the relative velocity expression for points A and B as

$$\mathbf{v}_B = \mathbf{v}_A + \omega_{rod} \times \mathbf{r}_{B/A} \qquad [a]$$

where

$$\mathbf{v}_A = -2\mathbf{j} \text{ m/s} \qquad \mathbf{v}_B = v_B\mathbf{i} \text{ m/s} \qquad \mathbf{r}_{B/A} = 6\mathbf{i} - 3\mathbf{j} + 4\mathbf{k} \text{ m} \qquad [b]$$

7.8 CONSTRAINED MOTION OF RIGID BODIES, INTERCONNECTIONS

With the unit vector **i** describing the direction along the collar B, we use Eq. [7.8.21] to find $\mathbf{k}'$, thus

$$\mathbf{k}' = \frac{\mathbf{i} \times \mathbf{r}_{A/B}}{|\mathbf{i} \times \mathbf{r}_{A/B}|} = \frac{3\mathbf{k} + 4\mathbf{j}}{5} \quad [\mathbf{c}]$$

so that the angular velocity of the rod can be expressed as

$$\boldsymbol{\omega}_{\text{rod}} = \omega_{\text{collar}}\mathbf{i} + \omega_{\text{rod/collar}}\mathbf{k}' = \omega_{\text{collar}}\mathbf{i} + \omega_{\text{rod/collar}}\frac{3\mathbf{k} + 4\mathbf{j}}{5} \quad [\mathbf{d}]$$

Comparing the elements of $\mathbf{k}'$ and Fig. 7.27 we conclude that ϕ has a negative value (clockwise rotation) for this problem. So does the angular velocity of the collar. We introduce Eqs. [b] and [d] into Eq. [a], which yields

$$v_B \mathbf{i} = -2\mathbf{j} + [\omega_{\text{collar}}\mathbf{i} + \omega_{\text{rod/collar}}\frac{1}{5}(3\mathbf{k} + 4\mathbf{j})] \times (6\mathbf{i} - 3\mathbf{j} + 4\mathbf{k})$$

$$= -2\mathbf{j} + \omega_{\text{collar}}(-3\mathbf{k} - 4\mathbf{j}) + \omega_{\text{rod/collar}}\frac{1}{5}(18\mathbf{j} + 9\mathbf{i} - 24\mathbf{k} + 16\mathbf{i}) \quad [\mathbf{e}]$$

Equating the components of Eq. [e] in the x, y, and z directions, we have a set of three equations and three unknowns ω_{collar}, $\omega_{\text{rod/collar}}$, and v_B, in the form

$$\mathbf{i} \text{ components} \rightarrow v_B = 5\omega_{\text{rod/collar}} \quad [\mathbf{f}]$$

$$\mathbf{j} \text{ components} \rightarrow 0 = -2 - 4\omega_{\text{collar}} + \frac{18}{5}\omega_{\text{rod/collar}} \quad [\mathbf{g}]$$

$$\mathbf{k} \text{ components} \rightarrow 0 = -3\omega_{\text{collar}} - \frac{24}{5}\omega_{\text{rod/collar}} \quad [\mathbf{h}]$$

Solving Eqs. [f]–[h], we obtain

$$\omega_{\text{rod/collar}} = \frac{1}{5} \text{ rad/s} \qquad \omega_{\text{collar}} = -\frac{8}{25} \text{ rad/s} \qquad v_B = 1 \text{ m/s} \quad [\mathbf{i}]$$

Note that we have three equations for the three unknowns, even though the system has a single degree of freedom.

Next, consider finding the angular velocity of the collar A. To accomplish this, we write the angular velocity of the rod in terms of the angular velocity of the collar B as

$$\boldsymbol{\omega}_{\text{rod}} = \omega_{\text{collar}B}\mathbf{i} + \omega_{\text{rod/collar}B}\frac{1}{5}(3\mathbf{k} + 4\mathbf{j}) = \frac{1}{25}(-8\mathbf{i} + 4\mathbf{j} + 3\mathbf{k}) \text{ rad/s} \quad [\mathbf{j}]$$

The angular velocity can also be expressed by noting that the angular velocity of collar A is in the y direction

$$\boldsymbol{\omega}_{\text{rod}} = \omega_{\text{collar}A}\mathbf{j} + \omega_{\text{rod/collar}A} = \omega_{\text{collar}A}\mathbf{j} + \frac{1}{25}[-8\mathbf{i} + (4 - 25\omega_{\text{collar}A})\mathbf{j} + 3\mathbf{k}] \text{ rad/s} \quad [\mathbf{k}]$$

Use of the relative velocity relation will not lend any more information, so we need to look at the nature of the angular velocity. Previously, we used a 1-3-1 Euler angle transformation to quantify the angular velocity and said that there is no spin associated with an axis along the length of the rod. When visualizing the angular motion with respect to collar A or collar B, we can then quantify this as the constraint $\boldsymbol{\omega}_{\text{rod/collar}A}$ and $\boldsymbol{\omega}_{\text{rod/collar}B}$ not having a component along the axis of rod AB. We can then write

$$\omega_{rod/collarA} \cdot \mathbf{r}_{B/A} = 0 \qquad \omega_{rod/collarB} \cdot \mathbf{r}_{B/A} = 0 \qquad \text{[l]}$$

The second of these equations is automatically satisfied. The first equation can be solved for the angular velocity of the collar. Carrying out the calculations, we obtain

$$[-8\mathbf{i} + (4 - 25\omega_{collarA})\mathbf{j} + 3\mathbf{k}] \cdot [6\mathbf{i} - 3\mathbf{j} + 4\mathbf{k}] = -48 - 12 + 75\omega_{collarA} + 12 = 0$$

$$\rightarrow \omega_{collarA} = \frac{16}{25} \text{ rad/s} \qquad \text{[m]}$$

7.9 ROLLING

An interesting case of constrained motion of rigid bodies is rolling. A body can roll over a fixed surface or over another body. For rolling to take place between two bodies, a continuous sequence of points on one of the bodies must be in continuous contact with a continuous sequence of points on the other body. A necessary condition for the continuity is that the contacting bodies must have smooth contours, so that the radii of curvature exist at each contact point on both bodies.[3] The contact that takes place is point contact (for thin disks and spheres) or line contact (for cylinders and thick disks, such as rigid wheels). The contour of a body is considered smooth enough to permit rolling if the peaks and dips on its surface are exceedingly small with respect to the overall dimensions of the body. The motion of bodies with sharp edges or with rough surfaces cannot be classified as rolling.

Rolling can occur in a variety of ways, as depicted in Figs. 7.28 and 7.29. The most common form is for a body to roll over a fixed surface; the surface can be planar or curved. A second form of rolling is for two bodies to roll together. Examples of this are gear mechanisms and a sphere rolling over another sphere. The contacting points define the *plane of contact*, which is tangent to both contacting bodies. The rolling constraint is defined as no relative velocity of the contacting points perpendicular to the plane of contact. At the point of contact, the radii of curvature of

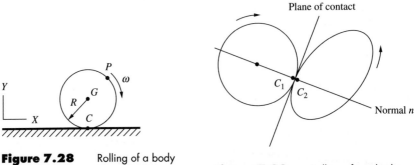

Figure 7.28 Rolling of a body over a surface

Figure 7.29 Rolling of two bodies

[3] Note that for a straight line the radius of curvature exists and it is at infinity.

the contours of the rolling bodies are along the same line. This last statement is a geometric constraint that determines whether one body *can* roll over another.

Consider two bodies rolling over one another, as shown in Fig. 7.29. The contacting points are denoted by C_1 and C_2, with velocities of $\mathbf{v}_{C_1}$ and $\mathbf{v}_{C_2}$, respectively. Let $\mathbf{n}$ denote the unit vector perpendicular to the plane of contact. The relative velocity of one contacting point with respect to the other is along the plane of contact. The rolling constraint can then be expressed as

$$(\mathbf{v}_{C_2} - \mathbf{v}_{C_1}) \cdot \mathbf{n} = 0 \qquad [7.9.1]$$

A rolling body is a five degree of freedom system. If the contacting points have the same velocity, the motion is referred to as *roll without slip*. If the contacting points have different velocities, the motion is referred to as *roll with slip*. Whether slipping exists or not depends on the forces acting on the rolling bodies, as well as on the amount of friction between the rolling bodies. The no-slip condition can be expressed as

$$\mathbf{v}_{C_1} = \mathbf{v}_{C_2} \qquad [7.9.2]$$

A body rolling without slipping has three degrees of freedom. An interesting point to note about rolling contact is that the accelerations of the contacting points are different, whether there is slipping or not. The contacting points approach each other, they have contact, and then they separate and move away from each other. The constraint force is applied to the contacting points only for one instant. Also, the accelerations of the contacting points usually have components along the contact plane. Hence, there is no direct constraint relation associated with the acceleration of rolling bodies. This can be remedied by taking into consideration that the shapes of the bodies are fixed and that the radius of curvature exists for all contacting points, which permits one to write the acceleration expressions for each body.

Let us first consider plane motion of an axisymmetric body—such as a disk or cylinder—of radius R that rolls on a fixed flat surface as shown in Fig. 7.28. We use as generalized coordinates the position of the center of mass, X and Y, and the rotation angle θ. The angular velocity of the body is $\boldsymbol{\omega} = -\omega \mathbf{K}$, where $\omega = \dot{\theta}$. We can express the velocity of a point P on the body as

$$\mathbf{v}_P = \mathbf{v}_G + \boldsymbol{\omega} \times \mathbf{r}_{P/G} = \dot{X}\mathbf{I} + \dot{Y}\mathbf{J} + \boldsymbol{\omega} \times \mathbf{r}_{P/G} \qquad [7.9.3]$$

where $\mathbf{r}_{P/G}$ is the position vector from the center of mass to P. The path followed by a point on the circumference of the rolling body is called a *cycloid*. Fig. 7.30 shows a cycloidal path. To find the velocity of the contact point C, we write Eq. [7.9.3] in terms of C. The no-slip condition can be expressed as $\mathbf{v}_C = \mathbf{0}$. Introducing this into Eq. [7.9.3], we obtain

$$\mathbf{v}_C = \mathbf{0} = \mathbf{v}_G + \boldsymbol{\omega} \times \mathbf{r}_{C/G} \qquad [7.9.4]$$

which we can conveniently write as

$$\mathbf{v}_G = \boldsymbol{\omega} \times \mathbf{r}_{G/C} = -\omega \mathbf{K} \times R\mathbf{J} = R\omega \mathbf{I} = R\dot{\theta}\mathbf{I} \qquad [7.9.5]$$

so that $\dot{X} = R\omega$, $\dot{Y} = 0$, and so roll without slip for plane motion is a single degree of freedom problem. Because there is no velocity of the point of contact C, the body

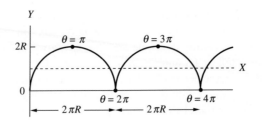

Figure 7.30   Cycloidal path

can be visualized as rotating about C at that instant. For roll without slip the point of contact is the instant center.

The acceleration of a point P on the rolling body can be obtained by differentiating Eq. [7.9.3],

$$\mathbf{a}_P = \mathbf{a}_G + \boldsymbol{\alpha} \times \mathbf{r}_{P/G} + \boldsymbol{\omega} \times (\boldsymbol{\omega} \times \mathbf{r}_{P/G}) \qquad [7.9.6]$$

The acceleration of the center of mass can be expressed as $\mathbf{a}_G = d(R\omega\mathbf{I})/dt = R\alpha\mathbf{I}$. The angular acceleration is $\boldsymbol{\alpha} = -\dot{\omega}\mathbf{K} = -\alpha\mathbf{K}$. The acceleration of a point on the rolling body is

$$\mathbf{a}_P = R\alpha\mathbf{I} + (-\alpha\mathbf{K} \times \mathbf{r}_{P/G}) - \omega^2 \mathbf{r}_{P/G} \qquad [7.9.7]$$

If we select point P as the contact point C $\mathbf{r}_{P/G} = -R\mathbf{J}$ and we obtain

$$\mathbf{a}_C = R\alpha\mathbf{I} + (-\dot{\omega}\mathbf{K} \times -R\mathbf{J}) - \omega^2(-R\mathbf{J}) = R\omega^2\mathbf{J} \qquad [7.9.8]$$

Now consider roll without slip on a fixed, concave-upward planar curve as shown in Fig. 7.31. We use a set of normal and tangential coordinates. It follows that we can write the velocity of the center of mass as

$$\mathbf{v}_G = v\mathbf{e}_t = R\omega\mathbf{e}_t \qquad [7.9.9]$$

The acceleration of the center of mass thus becomes

$$\mathbf{a}_G = \dot{v}\mathbf{e}_t + \frac{v^2}{\rho}\mathbf{e}_n = R\alpha\mathbf{e}_t + \frac{R^2\omega^2}{\rho}\mathbf{e}_n \qquad [7.9.10]$$

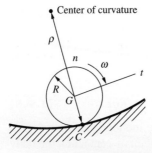

Figure 7.31   Rolling inside a concave surface

Substituting this expression into the relative velocity expression gives the acceleration of any point on the body. For the special case of the instant center, we have $\mathbf{r}_{C/G} = -R\mathbf{e}_n$ and

$$\mathbf{a}_C = \mathbf{a}_G + \boldsymbol{\alpha} \times \mathbf{r}_{C/G} - \omega^2 \mathbf{r}_{C/G}$$
$$= R\dot{\omega}\mathbf{e}_t + \frac{R^2\omega^2}{\rho}\mathbf{e}_n + (-\alpha\mathbf{K} \times -R\mathbf{e}_n) - \omega^2(-R\mathbf{e}_n) = R\omega^2\left(1 + \frac{R}{\rho}\right)\mathbf{e}_n \quad [7.9.11]$$

If the curve on which the disk is rolling is convex, the direction of the unit vector in the normal direction is reversed and $\mathbf{r}_{C/G} = R\mathbf{e}_n$. The acceleration of the instant center C becomes

$$\mathbf{a}_C = R\omega^2\left(-1 + \frac{R}{\rho}\right)\mathbf{e}_n \quad [7.9.12]$$

Next consider rolling in three dimensions. We make use of a 3-1-3 transformation sequence to describe the rolling of a flat axisymmetric body, such as a thin disk or a coin. We begin with an inertial set of axes XYZ, with the initial position of the coin as lying flat on the XY plane, as shown in Fig. 7.32. Figures 7.33 and 7.34 depict the disk after the first two Euler angle transformations have been carried out. The generalized coordinates used to describe the general motion of the disk are the coordinates of the center of mass X, Y, Z, and the three Euler angles ϕ, θ, and ψ. We use the F frame as the relative frame. From Eq. [7.5.3], the axes xyz and XYZ and unit vectors $\mathbf{ijk}$ and $\mathbf{IJK}$ are related by

$$\mathbf{i} = c\phi\mathbf{I} + s\phi\mathbf{J} \qquad \mathbf{j} = -s\phi c\theta\mathbf{I} + c\phi c\theta\mathbf{J} + s\theta\mathbf{K}$$
$$\mathbf{k} = s\phi s\theta\mathbf{I} - c\phi s\theta\mathbf{J} + c\theta\mathbf{K} \quad [7.9.13]$$

It should also be noted that the contact point between the disk and the surface is along the y axis. The angular velocity of the disk is

$$\boldsymbol{\omega} = \boldsymbol{\omega}_b = \omega_x\mathbf{i} + \omega_y\mathbf{j} + \omega_z\mathbf{k} = \dot{\theta}\mathbf{i} + \dot{\phi}\sin\theta\mathbf{j} + (\dot{\phi}\cos\theta + \dot{\psi})\mathbf{k} \quad [7.9.14]$$

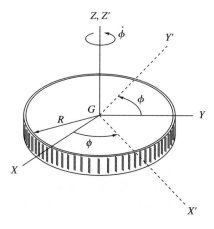

Figure 7.32 Initial position and precession

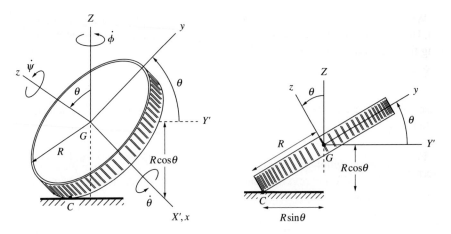

Figure 7.33 Nutation   **Figure 7.34** Side view

so that $\omega_x = \dot{\theta}$, $\omega_y = \dot{\phi}\sin\theta$, and $\omega_z = \dot{\phi}\cos\theta + \dot{\psi}$. The angular velocity of the reference frame is

$$\boldsymbol{\omega}_f = \dot{\theta}\mathbf{i} + \dot{\phi}\sin\theta\mathbf{j} + \dot{\phi}\cos\theta\mathbf{k} = \omega_x\mathbf{i} + \omega_y\mathbf{j} + \frac{\omega_y}{\tan\theta}\mathbf{k} \qquad [7.9.15]$$

We thus have

$$\omega_{fx} = \omega_x = \dot{\theta} \qquad \omega_{fy} = \omega_y = \dot{\phi}\sin\theta \qquad \omega_{fz} = \frac{\omega_y}{\tan\theta} = \dot{\phi}\cos\theta \qquad [7.9.16]$$

First, consider roll with slip. When the disk is rolling and slipping, the point of contact C has a nonzero velocity along the roll surface. Let us write the velocity of point C using the relative velocity expression between C and the center of mass, and express the result in terms of the inertial coordinates. Thus

$$\begin{aligned}\mathbf{v}_C &= \mathbf{v}_G + \boldsymbol{\omega}_b \times \mathbf{r}_{C/G} \\ &= \dot{X}\mathbf{I} + \dot{Y}\mathbf{J} + \dot{Z}\mathbf{K} + (\dot{\theta}\mathbf{i} + \dot{\phi}s\theta\mathbf{j} + (\dot{\phi}c\theta + \dot{\psi})\mathbf{k}) \times -R\mathbf{j} \\ &= (\dot{X} - R\dot{\theta}s\phi s\theta + R(\dot{\phi}c\theta + \dot{\psi})c\phi)\mathbf{I} \\ &\quad + (\dot{Y} + R\dot{\theta}c\phi s\theta + R(\dot{\phi}c\theta + \dot{\psi})s\phi)\mathbf{J} + (\dot{Z} - R\dot{\theta}c\theta)\mathbf{K}\end{aligned} \qquad [7.9.17]$$

The constraint associated with roll with slip states that the velocity of point C has no component in the vertical, that is, $\mathbf{v}_C \cdot \mathbf{K} = 0$. Setting the components of $\mathbf{v}_C$ in the Z direction to zero, we obtain

$$\dot{Z} = R\dot{\theta}\cos\theta \qquad [7.9.18]$$

This constraint is recognized as being the time derivative of

$$Z = R\sin\theta \qquad [7.9.19]$$

so that the constraint associated with roll with slip is holonomic. The constraint has a physical explanation: it represents the height of the center of mass from the roll surface.

We next consider roll without slip, where the velocity of the contact point C is zero. Hence, there are three constraints. Writing $\mathbf{v}_C = \mathbf{0}$ in terms of the components along the inertial axes X and Y, we obtain from Eq. [7.9.17]

$$\dot{X} = R\dot{\theta}s\phi s\theta - R(\dot{\phi}c\theta + \dot{\psi})c\phi \qquad \dot{Y} = -R\dot{\theta}c\phi s\theta - R(\dot{\phi}c\theta + \dot{\psi})s\phi \quad [7.9.20]$$

These equations are not perfect differentials and cannot be integrated to a form similar to Eq. [7.9.19]. Hence, constraints associated with the no-slip condition are nonholonomic.

For rolling without slipping, the velocity of the center of mass can be expressed in terms of the inertial coordinates as

$$\begin{aligned}\mathbf{v}_G &= \dot{X}\mathbf{I} + \dot{Y}\mathbf{J} + \dot{Z}\mathbf{K} \\ &= (R\dot{\theta}s\phi s\theta - R(\dot{\phi}c\theta + \dot{\psi})c\phi)\mathbf{I} \\ &\quad - (R\dot{\theta}c\phi s\theta + R(\dot{\phi}c\theta + \dot{\psi})s\phi)\mathbf{J} + R\dot{\theta}c\theta\mathbf{K} \quad [7.9.21]\end{aligned}$$

We can also express $\mathbf{v}_G$ by using the relative velocity expression with $\mathbf{v}_C = \mathbf{0}$, which yields

$$\mathbf{v}_G = \boldsymbol{\omega}_b \times \mathbf{r}_{G/C} = (\dot{\theta}\mathbf{i} + \dot{\phi}s\theta\mathbf{j} + (\dot{\phi}c\theta + \dot{\psi})\mathbf{k}) \times R\mathbf{j} = R\dot{\theta}\mathbf{k} - R(\dot{\phi}c\theta + \dot{\psi})\mathbf{i} \quad [7.9.22]$$

In terms of the angular velocity components along the F frame

$$\mathbf{v}_G = R\omega_x\mathbf{k} - R\omega_z\mathbf{i} \quad [7.9.23]$$

Consider the acceleration of the center of the disk. One can calculate $\mathbf{a}_G$ in a number of ways. The most cumbersome is to differentiate Eq. [7.9.21]. This approach does not yield any insight into the nature of the problem, except for the component of the motion in the Z direction. A more convenient approach is to use the F frame and to apply the transport theorem to Eq. [7.9.23], which yields

$$^A\mathbf{a}_G = {^F\!\frac{d}{dt}}{^A\mathbf{v}_G} + {^A\boldsymbol{\omega}^F} \times {^A\mathbf{v}_G} \quad [7.9.24]$$

Substituting the values for $\mathbf{v}_G$ and $\boldsymbol{\omega}_f$ into Eq. [7.9.24], we obtain

$$\begin{aligned}\mathbf{a}_G &= R\dot{\omega}_x\mathbf{k} - R\dot{\omega}_z\mathbf{i} + \left(\omega_x\mathbf{i} + \omega_y\mathbf{j} + \frac{\omega_y}{\tan\theta}\mathbf{k}\right) \times (R\omega_x\mathbf{k} - R\omega_z\mathbf{i}) \\ &= R(-\dot{\omega}_z + \omega_x\omega_y)\mathbf{i} + R\left(-\omega_x^2 - \frac{\omega_y\omega_z}{\tan\theta}\right)\mathbf{j} + R(\dot{\omega}_x + \omega_y\omega_z)\mathbf{k} \quad [7.9.25]\end{aligned}$$

In terms of the Euler angles, we write the expression for the acceleration by substituting Eq. [7.9.14] and its derivatives into Eq. [7.9.25], which yields

$$\begin{aligned}\mathbf{a}_G &= R(-\ddot{\phi}c\theta + 2\dot{\phi}\dot{\theta}s\theta - \ddot{\psi})\mathbf{i} + R(-\dot{\phi}c\theta(\dot{\phi}c\theta + \dot{\psi}) - \dot{\theta}^2)\mathbf{j} \\ &\quad + R(\ddot{\theta} + \dot{\phi}s\theta(\dot{\phi}c\theta + \dot{\psi}))\mathbf{k} \quad [7.9.26]\end{aligned}$$

The advantages of using a reference frame not attached to the body are obvious. Attaching the relative frame to the body would necessitate use of the spin angle ψ in all calculations. One would not be able to use the relation $\mathbf{a}_G = \boldsymbol{\alpha}_b \times \mathbf{r}_{G/C} + \boldsymbol{\omega}_b \times (\boldsymbol{\omega}_b \times \mathbf{r}_{G/C})$ to find the acceleration of the center of mass. Point C has a nonzero acceleration. Another reason for not using a body-fixed frame is that the position of the contact point between the disk and the surface changes with respect to a body-fixed frame as the disk rolls. One would have to take into consideration the motion of the point of contact. By contrast, viewed from the F frame, the point of contact always lies on the XY plane and ${}^F\mathbf{v}_C \cdot \mathbf{K} = 0$.

Now consider the acceleration of the contact point C. We first calculate the angular acceleration of the body using the transport theorem. Noting that $\boldsymbol{\omega}_s = \boldsymbol{\omega}_b - \boldsymbol{\omega}_f = (\omega_z - \omega_y/\tan\theta)\mathbf{k}$, we write

$$\boldsymbol{\alpha}_b = \frac{d}{dt}(\boldsymbol{\omega}_b)_{\text{rel}} + \boldsymbol{\omega}_f \times \boldsymbol{\omega}_s \qquad [7.9.27]$$

$$= \dot{\omega}_x\mathbf{i} + \dot{\omega}_y\mathbf{j} + \dot{\omega}_z\mathbf{k} + \left(\omega_x\mathbf{i} + \omega_y\mathbf{j} + \frac{\omega_y}{\tan\theta}\mathbf{k}\right) \times \left(\omega_z - \frac{\omega_y}{\tan\theta}\right)\mathbf{k}$$

$$= \left(\dot{\omega}_x + \omega_y\omega_z - \frac{\omega_y^2}{\tan\theta}\right)\mathbf{i} + \left(\dot{\omega}_y - \omega_x\omega_z + \frac{\omega_x\omega_y}{\tan\theta}\right)\mathbf{j} + \dot{\omega}_z\mathbf{k}$$

The relative acceleration equation between C and G is

$$\mathbf{a}_C = \mathbf{a}_G + \boldsymbol{\alpha}_b \times \mathbf{r}_{C/G} + \boldsymbol{\omega}_b \times (\boldsymbol{\omega}_b \times \mathbf{r}_{C/G}) \qquad [7.9.28]$$

Substituting in the values of $\mathbf{r}_{C/G} = -R\mathbf{j}$ and the values for $\mathbf{a}_G$, $\boldsymbol{\omega}_b$, and $\boldsymbol{\alpha}_b$ from Eqs. [7.9.26], [7.9.14] and [7.9.27], respectively, and carrying out the algebra, we obtain

$$\mathbf{a}_C = R\left(\omega_z^2 - \frac{\omega_x\omega_y}{\tan\theta}\right)\mathbf{j} + R\left(\omega_y\omega_z - \frac{\omega_y^2}{\tan\theta}\right)\mathbf{k} \qquad [7.9.29]$$

Example 7.9

The cone shown in Fig. 7.35 is of height L and apex angle β. It rolls on a smooth surface without slipping. The center of the base has a constant speed v. Find the angular velocity and angular acceleration of the cone.

Solution

To find the angular velocity of the cone we make use of instant centers. Figure 7.36 shows the geometry. The $h_1 h_2 h_3$ axes move with the line of contact between the cone and the horizontal surface. They are similar to the F frame. The vertical distance between point B and the surface

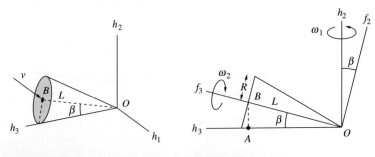

Figure 7.35   **Figure 7.36**

is $L \sin \beta$. Because of the no-slip condition, A is an instant center and the velocity of B is equal to the angular velocity times the height. The total angular velocity of the cone is then

$$\omega = \frac{v}{L \sin \beta} \qquad [a]$$

We express the angular velocity in vector form as $\boldsymbol{\omega} = -v/(L \sin \beta)\mathbf{h}_3$. To see this, we note that $\mathbf{v}_B = v\mathbf{h}_1$ and $\mathbf{r}_{B/A} = L \sin \beta \mathbf{h}_2$ and $\mathbf{v}_B = \boldsymbol{\omega} \times \mathbf{r}_{B/A}$.

To find the angular acceleration, we visualize the motion of the cone as the superposition of two angular velocities. The first angular velocity ω_1 is that of a cone which is only rotating in the horizontal plane. The second angular velocity, ω_2, is the angular velocity of the cone about its symmetry axis. We note that the $h_1 h_2 h_3$ coordinates are rotating with angular velocity ω_1. We thus have

$$\boldsymbol{\omega}_1 = \omega_1 \mathbf{h}_2 \qquad [b]$$

From Fig. 7.36 we find $\boldsymbol{\omega}_1$ as

$$\boldsymbol{\omega}_1 = \frac{v}{L \cos \beta} \mathbf{h}_2 \qquad [c]$$

This can be explained by considering the velocity of point B and that point B is rotating about the h_2 axis. The distance between B and the h_2 axis is $L \cos \beta$. The total angular velocity is

$$\boldsymbol{\omega} = \boldsymbol{\omega}_1 + \boldsymbol{\omega}_2 \qquad [d]$$

where ω_2 is the spin velocity of the cone. As the direction of this spin is along the symmetry axis of the cone, $\boldsymbol{\omega}_2$ can be expressed as

$$\boldsymbol{\omega}_2 = \omega_2 \mathbf{f}_3 = \omega_2 \sin \beta \mathbf{h}_2 + \omega_2 \cos \beta \mathbf{h}_3 \qquad [e]$$

We then use Eqs. [c] to [e] to relate the angular velocities as

$$\boldsymbol{\omega} = -\frac{v}{L \sin \beta} \mathbf{h}_3 = \boldsymbol{\omega}_1 + \boldsymbol{\omega}_2 = \frac{v}{L \cos \beta} \mathbf{h}_2 + \omega_2 \sin \beta \mathbf{h}_2 + \omega_2 \cos \beta \mathbf{h}_3 \qquad [f]$$

which can be solved for ω_2

$$\omega_2 = -\frac{v}{L \cos \beta \sin \beta} = -\frac{\omega}{\cos \beta} \qquad [g]$$

Noting that $\mathbf{h}_3 = \cos \beta \mathbf{f}_3 - \sin \beta \mathbf{f}_2$, and $R = L \tan \beta$, we can express the angular velocity as

$$\boldsymbol{\omega} = -\frac{v}{L \sin \beta} \mathbf{h}_3 = \frac{v}{L} \mathbf{f}_2 - \frac{v \cos \beta}{L \sin \beta} \mathbf{f}_3 = \frac{v}{L} \mathbf{f}_2 - \frac{v}{R} \mathbf{f}_3 \qquad [h]$$

The angular acceleration of the cone can be expressed more easily using the H frame, thus

$$\boldsymbol{\alpha} = \boldsymbol{\omega}_1 \times \boldsymbol{\omega}_2 = \frac{v}{L \cos \beta} \mathbf{h}_2 \times -\frac{v}{L \cos \beta \sin \beta} (\sin \beta \mathbf{h}_2 + \cos \beta \mathbf{h}_3)$$

$$= -\frac{v^2}{L^2 \cos \beta \sin \beta} \mathbf{h}_1 \qquad [i]$$

This problem can also be viewed from a Euler angle point of view, with ω_1 denoting the precession rate, constant nutation angle β, and spin rate of ω_2.

Example 7.10

A disk rolls over a horizontal surface while it rotates about a shaft, which is attached by a revolute joint to a bar with an arm, as shown in Fig. 7.37. The shaft is rotating with speed ω_1. Find the angular velocity and angular acceleration of the disk.

Solution

The reference frames are shown in Fig. 7.38. We obtain the relative frame $f_1 f_2 f_3$ by rotating the reference frame attached to the shaft (rotating with $\omega_1 = \dot{\phi}$ about the fixed a_3 axis) by an angle of $270° - \beta$ about the d_2 axis. We write

$$\mathbf{f}_1 = -\sin\beta \mathbf{d}_1 + \cos\beta \mathbf{d}_3 \qquad \mathbf{f}_2 = \mathbf{d}_2 \qquad \mathbf{f}_3 = -\cos\beta \mathbf{d}_1 - \sin\beta \mathbf{d}_3 \qquad [\mathbf{a}]$$

The angle β, which is fixed, is related to the height of the arm by

$$h = L\sin\beta + R\cos\beta \qquad [\mathbf{b}]$$

The angular velocity of the disk can be written as the sum of the angular velocities of the shaft and of the disk with respect to the shaft, or

$$\boldsymbol{\omega} = \boldsymbol{\omega}_1 + \boldsymbol{\omega}_2 \qquad [\mathbf{c}]$$

in which

$$\boldsymbol{\omega}_1 = \omega_1 \mathbf{d}_3 = \omega_1 \cos\beta \mathbf{f}_1 - \omega_1 \sin\beta \mathbf{f}_3 \qquad \boldsymbol{\omega}_2 = -\omega_2 \mathbf{f}_3 \qquad [\mathbf{d}]$$

are the respective angular velocities. Note that ω_2 is not known, and it will be found by using the rolling constraint. To do this, we obtain an expression for the velocity for the contact point as

$$\mathbf{v}_C = \mathbf{v}_G + \boldsymbol{\omega} \times \mathbf{r}_{C/G} = \mathbf{0} \qquad [\mathbf{e}]$$

where

$$\mathbf{v}_G = -(L + L\cos\beta)\omega_1 \mathbf{f}_2 \qquad \mathbf{r}_{C/G} = -R\mathbf{f}_1 \qquad [\mathbf{f}]$$

so that we have

$$\mathbf{v}_C = -(L + L\cos\beta)\omega_1 \mathbf{f}_2 + (\omega_1 \cos\beta \mathbf{f}_1 - (\omega_1 \sin\beta + \omega_2)\mathbf{f}_3) \times -R\mathbf{f}_1$$
$$= (-L\omega_1 - L\cos\beta\omega_1 + R\sin\beta\omega_1 + R\omega_2)\mathbf{f}_2 = \mathbf{0} \qquad [\mathbf{g}]$$

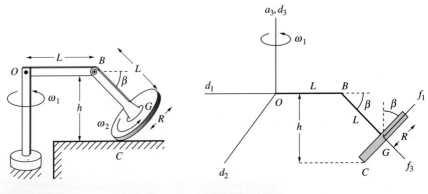

Figure 7.37 **Figure 7.38**

Solving the above, we obtain ω_2 in terms of ω_1 as

$$\omega_2 = \omega_1 \frac{(L + L\cos\beta - R\sin\beta)}{R} \quad \text{[h]}$$

We recognize the term in the brackets above to be the horizontal distance between O and C. Introducing the ratio $r = L/R$, we can now write the angular velocity of the disk as

$$\boldsymbol{\omega} = \omega_1 \mathbf{d}_3 - \omega_2 \mathbf{f}_3 = \omega_1 \cos\beta \mathbf{f}_1 - \omega_1(r + r\cos\beta - \sin\beta + \sin\beta)\mathbf{f}_3$$
$$= \omega_1 \cos\beta \mathbf{f}_1 - \omega_1 r(1 + \cos\beta)\mathbf{f}_3 \quad \text{[i]}$$

To find the angular acceleration we differentiate Eq. [i] with the result

$$\boldsymbol{\alpha} = \dot{\boldsymbol{\omega}}_{\text{rel}} + \boldsymbol{\omega}_f \times \boldsymbol{\omega}_s = \dot{\omega}_1 \mathbf{d}_3 - \dot{\omega}_2 \mathbf{f}_3 + (\omega_1 \cos\beta \mathbf{f}_1 - \omega_1 \sin\beta \mathbf{f}_3) \times (-\omega_2 \mathbf{f}_3) \quad \text{[j]}$$

Note that $\dot{\omega}_2$ is obtained by direct differentiation of Eq. [h]. Substituting in all the values and carrying out the algebra we obtain

$$\boldsymbol{\alpha} = \dot{\omega}_1 \cos\beta \mathbf{f}_1 + \omega_1^2(r + r\cos\beta - \sin\beta)\cos\beta \mathbf{f}_2 - \dot{\omega}_1 r(1 + \cos\beta)\mathbf{f}_3 \quad \text{[k]}$$

We cannot obtain the expression for the angular acceleration by direct differentiation of Eq. [i] because $f_1 f_2 f_3$ is not attached to the disk.

Figures 7.39 and 7.40 show the top and side views of a tricycle. Do a kinematic modeling on the tricycle and determine the number of degrees of freedom, assuming that all wheels roll without slip and the tricycle does not tip or roll.

| **Example 7.11** |

Solution

We select the inertial coordinates XYZ such that the tricycle moves on the XY plane, and we attach an xyz coordinate system to the tricycle at point G. The xyz axes are obtained by rotating the XYZ coordinates by an angle of ψ about the Z axis, as shown in Fig. 7.41.

The velocity of point G and the angular velocity of the main body are

$$\mathbf{v}_G = \dot{X}\mathbf{I} + \dot{Y}\mathbf{J} \qquad \boldsymbol{\Omega}_1 = \dot{\psi}\mathbf{K} = \dot{\psi}\mathbf{k} \quad \text{[a]}$$

and considering that $\mathbf{I} = \cos\psi \mathbf{i} - \sin\psi \mathbf{j}$, $\mathbf{J} = \cos\psi \mathbf{j} + \sin\psi \mathbf{i}$, we write the velocity of G in the xyz coordinate system as

$$\mathbf{v}_G = (\dot{X}\cos\psi + \dot{Y}\sin\psi)\mathbf{i} + (\dot{Y}\cos\psi - \dot{X}\sin\psi)\mathbf{j} \quad \text{[b]}$$

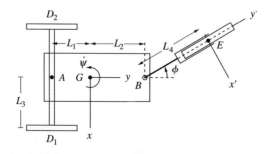

Figure 7.39 Top view of tricycle

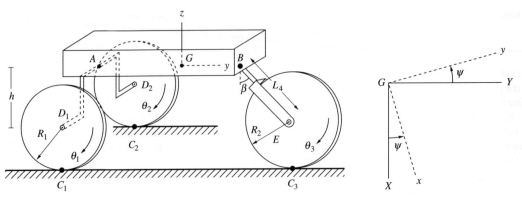

Figure 7.40 Side view

Figure 7.41

We next calculate the velocity of point D_1, which is the point to which the right rear wheel is connected. Noting that $\mathbf{r}_{D_1/G} = -L_1\mathbf{j} - h\mathbf{k} + L_3\mathbf{i}$, we obtain

$$\mathbf{\Omega}_1 \times \mathbf{r}_{D_1/G} = \dot{\psi}\mathbf{k} \times (-L_1\mathbf{j} - h\mathbf{k} + L_3\mathbf{i}) = L_1\dot{\psi}\mathbf{i} + L_3\dot{\psi}\mathbf{j} \qquad \text{[c]}$$

and

$$\mathbf{v}_{D_1} = \mathbf{v}_G + \mathbf{\Omega}_1 \times \mathbf{r}_{D_1/G} = (\dot{X}\cos\psi + \dot{Y}\sin\psi + L_1\dot{\psi})\mathbf{i} + (\dot{Y}\cos\psi - \dot{X}\sin\psi + L_3\dot{\psi})\mathbf{j} \qquad \text{[d]}$$

The point of contact between back wheel on the right and the ground is denoted by C_1. The angular velocity of that wheel is expressed as

$$\boldsymbol{\omega}_1 = \mathbf{\Omega}_1 - \dot{\theta}_1\mathbf{i} = \dot{\psi}\mathbf{k} - \dot{\theta}_1\mathbf{i} \qquad \text{[e]}$$

and, noting that $\mathbf{r}_{C_1/D_1} = -R_1\mathbf{k}$, the velocity of C_1 becomes

$$\mathbf{v}_{C_1} = \mathbf{v}_{D_1} + \boldsymbol{\omega}_1 \times \mathbf{r}_{C_1/D_1}$$
$$= (\dot{X}\cos\psi + \dot{Y}\sin\psi + L_1\dot{\psi})\mathbf{i} + (\dot{Y}\cos\psi - \dot{X}\sin\psi + L_3\dot{\psi} - R_1\dot{\theta}_1)\mathbf{j} \qquad \text{[f]}$$

Because of the roll without slip assumption, the velocity of C_1 is zero, $\mathbf{v}_{C_1} = \mathbf{0}$. This is tantamount to two constraint equations

$$\dot{X}\cos\psi + \dot{Y}\sin\psi + L_1\dot{\psi} = 0 \qquad \dot{Y}\cos\psi - \dot{X}\sin\psi + L_3\dot{\psi} - R_1\dot{\theta}_1 = 0 \qquad \text{[g,h]}$$

and we note that both constraints are nonholonomic. We can obtain the expression for the velocity of the point of contact for the back wheel on the left, $\mathbf{v}_{C_2}$, by substituting $-L_3$ for L_3 and $\dot{\theta}_2$ for $\dot{\theta}_1$ in Eq. [f]. Since that wheel also rolls without slipping, we set $\mathbf{v}_{C_2}$ to zero, which yields one additional constraint, namely the component of $\mathbf{v}_{C_2}$ in the y direction

$$\dot{Y}\cos\psi - \dot{X}\sin\psi - L_3\dot{\psi} - R_1\dot{\theta}_2 = 0 \qquad \text{[i]}$$

The component of the velocity of C_2 in the x direction is the same as its counterpart for C_1. It is interesting to note the velocity of point A, which is obtained by setting $L_3 = 0$ in Eq. [d] as

$$\mathbf{v}_A = \mathbf{v}_G + \mathbf{\Omega}_1 \times \mathbf{r}_{A/G} = (\dot{X}\cos\psi + \dot{Y}\sin\psi + L_1\dot{\psi})\mathbf{i} + (\dot{Y}\cos\psi - \dot{X}\sin\psi)\mathbf{j} \qquad \text{[j]}$$

However, from Eq. [g], the x component of $\mathbf{v}_A$ is zero, leading to the conclusion that

$$\mathbf{v}_A = (\dot{Y}\cos\psi - \dot{X}\sin\psi)\mathbf{j} \qquad [\mathbf{k}]$$

The velocity of point A is always only in the y direction, that is, in the direction of the body of the tricycle. This is the constraint mentioned in Chapter 4, when discussing vehicle dynamics problems. By virtue of the roll without slip approximation, we end up with this constraint.

Next, consider the steering mechanism. The turning of the wheel is denoted by ϕ and the rotation of the wheel by θ_3. We attach a coordinate system $x'y'z'$ to the wheel (obtained by rotating the xyz axes about the z axis by an angle ϕ). The velocity of point E, about which the front wheel turns, is obtained by

$$\mathbf{v}_E = \mathbf{v}_B + \mathbf{\Omega}_2 \times \mathbf{r}_{E/B} = \mathbf{v}_G + \mathbf{\Omega}_1 \times \mathbf{r}_{B/G} + \mathbf{\Omega}_2 \times \mathbf{r}_{E/B} \qquad [\mathbf{l}]$$

in which

$$\mathbf{\Omega}_2 = (\dot\psi + \dot\phi)\mathbf{k} \qquad \mathbf{r}_{B/G} = L_2\mathbf{j} \qquad \mathbf{r}_{E/B} = L_4\sin\beta\mathbf{j}' - L_4\cos\beta\mathbf{k} \qquad \mathbf{j}' = \cos\phi\mathbf{j} - \sin\phi\mathbf{i} \qquad [\mathbf{m}]$$

Substitution of all these values into Eq. [l] yields

$$\mathbf{v}_E = (\dot{X}\cos\psi + \dot{Y}\sin\psi - (L_2 + L_4\sin\beta\cos\phi)\dot\psi - L_4\sin\beta\cos\phi\dot\phi)\mathbf{i}$$
$$+ (\dot{Y}\cos\psi - \dot{X}\sin\psi - L_4\sin\beta\sin\phi(\dot\psi + \dot\phi))\mathbf{j} \qquad [\mathbf{n}]$$

To find the velocity of the contact point between the front wheel and the ground, we first develop an expression for the angular velocity of the front wheel. The wheel is at an angle of ϕ with the body of the tricycle, so that the angular velocity of the wheel becomes

$$\boldsymbol{\omega}_3 = \mathbf{\Omega}_2 - \dot\theta_3\mathbf{i}' = (\dot\psi + \dot\phi)\mathbf{k} - \dot\theta_3\mathbf{i}' = (\dot\psi + \dot\phi)\mathbf{k} - \dot\theta_3\cos\phi\mathbf{i} - \dot\theta_3\sin\phi\mathbf{j} \qquad [\mathbf{o}]$$

Noting that $\mathbf{r}_{C_3/E} = -R_3\mathbf{k}$, we have

$$\boldsymbol{\omega}_3 \times \mathbf{r}_{C_3/E} = ((\dot\psi + \dot\phi)\mathbf{k} - \dot\theta_3\cos\phi\mathbf{i} - \dot\theta_3\sin\phi\mathbf{j}) \times (-R_3)\mathbf{k}$$
$$= R_3\dot\theta_3\sin\phi\mathbf{i} - R_3\dot\theta_3\cos\phi\mathbf{j} \qquad [\mathbf{p}]$$

Combining Eqs. [n] and [p] we obtain for the velocity of C_3

$$\mathbf{v}_{C_3} = \mathbf{v}_E + \boldsymbol{\omega}_3 \times \mathbf{r}_{C_3/E}$$
$$= [\dot{X}\cos\psi + \dot{Y}\sin\psi - (L_2 + L_4\sin\beta\cos\phi)\dot\psi - L_4\sin\beta\cos\phi\dot\phi + R_3\dot\theta_3\sin\phi]\mathbf{i}$$
$$+ [\dot{Y}\cos\psi - \dot{X}\sin\psi - R_3\dot\theta_3\cos\phi - L_4\sin\beta\sin\phi(\dot\psi + \dot\phi)]\mathbf{j} \qquad [\mathbf{q}]$$

The roll without slip condition states that the velocity of C_3 is zero, which yields two more constraints that are also nonholonomic, which we write

$$\dot{X}\cos\psi + \dot{Y}\sin\psi - (L_2 + L_4\sin\beta\cos\phi)\dot\psi - L_4\sin\beta\cos\phi\dot\phi + R_3\dot\theta_3\sin\phi = 0 \qquad [\mathbf{r}]$$
$$\dot{Y}\cos\psi - \dot{X}\sin\psi - R_3\dot\theta_3\cos\phi - L_4\sin\beta\sin\phi(\dot\psi + \dot\phi) = 0 \qquad [\mathbf{s}]$$

Let us now calculate the number of degrees of freedom for this system. We introduced seven generalized coordinates: X, Y, ψ, θ_1, θ_2, ϕ, and θ_3. We generated five constraints, Eqs. [g], [h], [i], [r], and [s]. Hence, the tricycle is a two degree of freedom system.

References

Ginsberg, J. H. *Advanced Engineering Dynamics*. New York: Harper & Row, 1988.
Goldstein, H. *Classical Mechanics*. 2nd ed. Reading, MA: Addison-Wesley, 1981.
Junkins, J. L., and J. D. Turner. *Optimal Spacecraft Rotational Maneuvers*. Amsterdam–New York: Elsevier, 1986.
Kane, T. R., P. W. Likins, and D. A. Levinson. *Spacecraft Dynamics*. New York: McGraw-Hill, 1983.
Meirovitch, L. *Methods of Analytical Dynamics*. New York: McGraw-Hill, 1970.
Orthwein, W. *Machine Component Design*. St. Paul, MN: West Publishing, 1990.
Sandor, G. N., and A. G. Erdman. *Advanced Mechanism Design*. Vol. 2. Englewood Cliffs, NJ: Prentice-Hall, 1984.
Shigley, J. E. *Kinematic Analysis of Mechanisms*. 2nd ed. New York: McGraw-Hill, 1969.
Whittaker, E. T. *Analytical Dynamics of Particles and Rigid Bodies*. Cambridge: Cambridge University Press, 1933.

Homework Exercises

Section 7.2

1. The bar in Fig. 7.42 is always in contact with the corner D and the curved path. Find the angular velocity and angular acceleration of the bar as a function of the angle θ.

2. For the system shown in Fig. 7.43, find the instantaneous center of the rod AB and the angular velocity of the rod.

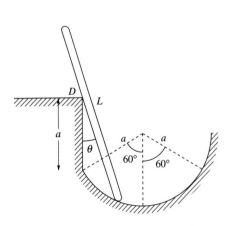

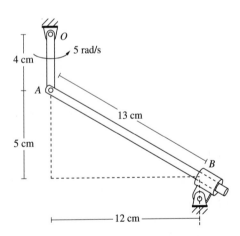

Figure 7.42   **Figure 7.43**

Section 7.3

3. A coordinate frame B is obtained by rotating frame A by an angle of 75° about a line joining the origin and the point (0.5, −0.7, 0.4). Find the direction cosine matrix [c] between the two frames. Then, calculate the direction cosines of the principal line from [c] and compare your result with the given information.

Section 7.4

4. A reference frame A is transformed into a reference frame B using a transformation consisting of two rotations. First a rotation about the a_1 axis, by an angle $\psi(t) = 0.3t$, and then a transformation about the resulting a_2' axis by an angle $\theta(t) = 0.6 \sin \pi t/4$. Using Eq. [7.4.13] find the angular velocity of the frame B with reference to frame A at $t = 3$ seconds.

Section 7.5

5. Solve the previous problem using Euler angles.
6. A disk of radius R is attached to a rod of length L_2, and it rotates about it with angular velocity ω (Fig. 7.44). The rod pivots about an arm of length L_1 with angle θ. This arm is attached to a vertical shaft that is rotating with the constant angular velocity of Ω. Find the angular velocity and angular acceleration of the disk, and the velocity and acceleration of the center of the disk.
7. Figure 7.45 depicts a spacecraft undergoing general motion. There is a point mass sliding inside a frictionless tube along the b_2 axis. Using a 3-1-3 Euler angle transformation (ϕ, θ, ψ) and denoting the position, velocity, and acceleration of the point mass with respect to the symmetry axis of the spacecraft by r, v, and a, respectively, derive expressions for the absolute velocity and acceleration of the mass. First write these expressions in terms of angular velocity parameters ω_i ($i = 1, 2, 3$). Then express them in terms of the Euler angles.

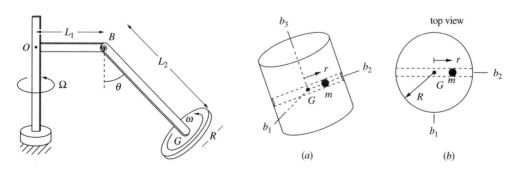

Figure 7.44   **Figure 7.45**

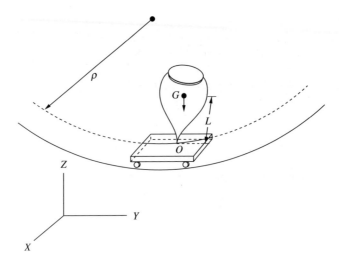

Figure 7.46

8. Figure 7.46 depicts a spinning top on a cart. The cart is moving along a curved track on the XY plane, the radius of curvature of the apex is ρ, and the constant speed of the apex is v. Find the velocity and acceleration of the center of mass of the top using a 3-2-3 transformation (ψ, θ, ϕ), with the inertial axes as shown.

Section 7.6

9. Consider a 3-2-1 transformation and express the partial angular velocities associated with the angular velocity vector, using the components of the angular velocity along the body axes as the generalized speeds.

Section 7.7

10. A coordinate frame B is obtained by rotating frame A by an angle of 75° about a line joining the origin and the point $(0.5, -0.7, 0.4)$. Find the Euler parameters associated with this rotation. From the Euler parameters, calculate the transformation matrix $[R]$ between the two coordinate frames.

11. The orientation of a rigid body is described by a 3-2-3 Euler angle rotation, with the rotation angles ψ, θ, and ϕ. At a certain instant, the values of the Euler angles and their rates of change are $\psi = 45°$, $\dot\psi = 0.2$ rad/s, $\theta = 30°$, $\dot\theta = 0.1$ rad/s, $\phi = 30°$, $\dot\phi = 1.0$ rad/s. Find the values of the Euler parameters and their derivatives for this instant.

12. Find the counterpart of Eq. [7.7.40] for a 3-2-1 transformation and for a 3-2-3 transformation.

13. Solve Problem 2.10 using Eq. [7.7.46].

14. Solve Problem 2.11 using Eq. [7.7.46].

15. Rodrigues parameters are defined as $\rho_i = e_i/e_0$ ($i = 1, 2, 3$). The Rodrigues vector is defined as $\boldsymbol{\rho} = \mathbf{n}\tan(\Phi/2)$, where $\mathbf{n}$ is the unit vector associated with the principal line. Derive an expression for the direction cosine matrix $[c]$ in terms of the Rodriguez parameters.

16. Show using the definition of the Rodrigues vector from Problem 15 that the vectors $\mathbf{r}$ and $\mathbf{r}'$ in Eq. [7.7.46] are related by $\mathbf{r}' = \mathbf{r} - (\mathbf{r} + \mathbf{r}') \times \boldsymbol{\rho}$.

Section 7.8

17. Consider the Cardan joint in Fig. 7.21. Plot Eq. [7.8.9] for different values of β and identify the range in which the Cardan joint can be assumed as a constant velocity joint (specified here as less than 3 percent variation in angular velocity).

18. The collars A and B are attached by a rod of length 30 cm, as shown in Fig. 7.47. The joint at A is a ball and socket joint and at B a pin joint. Collar A moves in the Z direction, while the guide bar for collar B is on the XY plane. At the point shown collar A is at a height of 24 cm and collar B is moving with a speed of 10 cm/s. Find the angular velocity of the rod.

19. The collars A and B are attached by a rod of length 30 in. The joint at A is a pin joint and at B a ball and socket joint, as shown in Fig. 7.48. The motion of collar B is restricted to the XY plane. At the instant shown, the rod is at a height of 25 in and collar A is moving down with the speed of 10 in/sec. Find the velocity of collar B and the angular velocity of the rod.

20. The two guide bars are on the same plane, as shown in Fig. 7.49. A rod of length 20 in. is attached to two collars, one at each end, by pin joints. Find the velocity of collar B if collar A has a speed of 3 in./sec at the position shown.

21. A sketch of a centrifugal governor is shown in Fig. 7.50. As the angular speed Ω increases, point D moves upward, forcing the point C to move up as well. We are given that $\theta = 40°$, point D is moving up with speed 0.2 m/s, and the

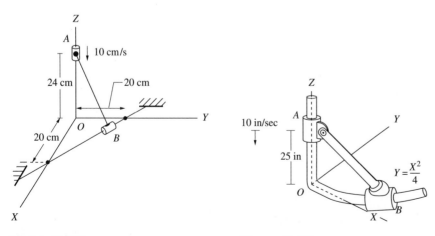

Figure 7.47   **Figure 7.48**

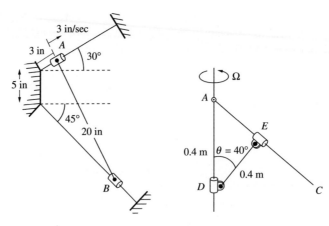

Figure 7.49 **Figure 7.50**

governor is rotating with $\Omega = 200$ rpm. Find the angular velocity of the arm DE and the velocity of the collar E with respect to the shaft AC.

Section 7.9

22. A cylinder of radius R rolls inside a parabolic surface defined by the relation $y = kx^2$, as shown in Fig. 7.51. Find the critical value of k that permits the cylinder to roll continuously inside the parabola.
23. Consider the disk in Fig. 7.33. Show that the velocity of the contact point, as viewed in the F frame, that is, $^F\mathbf{v}_C$, always lies on the XY plane.
24. Consider the gyropendulum of Fig. 7.14. Here, the disk is rotating without slipping inside a cylinder of radius $D = (\sqrt{3}L + R)/2$, as shown in Fig. 7.52. Given a constant precession rate of 1.8 rad/s, find the corresponding spin rate, as well as the angular velocity and angular acceleration of the disk.

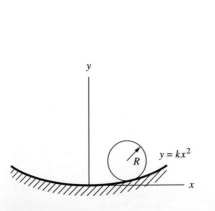
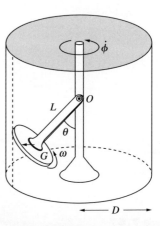

Figure 7.51 **Figure 7.52**

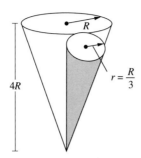

Figure 7.53

25. Consider the disk in Example 7.10. Now, the shaft arm is moving upward with respect to the horizontal surface with the relationship $\dot{h}$ = constant. This causes the point of contact between the disk and the surface to slip in the d_1 direction, while the rolling motion is still without slipping. Given that ω_1 = constant, find the angular velocity and angular acceleration of the disk for $R = L/4$.

26. The disk in Fig. 7.33 is rolling without slipping. Consider that $R = 8$ in, and that when the nutation angle $\theta = 75°$ and is constant, the precession rate is observed to be 0.4 rad/s and the angular velocity of the disk about its symmetry axis is 15 rad/s. Find the angular velocity of the disk and the velocity of the center of the disk.

27. The small cone in Fig. 7.53 rolls on the large cone (both cones are right-angled) and makes a trip every four seconds. Find the angular velocity and acceleration of the small cone.

28. Consider the unicycle in Fig. 7.54. The wheel is of mass m and radius R and is approximated as a disk, while the rider is modeled as a slender rod of mass $3m$ and length $3R$. Find the velocity and acceleration of the center of mass of the rider in terms of the Euler angles ψ, θ, and ϕ and a 3-2-3 transformation. Assume roll without slip and that the rider is upright.

29. Consider the axle in Fig. 7.55 to which two wheels of identical size are attached. The wheels roll about the axle without slipping. Perform a kinematic analysis

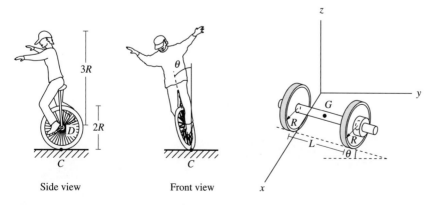

Side view   Front view

Figure 7.54   **Figure 7.55**

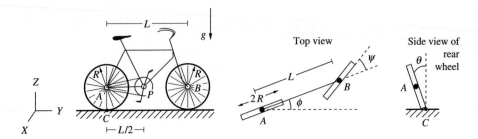

Figure 7.56

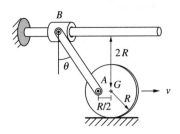

Figure 7.57

and determine the number of degrees of freedom. How would you select quasi-velocities for this problem?

30. Consider the bicycle in Fig. 7.56. Considering roll without slip of both wheels, determine how many degrees of freedom the bicycle has, and write an expression for the velocity of point B. Identify whether the constraints are holonomic or not. Begin with the frame and the coordinates of point A. The angle θ is with respect to the XY plane and it represents the tilting of the bicycle.

31. The wheel in Fig. 7.57 rolls without slipping, with its center having the constant speed v. Find the velocity and acceleration of the collar as a function of θ.

Table 7.1 Summary of common Euler angle transformations ($a_1a_2a_3$ to $b_1b_2b_3$, or XYZ to xyz)

	Name of Sequence		
	3-1-3 (Historically Significant)	**3-2-3 (NASA Standard Aerospace)**	**3-2-1 (NASA Standard Airplane)**
Sequence of Angles	ϕ (precession), θ (nutation), ψ (spin)	ψ (precession), θ (nutation), ϕ (spin)	ψ (heading), θ (attitude), ϕ (bank)
Application	Spinning or rolling bodies, orbital parameters	Spinning or rolling bodies	Vehicle motion, attitude dynamics
Names of Angular Velocities	ω_3: spin rate	ω_3: spin rate	ω_1: pitch, ω_2: roll, ω_3: yaw
Transformation Matrices	$[R_1] = \begin{bmatrix} c\phi & s\phi & 0 \\ -s\phi & c\phi & 0 \\ 0 & 0 & 1 \end{bmatrix}$ $[R_2] = \begin{bmatrix} 1 & 0 & 0 \\ 0 & c\theta & s\theta \\ 0 & -s\theta & c\theta \end{bmatrix}$ $[R_3] = \begin{bmatrix} c\psi & s\psi & 0 \\ -s\psi & c\psi & 0 \\ 0 & 0 & 1 \end{bmatrix}$	$[R_1] = \begin{bmatrix} c\psi & s\psi & 0 \\ -s\psi & c\psi & 0 \\ 0 & 0 & 1 \end{bmatrix}$ $[R_2] = \begin{bmatrix} c\theta & 0 & -s\theta \\ 0 & 1 & 0 \\ s\theta & 0 & c\theta \end{bmatrix}$ $[R_3] = \begin{bmatrix} c\phi & s\phi & 0 \\ -s\phi & c\phi & 0 \\ 0 & 0 & 1 \end{bmatrix}$	$[R_1] = \begin{bmatrix} c\psi & s\psi & 0 \\ -s\psi & c\psi & 0 \\ 0 & 0 & 1 \end{bmatrix}$ $[R_2] = \begin{bmatrix} c\theta & 0 & -s\theta \\ 0 & 1 & 0 \\ s\theta & 0 & c\theta \end{bmatrix}$ $[R_3] = \begin{bmatrix} 1 & 0 & 0 \\ 0 & c\phi & s\phi \\ 0 & -s\phi & c\phi \end{bmatrix}$
Combined Transformation Matrix $[R] = [R_3][R_2][R_1]$	$\begin{bmatrix} c\phi c\psi - s\phi c\theta s\psi & s\phi c\psi + c\phi c\theta s\psi & s\theta s\psi \\ -c\phi s\psi - s\phi c\theta c\psi & -s\phi s\psi + c\phi c\theta c\psi & s\theta c\psi \\ s\phi s\theta & -c\phi s\theta & c\theta \end{bmatrix}$	$\begin{bmatrix} c\psi c\theta c\phi - s\psi s\phi & s\psi c\theta c\phi + c\psi s\phi & -s\theta c\phi \\ -c\psi c\theta s\phi - s\psi c\phi & -s\psi c\theta s\phi + c\psi c\phi & s\theta s\phi \\ c\psi s\theta & s\psi s\theta & c\theta \end{bmatrix}$	$\begin{bmatrix} c\psi c\theta & s\psi c\theta & -s\theta \\ -s\psi c\phi + c\psi s\theta s\phi & c\psi c\phi + s\psi s\theta s\phi & c\theta s\phi \\ s\psi s\phi + c\psi s\theta c\phi & -c\psi s\phi + s\psi s\theta c\phi & c\theta c\phi \end{bmatrix}$
$[B]$ Matrix	$\begin{bmatrix} s\theta s\psi & c\psi & 0 \\ s\theta c\psi & -s\psi & 0 \\ c\theta & 0 & 1 \end{bmatrix}$	$\begin{bmatrix} -s\theta c\phi & s\phi & 0 \\ s\theta s\phi & c\phi & 0 \\ c\theta & 0 & 1 \end{bmatrix}$	$\begin{bmatrix} -s\theta & 0 & 1 \\ c\theta s\phi & c\phi & 0 \\ c\theta c\phi & -s\phi & 0 \end{bmatrix}$
Singularity at	$\theta = 0, \pm\pi$	$\theta = 0, \pm\pi$	$\theta = \pm\pi/2$
Angular Velocity in Body-Fixed Reference Frame	$\boldsymbol{\omega} = \dot{\phi}\mathbf{a}_3 + \dot{\theta}\mathbf{a}_1' + \dot{\psi}\mathbf{b}_3$ $= (\dot{\phi}s\theta s\psi + \dot{\theta}c\psi)\mathbf{b}_1 + (\dot{\phi}s\theta c\psi - \dot{\theta}s\psi)\mathbf{b}_2 + (\dot{\phi}c\theta + \dot{\psi})\mathbf{b}_3$	$\boldsymbol{\omega} = \dot{\psi}\mathbf{a}_3 + \dot{\theta}\mathbf{a}_2' + \dot{\phi}\mathbf{b}_3$ $= (-\dot{\psi}s\theta c\phi + \dot{\theta}s\phi)\mathbf{b}_1 + (\dot{\psi}s\theta s\phi + \dot{\theta}c\phi)\mathbf{b}_2 + (\dot{\psi}c\theta + \dot{\phi})\mathbf{b}_3$	$\boldsymbol{\omega} = \dot{\psi}\mathbf{a}_3 + \dot{\theta}\mathbf{a}_2' + \dot{\phi}\mathbf{b}_1$ $= (-\dot{\psi}s\theta + \dot{\phi})\mathbf{b}_1 + (\dot{\psi}c\theta s\phi + \dot{\theta}c\phi)\mathbf{b}_2 + (\dot{\psi}c\theta c\phi - \dot{\theta}s\phi)\mathbf{b}_3$
Derivatives of Euler Angles in Terms of Angular Velocities	$\dot{\phi} = \dfrac{1}{\sin\theta}(\omega_1 s\psi + \omega_2 c\psi)$ $\dot{\theta} = \omega_1 c\psi - \omega_2 s\psi$ $\dot{\psi} = \dfrac{1}{\sin\theta}(-\omega_1 c\theta s\psi - \omega_2 c\theta c\psi) + \omega_3$	$\dot{\psi} = \dfrac{1}{\sin\theta}(-\omega_1 c\phi + \omega_2 s\phi)$ $\dot{\theta} = \omega_1 s\phi + \omega_2 c\phi$ $\dot{\phi} = \dfrac{1}{\sin\theta}(\omega_1 c\theta c\phi - \omega_2 c\theta s\phi) + \omega_3$	$\dot{\psi} = \dfrac{1}{\cos\theta}(\omega_2 s\phi + \omega_3 c\phi)$ $\dot{\theta} = \omega_2 c\phi - \omega_3 s\phi$ $\dot{\phi} = \dfrac{1}{\cos\theta}(\omega_2 s\theta s\phi + \omega_3 s\theta c\phi) + \omega_1$

Table 7.1 (Continued)

	Name of Sequence		
	3-1-3 (Historically Significant)	3-2-3 (NASA Standard Aerospace)	3-2-1 (NASA Standard Airplane)
Transformation Matrix for F Frame	$\begin{bmatrix} c\phi & s\phi & 0 \\ -s\phi c\theta & c\phi c\theta & s\theta \\ s\phi s\theta & -c\phi s\theta & c\theta \end{bmatrix}$	$\begin{bmatrix} c\psi c\theta & s\psi c\theta & -s\theta \\ -s\psi & c\psi & 0 \\ c\psi s\theta & s\psi s\theta & c\theta \end{bmatrix}$	Not of any significant use
Angular Velocities in F Frame (for Inertially Symmetric Bodies)	$\boldsymbol{\omega}_b = \dot{\phi}\mathbf{a}_3 + \dot{\theta}\mathbf{a}_1' + \dot{\psi}\mathbf{f}_3$ $\boldsymbol{\omega}_f = \dot{\theta}\mathbf{f}_1 + \dot{\phi}s\theta\mathbf{f}_2 + [\dot{\phi}c\theta + \dot{\psi}]\mathbf{f}_3$ $\boldsymbol{\omega}_s = \dot{\psi}\mathbf{f}_3$	$\boldsymbol{\omega}_b = \dot{\psi}\mathbf{a}_3 + \dot{\theta}\mathbf{a}_2' + \dot{\phi}\mathbf{f}_3$ $= \dot{\theta}\mathbf{f}_1 + \dot{\phi}s\theta\mathbf{f}_2 + [\dot{\psi}c\theta + \dot{\phi}]\mathbf{f}_3$ $\boldsymbol{\omega}_f = \dot{\theta}\mathbf{f}_1 + \dot{\psi}s\theta\mathbf{f}_2 + \dot{\psi}c\theta\mathbf{f}_3$ $\boldsymbol{\omega}_s = \dot{\phi}\mathbf{f}_3$	Not of any significant use
Derivatives of Euler Angles in Terms of Angular Velocities Expressed in F Frame	$\dot{\phi} = \dfrac{\omega_2}{\sin\theta}$ $\dot{\theta} = \omega_1$ $\dot{\psi} = -\dfrac{\omega_2 \cos\theta}{\sin\theta} + \omega_3$	$\dot{\psi} = -\dfrac{\omega_1}{\sin\theta}$ $\dot{\theta} = \omega_2$ $\dot{\phi} = \dfrac{\omega_1 \cos\theta}{\sin\theta} + \omega_3$	Not of any significant use
Angular Acceleration of Body in Body Frame	$\alpha_1 = \ddot{\phi}s\theta s\psi + \dot{\phi}\dot{\theta}c\theta s\psi$ $\quad + \dot{\phi}\dot{\psi}s\theta c\psi + \ddot{\theta}c\psi - \dot{\theta}\dot{\psi}s\psi$ $\alpha_2 = \ddot{\phi}s\theta c\psi + \dot{\phi}\dot{\theta}c\theta c\psi$ $\quad - \dot{\phi}\dot{\psi}s\theta s\psi - \ddot{\theta}s\psi - \dot{\theta}\dot{\psi}c\psi$ $\alpha_3 = \ddot{\phi}c\theta - \dot{\phi}\dot{\theta}s\theta + \ddot{\psi}$	$\alpha_1 = -\ddot{\psi}s\theta c\phi - \dot{\psi}\dot{\theta}c\theta c\phi$ $\quad + \dot{\psi}\dot{\phi}s\theta s\phi + \ddot{\theta}s\phi + \dot{\theta}\dot{\phi}c\phi$ $\alpha_2 = \ddot{\psi}s\theta s\phi + \dot{\psi}\dot{\theta}c\theta s\phi$ $\quad + \dot{\psi}\dot{\phi}s\theta c\phi + \ddot{\theta}c\phi - \dot{\theta}\dot{\phi}s\phi$ $\alpha_3 = \ddot{\psi}c\theta - \dot{\psi}\dot{\theta}s\theta + \ddot{\phi}$	$\alpha_1 = -\ddot{\psi}s\theta - \dot{\psi}\dot{\theta}c\theta + \ddot{\phi}$ $\alpha_2 = \ddot{\psi}c\theta s\phi - \dot{\psi}\dot{\theta}s\theta s\phi$ $\quad + \dot{\psi}\dot{\phi}c\theta c\phi + \ddot{\theta}c\phi - \dot{\theta}\dot{\phi}s\phi$ $\alpha_3 = \ddot{\psi}c\theta c\phi - \dot{\psi}\dot{\theta}s\theta c\phi$ $\quad - \dot{\psi}\dot{\phi}c\theta s\phi - \ddot{\theta}s\phi - \dot{\theta}\dot{\phi}c\phi$
Angular Acceleration of Body in F Frame	$\boldsymbol{\alpha}_b = [\ddot{\theta} + \dot{\phi}\dot{\psi}s\theta]\mathbf{f}_1$ $\quad + [\ddot{\phi}s\theta + \dot{\phi}\dot{\theta}c\theta - \dot{\theta}\dot{\psi}]\mathbf{f}_2$ $\quad + [\ddot{\phi}c\theta - \dot{\phi}\dot{\theta}s\theta + \ddot{\psi}]\mathbf{f}_3$	$\boldsymbol{\alpha}_b = [-\ddot{\psi}s\theta - \dot{\psi}\dot{\theta}c\theta + \ddot{\theta}\phi]\mathbf{f}_1$ $\quad + [\ddot{\theta} + \dot{\psi}\dot{\phi}s\theta]\mathbf{f}_2$ $\quad + [\ddot{\psi}c\theta - \dot{\psi}\dot{\theta}s\theta + \ddot{\phi}]\mathbf{f}_3$	Not of any significant use
Euler Parameters in Terms of Euler Angles	$e_0 = c\dfrac{\phi+\psi}{2}c\dfrac{\theta}{2}$ $e_1 = c\dfrac{\phi-\psi}{2}s\dfrac{\theta}{2}$ $e_2 = s\dfrac{\phi-\psi}{2}s\dfrac{\theta}{2}$ $e_3 = s\dfrac{\phi+\psi}{2}c\dfrac{\theta}{2}$	$e_0 = c\dfrac{\phi+\psi}{2}c\dfrac{\theta}{2}$ $e_1 = s\dfrac{\phi-\psi}{2}s\dfrac{\theta}{2}$ $e_2 = c\dfrac{\phi-\psi}{2}s\dfrac{\theta}{2}$ $e_3 = s\dfrac{\phi+\psi}{2}c\dfrac{\theta}{2}$	$e_0 = c\dfrac{\psi}{2}c\dfrac{\theta}{2}c\dfrac{\phi}{2} + s\dfrac{\psi}{2}s\dfrac{\theta}{2}s\dfrac{\phi}{2}$ $e_1 = c\dfrac{\psi}{2}c\dfrac{\theta}{2}s\dfrac{\phi}{2} - s\dfrac{\psi}{2}s\dfrac{\theta}{2}c\dfrac{\phi}{2}$ $e_2 = c\dfrac{\psi}{2}s\dfrac{\theta}{2}c\dfrac{\phi}{2} + s\dfrac{\psi}{2}c\dfrac{\theta}{2}s\dfrac{\phi}{2}$ $e_3 = s\dfrac{\psi}{2}c\dfrac{\theta}{2}c\dfrac{\phi}{2} - c\dfrac{\psi}{2}s\dfrac{\theta}{2}s\dfrac{\phi}{2}$

chapter

8

RIGID BODY DYNAMICS: BASIC CONCEPTS

8.1 INTRODUCTION

This chapter presents basic concepts associated with the kinetics of rigid bodies. We develop the linear and angular momentum expressions. The Newtonian and Eulerian laws of motion are introduced as force and moment balances. We discuss analytical methods of obtaining equations of motion, such as Lagrange's equations and the direct application of D'Alembert's principle. We analyze qualitative and quantitative integration of the equations of motion, together with motion integrals. The developments in this chapter make extensive use of the geometric and kinematic properties developed in the previous two chapters.

When describing the motion of a particle three coordinates are used, one for each direction of the motion. For analyzing the motion of a rigid body one needs to use angular variables. The kinematics of a rigid body is studied using both inertial coordinates and relative coordinates. As we saw in the previous chapter, in most cases using a relative frame attached to the body gives better insight; this continues to be the case for the kinetics of rigid bodies. The use of body-fixed coordinates is attractive, as the moments of inertia of the body with respect to a set of body-fixed coordinates remains constant, and it becomes a simple task to differentiate the angular velocity. Thus, we derive several forms of the equations of motion, analogous to the concept of using different generalized coordinates. We extend the impulse-momentum and work-energy relationships to three-dimensional motion. This leads to a further study of impact problems, first introduced in Chapter 3.

8.2 Linear and Angular Momentum

Consider a rigid body, such as the one in Fig. 8.1, with center of mass G. Also consider an arbitrary point B, located on or off the body. The linear momentum of the body, denoted by $\mathbf{p}$, is defined as

$$\mathbf{p} = \int_{body} \mathbf{v}\, dm = \int_{body} \frac{d\mathbf{r}}{dt}\, dm \qquad [8.2.1]$$

The position vector of the differential element is

$$\mathbf{r} = \mathbf{r}_G + \boldsymbol{\rho} \qquad [8.2.2]$$

where $\boldsymbol{\rho}$ is the vector connecting the center of mass with the differential element. Differentiating the above expression, we obtain the velocity of the differential element as

$$\mathbf{v} = \frac{d\mathbf{r}}{dt} = \mathbf{v}_G + \boldsymbol{\omega} \times \boldsymbol{\rho} \qquad [8.2.3]$$

where $\boldsymbol{\omega}$ is the angular velocity of the body. Substituting Eq. [8.2.3] into Eq. [8.2.1], we obtain

$$\mathbf{p} = \int_{body} (\mathbf{v}_G + \boldsymbol{\omega} \times \boldsymbol{\rho})\, dm$$
$$= \mathbf{v}_G \int_{body} dm + \boldsymbol{\omega} \times \int_{body} \boldsymbol{\rho}\, dm = m\mathbf{v}_G + \boldsymbol{\omega} \times \int_{body} \boldsymbol{\rho}\, dm \qquad [8.2.4]$$

Recalling the definition of center of mass from Chapter 6, we write

$$\int_{body} \boldsymbol{\rho}\, dm = 0 \qquad [8.2.5]$$

so that introducing Eq. [8.2.5] into Eq. [8.2.4], we obtain the linear momentum as

$$\mathbf{p} = m\mathbf{v}_G \qquad [8.2.6]$$

As expected, the linear momentum of a rigid body is equal to its mass multiplied by the velocity of its center of mass. Hence, the translational motion of a rigid body

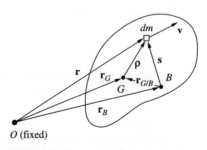

Figure 8.1

can be treated the same way as the motion of a particle having the same mass as the rigid body and where the entire mass of the body is concentrated at the center of mass.

The angular momentum, or moment of linear momentum, of the body about the center of mass, denoted by $\mathbf{H}_G$, is defined as

$$\mathbf{H}_G = \int_{body} \boldsymbol{\rho} \times \mathbf{v} \, dm \qquad [\mathbf{8.2.7}]$$

If we introduce Eq. [8.2.3] into this equation and make use of Eq. [8.2.5], we obtain

$$\mathbf{H}_G = \int_{body} \boldsymbol{\rho} \times (\mathbf{v}_G + \boldsymbol{\omega} \times \boldsymbol{\rho}) \, dm = \int_{body} \boldsymbol{\rho} \times (\boldsymbol{\omega} \times \boldsymbol{\rho}) \, dm \qquad [\mathbf{8.2.8}]$$

We express the vectors $\boldsymbol{\rho}$ and $\boldsymbol{\omega}$ in terms of their components in the body-fixed coordinate system xyz as

$$\boldsymbol{\rho} = x\mathbf{i} + y\mathbf{j} + z\mathbf{k} \qquad \boldsymbol{\omega} = \omega_x\mathbf{i} + \omega_y\mathbf{j} + \omega_z\mathbf{k} \qquad [\mathbf{8.2.9}]$$

Using the vector identity

$$\boldsymbol{\rho} \times (\boldsymbol{\omega} \times \boldsymbol{\rho}) = (\boldsymbol{\rho} \cdot \boldsymbol{\rho})\boldsymbol{\omega} - (\boldsymbol{\rho} \cdot \boldsymbol{\omega})\boldsymbol{\rho} \qquad [\mathbf{8.2.10}]$$

and expressing $\mathbf{H}_G$ in terms of its components as

$$\mathbf{H}_G = H_{Gx}\mathbf{i} + H_{Gy}\mathbf{j} + H_{Gz}\mathbf{k} \qquad [\mathbf{8.2.11}]$$

and carrying out the algebra, we obtain

$$H_{Gx} = \int_{body} [(y^2 + z^2)\omega_x - xy\omega_y - xz\omega_z] \, dm$$

$$H_{Gy} = \int_{body} [(x^2 + z^2)\omega_y - xy\omega_x - yz\omega_z] \, dm$$

$$H_{Gz} = \int_{body} [(x^2 + y^2)\omega_z - xz\omega_x - yz\omega_y] \, dm \qquad [\mathbf{8.2.12}]$$

Recalling the definitions for mass moments of inertia derived in Chapter 6, Eq. [8.2.12] becomes

$$H_{Gx} = I_{Gxx}\omega_x - I_{Gxy}\omega_y - I_{Gxz}\omega_z$$

$$H_{Gy} = I_{Gyy}\omega_y - I_{Gxy}\omega_x - I_{Gyz}\omega_z$$

$$H_{Gz} = I_{Gzz}\omega_z - I_{Gxz}\omega_x - I_{Gyz}\omega_y \qquad [\mathbf{8.2.13}]$$

or, in column vector format,

$$\{H_G\} = [I_G]\{\omega\} \qquad [\mathbf{8.2.14}]$$

where

$$\{H_G\} = [H_{Gx} \; H_{Gy} \; H_{Gz}]^T \qquad \{\omega\} = [\omega_x \; \omega_y \; \omega_z]^T \qquad [\mathbf{8.2.15}]$$

and $[I_G]$ is the inertia matrix defined in Eq. [6.3.4].

The expression for the angular momentum about the center of mass can also be derived conveniently in column vector format. Writing Eq. [8.2.8] as

$$\mathbf{H}_G = \int_{body} \boldsymbol{\rho} \times (\boldsymbol{\omega} \times \boldsymbol{\rho}) \, dm = \int_{body} -\boldsymbol{\rho} \times (\boldsymbol{\rho} \times \boldsymbol{\omega}) \, dm \qquad [\mathbf{8.2.16}]$$

and using Eqs. [2.3.7], we can express Eq. [8.2.16] in column vector format as

$$\{H_G\} = \int_{body} -[\tilde{\rho}][\tilde{\rho}]\{\omega\} \, dm \qquad [\mathbf{8.2.17}]$$

Noting that $[\tilde{\rho}]$ is a skew-symmetric matrix and using Eq. [6.4.2], we obtain

$$[I_G] = \int_{body} [\tilde{\rho}]^T [\tilde{\rho}] \, dm \qquad [\mathbf{8.2.18}]$$

which leads to Eq. [8.2.14].

The expression for angular momentum is simplified considerably when the coordinate axes used coincide with the principal axes or, for axisymmetric bodies, the F frame. The products of inertia vanish, and the angular momentum expression about the center of mass becomes

$$\mathbf{H}_G = I_{Gxx}\omega_x \mathbf{i} + I_{Gyy}\omega_y \mathbf{j} + I_{Gzz}\omega_z \mathbf{k} \qquad [\mathbf{8.2.19}]$$

or, using $b_1 b_2 b_3$ as the coordinate frame,

$$\mathbf{H}_G = I_1 \omega_1 \mathbf{b}_1 + I_2 \omega_2 \mathbf{b}_2 + I_3 \omega_3 \mathbf{b}_3 \qquad [\mathbf{8.2.20}]$$

where the body-fixed axes xyz or $b_1 b_2 b_3$ are aligned with the principal axes.

A special case of motion is rotation about a fixed axis. For example, for rotation about the z axis, $\boldsymbol{\omega} = \omega \mathbf{k}$, so

$$\mathbf{H}_G = -I_{Gxz}\omega \mathbf{i} - I_{Gyz}\omega \mathbf{j} + I_{Gzz}\omega \mathbf{k} \qquad [\mathbf{8.2.21}]$$

Example 8.1

Find the angular momentum about the centroid of the rolling cone shown in Fig. 7.35 using a set of centroidal axes that go through the symmetry axis of the cone.

Solution

This cone is the one used in Example 7.9. Noting from Fig. 7.36 that $\mathbf{h}_3 = \cos\beta \mathbf{f}_3 - \sin\beta \mathbf{f}_2$ and $R = L \tan\beta$, we can express the angular velocity as

$$\boldsymbol{\omega} = -\frac{v}{L \sin\beta}\mathbf{h}_3 = \frac{v}{L}\mathbf{f}_2 - \frac{v \cos\beta}{L \sin\beta}\mathbf{f}_3 = \frac{v}{L}\mathbf{f}_2 - \frac{v}{R}\mathbf{f}_3 \qquad [\mathbf{a}]$$

The centroidal moments of inertia of a right circular cone are $I_3 = \frac{3}{10}mR^2$, $I_1 = I_2 = \frac{3}{80}m(4R^2 + L^2)$. The angular momentum then becomes

$$\mathbf{H}_G = I_2 \omega_2 \mathbf{f}_2 + I_3 \omega_3 \mathbf{f}_3 = \frac{3}{80}m(4R^2 + L^2)\frac{v}{L}\mathbf{f}_2 - \frac{3}{10}mRv\mathbf{f}_3 \qquad [\mathbf{b}]$$

It is clear that the angular momentum and angular velocity are in different directions.

8.3 Transformation Properties of Angular Momentum

The transformation properties associated with angular momentum for a rigid body are very similar to the transformation properties associated with mass moments of inertia. We will see both the translational and rotational properties in sequence.

8.3.1 Translation of Coordinates

Consider angular momentum expressions about the center of mass and about an arbitrary point B. From Fig. 8.1 we express the distance from B to the differential element as $\mathbf{s}$, where

$$\mathbf{s} = \mathbf{r}_{G/B} + \boldsymbol{\rho} \qquad [8.3.1]$$

Introducing Eq. [8.3.1] into the definition of the angular momentum about B, we obtain

$$\mathbf{H}_B = \int_{\text{body}} \mathbf{s} \times \mathbf{v}\, dm = \int_{\text{body}} (\mathbf{r}_{G/B} + \boldsymbol{\rho}) \times (\mathbf{v}_G + \boldsymbol{\omega} \times \boldsymbol{\rho})\, dm \qquad [8.3.2]$$

which, when expanded, yields

$$\mathbf{H}_B = \int_{\text{body}} (\mathbf{r}_{G/B} \times \mathbf{v}_G + \boldsymbol{\rho} \times \mathbf{v}_G + \mathbf{r}_{G/B} \times (\boldsymbol{\omega} \times \boldsymbol{\rho}) + \boldsymbol{\rho} \times (\boldsymbol{\omega} \times \boldsymbol{\rho}))\, dm \qquad [8.3.3]$$

The second and third terms on the right side of this equation vanish due to the definition of the center of mass. The last term is recognized as the angular momentum about the center of mass. The angular momentum about point B can then be written as

$$\mathbf{H}_B = \mathbf{H}_G + m\mathbf{r}_{G/B} \times \mathbf{v}_G \qquad [8.3.4]$$

If the body is rotating about a fixed point C *on the body,* one can express the velocity of the center of mass as $\mathbf{v}_G = \boldsymbol{\omega} \times \mathbf{r}_{G/C}$, which, when introduced into Eq. [8.3.3], yields

$$\mathbf{H}_C = \int_{\text{body}} (\mathbf{r}_{G/C} + \boldsymbol{\rho}) \times (\boldsymbol{\omega} \times \mathbf{r}_{G/C} + \boldsymbol{\omega} \times \boldsymbol{\rho})\, dm$$

$$= \int_{\text{body}} (\mathbf{r}_{G/C} + \boldsymbol{\rho}) \times (\boldsymbol{\omega} \times (\mathbf{r}_{G/C} + \boldsymbol{\rho}))\, dm \qquad [8.3.5]$$

The column vector representation of Eq. [8.3.5] is recognized as the inertia matrix about point C times the angular velocity vector, or

$$\{H_C\} = [I_C]\{\omega\} \qquad [8.3.6]$$

Note that if point C is not attached to the body, Eq. [8.3.6] does not hold (except when point C can be viewed as an extension of the body) and one should use Eq. [8.3.4] to calculate the angular momentum.

8.3.2 ROTATION OF COORDINATES

While the translation theorem for angular momentum is valid only when one is considering the center of mass, the rotation relations are valid for angular momentum calculated about any point. Finding mass moments of inertia by rotation of axes can be carried out without any restriction on the origin of the coordinate system. We use a point B to which a set of coordinates is attached. We are given the angular momentum of the rigid body about B, and using this set of coordinates we are asked to find the components of the angular momentum vector about a rotated set of coordinates. This is essentially the same question asked in Chapter 2: Given a vector in a certain coordinate system, what are the components of that vector in a transformed coordinate system?

Given the angular momentum vector in the xyz coordinates as

$$\mathbf{H}_B = H_{Bx}\mathbf{i} + H_{By}\mathbf{j} + H_{Bz}\mathbf{k} \qquad [8.3.7]$$

we wish to compute its components in the $x'y'z'$ frame, obtained by rotating the xyz frame. We express the angular momentum in the $x'y'z'$ frame as

$$\mathbf{H}'_B = H_{Bx'}\mathbf{i}' + H_{By'}\mathbf{j}' + H_{Bz'}\mathbf{k}' \qquad [8.3.8]$$

Given the direction cosine matrix between the two frames as $[c] = [R]^T$, we can write in column vector format that

$$\{H'_B\} = [c]^T\{H_B\} = [R]\{H_B\} \qquad [8.3.9]$$

in which

$$\{H'_B\} = [H_{Bx'} \quad H_{By'} \quad H_{Bz'}]^T \qquad [8.3.10]$$

are the components of the angular momentum in the coordinate system $x'y'z'$.

If the point B coincides with the center of mass or if the point B is fixed in rotation, we have an interesting result. Denoting, as we did in Chapter 3, such a point by D, and considering that $\{H_D\} = [I_D]\{\omega\}$, the expression for the angular momentum in the $x'y'z'$ frame should have the form

$$\{H'_D\} = [I'_D]\{\omega'\} \qquad [8.3.11]$$

where

$$\{\omega'\} = [\omega_{x'} \quad \omega_{y'} \quad \omega_{z'}]^T \qquad [8.3.12]$$

are the components of the angular velocity vector expressed in the transformed coordinates and $[I'_D]$ is the inertia matrix in terms of the primed coordinates. We proceed to show that this is indeed the case. One can relate the angular velocities in the two frames as

$$\{\omega'\} = [c]^T\{\omega\} \qquad \{\omega\} = [c]\{\omega'\} \qquad [8.3.13]$$

Introducing Eq. [8.3.13] into Eq. [8.3.9], we obtain

$$\{H'_D\} = [c]^T\{H_D\} = [c]^T[I_D]\{\omega\} = [c]^T[I_D][c]\{\omega'\} \qquad [8.3.14]$$

Comparing Eq. [8.3.14] with Eq. [8.3.11] we recognize that

$$[I'_D] = [c]^T [I_D][c] \qquad \text{[8.3.15]}$$

which is the same relation obtained in Eq. [6.4.13] when discussing transformation of coordinates.

Considering the complicated nature of expressions associated with moments of inertia and angular momentum, the advantages of using the principal axes of a body as reference axes become more obvious. When finding the angular momentum of a body about a set of axes not coinciding with the principal axes, it is often more desirable to find the angular momentum about the principal axes and to then use Eq. [8.3.9] to find the angular momentum about the desired axes.

Example 8.2

Find the angular momentum of the disk in Fig. 2.21 about point O. The distance from O to B is L, from B to C is $L/3$ and the disk is of radius R.

Solution

We will use Eq. [8.3.4] and the F frame to calculate the angular momentum. The reference frame is shown in Figs. 2.22–2.23. From Example 2.7 the angular velocity has the form

$$\boldsymbol{\omega} = \omega_1 \mathbf{K} + \dot{\theta} \mathbf{j}' + \omega_3 \mathbf{i} \qquad \text{[a]}$$

where

$$\mathbf{K} = \sin 40° \mathbf{k}' - \cos 40° \mathbf{j}' \qquad \mathbf{k}' = \cos\theta \mathbf{k} - \sin\theta \mathbf{i} \qquad \mathbf{j}' = \mathbf{j} \qquad \text{[b]}$$

so that the angular velocity has the form

$$\boldsymbol{\omega} = (\omega_3 - \omega_1 \sin 40° \sin\theta)\mathbf{i} + (\dot{\theta} - \omega_1 \cos 40°)\mathbf{j} + \omega_1 \sin 40° \cos\theta \mathbf{k} \qquad \text{[c]}$$

The angular momentum about the center of mass (denoted by C in Fig. 2.21) is

$$\mathbf{H}_C = \mathbf{H}_G = I_{Gxx}\omega_x \mathbf{i} + I_{Gyy}\omega_y \mathbf{j} + I_{Gzz}\omega_z \mathbf{k} \qquad \text{[d]}$$

in which $I_{Gxx} = mR^2/2$, $I_{Gyy} = I_{Gzz} = mR^2/4$. Hence, we write the angular momentum about the center of mass as

$$\mathbf{H}_C = \frac{mR^2}{4}[2(\omega_3 - \omega_1 \sin 40° \sin\theta)\mathbf{i} + (\dot{\theta} - \omega_1 \cos 40°)\mathbf{j} + \omega_1 \sin 40° \cos\theta \mathbf{k}] \qquad \text{[e]}$$

We next calculate the velocity of C. We first find the velocity of point B as

$$\mathbf{v}_B = \omega_1 \mathbf{K} \times L\mathbf{j}' = \omega_1(\sin 40° \mathbf{k}' - \cos 40° \mathbf{j}) \times L\mathbf{j}' = -L\omega_1 \sin 40° \mathbf{i}' \qquad \text{[f]}$$

From Fig. 2.23 we have $\mathbf{i}' = \cos\theta \mathbf{i} + \sin\theta \mathbf{k}$, so that

$$\mathbf{v}_B = -L\omega_1 \sin 40° \cos\theta \mathbf{i} - L\omega_1 \sin 40° \sin\theta \mathbf{k} \qquad \text{[g]}$$

We find the velocity of the center of the disk as

$$\mathbf{v}_C = \mathbf{v}_B + \boldsymbol{\omega}_f \times \mathbf{r}_{C/B} \qquad \text{[h]}$$

in which $\mathbf{r}_{C/B} = L/3\mathbf{i}$, so that

$$\boldsymbol{\omega}_f \times \mathbf{r}_{C/B} = [-\omega_1 \sin 40° \sin\theta\mathbf{i} + (\dot\theta - \omega_1 \cos 40°)\mathbf{j} + \omega_1 \sin 40° \cos\theta\mathbf{k}] \times \frac{L}{3}\mathbf{i}$$

$$= \frac{1}{3}L\omega_1 \sin 40° \cos\theta\mathbf{j} - \frac{1}{3}L(\dot\theta - \omega_1 \cos 40°)\mathbf{k} \qquad [\mathbf{i}]$$

Hence, the velocity of C becomes

$$\mathbf{v}_C = -L\omega_1 \sin 40° \cos\theta\mathbf{i} + \frac{1}{3}L\omega_1 \sin 40° \cos\theta\mathbf{j} - \frac{1}{3}L\big(\dot\theta - \omega_1(\cos 40° - 3\sin 40° \sin\theta)\big)\mathbf{k}$$
$$[\mathbf{j}]$$

We now calculate the second term in Eq. [8.3.4]. We note that $\mathbf{r}_{G/O} = L\mathbf{j} + L/3\mathbf{i}$ so that

$$\mathbf{r}_{G/O} \times \mathbf{v}_G = -\frac{L^2}{3}\big(\dot\theta - \omega_1(3\sin 40° \sin\theta - \cos 40°)\big)\mathbf{i}$$

$$+ \frac{L^2}{9}\big(\dot\theta - \omega_1(3\sin 40° \sin\theta - \cos 40°)\big)\mathbf{j} + \frac{10}{9}L^2\omega_1 \sin 40° \cos\theta\mathbf{k} \qquad [\mathbf{k}]$$

Multiplying Eq. [k] by m and adding it to Eq. [e] gives the angular momentum of the disk about point O.

Note that Eq. [k] does not have any terms involving ω_3, the spin of the disk. It is clear that using Eq. [8.3.6] would lead to an incorrect result.

Example 8.3

Find the angular momentum associated with the rolling cone in Fig. 7.35 about the pivot point O, using the set of coordinates $h_1 h_2 h_3$.

Solution

We can solve this problem in a number of ways. We can calculate the angular momentum directly from Eq. [8.3.9] using the direction cosine matrix. We can find by direct integration the inertia elements and use Eq. [8.3.15]. Also, because the cone is rotating about point O, we can find the mass moment of inertia and angular momentum components about O using the parallel axis theorem.

We begin with the third approach. First, we use the frame $f_1 f_2 f_3$ and express the angular momentum about point O using these axes. Given the centroidal moments of inertia in Example 8.1, and that the center of mass lies on the f_3 axis and is a distance of $3L/4$ away from O,

$$I_{O1} = I_1 + m\left(\frac{3L}{4}\right)^2 = \frac{3}{80}m(4R^2 + 16L^2) \qquad I_{O2} = I_{O1} \qquad I_{O3} = I_3 = \frac{3}{10}mR^2 \qquad [\mathbf{a}]$$

with all the products of inertia zero. We proceed to find the inertia matrix about the $h_1 h_2 h_3$ axes. Expressing the unit vectors as

$$\mathbf{h}_1 = \mathbf{f}_1 \qquad \mathbf{h}_2 = \cos\beta\mathbf{f}_2 + \sin\beta\mathbf{f}_3 \qquad \mathbf{h}_3 = -\sin\beta\mathbf{f}_2 + \cos\beta\mathbf{f}_3 \qquad [\mathbf{b}]$$

the direction cosine matrix has the form

$$[c] = \begin{bmatrix} 1 & 0 & 0 \\ 0 & \cos\beta & -\sin\beta \\ 0 & \sin\beta & \cos\beta \end{bmatrix} \qquad [\mathbf{c}]$$

Using Eq. [8.3.15] we obtain the inertia matrix in terms of the H frame as

$$[_H I_O] = [c]^T [_F I_O][c] \qquad \textbf{[d]}$$

Note that we are using the notation introduced in Chapter 7 to denote the coordinate axes in which quantities of interest are resolved by a subscript on the left. Substituting in the values for $[c]$ and carrying out the calculations, we obtain

$$[_H I_O] = \begin{bmatrix} I_{O1} & 0 & 0 \\ 0 & I_{O1} \cos^2 \beta + I_{O3} \sin^2 \beta & (I_{O3} - I_{O1}) \sin \beta \cos \beta \\ 0 & (I_{O3} - I_{O1}) \sin \beta \cos \beta & I_{O3} \cos^2 \beta + I_{O1} \sin^2 \beta \end{bmatrix} \qquad \textbf{[e]}$$

The angular velocity vector can be expressed in the $h_1 h_2 h_3$ frame as

$$\boldsymbol{\omega} = -\frac{v}{L \sin \beta} \mathbf{h}_3 \qquad \textbf{[f]}$$

The angular momentum can now be written as

$$\mathbf{H}_O = -(I_{O3} - I_{O1}) \sin \beta \cos \beta \frac{v}{L \sin \beta} \mathbf{h}_2 - (I_{O3} \cos^2 \beta + I_{O1} \sin^2 \beta) \frac{v}{L \sin \beta} \mathbf{h}_3$$

$$= -(I_{O3} - I_{O1}) \frac{v \cos \beta}{L} \mathbf{h}_2 - (I_{O3} \cos^2 \beta + I_{O1} \sin^2 \beta) \frac{v}{L \sin \beta} \mathbf{h}_3 \qquad \textbf{[g]}$$

Let us compare this result with the direct use of Eq. [8.3.9], which yields

$$\{_H H_O\} = [c]^T \{_F H_O\} = [c]^T [_F I_O]\{\omega\}$$

$$= \begin{bmatrix} 1 & 0 & 0 \\ 0 & \cos \beta & \sin \beta \\ 0 & -\sin \beta & \cos \beta \end{bmatrix} \begin{bmatrix} I_{O1} & 0 & 0 \\ 0 & I_{O1} & 0 \\ 0 & 0 & I_{O3} \end{bmatrix} \begin{bmatrix} 0 \\ v/L \\ -v \cos \beta / L \sin \beta \end{bmatrix}$$

$$= \begin{bmatrix} 0 \\ (I_{O1} - I_{O3}) v \cos \beta / L \\ -I_{O1} v \sin \beta / L - I_{O3} v \cos^2 \beta / L \sin \beta \end{bmatrix} \qquad \textbf{[h]}$$

which, of course, is the same answer as Eq. [g]. We note that the use of Eq. [8.3.9] is simpler, but that the use of Eq. [8.3.15] gives more insight. This is because one sees what the inertia matrix associated with the transformed coordinates consists of.

8.4 RESULTANT FORCE AND MOMENT

In many cases, a body is subjected to a multitude of forces and moments, and it becomes convenient to express the combined effects of these external excitations. These combined effects are called *resultant force* and *resultant moment*.

Consider a system of N particles, in which an external force $\mathbf{F}_i$ and internal force $\mathbf{F}'_i$ act on each particle. The resultant of all forces is defined by $\mathbf{F}$ and given as

$$\mathbf{F} = \sum_{i=1}^{N} (\mathbf{F}_i + \mathbf{F}'_i) = \sum_{i=1}^{N} \mathbf{F}_i \qquad \textbf{[8.4.1]}$$

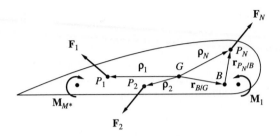

Figure 8.2 Applied forces and moments

The internal forces cancel each other. For a rigid body, we use the above relationship for N discrete forces. If there are distributed forces acting on the body, we consider a differential mass element with an external force $d\mathbf{F}$ acting on it. The resultant force is then defined as the sum of all external forces as

$$\mathbf{F} = \sum_{i=1}^{N} \mathbf{F}_i + \int_{\text{body}} d\mathbf{F} \qquad [8.4.2]$$

Consider now a body acted upon by N forces $\mathbf{F}_i$ ($i = 1, 2, \ldots, N$) through points P_i and M^* external moments $\mathbf{M}_i$ ($i = 1, 2, \ldots, M^*$), as shown in Fig. 8.2. The resultant moment about point B is defined as

$$\mathbf{M}_B = \sum_{i=1}^{N} \mathbf{r}_{P_i/B} \times \mathbf{F}_i + \sum_{i=1}^{M^*} \mathbf{M}_i = \sum_{i=1}^{N} \mathbf{r}_{P_i/B} \times \mathbf{F}_i + \mathbf{M}^* \qquad [8.4.3]$$

in which

$$\mathbf{M}^* = \sum_{i=1}^{M^*} \mathbf{M}_i \qquad [8.4.4]$$

is the sum of all external moments acting on the body. Note that because we are dealing with bodies that we assume to be rigid, the location of the individual moments $\mathbf{M}_i$ is immaterial when generating Eq. [8.4.3]. For flexible bodies, this is not the case; the location of the applied forces as well as moments becomes important.

The combined effects of all external excitations are usually expressed as a resultant force applied through a certain point and the resultant moment about that point. A common choice is the center of mass, as shown in Fig. 8.3. In this case, we write Eq. [8.4.3] in terms of the center of mass as

$$\mathbf{M}_G = \sum_{i=1}^{N} \boldsymbol{\rho}_i \times \mathbf{F}_i + \mathbf{M}^* \qquad [8.4.5]$$

When the resultant is expressed using a point other than the center of mass, say, B, the resultant moment becomes

$$\mathbf{M}_B = \mathbf{M}_G + \mathbf{r}_{G/B} \times \mathbf{F} \qquad [8.4.6]$$

The concept is illustrated in Fig. 8.4.

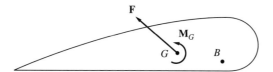

Figure 8.3   Resultant force and moment

Figure 8.4   Resultant force and moment about B

8.5 GENERAL EQUATIONS OF MOTION

In this section we consider a Newtonian approach and write the laws of motion for translation as well as rotation. Consider the translational equations first. For a system of N particles, the force balance for each particle has the form

$$m_i \mathbf{a}_i = \mathbf{F}_i + \mathbf{F}'_i \qquad i = 1, 2, \ldots, N \qquad \text{[8.5.1]}$$

where m_i is the mass of the ith particle and $\mathbf{F}_i$ is the external force, and $\mathbf{F}'_i$ is the internal force. For a rigid body, we sum the individual equations of motion, and take the limit as $N \to \infty$. We replace m_i by dm, $\mathbf{a}_i$ by $\mathbf{a}$, and the summation with integration. The left side of Eq. [8.5.1] then becomes

$$\lim_{N \to \infty} \sum_{i=1}^{N} m_i \mathbf{a}_i = \int_{\text{body}} \mathbf{a} \, dm = m\mathbf{a}_G = \frac{d}{dt}\mathbf{p} \qquad \text{[8.5.2]}$$

The right side of Eq. [8.5.1] becomes Eq. [8.4.1]. Thus, for a rigid body, Newton's second law is expressed as

$$\frac{d}{dt}\mathbf{p} = m\mathbf{a}_G = \mathbf{F} \qquad \text{[8.5.3]}$$

From a historical perspective, Newton developed his laws for the motion of rigid bodies, even though we first study them within the context of particles. Defining the *inertia force* acting on the body as $-m\mathbf{a}_G$, Newton's second law can be described as the inertia force being equal and opposite to the applied force.

The law governing rotational motion was formally stated by Euler in 1775, together with Eq. [8.5.3], in its form above. The law states that the rate of change of the angular momentum about the center of mass of a rigid body is equal to the sum of all applied moments about the center of mass, or

$$\frac{d}{dt}\mathbf{H}_G = \mathbf{M}_G \qquad \text{[8.5.4]}$$

Equations [8.5.3] and [8.5.4] provide a complete description of the governing equations of a body. They constitute the basis of *Newtonian mechanics,* also known as the *Newton-Euler formulation.*

Equation [8.5.4] can also be derived from Newton's second law using a differential element or particle formulation. Indeed, for a system of N particles, taking the cross product of Eq. [8.5.1] with $\boldsymbol{\rho}_i$, where $\boldsymbol{\rho}_i$ is the distance from the center of mass to the ith particle, we obtain

$$\boldsymbol{\rho}_i \times m_i \mathbf{a}_i = \boldsymbol{\rho} \times \mathbf{F}_i \qquad i = 1, 2, \ldots, N \qquad [8.5.5]$$

where the internal forces are not included in the formulation because they drop out. The term on the left side of this equation can be written as

$$\boldsymbol{\rho}_i \times m_i \mathbf{a}_i = \boldsymbol{\rho}_i \times m_i \frac{d\mathbf{v}_i}{dt} + \mathbf{v}_i \times m_i \mathbf{v}_i = \frac{d}{dt}(\boldsymbol{\rho}_i \times m_i \mathbf{v}_i) = \frac{d}{dt}\mathbf{H}_{Gi} \qquad [8.5.6]$$

where $\mathbf{H}_{Gi}$ is the angular momentum about the center of mass associated with the ith particle. If we sum Eq. [8.5.6] over N, we obtain the angular momentum of all the particles. For a rigid body, if we take the limit as N approaches infinity and add to it the sum of all applied moments, we obtain Eq. [8.5.4].

As discussed in Chapter 1, there is debate in the literature on whether Eq. [8.5.4] is a derived law or a stated law. One can argue that Eq. [8.5.4] is a stated principle, that is a fundamental law of mechanics, because of two major points: The internal forces of a rigid body are unknown, and for a deformable body the moment balance is a fundamental equation, as it leads to the symmetry of the stress tensor. (For a compelling argument in support of this issue, see pp. 259–271 in the text by Truesdell.)

The resultant of the applied forces and moments can be viewed in terms of the changes in the linear and angular momentum, as shown in Fig. 8.5. Hence, Figs. 8.2, 8.3, 8.4, and 8.5 are equivalent.

We next discuss two issues associated with writing the equations of motion:

- Whether one can write the rotational equations of motion about points other than the center of mass.
- What types of variables and reference frames one can use to express the equations of motion.

The translational equations of motion are not about a particular point. Recall that linear momentum is an absolute quantity. To obtain the rotational equations of motion about a point other than the center of mass, consider the angular momentum

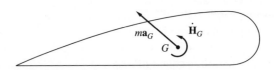

Figure 8.5   Resultant changes in linear and angular momentum

about an arbitrary point B, given by Eq. [8.3.4]. Recall that B is an arbitrary point and not necessarily attached to the body. Differentiating Eq. [8.3.4] we obtain

$$\frac{d}{dt}\mathbf{H}_B = \frac{d}{dt}\mathbf{H}_G + \frac{d}{dt}m(\mathbf{r}_{G/B} \times \mathbf{v}_G)$$
$$= \frac{d}{dt}\mathbf{H}_G + m\mathbf{r}_{G/B} \times \mathbf{a}_G + m(\mathbf{v}_{G/B}) \times \mathbf{v}_G \qquad [8.5.7]$$

The sum of moments about B is given by Eq. [8.4.6], which we can write as

$$\mathbf{M}_B = \mathbf{M}_G + \mathbf{r}_{G/B} \times \mathbf{F} = \frac{d}{dt}\mathbf{H}_G + m\mathbf{r}_{G/B} \times \mathbf{a}_G \qquad [8.5.8]$$

The last term on the right of Eq. [8.5.7] can also be expressed as

$$m(\mathbf{v}_{G/B}) \times \mathbf{v}_G = m(\mathbf{v}_G - \mathbf{v}_B) \times \mathbf{v}_G = m\mathbf{v}_G \times \mathbf{v}_B \qquad [8.5.9]$$

so that introducing Eq. [8.5.8] into Eq. [8.5.7] and considering Eq. [8.5.9] we obtain

$$\frac{d}{dt}\mathbf{H}_B = \mathbf{M}_B + m\mathbf{v}_G \times \mathbf{v}_B \qquad [8.5.10]$$

Equations [8.5.8] and [8.5.10] are the most general forms of the moment balance equations. When point B is attached to the body, one can show that

$$\frac{d}{dt}\mathbf{H}_B = \mathbf{M}_B + m(\boldsymbol{\omega} \times \mathbf{r}_{G/B}) \times \mathbf{v}_B \qquad [8.5.11]$$

When the moment balance is written about a point C (on or off the body) that is fixed in translation, we have

$$\frac{d}{dt}\mathbf{H}_C = \mathbf{M}_C \qquad [8.5.12]$$

We move on to the second issue: selection of the types of coordinates to express the equations of motion. For three-dimensional motion, using inertial parameters does not give much physical insight. For example, if we use an inertial coordinate system XYZ with unit vectors $\mathbf{IJK}$ and denote by X, Y, and Z the coordinates of the center of mass, the force balances become

$$m\ddot{X} = \mathbf{F} \cdot \mathbf{I} \qquad m\ddot{Y} = \mathbf{F} \cdot \mathbf{J} \qquad m\ddot{Z} = \mathbf{F} \cdot \mathbf{K} \qquad [8.5.13]$$

Unless the coordinates X, Y, and Z can be analyzed independently of each other, Eqs. [8.5.13] are not convenient to use by themselves. But consider writing these equations using a moving coordinate frame, which we will denote by xyz. Selecting the moving frame as a body-fixed frame, we write the components of the velocity, angular velocity, and force as

$$\mathbf{v}_G = v_x\mathbf{i} + v_y\mathbf{j} + v_z\mathbf{k}$$
$$\boldsymbol{\omega} = \omega_x\mathbf{i} + \omega_y\mathbf{j} + \omega_z\mathbf{k} \qquad \mathbf{F} = F_x\mathbf{i} + F_y\mathbf{j} + F_z\mathbf{k} \qquad [8.5.14]$$

To obtain the rates of change of the linear and angular velocities, we use the transport theorem. For the angular acceleration we have, as shown in Chapter 7,

$$\boldsymbol{\alpha} = \dot{\omega}_x\mathbf{i} + \dot{\omega}_y\mathbf{j} + \dot{\omega}_z\mathbf{k} + \boldsymbol{\omega} \times \boldsymbol{\omega} = \alpha_x\mathbf{i} + \alpha_y\mathbf{j} + \alpha_z\mathbf{k} \qquad [8.5.15]$$

so that $\alpha_x = \dot{\omega}_x$, $\alpha_y = \dot{\omega}_y$, $\alpha_z = \dot{\omega}_z$ for a set of body-fixed coordinates. Differentiating the expression for the velocity of the center of mass, we obtain

$$\frac{d}{dt}\mathbf{v}_G = \dot{v}_x\mathbf{i} + \dot{v}_y\mathbf{j} + \dot{v}_z\mathbf{k} + \boldsymbol{\omega} \times \mathbf{v}_G$$
$$= (\dot{v}_x + v_z\omega_y - v_y\omega_z)\mathbf{i} + (\dot{v}_y + v_x\omega_z - v_z\omega_x)\mathbf{j}$$
$$+ (\dot{v}_z + v_y\omega_x - v_x\omega_y)\mathbf{k} \qquad [8.5.16]$$

Introducing this expression into Newton's second law, we obtain for the translational equations

$$m(\dot{v}_x + v_z\omega_y - v_y\omega_z) = F_x$$
$$m(\dot{v}_y + v_x\omega_z - v_z\omega_x) = F_y$$
$$m(\dot{v}_z + v_y\omega_x - v_x\omega_y) = F_z \qquad [8.5.17]$$

In column vector form, the translational equations are written as

$$m\{\dot{v}_{Grel}\} + m[\tilde{\omega}]\{v_G\} = \{F\} \qquad [8.5.18]$$

or, using the notation for multiple reference frames, as

$$^B\frac{d}{dt}m\mathbf{v}_G + m{^A}\boldsymbol{\omega}^B \times \mathbf{v}_G = \mathbf{F} \qquad [8.5.19]$$

We obtain the rotational equations in a similar fashion. Consider Eq. [8.5.4]. We express the angular momentum vector as

$$\mathbf{H}_G = H_{Gx}\mathbf{i} + H_{Gy}\mathbf{j} + H_{Gz}\mathbf{k} \qquad [8.5.20]$$

in which H_{Gx}, H_{Gy}, and H_{Gz} are defined in Section 8.2. We write the resultant moment about the center of mass as

$$\mathbf{M}_G = M_{Gx}\mathbf{i} + M_{Gy}\mathbf{j} + M_{Gz}\mathbf{k} \qquad [8.5.21]$$

Using the transport theorem, the rate of change of the angular momentum vector becomes

$$\frac{d}{dt}\mathbf{H}_G = \dot{\mathbf{H}}_{Grel} + \boldsymbol{\omega} \times \mathbf{H}_G = \dot{H}_{Gx}\mathbf{i} + \dot{H}_{Gy}\mathbf{j} + \dot{H}_{Gz}\mathbf{k} + \boldsymbol{\omega} \times \mathbf{H}_G \qquad [8.5.22]$$

The *gyroscopic moment* is defined as $\boldsymbol{\omega} \times \mathbf{H}_G$. Hence, the rotational equations can be viewed as the gyroscopic moment plus the relative change in angular momentum being equal to the applied moment. In column vector form we have

$$\{\dot{H}_G\} = \{\dot{H}_{Grel}\} + [\tilde{\omega}]\{H_G\} \qquad [8.5.23]$$

Substitution of Eq. [8.5.22] into Eq. [8.5.4] yields the rotational equations in terms of the body-fixed angular velocities as

$$I_{xx}\alpha_x - I_{xy}(\alpha_y - \omega_x\omega_z) - I_{xz}(\alpha_z + \omega_x\omega_y) - (I_{yy} - I_{zz})\omega_y\omega_z - I_{yz}(\omega_y^2 - \omega_z^2) = M_{Gx}$$
$$I_{yy}\alpha_y - I_{yz}(\alpha_z - \omega_x\omega_y) - I_{xy}(\alpha_x + \omega_y\omega_z) - (I_{zz} - I_{xx})\omega_x\omega_z - I_{xz}(\omega_z^2 - \omega_x^2) = M_{Gy}$$
$$I_{zz}\alpha_z - I_{xz}(\alpha_x - \omega_y\omega_z) - I_{yz}(\alpha_y + \omega_x\omega_z) - (I_{xx} - I_{yy})\omega_x\omega_y - I_{xy}(\omega_x^2 - \omega_y^2) = M_{Gz}$$
$$[8.5.24]$$

In column vector format we have

$$[I_G]\{\alpha\} + [\tilde{\omega}][I_G]\{\omega\} = \{M_G\} \qquad [8.5.25]$$

and, in the notation for multiple reference frames, we have

$$^B\frac{d}{dt}\mathbf{H}_G + {}^A\boldsymbol{\omega}^B \times \mathbf{H}_G = \mathbf{M}_G \qquad [8.5.26]$$

If there is a fixed point C on the body about which the rigid body rotates, the rotational equations of motion have the form

$$[I_C]\{\alpha\} + [\tilde{\omega}][I_C]\{\omega\} = \{M_C\} \qquad [8.5.27]$$

The rotational equations are quite complicated even though there are no translational velocity terms in them. The complicated terms arise because both the angular momentum and the angular velocity change directions.

We can achieve significant simplification of the rotational equations of motion if we select the body axes as principal axes. In that case, all products of inertia vanish, and we have

$$I_1\alpha_1 - (I_2 - I_3)\omega_2\omega_3 = M_1$$
$$I_2\alpha_2 - (I_3 - I_1)\omega_1\omega_3 = M_2$$
$$I_3\alpha_3 - (I_1 - I_2)\omega_1\omega_2 = M_3 \qquad [8.5.28]$$

where the indices denote the principal directions. Equations [8.5.28] are known as *Euler's equations of motion*, stated originally by Euler in 1750. Except for certain special cases, such as some vehicle dynamics problems, Eqs. [8.5.28] are preferred over Eqs. [8.5.24]. Note also the order of the mass moments of inertia in this equation, which lends itself to an easy way of remembering these equations.

Euler's equations are for the general three-dimensional motion of a body. In many cases, part of the motion of the body is constrained; thus, the body has less than three rotational degrees of freedom. In such cases, Euler's equations yield the equation(s) of motion as well as expressions for the reactions.

The equations of motion expressed in terms of body-fixed coordinates give a lot of insight. Among their common uses is when conducting an instantaneous analysis, such as finding angular accelerations, finding the gyroscopic moment, or calculating the reactions at supports. Euler's equations can be integrated qualitatively to come up with integrals of the motion as well as for conducting a stability analysis. However, one cannot integrate these equations by themselves to find the orientation of the body at a given instant.

The above derivation of Euler's equations was based on using a set of rotating axes attached to the body. As discussed in Chapter 7, when dealing with the motion of an axisymmetric body, it is more convenient to make use of the F frame. The F frame is suitable for axisymmetric bodies, because the inertia matrix is constant in the F frame.

One can use the F frame in conjunction with Euler's equations in two ways: One can perform the differentiation in the F frame, or one can express the equations of motion using the components of the F frame. When performing the differentiation in the F frame, similar to Eq. [8.5.20] the translational equations of motion become

$$\frac{{}^F d}{dt} m\mathbf{v}_G + {}^A\boldsymbol{\omega}^F \times m\mathbf{v}_G = \mathbf{F} \quad \text{or} \quad m\{\dot{v}_{\text{Grel}}\} + m[\tilde{\omega}_f]\{v_G\} = \{F\} \quad [\mathbf{8.5.29}]$$

The rotational equations of motion become

$$\frac{{}^F d}{dt}\mathbf{H}_G + {}^A\boldsymbol{\omega}^F \times \mathbf{H}_G = \mathbf{M}_G \quad \text{or} \quad \{\dot{H}_G\} = [I_G]\{\dot{\omega}_{\text{brel}}\} + [\tilde{\omega}_f][I_G]\{\omega_b\} = \{M_G\}$$

[**8.5.30**]

Equations [8.5.30] are known as the *modified Euler equations*. We will study these equations in more detail in Chapter 9.

When resolving the equations of motion along the components of the F frame, one no longer has $\alpha_i = \dot{\omega}_i$ ($i = 1, 2, 3$), and one must calculate $\{\alpha\}$ using the transport theorem. We write

$$\{_f \dot{H}_G\} = [I_G]\{_f \alpha\} + [_f\tilde{\omega}][I_G]\{_F\omega\} = \{_F M_G\} \quad [\mathbf{8.5.31}]$$

in which

$$\{_F\alpha\} = \{_F\dot{\omega}\} + [_F\tilde{\omega}_f]\{_F\omega\} = \{_F\dot{\omega}\} + [_F\tilde{\omega}_f]\{\omega_s\} \quad [\mathbf{8.5.32}]$$

Example 8.4

A slender bar of mass m and length L is attached to a vertical shaft, which is rotating with constant angular speed Ω, as shown in Fig. 8.6. Find the equation of motion for the bar.

Solution

The only variable in this problem is the angle that the bar makes with the shaft, so that the three Euler's equations should yield one equation of motion and two other equations for the reactions. Because the bar is rotating about the fixed point B, it is preferable to write the

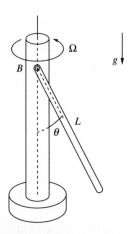

Figure 8.6

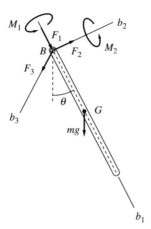

Figure 8.7

Euler's equations about the center of rotation. This way, we avoid calculation of the reaction forces at B.

The free-body diagram of the bar is shown in Fig. 8.7. We express the moment about B as

$$\mathbf{M}_B = M_1\mathbf{b}_1 + M_2\mathbf{b}_2 + \mathbf{r}_{G/B} \times \mathrm{mg}(\cos\theta\mathbf{b}_1 - \sin\theta\mathbf{b}_2) \qquad \text{[a]}$$

Noting that $\mathbf{r}_{G/B} = L/2\mathbf{b}_1$, we obtain

$$\mathbf{M}_B = M_1\mathbf{b}_1 + M_2\mathbf{b}_2 - \frac{1}{2}mgL\sin\theta\mathbf{b}_3 \qquad \text{[b]}$$

The angular velocity and angular acceleration have the form

$$\boldsymbol{\omega} = \omega_1\mathbf{b}_1 + \omega_2\mathbf{b}_2 + \omega_3\mathbf{b}_3 = -\Omega\cos\theta\mathbf{b}_1 + \Omega\sin\theta\mathbf{b}_2 + \dot\theta\mathbf{b}_3 \qquad \text{[c]}$$

$$\boldsymbol{\alpha} = \alpha_1\mathbf{b}_1 + \alpha_2\mathbf{b}_2 + \alpha_3\mathbf{b}_3 = \Omega\dot\theta\sin\theta\mathbf{b}_1 + \Omega\dot\theta\cos\theta\mathbf{b}_2 + \ddot\theta\mathbf{b}_3 \qquad \text{[d]}$$

The mass moments of inertia about B are

$$I_1 \approx 0 \qquad I_2 = I_3 = \frac{mL^2}{3} \qquad \text{[e]}$$

We now invoke Euler's equations, which yield

For b_1 $\qquad I_1\Omega\dot\theta\sin\theta = M_1 \qquad$ **[f]**

For b_2 $\qquad I_2\Omega\dot\theta\cos\theta + I_2\Omega\dot\theta\cos\theta = M_2 \qquad$ **[g]**

For b_3 $\qquad I_2\ddot\theta - I_2\Omega^2\sin\theta\cos\theta = -\dfrac{mgL}{2}\sin\theta \qquad$ **[h]**

Equation [h] is the equation of motion, while Eqs. [f] and [g] give expressions for the reactions. We simplify Eq. [h] to

$$\ddot\theta - \Omega^2\sin\theta\cos\theta + \frac{3g}{2L}\sin\theta = 0 \qquad \text{[i]}$$

The expressions for the moment reactions become

$$M_1 \approx 0 \qquad M_2 = 2I_2\Omega\dot{\theta}\cos\theta \qquad \text{[i]}$$

The reaction forces at B can be obtained from a force balance.

Example 8.5

Consider the gyropendulum in Example 7.5, repeated here as Fig. 8.8. Derive the equations of motion, assuming there is a moment M acting on the vertical shaft and that both shafts are massless.

Solution

We will solve this problem in two ways: First, we will use the modified Euler equations. Second, we will resolve Euler's equations along the coordinates of the F frame.

We split the gyropendulum into two parts: the vertical shaft and the disk and smaller shaft. The free-body diagram of the disk is shown in Fig. 8.9. Using the relative axes xyz, M_x and M_z are components of the reaction moments at point B in the x and z directions. The reaction forces are F_x, F_y, and F_z.

Because the assembly can be considered as rotating about point B, we can write the equations of motion about B. Observe that the xyz coordinate system describes the F frame, and it constitutes a set of principal axes, so that all products of inertia vanish. We have $\mathbf{K} = -\sin\theta\mathbf{i} + \cos\theta\mathbf{k}$, and

$$I_{Bzz} = \frac{1}{2}mR^2 \qquad I_{Byy} = I_{Bxx} = m\left(\frac{1}{4}R^2 + L^2\right) \qquad \text{[a]}$$

$$\boldsymbol{\omega}_b = \dot{\phi}\mathbf{K} + \dot{\theta}\mathbf{j}' + \dot{\psi}\mathbf{k} = -\dot{\phi}\sin\theta\mathbf{i} + \dot{\theta}\mathbf{j} + (\dot{\psi} + \dot{\phi}\cos\theta)\mathbf{k} \qquad \text{[b]}$$

$$\boldsymbol{\omega}_f = \dot{\phi}\mathbf{K} + \dot{\theta}\mathbf{j}' = -\dot{\phi}\sin\theta\mathbf{i} + \dot{\theta}\mathbf{j} + \dot{\phi}\cos\theta\mathbf{k} \qquad \text{[c]}$$

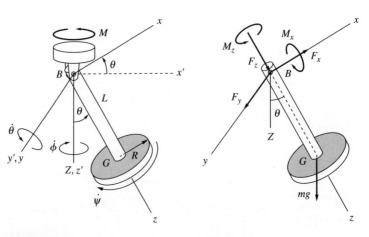

Figure 8.8 **Figure 8.9**

8.5 GENERAL EQUATIONS OF MOTION

Solution by Modified Euler Equations The angular momentum about B has the form

$$\mathbf{H}_B = -m\left(\frac{1}{4}R^2 + L^2\right)\dot{\phi}\sin\theta\,\mathbf{i} + m\left(\frac{1}{4}R^2 + L^2\right)\dot{\theta}\mathbf{j} + \frac{1}{2}mR^2(\dot{\psi} + \dot{\phi}\cos\theta)\mathbf{k} \quad [\mathbf{d}]$$

We next differentiate the expression for the angular momentum. The first term is the time derivative of $\mathbf{H}_B$ in the F frame and it is obtained by simple differentiation of Eq. [d] as

$$\frac{^Fd}{dt}\mathbf{H}_B = -m\left(\frac{1}{4}R^2 + L^2\right)(\ddot{\phi}\sin\theta + \dot{\phi}\dot{\theta}\cos\theta)\mathbf{i}$$

$$+ m\left(\frac{1}{4}R^2 + L^2\right)\ddot{\theta}\mathbf{j} + \frac{1}{2}mR^2(\ddot{\psi} + \ddot{\phi}\cos\theta - \dot{\phi}\dot{\theta}\sin\theta)\mathbf{k} \quad [\mathbf{e}]$$

The second term in the derivative is

$${}^A\boldsymbol{\omega}^F \times \mathbf{H}_B = \boldsymbol{\omega}_f \times \mathbf{H}_B$$

$$= (-\dot{\phi}\sin\theta\,\mathbf{i} + \dot{\theta}\mathbf{j} + \dot{\phi}\cos\theta\,\mathbf{k})$$

$$\times \left[-m\left(\frac{1}{4}R^2 + L^2\right)\dot{\phi}\sin\theta\,\mathbf{i} + m\left(\frac{1}{4}R^2 + L^2\right)\dot{\theta}\mathbf{j} + \frac{1}{2}mR^2(\dot{\psi} + \dot{\phi}\cos\theta)\mathbf{k}\right]$$

$$= \left[\frac{1}{2}mR^2(\dot{\psi} + \dot{\phi}\cos\theta)\dot{\theta} - m\left(\frac{1}{4}R^2 + L^2\right)\dot{\theta}\dot{\phi}\cos\theta\right]\mathbf{i}$$

$$+ \left[\frac{1}{2}mR^2(\dot{\psi} + \dot{\phi}\cos\theta)\dot{\phi}\sin\theta - m\left(\frac{1}{4}R^2 + L^2\right)\dot{\phi}^2\sin\theta\cos\theta\right]\mathbf{j} \quad [\mathbf{f}]$$

The external moment vector is

$$\mathbf{M}_B = M_x\mathbf{i} + M_z\mathbf{k} + L\mathbf{k} \times mg\mathbf{K} = M_x\mathbf{i} - mgL\sin\theta\,\mathbf{j} + M_z\mathbf{k} \quad [\mathbf{g}]$$

Combining Eqs. [e], [f], and [g] and separating the x, y, and z components, we obtain

$$\mathbf{i} \to -m\left(\frac{1}{4}R^2 + L^2\right)\ddot{\phi}\sin\theta + \frac{1}{2}mR^2\dot{\theta}\dot{\psi} - 2mL^2\dot{\phi}\dot{\theta}\cos\theta = M_x \quad [\mathbf{h}]$$

$$\mathbf{j} \to m\left(\frac{1}{4}R^2 + L^2\right)\ddot{\theta} + \frac{1}{2}mR^2\dot{\phi}\dot{\psi}\sin\theta + m\left(\frac{1}{4}R^2 - L^2\right)\dot{\phi}^2\sin\theta\cos\theta = -mgL\sin\theta \quad [\mathbf{i}]$$

$$\mathbf{k} \to \frac{1}{2}mR^2(\ddot{\psi} + \ddot{\phi}\cos\theta - \dot{\phi}\dot{\theta}\sin\theta) = M_z \quad [\mathbf{j}]$$

We have three degrees of freedom; thus, each one of Eqs. [h], [i], and [j] is an equation of motion.

Solution by Euler Equations Resolved Around F Frame To obtain the solution this way, we first need to obtain the expression for the angular acceleration. Given $\boldsymbol{\omega}_b$ and $\boldsymbol{\omega}_f$ in Eqs. [b], we obtain the angular acceleration as

$$\boldsymbol{\alpha}_b = \dot{\boldsymbol{\omega}}_{brel} + \boldsymbol{\omega}_f \times \boldsymbol{\omega}_b = \dot{\boldsymbol{\omega}}_{brel} + \boldsymbol{\omega}_f \times \boldsymbol{\omega}_s$$

$$= -(\ddot{\phi}\sin\theta + \dot{\phi}\dot{\theta}\cos\theta)\mathbf{i} + \ddot{\theta}\mathbf{j} + (\ddot{\psi} + \ddot{\phi}\cos\theta - \dot{\phi}\dot{\theta}\sin\theta)\mathbf{k}$$

$$+ (-\dot{\phi}\sin\theta\,\mathbf{i} + \dot{\theta}\mathbf{j} + \dot{\phi}\cos\theta\,\mathbf{k}) \times \dot{\psi}\mathbf{k}$$

$$= (-\ddot{\phi}\sin\theta - \dot{\phi}\dot{\theta}\cos\theta + \dot{\theta}\dot{\psi})\mathbf{i} + (\ddot{\theta} + \dot{\phi}\dot{\psi}\sin\theta)\mathbf{j} + (\ddot{\psi} + \ddot{\phi}\cos\theta - \dot{\phi}\dot{\theta}\sin\theta)\mathbf{k} \quad [\mathbf{k}]$$

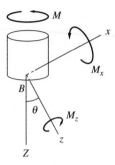

Figure 8.10

Substituting the values for the angular velocities in Eq. [b] and angular acceleration in Eq. [k] into the Euler's equations, we obtain

$$-m\left(\frac{1}{4}R^2 + L^2\right)(\ddot{\phi}\sin\theta + \dot{\phi}\dot{\theta}\cos\theta - \dot{\theta}\dot{\psi}) + m\left(\frac{1}{4}R^2 - L^2\right)\dot{\theta}(\dot{\psi} + \dot{\phi}\cos\theta) = M_x \quad [l]$$

$$m\left(\frac{1}{4}R^2 + L^2\right)(\ddot{\theta} + \dot{\phi}\dot{\psi}\sin\theta) + m\left(\frac{1}{4}R^2 - L^2\right)\dot{\phi}\sin\theta(\dot{\psi} + \dot{\phi}\cos\theta) = -mgL\sin\theta \quad [m]$$

$$\frac{1}{2}mR^2(\ddot{\psi} + \ddot{\phi}\cos\theta - \dot{\phi}\dot{\theta}\sin\theta) = M_z \quad [n]$$

which can be shown to be the same as Eqs. [h], [i], and [j].

Analysis of the Equations The equations of motion in Eqs. [h], [i], and [j] are in terms of the internal reaction moments M_x and M_z. Further, the external moment M does not appear. We relate the external moment M to the internal moments by conducting a moment balance for the vertical shaft about the Z axis, as shown in Fig. 8.10. Doing so, we obtain

$$-M + M_z\cos\theta - M_x\sin\theta = 0 \quad [o]$$

We still need another relationship for these moments. This added relationship can be obtained by separating the rod and the disk and analyzing the forces and moments these two members exert on each other. We will not pursue this, but rather, consider the case when there is a motor between the rod and disk, which makes the disk maintain a constant angular velocity with respect to the shaft. Such a motor is called a *servomotor*.

In the presence of a servomotor that keeps $\dot{\psi}$ constant so that it is no longer a variable, the system is reduced to two degrees of freedom. From Eq. [j] we get an expression for M_z as

$$M_z = \frac{1}{2}mR^2(\ddot{\phi}\cos\theta - \dot{\phi}\dot{\theta}\sin\theta) \quad [p]$$

which, when substituted into Eq. [o], gives an expression for M_x in terms of the external excitation M as

$$M_x = \frac{-M + M_z\cos\theta}{\sin\theta} = \frac{-M}{\sin\theta} + \frac{1}{2}mR^2\left(\frac{\ddot{\phi}\cos^2\theta}{\sin\theta} - \dot{\phi}\dot{\theta}\cos\theta\right) \quad [q]$$

Let us next calculate the value of the external torque needed in order to keep the precession $\dot\phi$ rate constant. Introducing Eq. [q] into Eq. [h] and setting $\ddot\phi$ to zero, we obtain

$$M = -\frac{1}{2}mR^2\dot\theta\dot\psi\sin\theta + 2mL^2\dot\phi\dot\theta\cos\theta\sin\theta - \frac{1}{2}mR^2\dot\phi\dot\theta\cos\theta\sin\theta \qquad [\mathbf{r}]$$

In this case we have a single degree of freedom system, and Eq. [i] is the equation of motion.

QUALITATIVE STABILITY ANALYSIS OF ROTATIONAL MOTION | Example 8.6

Consider a rigid body to which a set of body-fixed principal axes $b_1 b_2 b_3$ are attached at the center of mass. The body has an initial motion in the form of a rotation about one of the axes, say, b_1. We have $\omega_1 = \omega_0, \omega_2 = \omega_3 = 0$. No external moments act, so that from Eq. [8.5.28] ω_0 remains constant.

To investigate whether this motion is stable or not, a small moment is applied to the body, so that the angular velocities after the application of the moment become

$$\omega_1 = \omega_0 + \varepsilon_1 \qquad \omega_2 = \varepsilon_2 \qquad \omega_3 = \varepsilon_3 \qquad [\mathbf{a}]$$

where ε_i ($i = 1, 2, 3$) are small quantities. We wish to determine the evolution of these perturbed angular velocities in time. Introducing them into Euler's equations and noting that no further moments are applied, we obtain

$$I_1(\dot\omega_0 + \dot\varepsilon_1) - (I_2 - I_3)\varepsilon_2\varepsilon_3 = 0 \qquad I_2\dot\varepsilon_2 - (I_3 - I_1)(\omega_0 + \varepsilon_1)\varepsilon_3 = 0$$

$$I_3\dot\varepsilon_3 - (I_1 - I_2)(\omega_0 + \varepsilon_1)\varepsilon_2 = 0 \qquad [\mathbf{b}]$$

Because the change in the angular velocities is small, we linearize the above equations by eliminating quadratic and higher-order terms in ε_i ($i = 1, 2, 3$). Noting that ω_0 is constant, we get

$$I_1\dot\varepsilon_1 = 0 \qquad [\mathbf{c}]$$

$$I_2\dot\varepsilon_2 + (I_1 - I_3)\omega_0\varepsilon_3 = 0 \qquad [\mathbf{d}]$$

$$I_3\dot\varepsilon_3 + (I_2 - I_1)\omega_0\varepsilon_2 = 0 \qquad [\mathbf{e}]$$

Equation [c] states that ε_1 remains constant. To understand the behavior of the remaining two angular velocities, we conduct an eigenvalue analysis. Introducing the expansions

$$\varepsilon_2(t) = E_2 e^{\lambda t} \qquad \varepsilon_3(t) = E_3 e^{\lambda t} \qquad [\mathbf{f}]$$

into Eqs. [d] and [e] we obtain

$$\begin{bmatrix} I_2\lambda & (I_1 - I_3)\omega_0 \\ (I_2 - I_1)\omega_0 & I_3\lambda \end{bmatrix} \begin{bmatrix} E_2 \\ E_3 \end{bmatrix} e^{\lambda t} = \begin{bmatrix} 0 \\ 0 \end{bmatrix} \qquad [\mathbf{g}]$$

The solution of Eq. [g] requires that the determinant of the coefficient matrix be zero, which yields the characteristic equation

$$I_2 I_3 \lambda^2 - (I_2 - I_1)(I_1 - I_3)\omega_0^2 = 0 \qquad [\mathbf{h}]$$

The solution is

$$\lambda = \pm i \sqrt{\frac{(I_2 - I_1)(I_1 - I_3)\omega_0^2}{I_2 I_3}}$$ [i]

Two types of solutions are possible. If $I_2 > I_1$ and $I_3 > I_1$, or if $I_2 < I_1$ and $I_3 < I_1$, both roots of the characteristic equation are pure imaginary. Because there are no nonconservative forces, the system is critically stable. A higher-level stability analysis indicates that rotation about the axes representing rotation about the minimum and maximum moments of inertia are indeed stable. That is, if I_1 is the largest or smallest moment of inertia, rotation about the b_1 axis is stable.

On the other hand, if $I_2 > I_1$ and $I_3 < I_1$ or if $I_2 < I_1$ and $I_3 > I_1$, that is, if axis b_1 is the intermediate moment of inertia axis, the roots of the characteristic equation are real, one positive and one negative. Hence, rotation about the intermediate axis of inertia is unstable. One or both of the angular velocities grow exponentially to a level such that they can no longer be considered as perturbations. A set of equations other than the linearized equations must be used after that point.

Next, consider the special case when two of the mass moments of inertia are the same, such as in an axisymmetric body. We have two possibilities. In the first, b_1 is the symmetry axis and $I_2 = I_3$. From Eq. [i] all roots λ are pure imaginary, so there is critical stability. A higher-level stability analysis can be conducted which shows that perturbed angular motion of axisymmetric bodies spinning about the symmetry axis is indeed stable. Note that, unlike in the case of three distinct moments of inertia, the magnitudes of the moments of inertia did not affect this result.

The second case is descriptive of an axisymmetric body rotating about a transverse axis. That is, if b_1 is the symmetry axis, the rotation is about b_2 or b_3. This system leads to zero eigenvalues. The motion can be explained as a continuation of the tumbling motion and transfer of motion between the two transverse axes, with no rotation transferred to about the symmetry axis.

We can summarize the results as follows. For general bodies, rotation about the axes with the minimum and maximum moments of inertia is stable. Rotation about the intermediate axis of inertia is unstable. Any initial motion will turn itself into angular motion with components along every axis, resulting in some sort of wobbly motion. This phenomenon can readily be observed by taking a book or other object with three distinct principal moments of inertia and spinning it about its three principal axes, as illustrated in Fig. 8.11.

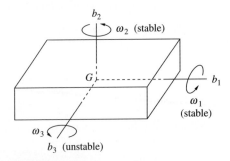

Figure 8.11

The instability phenomenon is not encountered in axisymmetric bodies. Another way of explaining this added stability due to axisymmetry (or, for the general case, inertial symmetry) is to recognize that for such a body, every principal axis represents a minimum or maximum moment of inertia. In engineering applications, one takes advantage of this stability feature and designs rotating components, such as engine parts, as axisymmetric. In Chapter 10, we will study stability issues associated with rotating bodies in more detail.

It turns out that in real-life applications, rotation is stable only around the axis of maximum moment of inertia. This is due to energy loss and transfer due to the flexibility of the body, and energy loss as a result of friction. In space mechanics, these issues were brought into the limelight after the Explorer satellite was launched in 1958, and nutational instabilities were observed. Ideally, the shape of a space vehicle would be like a disk. Because of practical considerations, satellites launched into space are spherical or slender cylindrical bodies.

8.6 Rotation about a Fixed Axis

An important special case is rotation about a fixed axis. Representation of such motion requires only one angular velocity component, making the angular velocity "simple."

Two types of rotation about a fixed axis are of interest. One is plane motion. Chapter 3 discusses the kinetics of plane motion. In the second type, one is primarily interested in bodies that spin in a fixed direction, but the axis of rotation is not a principal axis. In several situations a body spins about an axis that is fixed, for example a shaft used in power transmission, or the wheels of a vehicle. If the rotating body is not symmetric about the axis of rotation, the angular velocity of the body fluctuates or dynamic reactions are generated at the supports. Both of these are undesirable. Such behavior increases the loads on the supports, causing unnecessary vibration, noise, and possible damage.

We select the axis of rotation as one of the coordinate axes, say, z, and note that this axis is not necessarily a principal axis. That is, the products of inertia do not vanish. The components of the angular velocity vector are

$$\omega_x = \omega_y = 0 \qquad \omega_z = \omega \qquad \textbf{[8.6.1]}$$

Substituting this into the rotational equations of motion, Eqs. [8.5.24], we obtain

$$M_x = -I_{xz}\dot{\omega} + I_{yz}\omega^2 \qquad M_y = -I_{yz}\dot{\omega} - I_{xz}\omega^2 \qquad \textbf{[8.6.2]}$$

$$M_z = I_{zz}\dot{\omega} \qquad \textbf{[8.6.3]}$$

In this problem there is only one degree of freedom, so that Eqs. [8.6.2] represent relations for the reactions, while Eq. [8.6.3] is the equation of motion. It is interesting to note that the reaction moments M_x and M_y are influenced by the square of the angular velocity: this means that the problem of the axis of rotation not being a principal axis becomes more critical in high-speed situations.

In many applications, such as the balancing of a wheel or a shaft, weights are removed or added to a rotating body. If only static balancing is being done, one aligns

Example 8.7

A disk of mass m and radius R is attached to a rotating shaft, as shown in Fig. 8.12. Due to a manufacturing defect, the symmetry axis of the disk is not aligned with the shaft, but makes an angle of γ. The shaft rotates with the constant angular velocity Ω. Find the reactions at the bearings.

Solution

We use a set of inertial coordinates XYZ, with the Z axis along the shaft, and a set of xyz axes attached to the shaft, so that the z axis is along the shaft. The $x'y'z'$ axes are principal coordinates for the disk, and they are obtained by rotating the xyz axes clockwise by an angle of γ about the x axis. Fig. 8.13 shows the configuration and free-body diagram of the system.

The centroidal mass moments of inertia are $I_{x'x'} = I_{y'y'} = mR^2/4$, $I_{z'z'} = mR^2/2$, with all products of inertia zero. From Eqs. [8.6.2] and [8.6.3], because Ω is constant we only need to find I_{xz} and I_{yz}. It is easy to show that $I_{xz} = 0$, as the yz plane is a symmetry plane. To find I_{yz}, we use the definition of a product of inertia and express the y and z coordinates as

$$y = y'\cos\gamma + z'\sin\gamma \qquad z = -y'\sin\gamma + z'\cos\gamma \qquad \text{[a]}$$

so that I_{yz} becomes

$$I_{yz} = \int yz\,dm = \int (y'\cos\gamma + z'\sin\gamma)(-y'\sin\gamma + z'\cos\gamma)\,dm$$

$$= \sin\gamma\cos\gamma(I_{z'z'} - I_{y'y'}) = \frac{1}{4}mR^2\sin\gamma\cos\gamma \qquad \text{[b]}$$

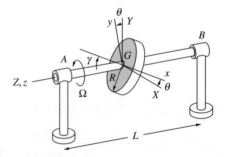

Figure 8.12

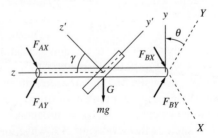

Figure 8.13

Introducing the values of I_{xz} and I_{yz} into Eqs. [8.6.2] yields the moments acting on the disk as

$$M_x = \frac{1}{4}mR^2\Omega^2 \sin\gamma \cos\gamma \qquad M_y = 0 \qquad \textbf{[c]}$$

Consequently, the disk exerts a moment $-M_x$ on the shaft. Consider now the reactions at the bearings. To this end, we will sum forces and moments for the shaft. To this end we express $-M_x$ in terms of its components along the inertial coordinates XYZ as

$$M_X = -M_x \cos\theta \qquad M_Y = -M_x \sin\theta \qquad \textbf{[d]}$$

where $\theta = \Omega t$ is the rotation angle between the XYZ and xyz frames. We are assuming that the center of the mass of the disk does not move. Hence, summing forces in the X and Y directions we obtain

$$F_{AX} = -F_{BX} \qquad F_{AY} + F_{BY} = mg \qquad \textbf{[e]}$$

Summing moments about G for the shaft gives

$$M_X = (F_{BY} - F_{AY})\frac{L}{2} \qquad M_Y = (F_{AX} - F_{BX})\frac{L}{2} \qquad \textbf{[f]}$$

Introducing Eq. [c] into Eq. [d] and solving Eqs. [e] and [f] for the reaction forces, we obtain

$$F_{BX} = -F_{AX} = \frac{1}{4}\frac{mR^2}{L}\Omega^2 \sin\gamma \cos\gamma \sin\Omega t$$

$$F_{AY} = \frac{1}{2}mg + \frac{1}{4}\frac{mR^2}{L}\Omega^2 \sin\gamma \cos\gamma \cos\Omega t$$

$$F_{BY} = \frac{1}{2}mg - \frac{1}{4}\frac{mR^2}{L}\Omega^2 \sin\gamma \cos\gamma \cos\Omega t \qquad \textbf{[g]}$$

The misalignment between the shaft and disk gives rise to bearing forces. These forces are related to the amount of misalignment and to the square of the angular velocity, and they are cyclic loads. For high-speed machinery, such misalignment can be detrimental, causing damage to the bearings as well as to the rotating parts.

8.7 EQUATIONS OF MOTION IN STATE FORM

In the previous two sections we derived the equations of motion. In order to extract more information from these equations, and to understand how the motion evolves, the equations of motion must be integrated. This integration can be carried out qualitatively or quantitatively. The qualitative integration leads to impulse-momentum and work-energy relationships, which we cover in Sections 8.8 and 8.9, as well as integrals of the motion. Here, we look into the quantitative integration of the equations of motion.

Integration of rotational equations of motion is usually difficult to carry out by hand, except for a few special cases. Computational techniques require that the equations of motion be expressed in state form.

Consider the rotational equations of motion independently from the translational equations for the time being, which is a valid assumption as long as the moments acting on the body are not functions of the translational velocities. We need to integrate Euler's equations, Eqs. [8.5.28], together with a set of equations that relate the body-fixed angular velocities ω_i ($i = 1, 2, 3$) to inertial quantities. We saw two sets of equations that accomplish this in Chapter 7: the Euler angles and the Euler parameters. The relationship between the Euler angles and body-fixed angular velocities is given in Eqs. [7.5.7]–[7.5.9] for a 3-1-3 transformation and at the table at the end of Chapter 7 for the 3-2-3 and 3-2-1 transformations. Let the body-fixed axes be the principal axes. One can then combine Eqs. [8.5.28] and [7.5.9] together, for a set of six first-order differential equations in terms of the variables ϕ, θ, ψ, ω_1, ω_2, ω_3 as

$$\dot{\omega}_1 = \frac{(I_2 - I_3)}{I_1}\omega_2\omega_3 + \frac{M_1}{I_1}$$

$$\dot{\omega}_2 = \frac{(I_3 - I_1)}{I_2}\omega_1\omega_3 + \frac{M_2}{I_2}$$

$$\dot{\omega}_3 = \frac{(I_1 - I_2)}{I_3}\omega_1\omega_2 + \frac{M_3}{I_3}$$

$$\dot{\phi} = \frac{1}{\sin\theta}(\omega_1 \sin\psi + \omega_2 \cos\psi)$$

$$\dot{\theta} = \omega_1 \cos\psi - \omega_2 \sin\psi$$

$$\dot{\psi} = -\frac{1}{\sin\theta}(\omega_1 \cos\theta \sin\psi + \omega_2 \cos\theta \cos\psi) + \omega_3 \quad \text{[8.7.1]}$$

subject to initial conditions $\omega_i(t_0)$ ($i = 1, 2, 3$), $\phi(t_0)$, $\theta(t_0)$, and $\psi(t_0)$, where t_0 is the initial time. If, instead of $\omega_i(t_0)$, the rates of the Euler angles are given at $t = t_0$, $\omega_i(t_0)$ can be found from $\dot{\phi}(t_0)$, $\dot{\theta}(t_0)$, and $\dot{\psi}(t_0)$ using Eqs. [7.5.7]. If the external excitations M_i ($i = 1, 2, 3$) are not explicit functions of the Euler angles, then the first three of the above equations, the Euler equations, can be integrated separately from the second three, the kinematic differential equations.

One problem associated with the above approach is that Euler angles have discontinuities at certain values of the second rotation angle. For example, a 3-1-3 or 3-2-3 transformation is discontinuous at $\theta = 0$ or $\theta = \pm\pi$. The 3-2-1 transformation has a singularity when the second transformation angle is $\pm\pi/2$. In the neighborhood of a singularity, one can switch to a different set of Euler angles, but the process is tedious.

By contrast, the Euler parameters discussed in Section 7.7 easily lend themselves to numerical computation. Rather than using the three relations in Eqs. [7.5.9], we use the four relations in Eqs. [7.7.27] and have a set of seven first-order differential equations in the form

$$\dot{\omega}_1 = \frac{(I_2 - I_3)}{I_1}\omega_2\omega_3 + \frac{M_1}{I_1}$$

$$\dot{\omega}_2 = \frac{(I_3 - I_1)}{I_2}\omega_1\omega_3 + \frac{M_2}{I_2} \qquad \dot{\omega}_3 = \frac{(I_1 - I_2)}{I_3}\omega_1\omega_2 + \frac{M_3}{I_3}$$

$$\dot{e}_0 = \frac{1}{2}(-\omega_1 e_1 - \omega_2 e_2 - \omega_3 e_3) \qquad \dot{e}_1 = \frac{1}{2}(\omega_1 e_0 + \omega_3 e_2 - \omega_2 e_3)$$

$$\dot{e}_2 = \frac{1}{2}(\omega_2 e_0 - \omega_3 e_1 + \omega_1 e_3) \qquad \dot{e}_3 = \frac{1}{2}(\omega_3 e_0 + \omega_2 e_1 - \omega_1 e_2) \quad \textbf{[8.7.2]}$$

The initial conditions for the Euler parameters, $e_j(t_0)$ ($j = 0, 1, 2, 3$) can be found from the initial values of the Euler angles using the relationships given in Eq. [7.7.40]. The initial conditions from the angular velocities can be input either as given quantities, or, if $\dot{\phi}(t_0)$, $\dot{\theta}(t_0)$ and $\dot{\psi}(t_0)$ are given instead, they can be converted to the angular velocities by using Eq. [7.5.7].

Note that even though the Euler parameters are related to each other by the equation

$$\{e\}^T\{e\} = e_0^2 + e_1^2 + e_2^2 + e_3^2 = 1 \qquad \textbf{[8.7.3]}$$

one does not need to solve Eqs. [8.7.2] together with Eq. [8.7.3]. Any accurate numerical solution that begins with an accurate description of $e_j(t_0)$ ($j = 0, 1, 2, 3$) should lead to the correct solution. In fact, Eq. [8.7.3] can be used to check the accuracy of the numerical solution. In this regard, the Rodrigues parameters (see Problems 7.15 and 7.16) are useful, as they result in six differential equations. Another set of constants that can be used to check accuracy are the entries of the matrix $[^A_B\tilde{\omega}^B]$, calculated by Eq. [7.7.30b]. Because this matrix is skew symmetric, its diagonal elements should be zero.

Consider next the translational equations. When the equations of motion are written in terms of the center of mass, the translational and rotational equations are independent of each other, unless coupling exists as a result of the applied external forces. Consider a 3-1-3 transformation and the translational equations of motion as six first-order equations. As variables, we will use the body-fixed translational velocities v_1, v_2, and v_3 and the coordinates of the center of mass in the inertial frame, which we will denote as A_1, A_2, and A_3.

The translational equations are given in Eq. [8.5.18]. To relate the rates of change of the inertial coordinates to the body-fixed velocities, we note that the relation between the velocity of the center of mass in inertial coordinates and body-fixed coordinates is

$$\{_A v_G\} = [R]^T\{_B v_G\} \qquad \textbf{[8.7.4]}$$

in which $\{_A v_G\} = [\dot{A}_1 \ \dot{A}_2 \ \dot{A}_3]^T$ and $\{_B v_G\} = [v_1 \ v_2 \ v_3]^T$ and $[R]$ is the transformation matrix, given in Eq. [7.5.3] for a 3-1-3 transformation. Introducing Eq. [7.5.3] into Eq. [8.7.4] and using Eq. [8.5.18], we obtain the translational equations of motion in state form as

$$\dot{v}_1 = -v_3\omega_2 + v_2\omega_3 + \frac{F_1}{m} \qquad \dot{v}_2 = -v_1\omega_3 + v_3\omega_1 + \frac{F_2}{m}$$

$$\dot{v}_3 = -v_2\omega_1 + v_1\omega_2 + \frac{F_3}{m}$$

$$\dot{A}_1 = (\cos\phi\cos\psi - \sin\phi\cos\theta\sin\psi)v_1 + (-\cos\phi\sin\psi - \sin\phi\cos\theta\cos\psi)v_2 + \sin\phi\sin\theta v_3$$

$$\dot{A}_2 = (\sin\phi\cos\psi + \cos\phi\cos\theta\sin\psi)v_1 + (-\sin\phi\sin\psi + \cos\phi\cos\psi\cos\theta)v_2 - \cos\phi\sin\theta v_3$$

$$\dot{A}_3 = \sin\theta\sin\psi v_1 + \sin\theta\cos\psi v_2 + \cos\theta v_3 \qquad [\mathbf{8.7.5}]$$

Equations [8.7.1] (or [8.7.2]) can be integrated independently of Eqs. [8.7.5] as long as the applied moments M_i ($i = 1, 2, 3$) are not functions of v_i or A_i ($i = 1, 2, 3$). Also, even though the latter three of Eqs. [8.7.5] are complicated expressions, they do not have the singularity problems that plague Eqs. [8.7.1]. The 12 equations [8.7.1] and [8.7.5] (or the 13 equations [8.7.2] and [8.7.5]) comprise the complete equations of an unrestrained rigid body in state form. In the presence of constraints, one must either obtain the equations of motion in terms of independent coordinates or include the constraint equations in the system description.

8.8 IMPULSE-MOMENTUM RELATIONSHIPS

In this section, we integrate the translational and rotational equations qualitatively with respect to time. Integration of Eqs. [8.5.3] and [8.5.4] over two points in time t_1 and t_2 yields

$$\int_{t_1}^{t_2} \frac{d}{dt}\mathbf{p}\,dt = \int_{t_1}^{t_2} \mathbf{F}\,dt \quad \Rightarrow \quad \mathbf{p}(t_2) - \mathbf{p}(t_1) = \int_{t_1}^{t_2} \mathbf{F}\,dt \qquad [\mathbf{8.8.1}]$$

$$\int_{t_1}^{t_2} \frac{d}{dt}\mathbf{H}_G\,dt = \int_{t_1}^{t_2} \mathbf{M}_G\,dt \quad \Rightarrow \quad \mathbf{H}_G(t_2) - \mathbf{H}_G(t_1) = \int_{t_1}^{t_2} \mathbf{M}_G\,dt \qquad [\mathbf{8.8.2}]$$

which constitute the linear and the angular impulse-momentum relationships for a rigid body, respectively. The angular impulse-momentum equation, Eq. [8.8.2], can be written in the same form when the rigid body is rotating about a fixed point.

The primary uses of Eqs. [8.8.1] and [8.8.2] are for cases when the excitation is a function of time, in the presence of impulsive forces and moments, or when momentum is conserved. The drawbacks associated with working with these relations are that they are vector relations and their manipulation often involves calculation of the reactions and constraint forces as well as integration of these forces over time. Also, when the forces and moments acting on the body are functions of time, Eqs. [8.5.3] and [8.5.4] cannot be integrated by themselves.

When the applied forces and moments are impulsive, the linear and angular impulse-momentum relations are extremely useful. One must be careful when dealing with bodies subjected to impulsive loads, as some of the reactions become impulsive as well. A special case of impulsive forcing is that of collisions, which we discuss in Section 8.12.

When the applied moment about the center of mass is zero, or its integral over time during a certain period of interest is zero, the angular momentum in the beginning of the period is the same as the angular momentum at the end of the period. The angular momentum is conserved. Similarly, if the applied forces or their integral over a period of time is zero, the linear momentum is conserved. In some cases, linear *and* angular momentum may be conserved about a certain direction, so that only a component of the relations $\mathbf{p}(t_2) = \mathbf{p}(t_1)$ or $\mathbf{H}_G(t_2) = \mathbf{H}_G(t_1)$ is used. Denoting the unit vector along which the linear momentum is conserved by $\mathbf{e}$ and the unit vector along which the angular momentum is conserved by $\mathbf{f}$, we can write for linear momentum

$$\mathbf{p}(t_2) \cdot \mathbf{e} = \mathbf{p}(t_1) \cdot \mathbf{e} \qquad [\mathbf{8.8.3}]$$

and for angular momentum conservation

$$\mathbf{H}_G(t_2) \cdot \mathbf{f} = \mathbf{H}_G(t_1) \cdot \mathbf{f} \qquad [\mathbf{8.8.4}]$$

The impulse and momentum conservation equations are very useful when more than one body is involved and motion is transferred from one body to another. A typical example is contact—or loss of contact—between two bodies. In the absence of external forces, the only forces acting on the bodies are internal for the system of two bodies. Hence, the linear momentum of the combined system and its angular momentum about the center of mass are preserved.

The axisymmetric spacecraft shown in Fig. 8.14 consists of a main body and a booster. The mass and centroidal radii of gyration of the main body and booster are given in the table. The xyz coordinates denote the F frame and point C is the center of mass of the combined system. Just before burnout, the spacecraft has acquired a velocity of 2500 m/s along the symmetry axis and an angular velocity about the symmetry axis of 0.1 rad/s. At burnout, the craft releases its booster. Immediately after the release, the center of mass of the booster has a velocity of 2000 m/s along the x direction and 2 m/s in the y direction. The angular

Example 8.8

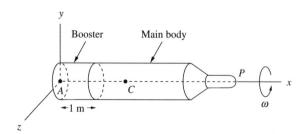

Figure 8.14

velocities of the booster are $\omega_x = 0.5$ rad/s, $\omega_y = -0.4$ rad/s, and $\omega_z = 0.9$ rad/s. Find the angular velocity of the main body and the velocity of its center of mass at this instant.

Part	Mass	Location of Center of Mass (on x axis)	κ_{xx}	$\kappa_{yy} = \kappa_{zz}$
Main body	1600 kg	3.5 m from A	0.35 m	2 m
Booster	400 kg	0.5 m from A	0.4 m	0.6 m

Solution

All forces and moments during separation are internal to the original system, which is the main body and booster together. Hence, during separation, the linear momentum and angular momentum about the center of mass of the spacecraft are conserved. Rather than calculate the centroidal mass moments of inertia of the system, we prefer to work with Eq. [8.3.4], which relates angular momentum about two points.

We denote quantities pertaining to the main body, booster, and system by subscripts M, B, and S, respectively. First we calculate the center of mass of the system. Using point A as the origin, we have

$$m_B \, 0.5 + m_M \, 3.5 = (m_B + m_M) r_S \qquad \text{[a]}$$

with the result

$$r_S = \frac{0.5(400) + 3.5(1600)}{2000} = 2.9 \text{ m} \qquad \text{[b]}$$

Let us first consider the linear momentum. Before separation the linear momentum is $\mathbf{p} = (m_B + m_M) 2500 \mathbf{i} = 5(10^6) \mathbf{i}$ kg·m/s. After separation we are given that $\mathbf{v}_B = 2000\mathbf{i} + 2\mathbf{j}$, so that the linear momentum after separation is

$$m_B(2000\mathbf{i} + 2\mathbf{j}) + m_M(v_{Mx}\mathbf{i} + v_{My}\mathbf{j} + v_{Mz}\mathbf{k}) = \mathbf{p} = 5(10^6)\mathbf{i} \text{ kg·m/s} \qquad \text{[c]}$$

Solving for the components of the velocity of the main body, we obtain

$$v_{Mx} = \frac{5(10^6) - m_B(v_{Bx})}{m_M} = \frac{5(10^6) - 400(2000)}{1600} = 2625 \text{ m/s}$$

$$v_{My} = \frac{-m_B(v_{By})}{m_M} = -\frac{2(400)}{1600} = -0.5 \text{ m/s} \qquad v_{Mz} = 0 \qquad \text{[d]}$$

Next we consider the angular momentum. We write the angular momentum of each body about the center of mass of the system, point C, as

$$\mathbf{H}_{CB} = \overline{\mathbf{H}}_B + m_B \mathbf{r}_{B/C} \times \mathbf{v}_B \qquad \mathbf{H}_{CM} = \overline{\mathbf{H}}_M + m_B \mathbf{r}_{M/C} \times \mathbf{v}_M \qquad \text{[e]}$$

where $\overline{\mathbf{H}}_B$ and $\overline{\mathbf{H}}_M$ denote the angular momenta of the booster and main body about their own center of mass, and $\mathbf{r}_{B/C} = -2.4\mathbf{i}$ m, $\mathbf{r}_{M/C} = 0.6\mathbf{i}$ m are vectors from the center of mass of the combined system to the centers of the mass of the booster and main body, respectively. Before separation, both components have a translational velocity in the x direction; thus, the cross products in the above equation vanish and we have

$$\mathbf{H}_C = \mathbf{H}_{CB} + \mathbf{H}_{CM} = (I_{Bxx} + I_{Mxx})\omega_x\mathbf{i} = (m_B\kappa_{Bxx}^2 + m_M\kappa_{Mxx}^2)\omega_x\mathbf{i}$$
$$= (400(0.4^2) + 1600(0.35^2)0.1)\mathbf{i} = 26\mathbf{i} \text{ kg} \cdot \text{m}^2/\text{s} \qquad [\mathbf{f}]$$

After separation, the angular momentum for the booster and main body are

$$\mathbf{H}_{CB} = (I_{Bxx}\omega_{Bx}\mathbf{i} + I_{Byy}\omega_{By}\mathbf{j} + I_{Bzz}\omega_{Bz}\mathbf{k}) + m_B(-2.4\mathbf{i}) \times (2000\mathbf{i} + 2\mathbf{j})$$
$$\mathbf{H}_{CM} = (I_{Mxx}\omega_{Mx}\mathbf{i} + I_{Myy}\omega_{My}\mathbf{j} + I_{Mzz}\omega_{Mz}\mathbf{k}) + m_M(0.6\mathbf{i}) \times (2625\mathbf{i} - 0.5\mathbf{j}) \qquad [\mathbf{g}]$$

where the mass moment of inertias components are

$$I_{Bxx} = m_B\kappa_{Bxx}^2 = 400(0.4^2) = 64 \text{ kg} \cdot \text{m}^2$$
$$I_{Byy} = m_B\kappa_{Byy}^2 = 400(0.6^2) = 144 \text{ kg} \cdot \text{m}^2$$
$$I_{Mxx} = m_M\kappa_{Mxx}^2 = 1600(0.35^2) = 196 \text{ kg} \cdot \text{m}^2$$
$$I_{Myy} = m_M\kappa_{Myy}^2 = 1600(2^2) = 6400 \text{ kg} \cdot \text{m}^2 \qquad [\mathbf{h}]$$

and $\omega_{Bx} = 0.5$ rad/s, $\omega_{By} = -0.4$ rad/s, and $\omega_{Bz} = 0.9$ rad/s. Hence, the angular momenta after separation are

$$\mathbf{H}_{CB} = 32\mathbf{i} - 57.6\mathbf{j} + (144(0.9) - 1920)\mathbf{k} \text{ kg} \cdot \text{m}^2/\text{s}$$
$$\mathbf{H}_{CM} = 196\omega_{Mx}\mathbf{i} + 6400\omega_{My}\mathbf{j} + (6400\omega_{Mz} - 480)\mathbf{k} \text{ kg} \cdot \text{m}^2/\text{s} \qquad [\mathbf{i}]$$

Summing the two angular momenta, equating to the angular momentum before separation, and solving for the unknown angular velocities, we obtain

$$\omega_{Mx} = \frac{26 - 32}{196} = -0.03061 \text{ rad/s} \qquad \omega_{My} = \frac{57.6}{6400} = 0.009 \text{ rad/s}$$
$$\omega_{Mz} = \frac{1920 + 480 - 129.6}{6400} = 0.3548 \text{ rad/s} \qquad [\mathbf{j}]$$

8.9 ENERGY AND WORK

In this section we examine the kinetic and potential energy for rigid bodies and develop expressions for the work done by forces and moments that act on rigid bodies. In essence, we extend the developments of Sec. 1.7 to rigid bodies. The kinetic energy of a rigid body is defined as

$$T = \frac{1}{2}\int_{body} \mathbf{v} \cdot \mathbf{v}\, dm \qquad [8.9.1]$$

where $\mathbf{v}$ is the absolute velocity of the differential element. Considering Fig. 8.1 we can express the velocity of the differential element in terms of the velocity of an arbitrary point B on the body as

$$\mathbf{v} = \mathbf{v}_B + \boldsymbol{\omega} \times \mathbf{s} \qquad [8.9.2]$$

Introducing this into Eq. [8.9.1] and performing the dot product gives

$$T = \frac{1}{2}\mathbf{v}_B \cdot \mathbf{v}_B \int_{body} dm + \frac{1}{2} \int_{body} (\boldsymbol{\omega} \times \mathbf{s}) \cdot (\boldsymbol{\omega} \times \mathbf{s}) \, dm + \mathbf{v}_B \cdot \boldsymbol{\omega} \times \int_{body} \mathbf{s} \, dm \quad [8.9.3]$$

The first term on the right side in this equation involves a translation, while the second term involves a rotation. The last term is a combination of translational and rotational terms, and it involves the first moment of the mass distribution. This term becomes zero if one of the following holds:

1. $\mathbf{v}_B = \mathbf{0}$, implying that the point B, the origin of the body-fixed coordinates, is fixed, so that the rigid body is rotating about point B.
2. $\boldsymbol{\omega} = \mathbf{0}$, implying that the body has no rotational motion.
3. B coincides with the center of mass G. If the origin of the coordinate system is selected as the center of mass, then the integral of $\mathbf{s} \, dm$ vanishes over the body.
4. If the vectors $\mathbf{v}_B$, $\boldsymbol{\omega}$, and $\mathbf{s}$ are such that the dot or cross product vanishes.

The fourth case is mathematically possible, but not practical. Hence, for all practical purposes, the kinetic energy expression is simplified if the body is rotating about a fixed point, if it is not rotating at all, or if it is expressed in terms of the center of mass.

Let us consider the kinetic energy in terms of the center of mass motion. The first term in Eq. [8.9.3] gives the translational part of the kinetic energy, denoted by T_{tran}, as

$$T_{\text{tran}} = \frac{1}{2} m \mathbf{v}_G \cdot \mathbf{v}_G \quad [8.9.4]$$

The second term in Eq. [8.9.3] gives the rotational component of the kinetic energy, denoted by T_{rot}. To evaluate it we make use of the vector relationship

$$(\mathbf{a} \times \mathbf{b}) \cdot \mathbf{c} = \mathbf{a} \cdot (\mathbf{b} \times \mathbf{c})$$

so that letting $\mathbf{a} = \boldsymbol{\omega}$, $\mathbf{b} = \boldsymbol{\rho}$, and $\mathbf{c} = \boldsymbol{\omega} \times \boldsymbol{\rho}$, we have

$$T_{\text{rot}} = \frac{1}{2} \int_{body} \boldsymbol{\omega} \cdot \boldsymbol{\rho} \times (\boldsymbol{\omega} \times \boldsymbol{\rho}) \, dm \quad [8.9.5]$$

Recalling from Eq. [8.2.8] the definition of angular momentum, we can express the rotational kinetic energy as

$$T_{\text{rot}} = \frac{1}{2} \boldsymbol{\omega} \cdot \mathbf{H}_G \quad [8.9.6]$$

This equation can also be obtained using the column vector representation of Eq. [8.9.5]. Indeed, noting that

$$(\boldsymbol{\omega} \times \boldsymbol{\rho}) \cdot (\boldsymbol{\omega} \times \boldsymbol{\rho}) = (\boldsymbol{\rho} \times \boldsymbol{\omega}) \cdot (\boldsymbol{\rho} \times \boldsymbol{\omega}) \quad [8.9.7]$$

whose column vector representation is

$$([\tilde{\rho}]\{\omega\})^T([\tilde{\rho}]\{\omega\}) = \{\omega\}^T[\tilde{\rho}]^T[\tilde{\rho}]\{\omega\} \qquad [\textbf{8.9.8}]$$

and recalling Eq. [6.4.2] we obtain the column vector representation of Eq. [8.9.7] as

$$T_{\text{rot}} = \frac{1}{2}\int_{\text{body}}\{\omega\}^T[\tilde{\rho}]^T[\tilde{\rho}]\{\omega\}\,dm = \frac{1}{2}\{\omega\}^T[I_G]\{\omega\} = \frac{1}{2}\{\omega\}^T\{H_G\} \qquad [\textbf{8.9.9}]$$

Let us summarize the kinetic energy expressions. For the general case of combined translational and rotational motion we have $T = T_{\text{tran}} + T_{\text{rot}}$, in which

$$T_{\text{tran}} = \frac{1}{2}m\mathbf{v}_G \cdot \mathbf{v}_G = \frac{1}{2}m\{v_G\}^T\{v_G\} \qquad T_{\text{rot}} = \boldsymbol{\omega} \cdot \mathbf{H}_G = \frac{1}{2}\{\omega\}^T[I_G]\{\omega\}$$
$$[\textbf{8.9.10}]$$

If the body is rotating about a fixed point C, we can write the kinetic energy as

$$T = T_{\text{rot}} = \frac{1}{2}\boldsymbol{\omega} \cdot \mathbf{H}_C = \frac{1}{2}\{\omega\}^T[I_C]\{\omega\} \qquad [\textbf{8.9.11}]$$

and if the body is only translating, with no rotational motion, we write

$$T = T_{\text{tran}} = \frac{1}{2}m\mathbf{v} \cdot \mathbf{v} \qquad T_{\text{rot}} = 0 \qquad [\textbf{8.9.12}]$$

as, for this case, all points on the body have the same velocity $\mathbf{v} = \mathbf{v}_G$. For plane motion, the kinetic energy can also be written about the instantaneous center of zero velocity. However, as the body moves the location of the instant center changes. Hence, the inertia matrix can also change, which creates difficulties if one needs to manipulate the angular momentum.

Taking the center of mass as the origin of the body-fixed reference frame, denoting $\{v_G\} = [v_x \quad v_y \quad v_z]^T$, and considering the inertia matrix and angular velocity vector, we get

$$T_{\text{tran}} = \frac{1}{2}m\{v_G\}^T\{v_G\} = \frac{1}{2}m(v_x^2 + v_y^2 + v_z^2)$$
$$T_{\text{rot}} = \frac{1}{2}\{\omega\}^T[I_G]\{\omega\}$$
$$= \frac{1}{2}(I_{xx}\omega_x^2 + I_{yy}\omega_y^2 + I_{zz}\omega_z^2) - I_{xy}\omega_x\omega_y - I_{xz}\omega_x\omega_z - I_{yz}\omega_y\omega_z \qquad [\textbf{8.9.13}]$$

If the coordinate axes are selected as the principal axes, the expression for the rotational kinetic energy further simplifies to

$$T_{\text{rot}} = \frac{1}{2}(I_{G_1}\omega_1^2 + I_{G_2}\omega_2^2 + I_{G_3}\omega_3^2) \qquad [\textbf{8.9.14}]$$

in which the indices $I_{G_i}(i = 1,2,3)$ denote the principal moments of inertia about the center of mass, and $\omega_i(i = 1,2,3)$ are the components of the angular velocity.

along the principal axes. Observe that Eq. [8.9.10] is another proof that the inertia matrix is positive definite. Kinetic energy is an absolute quantity; both the translational and rotational kinetic energy are always greater than or equal to zero. The only way for $\{\omega\}^T[I_G]\{\omega\}$ to be greater than zero for all nonzero values of the angular velocity $\{\omega\}$ is for $[I_G]$ to be a positive definite matrix.

Next, we write the rotational kinetic energy in terms of the Euler angles. Consider a 3-1-3 transformation. We recall that the angular velocity expression for a 3-1-3 transformation is

$$\boldsymbol{\omega} = \dot{\phi}\mathbf{a}_3 + \dot{\theta}\mathbf{a}_1' + \dot{\psi}\mathbf{b}_3$$
$$= (\dot{\phi}s\theta s\psi + \dot{\theta}c\psi)\mathbf{b}_1 + (\dot{\phi}s\theta c\psi - \dot{\theta}s\psi)\mathbf{b}_2 + (\dot{\phi}c\theta + \dot{\psi})\mathbf{b}_3 \qquad [\mathbf{8.9.15}]$$

The rotational kinetic energy then becomes

$$T_{\text{rot}} = \frac{1}{2}\left[I_1(\dot{\phi}s\theta s\psi + \dot{\theta}c\psi)^2 + I_2(\dot{\phi}s\theta c\psi - \dot{\theta}s\psi)^2 + I_3(\dot{\phi}c\theta + \dot{\psi})^2\right] \qquad [\mathbf{8.9.16}]$$

This relationship can be expressed in column vector form as

$$T_{\text{rot}} = \frac{1}{2}\{\omega\}^T[I_G]\{\omega\} = \frac{1}{2}\{\dot{\theta}\}^T[B]^T[I_G][B]\{\dot{\theta}\} \qquad [\mathbf{8.9.17}]$$

in which $\{\omega\} = [B]\{\dot{\theta}\}$, where $\{\dot{\theta}\} = [\dot{\phi}\ \dot{\theta}\ \dot{\psi}]^T$ denote the time derivatives of the Euler angles. The above form is not commonly used, because the corresponding inertia matrix, $[B]^T[I_G][B]$, is time dependent.

For inertially symmetric bodies, say, $I_1 = I_2$, the expression for rotational kinetic energy simplifies to

$$T_{\text{rot}} = \frac{1}{2}\left[I_1(\dot{\theta}^2 + \dot{\phi}^2\sin^2\theta) + I_3(\dot{\phi}\cos\theta + \dot{\psi})^2\right] \qquad [\mathbf{8.9.18}]$$

To describe the work done on a rigid body, consider Section 8.4 and assume that N forces $\mathbf{F}_i$ ($i = 1, 2, \ldots, N$) at points $\mathbf{r}_i$, and M^* moments $\mathbf{M}_i$ ($i = 1, 2, \ldots, M^*$) are acting on the body. Because the body is assumed to be rigid, the location of the applied moments is immaterial.

The incremental work performed by the forces and moments is defined as

$$dW = \sum_{i=1}^{N} \mathbf{F}_i \cdot d\mathbf{r}_i + \mathbf{M}^* \cdot d\boldsymbol{\theta} \qquad [\mathbf{8.9.19}]$$

in which $\mathbf{M}^*$ is the sum of all external moments acting on the body, and we have treated the incremental angular displacement as a vector quantity by considering $d\boldsymbol{\theta}$ as a vector. It is more convenient to divide the above equation by dt and work with power. Doing so we obtain

$$P = \frac{dW}{dt} = \sum_{i=1}^{N} \mathbf{F}_i \cdot \mathbf{v}_i + \mathbf{M}^* \cdot \boldsymbol{\omega} \qquad [\mathbf{8.9.20}]$$

where $\mathbf{v}_i$ is the velocity of the point to which the force $\mathbf{F}_i$ is being applied.

The expression for power can conveniently be expressed in terms of the motion of the center of mass. To this end, we write the velocity $\mathbf{v}_i$ in terms of the center of

mass as
$$\mathbf{v}_i = \mathbf{v}_G + \boldsymbol{\omega} \times \boldsymbol{\rho}_i \qquad i = 1, 2, \ldots, N \qquad [8.9.21]$$
and introduce it into $\sum \mathbf{F}_i \cdot \mathbf{v}_i$, which yields
$$\sum_{i=1}^{N} \mathbf{F}_i \cdot \mathbf{v}_i = \sum_{i=1}^{N} \mathbf{F}_i \cdot (\mathbf{v}_G + \boldsymbol{\omega} \times \boldsymbol{\rho}_i) = \mathbf{F} \cdot \mathbf{v}_G + \boldsymbol{\omega} \cdot \sum_{i=1}^{N} \boldsymbol{\rho}_i \times \mathbf{F}_i \qquad [8.9.22]$$

Introducing this expression into Eq. [8.9.20] and using, from Eq. [8.4.5], the expression for the resultant moment $\mathbf{M}_G$ about the center of mass we write the expression for power as
$$P = \mathbf{F} \cdot \mathbf{v}_G + \mathbf{M}_G \cdot \boldsymbol{\omega} \qquad [8.9.23]$$

The work done is the integral of power over time
$$W = \int_{t_1}^{t_2} P \, dt = \int_{t_1}^{t_2} (\mathbf{F} \cdot \mathbf{v}_G + \mathbf{M}_G \cdot \boldsymbol{\omega}) \, dt \qquad [8.9.24]$$

We can also write the expression for power in terms of the resultant of the forces and moments about an arbitrary point B as
$$P = \mathbf{F} \cdot \mathbf{v}_B + \mathbf{M}_B \cdot \boldsymbol{\omega} \qquad [8.9.25]$$

The general work-energy relation is
$$T_2 - T_1 = T_{\text{tran}2} + T_{\text{rot}2} - T_{\text{tran}1} - T_{\text{rot}1} = W_{1 \to 2} \qquad [8.9.26]$$

As we saw in Chapter 1, some of the forces and moments acting on a body may be conservative, that is, derivable from a potential function. It usually is more convenient to work with the potential energy associated with such forces and moments. Recalling the potential energy V, such that the infinitesimal work done by conservative forces and moments can be expressed as $dW_c = -dV$, we write
$$W_{c_{1 \to 2}} = V_1 - V_2 \qquad [8.9.27]$$
The reader is referred to Chapter 1 for further details. One can then separate the work expression into the potential energy and work of the nonconservative forces as
$$W_{1 \to 2} = -V_2 + V_1 + W_{nc1 \to 2} \qquad [8.9.28]$$
and write the work-energy theorem as
$$T_{\text{tran}1} + T_{\text{rot}1} + V_1 + W_{nc_{1 \to 2}} = T_{\text{tran}2} + T_{\text{rot}2} + V_2 \qquad [8.9.29]$$

The expressions for potential energy for rigid bodies do not lend themselves to any special form. The gravitational attraction between two bodies is a body force that is applied uniformly to every point on the body. Except for certain celestial mechanics or spacecraft dynamics problems, the gravitational attraction between two bodies is negligible compared to the gravitational attraction of the earth, and the force of gravity is conveniently represented as a single force acting through the center of mass. The gravitational potential energy becomes mg times the distance from a datum to the center of mass.

For nonnatural systems, the energy integral is $\hbar = T_2 - T_0 + V$ (note that here the subscripts denote parts of the kinetic energy), so that in conservation problems one should make use of the property that $\hbar$ is constant.

Example 8.9

Find the kinetic and potential energies of the rotating slender bar in Example 8.4 and identify the integrals of the motion.

Solution

The angular velocity of the bar is

$$\boldsymbol{\omega} = \omega_1 \mathbf{b}_1 + \omega_2 \mathbf{b}_2 + \omega_2 \mathbf{b}_3 = -\Omega \cos\theta \mathbf{b}_1 + \Omega \sin\theta \mathbf{b}_2 + \dot\theta \mathbf{b}_3 \qquad [a]$$

The mass moments of inertia about point B are

$$I_1 \approx 0 \qquad I_2 = I_3 = \frac{mL^2}{3} \qquad [b]$$

so that the kinetic energy is

$$T = T_{\text{rot}} = \frac{1}{2} I_2 (\Omega^2 \sin^2\theta + \dot\theta^2) = \frac{1}{6} mL^2 (\Omega^2 \sin^2\theta + \dot\theta^2) \qquad [c]$$

Using point B as datum, the potential energy has the form

$$V = -\frac{mgL}{2} \cos\theta \qquad [d]$$

We observe that this is a nonnatural system, and that the external forces and moments do not do any work, as they are all reaction moments. Hence, an integral of the motion is the Jacobi integral,

$$h = T_2 - T_0 + V = \frac{1}{6} mL^2 (\dot\theta^2 - \Omega^2 \sin^2\theta) - \frac{1}{2} mgL \cos\theta = \text{constant} \qquad [e]$$

If we are given the value of $\dot\theta$ when $\theta = \theta_1$ and are asked to find the value of $\dot\theta$ when $\theta = \theta_2$, we can solve for $\dot\theta_2$ using

$$\frac{1}{6} mL^2 (\dot\theta_1^2 - \Omega^2 \sin^2\theta_1) - \frac{1}{2} mgL \cos\theta_1 = \frac{1}{6} mL^2 (\dot\theta_2^2 - \Omega^2 \sin^2\theta_2) - \frac{1}{2} mgL \cos\theta_2 \qquad [f]$$

8.10 LAGRANGE'S EQUATIONS FOR RIGID BODIES

The extended Hamilton's principle and Lagrange's equations discussed in Chapter 4 have the same form whether they are written for particles or rigid bodies. In this section, we analyze the generalized forces, quantities suitable to be used as generalized coordinates, and the nature of the Lagrange's equations for rigid body motion. We also develop a physical interpretation of generalized momentum expressions.

8.10.1 VIRTUAL WORK AND GENERALIZED FORCES

One can obtain expressions for the virtual displacements by calculating the velocities and can replace time derivatives with the variational notation. To calculate the virtual work, we can take the expression for incremental work or for power, and make the appropriate substitution. For example, one way of expressing the incremental work done on a rigid body is Eq. [8.9.19], repeated here as

$$dW = \sum_{i=1}^{N} \mathbf{F}_i \cdot d\mathbf{r}_i + \mathbf{M}^* \cdot \mathbf{d\theta} \qquad [8.10.1]$$

The virtual work has the form

$$\delta W = \sum_{i=1}^{N} \mathbf{F}_i \cdot \delta\mathbf{r}_i + \mathbf{M}^* \cdot \delta\boldsymbol{\theta} \qquad [8.10.2]$$

We express the virtual work in terms of the generalized coordinates. For a system with n independent generalized coordinates, the variation of $\delta\mathbf{r}_i$ can be written as

$$\delta\mathbf{r}_i = \sum_{k=1}^{n} \frac{\partial \mathbf{r}_i}{\partial q_k} \delta q_k \qquad i = 1, 2, \ldots, N \qquad [8.10.3]$$

The problems of dealing with $\delta\boldsymbol{\theta}$ and its derivatives were discussed earlier. We also saw in Chapter 4 that if a vector $\mathbf{r}$ is a function of n generalized coordinates q_k ($k = 1, 2, \ldots, n$) and time t, we can write $\partial \mathbf{r}/\partial q_k = \partial \dot{\mathbf{r}}/\partial \dot{q}_k$. Introducing this substitution into the variation expressions, we obtain

$$\delta\mathbf{r}_i = \sum_{k=1}^{n} \frac{\partial \dot{\mathbf{r}}_i}{\partial \dot{q}_k} \delta q_k \qquad \delta\boldsymbol{\theta} = \sum_{k=1}^{n} \frac{\partial \boldsymbol{\omega}}{\partial \dot{q}_k} \delta q_k \qquad [8.10.4]$$

Substituting this equation into Eq. [8.10.2] gives

$$\delta W = \sum_{k=1}^{n} Q_k \delta q_k = \sum_{k=1}^{n} \left(\sum_{i=1}^{N} \mathbf{F}_i \cdot \frac{\partial \dot{\mathbf{r}}_i}{\partial \dot{q}_k} + \mathbf{M}^* \cdot \frac{\partial \boldsymbol{\omega}}{\partial \dot{q}_k} \right) \delta q_k \qquad [8.10.5]$$

so that the generalized forces have the form

$$Q_k = \sum_{i=1}^{N} \mathbf{F}_i \cdot \frac{\partial \dot{\mathbf{r}}_i}{\partial \dot{q}_k} + \mathbf{M}^* \cdot \frac{\partial \boldsymbol{\omega}}{\partial \dot{q}_k} \qquad k = 1, 2, \ldots, n \qquad [8.10.6]$$

The expression for the incremental work or power can be expressed in terms of the resultant of all forces $\mathbf{F}$ acting through the center of mass and the resultant moment $\mathbf{M}_G$ about the center of mass as Eq. [8.9.23]. The associated virtual work expression has the form

$$\delta W = \mathbf{F} \cdot \delta\mathbf{r}_G + \mathbf{M}_G \cdot \delta\boldsymbol{\theta} \qquad [8.10.7]$$

and the associated generalized forces become

$$Q_k = \mathbf{F} \cdot \frac{\partial \mathbf{v}_G}{\partial \dot{q}_k} + \mathbf{M}_G \cdot \frac{\partial \boldsymbol{\omega}}{\partial \dot{q}_k} \qquad [8.10.8]$$

When the resultant of forces and moments is expressed as the resultant force $\mathbf{F}$ about a point B and resultant moment $\mathbf{M}_B$ about B, as in Eq. [8.9.25], the virtual work can be expressed in terms of the motion of point B, and the generalized forces become

$$Q_k = \mathbf{F} \cdot \frac{\partial \mathbf{v}_B}{\partial \dot{q}_k} + \mathbf{M}_B \cdot \frac{\partial \boldsymbol{\omega}}{\partial \dot{q}_k} \qquad [8.10.9]$$

Depending on the problem at hand, one can use any one of Eqs. [8.10.6], [8.10.8], or [8.10.9] to calculate the generalized forces. For a system of N rigid bodies, Eq. [8.10.8] can be extended as

$$Q_k = \sum_{i=1}^{N} \left(\mathbf{F}_i \cdot \frac{\partial \mathbf{v}_{G_i}}{\partial \dot{q}_k} + \mathbf{M}_{G_i} \cdot \frac{\partial \boldsymbol{\omega}_i}{\partial \dot{q}_k} \right) \qquad [8.10.10]$$

in which $\mathbf{F}_i$ denotes the resultant of all external forces acting on the ith body, and $\mathbf{M}_{G_i}$ is the resultant moment about the center of mass of the ith body.

8.10.2 Generalized Coordinates

A set of generalized coordinates suitable to describe the translational motion of a rigid body are the displacements of its center of mass. However, the associated generalized velocities do not give too much insight; it is more suitable to use the components of the velocity in a moving coordinate frame, such as a set of body-fixed axes or the F frame. In essence, velocity and angular velocity components in a body frame are quasi-velocities (generalized speeds). Lagrange's equations, on the other hand, deal with generalized coordinates and generalized velocities. Using quasi-velocities is not feasible with the traditional form of Lagrange's equations.

Consider the Euler angles of precession (ϕ), nutation (θ), and spin (ψ) associated with a 3-1-3 Euler angle transformation. For the translational motion, we use the inertial coordinates A_1, A_2, and A_3 of the center of mass. Recall that the angular velocity expression for a 3-1-3 transformation is

$$\boldsymbol{\omega} = (\dot{\phi} s\theta s\psi + \dot{\theta} c\psi)\mathbf{b}_1 + (\dot{\phi} s\theta c\psi - \dot{\theta} s\psi)\mathbf{b}_2 + (\dot{\phi} c\theta + \dot{\psi})\mathbf{b}_3 \qquad [8.10.11]$$

The kinetic energy is

$$\begin{aligned} T &= \frac{1}{2}\left(I_1 \omega_1^2 + I_2 \omega_2^2 + I_3 \omega_3^2\right) + \frac{1}{2} m \left(\dot{A}_1^2 + \dot{A}_2^2 + \dot{A}_3^2\right) \\ &= \frac{1}{2}\left(I_1 (\dot{\phi} s\theta s\psi + \dot{\theta} c\psi)^2 + I_2(\dot{\phi} s\theta c\psi - \dot{\theta} s\psi)^2 + I_3(\dot{\phi} c\theta + \dot{\psi})^2\right) \\ &\quad + \frac{1}{2} m \left(\dot{A}_1^2 + \dot{A}_2^2 + \dot{A}_3^2\right) \end{aligned} \qquad [8.10.12]$$

Note that the precession angle ϕ is absent from T, so that ϕ can become a cyclic coordinate if it is not present in the potential energy and in the virtual work. Differentiation of the kinetic energy with respect to the Euler angles and their derivatives is cumbersome. An alternate way of dealing with the derivatives of the kinetic energy is to retain the angular velocity expression in the formulation as long as possible, as we will demonstrate. In the next chapter, we will see a modification to the

Lagrange's equations that makes it possible to derive the equations of motion in terms of the angular velocities.

Consider the physical interpretation of the generalized momenta terms associated with the Euler angles. The rotational kinetic energy is written as $T_{rot} = \boldsymbol{\omega} \cdot \mathbf{H}_G/2$. Note that because we are considering an unconstrained rigid body the translational kinetic energy has no terms involving the derivatives of the Euler angles. We first analyze the generalized momentum associated with the precession angle ϕ. Differentiating Eq. [8.10.12] with respect to $\dot\phi$ yields

$$\pi_\phi = \frac{\partial T}{\partial \dot\phi} = \frac{\partial T_{rot}}{\partial \dot\phi} \qquad [8.10.13]$$

$$= I_1(\dot\phi s\theta s\psi + \dot\theta c\psi)s\theta s\psi + I_2(\dot\phi s\theta c\psi - \dot\theta s\psi)s\theta c\psi + I_3(\dot\phi c\theta + \dot\psi)c\theta$$

Recalling the definition of the rotational kinetic energy, we can write π_ϕ also as

$$\pi_\phi = \frac{1}{2}\left[\frac{\partial \boldsymbol{\omega}}{\partial \dot\phi} \cdot \mathbf{H}_G + \boldsymbol{\omega} \cdot \frac{\partial \mathbf{H}_G}{\partial \dot\phi}\right] = \frac{\partial}{\partial \dot\phi}\left[\frac{1}{2}\{\omega\}^T[I_G]\{\omega\}\right] = \frac{\partial \omega^T}{\partial \dot\phi}[I_G]\{\omega\}$$

$$= \frac{\partial \boldsymbol{\omega}}{\partial \dot\phi} \cdot \mathbf{H}_G = I_1\omega_1\frac{\partial \omega_1}{\partial \dot\phi} + I_2\omega_2\frac{\partial \omega_2}{\partial \dot\phi} + I_3\omega_3\frac{\partial \omega_3}{\partial \dot\phi} \qquad [8.10.14]$$

The derivatives of the angular velocity vector with respect to the rates of change of the Euler angles can be obtained from Eq. [7.5.4] as

$$\frac{\partial \boldsymbol{\omega}}{\partial \dot\phi} = \mathbf{a}_3 \qquad \frac{\partial \boldsymbol{\omega}}{\partial \dot\theta} = \mathbf{a}_1' \qquad \frac{\partial \boldsymbol{\omega}}{\partial \dot\psi} = \mathbf{b}_3 \qquad [8.10.15]$$

In order to generalize Eq. [8.10.15] to any Euler angle sequence, we denote by $\mathbf{e}_\phi$ the unit vector about whose direction the Euler angle transformation with ϕ is conducted. Considering our coordinate system obtained by a 3-1-3 transformation, $\mathbf{e}_\phi = \mathbf{a}_3$. Similarly, we define $\mathbf{e}_\theta$ and $\mathbf{e}_\psi$ as the unit vectors about which the θ and ψ rotations are performed. For the 3-1-3 system under consideration

$$\mathbf{e}_\phi = \frac{\partial \boldsymbol{\omega}}{\partial \dot\phi} = \mathbf{a}_3 \qquad \mathbf{e}_\theta = \frac{\partial \boldsymbol{\omega}}{\partial \dot\theta} = \mathbf{a}_1' = \cos\psi\mathbf{b}_1 - \sin\psi\mathbf{b}_2 \qquad \mathbf{e}_\psi = \frac{\partial \boldsymbol{\omega}}{\partial \dot\psi} = \mathbf{b}_3$$

$$[8.10.16]$$

The unit vectors $\mathbf{e}_\phi$, $\mathbf{e}_\theta$, and $\mathbf{e}_\psi$ are in essence the partial velocities associated with the rates of the Euler angles. Also, from Eq. [7.5.5] we have $\mathbf{a}_3 = \sin\theta\sin\psi\mathbf{b}_1 + \sin\theta\cos\psi\mathbf{b}_2 + \cos\theta\mathbf{b}_3$, which, in light of Eqs. [8.10.13]–[8.10.15], can be expressed as

$$\mathbf{a}_3 = \mathbf{e}_\phi = \frac{\partial \omega_1}{\partial \dot\phi}\mathbf{b}_1 + \frac{\partial \omega_2}{\partial \dot\phi}\mathbf{b}_2 + \frac{\partial \omega_3}{\partial \dot\phi}\mathbf{b}_3 \qquad [8.10.17]$$

Physically, $\partial \omega_i/\partial \dot\phi$ ($i = 1, 2, 3$) is equivalent to the direction cosine, or, the cosine of the angle between the vectors $\mathbf{b}_i$ and $\mathbf{e}_\phi$.

Considering Eqs. [8.10.16] and Eq. [8.10.14], one can express the generalized momenta associated with the Euler angles as

$$\pi_\phi = \mathbf{e}_\phi \cdot \mathbf{H}_G \qquad \pi_\theta = \mathbf{e}_\theta \cdot \mathbf{H}_G \qquad \pi_\psi = \mathbf{e}_\psi \cdot \mathbf{H}_G \qquad [8.10.18]$$

Thus, *the generalized momenta associated with the Euler angles are the components of the angular momentum along the directions about which the Euler angle rotations have been performed.*

The rates of change of the unit vectors $\mathbf{e}_\phi$, $\mathbf{e}_\theta$, and $\mathbf{e}_\psi$ can be shown to be

$$\frac{d\mathbf{e}_\phi}{dt} = \frac{\partial \boldsymbol{\omega}}{\partial \phi} \qquad \frac{d\mathbf{e}_\theta}{dt} = \frac{\partial \boldsymbol{\omega}}{\partial \theta} \qquad \frac{d\mathbf{e}_\psi}{dt} = \frac{\partial \boldsymbol{\omega}}{\partial \psi} \qquad [8.10.19]$$

Equations [8.10.19] can be obtained by differentiation of Eqs. [8.10.16] and by making the proper substitutions.

We next examine the generalized forces associated with the Euler angles. The virtual work expression can be written as

$$\delta W = \sum_{k=1}^{n} Q_k \delta q_k = Q_\phi \delta \phi + Q_\theta \delta \theta + Q_\psi \delta \psi \qquad [8.10.20]$$

where Q_ϕ, Q_θ, and Q_ψ are the generalized forces. We can write the rotational equations of motion as

$$\frac{d}{dt}\pi_\phi - \frac{\partial T}{\partial \phi} = Q_\phi \qquad \frac{d}{dt}\pi_\theta - \frac{\partial T}{\partial \theta} = Q_\theta \qquad \frac{d}{dt}\pi_\psi - \frac{\partial T}{\partial \psi} = Q_\psi \qquad [8.10.21]$$

Now let us demonstrate that Lagrange's equations for an unconstrained rigid body are the angular momentum balances in the directions about which the Euler angle transformations are made.

We begin by showing that Q_ϕ, Q_θ, and Q_ψ are the components of applied external moment about the center of mass G (or center of rotation C) in the directions of the Euler angle rotations. Assuming N forces are applied to the body, we write the resultant moment as

$$\mathbf{M}_G = \sum_{i=1}^{N} \boldsymbol{\rho}_i \times \mathbf{F}_i \qquad [8.10.22]$$

where $\boldsymbol{\rho}_i$ ($i = 1, 2, \ldots, N$) are the position vectors of the points at which the forces are applied. Let us define the components of the external moment about the axes through which the Euler angles are transformed as

$$D_\phi = \mathbf{e}_\phi \cdot \mathbf{M}_G \qquad D_\theta = \mathbf{e}_\theta \cdot \mathbf{M}_G \qquad D_\psi = \mathbf{e}_\psi \cdot \mathbf{M}_G \qquad [8.10.23]$$

Take, for example, a 3-1-3 transformation and the spin angle ψ. Considering Eq. [8.10.22], we can write the component D_ψ as

$$D_\psi = \mathbf{e}_\psi \cdot \mathbf{M}_G = \sum_{i=1}^{N} \mathbf{e}_\psi \cdot (\boldsymbol{\rho}_i \times \mathbf{F}_i) \qquad [8.10.24]$$

Recalling the vector identity $\mathbf{a} \cdot (\mathbf{b} \times \mathbf{c}) = \mathbf{c} \cdot (\mathbf{a} \times \mathbf{b})$ we can write

$$D_\psi = \sum_{i=1}^N \mathbf{F}_i \cdot (\mathbf{e}_\psi \times \boldsymbol{\rho}_i) \qquad [\mathbf{8.10.25}]$$

We next evaluate the expression $\mathbf{e}_\psi \times \boldsymbol{\rho}_i$. Recalling the definition of the generalized force Q_k as

$$Q_k = \sum_{i=1}^N \mathbf{F}_i \cdot \frac{\partial \boldsymbol{\rho}_i}{\partial q_k} \qquad k = 1, 2, \ldots, n \qquad [\mathbf{8.10.26}]$$

we explore the relationship between $\mathbf{e}_\psi \times \boldsymbol{\rho}_i$ and $\partial \boldsymbol{\rho}_i / \partial \psi$. To this end, we write $\boldsymbol{\rho}_i$ in terms of the body coordinates and in column vector form as

$$\{_B\rho_i\} = [R_3(\psi)][R_2(\theta)][R_1(\phi)]\{_A\rho_i\} \qquad [\mathbf{8.10.27}]$$

so that differentiation of $\{_B\rho_i\}$ with respect to ψ yields

$$\frac{\partial \{_B\rho_i\}}{\partial \psi} = \frac{\partial [R_3(\psi)]}{\partial \psi} [R_2(\theta)][R_1(\phi)]\{_A\rho_i\} \qquad [\mathbf{8.10.28}]$$

It is easy to show that

$$\frac{\partial [R_3(\psi)]}{\partial \psi} = [\tilde{e}_\psi][R_3(\psi)] \qquad [\mathbf{8.10.29}]$$

and then, introducing Eq. [8.10.29] into Eq. [8.10.28], we obtain

$$\frac{\partial \{_B\rho_i\}}{\partial \psi} = [\tilde{e}_\psi][R_3(\psi)][R_2(\theta)][R_1(\phi)]\{_A\rho_i\} = [\tilde{e}_\psi]\{_B\rho_i\} \qquad [\mathbf{8.10.30}]$$

whose vector counterpart is

$$\frac{\partial \boldsymbol{\rho}_i}{\partial \psi} = \mathbf{e}_\psi \times \boldsymbol{\rho}_i \qquad i = 1, 2, \ldots N \qquad [\mathbf{8.10.31}]$$

Equation [8.10.31] is a very useful relationship. Introducing it into Eq. [8.10.25], we conclude that Q_ψ is indeed the same as D_ψ, and it has the form

$$Q_\psi = \sum_{i=1}^N \mathbf{F}_i \cdot \frac{\partial \boldsymbol{\rho}_i}{\partial \psi} = D_\psi = \mathbf{e}_\psi \cdot \mathbf{M}_G \qquad [\mathbf{8.10.32}]$$

In a similar fashion, one can show that Eq. [8.10.32] is applicable to the other Euler angles as well, regardless of which Euler angle transformation sequence is used, so that the generalized forces associated with the Euler angles are indeed the components of the applied moment in the direction of the Euler angle rotation, and

$$Q_\theta = \sum_{i=1}^N \mathbf{F}_i \cdot \frac{\partial \boldsymbol{\rho}_i}{\partial \theta} = \mathbf{e}_\theta \cdot \mathbf{M}_G \qquad Q_\phi = \sum_{i=1}^N \mathbf{F}_i \cdot \frac{\partial \boldsymbol{\rho}_i}{\partial \phi} = \mathbf{e}_\phi \cdot \mathbf{M}_G \qquad [\mathbf{8.10.33}]$$

8.10.3 Lagrange's Equations in Terms of the Euler Angles

Consider the dot product between the rate of change of the angular momentum and the unit vectors along the axes about which the Euler angle transformations are made. Take, for instance, $\dot{\mathbf{H}}_G \cdot \mathbf{e}_\theta$, which can be expressed as

$$\dot{\mathbf{H}}_G \cdot \mathbf{e}_\theta = \dot{\mathbf{H}}_G \cdot \mathbf{e}_\theta + \mathbf{H}_G \cdot \dot{\mathbf{e}}_\theta - \mathbf{H}_G \cdot \dot{\mathbf{e}}_\theta$$

$$= \frac{d}{dt}(\mathbf{H}_G \cdot \mathbf{e}_\theta) - \mathbf{H}_G \cdot \dot{\mathbf{e}}_\theta = \frac{d}{dt}\pi_\theta - \mathbf{H}_G \cdot \dot{\mathbf{e}}_\theta \qquad [8.10.34]$$

Considering Eq. [8.10.19] as well as the definitions of the rotational kinetic energy and angular momentum, we obtain

$$\mathbf{H}_G \cdot \dot{\mathbf{e}}_\theta = \left\{\frac{\partial \omega}{\partial \theta}\right\}^T [I_G]\{\omega\} = \frac{\partial}{\partial \theta}\left[\frac{1}{2}\{\omega\}^T[I_G]\{\omega\}\right] = \frac{\partial}{\partial \theta}T_{\text{rot}} = \frac{\partial}{\partial \theta}T \qquad [8.10.35]$$

Introducing Eq. [8.10.35] into Eq. [8.10.34], repeating the procedure for the other Euler angles, and considering Eqs. [8.10.32] and [8.10.33], we conclude that Lagrange's equations for an unconstrained rigid body can be recognized as the components of the moment balance along the directions about which the Euler angle rotations are performed:

$$\frac{d}{dt}\pi_\phi - \frac{\partial T}{\partial \phi} = Q_\phi \quad \text{or} \quad \dot{\mathbf{H}}_G \cdot \mathbf{e}_\phi = \mathbf{M}_G \cdot \mathbf{e}_\phi$$

$$\frac{d}{dt}\pi_\theta - \frac{\partial T}{\partial \theta} = Q_\theta \quad \text{or} \quad \dot{\mathbf{H}}_G \cdot \mathbf{e}_\theta = \mathbf{M}_G \cdot \mathbf{e}_\theta$$

$$\frac{d}{dt}\pi_\psi - \frac{\partial T}{\partial \psi} = Q_\psi \quad \text{or} \quad \dot{\mathbf{H}}_G \cdot \mathbf{e}_\psi = \mathbf{M}_G \cdot \mathbf{e}_\psi \qquad [8.10.36]$$

If the motion of the body is constrained in some fashion, Lagrange's equations for the rotational motion no longer have this form. Rather, one can only write them in their traditional form. Also, for rotation about a fixed point, Lagrange's equations are the moment balances about the center of rotation.

Unlike Euler's equations, Lagrange's equations are angular momentum balances about a set of nonorthogonal axes. The decision as to which form of the rotational equations to use depends on what type of explanation is sought. In general, if all three Euler angles need to be used and they are independent of each other and the general motion of the body is analyzed, it is easier to deal with Euler's equations, as they are in terms of the angular velocities and lead to simpler expressions.

The advantage of Lagrange's equations over Euler's equations becomes more apparent when one studies the motion of interconnected bodies. With Euler's equations, one must separate each component of the body and write the Euler equations for each body. Then, one eliminates the reaction forces by combining the individual equations. When using Lagrange's equations, one can circumvent part of the labor and avoid lengthy expressions by writing the kinetic energy in terms of the angular velocity components and then using the partial derivatives of the angular velocity with respect to the Euler angles.

8.10.4 LAGRANGE'S EQUATIONS IN TERMS OF EULER PARAMETERS

Another choice for generalized coordinates is the Euler parameters. In this case, we deal with a set of constrained generalized coordinates. Recall Eqs. [7.7.29] and [7.7.30c], repeated here as

$$\{\omega\} = 2[E]\{\dot{e}\} \qquad [E]\{e\} = \{0\} \qquad \textbf{[8.10.37a,b]}$$

The rotational kinetic energy and virtual work have the form

$$T_{\text{rot}} = \frac{1}{2}\{\omega\}^T[I_G]\{\omega\} \qquad \delta W = \{M_G\}^T\{\delta\theta\} \qquad \textbf{[8.10.38]}$$

We are not considering the translational kinetic energy, as it is not a function of the Euler parameters. Using Eq. [8.10.37a], we can write

$$\{\delta\theta\} = 2[E]\{\delta e\} \qquad \textbf{[8.10.39]}$$

Introduction of this and Eq. [8.10.37a] into Eqs. [8.10.38] yields the kinetic energy and virtual work in terms of the Euler parameters as

$$T_{\text{rot}} = 2\{\dot{e}\}^T[E]^T[I_G][E]\{\dot{e}\} \qquad \delta W = 2\{M_G\}^T[E]\{\delta e\} \qquad \textbf{[8.10.40]}$$

Because the Euler parameters are related to each other by $\{e\}^T\{e\} = 1$, this expression needs to be used as the constraint relation. We proceed with taking the partial derivatives of the kinetic energy as

$$\frac{\partial T_{\text{rot}}}{\partial \{\dot{e}\}} = 4\{\dot{e}\}^T[E]^T[I_G][E]$$

$$\frac{d}{dt}\left(\frac{\partial T_{\text{rot}}}{\partial \{\dot{e}\}}\right) = 4\{\ddot{e}\}^T[E]^T[I_G][E] + 4\{\dot{e}\}^T[\dot{E}]^T[I_G][E] + 4\{\dot{e}\}^T[E]^T[I_G][\dot{E}]$$

$$\textbf{[8.10.41]}$$

One can show, by examining the elements of $[\dot{E}]$, that $[\dot{E}]\{\dot{e}\} = \{0\}$, so the second term on the right side of this equation vanishes. To obtain the derivative of the kinetic energy with respect to $\{e\}$, we note from Eq. [8.10.37b] that

$$[E]\{\dot{e}\} = -[\dot{E}]\{e\} \qquad \textbf{[8.10.42]}$$

and then, introducing Eq. [8.10.42] into T_{rot} and differentiating, we obtain

$$\frac{\partial T_{\text{rot}}}{\partial \{e\}} = \frac{\partial}{\partial \{e\}} 2\{e\}^T[\dot{E}]^T[I_G][\dot{E}]\{e\} = 4\{e\}^T[\dot{E}]^T[I_G][\dot{E}] \qquad \textbf{[8.10.43]}$$

Using Eq. [8.10.42] we can also show that the last term on the right side of Eq. [8.10.41] can be written as

$$4\{\dot{e}\}^T[E]^T[I_G][\dot{E}] = -4\{e\}^T[\dot{E}]^T[I_G][\dot{E}] \qquad \textbf{[8.10.44]}$$

The modified virtual work, in the presence of the constraint, has the form

$$\delta\hat{W} = 2\{M_G\}^T[E]\{\delta e\} + 2\lambda\{e\}^T\{\delta e\} \qquad \textbf{[8.10.45]}$$

Combining Eqs. [8.10.41], [8.10.43], and [8.10.45], and taking the transpose of the resulting expression, we obtain Lagrange's equations in terms of the Euler parameters as

$$[E]^T[I_G][E]\{\ddot{e}\} - 2[\dot{E}]^T[I_G][\dot{E}]\{e\} = 0.5[E]^T\{M_G\} + 0.5\lambda\{e\} \quad \textbf{[8.10.46]}$$

subject to the constraint $\{e\}^T\{e\} = 1$.

8.10.5 LAGRANGE'S EQUATIONS AND CONSTRAINTS

The Lagrangian treatment of dynamical systems subjected to constraints was discussed in Section 4.10. The treatment of three-dimensional motion is essentially the same. One basically has to decide how to handle the constraints.

When the constraint is holonomic, one has a choice: generate a set of independent generalized coordinates that take into account the constraint, or deal with constrained generalized coordinates. When the constraint is nonholonomic, one must use constrained generalized coordinates.

When using constrained generalized coordinates, one has two options:

a. Introduce the kinematics of the constraint into the problem by means of Lagrange multipliers. When the constraint is expressed as a configuration constraint, it augments the Lagrangian. When the constraint is expressed in velocity form, it augments the virtual work. Both cases lead to a set of equations in terms of the Lagrange multipliers. After obtaining the equations, one can seek to eliminate the Lagrange multipliers and generate a set of independent equations of motion.

b. Introduce the constraint forces into the formulation. One does this by relaxing the constraints and accounting for the effect of the constraints by means of the constraint forces. These forces enter the equations of motion via the virtual work. Once the equations of motion are obtained, the kinematics of the constraints are introduced, and an expression is developed for the constraint force. Mathematically, this approach is equivalent to expressing the constraint in velocity form. The constraint force is essentially the Lagrange multiplier. While mathematically not any different from the others, this approach gives better physical insight. It is particularly useful when dealing with holonomic constraints and in the presence of friction forces. It also is a suitable approach to calculate the magnitudes of constraint forces that do no work.

Example 8.10

Derive the equations of motion for the spinning top shown in Fig. 7.13 using a 3-1-3 Euler angle transformation, and analyze the motion integrals. The centroidal moments of inertia are $I_1, I_2 = I_1$, and I_3. The center of mass is at a height of L from the bottom of the top. Assume that the point of contact between the top and the ground is stationary.

Solution

The motion of the top can be viewed as rotating about the fixed point O. The kinetic energy is

$$T = T_{\text{rot}} = \frac{1}{2}\left(I_{O1}\omega_1^2 + I_{O2}\omega_2^2 + I_{O3}\omega_3^2\right) \quad \textbf{[a]}$$

where the components of the mass moment of inertia about point O are

$$I_{O1} = I_1 + mL^2 \qquad I_{O2} = I_1 + mL^2 \qquad I_{O3} = I_3 \qquad \textbf{[b]}$$

Introducing the values for the angular velocities for a 3-1-3 transformation from Eq. [8.9.15], we obtain for the kinetic energy

$$T = \frac{1}{2}\left(I_{O1}(\dot{\phi}^2 \sin^2\theta + \dot{\theta}^2) + I_{O3}(\dot{\phi}\cos\theta + \dot{\psi})^2\right) \qquad \textbf{[c]}$$

The potential energy is

$$V = mgL\cos\theta \qquad \textbf{[d]}$$

The only other forces acting on the spinning top are those at the point of contact. They do no work, as they are being applied to a fixed point. It follows that the virtual work expression is zero.

Before obtaining the equations of motion, let us first analyze the integrals of the motion. The virtual work is zero, the Lagrangian is not an explicit function of time, and all terms in the kinetic energy are quadratic in the generalized coordinates. It follows that the first integral of the motion is the Jacobi integral, which for this case is the total energy

$$T + V = \text{constant} \qquad \textbf{[e]}$$

Examining Eqs. [c] and [d] closer, we note that the Lagrangian does not have any ϕ and ψ dependency. We conclude that the generalized momenta associated with these two coordinates are constant, which gives us two more integrals of the motion in the form

$$\pi_\phi = \frac{\partial T}{\partial \dot{\phi}} = I_{O1}\dot{\phi}\sin^2\theta + I_{O3}(\dot{\phi}\cos\theta + \dot{\psi})\cos\theta = \text{constant} \qquad \textbf{[f]}$$

$$\pi_\psi = \frac{\partial T}{\partial \dot{\psi}} = I_{O3}(\dot{\phi}\cos\theta + \dot{\psi}) = \text{constant} \qquad \textbf{[g]}$$

These first integrals are the components of the angular momentum along the a_3 and b_3 directions. We will discuss their physical interpretation further in Chapter 10. Equation [g] is recognized as $\omega_3 =$ constant. The first two equations of motion are then obtained by differentiating the generalized momenta associated with ϕ and ψ, and they have the form

$$\frac{d}{dt}\pi_\phi = 0 \qquad \frac{d}{dt}\pi_\psi = 0 \qquad \textbf{[h]}$$

We find the equation of motion associated with θ by invoking Lagrange's equations. We have

$$\frac{\partial T}{\partial \dot{\theta}} = (I_1 + mL^2)\dot{\theta} \qquad \frac{\partial T}{\partial \theta} = (I_1 + mL^2)\dot{\phi}^2 \sin\theta\cos\theta - I_3(\dot{\phi}\cos\theta + \dot{\psi})\dot{\phi}\sin\theta$$

$$\frac{\partial V}{\partial \theta} = -mgL\sin\theta \qquad \textbf{[i]}$$

leading to the equation of motion

$$(I_1 + mL^2)\ddot{\theta} - (I_1 + mL^2 - I_3)\dot{\phi}^2 \sin\theta\cos\theta + I_3\dot{\phi}\dot{\psi}\sin\theta - mgL\sin\theta = 0 \qquad \textbf{[j]}$$

Example 8.11

The slender bar of mass m and length L, shown in Fig. 8.15, is attached to the arm of a shaft which is rotating with a constant speed of Ω. The length of the arm is a. Find the equation of motion for the bar and the equilibrium position(s). Then, analyze the stability of the equilibrium.

Solution

We attach a $b_1 b_2 b_3$ coordinate system to the rod and write the angular velocity of the rod as the summation of the rate of change of θ and the angular velocity of the shaft as

$$\boldsymbol{\omega} = -\Omega \cos\theta \mathbf{b}_1 + \Omega \sin\theta \mathbf{b}_2 + \dot{\theta} \mathbf{b}_3 \qquad [a]$$

The mass moments of inertia about the center of mass are

$$I_1 = 0 \qquad I_2 = I_3 = \frac{1}{12} m L^2 \qquad [b]$$

To evaluate the kinetic energy, we need to find the velocity of the center of mass. Using the relative velocity relation

$$\mathbf{v}_G = \mathbf{v}_B + \boldsymbol{\omega} \times \mathbf{r}_{G/B} \qquad [c]$$

where

$$\mathbf{v}_B = -\Omega a \mathbf{b}_3 \qquad \mathbf{r}_{G/B} = \frac{L}{2} \mathbf{b}_1 \qquad [d]$$

we obtain

$$\mathbf{v}_G = -\Omega a \mathbf{b}_3 + (-\Omega \cos\theta \mathbf{b}_1 + \Omega \sin\theta \mathbf{b}_2 + \dot{\theta} \mathbf{b}_3) \times \frac{L}{2} \mathbf{b}_1$$

$$= \frac{L}{2} \dot{\theta} \mathbf{b}_2 - \Omega \left(\frac{L \sin\theta}{2} + a \right) \mathbf{b}_3 \qquad [e]$$

We write the kinetic energy as

$$T = \frac{1}{2}(I_1 \omega_1^2 + I_2 \omega_2^2 + I_3 \omega_3^2) + \frac{1}{2} m \mathbf{v}_G \cdot \mathbf{v}_G \qquad [f]$$

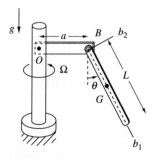

Figure 8.15

8.10 LAGRANGE'S EQUATIONS FOR RIGID BODIES

and, substituting the appropriate terms, we obtain

$$T = \frac{1}{24}mL^2(\dot{\theta}^2 + \Omega^2 \sin^2\theta) + \frac{1}{8}mL^2\dot{\theta}^2 + \frac{1}{2}m\Omega^2\left(\frac{L\sin\theta}{2} + a\right)^2 \qquad [\text{g}]$$

Note that we could not have written the kinetic energy as purely rotational about B because point B is not fixed.

The potential energy is

$$V = -mg\frac{L}{2}\cos\theta \qquad [\text{h}]$$

There are no nonconservative forces that do work. Application of Lagrange's equations yields the equation of motion as

$$\ddot{\theta} - \Omega^2 \sin\theta\cos\theta - \frac{3a}{2L}\Omega^2\cos\theta + \frac{3g}{2L}\sin\theta = 0 \qquad [\text{i}]$$

Finding the equilibrium position requires that we solve Eq. [i] with $\ddot{\theta} = 0$, for

$$-\Omega^2 \sin\theta\cos\theta - \frac{3a}{2L}\Omega^2\cos\theta + \frac{3g}{2L}\sin\theta = 0 \qquad [\text{j}]$$

Let us assume small motions and approximate $\sin\theta$ by θ and $\cos\theta$ by 1. Solving Eq. [j] we obtain for the equilibrium position θ_e

$$\theta_e = \frac{\dfrac{3a\Omega^2}{2L}}{\dfrac{3g}{2L} - \Omega^2} \qquad [\text{k}]$$

These equations also give a range of validity for the small angle assumption, namely that Ω^2 should be less than $3g/2L$. Comparing this result with the bead problem in Chapters 4 and 5, we observe that $\theta = 0$ is not an equilibrium point as long as the arm length a is not zero.

Next, we analyze the equilibrium point. From the above equation, as the rotational speed Ω gets larger, θ_e becomes larger. This can be explained by noting that the rod lifts up higher as the shaft spins faster. Considering the stability of the equilibrium point, Eqs. [g] and [h] indicate that the system is nonnatural, with $T_1 = 0$. The dynamic potential $U = V - T_0$ has the form

$$U = -mg\frac{L}{2}\cos\theta - \frac{1}{24}mL^2\Omega^2\sin^2\theta - \frac{1}{2}m\Omega^2\left(\frac{L\sin\theta}{2} + a\right)^2 \qquad [\text{l}]$$

The first derivative of U gives the equilibrium equation (Eq. [j] divided by 3). The second derivative of U is

$$\frac{\partial^2 U}{\partial \theta^2} = \frac{1}{2}mgL\cos\theta - \frac{1}{3}mL^2\Omega^2(\cos^2\theta - \sin^2\theta) + \frac{1}{2}mLa\Omega^2\sin\theta \qquad [\text{m}]$$

For small values of θ_e, if we introduce the small angle assumption to this equation, we obtain

$$\frac{\partial^2 U}{\partial \theta^2} = \frac{1}{2}mgL + mL\Omega^2\left(-\frac{1}{3}L + \frac{1}{2}a\theta_e\right) = \frac{1}{3}mL^2\left(\frac{3g}{2L} - \Omega^2\right) + \frac{1}{2}mLa\Omega^2\theta_e \qquad [\text{n}]$$

From Eq. [k], the first term on the right side of Eq. [n] is greater than zero; thus, the second derivative of U is larger than zero for all times when the small angle assumption is valid. We, of course, intuitively expected this to happen. Indeed, the stability property is valid also for large values of the equilibrium position.

When $a = 0$, we get that $\theta = 0$ is an equilibrium position for the bar as well, and that for small values of the rotation speed $\theta_e = 0$ represents a stable equilibrium position. To understand why $\theta = 0$ is not an equilibrium position when there is an arm, it is helpful to visualize the equilibrium position as a point where the centrifugal force and the gravitational force create moments that balance each other out. The instant an arm is placed on the shaft and the rod is suspended from that arm, the moment generated by the centrifugal force becomes much larger, and the equilibrium position shifts up.

Example 8.12

A spinning top of mass m_1, and centroidal moments of inertia I_1 and I_3, where I_3 is measured about the symmetry axis, is moving on a cart of mass m_2 which is constrained to move horizontally and in one direction, as shown in Fig 8.16. Find the equations of motion using a Lagrangian approach.

Solution

This problem is a four degree of freedom, holonomic problem. We select the translational coordinate for the cart and the three Euler angles as the generalized coordinates. We write the kinetic and potential energies of the cart and spinning top separately. We select a reference frame $a_1 a_2 a_3$ such that the cart always moves in the a_2 direction and use a 3-1-3 Euler angle transformation (ϕ, θ, ψ) to orient the top. We take advantage of the symmetry and use the F frame. The angular velocity of the top can be expressed as

$$\boldsymbol{\omega} = \omega_1 \mathbf{f}_1 + \omega_2 \mathbf{f}_2 + \omega_3 \mathbf{f}_3 = \dot{\theta} \mathbf{f}_1 + \dot{\phi} \sin\theta \mathbf{f}_2 + (\dot{\phi} \cos\theta + \dot{\psi}) \mathbf{f}_3 \qquad \textbf{[a]}$$

Next, let us write the kinetic and potential energies of the cart and top. For the cart we have

$$T_{\text{cart}} = \frac{1}{2} m_2 \dot{Y}^2 \qquad V_{\text{cart}} = 0 \qquad \textbf{[b]}$$

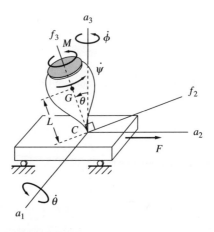

Figure 8.16

8.10 LAGRANGE'S EQUATIONS FOR RIGID BODIES

and for the top we have

$$T_{\text{top}} = \frac{1}{2}m_1\mathbf{v}_G\cdot\mathbf{v}_G + \frac{1}{2}\boldsymbol{\omega}\cdot\mathbf{H}_G = \frac{1}{2}m_1\mathbf{v}_G\cdot\mathbf{v}_G + \frac{1}{2}\left(I_1(\omega_1^2+\omega_2^2) + I_3\omega_3^2\right) \quad [\text{c}]$$

where the velocity of the center of mass of the top is

$$\mathbf{v}_G = \dot{Y}\mathbf{a}_2 + \boldsymbol{\omega}\times\mathbf{r}_{G/C} = \dot{Y}\mathbf{a}_2 + (\omega_1\mathbf{f}_1 + \omega_2\mathbf{f}_2 + \omega_3\mathbf{f}_3)\times L\mathbf{f}_3$$
$$= \dot{Y}\mathbf{a}_2 + \omega_2 L\mathbf{f}_1 - \omega_1 L\mathbf{f}_2 \quad [\text{d}]$$

We prefer to use (for notational purposes) ω_1, ω_2, and ω_3 to preserve the compact nature of the equations. We can express $\mathbf{a}_2$ in terms of the F frame by taking the second column of $[R]$ and setting $\psi = 0$ in Eq. [7.5.3] as

$$\mathbf{a}_2 = \sin\phi\mathbf{f}_1 + \cos\phi\cos\theta\mathbf{f}_2 - \cos\phi\sin\theta\mathbf{f}_3 \quad [\text{e}]$$

which results in the velocity expression for the center of mass of the top, written

$$\mathbf{v}_G = (\dot{Y}\sin\phi + \omega_2 L)\mathbf{f}_1 + (\dot{Y}\cos\phi\cos\theta - \omega_1 L)\mathbf{f}_2 - \dot{Y}\cos\phi\sin\theta\mathbf{f}_3 \quad [\text{f}]$$

Substituting the above equation in Eq. [c], carrying out the algebra, and combining with the kinetic energy of the cart yields

$$T = \frac{1}{2}(I_1 + m_1L^2)(\omega_1^2+\omega_2^2) + \frac{1}{2}I_3\omega_3^2 + \frac{1}{2}(m_1+m_2)\dot{Y}^2 + m_1L\dot{Y}(\omega_2\,\text{s}\,\phi - \omega_1\,\text{c}\,\phi\,\text{c}\,\theta)$$
$$= \frac{1}{2}(I_1 + m_1L^2)(\dot{\theta}^2 + \dot{\phi}^2\,\text{s}^2\,\theta) + \frac{1}{2}I_3(\dot{\phi}\,\text{c}\,\theta + \dot{\psi})^2 + \frac{1}{2}(m_1+m_2)\dot{Y}^2$$
$$+ m_1L\dot{Y}(\dot{\phi}\,\text{s}\,\phi\,\text{s}\,\theta - \dot{\theta}\,\text{c}\,\phi\,\text{c}\,\theta) \quad [\text{g}]$$

We notice that the last term in Eq. [g] provides the coupling between the translational and rotational motions. The external force that does work is F, which is along the a_2 direction, and the moment M, which is along the spin axis, f_3. The potential energy and virtual work are

$$V = m_1gL\cos\theta \qquad \delta W = F\delta Y + M(\cos\theta\delta\phi + \delta\psi) \quad [\text{h}]$$

To implement Lagrange's equations, we need to take the partial derivatives with respect to the generalized coordinates and generalized velocities. Once we write the angular velocities in terms of the Euler angles and start taking partial velocities, the equations become complex and lengthy. In addition, it becomes increasingly difficult to attribute a physical meaning to the equations of motion. We thus keep the angular velocities in the formulation as long as possible. We evaluate the partial derivatives of the angular velocities with respect to the Euler angles and their rates of change. Using Eq. [a] we have

$$\frac{\partial\omega_1}{\partial\dot{\phi}} = 0 \qquad \frac{\partial\omega_1}{\partial\dot{\theta}} = 1 \qquad \frac{\partial\omega_1}{\partial\dot{\psi}} = 0 \qquad \frac{\partial\omega_1}{\partial\theta} = 0$$

$$\frac{\partial\omega_2}{\partial\dot{\phi}} = \sin\theta \qquad \frac{\partial\omega_2}{\partial\dot{\theta}} = 0 \qquad \frac{\partial\omega_2}{\partial\dot{\psi}} = 0 \qquad \frac{\partial\omega_2}{\partial\theta} = \dot{\phi}\cos\theta$$

$$\frac{\partial\omega_3}{\partial\dot{\phi}} = \cos\theta \qquad \frac{\partial\omega_3}{\partial\dot{\theta}} = 0 \qquad \frac{\partial\omega_3}{\partial\dot{\psi}} = 1 \qquad \frac{\partial\omega_3}{\partial\theta} = -\dot{\phi}\sin\theta \quad [\text{i}]$$

470   CHAPTER 8 • RIGID BODY DYNAMICS: BASIC CONCEPTS

With respect to the precession angle, we have

$$\frac{\partial T}{\partial \dot{\phi}} = (I_1 + m_1 L^2)\left(\omega_1 \frac{\partial \omega_1}{\partial \dot{\phi}} + \omega_2 \frac{\partial \omega_2}{\partial \dot{\phi}}\right) + I_3 \omega_3 \frac{\partial \omega_3}{\partial \dot{\phi}} + m_1 L \dot{Y}\left(\frac{\partial \omega_2}{\partial \dot{\phi}} \sin\phi - \frac{\partial \omega_1}{\partial \dot{\phi}} \cos\phi \cos\theta\right)$$

$$= (I_1 + m_1 L^2)\omega_2 \sin\theta + I_3 \omega_3 \cos\theta + m_1 L \dot{Y} \sin\phi \sin\theta$$

$$\frac{\partial T}{\partial \phi} = m_1 L \dot{Y}(\omega_2 \cos\phi + \omega_1 \sin\phi \cos\theta) \qquad \frac{\partial V}{\partial \phi} = 0 \qquad \text{[j]}$$

so that, noting the relation $\omega_1 = \dot{\theta}$, the equation of motion associated with the precession angle becomes

$$(I_1 + m_1 L^2)(\dot{\omega}_2 \, s\,\theta + \omega_1 \omega_2 \, c\,\theta) + I_3 \dot{\omega}_3 \, c\,\theta + m_1 L \ddot{Y} s\,\phi s\,\theta - I_3 \omega_1 \omega_3 s\theta = M c\,\theta \qquad \text{[k]}$$

With respect to the nutation angle, we have

$$\frac{\partial T}{\partial \dot{\theta}} = (I_1 + m_1 L^2)\left(\omega_1 \frac{\partial \omega_1}{\partial \dot{\theta}} + \omega_2 \frac{\partial \omega_2}{\partial \dot{\theta}}\right) + I_3 \omega_3 \frac{\partial \omega_3}{\partial \dot{\theta}} + m_1 L \dot{Y}\left(\frac{\partial \omega_2}{\partial \dot{\theta}} \sin\phi - \frac{\partial \omega_1}{\partial \dot{\theta}} \cos\phi \cos\theta\right)$$

$$= (I_1 + m_1 L^2)\omega_1 - m_1 L \dot{Y} \cos\phi \cos\theta$$

$$\frac{\partial T}{\partial \theta} = (I_1 + m_1 L^2)\omega_2 \dot{\phi} \cos\theta + I_3 \omega_3 (-\dot{\phi}\sin\theta) + m_1 L \dot{Y}(\dot{\phi}\sin\phi\cos\theta + \omega_1 \cos\phi\sin\theta)$$

$$\frac{\partial V}{\partial \theta} = -m_1 g L \sin\theta \qquad \text{[l]}$$

Noting that $\omega_2 = \dot{\phi}\sin\theta$, we obtain the equation of motion for the nutation rate as

$$(I_1 + m_1 L^2)\dot{\omega}_1 - m_1 L \ddot{Y}\cos\phi\cos\theta - (I_1 + m_1 L^2)\frac{\omega_2^2}{\tan\theta} + I_3\omega_2\omega_3 - m_1 g L \sin\theta = 0 \qquad \text{[m]}$$

For the spin rate,

$$\frac{\partial T}{\partial \dot{\psi}} = I_3 \omega_3 \qquad \frac{\partial T}{\partial \psi} = 0 \qquad \frac{\partial V}{\partial \psi} = 0 \qquad \text{[n]}$$

so that the equation of motion has the form

$$I_3 \dot{\omega}_3 = M \qquad \text{[o]}$$

Comparing Eqs. [o] and [k], we observe that Eq. [o] is embedded in Eq. [k]. This is to be expected, because the equation of motion for the precession angle ϕ is nothing but the moment balance about the precession axis (a_3), and the equation for the spin angle is the moment balance about the b_3 axis. The angle between the two axes is θ. We multiply Eq. [o] by $\cos\theta$, subtract it from Eq. [k], and divide the result by $\sin\theta$, which gives

$$(I_1 + m_1 L^2)\left(\dot{\omega}_2 + \frac{\omega_1 \omega_2}{\tan\theta}\right) + m_1 L \ddot{Y}\sin\phi - I_3 \omega_1 \omega_3 = 0 \qquad \text{[p]}$$

Finally, with regards to the translational coordinate Y, we note that $\partial \omega_i/\partial \dot{Y} = 0$ ($i = 1, 2, 3$), thus

$$\frac{\partial T}{\partial \dot{Y}} = (m_1 + m_2)\dot{Y} + m_1 L(\omega_2 s\,\phi - \omega_1 c\,\phi c\,\theta) \qquad \frac{\partial T}{\partial Y} = 0 \qquad \frac{\partial V}{\partial Y} = 0 \qquad \text{[q]}$$

and the equation of motion becomes

$$(m_1 + m_2)\ddot{Y} + m_1 L(\dot{\omega}_2 s\phi - \dot{\omega}_1 c\phi c\theta) + m_1 L(\omega_2 \dot{\phi} c\phi + \omega_1 \dot{\phi} s\phi c\theta + \omega_1 \dot{\theta} c\phi s\theta) = F \quad [\mathbf{r}]$$

which we can rearrange as

$$(m_1 + m_2)\ddot{Y} + m_1 L(\dot{\omega}_2 s\phi - \dot{\omega}_1 c\phi c\theta) + m_1 L\left(\frac{\omega_2^2 c\phi}{\sin\theta} + \frac{\omega_1 \omega_2 s\phi}{\tan\theta} + \omega_1^2 c\phi s\theta\right) = F \quad [\mathbf{s}]$$

An interesting observation can be made with regards to the equations of motion. The final expressions do not contain a mixed product of $\dot{Y}$ and one of ω_1, ω_2, or ω_3. This is to be expected, for two reasons. First, we discussed in Chapter 4 that many of the $\partial T/\partial q_k$ terms, where q_k are generalized coordinates, cancel. Second, mixed velocity terms in the equations of motion are indicative of Coriolis effects and of motion with respect to a rotating frame. While we have mixed angular velocity terms in the equations of motion, there are no mixed velocity terms involving $\dot{Y}$, because Y is not a rotational coordinate. One can use this property when checking the accuracy of the equations of motion.

Let us now consider the spinning top without the cart. We should be able to derive its equations of motion by eliminating all the $m_2 \ddot{Y}$ terms from the equations of motion. Indeed, eliminating the $\ddot{Y}$ terms from Eqs. [m], [o], and [p] and rearranging, we obtain

$$(I_1 + m_1 L^2)\dot{\omega}_1 - \omega_2 \left[(I_1 + m_1 L^2)\frac{\omega_2}{\tan\theta} - I_3\omega_3\right] - m_1 g L \sin\theta = 0$$

$$(I_1 + m_1 L^2)\dot{\omega}_2 + \omega_1 \left[(I_1 + m_1 L^2)\frac{\omega_2}{\tan\theta} - I_3\omega_3\right] = 0$$

$$I_3 \dot{\omega}_3 = M \qquad [\mathbf{t}]$$

which can also be derived from Eqs. [8.5.30] or [8.5.31], as we did in Example 8.10.

The above example illustrates some of the difficulties encountered when using Lagrange's equations in conjunction with the general three-dimensional motion of rigid bodies. In this particular case, we alleviated some of the complexities in a roundabout way, by retaining ω_i ($i = 1, 2, 3$) in the equations of motion and taking their partial derivatives with respect to the Euler angles. If, on the other hand, we wanted to apply Euler's equations to this problem directly, we would have to separate the top and the cart into two separate systems. We would then write the individual equations of motion, which contain the reaction terms. After that, we would need to eliminate the reactions and obtain the equations of motion for the system. The question arises as to whether there are direct methods that permit the use of the body angular velocities in the equations of motion and also at the same time facilitate the solution for interconnected bodies. The answer to this question is discussed in Chapter 9.

Example 8.13

Derive the equations of motion for a disk of mass m and radius R that is rolling without slipping on a flat surface. Use Lagrange's equations.

Solution

The kinematics of the rolling disk is discussed in Section 7.9. Consider Figures 7.32–7.34, and use as generalized coordinates the translation of the center of mass X, Y, and Z, and the

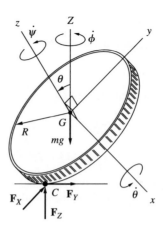

Figure 8.17

Euler angles ϕ, θ, and ψ in conjunction with a 3-1-3 sequence. The free-body diagram of the disk is shown in Fig. 8.17. The force F_Z is the normal force and F_Y and F_X are friction forces.

We obtained in Section 7.9 that the motion is governed by one holonomic and two nonholonomic constraints of the form

$$Z = R\sin\theta \qquad [\text{a}]$$

$$\dot{X} = R\dot{\theta}\sin\phi\sin\theta - R(\dot{\phi}\cos\theta + \dot{\psi})\cos\phi \qquad [\text{b}]$$

$$\dot{Y} = -R\dot{\theta}\cos\phi\sin\theta - R(\dot{\phi}\cos\theta + \dot{\psi})\sin\phi \qquad [\text{c}]$$

To find the equations of motion using Lagrange's equations, we write the expressions for the kinetic and potential energies. For a uniform thin disk we have

$$I_1 = I_2 = \frac{1}{4}mR^2 \qquad I_3 = \frac{1}{2}mR^2 \qquad [\text{d}]$$

We write the kinetic energy as

$$T = T_{\text{tran}} + T_{\text{rot}} = \frac{1}{2}m(\dot{X}^2 + \dot{Y}^2 + \dot{Z}^2) + \frac{1}{2}\left[I_1(\dot{\phi}^2\sin^2\theta + \dot{\theta}^2) + I_3(\dot{\phi}\cos\theta + \dot{\psi})^2\right]$$

$$= \frac{1}{2}m(\dot{X}^2 + \dot{Y}^2 + R^2\dot{\theta}^2\cos^2\theta) + \frac{1}{2}\left[I_1(\dot{\phi}^2\sin^2\theta + \dot{\theta}^2) + I_3(\dot{\phi}\cos\theta + \dot{\psi})^2\right] \qquad [\text{e}]$$

Note that in Eq. [e] we did not substitute for the values of $\dot{X}$ and $\dot{Y}$ from Eqs. [b] and [c], but we substituted the value of $\dot{Z}$. This is because the constraints [b] and [c] are nonholonomic and, while the expression for the kinetic energy would be correct if this substitution were made, the changes in the translational generalized coordinates would not be accounted for when the derivatives of the kinetic energy are taken. After deriving the equations of motion, one can perform the substitution. If Eqs. [b] and [c] are introduced into the kinetic energy and Lagrange's equations are invoked, we get an incorrect result.

The potential energy is

$$V = mgR\sin\theta \qquad [\text{f}]$$

The virtual work expression is due to the nonholonomic constraints [b] and [c], and it is expressed as

$$\delta W = \lambda_1[\delta X - R\sin\phi\sin\theta\delta\theta + R(\cos\theta\delta\phi + \delta\psi)\cos\phi]$$
$$+ \lambda_2[\delta Y + R\cos\phi\sin\theta\delta\theta + R(\cos\theta\delta\phi + \delta\psi)\sin\phi] \quad \textbf{[g]}$$

The generalized forces are obtained from the virtual work as

$$Q_X = \lambda_1 \qquad Q_Y = \lambda_2 \qquad Q_\phi = R\cos\phi\cos\theta\lambda_1 + R\sin\phi\cos\theta\lambda_2$$
$$Q_\theta = -R\sin\phi\sin\theta\lambda_1 + R\cos\phi\sin\theta\lambda_2 \qquad Q_\psi = R\cos\phi\lambda_1 + R\sin\phi\lambda_2 \quad \textbf{[h]}$$

Applying Lagrange's equations for a constrained system, we obtain five differential equations

$$m\ddot{X} = \lambda_1 \quad \textbf{[i]}$$

$$m\ddot{Y} = \lambda_2 \quad \textbf{[j]}$$

$$I_1\ddot{\phi}s^2\theta + I_3\ddot{\phi}c^2\theta + I_3\ddot{\psi}c\theta + 2I_1\dot{\phi}\dot{\theta}s\theta c\theta - I_3\dot{\theta}\dot{\psi}s\theta - 2I_3\dot{\phi}\dot{\theta}c\theta s\theta$$
$$= Rc\phi c\theta\lambda_1 + Rs\phi c\theta\lambda_2 \quad \textbf{[k]}$$

$$(I_1 + mR^2c^2\theta)\ddot{\theta} - mR^2\dot{\theta}^2 s\theta c\theta - I_1\dot{\phi}^2 s\theta c\theta - I_3(\dot{\phi}c\theta + \dot{\psi})\dot{\phi}s\theta + mgRc\theta$$
$$= -Rs\phi s\theta\lambda_1 + Rc\phi s\theta\lambda_2 \quad \textbf{[l]}$$

$$I_3(\ddot{\phi}c\theta - \dot{\phi}\dot{\theta}s\theta + \ddot{\psi}) = Rc\phi\lambda_1 + Rs\phi\lambda_2 \quad \textbf{[m]}$$

The five equations above and the two constraints [b] and [c] can be used together to solve for the seven unknowns: the five generalized coordinates and the two Lagrange multipliers.

One can reduce the number of equations by first introducing Eqs. [i] and [j] into Eqs. [k], [l], and [m]. This eliminates the Lagrange multipliers from the formulation. One can realize a further reduction by differentiating Eqs. [b] and [c], and substituting the results into Eqs. [k], [l], and [m], which are now in terms of $\ddot{X}$ and $\ddot{Y}$. As a result, we end up with three equations of motion, in terms of the Euler angles. The procedure is tedious; we outline here the procedure for the ψ equation only.

Introduction of Eqs. [i] and [j] into Eq. [m] yields

$$I_3(\ddot{\phi}\cos\theta - \dot{\phi}\dot{\theta}\sin\theta + \ddot{\psi}) = mR\ddot{X}\cos\phi + mR\ddot{Y}\sin\phi \quad \textbf{[n]}$$

Differentiation of Eqs. [b] and [c] gives

$$\ddot{X} = R\ddot{\theta}s\phi s\theta - R(\ddot{\phi}c\theta + \ddot{\psi})c\phi + 2R\dot{\phi}\dot{\theta}c\phi s\theta + R\dot{\theta}^2 s\phi c\theta + R\dot{\phi}^2 s\phi c\theta + R\dot{\phi}\dot{\psi}s\phi \quad \textbf{[o]}$$

$$\ddot{Y} = -R\ddot{\theta}c\phi s\theta - R(\ddot{\phi}c\theta + \ddot{\psi})s\phi + 2R\dot{\phi}\dot{\theta}s\phi s\theta - R\dot{\theta}^2 c\phi c\theta - R\dot{\phi}^2 c\phi c\theta - R\dot{\phi}\dot{\psi}c\phi \quad \textbf{[p]}$$

Introduction of Eqs. [o] and [p] into Eq. [n] and carrying out the algebra yields

$$(I_3 + mR^2)(\ddot{\phi}\cos\theta - \dot{\phi}\dot{\theta}\sin\theta + \ddot{\psi}) - mR^2\dot{\phi}\dot{\theta}\sin\theta = 0 \quad \textbf{[q]}$$

The other two equations of motion are obtained in the same way, and they have the form

$$(I_1 + mR^2)\ddot{\theta} + (I_3 + mR^2)\dot{\phi}s\theta(\dot{\phi}c\theta + \dot{\psi}) - I_1\dot{\phi}^2 s\theta c\theta + mgRc\theta = 0 \quad \textbf{[r]}$$

$$I_1\ddot{\phi}s\theta + 2I_1\dot{\phi}\dot{\theta}c\theta - I_3\dot{\theta}(\dot{\phi}c\theta + \dot{\psi}) = 0 \quad \textbf{[s]}$$

This example further illustrates the difficulties associated with obtaining the equations of motion of a system when all three Euler angles are present as generalized coordinates. Given the simplicity of the unconstrained equations, one wonders whether there is a more suitable way of obtaining the equations of motion.

Also, this example is more of a special case, where elimination of the Lagrange multipliers and redundant coordinates leads to a simpler set of equations. More often than not, one ends up with considerably more complicated equations after eliminating the Lagrange multipliers from the formulation.

Example 8.14

Consider Example 8.4 and calculate the moment acting on the shaft so that the shaft rotates with a constant angular velocity.

Solution

We will use the constraint relaxation method. We treat the rotation of the shaft as a variable $\dot{\phi}$, and consider a moment M acting on the shaft. From Example 8.9, the rotational kinetic energy has the form

$$T_{\text{rot}} = \frac{1}{6}mL^2(\dot{\phi}^2 \sin^2\theta + \dot{\theta}^2) + \frac{1}{2}I\dot{\phi}^2 \qquad \text{[a]}$$

where I is the mass moment of inertia of the shaft about the vertical. The virtual work is due to the moment M

$$\delta W = M\delta\phi \qquad \text{[b]}$$

and ϕ does not contribute to the potential energy. Invoking Lagrange's equations we obtain the equation of motion associated with ϕ as

$$\frac{1}{3}mL^2\ddot{\phi}\sin^2\theta + I\ddot{\phi} + \frac{2}{3}mL^2\dot{\phi}\dot{\theta}\sin\theta\cos\theta = M \qquad \text{[c]}$$

We next invoke the constraint that $\dot{\phi} = \Omega$ is constant, so that the expression for M becomes

$$M = \frac{2}{3}mL^2\Omega\dot{\theta}\sin\theta\cos\theta \qquad \text{[d]}$$

Let us compare this answer with Eq. [j] in Example 8.4, where we obtain the reaction moments. Considering Fig. 8.7, we recognize that

$$M = M_2 \sin\theta \qquad \text{[e]}$$

which we confirm upon comparing Eq. [d] with Eq. [j] in Example 8.4.

8.11 D'ALEMBERT'S PRINCIPLE FOR RIGID BODIES

In Section 5.13 we discussed certain shortcomings associated with Lagrange's equations and saw a way of writing the equations of motion for a system of particles directly from D'Alembert's principle. In this section, we do the same for rigid bodies.

8.11 D'Alembert's Principle for Rigid Bodies

For a system of N particles, D'Alembert's principle is written as

$$\sum_{i=1}^{N} m_i \mathbf{a}_i \cdot \delta \mathbf{r}_i = \sum_{i=1}^{N} \mathbf{F}_i \cdot \delta \mathbf{r}_i \qquad [8.11.1]$$

in which $\mathbf{F}_i$ denotes all the forces external to the ith particle. For a rigid body, we replace the summation by an integration, m_i by dm, $\mathbf{F}_i$ by $d\mathbf{F}$, and we drop the subscript i, with the result

$$\int_{\text{body}} \mathbf{a} \cdot \delta \mathbf{r}\, dm = \int_{\text{body}} d\mathbf{F} \cdot \delta \mathbf{r} \qquad [8.11.2]$$

Writing from Fig. 8.1 the position vector as $\mathbf{r} = \mathbf{r}_G + \boldsymbol{\rho}$, we expand $\mathbf{a}$ and $\delta \mathbf{r}$ in terms of the center of mass as

$$\delta \mathbf{r} = \delta \mathbf{r}_G + \delta \boldsymbol{\theta} \times \boldsymbol{\rho} \qquad \mathbf{a} = \mathbf{a}_G + \boldsymbol{\alpha} \times \boldsymbol{\rho} + \boldsymbol{\omega} \times (\boldsymbol{\omega} \times \boldsymbol{\rho}) \qquad [8.11.3]$$

Note that, as before, the variation of the rotation is denoted by $\delta \boldsymbol{\theta}$, indicating that this is not a derived quantity, but rather a defined one. Introduction of Eqs. [8.11.3] to the left side of Eq. [8.11.2] yields

$$\int_{\text{body}} \mathbf{a} \cdot \delta \mathbf{r}\, dm = \int_{\text{body}} \left(\mathbf{a}_G + \boldsymbol{\alpha} \times \boldsymbol{\rho} + \boldsymbol{\omega} \times (\boldsymbol{\omega} \times \boldsymbol{\rho}) \right) \cdot (\delta \mathbf{r}_G + \delta \boldsymbol{\theta} \times \boldsymbol{\rho})\, dm$$

$$= \int_{\text{body}} \left(\mathbf{a}_G \cdot \delta \mathbf{r}_G + (\boldsymbol{\alpha} \times \boldsymbol{\rho}) \cdot (\delta \boldsymbol{\theta} \times \boldsymbol{\rho}) + \right.$$

$$\left. \left(\boldsymbol{\omega} \times (\boldsymbol{\omega} \times \boldsymbol{\rho})\right) \cdot (\delta \boldsymbol{\theta} \times \boldsymbol{\rho}) \right) dm \qquad [8.11.4]$$

All other terms drop out due to the definition of the center of mass. The first term on the right side of the above equation is recognized as $m\mathbf{a}_G \cdot \delta \mathbf{r}_G$. To evaluate the second and third terms, we make use of column vector formulation to write

$$(\boldsymbol{\alpha} \times \boldsymbol{\rho}) \cdot (\delta \boldsymbol{\theta} \times \boldsymbol{\rho}) = \{\delta \boldsymbol{\theta}\}^T [\tilde{\rho}]^T [\tilde{\rho}] \{\alpha\} \qquad [8.11.5]$$

and realize from Eq. [8.2.18] that $\int [\tilde{\rho}]^T [\tilde{\rho}]\, dm = [I_G]$ and that $[I_G]\{\alpha\} = \{dH_G/dt\}_{\text{rel}}$. Manipulation of the last term on the right side of Eq. [8.11.4] is more complicated. After a number of manipulations one can show that

$$(\boldsymbol{\omega} \times \boldsymbol{\omega} \times \boldsymbol{\rho}) \cdot (\delta \boldsymbol{\theta} \times \boldsymbol{\rho}) = \delta \boldsymbol{\theta} \cdot [\boldsymbol{\omega} \times (\boldsymbol{\rho} \times (\boldsymbol{\omega} \times \boldsymbol{\rho}))] \qquad [8.11.6]$$

where we recognize the expression that leads to the angular momentum, $\boldsymbol{\rho} \times (\boldsymbol{\omega} \times \boldsymbol{\rho})$. It follows that the second and third terms in Eq. [8.11.4] reduce to

$$\int_{\text{body}} \left[(\boldsymbol{\alpha} \times \boldsymbol{\rho}) \cdot (\delta \boldsymbol{\theta} \times \boldsymbol{\rho}) + \left(\boldsymbol{\omega} \times (\boldsymbol{\omega} \times \boldsymbol{\rho})\right) \cdot (\delta \boldsymbol{\theta} \times \boldsymbol{\rho}) \right] dm$$

$$= \delta \boldsymbol{\theta} \cdot \left(\frac{d\mathbf{H}_G}{dt} \right)_{\text{rel}} + \delta \boldsymbol{\theta} \cdot (\boldsymbol{\omega} \times \mathbf{H}_G) = \dot{\mathbf{H}}_G \cdot \delta \boldsymbol{\theta} \qquad [8.11.7]$$

Introducing the expression for $\delta \mathbf{r}$ into the right side of Eq. [8.11.2] and using the results from Section 8.4, we obtain

$$\int_{\text{body}} d\mathbf{F} \cdot \delta \mathbf{r} = \int_{\text{body}} d\mathbf{F} \cdot (\delta \mathbf{r}_G + \delta \boldsymbol{\theta} \times \boldsymbol{\rho})\, dm = \mathbf{F} \cdot \delta \mathbf{r}_G + \mathbf{M}_G \cdot \delta \boldsymbol{\theta} \qquad [8.11.8]$$

in which **F** is the resultant force and $\mathbf{M}_G$ is the resultant moment about the center of mass. Substituting Eqs. [8.11.7] and [8.11.8] into Eq. [8.11.2], we obtain D'Alembert's principle for a rigid body as

$$m\mathbf{a}_G \cdot \delta \mathbf{r}_G + \dot{\mathbf{H}}_G \cdot \delta \boldsymbol{\theta} = \mathbf{F} \cdot \delta \mathbf{r}_G + \mathbf{M}_G \cdot \delta \boldsymbol{\theta} \qquad [8.11.9]$$

When there is more than one body in the system, D'Alembert's principle becomes

$$\sum_{i=1}^{N} (m\mathbf{a}_{G_i} \cdot \delta \mathbf{r}_{G_i} + \dot{\mathbf{H}}_{G_i} \cdot \delta \boldsymbol{\theta}_i) = \sum_{i=1}^{N} (\mathbf{F}_i \cdot d\mathbf{r}_{G_i} + \mathbf{M}_{G_i} \cdot \delta \boldsymbol{\theta}_i) \qquad [8.11.10]$$

where all the internal forces and moments that one body exerts on another cancel. Actually, Eq. [8.11.10] is the most general definition of D'Alembert's principle. For a system of particles, all one does is to eliminate the rotational terms from Eq. [8.11.10].

We next consider writing the equations of motion directly from D'Alembert's principle. To this end, we express the variation of the centers of mass and of the rotations in terms of independent generalized coordinates and velocities as

$$\delta \mathbf{r}_{G_i} = \sum_{k=1}^{n} \frac{\partial \mathbf{r}_{G_i}}{\partial q_k} \delta q_k = \sum_{k=1}^{n} \frac{\partial \mathbf{v}_{G_i}}{\partial \dot{q}_k} \delta q_k \qquad \delta \boldsymbol{\theta}_i = \sum_{k=1}^{n} \frac{\partial \boldsymbol{\omega}_i}{\partial \dot{q}_k} \delta q_k \qquad [8.11.11]$$

We introduce Eqs. [8.11.11] into Eq. [8.11.10], with the result

$$\sum_{k=1}^{n} \sum_{i=1}^{N} \left(m\mathbf{a}_{G_i} \cdot \frac{\partial \mathbf{v}_{G_i}}{\partial \dot{q}_k} + \dot{\mathbf{H}}_{G_i} \cdot \frac{\partial \boldsymbol{\omega}_i}{\partial \dot{q}_k} - \mathbf{F}_i \cdot \frac{\partial \mathbf{v}_{G_i}}{\partial \dot{q}_k} - \mathbf{M}_{G_i} \cdot \frac{\partial \boldsymbol{\omega}_i}{\partial \dot{q}_k} \right) \delta q_k = 0 \qquad [8.11.12]$$

We make use of the property that the variations of independent generalized coordinates are independent themselves. For Eq. [8.11.12] to hold, the coefficients of the variations of the generalized coordinates must vanish individually. We hence obtain the equations of motion, as the coefficients of the variations of the generalized coordinates, as

$$\sum_{i=1}^{N} \left(m\mathbf{a}_{G_i} \cdot \frac{\partial \mathbf{v}_{G_i}}{\partial \dot{q}_k} + \dot{\mathbf{H}}_{G_i} \cdot \frac{\partial \boldsymbol{\omega}_i}{\partial \dot{q}_k} \right) = Q_k \qquad k = 1, 2, \ldots, n \qquad [8.11.13]$$

in which the generalized forces are given by

$$Q_k = \sum_{i=1}^{N} \left(\mathbf{F}_i \cdot \frac{\partial \mathbf{v}_{G_i}}{\partial \dot{q}_k} + \mathbf{M}_{G_i} \cdot \frac{\partial \boldsymbol{\omega}_i}{\partial \dot{q}_k} \right) \qquad [8.11.14]$$

Note the similarity of the above equations with Lagrange's equations derived in the previous section. The right sides are identical. On the left sides, in Lagrange's equations we have derivatives of the kinetic energy, and in Eq. [8.11.13] we have vector products. Writing the rotational equations of motion in this form is usually more convenient than taking derivatives of the kinetic energy.

The disadvantage of Eq. [8.11.13] is that the acceleration and rate of change of the angular momentum need to be calculated, which involves many more operations than computation of the kinetic energy. In a sense, Eq. [8.11.13] is a compromise between Lagrange's equations and Euler's equations, combining aspects of both. In Chapter 9, we study additional ways of writing the equations of motion that are based on D'Alembert's principle.

8.12 IMPACT OF RIGID BODIES

An interesting application of the impulse-momentum relations is when two moving rigid bodies or a moving body and a stationary object collide. The duration of the collision is very important when analyzing the properties of the ensuing motion. When the collision takes place in an extremely short period of time, it can be considered as impulsive motion. The impulsive reaction forces that are generated form an impulsive constraint at the location of the collision. The velocities and angular velocities before and after the collision are related to each other by linear and angular momentum principles, as well as by *Poisson's hypothesis.*

A collision takes place along the *line of impact,* which is perpendicular to the surface tangent to both colliding bodies. This surface is defined the same way the rolling surface is defined in Chapter 7. For particles, smooth spheres, or disks, the line of impact joins the centers of mass of the colliding bodies.

For collisions of rigid bodies, the line of impact does not necessarily go through the centers of mass of the colliding bodies. Rather, the line of impact connects the impact point and centers of curvature of the contours of the colliding bodies at the point of impact, as shown in Fig 8.18. Also, one of the impacting bodies may not have a radius of curvature at the point of impact, because the impacting point is a

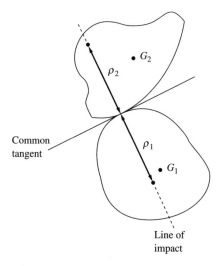

Figure 8.18

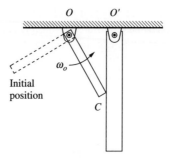

Figure 8.19

sharp edge. An example is given in Fig. 8.19. The rod on the left is given an initial motion by releasing it from an angle. The line of impact is perpendicular to the rod on the right. If, at the point of contact, the contours of both colliding bodies do not have a radius of curvature, one then has to make a reasonable assumption regarding the line of impact.

As a result of the property that the line of impact does not lie along the line joining the centers of mass of the colliding bodies, Poisson's hypothesis, which we studied in Chapter 3, does not always reduce to Eq. [3.5.10], repeated here as

$$\Delta v_2 = -e \Delta v_1 \qquad [8.12.1]$$

in which e is the coefficient of restitution and Δv is the difference in the components of the velocities of the colliding points along the line of impact, and the subscripts 1 and 2 denote before and after the impact, respectively. Also referred to as *Newton's experimental law*,[1] Eq. [8.12.1], which can be regarded as a kinematic relationship, holds for the component of the velocity along the line of impact when the contours of the colliding bodies are smooth at the point of impact and there is no friction involved. Recall that Poisson's hypothesis states that the impulse takes place in two stages: a compression stage, where the bodies compress each other until the relative velocity of the two bodies along the line of impact is zero, and a restitution stage, where the colliding points split from each other and the bodies regain their original shapes. Inherent in this hypothesis is the consideration that the colliding bodies have some elasticity in them to permit compression and restitution.

The coefficient of restitution e indicates the strengths of the two stages of impact. It varies between 0 and 1, its value depending on the material properties of the colliding bodies as well as the circumstances under which contact takes place. When $e = 1$, the collision is known as perfectly elastic impact and there is no energy loss. The strengths of the impact during the compression and restitution are the same. A value of e between 0 and 1 indicates that the strength of the impact has lessened in the restitution phase. This reduction in the strength is due to the elasticity and

---

[1] It was Newton who first formulated the velocity relations for impact. Poisson later on generalized what happens during impact to what is known as Poisson's hypothesis.

internal energy dissipation. When $e = 0$, there is plastic impact and no restitution. The colliding bodies do not separate from each other immediately after impact. The shapes of the bodies also affect the coefficient of restitution. When the collision takes place over a sharp edge, the value of e becomes smaller than its value for impact over a smooth surface.

It should be noted that the coefficient of restitution represents a gross simplification of what happens during impact. Another issue that needs consideration is the material damage that may ensue as a result of impact. Since the implicit assumption is that there is no damage to the colliding bodies as a result of impact, we are inherently considering low-velocity impact. This itself challenges the validity of the assumption that impact takes place during an extremely small period of time. Yet another issue is the presence of frictional forces during impact. Recalling that friction forces are obtained from normal forces through the multiplication of coefficients of friction and that the coefficients of friction themselves are largely simplifications, the issue of dealing with frictional impulsive forces becomes even more complicated. Take care to see if the assumptions used are correct and whether dealing with a frictional impulsive force is realistic.

Example 8.15

Consider a cube resting on a smooth surface, subjected to an impulsive force $\hat{F}$, as shown in Fig. 8.20. Find the velocity of the center of the mass of the cube and its angular velocity after impact as a function of the coefficient of restitution e.

Solution

We draw in Figs. 8.21–8.22 the free-body diagrams of the cube, during the compression and restitution, respectively. The impact of the cube with the smooth surface takes place at point C. We conclude this by noting that point C can move horizontally but it cannot move vertically because $\hat{F}$ causes a clockwise rotation of the cube. We denote the impulsive reaction during compression by $\hat{R}$.

We denote the components of the velocity of the center of mass and angular velocity at the end of the compression period by v_{xc}, v_{yc}, and ω_c, and after the collision is over, by v_x, v_y and ω, respectively. After compression, the linear impulse-momentum theorem yields in the x and y directions

$$mv_{xc} = \hat{F} \qquad mv_{yc} = \hat{R} \qquad \text{[a]}$$

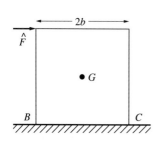

Figure 8.20

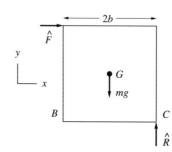

Figure 8.21   Compression

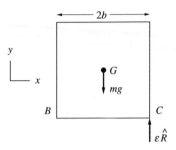

Figure 8.22   Restitution

The angular impulse-momentum relationship about the center of mass yields

$$I_G \omega_c = b(\hat{R} - \hat{F}) \qquad \text{[b]}$$

where $I_G = m((2b)^2 + (2b)^2)/12 = 2mb^2/3$ is the centroidal mass moment of inertia. Eqs. [a] and [b] constitute three equations that need to be solved for the four unknowns v_{xc}, v_{yc}, ω_c, and $\hat{R}$. The fourth equation comes from the kinematics and from the realization that point C has no vertical velocity at the end of the compression period. To this end, we write the relative velocity equation between the center of mass and point C as

$$\mathbf{v}_C = \mathbf{v}_G + \omega_c \mathbf{k} \times \mathbf{r}_{C/G} = (v_{xc} + \omega_c b)\mathbf{i} + (v_{yc} + \omega_c b)\mathbf{j} \qquad \text{[c]}$$

where $\mathbf{v}_G = v_{xc}\mathbf{i} + v_{yc}\mathbf{j}$ and $\mathbf{r}_{C/G} = b\mathbf{i} - b\mathbf{j}$. Noting that the vertical component of $\mathbf{v}_C$ is zero, we obtain for the component of the motion in the y direction

$$\mathbf{v}_C \cdot \mathbf{j} = v_{yc} + \omega_c b = 0 \qquad \text{[d]}$$

Solving Eqs. [a] through [d] simultaneously, we obtain

$$v_{xc} = \frac{\hat{F}}{m} \qquad v_{yc} = \frac{\hat{F}b^2}{I_G + mb^2} \qquad \omega_c = -\frac{\hat{F}b}{I_G + mb^2} \qquad \hat{R} = \frac{m\hat{F}b^2}{I_G + mb^2} \qquad \text{[e]}$$

Next, we write the impulse-momentum relationships for the restitution phase. From the free-body diagram, the velocity in the x direction does not change. In the y direction we have

$$mv_{yc} + e\hat{R} = mv_y \qquad \text{[f]}$$

Introducing the values for v_{yc} and $\hat{R}$ into this equation, we obtain

$$v_y = \frac{\hat{F}b^2}{I_G + mb^2}(1 + e) \qquad \text{[g]}$$

Similarly, for the angular impulse momentum about the center of mass we have

$$I_G \omega_c + e\hat{R}b = I_G \omega \qquad \text{[h]}$$

whose solution is

$$I_G \omega = \frac{(-I_G + emb^2)\hat{F}b}{I_G + mb^2} = \left(e - \frac{2}{3}\right)\frac{m\hat{F}b^3}{I_G + mb^2} \qquad \text{[i]}$$

The sign of the angular velocity after impact depends on the value of the coefficient of restitution. For perfectly elastic impact, the angular velocity is positive, counterclockwise by our sign convention. Because there is no energy loss, it is as if the cube bounces back. However, as the cube is resting horizontally, a counterclockwise rotation is impossible and a second impact is immediately generated by an impulsive force at point B. The cube rocks back and forth as it translates forward.

As the coefficient of restitution becomes smaller, the resistance of the cube to clockwise rotation becomes smaller. When $e = 2/3$, the angular velocity ω becomes zero and the cube just acquires a translational velocity with no rotation. When e is less than 2/3, the cube acquires a clockwise angular velocity. As the coefficient of restitution becomes even smaller, the angular velocity becomes large enough to tip the cube over.

Note that in order to solve this problem using Eq. [8.12.1], we would have to assume that the cube is originally resting an infinitesimal distance above the surface; then we would use the kinematic relations as the infinitesimal distance goes to zero.

REFERENCES

Etkin, B. *Dynamics of Flight; Stability and Control.* 2nd ed. New York: John Wiley, 1982.
Ginsberg, J. H. *Advanced Engineering Dynamics.* New York: Harper & Row, 1988.
Greenwood, D. T. *Principles of Dynamics.* 2nd ed. Englewood Cliffs, N.J.: Prentice-Hall, 1988.
Junkins, J. L., and J. D. Turner. *Optimal Spacecraft Rotational Maneuvers.* Amsterdam-New York: Elsevier, 1986.
Kilmister, C. W., and J. E. Reeve. *Rational Mechanics.* New York: American Elsevier, 1966.
Meirovitch, L. *Methods of Analytical Dynamics.* New York: McGraw-Hill, 1970.
Truesdell, C. T. *Essays in the History of Mechanics.* New York: Springer-Verlag, 1968.

HOMEWORK EXERCISES

SECTIONS 8.2 AND 8.3

1. Consider the slender bar in Fig. 8.15. Find the angular momentum of the bar about point B.
2. Replace the slender bar in the previous problem by a rectangular plate, as shown in Fig. 8.23. Find the angular momentum about point B.

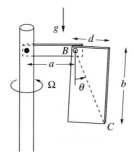

Figure 8.23

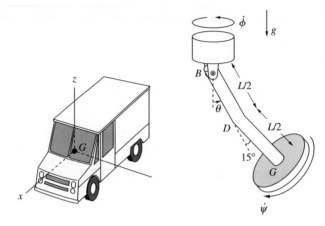

Figure 8.24   **Figure 8.25**

3. Figure 8.17 shows a disk of radius R rolling without slipping. Using a 3-1-3 Euler angle transformation, find its linear momentum and angular momentum about the center of mass and about the contact point.

4. The vehicle in Fig. 8.24 has an inertia matrix as given below. If for the instant considered the angular velocities of the vehicle are $\omega_x = 0.1$ rad/s, $\omega_y = -0.04$ rad/s, and $\omega_z = -0.15$ rad/s, find the angular momentum of the vehicle expressed in terms of principal axes.

$$[I_G] = \begin{bmatrix} 100 & 0 & 50 \\ 0 & 300 & 0 \\ 50 & 0 & 250 \end{bmatrix} \text{ kg} \cdot \text{m}^2$$

5. The light shaft of a spherical pendulum is bent as shown in Fig. 8.25. Find the angular momentum of the pendulum about B.

6. Find the angular momentum of the disk in Fig. 2.47 about point A.

Section 8.5

7. Find the equation of motion and magnitudes of the reaction forces for the spinning pendulum in Fig. 8.15 using Euler's equations.

8. Consider Example 8.5. Find the equation of motion when the rod connecting the shaft to the disk has a mass of $m/2$. Both $\dot{\phi}$ and $\dot{\psi}$ are kept constant by servomotors.

9. Consider the spinning top in Fig. 8.16. The cart is not moving, and at the instant shown $\theta = 15°$, $\dot{\theta} = 0$, $\ddot{\theta} = 0$, and $\dot{\phi} = 0.2$ rad/s and is constant. Find the spin rate $\dot{\psi}$ necessary to maintain this condition.

10. Consider the rolling cone in Fig. 7.35. Pivot O is fixed, and $R/L = 0.2$. Given that friction is sufficient to prevent slipping, find the necessary condition on the

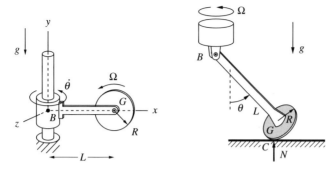

Figure 8.26   **Figure 8.27**

angular velocity of the cone that will prevent the cone from tipping. *Hint:* Model the normal force acting on the cone as a single force acting from a distance d from O find d and relate the angular velocity to d.

11. The disk of mass m and radius R in Fig. 8.26 is being held by a light bar. The disk rotates with constant angular velocity Ω. The other end of the bar is connected to a joint at B. The xyz axes are attached to the bar. Find the external moment necessary to have the bar rotate and the reaction moments when the joint permits motion about (*a*) the x axis, (*b*) the y axis, and (*c*) the z axis.

12. Consider the spinning top in Fig. 7.13. Find the equations of motion using Eqs. [8.5.31] and [8.5.32].

13. The disk in Fig. 8.27 rotates about the light rod BG and it rolls on a horizontal surface. The shaft to which point B is attached rotates with the constant angular velocity Ω. Find the reactions at B and the normal force at the point of contact C.

14. Find the equation of motion of the plate in Fig. 8.23 for $a = 0$, $b = d$. Calculate the reactions at B.

15. Consider the rolling disk in Example 8.13. Show that when the nutation rate is zero, the center of the disk traverses a circular path. Relate the radius of this path to the spin rate.

16. Consider the unicycle in Fig. 7.54. Assume roll without slip and find the normal and friction forces at the point of contact.

Section 8.6

17. The triangular plate in Fig 8.28 is attached to a massless rod. The rod spins with constant angular velocity Ω. Find the reactions at the supports and the moment that needs to be exerted on the shaft to maintain this angular velocity.

18. The rectangular plate shown in Fig 8.29 is attached to a massless rod. The rod spins with constant angular velocity Ω. Find the reactions at the supports A and B.

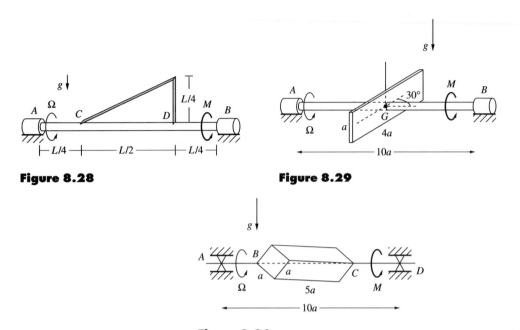

Figure 8.28

Figure 8.29

Figure 8.30

19. The rectangular prism of mass m is attached to a shaft that goes through points B and C (Fig. 8.30). Find the reactions at the supports and the moment necessary to exert on the shaft to keep the angular velocity Ω constant.

Section 8.8

20. The rectangular plate of mass 5 kg in Fig. 8.31 is resting on the xz plane on one of its edges, which is attached to a spherical joint. Suddenly, the plate is hit by an impulsive force of magnitude 300 N·s in the negative y direction. Find the angular velocities of the plate and the velocity of the center of mass after this impulse is applied.

21. The cylinder in Fig. 8.32 of length L and radius R initially has an angular velocity of $\omega_x = 0.4$ rad/s. The cylinder breaks into two equal parts at point C, and it is observed immediately after the break that the difference in velocities at C is $\mathbf{v}_{C_2} - \mathbf{v}_{C_1} = 0.2\,\mathbf{j}$ m/s, and the cylinder on the left has angular velocities of $\omega_x = 0.2$ rad/s, $\omega_y = -0.1$ rad/s. Find the angular velocity of the cylinder on the right.

22. The slab from Space Odyssey 2001, shown in Fig. 8.33, is tumbling freely in space with $\mathbf{v}_G = \mathbf{0}$. The slab is of mass 10 kg and has dimensions $a = 1$ m, $b = 0.2$ m, and $c = 0.5$ m. Suddenly, a rock of mass 0.5 kg traveling with speed 12 m/s in the negative y direction hits the slab and gets lodged in it. Right before impact, the slab has angular velocities of $\omega_x = 0.1$ rad/s, $\omega_y = 0.2$ rad/s,

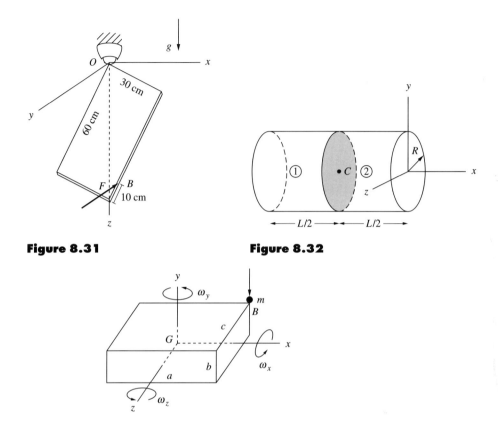

Figure 8.31

Figure 8.32

Figure 8.33

and $\omega_z = 0.4$ rad/s. Find the velocity of the center of mass of the slab and its angular velocity immediately after impact.

SECTION 8.9

23. The double pendulum in Fig. 8.34 is swinging in the local xz plane, which is rotating with the constant angular velocity of Ω. Write the kinetic and potential energies and the equilibrium equations, and identify the integral(s) of the motion.

24. Consider Fig. 8.15, with $a = 0.3$ m, $L = 0.6$ m, mass $m = 2$ kg. The shaft is rotating with constant angular velocity $\Omega_1 = 2$ rad/s. Initially, the pin joint at B is locked in the position $\theta = 90°$. The lock breaks and the rod begins to swing. Calculate the angular velocity of the rod when $\theta = 60°$.

25. Consider the previous problem, with $a = 0$, and in the absence of a motor that maintains a constant angular velocity of the shaft. Given the same initial conditions as above, find the angular velocity of the shaft and of the rod when $\theta = 60°$. *Hint:* What are the integrals of the motion?

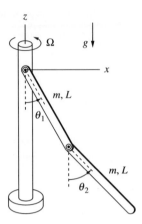

Figure 8.34

Section 8.10

26. Derive the equations of motion for a spinning top using a 3-2-3 Euler angle transformation and by using Lagrange's equations. Compare your answer with the results of Example 8.10.

27. Find the equation of motion of the gyropendulum in Example 8.5 using Lagrange's equations.

28. Find the equation of motion of the spinning plate in Fig. 8.23 using Lagrange's equations.

29. The spacecraft in Fig. 8.35 has the following centroidal inertias: $I_1 = I_2 = 5m\kappa^2$, $I_3 = m\kappa^2$, where κ is the radius of gyration. Along the b_2 axis there is a frictionless tube, inside which a point mass $m/10$ moves. The point mass

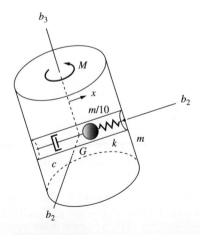

Figure 8.35

is connected to the ends of the tube by a spring k and a dashpot c. An external torque $\mathbf{M} = M\mathbf{b}_3$ is applied. Derive the equations of motion using a 3-1-3 Euler angle sequence. Ignore the translation of the spacecraft.

30. The shaft in Fig. 8.23 rotates with constant angular velocity Ω. Find the moment that must be exerted on the shaft to keep its angular velocity constant. Do this by treating the rotation of the shaft as a constrained generalized coordinate. Let $a = 0$.

31. Consider Example 8.11 and, using constrained generalized coordinates, find the moment on the shaft needed to keep Ω constant.

Section 8.11

32. Consider Example 8.10 and derive the equations of motion by using Eq. [8.11.13].

33. Find the equation of motion of the plate in Fig. 8.23 using Eq. [8.11.13].

Section 8.12

34. Consider the collision of the two rods in Fig. 8.19. Assume that the rod on the left is released from some initial position and calculate the angular velocities of the rods after impact as a function of the coefficient of restitution. The rod on the left is of mass m and length L, and the rod on the right of mass $4m$ and length $2L$. The distance between O and O' is $L/2$.

35. A beam of mass $m = 2$ kg and length $L = 0.6$ m hits a table as it is falling down (Fig. 8.36). At the time of impact the beam is horizontal, with its center of mass having a speed of 0.5 m/s and a clockwise angular velocity of 0.2 rad/s. Find the velocity of the center of mass and angular velocity immediately after impact for $e = 0$ and $e = 1$.

36. The plate in Fig. 8.37 has a mass of 3 kg and width $a = 20$ cm. It is dropped from a height and hits a stationary table at corner P. At the time of impact the plate is horizontal and it has a vertical velocity of 0.5 m/s, with no

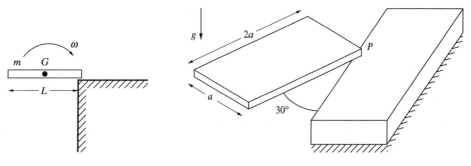

Figure 8.36 **Figure 8.37**

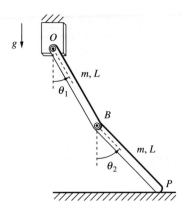

Figure 8.38

angular velocity. Find the velocity of point P and the angular velocity of the plate after impact for $e = 0.5$.

37. The double pendulum shown in Fig. 8.38 impacts the horizontal surface at the instant when $\theta_1 = 30°$, $\theta_2 = 45°$, $\dot{\theta}_1 = -0.2$ rad/s, and $\dot{\theta}_2 = -0.3$ rad/s. Find the angular velocities immediately after impact, assuming that there is no friction between point P and the surface, and $e = 0.8$.

38. Consider the slab in Fig. 8.33, except that the impact between the rock and slab has a coefficient of restitution of 0.9 and that the slab has no initial angular velocity. Find the velocity of the center of mass of the slab and its angular velocities immediately after impact.

39. The coin of weight 0.5 oz and radius 0.4 in, such as the one in Fig. 8.17, is dropped onto the ground. At the point of contact, the angular velocities of the coin are $\omega_x = 0.5$ rad/s, $\omega_y = 0$, $\omega_z = 5$ rad/s, the center of mass has a vertical speed of 9 in/sec, and $\theta = 60°$. Assuming perfectly elastic impact, calculate the angular velocities of the coin immediately after impact.

chapter 9

DYNAMICS OF RIGID BODIES: ADVANCED CONCEPTS

9.1 INTRODUCTION

This chapter presents additional methods for deriving the equations of motion of rigid bodies. These methods are particularly suitable for bodies subjected to nonholonomic constraints, and bodies that are interconnected. These additional equations are based on two concepts: First, the rotational equations of motion represent the moment balance of a body. The angular momentum of a body can be written in many ways, about the center of mass or about an arbitrary point. When differentiating the angular momentum, one can use a set of coordinates attached to the body, a set of coordinates that take advantage of symmetry of the body (if any), or a totally different set. Further, when expressing the moment balance, one can resolve the equations using the body frame or another frame. Each one of these choices leads to a different way of writing the equations of motion.

Second, analytical methods for obtaining the equations of motion, such as in Lagrange's equations, are derived based on D'Alembert's and Hamilton's principles. As discussed in Chapter 5, there are other variational principles that one can consider, and they lead to additional forms of the equations of motion.

The chapter begins with the modified Euler's equations, discussed briefly in Chapter 8 and particularly useful for inertially symmetric bodies. It moves on to a discussion of moment equations about an arbitrary point. There is an introduction to the equations of motion in terms of quasi-velocities (generalized speeds). The Gibbs-Appell and Kane's equations are derived and compared. We demonstrate that the two are indeed equivalent.

When studying the equations of motion in terms of quasi-velocities, one may question the reasoning behind introducing such coordinates in association with rigid bodies. After all, one can define such quantities associated with a system of particles or any other system. These variables and the forms of the equations that make use of them could have been introduced in Chapter 4. The reason for introducing them here is that generalized speeds are the most useful for nonholonomic or interconnected systems and that such properties are, for the most part, associated with three-dimensional rigid body motion.

9.2 MODIFIED EULER'S EQUATIONS

The modified Euler's equations are a variant of Euler's equations that are applicable to inertially symmetric bodies. They are based on writing the equations of rotational motion with respect to a reference frame other than one fixed to the body.

Consider the rotational motion equations about the center of mass

$$\frac{d}{dt}\{H_G\} = \{M_G\} \qquad [9.2.1]$$

If we view the angular momentum from a reference frame N, we can write the above equation as

$$^A\frac{d}{dt}\{H_G\} = {}^N\frac{d}{dt}\{H_G\} + [{}^A\tilde{\omega}^N]\{H_G\} = \{M_G\} \qquad [9.2.2]$$

where the A frame is inertial. When selecting frame N, the objective is to find a reference frame that will make the calculations simpler. Particularly appealing is a reference frame in which the elements of the inertia matrix remain time invariant. A reference frame that satisfies this criterion is one that is attached to the body. The rotational equations of motion can then be written as Eq. [8.5.25], or

$$[I_G]\{\dot{\omega}\} + [\tilde{\omega}][I_G]\{\omega\} = \{M_G\} \qquad [9.2.3]$$

For inertially symmetric bodies, when one uses the F frame, the elements of the inertia matrix are time invariant. As discussed in Chapter 8, one can use the F frame in conjunction with the moment balance equations in two ways: One can perform the differentiation in the F frame, or one can express Euler's equations using the components of the F frame. In this section, we discuss the first approach in more detail.

Consider an inertially symmetric body, with body axes $b_1b_2b_3$, where b_3 is the symmetry axis. It follows that $I_1 = I_2$, and the body axes are principal axes. Further, any rotation of coordinates about b_3 yields another set of principal axes.

We introduce the F frame to the formulation. Chapter 7 introduced the notations of ω_b, ω_f, and ω_s to denote the angular velocities of the body, frame, and spin, respectively. Mathematically, if the body-fixed frame is obtained from, say, a 3-1-3 transformation, the F frame is obtained after the 3-1 transformation. This type of transformation completely describes the orientation of the symmetry axis. For a 3-1-3 transformation, the angular velocities are

9.2 MODIFIED EULER'S EQUATIONS

$$^A\boldsymbol{\omega}^B = \boldsymbol{\omega}_b = \omega_1\mathbf{f}_1 + \omega_2\mathbf{f}_2 + \omega_3\mathbf{f}_3 = \dot{\theta}\mathbf{f}_1 + \dot{\phi}\sin\theta\mathbf{f}_2 + (\dot{\phi}\cos\theta + \dot{\psi})\mathbf{f}_3$$

$$^A\boldsymbol{\omega}^F = \boldsymbol{\omega}_f = \omega_{f1}\mathbf{f}_1 + \omega_{f2}\mathbf{f}_2 + \omega_{f3}\mathbf{f}_3 = \dot{\theta}\mathbf{f}_1 + \dot{\phi}\sin\theta\mathbf{f}_2 + \dot{\phi}\cos\theta\mathbf{f}_3$$

$$^F\boldsymbol{\omega}^B = \boldsymbol{\omega}_s = \dot{\psi}\mathbf{f}_3 \qquad ^A\boldsymbol{\omega}^B = {}^A\boldsymbol{\omega}^F + {}^F\boldsymbol{\omega}^B = \boldsymbol{\omega}_f + \boldsymbol{\omega}_s \qquad [9.2.4]$$

We then have

$$\omega_1 = \omega_{f1} = \dot{\theta} \qquad \omega_2 = \omega_{f2} = \dot{\phi}\sin\theta \qquad \omega_3 = \dot{\phi}\cos\theta + \dot{\psi}$$

$$\omega_{f3} = \dot{\phi}\cos\theta = \omega_2/\tan\theta \qquad [9.2.5]$$

For a 3-2-3 transformation, with the sequence ψ, θ, and ϕ, the angular velocities of the body and frame become

$$^A\boldsymbol{\omega}^B = -\dot{\psi}\sin\theta\mathbf{f}_1 + \dot{\theta}\mathbf{f}_2 + (\dot{\psi}\cos\theta + \dot{\phi})\mathbf{f}_3$$

$$^A\boldsymbol{\omega}^F = -\dot{\psi}\sin\theta\mathbf{f}_1 + \dot{\theta}\mathbf{f}_2 + \dot{\psi}\cos\theta\mathbf{f}_3 \qquad [9.2.6]$$

and we have

$$\omega_1 = \omega_{f1} = -\dot{\psi}\sin\theta \qquad \omega_2 = \omega_{f2} = \dot{\theta} \qquad \omega_3 = \dot{\psi}\cos\theta + \dot{\phi}$$

$$\omega_{f3} = \dot{\psi}\cos\theta = -\omega_1/\tan\theta \qquad [9.2.7]$$

We can write the rotational equations of motion using the angular velocity of the F frame as

$$^A\frac{d}{dt}\{H_G\} = {}^F\frac{d}{dt}\{H_G\} + [{}^A\tilde{\omega}^F]\{H_G\} = \{M_G\} \qquad [9.2.8]$$

where we note that all the column vectors are resolved in the F frame, and

$$\{H_G\} = [I_G]\{{}^A\omega^B\} \qquad [9.2.9]$$

The equations of motion become

$$[I_G]^F\frac{d}{dt}\{{}^A\omega^B\} + [{}^A\tilde{\omega}^F][I_G]\{{}^A\omega^B\} = \{M_G\} \quad \text{or} \quad [I_G]\{\dot{\omega}\}_{\text{rel}} + [\tilde{\omega}_f][I_G]\{\omega\} = \{M_G\}$$

$$[9.2.10]$$

Equations [9.2.10] are known as the *modified Euler's equations*. Denoting the inertia matrix by $[I_G] = \text{diag}(I_1\ I_1\ I_3)$ and carrying out the matrix algebra, we can express them as

$$I_1\dot{\omega}_1 + \omega_2(I_3\omega_3 - I_1\omega_{f3}) = M_1$$

$$I_2\dot{\omega}_2 - \omega_1(I_3\omega_3 - I_1\omega_{f3}) = M_2$$

$$I_3\dot{\omega}_3 = M_3 \qquad [9.2.11]$$

Also, for rotation about a fixed point C on the body, the modified Euler's equations can be written about C as well as the center of mass. Depending on the Euler angle transformation sequence used, the modified Euler's equations look slightly different when we substitute the value of ω_{f3} into them. Substituting for ω_{f3} from Eqs. [9.2.5] and [9.2.7], Table 9.1 gives the modified Euler's equations for the 3-1-3 and 3-2-3 transformations.

Table 9.1 Modified Euler equations for 3-1-3 and 3-2-3 transformations

3-1-3 Transformation Sequence	3-2-3 Transformation Sequence
$I_1\dot{\omega}_1 + \omega_2\left(I_3\omega_3 - \dfrac{I_1\omega_2}{\tan\theta}\right) = M_1$	$I_1\dot{\omega}_1 + \omega_2\left(I_3\omega_3 + \dfrac{I_1\omega_1}{\tan\theta}\right) = M_1$
$I_2\dot{\omega}_2 - \omega_1\left(I_3\omega_3 - \dfrac{I_1\omega_2}{\tan\theta}\right) = M_2$	$I_2\dot{\omega}_2 - \omega_1\left(I_3\omega_3 + \dfrac{I_1\omega_1}{\tan\theta}\right) = M_2$
$I_3\dot{\omega}_3 = M_3$	$I_3\dot{\omega}_3 = M_3$

Table 9.2 Kinematic differential equations for the modified Euler's equations

3-1-3 Transformation Sequence	3-2-3 Transformation Sequence
$\dot{\phi} = \dfrac{\omega_2}{\sin\theta}$	$\dot{\psi} = -\dfrac{\omega_1}{\sin\theta}$
$\dot{\theta} = \omega_1$	$\dot{\theta} = \omega_2$
$\dot{\psi} = -\dfrac{\omega_2}{\tan\theta} + \omega_3$	$\dot{\phi} = \dfrac{\omega_1}{\tan\theta} + \omega_3$

To write the associated kinematic differential equations, we solve Eqs. [9.2.5] and [9.2.7] for ω_i ($i = 1, 2, 3$). Table 9.2 gives the result.

Be aware that the equations of motion obtained by the modified Euler's equations are not a restricted form of what one would obtain using the original Euler's equations. Once one has solved Eqs. [9.2.11] together with the kinematic differential equations, one has the complete time history of $\psi(t)$. An interesting observation is that the singularity at $\sin\theta = 0$ is still present. Furthermore, the modified Euler's equations and corresponding kinematic differential equations are coupled through the $\tan\theta$ term. This coupling diminishes the usefulness of the equations in numerical analysis. The primary uses of the modified Euler's equations are to derive the equations of motion and to aid in a qualitative analysis.

The modified Euler's equations are for the rotational component of the motion. We next investigate the translational equations of motion in terms of a set of axes other than the body-fixed axes. Recall that the translational equations of motion are given in terms of the body axes by Eq. [8.5.19] as

$$^B\dfrac{d}{dt}\{mv_G\} + m[^A\tilde{\omega}^B]\{v_G\} = \{F\} \qquad [9.2.12]$$

The mass of the body does not change under a coordinate transformation, so that Eq. [9.2.12] can be written about any reference frame. Considering the F frame, we have

$$^F\dfrac{d}{dt}\{mv_G\} + [^A\tilde{\omega}^F]\{mv_G\} = \{F\} \qquad [9.2.13]$$

Equations [9.2.10] and [9.2.13] can be useful not only to simplify the equations of motion of an inertially symmetric body but also to solve nonholonomic problems.

We can write the above equations in terms of the kinetic energy. Expressing the kinetic energy as

$$T = \frac{1}{2}\{\omega\}^T[I_G]\{\omega\} + \frac{1}{2}m\{v_G\}^T\{v_G\} \qquad [9.2.14]$$

we can write

$$\{mv_G\} = \{p\} = \left(\frac{\partial T}{\partial\{v_G\}}\right)^T \qquad [I_G]\{\omega\} = \left(\frac{\partial T}{\partial\{\omega\}}\right)^T \qquad [9.2.15]$$

where the notation is obvious. The equations of motion are

$$\frac{d}{dt}\left(\frac{\partial T}{\partial\{v_G\}}\right)^T + [\tilde{\omega}_f]\left(\frac{\partial T}{\partial\{v_G\}}\right)^T = \{F\}$$

$$\frac{d}{dt}\left(\frac{\partial T}{\partial\{\omega\}}\right)^T + [\tilde{\omega}_f]\left(\frac{\partial T}{\partial\{\omega\}}\right)^T = \{M_G\} \qquad [9.2.16a,b]$$

Example 9.1

Consider the spinning top in Fig. 7.13. Write the rotational equations of motion about O.

Solution

The modified Euler's equations are particularly useful, as we have an axisymmetric body and the external moments are not a function of the spin angle.

The angular velocities are

$$\omega_1 = \dot{\theta} \qquad \omega_2 = \dot{\phi}\sin\theta \qquad \omega_3 = \dot{\phi}\cos\theta + \dot{\psi} \qquad \omega_{f3} = \frac{\omega_2}{\tan\theta} = \dot{\phi}\cos\theta \qquad [a]$$

The free-body diagram of the top is given in Fig. 9.1. The external moments about O are due to gravity and an applied moment about the f_3 axis, so that

$$M_{O1} = mgL\sin\theta \qquad M_{O2} = 0 \qquad M_{O3} = M \qquad [b]$$

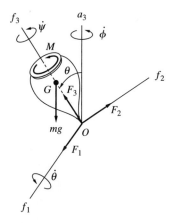

Figure 9.1

Introducing Eqs. [a] and [b] into the modified Euler's equations, written about the fixed point O, we obtain

$$I_{O1}\ddot{\theta} + \dot{\phi}\sin\theta(I_{O3}(\dot{\psi} + \dot{\phi}\cos\theta) - I_{O1}\dot{\phi}\cos\theta) = mgL\sin\theta \qquad [\text{c}]$$

$$I_{O1}\frac{d}{dt}(\dot{\phi}\sin\theta) + \dot{\theta}(I_{O1}\dot{\phi}\cos\theta - I_{O3}(\dot{\psi} + \dot{\phi}\cos\theta)) = 0 \qquad [\text{d}]$$

$$I_{O3}\frac{d}{dt}(\dot{\psi} + \dot{\phi}\cos\theta) = M \qquad [\text{e}]$$

These equations are, of course, identical to those obtained in Example 8.10. Using the modified Euler's equations is much simpler than using Lagrange's equations. Further, they are much simpler than Euler's equations. Had we used the regular Euler's equations, we would have obtained two equations that are combinations of Eqs. [c] and [d] multiplied by $\sin\psi$ and $\cos\psi$, and Eq. [e] would be unchanged.

Example 9.2

A *dual spin satellite* consists of a main body with an attached spinning disk (rotor). Such a satellite is also called a *gyrostat*.[1] The spinning disk can be placed outside or inside the main body, and it is there primarily for stability and maneuvering purposes. Derive the equations of motion for the dual spin satellite shown in Fig. 9.2. The disk is located at the center of mass of the main body. The spin axis of the disk is aligned with the principal axis b_1 of the main body. The disk is spinning with an angular velocity of Ω with respect to the main body, and it is powered by a motor that imparts a moment M along the b_1 axis.

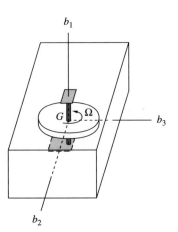

Figure 9.2 A dual spin satellite

---

[1] More sophisticated spacecraft have three rotors on nonparallel axes that permit rotational maneuvers in three dimensions.

9.2 MODIFIED EULER'S EQUATIONS

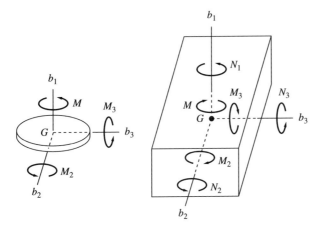

Figure 9.3 Free-body diagrams

Solution

We separate the disk and satellite into two bodies and draw the free-body diagrams, as shown in Fig. 9.3. A motor exerts a moment M on the disk to keep it rotating about the spin axis, and the two bodies exert moments M_2 and M_3 on each other about the transverse axes.

The equations of motion for the main body obtained by using Euler's equations have the standard form of

$$I_1\dot{\omega}_1 - (I_2 - I_3)\omega_2\omega_3 = -M + N_1 \qquad \text{[a]}$$

$$I_2\dot{\omega}_2 - (I_3 - I_1)\omega_1\omega_3 = -M_2 + N_2 \qquad \text{[b]}$$

$$I_3\dot{\omega}_3 - (I_1 - I_2)\omega_1\omega_2 = -M_3 + N_3 \qquad \text{[c]}$$

where I_1, I_2, and I_3 are the centroidal moments of inertia of the main body, ω_1, ω_2, and ω_3 are the corresponding body angular velocities, and N_1, N_2, and N_3 are the excitations external to the satellite.

To derive the equations of motion of the rotor, we view its motion from the main body. Hence, the angular velocities of the rotor in the b_2 and b_3 directions are the same as that of the main body. These two factors make the modified Euler's equations ideally suited for analyzing the motion with the $b_1b_2b_3$ axes as the F frame for the rotor. The angular velocity vector for the rotor is written as

$$\{\omega_R\} = [\omega_1 + \Omega \quad \omega_2 \quad \omega_3]^T \qquad \text{[d]}$$

and the angular velocity of the reference frame is

$$\{\omega_f\} = [\omega_1 \quad \omega_2 \quad \omega_3]^T \qquad \text{[e]}$$

Denoting the centroidal mass moments of inertia of the rotor by J_1, $J_2 = J_3$, the modified Euler's equations for the rotor become

$$\begin{bmatrix} J_1(\dot{\omega}_1 + \dot{\Omega}) \\ J_2\dot{\omega}_2 \\ J_2\dot{\omega}_3 \end{bmatrix} + \begin{bmatrix} 0 & -\omega_3 & \omega_2 \\ \omega_3 & 0 & -\omega_1 \\ -\omega_2 & \omega_1 & 0 \end{bmatrix} \begin{bmatrix} J_1(\omega_1 + \Omega) \\ J_2\omega_2 \\ J_2\omega_3 \end{bmatrix} = \begin{bmatrix} M \\ M_2 \\ M_3 \end{bmatrix} \qquad \text{[f]}$$

which reduce to

$$J_1(\dot{\omega}_1 + \dot{\Omega}) = M \quad \text{[g]}$$

$$J_2\dot{\omega}_2 + (J_1 - J_2)\omega_1\omega_3 + J_1\Omega\omega_3 = M_2 \quad \text{[h]}$$

$$J_2\dot{\omega}_3 + (J_2 - J_1)\omega_1\omega_2 - J_1\Omega\omega_2 = M_3 \quad \text{[i]}$$

Equations [a] and [g] are the equations of motion for ω_1 and Ω. To find the equations of motion for ω_2 and ω_3, we need to eliminate M_2 and M_3. To this end, we add Eq. [b] to Eq. [h] and Eq. [c] to Eq. [i], with the result

$$(I_2 + J_2)\dot{\omega}_2 + (I_1 + J_1 - I_3 - J_2)\omega_1\omega_3 + J_1\Omega\omega_3 = N_2 \quad \text{[j]}$$

$$(I_3 + J_2)\dot{\omega}_3 + (I_2 + J_2 - I_1 - J_1)\omega_1\omega_2 - J_1\Omega\omega_2 = N_3 \quad \text{[k]}$$

The four equations of motion are Eqs. [a], [g], [j], and [k]. One can substitute for $\dot{\omega}_1$ in Eq. [g] from Eq. [a] to make all the equations of motion have a single angular acceleration term, thereby putting them in state form. Doing so yields

$$J_1\dot{\Omega} - \frac{J_1(I_3 - I_2)\omega_2\omega_3}{I_1} = M\left(1 + \frac{J_1}{I_1}\right) - N_1\frac{J_1}{I_1} \quad \text{[l]}$$

Also, one can add Eqs. [a] and [g] to get an equation for ω_1 as

$$(I_1 + J_1)\dot{\omega}_1 + J_1\dot{\Omega} - (I_2 - I_3)\omega_2\omega_3 = N_1 \quad \text{[m]}$$

When a servomotor is used to keep Ω constant, Eqs. [j], [k], and [m] become the describing equations of the satellite.

Example 9.3

Consider the rolling (without slipping) disk of mass m and radius R shown in Fig. 9.4. We will make use of the modified Euler's equations and a 3-1-3 transformation to derive the equations of motion. Note that in Example 8.13, we derived the equations of motion for this system using Lagrange's equations for constrained coordinates.

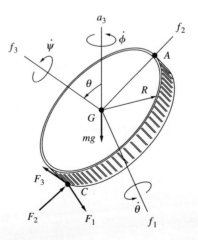

Figure 9.4

Solution

We will write both the translational and rotational equations of motion and then use the rolling constraint to eliminate three of the variables. This elimination of the nonholonomic constraint is possible because we are dealing with the angular velocities of the body. From Eq. [9.2.13] the translational equations of motion have the form

$$\dot{v}_1 + \omega_2 v_3 - \omega_{f3} v_2 = \frac{F_1}{m} \tag{a}$$

$$\dot{v}_2 + \omega_{f3} v_1 - \omega_1 v_3 = \frac{F_2}{m} - g\sin\theta \tag{b}$$

$$\dot{v}_3 + \omega_1 v_2 - \omega_2 v_1 = \frac{F_3}{m} - g\cos\theta \tag{c}$$

where v_1, v_2, and v_3 denote the velocities of the center of mass of the disk along the coordinate axes of the F frame. The modified Euler's equations about the center of mass are

$$I_1 \dot{\omega}_1 + \omega_2 (I_3 \omega_3 - I_1 \omega_{f3}) = -F_3 R \tag{d}$$

$$I_1 \dot{\omega}_2 - \omega_1 (I_3 \omega_3 - I_1 \omega_{f3}) = 0 \tag{e}$$

$$I_3 \dot{\omega}_3 = F_1 R \tag{f}$$

The rolling constraint can be expressed as

$$\mathbf{v}_G = \boldsymbol{\omega}_b \times \mathbf{r}_{G/C} \Rightarrow v_1 \mathbf{f}_1 + v_2 \mathbf{f}_2 + v_3 \mathbf{f}_3 = (\omega_1 \mathbf{f}_1 + \omega_2 \mathbf{f}_2 + \omega_3 \mathbf{f}_3) \times R\mathbf{f}_2$$
$$= \omega_1 R \mathbf{f}_3 - \omega_3 R \mathbf{f}_1 \tag{g}$$

leading to the three constraint equations

$$v_1 = -\omega_3 R \qquad v_2 = 0 \qquad v_3 = \omega_1 R \tag{h}$$

Introduction of Eqs. [h] into Eqs. [a] and [c] yields expressions for F_1 and F_3 as

$$F_1 = m(-\dot{\omega}_3 R + \omega_1 \omega_2 R) \qquad F_3 = m(\dot{\omega}_1 R + \omega_2 \omega_3 R) + mg\cos\theta \tag{i}$$

Substituting these values of F_1 and F_3 into Eqs. [d] and [f] and using the relation $\omega_{f3} = \omega_2/\tan\theta$, we obtain the equations of motion as

$$(I_1 + mR^2)\dot{\omega}_1 + (I_3 + mR^2)\omega_2 \omega_3 - \frac{I_1 \omega_2^2}{\tan\theta} = -mgR\cos\theta \tag{j}$$

$$I_1 \dot{\omega}_2 - I_3 \omega_1 \omega_3 + \frac{I_1 \omega_1 \omega_2}{\tan\theta} = 0 \tag{k}$$

$$(I_3 + mR^2)\dot{\omega}_3 - mR^2 \omega_1 \omega_2 = 0 \tag{l}$$

If desired, one can write the equations of motion in terms of the Euler angles by substituting the values of ω_1, ω_2, and ω_3 into the above equations, which yields

$$(I_1 + mR^2)\ddot{\theta} + (I_3 + mR^2)\dot{\phi}\sin\theta(\dot{\phi}\cos\theta + \dot{\psi}) - I_1\dot{\phi}^2\sin\theta\cos\theta = -mgR\cos\theta \tag{m}$$

$$I_1 \frac{d}{dt}(\dot{\phi}\sin\theta) - I_3\dot{\theta}(\dot{\phi}\cos\theta + \dot{\psi}) + I_1\dot{\theta}\dot{\phi}\cos\theta = 0 \tag{n}$$

$$(I_3 + mR^2)\frac{d}{dt}(\dot{\phi}\cos\theta + \dot{\psi}) - mR^2\dot{\theta}\dot{\phi}\sin\theta = 0 \tag{o}$$

9.3 Moment Equations about an Arbitrary Point

The Euler's equations and the modified Euler's equations that we have studied so far can be written about the center of mass, or, if the body is rotating about a fixed point, about that point. In this section, we examine the moment balance equations about an arbitrary point. Consider a body whose motion is being viewed from a reference frame. We make no restriction on the nature of the reference frame; it can be either fixed to the body or not fixed to the body. Furthermore, the origin of the coordinate frame, which we denote by B, can move with respect to the body. As discussed in Chapter 8, the rotational equations of motion about an arbitrary point can be written in several forms. We use the form given in Eq. [8.5.10] and write

$$\mathbf{M}_B = \dot{\mathbf{H}}_B + m\mathbf{v}_B \times \mathbf{v}_G \qquad [9.3.1]$$

where $\mathbf{M}_B$ is the sum of all external moments acting on point B, $\mathbf{H}_B$ is the angular momentum about point B, and $\mathbf{v}_B$ is the velocity of point B.

We wish to express Eq. [9.3.1] in column vector format. We use the angular velocity of the body and the velocity of point B, both expressed in terms of the reference frame attached to B, as the motion variables. The angular velocity of the reference frame is denoted by $\boldsymbol{\omega}_N$. It follows that the angular momentum about B can be related to the kinetic energy by

$$\{H_B\} = \left(\frac{\partial T}{\partial \{\omega\}}\right)^T \qquad [9.3.2]$$

We also express the velocity of the center of mass of the body as

$$\{v_G\} = \left(\frac{\partial T}{\partial \{v_G\}}\right)^T \qquad [9.3.3]$$

Introducing Eq. [9.3.2] and [9.3.3] into Eq. [9.3.1] we write the rotational equation of motion of a body about an arbitrary point B as

$$\frac{d}{dt}\left(\frac{\partial T}{\partial(\omega)}\right)^T + [\tilde{\omega}_N]\left(\frac{\partial T}{\partial\{\omega\}}\right)^T + m[\tilde{v}_B]\{v_G\} = \{M_B\} \qquad [9.3.4]$$

or

$$\frac{d}{dt}\left(\frac{\partial T}{\partial\{\omega\}}\right)^T + [\tilde{\omega}_N]\left(\frac{\partial T}{\partial\{\omega\}}\right)^T + [\tilde{v}_B]\left(\frac{\partial T}{\partial\{v_G\}}\right)^T = \{M_B\} \qquad [9.3.5]$$

The translational equation of motion is not dependent on the location of point B. It can be expressed in terms of ω_N as

$$m\frac{d}{dt}\{v_G\} + m[\tilde{\omega}_N]\{v_G\} = \{F\} \qquad [9.3.6]$$

or

$$\frac{d}{dt}\left(\frac{\partial T}{\partial\{v_G\}}\right)^T + [\tilde{\omega}_N]\left(\frac{\partial T}{\partial\{v_G\}}\right)^T = \{F\} \qquad [9.3.7]$$

9.3 MOMENT EQUATIONS ABOUT AN ARBITRARY POINT

Equations [9.3.4] and [9.3.6] are known as the *Boltzmann-Hamel equations*. Compare Eqs. [9.2.16b] and [9.3.4]. The added term in the moment equation is due to the point B not being the center of mass or center of rotation. Also, it should be reemphasized that the rate variables in the Boltzmann-Hamel equations are the components of the velocity of point B and not the components of the velocity of the center of mass, even though to write these equations in compact form we use the partial derivatives of the kinetic energy with respect to the velocity of the center of mass.

The moment equations about an arbitrary point make no restrictions on whether the body should be inertially symmetric or not. However, their usefulness diminishes if the inertia matrix is time varying. We thus identify the following major uses:

a. If there is no inertial symmetry, one should select the reference frame such that it is attached to the body. This implies that $\boldsymbol{\omega}_N = \boldsymbol{\omega}_b$.

b. If there is inertial symmetry, one can select the reference frame as the F frame. However, note that the moment center is now a point lying on the reference frame and not necessarily on the body and that the moment center can move with respect to the body.

Selection of the moment center is an important issue. In general, one selects the point B such that reaction forces do not create a moment around it, or as a point whose velocity is described by a simple expression. In this regard, the Boltzmann-Hamel equations come in handy for considering the motion of multiple connected bodies. Also, as with the Euler's or modified Euler's equations, nonholonomic constraints can be dealt with effectively.

Example 9.4

A spinning top of mass m_1 and centroidal moments of inertia I_1 and I_3, where I_3 is about the symmetry axis, is moving on a cart of mass m_2 which is constrained to move horizontally and in one direction. Find the equations of motion using the Boltzmann-Hamel equations.

Solution

This is the same problem considered in Example 8.12. We have four degrees of freedom, and we will use the velocity of the cart ($\dot{Y} = u$) and the angular velocities of the top observed from a 3-1-3 Euler angle transformation as the rate variables. Refer to Example 8.12 for details. The free-body diagram of the top is shown in Fig. 9.1 and of the cart in Fig. 9.5. Note that we are conveniently expressing the reaction forces at O in terms of both the inertial and F frames.

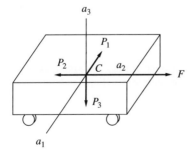

Figure 9.5 Free-body diagram of cart

CHAPTER 9 • DYNAMICS OF RIGID BODIES: ADVANCED CONCEPTS

The kinetic energy of the system has the form

$$T = \frac{1}{2}(I_1 + m_1 L^2)(\omega_1^2 + \omega_2^2) + \frac{1}{2}I_3\omega_3^2 + \frac{1}{2}(m_1 + m_2)u^2 + m_1 Lu(\omega_2 \sin\phi - \omega_1 \cos\phi \cos\theta)$$

[a]

Taking the partial derivatives, we obtain

$$\frac{\partial T}{\partial \omega_1} = (I_1 + m_1 L^2)\omega_1 - m_1 Lu \cos\phi \cos\theta$$

$$\frac{\partial T}{\partial \omega_2} = (I_1 + m_1 L^2)\omega_2 + m_1 Lu \sin\phi \qquad \frac{\partial T}{\partial \omega_3} = I_3 \omega_3 \qquad [b]$$

The angular velocity of the reference frame is

$$\omega_{f1} = \omega_1 = \dot\theta \qquad \omega_{f2} = \omega_2 = \dot\phi \sin\theta \qquad \omega_{f3} = \dot\phi \cos\theta = \frac{\omega_2}{\tan\theta} \qquad [c]$$

The time derivatives of Eqs. [b] are

$$\frac{d}{dt}\left(\frac{\partial T}{\partial \omega_1}\right) = (I_1 + m_1 L^2)\dot\omega_1 - m_1 L\dot u \cos\phi \cos\theta + m_1 Lu \frac{\omega_2 \sin\phi}{\tan\theta} + m_1 Lu \omega_1 \cos\phi \sin\theta$$

$$\frac{d}{dt}\left(\frac{\partial T}{\partial \omega_2}\right) = (I_1 + m_1 L^2)\dot\omega_2 + m_1 L\dot u \sin\phi + m_1 Lu \frac{\omega_2 \cos\phi}{\sin\theta}$$

$$\frac{d}{dt}\left(\frac{\partial T}{\partial \omega_3}\right) = I_3 \dot\omega_3 \qquad [d]$$

The term $[\tilde\omega_f]\{\partial T/\partial \omega\}$ becomes

$$[\tilde\omega_f]\left(\frac{\partial T}{\partial\{\omega\}}\right)^T = \begin{bmatrix} 0 & -\frac{\omega_2}{t\theta} & \omega_2 \\ \frac{\omega_2}{t\theta} & 0 & -\omega_1 \\ -\omega_2 & \omega_1 & 0 \end{bmatrix} \begin{bmatrix} (I_1 + m_1 L^2)\omega_1 - m_1 Lu\, c\phi c\theta \\ (I_1 + m_1 L^2)\omega_2 + m_1 Lu\, s\phi \\ I_3 \omega_3 \end{bmatrix}$$

$$= \begin{bmatrix} -(I_1 + m_1 L^2)\frac{\omega_2^2}{t\theta} - m_1 Lu \frac{\omega_2 s\phi}{t\theta} + I_3 \omega_2 \omega_3 \\ (I_1 + m_1 L^2)\frac{\omega_1 \omega_2}{t\theta} - m_1 Lu \frac{\omega_2 c\phi c\theta}{t\theta} - I_3 \omega_1 \omega_3 \\ m_1 Lu(\omega_2 c\phi c\theta + \omega_1 s\phi) \end{bmatrix} \qquad [e]$$

The velocity of the point of contact C is

$$\mathbf{v}_C = u\mathbf{a}_2 = us\phi\, \mathbf{f}_1 + uc\phi c\theta\, \mathbf{f}_2 - uc\phi s\theta\, \mathbf{f}_3 \qquad [f]$$

and, from Example 8.12, the velocity of the center of mass is

$$\mathbf{v}_G = (u \sin\phi + \omega_2 L)\mathbf{f}_1 + (u \cos\phi \cos\theta - \omega_1 L)\mathbf{f}_2 - u \cos\phi \sin\theta\, \mathbf{f}_3 \qquad [g]$$

Note that the velocity of the center of mass is expressed in terms of u and the angular velocities. The velocity term in the Boltzmann-Hamel equations yield

$$m_1[\tilde v_C]\{v_G\} = m_1 L \begin{bmatrix} -u\omega_1 \cos\phi \sin\theta \\ -u\omega_2 \cos\phi \sin\theta \\ -u\omega_1 \sin\phi - u\omega_2 \cos\phi \cos\theta \end{bmatrix} \qquad [h]$$

The moment about point C can be expressed as

$$\mathbf{M}_C = \mathbf{r}_{G/C} \times -m_1 g \mathbf{a}_3 + M \mathbf{f}_3 \qquad [\mathbf{i}]$$

where $\mathbf{r}_{G/C} = L\mathbf{f}_3$ and $\mathbf{a}_3 = \sin\theta \mathbf{f}_2 + \cos\theta \mathbf{f}_3$, so that in column vector format we have

$$\{M_C\} = \begin{bmatrix} m_1 g L \sin\theta \\ 0 \\ M \end{bmatrix} \qquad [\mathbf{j}]$$

We write the rotational components of the moment equations by combining Eqs. [d], [e], [h], and [j], which yields

$$(I_1 + m_1 L^2)\left(\dot{\omega}_1 - \frac{\omega_2^2}{\tan\theta}\right) - m_1 L \dot{u} \cos\phi \cos\theta + I_3 \omega_2 \omega_3 - m_1 g L \sin\theta = 0 \qquad [\mathbf{k}]$$

$$(I_1 + m_1 L^2)\left[\dot{\omega}_2 + \frac{\omega_1 \omega_2}{\tan\theta}\right] + m_1 L \dot{u} \sin\phi - I_3 \omega_1 \omega_3 = 0 \qquad [\mathbf{l}]$$

$$I_3 \dot{\omega}_3 = M \qquad [\mathbf{m}]$$

Next, we obtain the translational equation of motion using Eq. [9.3.6]. We note that there is only one degree of freedom involved, Y, which is in the $\mathbf{a}_2$ direction. To solve for the equations of motion, we can use the relative frame F and deal with all three equations, or we can express $[\tilde{\omega}_N]$, $\{v_G\}$, and $d\{v_G\}/dt$ in the $a_1 a_2 a_3$ coordinate system and use only the second row of Eq. (9.3.6). Another option is to write the sum of the forces in the a_2 direction for the combined system. Here, we opt for the first approach (primarily for illustrative purposes). Looking at the free-body diagram of the cart (Fig. 9.5), we can write the force balance in the a_2 direction as

$$F - P_2 = m_2 \dot{u} \qquad [\mathbf{n}]$$

from which we solve for the reaction P_2 as $P_2 = F - m_2 \dot{u}$. The forces acting on the top have the form

$$\mathbf{F} = P_1 \mathbf{a}_1 + P_2 \mathbf{a}_2 + P_3 \mathbf{a}_3$$
$$= (F - m_2 \dot{u})\sin\phi \mathbf{f}_1 + (F - m_2 \dot{u})\cos\phi\cos\theta \mathbf{f}_2 - (F - m_2 \dot{u})\cos\phi\sin\theta \mathbf{f}_3$$
$$+ P_1 \cos\phi \mathbf{f}_1 + (-P_1 \sin\phi\cos\theta + P_3 \sin\theta)\mathbf{f}_2 + (P_1 \sin\phi\sin\theta + P_3 \cos\theta)\mathbf{f}_3 \qquad [\mathbf{o}]$$

Introducing the expression for the force into Eq. [9.3.6] and carrying out the algebra yields the equations

$$m_1 \dot{u} s\phi + m_1 \dot{\omega}_2 L + \frac{m_1 \omega_1 \omega_2 L}{t\theta} = (F - m_2 \dot{u})s\phi + P_1 c\phi \qquad [\mathbf{p}]$$

$$m_1 \dot{u} c\phi c\theta - m_1 \dot{\omega}_1 L + \frac{m_1 \omega_2^2 L}{t\theta} = (F - m_2 \dot{u})c\phi c\theta - P_1 s\phi c\theta + P_3 s\theta \qquad [\mathbf{q}]$$

$$m_1 \dot{u} c\phi s\theta + m_1 L(\omega_1^2 + \omega_2^2) = (F - m_2 \dot{u})c\phi s\theta - P_1 s\phi s\theta - P_3 c\theta \qquad [\mathbf{r}]$$

We next eliminate P_1 and P_3. Multiplying Eq. [p] by $\sin\phi$, Eq. [q] by $\cos\phi\cos\theta$, and Eq. [r] by $\cos\phi\sin\theta$, and adding them, we obtain

$$(m_1 + m_2)\dot{u} + m_1 L(\dot{\omega}_2 s\phi - \dot{\omega}_1 c\phi c\theta) + m_1 L\left(\frac{\omega_2^2 c\phi}{s\theta} + \frac{\omega_1 \omega_2 s\phi}{t\theta} + \omega_1^2 c\phi s\theta\right) = F \quad \textbf{[s]}$$

Comparing Eqs. [k], [l], [m], and [s] with their counterparts in Example 8.12, we see that they are the same. Note that the actual values of the magnitudes of P_1, P_2, and P_3 were not calculated in this problem, due to selecting the moment center as point C.

Example 9.5

Derive the equations of motion of the rolling disk considered in Example 9.3 using the Boltzmann-Hamel equations.

Solution

The disk configuration is shown in Fig. 9.4. For roll without slip, the velocity of the contact point is zero, which constitutes the nonholonomic constraint. We use the 3-1-3 Euler angle transformation and F frame. The kinetic energy can be written as

$$T = \frac{1}{2}\{\omega\}^T[I_C]\{\omega\} = \frac{1}{2}\left(I_{C1}\omega_1^2 + I_{C2}\omega_2^2 + I_{C3}\omega_3^2\right) \quad \textbf{[a]}$$

where the components of the mass moments of inertia about the contact point and angular velocity are

$$I_{C1} = I_1 + mR^2 \qquad I_{C2} = I_1 \qquad I_{C3} = I_3 + mR^2$$
$$\omega_1 = \dot{\theta} \qquad \omega_2 = \dot{\phi}\sin\theta \qquad \omega_3 = \dot{\phi}\cos\theta + \dot{\psi} \quad \textbf{[b]}$$

Note that we have automatically incorporated the nonholonomic velocity constraint into the kinetic energy. We proceed with solving the problem as if it were a three degree of freedom unconstrained problem.

Because of roll without slip, the velocity of the contact point is zero. However, recalling the argument with regards to the components of the velocity of the moment center, we need to have the velocity of a point on the reference frame that coincides with the contact point. Denoting that point by C' and noting that C' is always in contact with the rolling surface, we write the relative velocity expression between the center of mass and point C' in the F frame. We have

$$^A\mathbf{v}_G = {^A\mathbf{v}_{C'}} + {^A\boldsymbol{\omega}^F} \times \mathbf{r}_{G/C'} \quad \textbf{[c]}$$

Using the values

$$^A\mathbf{v}_G = -R\omega_3\mathbf{f}_1 + R\omega_1\mathbf{f}_3 \qquad ^A\boldsymbol{\omega}^F = \omega_1\mathbf{f}_1 + \omega_2\mathbf{f}_2 + \frac{\omega_2}{\tan\theta}\mathbf{f}_3 \qquad \mathbf{r}_{G/C'} = \mathbf{r}_{G/C} = R\mathbf{f}_2$$
$$\textbf{[d]}$$

and substitution of these values into Eq. [c] yields the velocity of the moment center as

$$^A\mathbf{v}_{C'} = -R\left(\omega_3 - \frac{\omega_2}{\tan\theta}\right)\mathbf{f}_1 = -R\dot{\psi}\mathbf{f}_1 \quad \textbf{[e]}$$

This result may seem surprising at first. Imagine a coin rolling without slipping. If we look at the bottom point of the coin from a distance we think that point is moving. In reality, our eyes are fooled by the change in location of the contact point between the coin and the surface. What we do when looking at the motion of the coin is to follow the motion of the F frame. We do not visualize the spin of the coin unless there are special markings on the coin that draw attention.

The Boltzmann-Hamel equations are written about point C' as

$$\begin{bmatrix} (I_1 + mR^2)\dot{\omega}_1 \\ I_1\dot{\omega}_2 \\ (I_3 + mR^2)\dot{\omega}_3 \end{bmatrix} + \begin{bmatrix} 0 & -\frac{\omega_2}{\tan\theta} & \omega_2 \\ \frac{\omega_2}{\tan\theta} & 0 & -\omega_1 \\ -\omega_2 & \omega_1 & 0 \end{bmatrix} \begin{bmatrix} (I_1 + mR^2)\omega_1 \\ I_1\omega_2 \\ (I_3 + mR^2)\omega_3 \end{bmatrix}$$

$$+ mR \begin{bmatrix} 0 & 0 & 0 \\ 0 & 0 & \omega_3 - \frac{\omega_2}{\tan\theta} \\ 0 & -\omega_3 + \frac{\omega_2}{\tan\omega} & 0 \end{bmatrix} \begin{bmatrix} -R\omega_3 \\ 0 \\ R\omega_1 \end{bmatrix} = \begin{bmatrix} -mgR\cos\theta \\ 0 \\ 0 \end{bmatrix} \quad [\mathbf{f}]$$

Expansion of Eq. [f] gives the equations of motion as

$$(I_1 + mR^2)\dot{\omega}_1 + (I_3 + mR^2)\omega_2\omega_3 - \frac{I_1\omega_2^2}{\tan\theta} = -mgR\cos\theta$$

$$I_1\dot{\omega}_2 - I_3\omega_1\omega_3 + \frac{I_1\omega_1\omega_2}{\tan\theta} = 0$$

$$(I_3 + mR^2)\dot{\omega}_3 - mR^2\omega_1\omega_2 = 0 \quad [\mathbf{g}]$$

and we note that they are the same as Eqs. [j]–[l] in Example 9.3. Using the Boltzmann-Hamel equations gave the equations of motion much more rapidly than using the modified Euler's or Lagrange's equations. By selecting the moment center as coincident with the point of contact, we were able to eliminate the contributions of the forces acting there.

Had we used a body-fixed frame to write the equations of motion, the point of contact at a particular instant would be at different locations on the body throughout the motion. One would then derive an expression for the velocity of point C and incorporate that into the Boltzmann-Hamel equations. The end result would be more complicated expressions. This situation is common in rolling problems.

9.4 Classification of the Moment Equations

In Chapter 8 and in the earlier sections of this chapter we saw several ways of formulating the moment equations. In this section we outline the process of selecting the best form for a given problem. The principal choices are with respect to:

a. Selecting the point about which these equations are written.
b. Selecting the reference frame in which the differentiation is performed.
c. Selecting the coordinate axes to resolve these equations.

For arbitrary bodies one usually differentiates the moment equations and resolves them in a reference frame attached to the body. Otherwise, the inertia matrix will be time varying—in all probability, any advantages realized from using a

Table 9.3 Classification of the moment equations

Type of Body	Moment Equations about	Differentiation Performed in	Equations Resolved in	Name of Equations, Comments
Arbitrary	Center of mass or center of rotation	Body frame	Body frame	Euler's [8.5.28]
		Arbitrary	Arbitrary	Time-varying inertia matrix [9.2.2]
	An arbitrary point	Body frame	Body frame	Boltzmann-Hamel [9.3.4], [9.3.6]
		Arbitrary	Arbitrary	Time-varying inertia matrix [9.3.4], [9.3.6]
Inertially symmetric	Center of mass or center of rotation	Body frame	F frame	Euler's [8.5.31], [8.5.32]
		F frame	F frame	Modified Euler's [9.2.3]
	An arbitrary point	Body frame	F frame	Boltzmann-Hamel Angular acc. term lengthy
		F frame	F frame	Boltzmann-Hamel

reference frame different than one attached to the body will be counteracted by the time-varying inertia matrix. The choice of a different frame becomes attractive for bodies possessing inertial symmetry, in which case one can use the F frame to differentiate or resolve the moment balance equations. One should be careful, though, to correctly calculate the velocity of the contact point, especially if it is located along the F frame. Table 9.3 summarizes the different possibilities.

There are a few other possible cases, but those do not have common applications. When selecting the approach for writing the equations of motion, take into consideration a variety of factors, including the acting forces, the additional terms you need to calculate, and the complexity of the resulting equations. Of course, you should also compare these different approaches with the analytical techniques to derive the equations of motion. Experience with a variety of problems is the best way to develop guidelines for selecting the method to derive the equations of motion.

9.5 Quasi-Coordinates and Quasi-Velocities (Generalized Speeds)

In the previous sections, we saw additional ways of writing the equations of motion. What makes these equations so useful is that they are first-order equations and the angular velocity components along the body axes are used directly in the equations of motion. We also saw in Chapter 8 that in certain cases it is more convenient to retain the angular velocity components in the formulation. Angular velocities are not generalized velocities and they are *not* perfect differentials.

Recall Examples 4.14 and 5.13, which dealt with a nonholonomic constraint in conjunction with a vehicle dynamics problem. We saw that switching to a rate variable different than the generalized velocities simplified the equations of motion considerably.

The question arises as to whether one can define other such quantities and make use of them in deriving the equations of motion. As we saw in Chapter 7, it is possible to define a set of variables called *quasi-velocities* or *generalized speeds*.[2] These are quantities that are linear combinations of the generalized velocities but which cannot necessarily be integrated to the generalized coordinates.

We define *quasi-coordinates* as a set of coordinates whose derivatives are meaningful quantities (quasi-velocities), but they themselves do not necessarily have a physical meaning. We will denote the quasi-coordinates by $q'_1, q'_2, \ldots, q'_n$ and the quasi-velocities by $u_1, u_2, \ldots, u_n$. Using this definition, generalized velocities become a subset of the quasi-velocities. An alternate definition for quasi-velocities is variables that cannot always be expressed as derivatives of displacement coordinates.

Let us first consider a holonomic system having n degrees of freedom, with n independent generalized coordinates $q_1, q_2, \ldots, q_n$ and generalized velocities $\dot{q}_1, \dot{q}_2, \ldots, \dot{q}_n$. We defined earlier a set of independent generalized speeds $u_1, u_2, \ldots, u_n$ as

$$u_k = \sum_{r=1}^{n} Y_{kr} \dot{q}_r + Z_k \qquad k = 1, 2, \ldots, n \qquad [9.5.1]$$

where $Y_{kr} = Y_{kr}(q_1, q_2, \ldots, q_n, t)$, $Z_k = Z_k(q_1, q_2, \ldots, q_n, t)$ $(k, r = 1, 2, \ldots, n)$ are functions of the generalized coordinates and time. In column vector form we can write Eq. [9.5.1] as

$$\{u\} = [Y]\{\dot{q}\} + \{Z\} \qquad [9.5.2]$$

In order for the set of generalized speeds to completely describe the system under consideration, it must be defined such that $[Y]$ is nonsingular. For example, the relation between the body angular velocities and the Euler angles, Eq. [7.5.7], defines a valid set of quasi-velocities.[3] By contrast, as we saw in Example 7.6, the angular velocities of the F frame do not constitute a set of quasi-velocities, as the relation between them and the Euler angles is defined by a matrix that is singular at all times.

Inversion of Eq. [9.5.2] yields

$$\{\dot{q}\} = [Y]^{-1}(\{u\} - \{Z\}) = [W]\{u\} + \{X\} \qquad [9.5.3]$$

in which $[W] = [Y]^{-1}$ and $\{X\} = -[Y]^{-1}\{Z\}$. The above equations are referred to as *kinematic equations* or *kinematic differential equations* for the generalized speeds.

---

[2] Each of these names has been used in conjunction with one of two distinct ways to derive equations of motion, the Gibbs-Appell equations and Kane's equations. Because the two methods are equivalent, we use the terms interchangeably.

[3] Equation [7.5.7] has a singularity when the second Euler angle reaches a certain value. This singularity may be avoided by switching to a different set of Euler angles, or by using Euler parameters. Hence, the transformation between angular velocities and Euler angles is considered here as a valid transformation between the generalized velocities and the quasi-velocities.

Consider next a position vector **r**. For an n degree of freedom system **r** can be expressed as $\mathbf{r} = \mathbf{r}(q_1, q_2, \ldots, q_n, t)$. The time derivative of **r** has the form

$$\mathbf{v} = \dot{\mathbf{r}} = \sum_{k=1}^{n} \frac{\partial \mathbf{r}}{\partial q_k} \dot{q}_k + \frac{\partial \mathbf{r}}{\partial t} \qquad [9.5.4]$$

or, in column vector form,

$$\{v\} = [r_q]\{\dot{q}\} + \frac{\partial \{r\}}{\partial t} \qquad [9.5.5]$$

in which $[r_q]$ is a matrix of order $3 \times n$ whose elements are

$$[r_q] = \left[\frac{\partial \{r\}}{\partial q_1} \quad \frac{\partial \{r\}}{\partial q_2} \quad \cdots \quad \frac{\partial \{r\}}{\partial q_n} \right] \qquad [9.5.6]$$

Further, in Chapter 4 it was derived that

$$\frac{\partial \mathbf{r}}{\partial q_k} = \frac{\partial \mathbf{v}}{\partial \dot{q}_k} \qquad k = 1, 2, \ldots, n \qquad [9.5.7]$$

so that calculation of the matrix $[r_q]$ can be accomplished using the position vector **r** or the velocity **v**. In the latter case, we have

$$[r_q] = \left[\frac{\partial \{v\}}{\partial \dot{q}_1} \quad \frac{\partial \{v\}}{\partial \dot{q}_2} \quad \cdots \quad \frac{\partial \{v\}}{\partial \dot{q}_n} \right] \qquad [9.5.8]$$

This is a very important result, as it permits calculation of $[r_q]$ when only the velocity vector is available, such as in cases involving angular velocities. Introducing Eq. [9.5.3] into Eq. [9.5.5], we obtain

$$\{v\} = [r_q][W]\{u\} + [r_q]\{X\} + \frac{\partial \{r\}}{\partial t} \qquad [9.5.9]$$

We define the following quantities:

Partial velocity matrix: $[v^q] = [r_q][W]$, a matrix of order $3 \times n$

Time partial velocity vector: $\{v^t\} = [r_q]\{X\} + \frac{\partial \{r\}}{\partial t}$, a vector of order 3

Thus, the velocity can be expressed as

$$\{v\} = [v^q]\{u\} + \{v^t\} \qquad [9.5.10]$$

In algebraic vector format, we can express this relation as

$$\mathbf{v} = \sum_{k=1}^{n} \mathbf{v}^k u_k + \mathbf{v}^t \qquad [9.5.11]$$

in which $\mathbf{v}^k$ is the kth *partial velocity* of **v** and $\mathbf{v}^t$ is the time partial velocity of **v**.[4] It follows that with a properly defined set of generalized speeds, one can express the

---

[4] Note that this notation is slightly different than the notation used by Kane. We denote the indices of the partial velocities by superscripts, whereas Kane denotes these indices by subscripts.

velocity of a point in terms of the generalized speeds and the partial velocities of the system.

Observe that partial velocities can also be expressed as

$$\mathbf{v}^k = \frac{\partial \mathbf{v}}{\partial u_k} \qquad \mathbf{v}^t = \frac{\partial \mathbf{v}}{\partial t} \qquad [9.5.12]$$

This equation can be interpreted as follows: The kth partial velocity denotes the direction of the velocity along the direction affected by the kth generalized speed. The reader should compare this interpretation with the interpretation in Chapter 8 of $\mathbf{e}_\phi$, $\mathbf{e}_\theta$, and $\mathbf{e}_\psi$, unit vectors about which Euler angle transformations are taken.

One can also define partial velocities for angular velocities. The procedure is the same, where one first expresses the angular velocities in terms of the generalized velocities, and then relates the generalized velocities to the generalized speeds, with the result

$$\boldsymbol{\omega} = \sum_{k=1}^{n} \boldsymbol{\omega}^k u_k + \boldsymbol{\omega}^t \qquad [9.5.13]$$

in which $\boldsymbol{\omega}^k$ and $\boldsymbol{\omega}^t$ are partial angular velocities. Also, we have so far considered inertial reference frames and the associated partial velocities. It turns out that partial velocities can be calculated for velocities that are measured from relative frames. We can then write the above expressions in the more general form

$$^A\mathbf{v} = \sum_{k=1}^{n} {}^A\mathbf{v}^k u_k + {}^A\mathbf{v}^t \qquad {}^B\mathbf{v} = \sum_{k=1}^{n} {}^B\mathbf{v}^k u_k + {}^B\mathbf{v}^t \qquad {}^A\boldsymbol{\omega}^B = \sum_{k=1}^{n} {}^A\boldsymbol{\omega}^{B(k)} u_k + {}^A\boldsymbol{\omega}^{B(t)}$$

$$[9.5.14]$$

where A and B denote reference frames.

Let us now consider the virtual work. For a system of N particles, the virtual work is

$$\delta W = \sum_{i=1}^{N} \mathbf{F}_i \cdot \delta \mathbf{r}_i = \sum_{i=1}^{N} \{F_i\}^T \{\delta r_i\} \qquad [9.5.15]$$

In terms of the generalized coordinates, the expression for the virtual work is written as

$$\delta W = \sum_{k=1}^{n} Q_k \delta q_k = \{Q\}^T \{\delta q\} \qquad [9.5.16]$$

where the generalized forces are found from

$$Q_k = \sum_{i=1}^{N} \mathbf{F}_i \cdot \frac{\partial \mathbf{r}_i}{\partial q_k} = \sum_{i=1}^{N} \mathbf{F}_i \cdot \frac{\partial \mathbf{v}_i}{\partial \dot{q}_k} \qquad [9.5.17]$$

or, in column vector format,

$${Q} = \sum_{i=1}^{N} [r_{iq}]^T \{F_i\} \qquad [9.5.18]$$

where $\{F_i\}$ is a column vector of order 3 containing the components of $\mathbf{F}_i$.

We next express the generalized forces in terms of the generalized speeds. Left-multiplying the above equation by $[W]^T = [Y]^{-T}$ and defining by $\{U\} = [W]^T \{Q\} = [U_1 \ U_2 \ldots U_n]^T$, we obtain

$$\{U\} = [W]^T \{Q\} = \sum_{i=1}^{N} [W]^T [r_{iq}]^T \{F_i\} \qquad [9.5.19]$$

Introducing the definition of the partial velocity matrix to the equation, we have

$$\{U\} = \sum_{i=1}^{N} [v_i^q]^T \{F_i\} \qquad [9.5.20]$$

or, in algebraic vector form,

$$U_k = \sum_{i=1}^{N} \mathbf{v}_i^k \cdot \mathbf{F}_i \qquad [9.5.21]$$

where $\mathbf{v}_i^k$ is the kth partial velocity associated with the point to which F_i is applied. We refer to U_k as the *generalized force associated with the kth generalized speed* and to $\{U\}$ as the *generalized force vector associated with generalized speeds*. It follows that the virtual work can be written in terms of quasi-coordinates as

$$\delta W = \{Q\}^T [W] \{\delta q'\} = \{U\}^T \{\delta q'\} \qquad [9.5.22]$$

Note that we are using the variations of the quasi-coordinates for derivation purposes only. When the generalized speeds are selected as the generalized velocities, $u_k = \dot{q}_k$ ($k = 1, 2, \ldots, n$) the associated generalized forces U_k become the same as the generalized forces Q_k.

Next, we express the generalized forces associated with a rigid body. Consider a rigid body subjected to N forces $\mathbf{F}_i$ ($i = 1, 2, \ldots, N$) and M^* torques $\mathbf{M}_i$ ($i = 1, 2, \ldots, M^*$). The generalized force for the quasi-velocities takes the form

$$U_k = \sum_{i=1}^{N} \mathbf{v}_i^k \cdot \mathbf{F}_i + \sum_{i=1}^{M^*} \boldsymbol{\omega}^k \cdot \mathbf{M}_i \qquad k = 1, 2, \ldots, n \qquad [9.5.23]$$

We can express the generalized forces in terms of the resultant of all the forces and moments. Writing the resultant of all applied forces and moments as force $\mathbf{F}$ passing through a point B and a moment $\mathbf{M}_B$ about B, where

$$\mathbf{F} = \sum_{i=1}^{N} \mathbf{F}_i \qquad \mathbf{M}_B = \sum_{i=1}^{M^*} (\mathbf{M}_i + \mathbf{r}_i \times \mathbf{F}_i) \qquad [9.5.24]$$

in which $\mathbf{r}_i$ is the position vector from point B to the point where $\mathbf{F}_i$ is applied, we can express Eq. [9.5.23] as

9.5 QUASI-COORDINATES AND QUASI-VELOCITIES (GENERALIZED SPEEDS)

$$U_k = \mathbf{v}_B^k \cdot \mathbf{F} + \boldsymbol{\omega}^k \cdot \mathbf{M}_B \qquad [9.5.25]$$

When using the center of mass to express the resultant force and moment, we write

$$U_k = \mathbf{v}_G^k \cdot \mathbf{F} + \boldsymbol{\omega}^k \cdot \mathbf{M}_G \qquad [9.5.26]$$

The decision about which of Eqs. [9.5.23], [9.5.25], and [9.5.26] to use depends on the number and nature of the forces and moments. For a system of N rigid bodies, we extend Eq. [9.5.26] to

$$U_k = \sum_{i=1}^{N} \mathbf{v}_{G_i}^k \cdot \mathbf{F}_i + \sum_{i=1}^{N} \boldsymbol{\omega}_i^k \cdot \mathbf{M}_{G_i} \qquad [9.5.27]$$

in which G_i denotes the center of mass of the ith body, and $\mathbf{F}_i$ and $\mathbf{M}_{G_i}$ denote the resultant of all forces external to the ith body and the resultant moment about the center of mass of the ith body due to these external forces.

Example 9.6

Consider the spinning top on a cart in Example 9.4 and find the partial velocities associated with the cart, the center of mass of the top, and the angular velocity of the top. Then, calculate the associated generalized forces.

Solution

The free-body diagram of the top and cart are given in Figs. 9.1 and 9.5. The velocities of the cart and the center of the top, and the angular velocities of the top are

$$\mathbf{v}_C = \dot{Y}\mathbf{a}_2 = \dot{Y}\sin\phi\,\mathbf{f}_1 + \dot{Y}\cos\phi\cos\theta\,\mathbf{f}_2 - \dot{Y}\cos\phi\sin\theta\,\mathbf{f}_3 \qquad [a]$$

$$\mathbf{v}_G = (\dot{Y}\sin\phi + \omega_2 L)\mathbf{f}_1 + (\dot{Y}\cos\phi\cos\theta - \omega_1 L)\mathbf{f}_2 - \dot{Y}\cos\phi\sin\theta\,\mathbf{f}_3 \qquad [b]$$

$$\boldsymbol{\omega} = \omega_1\mathbf{f}_1 + \omega_2\mathbf{f}_2 + \omega_3\mathbf{f}_3 = \dot{\theta}\mathbf{f}_1 + \dot{\phi}\sin\theta\,\mathbf{f}_2 + (\dot{\phi}\cos\theta + \dot{\psi})\mathbf{f}_3 \qquad [c]$$

We define the generalized speeds as

$$u_1 = \omega_1 \qquad u_2 = \omega_2 \qquad u_3 = \omega_3 \qquad u_4 = \dot{Y} \qquad [d]$$

so that the velocity and angular velocity expressions can be written as

$$\mathbf{v}_C = u_4\mathbf{a}_2 = u_4(\sin\phi\,\mathbf{f}_1 + \cos\phi\cos\theta\,\mathbf{f}_2 - \cos\phi\sin\theta\,\mathbf{f}_3) \qquad [e]$$

$$\mathbf{v}_G = -Lu_1\mathbf{f}_2 + Lu_2\mathbf{f}_1 + u_4(\sin\phi\,\mathbf{f}_1 + \cos\phi\cos\theta\,\mathbf{f}_2 - \cos\phi\sin\theta\,\mathbf{f}_3) \qquad [f]$$

$$\boldsymbol{\omega} = u_1\mathbf{f}_1 + u_2\mathbf{f}_2 + u_3\mathbf{f}_3 \qquad [g]$$

from which we identify the partial velocities as

$$\mathbf{v}_C^1 = \mathbf{v}_C^2 = \mathbf{v}_C^3 = 0 \qquad \mathbf{v}_C^4 = \mathbf{a}_2 = (\sin\phi\,\mathbf{f}_1 + \cos\phi\cos\theta\,\mathbf{f}_2 - \cos\phi\sin\theta\,\mathbf{f}_3) \qquad [h]$$

$$\mathbf{v}_G^1 = -L\mathbf{f}_2 \qquad \mathbf{v}_G^2 = L\mathbf{f}_1 \qquad \mathbf{v}_G^3 = 0$$

$$\mathbf{v}_G^4 = \mathbf{a}_2 = (\sin\phi\,\mathbf{f}_1 + \cos\phi\cos\theta\,\mathbf{f}_2 - \cos\phi\sin\theta\,\mathbf{f}_3) \qquad [i]$$

$$\boldsymbol{\omega}^1 = \mathbf{f}_1 \qquad \boldsymbol{\omega}^2 = \mathbf{f}_2 \qquad \boldsymbol{\omega}^3 = \mathbf{f}_3 \qquad \boldsymbol{\omega}^4 = 0 \qquad [j]$$

The forces external to the system that do work are gravity and the applied force F. The cart exerts a moment M about the $\mathbf{f}_3$ direction to the top through a motor, and there is a counter moment of equal magnitude applied to the cart by the top. To find the generalized forces we use Eqs. [9.5.23] for each force and moment. We identify the external inputs as

$$\mathbf{F}_g = -mg\mathbf{a}_3 = -mg(\sin\theta\,\mathbf{f}_2 + \cos\theta\,\mathbf{f}_3), \text{ acting through } G$$

$$\mathbf{F}_C = F\mathbf{a}_2 = F(\sin\phi\,\mathbf{f}_1 + \cos\phi\cos\theta\,\mathbf{f}_2 - \cos\phi\sin\theta\,\mathbf{f}_3), \text{ acting through } C, \text{ and}$$

$$\mathbf{M} = M\mathbf{f}_3 \qquad \text{[k]}$$

The generalized forces are calculated as

$$U_k = \mathbf{v}_G^k \cdot \mathbf{F}_g + \mathbf{v}_C^k \cdot \mathbf{F}_C + \boldsymbol{\omega}^k \cdot \mathbf{M} \qquad k = 1, 2, 3, 4 \qquad \text{[l]}$$

Evaluating the individual expressions, we find

$$U_1 = \mathbf{v}_G^1 \cdot \mathbf{F}_g + \mathbf{v}_C^1 \cdot \mathbf{F}_C + \boldsymbol{\omega}^1 \cdot \mathbf{M}$$
$$= -mg(\sin\theta\,\mathbf{f}_2 + \cos\theta\,\mathbf{f}_3)\cdot(-L\mathbf{f}_2) + F\mathbf{a}_2\cdot 0 + M\mathbf{f}_3\cdot\mathbf{f}_1 = mgL\sin\theta$$

$$U_2 = \mathbf{v}_G^2 \cdot \mathbf{F}_g + \mathbf{v}_C^2 \cdot \mathbf{F}_C + \boldsymbol{\omega}^2 \cdot \mathbf{M}$$
$$= -mg(\sin\theta\,\mathbf{f}_2 + \cos\theta\,\mathbf{f}_3)\cdot L\mathbf{f}_1 + F\mathbf{a}_2\cdot 0 + M\mathbf{f}_3\cdot\mathbf{f}_2 = 0$$

$$U_3 = \mathbf{v}_G^3 \cdot \mathbf{F}_g + \mathbf{v}_C^3 \cdot \mathbf{F}_C + \boldsymbol{\omega}^3 \cdot \mathbf{M} = -mg\mathbf{a}_3\cdot 0 + F\mathbf{a}_2\cdot 0 + M\mathbf{f}_3\cdot\mathbf{f}_3 = M$$

$$U_4 = \mathbf{v}_G^4 \cdot \mathbf{F}_g + \mathbf{v}_C^4 \cdot \mathbf{F}_C + \boldsymbol{\omega}^4 \cdot \mathbf{M} = -mg\mathbf{a}_3\cdot\mathbf{a}_2 + F\mathbf{a}_2\cdot\mathbf{a}_2 + M\mathbf{f}_3\cdot 0 = F \qquad \text{[m]}$$

9.6 GENERALIZED SPEEDS AND CONSTRAINTS

Consider a system that originally has n degrees of freedom subjected to m equality constraints. We write the constraints in velocity form as

$$\sum_{k=1}^{n} a_{jk}\dot{q}_k + a_{0j} = 0 \qquad j = 1, 2, \ldots, m \qquad \text{[9.6.1]}$$

where the coefficients $a_{jk}(j = 1, 2, \ldots, m; k = 1, 2, \ldots, n)$ and $a_{0j}(j = 1, 2, \ldots, m)$ are functions of the generalized coordinates and of time. When the constraints are nonholonomic, one cannot readily eliminate them from the formulation, so we explore the use of generalized speeds. We write Eq. [9.6.1] in matrix form as

$$[a]\{\dot{q}\} + \{b\} = \{0\} \qquad \text{[9.6.2]}$$

Introducing Eq. [9.5.3] into Eq. [9.6.2], we express the constraint in terms of the quasi-velocities as

$$[a][W]\{u\} + [a]\{X\} + \{b\} \Rightarrow [A]\{u\} + \{B\} = \{0\} \qquad \text{[9.6.3]}$$

and we note that the matrix $[A] = [a][W]$ is of order $m \times n$. The generalized velocities and the generalized speeds are no longer independent. We now refer to $\{u\}$ as the *constrained generalized speeds*. Because we have $p = n - m$ degrees of freedom, it is of interest to generate a set of p unconstrained (independent) generalized speeds.

We order the generalized speeds such that the first p are coordinates from which the independent generalized speeds are constructed and collect them in a vector $\{u_p\} = [u_1 \; u_2 \; \ldots \; u_p]^T$. The remaining generalized speeds, called *surplus*

9.6 GENERALIZED SPEEDS AND CONSTRAINTS

generalized speeds, are expressed as $\{u_S\} = [u_{p+1} \; u_{p+2} \; \ldots \; u_n]^T$. We hence partition the generalized speeds into

$$\{u\} = [\{u_p\}^T \; \{u_S\}^T]^T \qquad [9.6.4]$$

We define a set of independent generalized speeds by $\{\check{u}\} = [\check{u}_1 \; \check{u}_2 \; \ldots \; \check{u}_p]^T$. The vector containing the p independent generalized speeds $\{\check{u}\}$ is related to $\{u_p\}$ by

$$\{u_p\} = [T]\{\check{u}\} \qquad \{\check{u}\} = [T]^{-1}\{u_p\} \qquad [9.6.5]$$

in which $[T]$ is a nonsingular transformation matrix of order $p \times p$. Without loss of generality, we assume here that $\{\check{u}\}$ and $\{u_p\}$ coincide, so that $[T] = [1]$. We next partition the constraint matrix $[A]$ as

$$[A] = [[A_p] \; [A_S]] \qquad [9.6.6]$$

where $[A_p]$ is of order $m \times p$ and $[A_S]$ is of order $m \times m$. Introduction of this partition into Eq. [9.6.3] gives

$$[A_p]\{u_p\} + [A_S]\{u_S\} + \{B\} = \{0\} \qquad [9.6.7]$$

Solving for $\{u_S\}$, we obtain

$$\{u_S\} = -[A_S]^{-1}[A_p]\{u_p\} - [A_S]^{-1}\{B\} = [A']\{\check{u}\} + \{B'\} \qquad [9.6.8]$$

in which

$$[A'] = -[A_S]^{-1}[A_p] \qquad \{B'\} = -[A_S]^{-1}\{B\} \qquad [9.6.9]$$

For Eq. [9.6.8] to hold, $[A_S]$ must be nonsingular, which is the requirement when partitioning the generalized speeds into independent and surplus sets.

Next, we write the generalized velocities in terms of the independent and surplus generalized speeds. The relation between the generalized velocities and the generalized speeds, Eq. [9.5.3], can be partitioned as

$$\{\dot{q}\} = [W]\{u\} + \{X\} = [W_p]\{u_p\} + [W_S]\{u_S\} + \{X\} \qquad [9.6.10]$$

where $[W_p]$ and $[W_S]$ are partitions of $[W]$ of orders $n \times p$ and $n \times m$, respectively, $[W] = [[W_p] \; [W_S]]$. Introducing the expression from Eq. [9.6.8] into the above equation, we obtain

$$\{\dot{q}\} = [W_p]\{u_p\} - [W_S][A_S]^{-1}([A_p]\{u_p\} + \{B\}) + \{X\} = [\check{W}]\{\check{u}\} + \{\check{X}\} \qquad [9.6.11]$$

in which

$$[\check{W}] = [W_p] - [W_S][A_S]^{-1}[A_p] = [W_p] + [W_S][A'] \qquad [9.6.12]$$

We next extend the concept of independent generalized speeds to the partial velocities. We rewrite Eq. [9.5.10] as

$$\{v\} = [V]\{u\} + \{v^t\} \qquad [9.6.13]$$

where $[V]$ is of order $3 \times n$ and is the partial velocity matrix associated with $\{v\}$. We partition $[V]$ in terms of the independent and surplus parts as $[V] = [[V_p][V_S]]$ and write

$$\{v\} = [V_p]\{u_p\} + [V_S]\{u_S\} + \{v^t\} \qquad [9.6.14]$$

Introducing the expression for $\{u_S\}$ from Eq. [9.6.8] to this equation and carrying out the algebra gives

$$\{v\} = [\check{V}]\{\check{u}\} + [V_S]\{B'\} + \{v'\} \qquad [9.6.15]$$

The partial velocity matrix associated with the independent generalized speeds can be expressed as

$$[\check{V}] = [V_p] + [V_S][A'] \qquad [9.6.16]$$

Note the similarity in form between Eqs. [9.6.12] and [9.6.16]. To express the partial velocities in algebraic vector form we note that $[V_p]$ contains the first p partial velocities, and $[V_S][A']$ the remaining m. Hence, the partial velocities associated with the independent partial velocities can be expressed as

$$\check{\mathbf{v}}^k = \mathbf{v}^k + \sum_{j=1}^{m} A'_{jk}\mathbf{v}^{j+p} \qquad k = 1, 2, \ldots, p \qquad \check{\mathbf{v}}^t = \mathbf{v}^t + \sum_{j=1}^{m} B'_j \mathbf{v}^{j+p}$$

$$[9.6.17]$$

where A'_{jk} and B'_j are the entries of $[A']$ and $\{B'\}$, respectively, which were defined in Eq. [9.6.9]. When the transformation matrix $[T]$ is not the identity matrix, we have

$$[A'] = -[A_S]^{-1}[A_p][T] \qquad [\check{W}] = [W_p][T] + [W_S][A'] \qquad [9.6.18]$$

$$\check{\mathbf{v}}^k = \sum_{r=1}^{p} T_{kr}\mathbf{v}^r + \sum_{j=1}^{m} A'_{jk}\mathbf{v}^{j+p} \qquad [9.6.19]$$

in which T_{kr} are the entries of $[T]$.

We repeat the procedure for independent partial angular velocities. The results turn out to be the same as Eqs. [9.6.17], with the $\mathbf{v}$ terms replaced by $\boldsymbol{\omega}$:

$$\check{\boldsymbol{\omega}}^k = \boldsymbol{\omega}^k + \sum_{j=1}^{m} A'_{jk}\boldsymbol{\omega}^{j+p} \qquad \check{\boldsymbol{\omega}}^t = \boldsymbol{\omega}^t + \sum_{j=1}^{m} B'_j \boldsymbol{\omega}^{j+p} \qquad [9.6.20]$$

We now calculate the generalized forces associated with the independent generalized speeds. To this end, we make use of the quasi-coordinates. From Eq. [9.6.11] we write the virtual displacements as $\delta\{q\} = [\check{W}]\delta\{\check{q}'\}$, where $\delta\{\check{q}'\}$ are the virtual displacements associated with the independent quasi-coordinates. As before, we are using this expression only for derivation purposes. Introducing the virtual displacements into the virtual work, we can write

$$\delta W = \{Q\}^T \delta\{q\} = \{Q\}^T [\check{W}] \delta\{\check{q}'\} = \{\check{U}\}^T \delta\{\check{q}'\} \qquad [9.6.21]$$

where

$$\{\check{U}\}^T = \{Q\}^T [\check{W}] = \{Q\}^T [W_p] + \{Q\}^T [W_S][A'] \qquad [9.6.22]$$

is a vector of order p in the form $\{\check{U}\}^T = [\check{U}_1 \ \check{U}_2 \ \ldots \ \check{U}_p]$. To express the generalized forces associated with the independent generalized speeds, we partition the generalized forces as

$$\{U\}^T = \{Q\}^T [[W_p] \ [W_S]] = \{\{U_p\}^T \ \{U_S\}^T\} \qquad [9.6.23]$$

Introducing Eq. [9.6.23] into Eq. [9.6.22] gives

$$\{\check{U}\}^T = \{Q\}^T[\check{W}] = \{U_p\}^T + \{U_S\}^T[A'] \qquad [9.6.24]$$

which can be expressed in scalar form as

$$\check{U}_k = U_k + \sum_{j=1}^{m} U_{j+p} A'_{jk} \qquad k = 1, 2, \ldots, p \qquad [9.6.25]$$

We have eliminated the surplus generalized speeds from the formulation and arrived at a set of independent generalized speeds, partial velocities, and generalized forces. This is a significant advantage of using generalized speeds rather than generalized velocities. Recall that with Lagrangian mechanics, nonholonomic constraints and associated Lagrange multipliers cannot be eliminated from the formulation before the equations of motion are derived. Only *after* deriving the equations of motion in constrained form can one use a series of substitutions to eliminate the Lagrange multipliers and surplus coordinates.

One can arrive at a set of independent generalized speeds, partial velocities, and generalized forces in two ways. One is by observing the kinematics of the system and directly identifying a set of independent generalized speeds. The other is by beginning with a set of constrained generalized speeds and then applying the constraint equations derived above. The latter approach is most often a complicated task; one is usually better off generating a set of independent generalized speeds directly.

Under certain circumstances, such as in problems involving Coulomb friction, it may be more desirable to deal with constrained quasi-velocities than with unconstrained ones. In such cases, one deliberately leaves a constrained coordinate in the system formulation. This procedure is similar to the relaxation of constraints using Lagrange's equations.

Example 9.7

Consider the rolling disk problem. First consider the case when the disk is rolling and slipping, and obtain the partial velocities associated with the center of mass and the partial angular velocities. Then, impose the rolling constraint and obtain the corresponding partial velocities and partial angular velocities.

Solution

We use the 3-1-3 Euler angle transformation and the F frame. We select the inertial frame $a_1 a_2 a_3$ such that the $a_1 a_2$ plane is the rolling surface, and the a_3 direction is the vertical. We invoke the constraint that the disk is always in contact with the surface, so that for roll with slip we have five degrees of freedom. We select the quasi-velocities as

$$u_1 = \omega_1 = \dot{\theta} \qquad u_2 = \omega_2 = \dot{\phi} \sin\theta$$

$$u_3 = \omega_3 = \dot{\phi} \cos\theta + \dot{\psi} \qquad u_4 = \dot{X} \qquad u_5 = \dot{Y} \qquad [a]$$

where X and Y denote the coordinates of the center of mass in the a_1 and a_2 directions, respectively. Noting that the height of the center of mass is given by the holonomic constraint relation $Z = R\sin\theta$, the component of the velocity of the center of mass in the a_3 direction is $R\dot{\theta}\cos\theta = Ru_1\cos\theta$. The velocity of the center of mass and the angular velocity of the body can be written as

$$\mathbf{v}_G = \dot{X}\mathbf{a}_1 + \dot{Y}\mathbf{a}_2 + \dot{Z}\mathbf{a}_3 = u_4\mathbf{a}_1 + u_5\mathbf{a}_2 + Ru_1\cos\theta\,\mathbf{a}_3 \quad\text{[b]}$$

$$\boldsymbol{\omega} = u_1\mathbf{f}_1 + u_2\mathbf{f}_2 + u_3\mathbf{f}_3 \quad\text{[c]}$$

The partial velocities and partial angular velocities can now be written as

$$\mathbf{v}_G^1 = R\cos\theta\,\mathbf{a}_3 \quad \mathbf{v}_G^2 = \mathbf{v}_G^3 = 0 \quad \mathbf{v}_G^4 = \mathbf{a}_1 \quad \mathbf{v}_G^5 = \mathbf{a}_2$$

$$\boldsymbol{\omega}^1 = \mathbf{f}_1 \quad \boldsymbol{\omega}^2 = \mathbf{f}_2 \quad \boldsymbol{\omega}^3 = \mathbf{f}_3 \quad \boldsymbol{\omega}^4 = 0 \quad \boldsymbol{\omega}^5 = 0 \quad\text{[d]}$$

To find the partial velocities associated with the constrained system, we first calculate the velocity of the contact point C and set it equal to zero, thus

$$\mathbf{v}_C = \mathbf{v}_G + \boldsymbol{\omega}\times\mathbf{r}_{C/G} = u_4\mathbf{a}_1 + u_5\mathbf{a}_2 + Ru_1c\theta\mathbf{a}_3 + (u_1\mathbf{f}_1 + u_2\mathbf{f}_2 + u_3\mathbf{f}_3)\times(-R\mathbf{f}_2)$$

$$= (u_4 - Ru_1s\phi s\theta + Ru_3c\phi)\mathbf{a}_1 + (u_5 + Ru_1c\phi s\theta + Ru_3s\phi)\mathbf{a}_2 = 0 \quad\text{[e]}$$

As expected, components of the motion in the a_3 direction canceled each other.

We select the independent generalized speeds as $\check{u}_1 = u_1$, $\check{u}_2 = u_2$, and $\check{u}_3 = u_3$, so that the surplus coordinates are $\{u_S\} = [u_4\ u_5]^T$. We are retaining the angular velocities in the formulation. Comparing Eq. [e] with Eqs. [9.6.7] and [9.6.8], we have

$$[A_S] = \begin{bmatrix} 1 & 0 \\ 0 & 1 \end{bmatrix} \quad [A_p] = R\begin{bmatrix} -\sin\phi\sin\theta & 0 & \cos\phi \\ \cos\phi\sin\theta & 0 & \sin\phi \end{bmatrix} \quad \{B\} = \begin{bmatrix} 0 \\ 0 \end{bmatrix}$$

$$[A'] = -[A_S]^{-1}[A_p] = -[A_p] = -R\begin{bmatrix} -\sin\phi\sin\theta & 0 & \cos\phi \\ \cos\phi\sin\theta & 0 & \sin\phi \end{bmatrix} \quad\text{[f]}$$

The matrix $[A_S]$ is nonsingular; thus, our choice of independent generalized speeds is a valid one. Next, we introduce the constraints into the expression for the velocity of the center of mass. Introducing the expressions for u_4 and u_5 from the right side of Eq. [e] into Eq. [b] and expressing the result in terms of the F frame, we obtain

$$\mathbf{v}_G = (Ru_1\sin\phi\sin\theta - Ru_3\cos\phi)\mathbf{a}_1$$
$$- (Ru_1\cos\phi\sin\theta + Ru_3\sin\phi)\mathbf{a}_2 + Ru_1\cos\theta\,\mathbf{a}_3$$
$$= -Ru_3\mathbf{f}_1 + Ru_1\mathbf{f}_3 \quad\text{[g]}$$

which is the well-known result for the velocity of the center of mass. The partial velocities associated with the velocity of the center of mass in terms of the independent generalized speeds are

$$\check{\mathbf{v}}_G^1 = R\mathbf{f}_3 \quad \check{\mathbf{v}}_G^2 = 0 \quad \check{\mathbf{v}}_G^3 = -R\mathbf{f}_1 \quad\text{[h]}$$

The independent partial angular velocities remain the same as before and have the form

$$\check{\boldsymbol{\omega}}^1 = \mathbf{f}_1 \quad \check{\boldsymbol{\omega}}^2 = \mathbf{f}_2 \quad \check{\boldsymbol{\omega}}^3 = \mathbf{f}_3 \quad\text{[i]}$$

We next calculate the partial velocities associated with the independent generalized speeds using Eqs. [9.6.17]. We find

$$\check{\mathbf{v}}_G^1 = \mathbf{v}_G^1 + A'_{11}\mathbf{v}_G^4 + A'_{21}\mathbf{v}_G^5 = R\cos\theta\,\mathbf{a}_3 + R\sin\phi\sin\theta\,\mathbf{a}_1 - R\cos\phi\sin\theta\,\mathbf{a}_2 = R\mathbf{f}_3$$

$$\check{\mathbf{v}}_G^2 = \mathbf{v}_G^2 + A'_{12}\mathbf{v}_G^4 + A'_{22}\mathbf{v}_G^5 = 0$$

$$\check{\mathbf{v}}_G^3 = \mathbf{v}_G^3 + A'_{13}\mathbf{v}_G^4 + A'_{23}\mathbf{v}_G^5 = 0 - R\cos\phi\,\mathbf{a}_1 - R\sin\phi\,\mathbf{a}_2 = -R\mathbf{f}_1 \quad\text{[j]}$$

Consider the rolling disk. First analyze the case when the disk is rolling and slipping and obtain the generalized forces. Then impose the rolling constraint and obtain the associated generalized forces.

Example 9.8

Solution

We will use the expressions generated for the partial velocities in Example 9.7. For roll with slip, there are five degrees of freedom. We select the quasi-velocities as in the previous example as

$$u_1 = \omega_1 = \dot{\theta} \qquad u_2 = \omega_2 = \dot{\phi}\sin\theta$$
$$u_3 = \omega_3 = \dot{\phi}\cos\theta + \dot{\psi} \qquad u_4 = \dot{X} \qquad u_5 = \dot{Y} \qquad \text{[a]}$$

and can write the velocity of the center of mass and the angular velocity vector as

$$\mathbf{v}_G = u_4\mathbf{a}_1 + u_5\mathbf{a}_2 + Ru_1\cos\theta\,\mathbf{a}_3 \qquad \boldsymbol{\omega} = u_1\mathbf{f}_1 + u_2\mathbf{f}_2 + u_3\mathbf{f}_3 \qquad \text{[b]}$$

The associated partial velocities are

$$\mathbf{v}_G^1 = R\cos\theta\,\mathbf{a}_3 \qquad \mathbf{v}_G^2 = \mathbf{v}_G^3 = 0 \qquad \mathbf{v}_G^4 = \mathbf{a}_1 \qquad \mathbf{v}_G^5 = \mathbf{a}_2$$
$$\boldsymbol{\omega}^1 = \mathbf{f}_1 \qquad \boldsymbol{\omega}^2 = \mathbf{f}_2 \qquad \boldsymbol{\omega}^3 = \mathbf{f}_3 \qquad \boldsymbol{\omega}^4 = 0 \qquad \boldsymbol{\omega}^5 = 0 \qquad \text{[c]}$$

The forces acting on the disk are the force of gravity, $\mathbf{F}_g = -mg\mathbf{a}_3$, and three reaction forces at the point of contact, $\mathbf{F}_C = P_1\mathbf{a}_1 + P_2\mathbf{a}_2 + P_3\mathbf{a}_3 = F_1\mathbf{f}_1 + F_2\mathbf{f}_2 + F_3\mathbf{f}_3$, where P_i and F_i are their components in the A and F frames. We can use two approaches to find the generalized forces associated with the quasi-coordinates. One approach is to treat each force separately and use Eq. [9.5.23], which requires the partial velocities associated with the individual points at which the forces are applied. For the problem under consideration, this requires calculation of the velocity of point C and of the center of mass G. As there are no applied torques, there is no need for partial angular velocities.

The second approach is to take the resultant and use Eq. [9.5.26]. Here, one needs to calculate the resultant moment, the partial velocities associated with the center of mass, and the partial angular velocities. To demonstrate the merits of the two approaches, we will use both.

In the resultant approach, we first calculate the resultant force, $\mathbf{F}$, which has the form

$$\mathbf{F} = \mathbf{F}_C + \mathbf{F}_g = P_1\mathbf{a}_1 + P_2\mathbf{a}_2 + (P_3 - mg)\mathbf{a}_3$$
$$= F_1\mathbf{f}_1 + (F_2 - mg\sin\theta)\mathbf{f}_2 + (F_3 - mg\cos\theta)\mathbf{f}_3 \qquad \text{[d]}$$

The resultant moment about the center of mass is

$$\mathbf{M}_G = \mathbf{r}_{C/G} \times \mathbf{F}_C = -R\mathbf{f}_2 \times (F_1\mathbf{f}_1 + F_2\mathbf{f}_2 + F_3\mathbf{f}_3) = -F_3 R\mathbf{f}_1 + F_1 R\mathbf{f}_3 \qquad \text{[e]}$$

Note that we are conveniently using unit vectors from both the A and F frames to simplify the expressions. The associated generalized forces become

$$U_1 = \mathbf{v}_G^1 \cdot \mathbf{F} + \boldsymbol{\omega}^1 \cdot \mathbf{M}_G = (P_3 - mg)R\cos\theta - F_3 R$$
$$U_2 = \mathbf{v}_G^2 \cdot \mathbf{F} + \boldsymbol{\omega}^2 \cdot \mathbf{M}_G = 0$$
$$U_3 = \mathbf{v}_G^3 \cdot \mathbf{F} + \boldsymbol{\omega}^3 \cdot \mathbf{M}_G = F_1 R$$
$$U_4 = \mathbf{v}_G^4 \cdot \mathbf{F} + \boldsymbol{\omega}^4 \cdot \mathbf{M}_G = P_1$$
$$U_5 = \mathbf{v}_G^5 \cdot \mathbf{F} + \boldsymbol{\omega}^5 \cdot \mathbf{M}_G = P_2 \qquad \text{[f]}$$

When treating each force separately, we first find the velocity of the contact point C, which is

$$\mathbf{v}_C = \mathbf{v}_G + \boldsymbol{\omega} \times \mathbf{r}_{C/G}$$
$$= u_4\mathbf{a}_1 + u_5\mathbf{a}_2 + Ru_1\cos\theta\,\mathbf{a}_3 + (u_1\mathbf{f}_1 + u_2\mathbf{f}_2 + u_3\mathbf{f}_3) \times (-R\mathbf{f}_2)$$
$$= u_4\mathbf{a}_1 + u_5\mathbf{a}_2 + Ru_1\cos\theta\,\mathbf{a}_3 + Ru_3\mathbf{f}_1 - Ru_1\mathbf{f}_3 \qquad \text{[g]}$$

so that the partial velocities associated with the velocity of point C have the form

$$\mathbf{v}_C^1 = -R\mathbf{f}_3 + R\cos\theta\,\mathbf{a}_3 \qquad \mathbf{v}_C^2 = 0 \qquad \mathbf{v}_C^3 = R\mathbf{f}_1 \qquad \mathbf{v}_C^4 = \mathbf{a}_1 \qquad \mathbf{v}_C^5 = \mathbf{a}_2 \qquad \text{[h]}$$

We proceed to find the generalized forces. Noting that there are no applied torques, we write the generalized forces as

$$U_k = \mathbf{v}_G^k \cdot \mathbf{F}_g + \mathbf{v}_C^k \cdot \mathbf{F}_C \qquad k = 1, 2, \ldots, 5 \qquad \text{[i]}$$

Carrying out the dot products, we obtain

$$U_1 = -mg\mathbf{a}_3 \cdot R\cos\theta\,\mathbf{a}_3 + (P_1\mathbf{a}_1 + P_2\mathbf{a}_2 + P_3\mathbf{a}_3) \cdot (-R\mathbf{f}_3 + R\cos\theta\,\mathbf{a}_3)$$
$$= -mgR\cos\theta - F_3R + P_3R\cos\theta$$

$$U_2 = 0$$

$$U_3 = -mg\mathbf{a}_3 \cdot 0 + (F_1\mathbf{f}_1 + F_2\mathbf{f}_2 + F_3\mathbf{f}_3) \cdot R\mathbf{f}_1 = F_1 R$$

$$U_4 = -mg\mathbf{a}_3 \cdot \mathbf{a}_1 + (P_1\mathbf{a}_1 + P_2\mathbf{a}_2 + P_3\mathbf{a}_3) \cdot \mathbf{a}_1 = P_1$$

$$U_5 = -mg\mathbf{a}_3 \cdot \mathbf{a}_2 + (P_1\mathbf{a}_1 + P_2\mathbf{a}_2 + P_3\mathbf{a}_3) \cdot \mathbf{a}_2 = P_2 \qquad \text{[j]}$$

which is identical to Eqs. [f].

Now we impose the no-slip constraint and obtain the generalized forces associated with the independent generalized speeds. Here, we use the resultant approach. From Example 9.7, the independent partial velocities are

$$\check{\mathbf{v}}_G^1 = R\mathbf{f}_3 \qquad \check{\mathbf{v}}_G^2 = 0 \qquad \check{\mathbf{v}}_G^3 = -R\mathbf{f}_1 \qquad \check{\boldsymbol{\omega}}^1 = \mathbf{f}_1 \qquad \check{\boldsymbol{\omega}}^2 = \mathbf{f}_2 \qquad \check{\boldsymbol{\omega}}^3 = \mathbf{f}_3 \qquad \text{[k]}$$

so that the generalized forces become

$$\check{U}_1 = \check{\mathbf{v}}_G^1 \cdot \mathbf{F} + \check{\boldsymbol{\omega}}^1 \cdot \mathbf{M}_G = (F_3 - mg\cos\theta)R - F_3R = -mgR\cos\theta$$

$$\check{U}_2 = \check{\mathbf{v}}_G^2 \cdot \mathbf{F} + \check{\boldsymbol{\omega}}^2 \cdot \mathbf{M}_G = 0$$

$$\check{U}_3 = \check{\mathbf{v}}_G^3 \cdot \mathbf{F} + \check{\boldsymbol{\omega}}^3 \cdot \mathbf{M}_G = -F_1R + F_1R = 0 \qquad \text{[l]}$$

As expected, the reaction forces at the contact point do not appear in the generalized forces. Let us now calculate the generalized forces associated with the independent generalized speeds using the constraint equation, Eq. [9.6.25]. From Example 9.7 we have

$$[A'] = -R\begin{bmatrix} -\sin\phi\sin\theta & 0 & \cos\phi \\ \cos\phi\sin\theta & 0 & \sin\phi \end{bmatrix} \qquad \text{[m]}$$

thus, we can write Eq. [9.6.25] as

$$\check{U}_1 = U_1 + A'_{11}U_4 + A'_{21}U_5 = U_1 + U_4R\sin\phi\sin\theta - U_5R\cos\phi\sin\theta$$
$$= -mgR\cos\theta - F_3R + P_3R\cos\theta + P_1R\sin\phi\sin\theta - P_2R\cos\phi\sin\theta$$

$$\check{U}_2 = U_2 + A'_{12}U_4 + A'_{22}U_5 = U_2 = 0$$
$$\check{U}_3 = U_3 + A'_{13}U_4 + A'_{23}U_5 = U_3 - U_4 R\cos\phi - U_5 R\sin\phi$$
$$= F_1 R - P_1 R\cos\phi - P_2 R\sin\phi \qquad \text{[n]}$$

Noting that $F_1 = P_1\cos\phi + P_2\sin\phi$ and $F_3 = P_1\sin\phi\sin\theta - P_2\cos\phi\sin\theta + P_3\cos\theta$, we observe that all the P_i and F_i ($i = 1, 2, 3$) expressions cancel from Eqs. [n], and we obtain Eq. [l].

9.7 GIBBS-APPELL EQUATIONS[5]

The Gibbs-Appell method makes use of a scalar function in terms of accelerations to derive the equations of motion, analogous to the concept of using kinetic energy in Lagrange's equations. We describe the method first for a system of particles and then for rigid bodies. Consider a system of N particles that has n degrees of freedom. We will initially consider independent quasi-velocities. The force balance for each particle can be written as

$$m_i \mathbf{a}_i = \mathbf{F}_i + \mathbf{F}'_i \qquad i = 1, 2, \ldots, N \qquad \textbf{[9.7.1]}$$

In order to simplify the derivation and to establish that the Gibbs-Appell and Kane's equations are indeed identical, we will carry out the derivation of the Gibbs-Appell equations using partial velocities. This way of deriving the Gibbs-Appell equations is not the traditional derivation.

We select a set of n independent quasi-velocities $u_1, u_2, \ldots, u_n$ and write the velocity and acceleration of each particle. The velocity of each particle can be written as

$$\mathbf{v}_i = \sum_{k=1}^{n} \mathbf{v}_i^k u_k + \mathbf{v}_i^t \qquad \textbf{[9.7.2]}$$

The acceleration of each particle can be written as

$$\mathbf{a}_i = \sum_{k=1}^{n} \mathbf{v}_i^k \dot{u}_k + \sum_{k=1}^{n} \dot{\mathbf{v}}_i^k u_k + \frac{d}{dt}\mathbf{v}_i^t \qquad \textbf{[9.7.3]}$$

Define the *Gibbs-Appell function,* denoted by S, as having the form

$$S = \frac{1}{2}\sum_{i=1}^{N} m_i \mathbf{a}_i \cdot \mathbf{a}_i \qquad \textbf{[9.7.4]}$$

The Gibbs-Appell function is also referred to as the *energy of acceleration* or the *Gibbs function.* Let us next differentiate S with respect to the time derivative of the kth quasi-velocity, which yields

---

[5] There is debate on whether these equations should be called the Gibbs-Appell or the Appell equations. The name Gibbs-Appell is attributed to Pars.

$$\frac{\partial S}{\partial \dot{u}_k} = \sum_{i=1}^{N} m_i \mathbf{a}_i \cdot \frac{\partial \mathbf{a}_i}{\partial \dot{u}_k} \qquad [9.7.5]$$

Noting from Eq. [9.7.3] that $\dot{\mathbf{v}}_i^k$ and $\dot{\mathbf{v}}_i^t$ do not contain any derivatives of the generalized speeds, we can write

$$\frac{\partial \mathbf{a}_i}{\partial \dot{u}_k} = \frac{\partial \mathbf{v}_i}{\partial u_k} = \mathbf{v}_i^k \qquad k = 1, 2, \ldots, n \qquad [9.7.6]$$

so that, substituting this result into Eq. [9.7.5] and using the definition of the generalized forces associated with the quasi-velocities, we obtain

$$\frac{\partial S}{\partial \dot{u}_k} = \sum_{i=1}^{N} \mathbf{v}_i^k \cdot m_i \mathbf{a}_i = \sum_{i=1}^{N} \mathbf{v}_i^k \cdot \mathbf{F}_i = U_k \qquad [9.7.7]$$

where the contributions of the reaction forces have canceled out. Equation [9.7.7] can be rewritten as

$$\frac{\partial S}{\partial \dot{u}_k} = U_k \qquad k = 1, 2, \ldots, n \qquad [9.7.8]$$

Equation [9.7.8] represents the *Gibbs-Appell equations* or *Appell equations* in terms of independent quasi-velocities. They are n first-order equations. When combined with the kinematic differential equations relating the generalized velocities to the generalized speeds, they form a set of $2n$ first-order equations that describe the evolution of the dynamical system completely.

To write the Gibbs-Appell equations, we first determine the number of degrees of freedom of the system and select an appropriate set of independent generalized coordinates and quasi-velocities. We proceed with the kinematics of the problem, developing expressions for the velocity and for the acceleration of each particle. From the velocity terms, we identify the partial velocities. From the acceleration terms, we generate the function S. Because we will be taking partial derivatives of S with respect to the rates of change of the generalized speeds, we can ignore any term in S that does not contain derivatives of generalized speeds. After obtaining S, we take the partial derivatives. We obtain the generalized forces U_k after calculating the forces acting on each particle. We then invoke Eq. [9.7.8].

The term $\partial S/\partial \dot{u}_k$ can also be calculated by using the second term on the right side in Eq. [9.7.7], without the need to calculate S explicitly. Doing so, as we will see in the next section, is tantamount to using Kane's equations.

Let us compare the Gibbs-Appell equations and Lagrange's equations. In Lagrange's equations, first the kinetic energy is calculated, which is a function of the velocities. Lagrange's equations can handle holonomic constraints as—at least in theory—one can find a set of independent generalized coordinates that take into account the constraints. In the presence of nonholonomic constraints, one cannot eliminate the constraints before deriving the equations of motion. But the Gibbs-Appell method uses an acceleration function. Hence, for systems acted upon by nonholonomic constraints, it becomes possible to find a set of independent quasi-velocities that are compatible with the constraints. In such systems, one ends up with $n - m = p$ independent equations and n kinematic differential equations, for a total of $n + p$ first-order equations.

Lagrange's equations can be shown to be a special case of the Gibbs-Appell equations. Selecting the quasi-velocities to be the same as the generalized velocities, $u_k = \dot{q}_k$ ($k = 1, 2, \ldots, n$), we observe that $Q_k = U_k$, and by manipulating S one can show that

$$\frac{\partial S}{\partial \dot{u}_k} = \frac{\partial S}{\partial \ddot{q}_k} = \frac{d}{dt}\frac{\partial T}{\partial \dot{q}_k} - \frac{\partial T}{\partial q_k} \qquad k = 1, 2, \ldots, n \qquad [9.7.9]$$

It is for these reasons that the function S is also referred to as the *fundamental function*.

One might wonder at this point why we bothered to learn about Lagrange's equations, if the Gibbs-Appell equations represent the more general case and can handle holonomic as well as nonholonomic constraints, as well as leading to equations of motion that can be put into state form with greater ease. There are several answers. First of all, for holonomic systems, the Gibbs-Appell equations have no advantage over Lagrange's equations. The Gibbs-Appell equations are more cumbersome; they require the calculation of acceleration terms, as opposed to the velocity terms needed for Lagrange's equations.

From a physical perspective, velocities are easier to visualize than accelerations, so that dealing with velocities gives a greater physical insight. Following this line of argument, the kinetic and potential energy are quantities that are much easier to visualize than the function S, which looks much more abstract. Dealing with energy expressions helps one differentiate between natural and nonnatural systems easier, and it simplifies the solution of equilibrium problems. Finally, integrals of the motion are much more easily derived from the kinetic and potential energy than they are from the fundamental function S.

The definition of S is motivated by Gauss's principle for virtual accelerations, which is obtained by considering a special class of variation (see Section 5.13). By contrast, Lagrange's equations are based on D'Alembert's principle and its extension to scalar functions, the extended Hamilton's principle.

Another shortcoming of the Gibbs-Appell equations is when deriving the equations of motion associated with spatially continuous systems. As we will see in Chapter 11, for deformable bodies the extended Hamilton's principle leads to both the equations of motion and boundary conditions. By contrast, the Gibbs-Appell equations can be used only after the flexible motion is discretized.

We are ready to extend the Gibbs-Appell equations to rigid bodies. The general form of the equations of motion is still Eq. [9.7.8], so we develop the expression for the fundamental function. Intuitively, we expect it to contain expressions associated with the acceleration of the center of mass and with the rate of change of angular momentum. From Fig. 8.1, the acceleration of a differential mass element on the body is written in terms of the center of mass motion as

$$\mathbf{a} = \mathbf{a}_G + \boldsymbol{\alpha} \times \boldsymbol{\rho} + \boldsymbol{\omega} \times (\boldsymbol{\omega} \times \boldsymbol{\rho}) \qquad [9.7.10]$$

where $\boldsymbol{\omega}$ and $\boldsymbol{\alpha}$ are the angular velocity and acceleration of the body, respectively, and $\boldsymbol{\rho}$ is the position vector from the center of mass to the differential element. It follows that $\int \boldsymbol{\rho}\, dm = \mathbf{0}$. The Gibbs-Appell function can be written as an integral over the entire body as

$$S = \frac{1}{2}\int \mathbf{a} \cdot \mathbf{a}\, dm \qquad [9.7.11]$$

We introduce Eq. [9.7.10] into Eq. [9.7.11] and integrate over the body. All of the $\int \boldsymbol{\rho}\, dm$ terms vanish and terms that do not contain any accelerations or angular accelerations can be ignored, as they do not contribute to $\partial S/\partial \dot{u}_k$. The expression for S reduces to

$$S = \frac{1}{2}m\mathbf{a}_G \cdot \mathbf{a}_G + \frac{1}{2}\int (\boldsymbol{\alpha} \times \boldsymbol{\rho}) \cdot (\boldsymbol{\alpha} \times \boldsymbol{\rho})\, dm + \int (\boldsymbol{\alpha} \times \boldsymbol{\rho}) \cdot (\boldsymbol{\omega} \times (\boldsymbol{\omega} \times \boldsymbol{\rho}))\, dm \qquad [9.7.12]$$

It is convenient to express S in terms of the angular momentum. Here, we follow an approach similar to the one in Section 8.11. The second term on the right side of Eq. [9.7.12] can be expressed as

$$(\boldsymbol{\alpha} \times \boldsymbol{\rho}) \cdot (\boldsymbol{\alpha} \times \boldsymbol{\rho}) \Rightarrow \{\alpha\}^T [\tilde{\rho}]^T [\tilde{\rho}]\{\alpha\} \qquad [9.7.13]$$

and we realize that $\int [\tilde{\rho}]^T [\tilde{\rho}]\, dm = [I_G]$ and that $[I_G]\{\alpha\} = \{\dot{H}_G\}_{\text{rel}}$. Manipulation of the last term on the right side of Eq. [9.7.12] is more complicated. After a number of manipulations one can show that

$$(\boldsymbol{\alpha} \times \boldsymbol{\rho}) \cdot (\boldsymbol{\omega} \times (\boldsymbol{\omega} \times \boldsymbol{\rho})) = \boldsymbol{\alpha} \cdot (\boldsymbol{\omega} \times (\boldsymbol{\rho} \times (\boldsymbol{\omega} \times \boldsymbol{\rho}))) \qquad [9.7.14]$$

where we recognize the expression that leads to the angular momentum term, $\boldsymbol{\rho} \times (\boldsymbol{\omega} \times \boldsymbol{\rho})$. Carrying out the integration, we can write the Gibbs-Appell function for a rigid body as

$$S = \frac{1}{2}m\mathbf{a}_G \cdot \mathbf{a}_G + \frac{1}{2}\boldsymbol{\alpha} \cdot \dot{\mathbf{H}}_{G\,\text{rel}} + \boldsymbol{\alpha} \cdot (\boldsymbol{\omega} \times \mathbf{H}_G) \qquad [9.7.15]$$

The last two terms in this equation indicate that S is related to the rate of change of the angular momentum. In column vector format we have

$$S = \frac{1}{2}m\{a_G\}^T\{a_G\} + \frac{1}{2}\{\alpha\}^T[I_G]\{\alpha\} + \{\alpha\}^T[\tilde{\omega}][I_G]\{\omega\} \qquad [9.7.16]$$

When taking the partial derivatives of S with respect to the derivatives of the generalized speeds, a simplification takes place when we recall Eq. [9.7.6]. Indeed, for the acceleration of the center of mass, one can write

$$\frac{\partial \mathbf{a}_G}{\partial \dot{u}_k} = \frac{\partial \mathbf{v}_G}{\partial u_k} = \mathbf{v}_G^k \qquad k = 1, 2, \ldots, n \qquad [9.7.17]$$

and, using the same argument for the angular acceleration vector, write

$$\frac{\partial \boldsymbol{\alpha}}{\partial \dot{u}_k} = \frac{\partial \boldsymbol{\omega}}{\partial u_k} = \boldsymbol{\omega}^k \qquad [9.7.18]$$

Taking the partial derivative of S with respect to $\dot{u}_k$ and introducing the above two equations into these partial derivatives, we obtain

9.7 GIBBS-APPELL EQUATIONS

$$\frac{\partial S}{\partial \ddot{u}_k} = m\{a_G\}^T \frac{\partial \{a_G\}}{\partial \ddot{u}_k} + \left(\frac{\partial \{\alpha\}}{\partial \ddot{u}_k}\right)^T [I_G]\{\alpha\} + \left(\frac{\partial \{\alpha\}}{\partial \ddot{u}_k}\right)^T [\tilde{\omega}][I_G]\{\omega\}$$

$$= \mathbf{v}_G^k \cdot m\mathbf{a}_G + \boldsymbol{\omega}^k \cdot (\dot{\mathbf{H}}_{G\,\mathrm{rel}} + \boldsymbol{\omega} \times \mathbf{H}_G) = \mathbf{v}_G^k \cdot m\mathbf{a}_G + \boldsymbol{\omega}^k \cdot \dot{\mathbf{H}}_G \qquad \textbf{[9.7.19]}$$

so that, as with particles, the partial derivatives of S can be calculated without evaluating S itself. Considering that the expressions for the generalized forces are given by Eq. [9.5.26], we can write the Gibbs-Appell equations for a rigid body as

$$\mathbf{v}_G^k \cdot m\mathbf{a}_G + \boldsymbol{\omega}^k \cdot \dot{\mathbf{H}}_G = \mathbf{v}_G^k \cdot \mathbf{F} + \boldsymbol{\omega}^k \cdot \mathbf{M}_G \qquad k = 1, 2, \ldots, n \qquad \textbf{[9.7.20]}$$

For a system consisting of several bodies, the Gibbs-Appell equations can be written as

$$\sum_{i=1}^{N} (\mathbf{v}_{G_i}^k \cdot m_i \mathbf{a}_{G_i} + \boldsymbol{\omega}_i^k \cdot \dot{\mathbf{H}}_{G_i}) = \sum_{i=1}^{N} (\mathbf{v}_{G_i}^k \cdot \mathbf{F}_i + \boldsymbol{\omega}_i^k \cdot \mathbf{M}_{G_i}) \qquad k = 1, 2, \ldots, n$$

$$\textbf{[9.7.21]}$$

When the body is rotating about a fixed point C, the Gibbs-Appell function can be written about that point. The Gibbs-Appell function takes the form

$$S = \frac{1}{2}\boldsymbol{\alpha} \cdot \dot{\mathbf{H}}_{C\,\mathrm{rel}} + \boldsymbol{\alpha} \cdot (\boldsymbol{\omega} \times \mathbf{H}_C) \qquad \textbf{[9.7.22]}$$

On the issue of the selection of the quasi-velocities: A good choice of these should simplify the equations of motion and, if possible, give the generalized speeds a physical interpretation. If there are nonholonomic quantities, such as constraints, angular velocities, or body-fixed velocity components, the selection should reflect those quantities.

Example 9.9

Derive the translational and rotational equations of motion for a rigid body, using the Gibbs-Appell equations.

Solution

We will use the column vector notation. We select the quasi-velocities as the three angular velocities ω_1, ω_2, and ω_3 of the body and three components of the velocity of the center of mass along body-fixed axes, denoted by v_1, v_2, and v_3. Hence,

$$u_k = \omega_k \qquad u_{3+k} = v_k \qquad k = 1, 2, 3 \qquad \textbf{[a]}$$

To obtain the rates of change of the quasi-velocities, we note that

$$\{v_G\} = [v_1 \quad v_2 \quad v_3]^T = [u_4 \quad u_5 \quad u_6]^T \qquad \{\omega\} = [\omega_1 \quad \omega_2 \quad \omega_3]^T = [u_1 \quad u_2 \quad u_3]^T$$

$$\textbf{[b]}$$

and that

$$\{a_G\} = \frac{d}{dt}\{v_G\}_{\mathrm{rel}} + [\tilde{\omega}]\{v_G\} = \{a_G\}_{\mathrm{rel}} + [\tilde{\omega}]\{v_G\} = [\dot{v}_1 \quad \dot{v}_2 \quad \dot{v}_3]^T + [\tilde{\omega}]\{v_G\} \qquad \textbf{[c]}$$

$$\{\alpha\} = \frac{d}{dt}\{\omega\} = [\alpha_1 \quad \alpha_2 \quad \alpha_3]^T = [\dot{u}_1 \quad \dot{u}_2 \quad \dot{u}_3]^T \qquad \textbf{[d]}$$

The partial velocities of the center of mass and the partial angular velocities have the form

$$\{v_G^k\} = \{0\} \qquad k = 1, 2, 3$$

$$\{v_G^4\} = [1 \ 0 \ 0]^T \qquad \{v_G^5\} = [0 \ 1 \ 0]^T \qquad \{v_G^6\} = [0 \ 0 \ 1]^T$$

$$\{\omega^1\} = [1 \ 0 \ 0]^T \qquad \{\omega^2\} = [0 \ 1 \ 0]^T \qquad \{\omega^3\} = [0 \ 0 \ 1]^T$$

$$\{\omega^k\} = \{0\} \qquad k = 4, 5, 6 \qquad \textbf{[e]}$$

We write the Gibbs-Appell function in column vector format. For a set of body-fixed coordinates,

$$S = \frac{1}{2}m\{a_G\}^T\{a_G\} + \frac{1}{2}\{\alpha\}^T[I_G]\{\alpha\} + \{\alpha\}^T[\tilde{\omega}][I_G]\{\omega\} \qquad \textbf{[f]}$$

Consider a resultant force vector $\{F\} = [F_1 \ F_2 \ F_3]^T$ and a resultant moment vector $\{M_G\} = [M_1 \ M_2 \ M_3]^T$ about the center of mass. We calculate the generalized forces, using Eq. [9.5.26], as

$$U_k = \{v_G^k\}^T\{F\} + \{\omega^k\}^T\{M_G\} \qquad k = 1, 2, \ldots, 6 \qquad \textbf{[g]}$$

Evaluating the expressions, we obtain

$$U_k = M_k \qquad U_{3+k} = F_k \qquad k = 1, 2, 3 \qquad \textbf{[h]}$$

Consider the rotational equations of motion first. They correspond to the first three Gibbs-Appell equations associated with u_1, u_2, and u_3. Introduce the notation

$$\{u'\} = [u_1 \ u_2 \ u_3]^T = \{\omega\} \qquad \{u''\} = [u_4 \ u_5 \ u_6]^T = [v_1 \ v_2 \ v_3]^T = \{v_G\} \qquad \textbf{[i]}$$

It follows that

$$\{\dot{u}'\} = [\dot{u}_1 \ \dot{u}_2 \ \dot{u}_3]^T = \{\alpha\} \qquad \{\dot{u}''\} = [\dot{u}_4 \ \dot{u}_5 \ \dot{u}_6]^T = [\dot{v}_1 \ \dot{v}_2 \ \dot{v}_3]^T \qquad \textbf{[j]}$$

To obtain the left side of the Gibbs-Appell equations, we differentiate S with respect to the derivatives of the first three generalized speeds. From Eq. [d] this corresponds to differentiating S with respect to $\{\alpha\}$. Doing so, we obtain

$$\left(\frac{\partial S}{\partial \{\dot{u}'\}}\right)^T = \left(\frac{\partial S}{\partial \{\alpha\}}\right)^T = [I_G]\{\alpha\} + [\tilde{\omega}][I_G]\{\omega\} \qquad \textbf{[k]}$$

From Eq. [h], the right side of the Gibbs-Appell equations associated with the first three and last three generalized speeds have the form

$$[U_1 \ U_2 \ U_3]^T = [M_1 \ M_2 \ M_3]^T = \{M\}$$

$$[U_4 \ U_5 \ U_6]^T = [F_1 \ F_2 \ F_3]^T = \{F\} \qquad \textbf{[l]}$$

Combining Eq. [k] and the first of Eqs. [l] yields the rotational equations of motion as

$$[I_G]\{\alpha\} + [\tilde{\omega}][I_G]\{\omega\} = \{M\} \qquad \textbf{[m]}$$

We next consider the translational equations. Differentiating S with respect to the generalized speeds associated with translation, we obtain

$$\frac{\partial S}{\partial \{\dot{u}''\}} = m\{a_G\}^T \frac{\partial \{a_G\}}{\partial \{\dot{u}''\}} = m\{a_G\}^T \qquad \textbf{[n]}$$

because $\frac{\partial\{a_G\}}{\partial\{\ddot{u}''\}}$ is equal to the identity matrix. Combining Eq. [n] and the second of Eqs. [l] yields

$$m\{a_G\} = m\{a_G\}_{\text{rel}} + m[\tilde{\omega}]\{v_G\} = \{F\} \qquad \text{[o]}$$

which are the translational motion equations.

Consider the vehicle in Example 4.14 and derive the equations of motion using the Gibbs-Appell equations. It is given that the velocity of point A is always along the x-axis.

Example 9.10

Solution

The vehicle configuration is shown in Fig. 4.8. The orientation of the vehicle can be described by three generalized coordinates, X, Y, and θ, with X and Y denoting the coordinates of the center of mass. (Another reasonable choice for the generalized coordinates would be the coordinates of point A: X_A, Y_A, and θ.) The nonholonomic constraint indicates that the velocity of point A is always in the direction of the vehicle, or

$$\mathbf{v}_A \cdot \mathbf{j} = 0 \qquad \text{[a]}$$

We select the generalized speeds as $u_1 = v_A$ and $u_2 = \dot{\theta}$, so that $\mathbf{v}_A = u_1\mathbf{i}$, $\boldsymbol{\omega} = u_2\mathbf{k}$. In order to write the generalized forces, we need to calculate the partial velocities associated with points of application of the forces, C and D. To derive the fundamental equations without calculating S explicitly, we also need the partial velocities of G. The velocities of G, C, and D are

$$\mathbf{v}_G = \mathbf{v}_A + \boldsymbol{\omega} \times \mathbf{r}_{G/A} = u_1\mathbf{i} + u_2\mathbf{k} \times L\mathbf{i} = u_1\mathbf{i} + Lu_2\mathbf{j}$$

$$\mathbf{v}_C = \mathbf{v}_A + \boldsymbol{\omega} \times \mathbf{r}_{C/A} = u_1\mathbf{i} + u_2\mathbf{k} \times h\mathbf{j} = (u_1 - hu_2)\mathbf{i}$$

$$\mathbf{v}_D = \mathbf{v}_A + \boldsymbol{\omega} \times \mathbf{r}_{D/A} = u_1\mathbf{i} + u_2\mathbf{k} \times -h\mathbf{j} = (u_1 + hu_2)\mathbf{i} \qquad \text{[b]}$$

The partial velocities are

$$\mathbf{v}_G^1 = \mathbf{i} \qquad \mathbf{v}_G^2 = L\mathbf{j} \qquad \mathbf{v}_C^1 = \mathbf{i} \qquad \mathbf{v}_C^2 = -h\mathbf{i}$$

$$\mathbf{v}_D^1 = \mathbf{i} \qquad \mathbf{v}_D^2 = h\mathbf{i} \qquad \boldsymbol{\omega}^1 = 0 \qquad \boldsymbol{\omega}^2 = \mathbf{k} \qquad \text{[c]}$$

The acceleration of the center of mass is

$$\mathbf{a}_G = \dot{u}_1\mathbf{i} + L\dot{u}_2\mathbf{j} + u_1\boldsymbol{\omega} \times \mathbf{i} + Lu_2\boldsymbol{\omega} \times \mathbf{j} = (\dot{u}_1 - Lu_2^2)\mathbf{i} + (L\dot{u}_2 + u_1u_2)\mathbf{j} \qquad \text{[d]}$$

The angular momentum is $\mathbf{H}_G = I_G\dot{u}_2\mathbf{k}$, in which I_G is the mass moment of inertia about the center of mass. The rate of change of angular momentum is simply $\dot{\mathbf{H}}_G = I_G\dot{u}_2\mathbf{k}$. Because this is a plane motion problem, the angular velocity and angular momentum are in the same direction; thus, the Gibbs function reduces to

$$S = \frac{1}{2}m\mathbf{a}_G \cdot \mathbf{a}_G + \frac{1}{2}\boldsymbol{\alpha} \cdot \dot{\mathbf{H}}_G = \frac{1}{2}\left(m(\dot{u}_1 - Lu_2^2)^2 + m(L\dot{u}_2 + u_1u_2)^2 + I_G\dot{u}_2^2\right) \qquad \text{[e]}$$

Taking partial derivatives of S we obtain

$$\frac{\partial S}{\partial \dot{u}_1} = \mathbf{v}_G^1 \cdot m\mathbf{a}_G + \boldsymbol{\omega}^1 \cdot \dot{\mathbf{H}}_G = m(\dot{u}_1 - Lu_2^2)$$

$$\frac{\partial S}{\partial \dot{u}_2} = \mathbf{v}_G^2 \cdot m\mathbf{a}_G + \boldsymbol{\omega}^2 \cdot \dot{\mathbf{H}}_G = mL(L\dot{u}_2 + u_1u_2) + I_G\dot{u}_2 = (I_G + mL^2)\dot{u}_2 + mLu_1u_2 \qquad \text{[f]}$$

The external forces are $\mathbf{F}_C = F_C\mathbf{i}$, $\mathbf{F}_D = F_D\mathbf{i}$, so that the right sides of the fundamental equations become

$$U_1 = \mathbf{v}_C^1 \cdot \mathbf{F}_C + \mathbf{v}_D^1 \cdot \mathbf{F}_D = F_C + F_D$$

$$U_2 = \mathbf{v}_C^2 \cdot \mathbf{F}_C + \mathbf{v}_D^2 \cdot \mathbf{F}_D = h(F_D - F_C) \qquad [\mathbf{g}]$$

Equating Eqs. [f] and [g] gives the equations of motion as

$$m(\dot{u}_1 - Lu_2^2) = F_C + F_D \qquad [\mathbf{h}]$$

$$(I_G + mL^2)\dot{u}_2 + mLu_1u_2 = h(F_D - F_C) \qquad [\mathbf{i}]$$

One can assign a physical interpretation to the equations of motion. Equation [h] is the force balance in the x direction. Equation [i] represents the moment balance about A.

Let us compare the effort to obtain the equations of motion with Example 4.14, where Lagrange's equations are used and with Example 5.13, where Jourdain's variational principle is used. In both these examples we begin with a set of constrained coordinates and impose the constraint. By contrast, when using the Gibbs-Appell equations, we begin with a set of independent variables. It is clear that we obtained the equations of motion more easily using the Gibbs-Appell equations.

The kinematic differential equations are obtained by expressing the rates of change of the generalized coordinates X, Y, and θ in terms of the generalized speeds. To this end, we write the velocity of the center of mass as

$$\mathbf{v}_G = \dot{X}\mathbf{I} + \dot{Y}\mathbf{J} = (\dot{X}\cos\theta + \dot{Y}\sin\theta)\mathbf{i} + (-\dot{X}\sin\theta + \dot{Y}\cos\theta)\mathbf{j} \qquad [\mathbf{j}]$$

and equating this equation to the expression of $\mathbf{v}_G$ in terms of the generalized speeds in Eq. [b] we obtain

$$\dot{X}\cos\theta + \dot{Y}\sin\theta = u_1 \qquad -\dot{X}\sin\theta + \dot{Y}\cos\theta = Lu_2 \qquad [\mathbf{k}]$$

Solving these equations for $\dot{X}$ and $\dot{Y}$, we reach

$$\dot{X} = u_1\cos\theta - Lu_2\sin\theta \qquad [\mathbf{l}]$$

$$\dot{Y} = u_1\sin\theta + Lu_2\cos\theta \qquad [\mathbf{m}]$$

The kinematic differential equation for θ is simply

$$\dot{\theta} = u_2 \qquad [\mathbf{n}]$$

The five equations [h], [i], [l], [m], and [n] constitute the complete set of differential equations that describe the motion of this system.

On the selection of the quasi-velocities for Example 9.10: We selected u_1 as the velocity of A in view of the constraint. Had we selected one of the quasi-velocities as $\dot{X}$ or $\dot{Y}$, the resulting equations of motion would be quite complicated. Also, components of the velocity along a body-fixed frame provide physical insight.

9.8 KANE'S EQUATIONS

Kane's equations can be derived in a variety of ways. One can take a system of particles and use the equation of motion of each particle, or one can begin with

D'Alembert's principle and use the transformation from the generalized velocities to the quasi-velocities. In order to establish the equivalence between the Gibbs-Appell and Kane's equations, we begin with the first approach.

Consider a system of N particles that has n degrees of freedom. The equation of motion for each particle can be written as Eq. [9.7.1]. We select an appropriate set of independent generalized coordinates and quasi-velocities. Associated with each particle and each quasi-velocity there is a partial velocity vector $\mathbf{v}_i^k$ ($k = 1, 2, \ldots, n; i = 1, 2, \ldots, N$). To obtain Kane's equations, one takes the dot product of each of the N equations of motion with the partial velocities and adds the resulting equations, which yields

$$\sum_{i=1}^{N} m_i \mathbf{a}_i \cdot \mathbf{v}_i^k = \sum_{i=1}^{N} \mathbf{F}_i \cdot \mathbf{v}_i^k \qquad k = 1, 2, \ldots, n \qquad [9.8.1]$$

The right side of this equation is recognized as the generalized force U_k. We define by $\mathbf{F}_i^* = -m_i \mathbf{a}_i$ the *inertia force* associated with the ith mass, and the sum of the inertia forces multiplied by the partial velocities as U_k^*, the *generalized inertia force*, is defined as[6]

$$U_k^* = -\sum_{i=1}^{N} \mathbf{v}_i^k \cdot \mathbf{F}_i^* = \sum_{i=1}^{N} \mathbf{v}_i^k \cdot m_i \mathbf{a}_i \qquad [9.8.2]$$

Introducing this expression into Eq. [9.8.1] yields

$$U_k^* = U_k \qquad k = 1, 2, \ldots, n \qquad [9.8.3]$$

which is known as *Kane's equations*. To establish the equivalence between the Kane's equations and the Gibbs-Appell equations, we focus on Eqs. [9.7.7] and [9.8.2], from which

$$\sum_{i=1}^{N} m_i \mathbf{a}_i \cdot \mathbf{v}_i^k = U_k^* = \frac{\partial S}{\partial \dot{u}_k} \qquad [9.8.4]$$

It is clear that the Gibbs-Appell and Kane's equations are identical.

For rigid bodies, the generalized inertia forces can be found in a way similar to finding the generalized forces. Indeed, we define a resultant inertia force $\mathbf{F}^* = -m\mathbf{a}_G$, acting through the center of mass, and a resultant *inertia torque* $\mathbf{M}^* = -d\mathbf{H}_G/dt$, similar to the negative of the gyroscopic moment. The generalized inertia force has the form

$$U_k^* = -\mathbf{v}_G^k \cdot \mathbf{F}^* - \boldsymbol{\omega}^k \cdot \mathbf{M}^* = \mathbf{v}_G^k \cdot m\mathbf{a}_G + \boldsymbol{\omega} \cdot \dot{\mathbf{H}}_G \qquad [9.8.5]$$

Compare this equation with Eq. [9.7.19]. For interconnected bodies, one uses the resultant acceleration and change in angular momentum for each body and sums, as we did for particles, to obtain Eq. [9.7.21].

Even though we performed near-identical manipulations to obtain the Gibbs-Appell and Kane's equations, the two are inspired from very different viewpoints.

---

[6]This definition is different than Kane's definition of generalized inertia force.

The motivation behind the development of the Gibbs-Appell equations was to develop a fundamental function, in a sense similar to the extended Hamilton's principle but one that can handle nonholonomic constraints. These equations can be derived without using partial velocities. On the other hand, Kane's equations are based on Kane's initial work with D'Alembert's principle, and on Lagrange's form of D'Alembert's principle.

As discussed before, Kane's equations can be derived from D'Alembert's principle is as follows: We invoke D'Alembert's principle for rigid bodies, derived in Section 8.11. For a system of N bodies, the principle is written in terms of the generalized velocities as

$$\sum_{i=1}^{N} \left(m_i \mathbf{a}_{G_i} \cdot \frac{\partial \mathbf{v}_{G_i}}{\partial \dot{q}_k} + \dot{\mathbf{H}}_{G_i} \cdot \frac{\partial \boldsymbol{\omega}_i}{\partial \dot{q}_k} \right) = Q_k \qquad k = 1, 2, \ldots, n \qquad [9.8.6]$$

in which the generalized forces are given by

$$Q_k = \sum_{i=1}^{N} \left(\mathbf{F}_i \cdot \frac{\partial \mathbf{v}_{G_i}}{\partial \dot{q}_k} + \mathbf{M}_{G_i} \cdot \frac{\partial \boldsymbol{\omega}_i}{\partial \dot{q}_k} \right) \qquad [9.8.7]$$

We introduce to these two equations the relationships between the generalized velocities and the generalized speeds. Making use of column vector notation and the partial velocity matrix, we can express the partial derivatives of $\mathbf{v}_{G_i}$ and $\mathbf{w}_i$ with respect to the generalized velocities as

$$\frac{\partial \{v_{G_i}\}}{\partial \{\dot{q}\}} = \frac{\partial \{v_{G_i}\}}{\partial \{u\}} \frac{\partial \{u\}}{\partial \{\dot{q}\}} = \frac{\partial \{v_{G_i}\}}{\partial \{u\}} [W]^{-1} = [v_{G_i}^q][W]^{-1}$$

$$\frac{\partial \{\omega_i\}}{\partial \{\dot{q}\}} = \frac{\partial \{\omega_i\}}{\partial \{u\}} \frac{\partial \{u\}}{\partial \{\dot{q}\}} = \frac{\partial \{\omega_i\}}{\partial \{u\}} [W]^{-1} = [\omega_i^q][W]^{-1} \qquad [9.8.8]$$

so that Eq. [9.8.6] can be expressed as

$$\sum_{i=1}^{N} \left(m_i \{a_{G_i}\}^T [v_{G_i}^q][W]^{-1} + \{\dot{H}_{G_i}\}^T [v_{G_i}^q][W]^{-1} \right) = \{Q\}^T \qquad [9.8.9]$$

We right-multiply both sides of this equation by $[W]$, for

$$\sum_{i=1}^{N} \left(m_i \{a_{G_i}\}^T [v_{G_i}^q] + \{\dot{H}_{G_i}\}^T [v_{G_i}^q] \right) = \{Q\}^T [W] = \{U\}^T \qquad [9.8.10]$$

The right side of this equation is recognized as the generalized force vector associated with generalized speeds. The left side is recognized as the column vector representation of the generalized inertia forces. Hence, Kane's equations can be derived directly from D'Alembert's principle for rigid bodies, by a simple change of variables from the generalized velocities to the generalized speeds.

Derivation of Kane's equations can also be accomplished without the use of any variational principle, such as D'Alembert's or Hamilton's or of variational calculus. Nor do they require the development of scalar functions such as energy and S. In this regard, they are different than the others that we have studied.

Another advantage of Kane's equations is that the approach, especially the use of partial velocities, is desirable for interconnected and large-order systems. This is because the procedure of obtaining the partial velocities and the generalized forces can be mechanized and made suitable for computer implementation.

Rather than adding to the controversy on differences and similarities between the Gibbs-Appell and Kane's equations, we will refer to the equations of motion derived in this and the previous section as the *fundamental equations of motion*. We will differentiate the forms by referring to the *Kane's form of the fundamental equations of motion* and the *Gibbs-Appell form of the fundamental equations of motion*.

Another interesting interpretation of the fundamental equations is as follows. Consider that we have one body, whose translational and rotational equations of motion are

$$m\mathbf{a}_G - \mathbf{F} = 0 \qquad \dot{\mathbf{H}}_G - \mathbf{M}_G = 0 \qquad [9.8.11]$$

The fundamental equations are obtained by taking the dot product of the first of Eq. [9.8.11] with $\mathbf{v}_G^k$ and of the second Eq. [9.8.11] with $\boldsymbol{\omega}^k$ and summing the two expressions. This results in the equation of motion for the kth generalized speed. In essence, the kth fundamental equation is simply the sum of the components of the force and moment balance along the directions of the partial velocities. This interpretation should be compared with the interpretation of Lagrange's equations in terms of the Euler angles.

Example 9.11

Obtain the equations of motion of a disk rolling without slipping using the fundamental equations.

Solution

We discussed this problem in Examples 9.3 and 9.5. We begin with a set of independent generalized speeds $u_k = \omega_k$ ($k = 1, 2, 3$) and use the F frame. First, we calculate the acceleration of the center of mass. The velocity of the center of mass is given by

$$\mathbf{v}_G = -Ru_3\mathbf{f}_1 + Ru_1\mathbf{f}_3 \qquad [\mathbf{a}]$$

so that the acceleration becomes

$$\mathbf{a}_G = \frac{d}{dt}(\mathbf{v}_G)_{\text{rel}} + \boldsymbol{\omega}_f \times \mathbf{v}_G$$

$$= (-R\dot{u}_3 + Ru_1u_2)\mathbf{f}_1 - \left(Ru_1^2 + \frac{Ru_2u_3}{\tan\theta}\right)\mathbf{f}_2 + (R\dot{u}_1 + Ru_2u_3)\mathbf{f}_3 \qquad [\mathbf{b}]$$

At this point, we can either invoke the Gibbs-Appell or the Kane's forms of the fundamental equations. Let us use the Gibbs-Appell form and write the Gibbs function

$$S = \frac{1}{2}m\{a_G\}^T\{a_G\} + \frac{1}{2}\{\alpha\}^T[I_G]\{\alpha\} + \{\alpha\}^T[\tilde{\omega}][I_G]\{\omega\} \qquad [\mathbf{c}]$$

We will take partial derivatives of each term with respect to the quasi-velocities. We note that

$$\frac{\partial \mathbf{a}_G}{\partial \dot{u}_1} = \mathbf{v}_G^1 = R\mathbf{f}_3 \qquad \frac{\partial \mathbf{a}_G}{\partial \dot{u}_2} = \mathbf{v}_G^2 = 0 \qquad \frac{\partial \mathbf{a}_G}{\partial \dot{u}_3} = \mathbf{v}_G^3 = -R\mathbf{f}_1 \qquad [\mathbf{d}]$$

For the first term of S

$$\frac{\partial\left(\frac{1}{2}m\{a_G\}^T\{a_G\}\right)}{\partial \dot{u}_k} = m\{a_G\}^T\{v_G^k\} \qquad k = 1, 2, 3 \qquad \text{[e]}$$

and carrying out the algebra we obtain

$$\frac{\partial\left(\frac{1}{2}m\{a_G\}\{a_G\}\right)}{\partial\{\dot{u}\}} = mR^2 \begin{bmatrix} \dot{u}_1 + u_2 u_3 \\ 0 \\ \dot{u}_3 - u_1 u_2 \end{bmatrix}^T \qquad \text{[f]}$$

To differentiate the second and third terms of S, we note that $\partial\{\alpha\}/\partial\{\dot{u}\} = [1]$. The derivatives of the second and third terms in S yield Euler's equations expressed in the F frame, Eqs. [8.5.31]. These equations are the same as the modified Euler's equations, as given in the left side of Table 9.1. Adding these equations to Eq. [f], and noting that the generalized forces are given by Eq. [l] of Example 9.8, we arrive at the equations of motion as

$$(I_1 + mR^2)\dot{u}_1 - \frac{I_1 u_2^2}{\tan\theta} + (I_3 + mR^2)u_2 u_3 = -mgR\cos\theta$$

$$I_1 \dot{u}_2 + \frac{I_1 u_1 u_2}{\tan\theta} - I_3 u_1 u_3 = 0$$

$$(I_3 + mR^2)\dot{u}_3 - mR^2 u_1 u_2 = 0 \qquad \text{[g]}$$

Use of the Gibbs-Appell form of the fundamental equations led to a physical interpretation. The equations of motion are basically the modified Euler's equations with an additional moment term. This additional expression, given in Eq. [f], is in essence the contribution of the moment generated about the center of mass by the forces acting on the contact point.

We have now solved the problem of the rolling disk four times. We first studied this problem in Chapter 8, using Lagrange's equations with constraints. Next we used the modified Euler's equations, where we discussed both a force and moment balance. Then we used the Boltzmann-Hamel equations and summed moments about the contact point. Finally, in Example 9.11, we used the fundamental equations. The reader is urged to compare all these methods of solution.

Example 9.12

Obtain the equations of motion for the spinning top on a cart using Kane's form of the fundamental equations.

Solution

We saw this problem in Examples 8.12 and 9.4. The forces and moments that do work on the combined system are

$$\mathbf{F}_1 = -mg\mathbf{a}_3 \quad \text{applied at } G \qquad \mathbf{F}_2 = F\mathbf{a}_2 \quad \text{applied at } C$$

$$\mathbf{M} = M\mathbf{f}_3 \quad \text{at } C \qquad \text{[a]}$$

Note that the top exerts an equal and opposite moment on the cart. The other forces that the cart and top exert on each other do not enter the formulation, as they are internal to the system.

We need to calculate the following partial velocities and partial angular velocities:

$\mathbf{v}_G^k$, to evaluate U_k^* and to find the component of U_k due to $\mathbf{F}_1$

$\boldsymbol{\omega}^k$, to evaluate U_k^* and to find the component of U_k due to $\mathbf{M}$

$\mathbf{v}_C^k$, to evaluate U_k^* and to find the component of U_k due to $\mathbf{F}_2$ [b]

All these partial velocities were calculated in Example 9.6. We summarize the results as

Generalized speeds: $\quad u_1 = \omega_1 \quad u_2 = \omega_2 \quad u_3 = \omega_3 \quad u_4 = \dot{Y}$ [c]

Angular velocity: $\quad \boldsymbol{\omega} = \omega_1 \mathbf{f}_1 + \omega_2 \mathbf{f}_2 + \omega_3 \mathbf{f}_3 = u_1 \mathbf{f}_1 + u_2 \mathbf{f}_2 + u_3 \mathbf{f}_3$ [d]

Partial angular
velocities: $\quad \boldsymbol{\omega}^1 = \mathbf{f}_1 \quad \boldsymbol{\omega}^2 = \mathbf{f}_2 \quad \boldsymbol{\omega}^3 = \mathbf{f}_3 \quad \boldsymbol{\omega}^4 = 0$ [e]

Partial velocities for C: $\quad \mathbf{v}_C^1 = \mathbf{v}_C^2 = \mathbf{v}_C^3 = 0$

$$\mathbf{v}_C^4 = \mathbf{a}_2 = (\sin\phi \mathbf{f}_1 + \cos\phi\cos\theta \mathbf{f}_2 - \cos\phi\sin\theta \mathbf{f}_3)$$ [f]

Velocity of G: $\quad \mathbf{v}_G = -L\mathbf{f}_2 u_1 + L\mathbf{f}_1 u_2$
$\quad\quad\quad\quad\quad\quad + (\sin\phi \mathbf{f}_1 + \cos\phi\cos\theta \mathbf{f}_2 - \cos\phi\sin\theta \mathbf{f}_3)u_4$ [g]

Partial velocities for G: $\quad \mathbf{v}_G^1 = -L\mathbf{f}_2 \quad \mathbf{v}_G^2 = L\mathbf{f}_1 \quad \mathbf{v}_G^3 = 0$

$$\mathbf{v}_G^4 = \mathbf{a}_2 = (\sin\phi \mathbf{f}_1 + \cos\phi\cos\theta \mathbf{f}_2 - \cos\phi\sin\theta \mathbf{f}_3)$$ [h]

Generalized forces: $\quad U_1 = mgL\sin\theta \quad U_2 = 0 \quad U_3 = M \quad U_4 = F$ [i]

To derive the equations of motion, we need to obtain the accelerations of G and C, and the rate of change of angular momentum. The acceleration of C is simply

$$\mathbf{a}_C = \dot{u}_4 \mathbf{a}_2$$ [j]

We find the acceleration of G using the relation

$$\mathbf{a}_G = \mathbf{a}_C + \boldsymbol{\alpha} \times \mathbf{r}_{G/C} + \boldsymbol{\omega} \times (\boldsymbol{\omega} \times \mathbf{r}_{G/C})$$ [k]

in which $\mathbf{r}_{G/C} = L\mathbf{f}_3$ and

$$\boldsymbol{\alpha} = \dot{\boldsymbol{\omega}}_{\text{rel}} + \boldsymbol{\omega}_f \times \boldsymbol{\omega}_s = (\dot{\omega}_1 \mathbf{f}_1 + \dot{\omega}_2 \mathbf{f}_2 + \dot{\omega}_3 \mathbf{f}_3) + \left(\omega_1 \mathbf{f}_1 + \omega_2 \mathbf{f}_2 + \frac{\omega_2}{\tan\theta}\mathbf{f}_3\right) \times \left(\omega_3 - \frac{\omega_2}{\tan\theta}\right)\mathbf{f}_3$$

$$= \left(\dot{u}_1 - \frac{u_2^2}{\tan\theta} + u_2 u_3\right)\mathbf{f}_1 + \left(\dot{u}_2 + \frac{u_1 u_2}{\tan\theta} - u_1 u_3\right)\mathbf{f}_2 + \dot{u}_3 \mathbf{f}_3$$ [l]

As mentioned before, this differentiation is necessary because the angular velocity is being expressed in the F frame and not a body frame. Carrying out the necessary steps, the acceleration of point G becomes

$$\mathbf{a}_G = \left(\dot{u}_4 \sin\phi + L\dot{u}_2 + \frac{L u_1 u_2}{\tan\theta}\right)\mathbf{f}_1 + \left(\dot{u}_4 \cos\phi\cos\theta - L\dot{u}_1 + \frac{L u_2^2}{\tan\theta}\right)\mathbf{f}_2$$
$$+ (-\dot{u}_4 \cos\phi\sin\theta - L u_1^2 - L u_2^2)\mathbf{f}_3$$ [m]

As expected, all $u_3 = \omega_3$ terms drop out of the acceleration expression. Taking the dot products between the accelerations and the associated partial velocities yields

$$\mathbf{v}_C^1 \cdot \mathbf{a}_C = 0 \qquad \mathbf{v}_C^2 \cdot \mathbf{a}_C = 0 \qquad \mathbf{v}_C^3 \cdot \mathbf{a}_C = 0 \qquad \mathbf{v}_C^4 \cdot \mathbf{a}_C = \dot{u}_4 \qquad [\mathbf{n}]$$

$$\mathbf{v}_G^1 \cdot \mathbf{a}_G = -L\mathbf{f}_2 \cdot \mathbf{a}_G = L^2 \dot{u}_1 - L\dot{u}_4 \cos\phi \cos\theta - \frac{L^2 u_2^2}{\tan\theta}$$

$$\mathbf{v}_G^2 \cdot \mathbf{a}_G = L\mathbf{f}_1 \cdot \mathbf{a}_G = L^2 \dot{u}_2 + L\dot{u}_4 \sin\phi + \frac{L^2 u_1 u_2}{\tan\theta}$$

$$\mathbf{v}_G^3 \cdot \mathbf{a}_G = 0$$

$$\mathbf{v}_G^4 \cdot \mathbf{a}_G = (\sin\phi \mathbf{f}_1 + \cos\phi\cos\theta \mathbf{f}_2 - \cos\phi\sin\theta \mathbf{f}_3) \cdot \mathbf{a}_G$$

$$= \dot{u}_4 + L\dot{u}_2 \sin\phi + \frac{L u_1 u_2 \sin\phi}{\tan\theta}$$

$$- L\dot{u}_1 \cos\phi\cos\theta + L u_1^2 \cos\phi\sin\theta + \frac{L u_2^2 \cos\phi}{\sin\theta} \qquad [\mathbf{o}]$$

We next find the rate of change of the angular momentum. We derived a general expression for this earlier in this chapter, when studying the modified Euler's equations. Using Table 9.1, for a 3-1-3 transformation the rate of change of angular momentum of the top is

$$\dot{\mathbf{H}}_G = [I_1 \dot{\omega}_1 + \omega_2 \left(I_3 \omega_3 - \frac{I_1 \omega_2}{\tan\theta}\right)]\mathbf{f}_1 + \left[I_2 \dot{\omega}_2 - \omega_1 \left(I_3 \omega_3 - \frac{I_1 \omega_2}{\tan\theta}\right)\right]\mathbf{f}_2 + I_3 \dot{\omega}_3 \mathbf{f}_3 \qquad [\mathbf{p}]$$

The contribution from angular momentum of the cart is zero, as the cart has zero angular velocity. Taking the dot product between $\dot{\mathbf{H}}_G$ and the partial angular velocities from Eqs. [e], we obtain

$$\boldsymbol{\omega}^1 \cdot \dot{\mathbf{H}}_G = \mathbf{f}_1 \cdot \dot{\mathbf{H}}_G = I_1 \dot{u}_1 + u_2 \left(I_3 u_3 - \frac{I_1 u_2}{\tan\theta}\right)$$

$$\boldsymbol{\omega}^2 \cdot \dot{\mathbf{H}}_G = \mathbf{f}_2 \cdot \dot{\mathbf{H}}_G = I_1 \dot{u}_2 - u_1 \left(I_3 u_3 - \frac{I_1 u_2}{\tan\theta}\right)$$

$$\boldsymbol{\omega}^3 \cdot \dot{\mathbf{H}}_G = \mathbf{f}_3 \cdot \dot{\mathbf{H}}_G = I_3 \dot{u}_3 \qquad \boldsymbol{\omega}^4 \cdot \dot{\mathbf{H}}_G = 0 \qquad [\mathbf{q}]$$

The generalized inertia forces U_k^* are calculated as

$$U_k^* = \mathbf{v}_C^k \cdot m_2 \mathbf{a}_C + \mathbf{v}_G^k \cdot m_1 \mathbf{a}_G + \boldsymbol{\omega}^k \cdot \dot{\mathbf{H}}_G \qquad k = 1, 2, 3, 4 \qquad [\mathbf{r}]$$

Combining Eqs. [n], [o] and [q] with the generalized forces in Eq. [i], we obtain the equations of motion as

$$U_1^* = U_1 \rightarrow 0 + m_1 \left(L^2 \dot{u}_1 - L\dot{u}_4 c\phi c\theta - \frac{L^2 u_2^2}{t\theta}\right) + I_1 \dot{u}_1 + u_2 \left(I_3 u_3 - \frac{I_1 u_2}{t\theta}\right) = mgL s\theta$$

$$\rightarrow (I_1 + m_1 L^2) \dot{u}_1 - m_1 L \dot{u}_4 c\phi c\theta - (I_1 + m_1 L^2) \frac{u_2^2}{t\theta} + I_3 u_2 u_3 - mgL s\theta = 0 \qquad [\mathbf{s}]$$

$$U_2^* = U_2 \rightarrow 0 + m_1 \left[L^2 \dot{u}_2 + L\dot{u}_4 s\phi + \frac{L^2 u_1 u_2}{t\theta}\right] + I_1 \dot{u}_2 - u_1 \left(I_3 u_3 - \frac{I_1 u_2}{t\theta}\right) = 0$$

$$\rightarrow (I_1 + m_1 L^2)\left(\dot{u}_2 + \frac{u_1 u_2}{t\theta}\right) + m_1 L \dot{u}_4 s\phi - I_3 u_1 u_3 = 0 \qquad [\mathbf{t}]$$

$$U_3^* = U_3 \rightarrow 0 + 0 + I_3 \dot{u}_3 = M \rightarrow I_3 \dot{u}_3 = M \qquad [\mathsf{u}]$$

$$U_4^* = U_4 \rightarrow m_2 \dot{u}_4$$
$$+ m_1 \left(\dot{u}_4 + L\dot{u}_2 \, \mathsf{s}\phi + \frac{L u_1 u_2 \, \mathsf{s}\phi}{\mathsf{t}\theta} - L\dot{u}_1 \, \mathsf{c}\phi \mathsf{c}\theta + L u_1^2 \, \mathsf{c}\phi \mathsf{s}\theta + \frac{L u_2^2 \, \mathsf{c}\phi}{\mathsf{s}\theta} \right) + 0 = F$$
$$\rightarrow (m_1 + m_2) \dot{u}_4 - m_1 L \dot{u}_1 \, \mathsf{c}\phi \mathsf{c}\theta + m_1 L \dot{u}_2 \, \mathsf{s}\phi$$
$$+ \frac{m_1 L u_1 u_2 \, \mathsf{s}\phi}{\mathsf{t}\theta} + m_1 L u_1^2 \, \mathsf{c}\phi \mathsf{s}\theta + \frac{m_1 L u_2^2 \, \mathsf{c}\phi}{\mathsf{s}\theta} = F \qquad [\mathsf{v}]$$

9.9 THE FUNDAMENTAL EQUATIONS AND CONSTRAINTS

As we saw in the previous sections, the fundamental equations have the advantage that by a judicious choice of the generalized speeds one can account for nonholonomic constraints in the formulation and obtain equations of motion in terms of independent generalized speeds. In certain cases, however, it is desirable or necessary to leave the problem formulation in terms of constrained coordinates.

Consider a system with n degrees of freedom described in terms of n generalized coordinates q_k ($k = 1, 2, \ldots, n$) and n generalized speeds u_k ($k = 1, 2, \ldots, n$). We now apply m constraints to the system of the form

$$[A]\{u\} + \{B\} = \{0\} \qquad [9.9.1]$$

where $[A]$ and $\{B\}$ are defined as in Section 9.6. From Eq. [9.5.22], the expression for virtual work in terms of the quasi-coordinates is

$$\delta W = \{Q\}^T [W]\{\delta q'\} = \{U\}^T \{\delta q'\} \qquad [9.9.2]$$

As we did in Lagrangian mechanics, we take the variation of the constraint in Eq. [9.9.1] and left-multiply it by the Lagrange multipliers $\{\lambda\}^T = \{\lambda_1 \quad \lambda_2 \quad \ldots \quad \lambda_m\}$, with the result

$$\{\lambda\}^T [A]\{\delta q'\} = 0 \qquad [9.9.3]$$

We then add the above expression to the virtual work and create an augmented virtual work function

$$\delta \hat{W} = \{U\}^T \{\delta q'\} + \{\lambda\}^T [A]\{\delta q'\} = \{\hat{U}\}^T \{\delta q'\} \qquad [9.9.4]$$

in which

$$\{\hat{U}\} = \{U\} + [A]^T \{\lambda\} \qquad [9.9.5]$$

is a set of generalized forces associated with the constrained generalized coordinates. Individually, the elements of $\{\hat{U}\}$ have the form

$$\hat{U}_k = U_k + \sum_{j=1}^{m} A_{jk} \lambda_j \qquad k = 1, 2, \ldots, n \qquad [9.9.6]$$

This expression for the generalized forces is then used in the Gibbs-Appell equations, with the result

$$\frac{\partial S}{\partial \dot{u}_k} = U_k + \sum_{j=1}^{m} A_{jk}\lambda_j \qquad [9.9.7]$$

or

$$U_k^* = U_k + \sum_{j=1}^{m} A_{jk}\lambda_j \qquad [9.9.8]$$

The n equations [9.9.8], the m constraint equations [9.9.1], and the n kinematic differential equations [9.5.3] are used to solve for the $2n + m$ variables $q_1, q_2, \ldots, q_n$, $u_1, u_2, \ldots, u_n, \lambda_1, \lambda_2, \ldots, \lambda_m$.

9.10 Relationships between the Fundamental Equations and Lagrange's Equations

The relationship between Lagrange's equations and the fundamental equations is given in Eq. [9.7.9], for when the system is unconstrained and the generalized speeds are selected as the generalized velocities. In this case the generalized forces Q_k and U_k ($k = 1, 2, \ldots, n$) coincide. If the applied forces are conservative, one can take advantage of the potential energy to calculate the generalized forces

$$U_k = Q_k = -\frac{\partial V}{\partial q_k} \qquad k = 1, 2, \ldots, n \qquad [9.10.1]$$

Consider now the case when the generalized speeds and generalized velocities are related by Eq. [9.5.3] as

$$\{\dot{q}\} = [W]\{u\} + \{X\} \qquad [9.10.2]$$

We make use of Eq. [9.5.19], which relates the generalized forces by

$$\{U\} = [W]^T\{Q\} \qquad [9.10.3]$$

and separate the generalized forces into those that can be derived from a potential and those that cannot, as

$$\{U\} = \{U_c\} + \{U_{nc}\} \qquad [9.10.4]$$

It follows that

$$\{U_c\} = [W]^T\{Q_c\} = -[W]^T \left(\frac{\partial V}{\partial \{q\}}\right)^T \qquad [9.10.5]$$

or, in scalar form,

$$U_{ck} = -\sum_{s=1}^{n} W_{sk}\frac{\partial V}{\partial q_s} \qquad k = 1, 2, \ldots, n \qquad [9.10.6]$$

We write the Lagrange's equations in matrix form as

$$\frac{d}{dt}\frac{\partial T}{\partial \{\dot{q}\}} - \frac{\partial T}{\partial \{q\}} + \frac{\partial V}{\partial \{q\}} = \{Q_{nc}\}^T \qquad [9.10.7]$$

9.10 RELATIONSHIPS BETWEEN THE FUNDAMENTAL EQUATIONS AND LAGRANGE'S EQUATIONS

in which $\{Q_{nc}\}$ contains the contribution of all external forces not derivable from a potential. Right-multiplying the above equation by $[W]$ and using Eq. [9.10.3], we obtain

$$\left(\frac{d}{dt}\frac{\partial T}{\partial \{\dot{q}\}} - \frac{\partial T}{\partial \{q\}} + \frac{\partial V}{\partial \{q\}}\right)[W] = \{Q_{nc}\}^T[W] = \{U_{nc}\}^T \qquad [9.10.8]$$

which can be expressed as

$$\sum_{s=1}^{n}\left(\frac{d}{dt}\frac{\partial T}{\partial \dot{q}_s} - \frac{\partial T}{\partial q_s} + \frac{\partial V}{\partial q_s}\right)W_{sk} = U_{nck} \qquad [9.10.9]$$

where $\{U_{nc}\}^T = [U_{nc1}\ U_{nc2}\ \ldots\ U_{ncn}]$ is the generalized force vector. If one does not wish to make use of the potential energy formulation, the equation becomes

$$\sum_{s=1}^{n}\left(\frac{d}{dt}\frac{\partial T}{\partial \dot{q}_s} - \frac{\partial T}{\partial q_s}\right)W_{sk} = U_k \qquad [9.10.10]$$

In the presence of constraints, the relationship between the generalized forces is given by Eq. [9.6.22] as

$$\{\check{U}\} = [\check{W}]^T\{Q\} \qquad [9.10.11]$$

Let us right-multiply Eq. [9.10.7] by the $[\check{W}]$ matrix. From Eq. [9.6.12] we have

$$[\check{W}] = [W_P] + [W_S][A'] \qquad [9.10.12]$$

in which $[W_P]$ and $[W_S]$ are the partitions of $[W]$. Hence, we obtain for the equations of motion

$$\sum_{s=1}^{n}\left(\frac{d}{dt}\frac{\partial T}{\partial \dot{q}_s} - \frac{\partial T}{\partial q_s} + \frac{\partial V}{\partial q_s}\right)\left(W_{sk} + \sum_{j=1}^{m}W_{s,p+j}A'_{jk}\right) = \check{U}_{nck} \qquad k = 1, 2, \ldots, p$$

$$[9.10.13]$$

A special case of Eq. [9.10.6] arises when the generalized speeds are selected as the same as the independent generalized velocities, $\check{u}_k = \dot{q}_k$, $k = 1, 2, \ldots, p$. In this case, $\check{U}_{nck}$ coincide with Q_k, and the first $p \times p$ partition of the $[W]$ matrix becomes an identity matrix, or

$$W_{sk} = \delta_{sk} \qquad s, k = 1, 2, \ldots, p \qquad [9.10.14]$$

and, as a result, we write Eq. [9.10.8] as

$$\left(\frac{d}{dt}\frac{\partial T}{\partial \dot{q}_k} - \frac{\partial T}{\partial q_k} + \frac{\partial V}{\partial q_k}\right)\left(1 + \sum_{s=1}^{n}\sum_{j=1}^{m}W_{s,p+j}A'_{jk}\right) = \check{U}_{nck} \qquad [9.10.15]$$

Equation [9.10.15] in essence represents the removal of the Lagrange multiplier from Lagrange's equations. This procedure, while relatively straightforward in derivation, often involves several complicated steps for actual problems and is not commonly carried out. Equations [9.10.10] and [9.10.15] are known as the *Passarello-Huston equations*.

Note that we could have derived Eq. [9.10.15] in Section 4.10, when we discussed Lagrange's equations for constrained systems. Instead, we left it for here, so that it could be done for both Lagrange's as well as the fundamental equations.

We next consider Lagrange's equations in terms of quasi-velocities. By substituting the quasi-velocities into the kinetic energy, we write the kinetic energy as

$$T = \overline{T}(\{q\}, \{u\}) \qquad [9.10.16]$$

where the overbar denotes that a change of variables has occurred. We also recall the relationships

$$\{u\} = [Y]\{\dot{q}\} + \{Z\} \qquad \{\dot{q}\} = [W]\{u\} + \{X\} \qquad [9.10.17a,b]$$

Using Eq. [9.10.17a], we can express the derivative of the Lagrangian with respect to the generalized velocities in terms of the generalized speeds as

$$\frac{\partial T}{\partial \{\dot{q}\}} = \frac{\partial \overline{T}}{\partial \{u\}} \frac{\partial \{u\}}{\partial \{\dot{q}\}} = \frac{\partial \overline{T}}{\partial \{u\}}[Y] \qquad [9.10.18]$$

which, upon differentiation with respect to time, becomes

$$\frac{d}{dt}\frac{\partial T}{\partial \{\dot{q}\}} = \frac{d}{dt}\frac{\partial \overline{T}}{\partial \{u\}}[Y] + \frac{\partial \overline{T}}{\partial \{u\}}[\dot{Y}] \qquad [9.10.19]$$

Noting that the generalized speeds are functions of both the generalized coordinates and generalized velocities, the derivative of the kinetic energy with respect to the generalized coordinates becomes

$$\frac{\partial T}{\partial \{q\}} = \frac{\partial \overline{T}}{\partial \{q\}} + \frac{\partial \overline{T}}{\partial \{u\}}\frac{\partial \{u\}}{\partial \{q\}} \qquad [9.10.20]$$

In this equation, the first term is the explicit derivative, and the second term uses the chain rule to find the contribution from the generalized speeds. Introducing Eqs. [9.10.19] and [9.10.20] into Lagrange's equations, we obtain

$$\frac{d}{dt}\frac{\partial \overline{T}}{\partial \{u\}}[Y] + \frac{\partial \overline{T}}{\partial \{u\}}\left([\dot{Y}] - \frac{\partial \{u\}}{\partial \{q\}}\right) - \frac{\partial \overline{T}}{\partial \{q\}} = \{Q\}^T \qquad [9.10.21]$$

We right-multiply this equation by $[W]$ and make use of the relationship for the generalized forces $\{U\} = [W]^T\{Q\}$, with the result

$$\frac{d}{dt}\frac{\partial \overline{T}}{\partial \{u\}} + \frac{\partial \overline{T}}{\partial \{u\}}[\mathscr{L}] - \frac{\partial \overline{T}}{\partial \{q\}}[W] = \{U\}^T \qquad [9.10.22]$$

in which

$$[\mathscr{L}] = \left([\dot{Y}] - \frac{\partial \{u\}}{\partial \{q\}}\right)[W] = \left(\frac{d}{dt}\frac{\partial \{u\}}{\partial \{\dot{q}\}} - \frac{\partial \{u\}}{\partial \{q\}}\right)[W] \qquad [9.10.23]$$

Equation [9.10.22] is *Lagrange's equations for quasi-coordinates*. Their primary difference from the traditional Lagrange's equations is the calculation of the coefficient matrix $[\mathscr{L}]$. One can make use of the several different ways to express $[\mathscr{L}]$. One way is to use the preceding definition directly. Once $[\mathscr{L}]$ is calculated, if it is not in terms of generalized speeds, it can be expressed in terms of them by

9.10 RELATIONSHIPS BETWEEN THE FUNDAMENTAL EQUATIONS AND LAGRANGE'S EQUATIONS

simple substitutions. The same procedure can be followed when evaluating the time derivative of $\partial \overline{T}/\partial\{u\}$.

Another way to evaluate $[\mathscr{L}]$ is as follows. $[\mathscr{L}]$ can be written entirely in terms of the generalized coordinates and generalized speeds. To demonstrate this, differentiate an element of $[Y]$, say, Y_{ij}, as

$$\dot{Y}_{ij} = \sum_{k=1}^{n} \frac{\partial Y_{ij}}{\partial q_k} \dot{q}_k = \frac{\partial Y_{ij}}{\partial \{q\}} \{\dot{q}\} = \frac{\partial Y_{ij}}{\partial \{q\}}[W]\{u\} + \frac{\partial Y_{ij}}{\partial \{q\}}\{X\} \qquad i,j = 1,2,\ldots,n \quad [9.10.24]$$

and note that this operation is performed for each component of $[Y]$. We expand $\partial\{u\}/\partial\{q\}$ as

$$\frac{\partial \{u\}}{\partial \{q\}} = \left[\frac{\partial \{u\}}{\partial q_1} \quad \frac{\partial \{u\}}{\partial q_2} \quad \cdots \quad \frac{\partial \{u\}}{\partial q_n} \right] \qquad [9.10.25]$$

where each of the column vectors can be expressed as

$$\frac{\partial \{u\}}{\partial q_k} = \frac{(\partial [Y]\{\dot{q}\} + \{Z\})}{\partial q_k} = \frac{\partial [Y]}{\partial q_k}\{\dot{q}\} + \frac{\partial \{Z\}}{\partial q_k}$$

$$= \frac{\partial [Y]}{\partial q_k}[W]\{u\} + \frac{\partial \{X\}}{\partial q_k} + \frac{\partial \{Z\}}{\partial q_k} \qquad k = 1,2,\ldots,n \quad [9.10.26]$$

In the presence of forces that can be derived from a potential, one can express the contribution of these forces using the potential energy V. In this case, Lagrange's equations for quasi-coordinates have the form

$$\frac{d}{dt}\frac{\partial \overline{T}}{\partial \{u\}} + \frac{\partial \overline{T}}{\partial \{u\}}[\mathscr{L}] - \frac{\partial \overline{T}}{\partial \{q\}}[W] + \frac{\partial V}{\partial \{q\}}[W] = \{U_{nc}\}^T \qquad [9.10.27]$$

For constrained systems, once a set of independent generalized speeds are selected, Eq. [9.10.23] can still be used, noting that both $[Y]$ and $[W]$ are rectangular matrices of orders $p \times n$ and $n \times p$, respectively. A Lagrange multiplier matrix enters the formulation in the same way that it does in the traditional Lagrange's equations.

Let us compare Lagrange's equations for quasi-coordinates with the fundamental equations. Both sets of equations are fundamental, and they can handle nonholonomic systems. Other forms of the equations of motion, such as Euler's equations, the Boltzmann-Hamel equations, and the traditional form of Lagrange's equations, can be derived from them. The right sides of both equations are identical. The advantage of Eq. [9.10.22] is that it does not involve calculation of acceleration terms or of rate of change of angular momenta. Its disadvantage is the effort required in the calculation of $[\mathscr{L}]$. The fundamental equations, especially in the Kane's form, lend themselves to efficient calculation of partial velocities, making it more convenient to derive the equations of motion in many cases. On the other hand, Eq. [9.10.22] retains the Lagrangian formulation and has more physical insight. It utilizes the kinetic and potential energies and eases the generation of motion integrals.

One can consider Lagrange's equations for quasi-coordinates as a set of equations derived from the extended Hamilton's principle, which is an integral principle. By contrast, the fundamental equations are not based on an integral principle.

Another advantage of the Lagrangian formulation will become evident in Chapter 11, where we study the dynamics of deformable bodies.

Similar to the debate about the difference between the Gibbs-Appell and Kane's forms of the fundamental equations, there is debate about whether the fundamental equations or Lagrange's equations for quasi-coordinates are better and easier to use, or more fundamental themselves.

Example 9.13

Again consider the spinning top on cart problem. Obtain the equations of motion using Lagrange's equations for quasi-coordinates.

Solution

Because the right sides of Lagrange's equations and the fundamental equations are the same, we will only calculate the left sides of the equation of motion. The kinetic energy for this problem is calculated in Example 8.12 and it has the form (using a 3-1-3 transformation and F frame)

$$T = \frac{1}{2}(I_1 + m_1 L^2)(\omega_1^2 + \omega_2^2) + \frac{1}{2}I_3 \omega_3^2 + \frac{1}{2}(m_1 + m_2)\dot{Y}^2 + m_1 L \dot{Y}(\omega_2 s\phi - \omega_1 c\phi c\theta) \quad \text{[a]}$$

We select the generalized coordinates and speeds as

$$q_1 = \phi \qquad q_2 = \theta \qquad q_3 = \psi \qquad q_4 = Y \qquad u_i = \omega_i \qquad i = 1, 2, 3 \qquad u_4 = \dot{Y} \quad \text{[b]}$$

so that the kinetic energy is written as

$$\overline{T} = \frac{1}{2}(I_1 + m_1 L^2)(u_1^2 + u_2^2) + \frac{1}{2}I_3 u_3^2 + \frac{1}{2}(m_1 + m_2)u_4^2 + m_1 L u_4(u_2 s q_1 - u_1 c q_1 c q_2) \quad \text{[c]}$$

Noting that the angular velocity vector has the form $\boldsymbol{\omega} = \dot{\theta}\mathbf{f}_1 + \dot{\phi}s\theta\mathbf{f}_2 + (\dot{\phi}c\theta + \dot{\psi})\mathbf{f}_3$, we write the relationship between the generalized speeds and generalized velocities as

$$\begin{bmatrix} u_1 \\ u_2 \\ u_3 \\ u_4 \end{bmatrix} = \begin{bmatrix} 0 & 1 & 0 & 0 \\ s q_2 & 0 & 0 & 0 \\ c q_2 & 0 & 1 & 0 \\ 0 & 0 & 0 & 1 \end{bmatrix} \begin{bmatrix} \dot{q}_1 \\ \dot{q}_2 \\ \dot{q}_3 \\ \dot{q}_4 \end{bmatrix} \quad \text{[d]}$$

and the inverse relationship as

$$\begin{bmatrix} \dot{q}_1 \\ \dot{q}_2 \\ \dot{q}_3 \\ \dot{q}_4 \end{bmatrix} = \begin{bmatrix} 0 & \frac{1}{s q_2} & 0 & 0 \\ 1 & 0 & 0 & 0 \\ 0 & -\frac{1}{t q_2} & 1 & 0 \\ 0 & 0 & 0 & 1 \end{bmatrix} \begin{bmatrix} u_1 \\ u_2 \\ u_3 \\ u_4 \end{bmatrix} \quad \text{[e]}$$

The coefficient matrices in Eqs. [d] and [e] are recognized as $[Y]$ and $[W]$, with $\{Z\} = \{X\} = \{0\}$.

The derivatives of the kinetic energy with respect to the generalized speeds is

$$\frac{\partial \overline{T}}{\partial u_1} = (I_1 + m_1 L^2)u_1 - m_1 L u_4 c q_1 c q_2 \qquad \frac{\partial \overline{T}}{\partial u_2} = (I_1 + m_1 L^2)u_2 + m_1 L u_4 s q_1$$

$$\frac{\partial \overline{T}}{\partial u_3} = I_3 u_3 \qquad \frac{\partial \overline{T}}{\partial u_4} = (m_1 + m_2)u_4 + m_1 L(u_2 s q_1 - u_1 c q_1 c q_2) \quad \text{[f]}$$

9.10 RELATIONSHIPS BETWEEN THE FUNDAMENTAL EQUATIONS AND LAGRANGE'S EQUATIONS

and the explicit derivatives with respect to the generalized coordinates are

$$\frac{\partial T}{\partial q_1} = m_1 L u_4 (u_2\, c\, q_1 + u_1\, s\, q_1\, c\, q_2) \qquad \frac{\partial T}{\partial q_2} = m_1 L u_4 u_1\, c\, q_1\, s\, q_2$$

$$\frac{\partial T}{\partial q_3} = 0 \qquad \frac{\partial T}{\partial q_4} = 0 \qquad \text{[g]}$$

We next calculate the $[\mathscr{L}]$ matrix. The time derivative of $[Y]$ multiplied by $[W]$ is

$$[\dot{Y}][W] = \begin{bmatrix} 0 & 0 & 0 & 0 \\ u_1\, c\, q_2 & 0 & 0 & 0 \\ -u_1\, s\, q_2 & 0 & 0 & 0 \\ 0 & 0 & 0 & 0 \end{bmatrix} \begin{bmatrix} 0 & \dfrac{1}{s\, q_2} & 0 & 0 \\ 1 & 0 & 0 & 0 \\ 0 & -\dfrac{1}{t\, q_2} & 1 & 0 \\ 0 & 0 & 0 & 1 \end{bmatrix} = \begin{bmatrix} 0 & 0 & 0 & 0 \\ 0 & \dfrac{u_1}{t\, q_2} & 0 & 0 \\ 0 & -u_1 & 0 & 0 \\ 0 & 0 & 0 & 0 \end{bmatrix} \qquad \text{[h]}$$

To evaluate the second term in Eq. [9.10.23] we note that $[Y]$ contains contributions only from the second generalized coordinate, so that

$$\frac{\partial \{u\}}{\partial q_k} = \{0\} \quad k = 1, 3, 4 \qquad \frac{\partial \{u\}}{\partial q_2} = \frac{\partial [Y]}{\partial q_2}[W]\{u\} \qquad \text{[i]}$$

Also, evaluation of the derivative of $[Y]$ with respect to q_2 is similar to the time derivative but without the u_1 terms, so that we obtain

$$\frac{\partial \{u\}}{\partial q_2} = \begin{bmatrix} 0 & 0 & 0 & 0 \\ c\, q_2 & 0 & 0 & 0 \\ -s\, q_2 & 0 & 0 & 0 \\ 0 & 0 & 0 & 0 \end{bmatrix} \begin{bmatrix} 0 & \dfrac{1}{s\, q_2} & 0 & 0 \\ 1 & 0 & 0 & 0 \\ 0 & -\dfrac{1}{t\, q_2} & 1 & 0 \\ 0 & 0 & 0 & 1 \end{bmatrix} \{u\} = \begin{bmatrix} 0 & 0 & 0 & 0 \\ 0 & \dfrac{1}{t\, q_2} & 0 & 0 \\ 0 & -1 & 0 & 0 \\ 0 & 0 & 0 & 0 \end{bmatrix} \begin{bmatrix} u_1 \\ u_2 \\ u_3 \\ u_4 \end{bmatrix} = \begin{bmatrix} 0 \\ \dfrac{u_2}{t\, q_2} \\ -u_2 \\ 0 \end{bmatrix} \qquad \text{[j]}$$

Thus, combining all generalized coordinates, we obtain

$$\frac{\partial \{u\}}{\partial \{q\}}[W] = \begin{bmatrix} 0 & 0 & 0 & 0 \\ 0 & \dfrac{u_2}{t\, q_2} & 0 & 0 \\ 0 & -u_2 & 0 & 0 \\ 0 & 0 & 0 & 0 \end{bmatrix} \begin{bmatrix} 0 & \dfrac{1}{s\, q_2} & 0 & 0 \\ 1 & 0 & 0 & 0 \\ 0 & -\dfrac{1}{t\, q_2} & 1 & 0 \\ 0 & 0 & 0 & 1 \end{bmatrix} = \begin{bmatrix} 0 & 0 & 0 & 0 \\ \dfrac{u_2}{t\, q_2} & 0 & 0 & 0 \\ -u_2 & 0 & 0 & 0 \\ 0 & 0 & 0 & 0 \end{bmatrix} \qquad \text{[k]}$$

We can now write the $[\mathscr{L}]$ matrix as

$$[\mathscr{L}] = [\dot{Y}][W] - \frac{\partial \{u\}}{\partial \{q\}}[W] = \begin{bmatrix} 0 & 0 & 0 & 0 \\ -\dfrac{u_2}{t\, q_2} & \dfrac{u_1}{t\, q_2} & 0 & 0 \\ u_2 & -u_1 & 0 & 0 \\ 0 & 0 & 0 & 0 \end{bmatrix} \qquad \text{[l]}$$

Next, we evaluate the individual expressions in Lagrange's equations. The time derivatives of the $\partial T/\partial u_k$ are

$$\frac{d}{dt}\frac{\partial \overline{T}}{\partial u_1} = (I_1 + m_1 L^2)\dot{u}_1 - m_1 L \dot{u}_4 \, cq_1 \, cq_2 + \frac{m_1 L u_2 u_4 \, sq_1}{tq_2} + m_1 L u_1 u_4 \, cq_1 \, sq_2$$

$$\frac{d}{dt}\frac{\partial \overline{T}}{\partial u_2} = (I_1 + m_1 L^2)\dot{u}_2 + m_1 L \dot{u}_4 \, sq_1 + \frac{m_1 L u_2 u_4 \, c \, q_1}{s q_2}$$

$$\frac{d}{dt}\frac{\partial T}{\partial u_3} = I_3 \dot{u}_3$$

$$\frac{d}{dt}\frac{\partial \overline{T}}{\partial u_4} = (m_1 + m_2)\dot{u}_4 + m_1 L \left(\dot{u}_2 sq_1 - \dot{u}_1 cq_1 cq_2 + \frac{u_2^2 cq_1}{sq_2} + u_1^2 cq_1 sq_2 + \frac{u_1 u_2 sq_1}{tq_2} \right) \quad [\mathbf{m}]$$

The transpose of second term on the left side of Lagrange's equations becomes

$$[\mathscr{L}]^T \left(\frac{\partial \overline{T}}{\partial \{u\}}\right)^T = \begin{bmatrix} 0 & -\frac{u_2}{tq_2} & u_2 & 0 \\ 0 & \frac{u_1}{tq_2} & -u_1 & 0 \\ 0 & 0 & 0 & 0 \\ 0 & 0 & 0 & 0 \end{bmatrix} \begin{bmatrix} (I_1 + m_1 L^2)u_1 - m_1 L u_4 \, c \, q_1 \, c \, q_2 \\ (I_1 + m_1 L^2)u_2 + m_1 L u_4 \, s \, q_1 \\ I_3 u_3 \\ (m_1 + m_2)u_4 + m_1 L(u_2 \, s \, q_1 - u_1 \, c \, q_1 \, c \, q_2) \end{bmatrix}$$

$$= \begin{bmatrix} -(I_1 + m_1 L^2)\frac{u_2^2}{tq_2} - \frac{m_1 L u_2 u_4 \, s \, q_1}{tq_2} + I_3 u_2 u_3 \\ (I_1 + m_1 L^2)\frac{u_1 u_2}{tq_2} + \frac{m_1 L u_1 u_4 \, s \, q_1}{tq_2} - I_3 u_1 u_3 \\ 0 \\ 0 \end{bmatrix} \quad [\mathbf{n}]$$

and the transpose of third term is

$$-[W]^T \left(\frac{\partial \overline{T}}{\partial \{q\}}\right)^T = - \begin{bmatrix} 0 & 1 & 0 & 0 \\ \frac{1}{sq_2} & 0 & -\frac{1}{tq_2} & 0 \\ 0 & 0 & 1 & 0 \\ 0 & 0 & 0 & 1 \end{bmatrix} \begin{bmatrix} m_1 L u_4 (u_2 cq_1 + u_1 sq_1 cq_2) \\ m_1 L u_4 u_1 cq_1 sq_2 \\ 0 \\ 0 \end{bmatrix}$$

$$= \begin{bmatrix} -m_1 L u_4 u_1 cq_1 sq_2 \\ -\frac{m_1 L u_4 (u_2 cq_1 + u_1 sq_1 cq_2)}{sq_2} \\ 0 \\ 0 \end{bmatrix} \quad [\mathbf{o}]$$

The potential energy and virtual work are

$$V = m_1 g L \cos \theta = m_1 g L \cos q_2 \qquad \delta W = M \delta q_3' + F \delta q_4' \quad [\mathbf{p}]$$

so that the generalized force vector becomes

$$\{U\} = [m_1 gL \sin q_2 \quad 0 \quad M \quad F]^T \qquad \textbf{[q]}$$

Combining Eqs. [m], [n], [o], and [q] leads to the equations of motion, for the fourth time. The reader should compare these methods of solution and judge, at least for this problem, the merits and disadvantages of each method. Note that, as encountered in many cases when using the traditional Lagrange's equations, certain terms in Eqs. [m], [n], and [o] canceled each other. This is one of the disadvantages of using Lagrange's equations; in many cases they lead to wasted manipulations. However, considering the proliferation of symbolic manipulation software, this disadvantage loses its importance.

9.11 IMPULSE-MOMENTUM RELATIONSHIPS FOR GENERALIZED SPEEDS

The impulse and momentum relationships for systems described by generalized speeds are very similar to those obtained using generalized velocities. In Section 5.8, we defined the generalized momentum π_k ($k = 1, 2, \ldots, n$) associated with the generalized coordinate q_k as

$$\pi_k = \frac{\partial T}{\partial \dot{q}_k} \qquad k = 1, 2, \ldots, n \qquad \textbf{[9.11.1]}$$

In terms of independent generalized speeds, generalized momenta are defined similarly as

$$\pi'_k = \frac{\partial T}{\partial u_k} \qquad \textbf{[9.11.2]}$$

For a system of N particles the generalized momenta can be shown to be

$$\pi'_k = \sum_{i=1}^{N} m_i \mathbf{v}_i \cdot \mathbf{v}_i^k \qquad \textbf{[9.11.3]}$$

We relate the two generalized momenta by noting from Eq. [9.5.3] that $\{\dot{q}\} = [W]\{u\} + \{X\}$, so that $\partial\{\dot{q}\}/\partial\{u\} = [W]$. Hence, writing Eq. [9.11.1] in column vector format as $\{\pi\}^T = \partial T/\partial\{\dot{q}\}$, we can invoke the chain rule of differentiation and have

$$\{\pi'\}^T = \frac{\partial T}{\partial \{u\}} = \frac{\partial T}{\partial \{\dot{q}\}} \frac{\partial \{\dot{q}\}}{\partial \{u\}} = \{\pi\}^T [W] \qquad \textbf{[9.11.4]}$$

so that

$$\{\pi'\} = [W]^T \{\pi\} \qquad \textbf{[9.11.5]}$$

The generalized impulse associated with a generalized coordinate was defined in Section 5.10 as the integral of the generalized force over time, or

$$\hat{Q}_k = \int_{t_1}^{t_2} Q_k \, dt \qquad k = 1, 2, \ldots, n \qquad \textbf{[9.11.6]}$$

For generalized forces associated with generalized speeds we define the generalized impulse as

$$\hat{U}_k = \int_{t_1}^{t_2} U_k \, dt \qquad [9.11.7]$$

From Eq. [9.5.19] we relate the generalized impulses associated with generalized coordinates to those associated with generalized speeds by

$$\{\hat{U}\} = [W]^T \{\hat{Q}\} \qquad [9.11.8]$$

For generalized coordinates, we saw in Sec. 5.10 that for a true impulsive force—that is, for a very large force applied over an infinitesimal time instant Δt—the difference in the generalized momenta between any two time instances is equal to the generalized impulse, or

$$\lim_{\Delta t \to 0} \int_{t_1}^{t_1+\Delta t} Q_k \, dt = \Delta \pi_k \qquad [9.11.9]$$

To see the corresponding relationship for generalized speeds, we write this equation in column vector form as

$$\lim_{\Delta t \to 0} \int_{t_1}^{t_1+\Delta t} \{Q\} \, dt = \{\Delta \pi\} \qquad [9.11.10]$$

Left-multiplying both sides by $[W]^T$ and considering Eqs. [9.11.5] and [9.11.8], we obtain

$$\lim_{\Delta t \to 0} \int_{t_1}^{t_1+\Delta t} \{U\} \, dt = \{\Delta \pi'\} \qquad [9.11.11]$$

indicating that when acted upon an impulsive force, the generalized momenta associated with generalized speeds behave in the same way as the generalized momenta associated with generalized coordinates. Hence, when nonholonomic variables such as angular velocities are involved, it is more convenient to use Eq. [9.11.11] over Eq. [9.11.10].

Example 9.14

Consider the spinning top on a cart once more. It is given that at a certain instant the top is spinning while the cart is moving. An impulsive force $\hat{F}$ is applied to the cart in the a_2 direction. Find the change in the velocity of the cart and in the angular velocities of the top immediately after the impulse.

Solution

It is more convenient to use generalized speeds in this problem. From Example 9.13, the kinetic energy is

$$T = \frac{1}{2}(I_1 + m_1 L^2)(u_1^2 + u_2^2) + \frac{1}{2} I_3 u_3^2 + \frac{1}{2}(m_1 + m_2) u_4^2$$
$$+ m_1 L u_4 (u_2 \sin \phi - u_1 \cos \phi \cos \theta) \qquad [a]$$

so that the generalized momenta associated with the generalized speeds have the form

$$\pi'_1 = \frac{\partial T}{\partial u_1} = (I_1 + m_1 L^2)u_1 - m_1 L u_4 \cos\phi \cos\theta$$

$$\pi'_2 = \frac{\partial T}{\partial u_2} = (I_1 + m_1 L^2)u_2 + m_1 L u_4 \sin\phi$$

$$\pi'_3 = \frac{\partial T}{\partial u_3} = I_3 u_3$$

$$\pi'_4 = \frac{\partial T}{\partial u_4} = (m_1 + m_2)u_4 + m_1 L(u_2 \sin\phi - u_1 \cos\phi \cos\theta) \quad \text{[b]}$$

From Eq. [m] in Example 9.6, the generalized forces have the form

$$U_1 = m_1 g L \sin\theta \quad U_2 = 0 \quad U_3 = M \quad U_4 = F \quad \text{[c]}$$

Of the applied forces, only F is impulsive, so that only it contributes to the generalized forces during the impulse, with the result

$$\hat{U}_1 = 0 \quad \hat{U}_2 = 0 \quad \hat{U}_3 = 0 \quad \hat{U}_4 = \hat{F} \quad \text{[d]}$$

We equate Eqs. [b] and [d] to solve for the generalized speeds after the impulse. This gives four equations for the four unknowns $\Delta\pi'_k$ ($k = 1, 2, 3, 4$). We note that u_3 only appears in one of the equations, so that it can be solved for independently of the others, with the result $\Delta\pi'_3 = 0$. This indicates that the spin of the top is not affected by the impulse, and ω_3 is unchanged.

To solve for the remaining unknowns, we denote the changes in the generalized speeds by Δu_k and write the balance equations as

$$\begin{bmatrix} I_1 + m_1 L^2 & 0 & -m_1 L \cos\phi \cos\theta \\ 0 & I_1 + m_1 L^2 & m_1 L \sin\phi \\ -m_1 L \cos\phi \cos\theta & m_1 L \sin\phi & m_1 + m_2 \end{bmatrix} \begin{bmatrix} \Delta u_1 \\ \Delta u_2 \\ \Delta u_4 \end{bmatrix} = \begin{bmatrix} 0 \\ 0 \\ \hat{F} \end{bmatrix} \quad \text{[e]}$$

From the first and second of these equations we have

$$\Delta u_1 = \frac{m_1 L \cos\phi \cos\theta}{I_1 + m_1 L^2} \Delta u_4 \qquad \Delta u_2 = -\frac{m_1 L \sin\phi}{I_1 + m_1 L^2} \Delta u_4 \quad \text{[f]}$$

which, when introduced into the third equation yield

$$\left[-\frac{m_1^2 L^2 \cos^2\phi \cos^2\theta}{I_1 + m_1 L^2} - \frac{m_1^2 L^2 \sin^2\phi}{I_1 + m_1 L^2} + m_1 + m_2 \right] \Delta u_4 = \hat{F} \quad \text{[g]}$$

The change in the horizontal speed of the cart is dependent on the angles ϕ and θ, so the position of the top influences the velocities and angular velocities after the impulse.

REFERENCES

Ginsberg, J. H. *Advanced Engineering Dynamics*. New York: Harper & Row, 1988.
Goldstein, H. *Classical Mechanics*. 2nd ed. Reading, MA: Addison-Wesley, 1981.
Greenwood, D. T. *Classical Dynamics*. Englewood Cliffs, NJ: Prentice-Hall, 1977.
Kane, T. R., and D. A. Levinson. *Dynamics: Theory and Applications*. New York: McGraw-Hill, 1985.
Meirovitch, L. *Methods of Analytical Dynamics*. New York: McGraw-Hill, 1970.
Pars, L. A. *A Treatise of Analytical Dynamics*. New York: Wiley, 1965.

HOMEWORK EXERCISES

SECTION 9.2

1. Using the modified Euler's equations, derive the equations of motion of a disk rolling without slipping on a surface that is like a wedge, as shown in Fig. 9.6.
2. Using the modified Euler's equations, derive the equations of motion for the disk in Fig. 8.26. The xyz axes also constitute principal axes for the fork, which has a mass $m/2$ and can be modeled as a slender rod.

SECTION 9.3

3. Solve Problem 1 using the Boltzmann-Hamel equations.
4. Obtain the equations of motion of the dual spin spacecraft shown in Fig. 9.7. The centers of mass of the rotor and the main body are not at the same point, but along the b_3 axis, which is a principal axis. Use the Boltzmann-Hamel equations and consider only the rotational motion of the main body. The shaft connecting the main body to the rotor is light.
5. Find the equation of motion of the gyropendulum in Example 8.5 using the Boltzmann-Hamel equations. Assume that both $\dot{\phi}$ and $\dot{\psi}$ are constant.

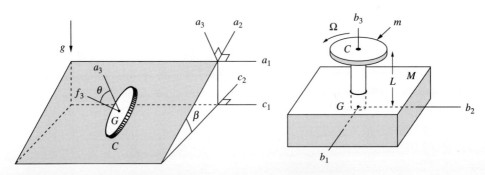

Figure 9.6 **Figure 9.7**

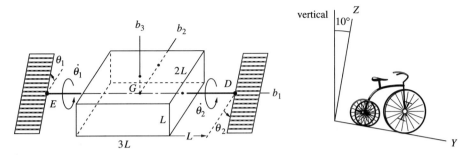

Figure 9.8 **Figure 9.9**

SECTIONS 9.5 AND 9.6

6. Consider the satellite in Fig. 9.8 with the solar panels. The panels are operated by motors that exert moments N_1 and N_2 about the b_1 axis, and moments M_1, M_2, and M_3 act on the main body. Select a set of generalized speeds, and write the partial velocities. Consider only rotational motion of the satellite.

7. Generate the generalized forces and the generalized inertia forces for the preceding problem. The main body has mass moments of inertia I_1, I_2, and I_3 about the $b_1 b_2 b_3$ axes, which are principal axes, and the solar panels can be treated as plates.

8. Consider the vehicle in Fig. 4.8 and consider that the constraint is that the tip of the vehicle moves in the x axis only. Select a set of dependent generalized speeds, invoke the constraint, and arrive at a set of independent generalized speeds.

9. Consider the tricycle in Figs. 7.39–7.40. Select as independent generalized speeds the speed of point A and $\dot{\phi}$, and express the partial velocities associated with the centers of masses of all components (main body, wheels, and handlebars) as well as all the partial angular velocities.

10. For the preceding problem, can you think of a better choice of generalized speeds? Compare with the results of selecting the orientation and speed of the front wheel as the variables.

11. Reconsider problem 10. This time the tricycle is on an incline as shown in Fig. 9.9. Calculate the generalized forces. Assume that the handlebar is massless, the main body of the tricycle has mass m, the rear wheels $m/10$ each, and the front wheel $m/5$.

12. Consider the rod to which two wheels are attached in Fig. 7.55. The rod is of mass m and the wheels are each of mass $2m$. Consider the following sets of generalized speeds: (a) the angular velocities $\dot{\theta}_1$ and $\dot{\theta}_2$ of the wheels, (b) the speed of the center of the rod, and the turning rate of the rod. Consider that the system is on an incline, like the one in Fig. 9.9. Find the generalized forces for both sets of quasi-velocities.

13. Consider Example 9.7 and express the translation of the disk and the constraint equations by beginning with the contact point of the disk.

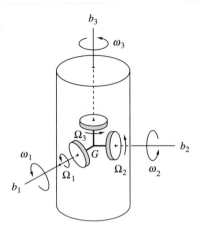

Figure 9.10

Sections 9.7 and 9.8

14. Generate the Gibbs-Appell function for Problem 4.
15. Generate the Gibbs-Appell function for Problem 6.
16. Solve Problem 2 using the Kane's form of the fundamental equations.
17. Solve Problem 5 using the Kane's form of the fundamental equations.
18. Derive Hamilton's equations from the fundamental equations.
19. Solve the dual spin spacecraft problem in Fig. 9.7 using the fundamental equations. Consider rotational motion only.
20. Figure 9.10 shows a spacecraft that has three rotors of identical size and shape, used primarily for attitude maneuvers. Each rotor is driven by its own motor and is at a distance L from the center of mass. Derive the equations of motion using the fundamental equations.
21. Consider the spacecraft in Fig. 8.35. Obtain the equations of motion using the fundamental equations.
22. Consider the rod with the two wheels attached in Fig. 7.55. Derive the equations of motion, using the two sets of quasi-velocities in Problem 13.

Section 9.10

23. Consider the rotational motion of an axisymmetric rigid body. Show that the modified Euler's equations can be derived from Lagrange's equations for quasi-coordinates.
24. Solve for the equations of motion of the vehicle in Example 9.10 using Lagrange's equations for quasi-coordinates.
25. Solve Problem 2 using Lagrange's equations for quasi-coordinates.
26. Solve Problem 4 using Lagrange's equations for quasi-coordinates.

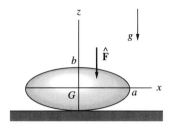

Figure 9.11

SECTION 9.11

27. Consider the tricycle in Figs. 7.39–40. The tricycle is at rest on a flat surface, with $\phi = 15°$. An impulsive force $\hat{F}$ is applied at point A along the y axis. Find the resulting angular velocities of the wheels, and the velocity of G. Treat the wheels as disks and the main body of the tricycle as a plate, and ignore the weight of the front handle and connecting rods between the main body and wheels.

28. Consider the disk in Fig. 9.4, rolling without slipping with a constant nutation angle of 45°. An impulsive force is applied to the disk in the f_3 direction at point A. Calculate the resulting precession, nutation, and spin rates immediately after the impulse, assuming that the disk is still rolling without slipping after the impulsive force is applied.

29. The ellipsoid of revolution (about the x axis) in Fig. 9.11 is at rest on a horizontal surface. It is struck by an impulsive force $\hat{F}$ in the z direction and applied through the coordinates $x = a/2$, $y = b/2$. Find the velocity of the center of mass and the angular velocity of the ellipsoid, assuming that the ellipsoid rolls without slipping on the horizontal surface.

chapter

10

QUALITATIVE ANALYSIS OF RIGID BODY MOTION

10.1 INTRODUCTION

In this chapter, we analyze the motion of rigid bodies from a qualitative perspective. We study the nature of the motion and make use of motion integrals. The main interest is in gyroscopic motion and on the effects of the gyroscopic moment. Gyroscopic motion is commonly described as motion in which the angular momentum and angular velocity vectors have different directions and one of the components of the angular velocity is much larger than the other two. This type of motion is commonly encountered in bodies that spin and the direction of the spin axis is not stationary. The gyroscopic moment becomes large and this enables tops, disks, and gyroscopes to continue their motion and prevents them from falling down. In essence, gyroscopic motion is a battle between gravitational moment and gyroscopic moment. A specific spin rate is required to overcome the effects of gravity.

We begin the analysis in this chapter with the simplest case of moment-free motion. We study axisymmetric as well as nonsymmetric bodies. We focus on bodies under the action of forces and moments; the rolling disk and spinning top are studied, as well as gyroscopes.

While the qualitative analysis of gyroscopic motion and of spinning bodies is a very useful and important tool, it becomes even more interesting to supplement the qualitative analysis with quantitative results. Numerical integration of the equations of motion was discussed in Chapter 1. The reader is urged to take advantage of this. By numerically integrating the equations of motion one can quantitatively observe many of the interesting results that are presented in this chapter.

This chapter does not have many examples. In essence, each section can be considered as an example analyzing a specific situation.

10.2 Torque-Free Motion of Inertially Symmetric Bodies

We begin with the motion of a rigid body that is not subjected to any external moments. Since the translational equations are straightforward, we will only analyze the rotational equations. Selecting the principal axes as the body coordinates and in the absence of external moments, Euler's equations, Eqs. [8.5.28], simplify to

$$I_1\dot{\omega}_1 - (I_2 - I_3)\omega_2\omega_3 = M_1 = 0$$
$$I_2\dot{\omega}_2 - (I_3 - I_1)\omega_1\omega_3 = M_2 = 0$$
$$I_3\dot{\omega}_3 - (I_1 - I_2)\omega_1\omega_2 = M_3 = 0 \qquad [\mathbf{10.2.1}]$$

We next consider the special case where there is inertial symmetry in the structure and that two of the principal moments of inertia are the same, say $I_1 = I_2$. This assumption not only simplifies the equations of motion but increases the stability margin as well. (Example 8.6 analyzed this increased stability.) From a mathematical standpoint, an arbitrarily shaped body with two equal principal moments of inertia will behave in the same way as an axisymmetric body. However, one does not frequently encounter this situation. Many times, air flow or resistance by another fluid around a body has a substantial effect on the nature of the motion; the shape of the body then becomes a significant factor and there is a difference in the response of axisymmetric and inertially symmetric bodies.

Several bodies that undergo rotational motion are specifically designed as axisymmetric. Examples of this include rockets and satellites, machinery parts, frisbees, and balls such as American footballs and rugby balls. Soccer and ping pong balls have all three principal moments of inertia as the same.

Setting $I_1 = I_2$ in Eqs. [10.2.1] yields

$$I_1\dot{\omega}_1 - (I_1 - I_3)\omega_2\omega_3 = 0 \qquad I_1\dot{\omega}_2 - (I_3 - I_1)\omega_1\omega_3 = 0 \qquad I_3\dot{\omega}_3 = 0$$
$$[\mathbf{10.2.2a,b,c}]$$

Equation [10.2.2c] indicates that ω_3 is constant, so it is an integral of the motion. Introducing the rotation constant Ω

$$\Omega = \frac{(I_3 - I_1)\omega_3}{I_1} \qquad [\mathbf{10.2.3}]$$

Eqs. [10.2.2a] and [10.2.2b] can be expressed as

$$\dot{\omega}_1 + \Omega\omega_2 = 0 \qquad \dot{\omega}_2 - \Omega\omega_1 = 0 \qquad [\mathbf{10.2.4a,b}]$$

Equations [10.2.4] describe a gyroscopic system, with Ω being the parameter describing the gyroscopic properties. They can be written in column vector format as

$$\begin{bmatrix} \dot{\omega}_1 \\ \dot{\omega}_2 \end{bmatrix} = -\begin{bmatrix} 0 & \Omega \\ -\Omega & 0 \end{bmatrix}\begin{bmatrix} \omega_1 \\ \omega_2 \end{bmatrix} \qquad [\mathbf{10.2.5}]$$

10.2 TORQUE-FREE MOTION OF INERTIALLY SYMMETRIC BODIES

where we notice that the coefficient matrix is skew symmetric. A skew-symmetric coefficient matrix is usually the sign of gyroscopic behavior.[1] The rotation constant Ω describes the rate at which the gyroscopic motion unfolds. Multiplying Eq. [10.2.4a] by ω_1 and Eq. [10.2.4b] by ω_2 and adding the two equations leads to

$$\omega_1 \dot{\omega}_1 + \omega_2 \dot{\omega}_2 = \frac{1}{2} \frac{d}{dt}(\omega_1^2 + \omega_2^2) = 0 \quad [\textbf{10.2.6}]$$

from which it is realized that $\omega_1^2 + \omega_2^2$ is constant. We define by $\omega_{12} = \sqrt{\omega_1^2 + \omega_2^2}$ the magnitude of the projection of the angular velocity vector on the $b_1 b_2$ plane, as shown in Fig. 10.1. Hence, we have a second integral of the motion. The two first integrals can be combined to yield a third integral of the motion, thus

$$\omega_1^2 + \omega_2^2 + \omega_3^2 = \text{constant} = |\boldsymbol{\omega}|^2 = \omega^2 \quad [\textbf{10.2.7}]$$

indicating that the magnitude of the angular velocity vector is also constant.

Because the body is torque free, both the angular momentum about the center of mass and the rotational kinetic energy are conserved, pointing to two other integrals of the motion as

$$\mathbf{H}_G = I_1 \omega_1 \mathbf{b}_1 + I_1 \omega_2 \mathbf{b}_2 + I_3 \omega_3 \mathbf{b}_3 = \text{constant}$$
$$2T_{\text{rot}} = \boldsymbol{\omega} \cdot \mathbf{H}_G = I_1 \omega_1^2 + I_1 \omega_2^2 + I_3 \omega_3^2 = \text{constant} \quad [\textbf{10.2.8}]$$

in which $\mathbf{b}_i$ ($i = 1, 2, 3$) denote the unit vectors along the principal axes. We use the integrals of the motion to explain the motion of the body. Defining the projection of the angular momentum vector onto the $b_1 b_2$ plane by $\mathbf{H}_{12}$, we observe that the magnitude of $\mathbf{H}_{12}$ is

$$H_{12} = |\mathbf{H}_{12}| = I_1 \sqrt{\omega_1^2 + \omega_2^2} = I_1 \omega_{12} \quad [\textbf{10.2.9}]$$

so that the magnitude of $\mathbf{H}_{12}$ is also constant. The above equations imply that the projections of $\mathbf{H}_G$ and $\boldsymbol{\omega}$ onto the $b_1 b_2$ plane lie along the same line and hence,

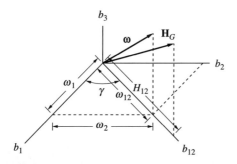

Figure 10.1

---

[1] The reader is urged to compare Eq. [10.2.5] with the linearized equations for the Foucault's pendulum, Example 2.16, Eq. [e].

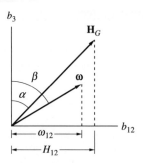

Figure 10.2

as described in Fig. 10.1, the vectors $\mathbf{b}_3$, $\mathbf{H}_G$, and $\boldsymbol{\omega}$ must lie on the same plane. Defining the axis b_{12} such that $\mathbf{H}_{12} = H_{12}\mathbf{b}_{12}$, where $\mathbf{b}_{12}$ is the unit vector in the b_{12} direction, this plane is defined by the b_3 and b_{12} axes. Fig. 10.2 depicts the plane.

The motion can be interpreted as the rotation of this plane. The rotation is about the angular momentum vector $\mathbf{H}_G$, as both the magnitude and direction of this vector are constant. The rotation of the angular velocity vector can also be described as the movement of two imaginary cones, called the *body and space cones*, on top of each other. The body cone is fixed to the body. It is generated by the rotation of the angular velocity vector about the b_3 axis (the spin and symmetry axis). The space cone is fixed in the inertial space and is generated by the rotation of the angular velocity vector about the angular momentum vector. In Chapter 7, we saw that the body and space cones can take arbitrary shapes. Because of the inertial symmetry, the cones generated here are right circular cones. Figures 10.3 and 10.4 show the body and space cones, as well as the orientation of $\mathbf{H}_G$ and $\boldsymbol{\omega}$ along the b_3 and b_{12} directions, depending on the angles α and β. We select the inertial frame such that the angular momentum vector $\mathbf{H}_G$ (which is constant in both magnitude and direction) coincides with the a_3 direction. We thus have

$$\mathbf{H}_G = H_G \mathbf{a}_3 \qquad [10.2.10]$$

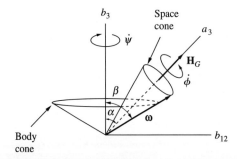

Figure 10.3 Flat body

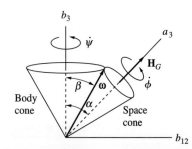

Figure 10.4 Slender body

where H_G is the magnitude of the angular momentum vector. From Fig. 10.2 the angles α and β are constants, and they can be expressed as

$$\tan\alpha = \frac{H_{12}}{H_3} = \frac{I_1\omega_{12}}{I_3\omega_3} = \text{constant} \qquad \tan\beta = \frac{\omega_{12}}{\omega_3} = \text{constant} \qquad [\mathbf{10.2.11}]$$

in which H_{12} and H_3 are components of the angular momentum along the b_{12} and b_3 axes, respectively. The relationship between α and β depends on the magnitude of the mass moments of inertia I_1 and I_3. We have two possible cases:

1. If $I_1 < I_3$, then $\alpha < \beta$. This case corresponds to the motion of a flat body, such as a disk.
2. If $I_1 > I_3$, then $\alpha > \beta$. This case corresponds to the motion of a slender body, such as a rod or a football.

When viewed from the body-fixed frame, $\boldsymbol{\omega}$ generates the body cone, which has an apex angle of β around b_3. When viewed from the inertial frame, $\boldsymbol{\omega}$ generates the space cone with an apex angle of $|\beta - \alpha|$. For a flat body with $I_1 < I_3$, the space cone rolls inside the body cone, and for a slender body, $I_1 > I_3$, the body and space cones lie outside of each other, as shown in Figs. 10.3–10.4. The space cone is easier to visualize, because it shows the direction of the angular velocity vector from an inertial frame. The visualization is also simpler for slender bodies.

Consider, for example, a football being thrown. Under ideal circumstances, the ball would just have a spin along its longitudinal axis, implying an angular velocity vector for the ball that has constant direction in addition to constant magnitude.[2] However, if there is a wobble associated with the throw, the angular velocity now has two components: the wobble and the spin. Consequently, the angular velocity vector no longer has a constant direction, but a direction that rotates itself.

We next address the issue of the rate with which ω_{12}, the component of the angular velocity vector along the b_1b_2 plane, rotates. This is, of course, the same as the rate with which the plane defined by b_3, $\mathbf{H}_{12}$, and ω_{12} rotates, or how fast a body cone is swept. Considering Eqs. [10.2.4], we intuitively expect this rate to be the rotation constant $\Omega = \omega_3(I_3 - I_1)/I_1$. To verify this, consider Fig. 10.1 and the angle γ, which is the angle ω_{12} makes with the b_1 axis. The rotation rate is then $\dot{\gamma}$. Simple geometry indicates that

$$\frac{\sin\gamma}{\cos\gamma} = \frac{\omega_2}{\omega_1} \qquad [\mathbf{10.2.12}]$$

Differentiation of both sides of the above equation with respect to time yields

$$\frac{\cos\gamma(\dot{\gamma}\cos\gamma) - \sin\gamma(-\dot{\gamma}\sin\gamma)}{\cos^2\gamma} = \frac{\omega_1\dot{\omega}_2 - \omega_2\dot{\omega}_1}{\omega_1^2} \qquad [\mathbf{10.2.13}]$$

---

[2] The motion of the football is actually a much more complex phenomenon, due to the aerodynamics and gravity moment.

Introducing Eqs. [10.2.4] into Eq. [10.2.13], we obtain

$$\frac{\dot{\gamma}}{\cos^2 \gamma} = \frac{\Omega(\omega_2^2 + \omega_1^2)}{\omega_1^2} \qquad [10.2.14]$$

We observe from Fig. 10.1 that $\cos^2 \gamma = \omega_1^2/(\omega_2^2 + \omega_1^2)$, which leads to the expected result that $\dot{\gamma} = \Omega$.

The total motion of the body, as stated earlier, is some kind of wobbling, unless motion is initiated with a perfect spin about one of the principal axes. The amount of wobble, as observed from the value for Ω, depends on how the principal moments of inertia are different from each other. However, the wobble remains constant, because the apex angles of the body and space cones are the same.

An example of wobbling motion is the rotation of the earth. The earth is not a perfect sphere—it is an oblate spheroid. If we attach a set of body axes to the earth and select the b_3 axis as the polar axis, we have (from measurements that have been taken)

$$\frac{I_3 - I_1}{I_1} = 0.0033 \qquad [10.2.15]$$

and, taking the angular velocity of the earth as 1 rev/day, we get $\Omega = 0.0033$ rev/day. This implies that the earth should sweep one body cone in $1/0.0033 = 300$ days.

Astronomers have observed that the polar axis of the Earth indeed has some wobbling motion, a phenomenon called *variation in latitude*. The period of the wobble has been measured as 433 days and the radius swept by the poles in one body cone as about 4 meters. The difference from the predicted period of 300 days is attributed to two reasons. The earth is not rigid and not entirely torque free. Because it is an oblate spheroid and not a perfect sphere, the gravitational attraction of the sun and moon creates a gravity gradient torque. Also, the lack of rigidity causes some energy transfer from one component of the rotation to another.

Let us introduce the Euler angles to the formulation. As before, we consider a 3-1-3 sequence with the rotations of ϕ (precession), θ (nutation), and ψ (spin). The transformation matrix $[R]$ is such that

$$\begin{bmatrix} b_1 \\ b_2 \\ b_3 \end{bmatrix} = [R] \begin{bmatrix} a_1 \\ a_2 \\ a_3 \end{bmatrix} \qquad [10.2.16]$$

as given in Eq. [7.5.3]. From this we have

$$\mathbf{a}_3 = \sin\theta \sin\psi \mathbf{b}_1 + \sin\theta \cos\psi \mathbf{b}_2 + \cos\theta \mathbf{b}_3 \qquad [10.2.17]$$

We express the angular momentum in terms of the body-fixed coordinates as

$$\mathbf{H}_G = H_G \mathbf{a}_3 = H_G(\sin\theta \sin\psi \mathbf{b}_1 + \sin\theta \cos\psi \mathbf{b}_2 + \cos\theta \mathbf{b}_3)$$
$$= I_1 \omega_1 \mathbf{b}_1 + I_1 \omega_2 \mathbf{b}_2 + I_3 \omega_3 \mathbf{b}_3 \qquad [10.2.18]$$

10.2 TORQUE-FREE MOTION OF INERTIALLY SYMMETRIC BODIES

from which we obtain

$$I_1\omega_1 = H_G \sin\theta \sin\psi \qquad I_1\omega_2 = H_G \sin\theta \cos\psi \qquad I_3\omega_3 = H_G \cos\theta \qquad [10.2.19]$$

Notice from the last of these equations that the nutation angle θ is constant and hence the nutation rate is zero. Indeed, we have

$$\cos\theta = \frac{I_3\omega_3}{H_G} = \text{constant} \qquad [10.2.20]$$

In addition, we realize that the angle between the a_3 and b_3 axes is θ, so that $\theta = \alpha$. To obtain expressions for the precession and spin rates, we make use of the relationship between the body-fixed angular velocities and the Euler angles, Eq. [7.5.7]

$$\begin{bmatrix} \omega_1 \\ \omega_2 \\ \omega_3 \end{bmatrix} = \begin{bmatrix} s\theta s\psi & c\psi & 0 \\ s\theta c\psi & -s\psi & 0 \\ c\theta & 0 & 1 \end{bmatrix} \begin{bmatrix} \dot\phi \\ \dot\theta \\ \dot\psi \end{bmatrix} \qquad [10.2.21]$$

In view of the nutation rate being zero, these reduce to

$$\omega_1 = \dot\phi s\theta s\psi \qquad \omega_2 = \dot\phi s\theta c\psi \qquad \omega_3 = \dot\phi c\theta + \dot\psi \qquad [10.2.22]$$

Comparing Eqs. [10.2.19] and [10.2.22], we conclude that

$$\dot\phi = \frac{H_G}{I_1} = \text{constant}$$

$$\dot\psi = -\dot\phi c\theta + \frac{H_G \cos\theta}{I_3} = \left(\frac{H_G}{I_3} - \frac{H_G}{I_1}\right)\cos\theta = \omega_3\left(1 - \frac{I_3}{I_1}\right) = \text{constant}$$

$$[10.2.23]$$

so that the precession and spin rates are constant as well. Combining Eqs. [10.2.23], we arrive at the relation between the precession and spin rates as

$$\dot\phi = \frac{I_3}{(I_1 - I_3)\cos\theta}\dot\psi \qquad [10.2.24]$$

This equation states that for torque-free motion of a body with two equal moments of inertia, the precession rate $\dot\phi$ and spin rate $\dot\psi$ are proportional to each other. The nutation rate is zero. For example, if a football is thrown with a wobble, the magnitude and angle of the wobble should remain as it is and not change.[3] Furthermore, the direction of the wobble remains the same. From Eq. [10.2.24] and Figs. 10.3 and 10.4 we observe that again two cases are possible, depending on the magnitudes of the moments of inertia:

1. When $I_1 > I_3$, such as in a slender body, the precession and spin have the same sense. This type of motion is called *direct precession*. The space cone rolls

---

[3] In reality, the wobble changes due to aerodynamic and gravitational effects.

outside the body cone as shown in Fig. 10.4. The wobble is in the same direction as the spin.

2. When $I_1 < I_3$, such as in a flat body, the precession and spin have opposite senses. The motion is called *retrograde precession*. The space cone rolls inside the body cone as shown in Fig. 10.3.

Direct precession is relatively easy to observe, as for a slender body the ratio of the largest to smallest moment of inertia can take on any value. Making I_1/I_3 a large quantity, such as 4 or 5, one can easily view direct precession. On the other hand, because in flat bodies the ratios of the moments of inertia about the principal axes are closer to each other than in slender bodies, it is usually more difficult to visually observe retrograde precession. For a circular cylinder $I_3 = mR^2/2$ and $I_1 = mR^2/4 + mL^2/12$. The maximum value of the inertia ratio is for a thin disk, $I_3/I_1 = 2$. As the disk thickness increases, the two moments of inertia become closer.

We now relate the values of the body and space cone angles to the Euler angles. As stated above, α, the angle between the angular momentum vector and b_3 is the same as θ, so $\alpha = \theta$. From Eqs. [10.2.11] and [10.2.19] we can write a relationship for the angle β as

$$\tan \beta = \frac{\omega_{12}}{\omega_3} = \frac{H_G/I_1 \sin \theta}{H_G/I_3 \cos \theta} = \frac{I_3}{I_1} \tan \theta \qquad [\mathbf{10.2.25}]$$

which verifies that whether the angle β is larger than α depends on the shape of the body.

We now relate the Euler angles and body angular velocities to each other. Considering Eqs. [10.2.10], [10.2.23], and [10.2.25], one can show that

$$H_G = \omega \sqrt{I_1^2 \sin^2 \beta + I_3^2 \cos^2 \beta}$$

$$\dot{\phi} = \omega \sqrt{\sin^2 \beta + \left(\frac{I_3}{I_1}\right)^2 \cos^2 \beta} \qquad \dot{\psi} = \left(1 - \frac{I_3}{I_1}\right) \omega \cos \beta \qquad \theta = \tan^{-1}\left(\frac{I_1 \tan \beta}{I_3}\right)$$

$$[\mathbf{10.2.26}]$$

thus, given the total angular velocity ω and the angle β, one can determine the angular momentum, the precession and spin rates, and the nutation angle.

Torque-free motion of axisymmetric bodies lends itself to an interesting Euler parameter description. Because precession and spin rates and the nutation angle are constants, one can find closed-form expressions for the Euler parameters in terms of the initial conditions. The Euler parameters are expressed in terms of the Euler angles in Eq. [7.7.40]. Hence, we obtain for a 3-1-3 transformation

$$e_0 = \cos\left(\frac{\dot{\phi} + \dot{\psi}}{2} t\right) \cos \frac{\theta}{2} \qquad e_1 = \cos\left(\frac{\dot{\phi} - \dot{\psi}}{2} t\right) \sin \frac{\theta}{2}$$

$$e_2 = \sin\left(\frac{\dot{\phi} - \dot{\psi}}{2} t\right) \sin \frac{\theta}{2} \qquad e_3 = \sin\left(\frac{\dot{\phi} + \dot{\psi}}{2} t\right) \cos \frac{\theta}{2} \qquad [\mathbf{10.2.27}]$$

10.2 TORQUE-FREE MOTION OF INERTIALLY SYMMETRIC BODIES

Example 10.1

An American football (axially symmetric body) with a mass moment of inertia ratio of $I_1 = 4I_3$ is thrown with a spin Ω about its axis of symmetry. A player tips the football, and in doing so exerts an angular impulse of $\varepsilon I_1 \Omega$ in a transverse direction, where ε is a small number. The change in the linear momentum is assumed to be negligible. Obtain the precession and spin rates after the impulse is applied.

Solution

Without loss of generality, we assume that the angular impulse is applied about the b_1 axis. The angular momentum before the impulse is

$$\mathbf{H}_G(\text{before}) = I_3 \Omega \mathbf{b}_3 \quad \text{[a]}$$

After the impulse, using the angular impulse-momentum theorem, the angular momentum becomes

$$\mathbf{H}_G(\text{after}) = \varepsilon I_1 \Omega \mathbf{b}_1 + I_3 \Omega \mathbf{b}_3 \quad \text{[b]}$$

so that the angular velocity after the impulse is

$$\boldsymbol{\omega} = \varepsilon \Omega \mathbf{b}_1 + \Omega \mathbf{b}_3 \quad \text{[c]}$$

We now construct the body and space cones. Because $I_1 > I_3$, we have a case of direct precession; the body cone lies outside the space cone, α is the angle between $\mathbf{b}_3$ and $\mathbf{H}_G$, and β is the angle between $\mathbf{b}_3$ and $\boldsymbol{\omega}$. Considering Fig. 10.4 we can write

$$\tan \alpha = \frac{\varepsilon I_1 \Omega}{I_3 \Omega} = 4\varepsilon \qquad \tan \beta = \frac{\varepsilon \Omega}{\Omega} = \varepsilon \quad \text{[d]}$$

so that the body and space cones depend on the magnitude of the impulse. From Eqs. [10.2.23] we have the precession and spin rates

$$\dot{\phi} = \frac{H_G}{I_1} \qquad \dot{\theta} = 0 \qquad \dot{\psi} = \omega_3 \left(1 - \frac{I_3}{I_1}\right) \quad \text{[e]}$$

and that the nutation angle $\theta = \alpha = \tan^{-1}(4\varepsilon)$. From Eq. [b], we obtain

$$H_G = I_3 \Omega \sqrt{1 + 16\varepsilon^2} \quad \text{[f]}$$

which then leads to the precession and spin rates after the impulse as

$$\dot{\phi} = \frac{H_G}{I_1} = \frac{I_3 \Omega \sqrt{1 + 16\varepsilon^2}}{4 I_3} = \frac{\Omega \sqrt{1 + 16\varepsilon^2}}{4} \quad \text{[g]}$$

$$\dot{\psi} = \omega_3 \left(1 - \frac{I_3}{I_1}\right) = \frac{3\Omega}{4} \quad \text{[h]}$$

It is interesting to note that the spin rate $\dot{\psi}$ is independent of the magnitude of the angular impulse. The nutation angle and the precession rate depend on the strength of the impulse. For very small values of ε, the precession rate can be approximated as

$$\dot{\phi} \approx \frac{\Omega}{4}(1 + 8\varepsilon^2) \quad \text{[i]}$$

indicating that the minimum value of the precession rate after the impulse is $\Omega/4$.

From Eqs. [g] and [h] we obtain two additional results: First, the precession and spin rates depend on the ratios of the mass moments of inertia. Secondly, the spin rate after the angular impulse is always less than the spin rate before the impulse, known intuitively to players and spectators all around.

10.3 General Case of Torque-Free Motion

We next analyze the torque-free motion of an arbitrary rigid body whose principal moments of inertia are different from each other. Using a 3-1-3 Euler angle transformation, we write the rotational kinetic energy as

$$T_{\text{rot}} = \frac{1}{2}\left[I_1(\dot{\phi}\,s\,\theta\,s\,\psi + \dot{\theta}\,c\,\psi)^2 + I_2(\dot{\phi}\,s\,\theta\,c\,\psi - \dot{\theta}\,s\,\psi)^2 + I_3(\dot{\phi}\,c\,\theta + \dot{\psi})^2\right]$$

[10.3.1]

where we note that ϕ is absent from the Lagrangian, so that it is a cyclic coordinate. Therefore, in the absence of moments about the a_3 axis, which is the case considered here, π_ϕ is an integral of the motion and has the form

$$\pi_\phi = \frac{\partial L}{\partial \dot{\phi}}$$
$$= I_1(\dot{\phi}\,s\,\theta\,s\,\psi + \dot{\theta}\,c\,\psi)s\,\theta\,s\,\psi + I_2(\dot{\phi}\,s\,\theta\,c\,\psi - \dot{\theta}\,s\,\psi)s\,\theta\,c\,\psi + I_3(\dot{\phi}\,c\,\theta + \dot{\psi})c\,\theta$$
$$= \text{constant}$$

[10.3.2]

Unlike the axially symmetric bodies studied in the previous section, π_ψ, the component of the angular momentum along the spin axis, is not constant. As a result, the precession, nutation, and spin rates are no longer constant, either. This makes it difficult to visualize the rotational motion of arbitrary rigid bodies.

Because the motion is torque free, the angular momentum about the center of mass is conserved, and it provides three additional integrals of the motion. These integrals can be used to obtain relations between the precession, nutation, and spin rates. Combining Eqs. [10.2.19] and [10.2.21] gives

$$I_1\omega_1 = I_1(\dot{\phi}\sin\theta\sin\psi + \dot{\theta}\cos\psi) = H_G\sin\theta\sin\psi$$
$$I_2\omega_2 = I_2(\dot{\phi}\sin\theta\cos\psi - \dot{\theta}\sin\psi) = H_G\sin\theta\cos\psi$$
$$I_3\omega_3 = I_3(\dot{\phi}\cos\theta + \dot{\psi}) = H_G\cos\theta$$

[10.3.3]

which can be solved for the precession, nutation, and spin rates. The result is

$$\dot{\phi} = \frac{H_G}{I_1}\sin^2\psi + \frac{H_G}{I_2}\cos^2\psi$$

$$\dot{\theta} = \frac{1}{2}\left(\frac{H_G}{I_1} - \frac{H_G}{I_2}\right)\sin\theta\sin 2\psi$$

$$\dot{\psi} = \frac{H_G}{I_3}\cos\theta - \left(\frac{H_G}{I_1}\sin^2\psi + \frac{H_G}{I_2}\cos^2\psi\right)\cos\theta = \frac{H_G}{I_3}\cos\theta - \dot{\phi}\cos\theta$$

[10.3.4]

These represent three first-order differential equations. They are expressions for the Euler angles in terms of the integrals of the motion. Because they are in state form, they can be integrated once the initial conditions are specified.

The precession rate is always positive, indicating that even though the precession rate of a body may change during motion, the direction of precession does not change. The same cannot be said of the nutation and spin rates. These rates can oscillate, depending on the initial conditions. Note that we observed this same relationship when we conducted a stability analysis in Example 8.6 with regards to the stability of rotational motion about different principal axes.

The nutation rate is dependent on the ratios of the two moments of inertia I_1 and I_2. For the same initial conditions the nutation rate will be different depending on whether I_1 is larger than I_2. On the other hand, the spin equation is the same for general as well as axially symmetric bodies.

Considering the special case of axisymmetric bodies with $I_2 = I_1$, Eqs. [10.3.4] reduce to

$$\dot{\phi} = \frac{H_G}{I_1} \qquad \dot{\theta} = 0 \qquad \dot{\psi} = \left(\frac{H_G}{I_3} - \frac{H_G}{I_1}\right)\cos\theta \qquad [\mathbf{10.3.5}]$$

which are the same as Eqs. [10.2.23]. The nutation angle is constant for axisymmetric bodies, constituting one of the biggest differences between the torque-free motions of general and inertially symmetric bodies.

Observe from this and the previous sections that in order to understand the nature of the motion qualitatively, one needs to analyze the motion using variables from both the inertial and body-fixed frames. Each of these reference frames provides a different kind of insight into the problem.

10.4 POLHODES

An interesting qualitative interpretation of the torque-free rotational motion of arbitrary rigid bodies is due to Poinsot's construction. The approach makes use of the energy and angular momentum integrals, as well as the inertia ellipsoid.

Consider a set of body-fixed coordinates. These coordinates have a general configuration and they do not have to be principal coordinates. The rotational kinetic energy and angular momentum about the center of mass are

$$T_{\text{rot}} = \frac{1}{2}\{\omega\}^T[I_G]\{\omega\} \qquad \{H_G\} = [I_G]\{\omega\} \qquad [\mathbf{10.4.1}]$$

We express the angular velocity vector using the unit vector along which it is directed as

$$\boldsymbol{\omega} = \omega \mathbf{e}_\omega \qquad \text{or} \qquad \{\omega\} = \omega\{e_\omega\} \qquad [\mathbf{10.4.2}]$$

where $\mathbf{e}_\omega$ is the unit vector along the instantaneous axis of rotation. The rotational kinetic energy can be expressed as

$$T_{\text{rot}} = \frac{1}{2}\{\omega\}^T[I_G]\{\omega\} = \frac{1}{2}\omega^2\{e_\omega\}^T[I_G]\{e_\omega\} = \frac{1}{2}I^*\omega^2 \qquad [\mathbf{10.4.3}]$$

where $I^* = \{e_\omega\}^T[I_G]\{e_\omega\}$ is the mass moment of inertia about the instantaneous axis of rotation. Note that I^* is not constant, and its value changes as the angular velocity changes.

We next compare the expression for the kinetic energy with the inertia ellipsoid relations defined in Chapter 6. We defined the quadratic expression $\{u\}^T[I_G]\{u\}$, where $\{u\} = [u_1 \ u_2 \ u_3]^T$ was a vector whose magnitude was 1, $\{u\}^T\{u\} = 1$. We saw that the expression

$$\frac{1}{\lambda}\{u\}^T[I_G]\{u\} = 1 \qquad [10.4.4]$$

defined the *inertia ellipsoid*, a closed quadratic surface (Fig. 6.16). The value of λ depends on the orientation of the vector $\{u\}$. When $\{u\}$ coincides with the unit vector along a principal axis, λ becomes the principal moment of inertia corresponding to the principal axis. The maximum and minimum moments of inertia are the major and minor axes of the ellipsoid of inertia.

In the expression for kinetic energy in Eq. [10.4.3], $\{e_\omega\}$ is analogous to $\{u\}$ in Eq. [10.4.4]. Similarly, the term λ is analogous to T_{rot}. It follows that the expression for kinetic energy can be described in terms of an ellipsoid of inertia. For a given level of kinetic energy, the angular velocity will be smallest about the axis with the largest moment of inertia.

It is useful to express the inertia ellipsoid in terms of normalized quantities. To this end, we define a vector $\boldsymbol{\rho} = \mathbf{e}_\omega/\sqrt{I^*}$, or $\mathbf{e}_\omega = \boldsymbol{\rho}\sqrt{I^*}$. Substituting these expressions for the kinetic energy yields

$$T_{\text{rot}} = \frac{1}{2}\omega^2\{e_\omega\}^T[I_G]\{e_\omega\} = \frac{1}{2}\omega^2 I^*\{\rho\}^T[I_G]\{\rho\} \qquad [10.4.5]$$

from which we conclude that

$$\{\rho\}^T[I_G]\{\rho\} = 1 \qquad [10.4.6]$$

From now on we will use Eq. [10.4.6] to describe the inertia ellipsoid associated with the kinetic energy. Note that the inertia ellipsoid is fixed on the body. This ellipsoid is illustrated in Fig. 6.17.

We next explore the relationship between the angular momentum about the center of mass and the inertia ellipsoid. We define the intersection between the inertia ellipsoid and the angular velocity vector by P. We denote the unit normal vector to the inertia ellipsoid at point P by $\mathbf{n}$, as shown in Fig. 10.5. In Chapter 4, we saw that given a surface defined by $f(x, y, z) = 0$, we can find the unit vector normal to that surface by means of the del operation and by then normalizing the magnitude of the found vector. For example, given the equation of a circle of $x^2 + y^2 = R^2$, the normal to the circle is found by $\nabla f = \nabla(x^2 + y^2 - R^2) = \partial f/\partial x \mathbf{i} + \partial f/\partial y \mathbf{j} = 2x\mathbf{i} + 2y\mathbf{j}$. Extending this case to the inertia ellipsoid and the column vector representation, we can express the normal to the ellipsoid as

$$K\{n\}^T = \nabla(\{\rho\}^T[I_G]\{\rho\} - 1) = \frac{d(\{\rho\}^T[I_G]\{\rho\} - 1)}{d\{\rho\}} = 2\{\rho\}^T[I_G] \qquad [10.4.7]$$

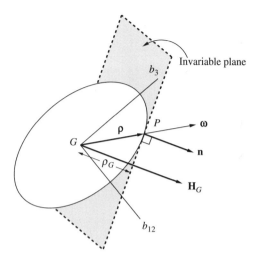

Figure 10.5 Inertia ellipsoid and the invariable plane

where K is a normalization constant. However, the vectors $\{\omega\}$ and $\{\rho\}$ are parallel, and $\{H_G\} = [I_G]\{\omega\}$. We can then write

$$K^*\{n\} = \{H_G\} \qquad [10.4.8]$$

where K^* is yet another constant, showing that the *normal to the inertia ellipsoid at point* P *is always parallel to the angular momentum vector.*

Let us analyze the physical interpretation of this result. The vector $\{\rho\}$ lies on the instantaneous axis of rotation, so that the rotational motion can be viewed as the rolling (without slipping) of the inertia ellipsoid, with the normal to the ellipsoid at point P always perpendicular to a plane whose orientation is fixed. This plane is called the *invariable plane*. While this plane can translate, it cannot rotate.

As discussed before, the point of tangency (contact) between the inertia ellipsoid and the invariable plane is on the instantaneous axis of rotation. As the angular velocity changes, so does the orientation of the inertia ellipsoid and of the point of contact with the invariable plane. The loci of point P, that is, of the tangent points on the ellipsoid as it rolls on the invariable plane, are called *polhodes*, and the loci of the tangent points on the invariable plane are called *herpolhodes*. The polhodes are closed curves because the kinetic energy is constant.

It is of interest to look at the component of $\boldsymbol{\rho}$ along the fixed direction of the angular momentum. From Fig. 10.5, denoting this distance by ρ_G we can write

$$\rho_G = \boldsymbol{\rho} \cdot \mathbf{n} \qquad [10.4.9]$$

where $\mathbf{n}$ is the unit vector along the direction of the angular momentum, $\mathbf{n} = \mathbf{H}_G/H_G$. Noting that the vector $\boldsymbol{\rho}$ can be written as $\boldsymbol{\rho} = \boldsymbol{\omega}/(\omega\sqrt{I^*})$, and that $2T_{\text{rot}} = I^*\omega^2$, we obtain

$$\rho_G = \frac{1}{\sqrt{I^*}}\left(\frac{\boldsymbol{\omega} \cdot \mathbf{H}_G}{\omega H_G}\right) = \frac{2T_{\text{rot}}}{H_G\sqrt{2T_{\text{rot}}}} = \frac{\sqrt{2T_{\text{rot}}}}{H_G} = \text{constant} \quad [\mathbf{10.4.10}]$$

so that the perpendicular distance from the center of mass to the invariable plane is constant. Equation [10.4.10] is yet another integral of the motion and, as other integrals of the motion, it depends on the initial conditions with which the body is released.

Up to now, the orientation of the body-fixed coordinate axes was not specified. Without loss of generality, and in order to simplify the derivations, we now consider the body axes to be the principal axes. The inertia matrix is now diagonal, and

$$H_G^2 = \{\omega\}^T [I_G][I_G]\{\omega\} = \sum_{i=1}^{3} I_i^2 \omega_i^2 \quad [\mathbf{10.4.11}]$$

and, using the relation $\{\omega\} = \omega\sqrt{I^*}\{\rho\}$, or $\omega_i = \omega \rho_i \sqrt{I^*}$ ($i = 1, 2, 3$), we can write

$$H_G^2 = I^* \omega^2 \sum_{i=1}^{3} I_i^2 \rho_i^2 \quad [\mathbf{10.4.12}]$$

which can be rearranged as

$$I_1^2 \rho_1^2 + I_2^2 \rho_2^2 + I_3^2 \rho_3^2 = \{\rho\}^T [I_G][I_G]\{\rho\} = \frac{H_G^2}{2T_{\text{rot}}} = \text{constant} \quad [\mathbf{10.4.13}]$$

This equation defines yet another closed quadratic surface, that is, another ellipsoid of inertia that is attached to the body. Comparing Eqs. [10.4.10] and [10.4.13], we conclude that

$$\frac{H_G^2}{2T_{\text{rot}}} = \frac{1}{\rho_G^2} \quad [\mathbf{10.4.14}]$$

We next relate the motion of the ellipsoid defined in Eq. [10.4.13] to the inertia ellipsoid defined in Eq. [10.4.6]. The two ellipsoids are in contact with each other at point P, leading to the conclusion that as the body rotates the intersection of the two ellipsoids describes the polhodes. We can construct the polhode curves by solving Eqs. [10.4.6] and [10.4.13] simultaneously. Note that if we consider Eq. [10.4.6] as describing a closed surface and Eq. [10.4.13] as a constraint relation, the result becomes a set of curves on the closed surface. We get different curves for different starting points.

One can simplify the construction and interpretation of the polhode curves by looking at their projections along the principal axes. To this end, we define the variable

$$D = \frac{H_G^2}{2T_{\text{rot}}} = \frac{1}{\rho_G^2} \quad [\mathbf{10.4.15}]$$

10.4 POLHODES

whose value depends on the initial conditions. It follows from the above discussion that $I_{\min} \leq D \leq I_{\max}$ and that when $D = I_i$ ($i = 1, 2, 3$) this corresponds to an angular motion about the axis b_i. Without loss of generality, we select our body axes $b_1 b_2 b_3$ such that $I_1 < I_2 < I_3$. This implies that the inertia ellipsoid is longest in the b_1 direction.

We next find the projections of the polhode curves on the $b_1 b_2$, $b_1 b_3$, and $b_2 b_3$ planes. Recall that the polhode equations are

$$I_1 \rho_1^2 + I_2 \rho_2^2 + I_3 \rho_3^2 = 1 \qquad I_1^2 \rho_1^2 + I_2^2 \rho_2^2 + I_3^2 \rho_3^2 = D \quad \textbf{[10.4.16]}$$

Multiplying the first of Eqs. [10.4.16] by I_3 and subtracting the second from it yields the projections of the curves on the $b_1 b_2$ plane as

$$I_1(I_3 - I_1)\rho_1^2 + I_2(I_3 - I_2)\rho_2^2 = I_3 - D \qquad \textbf{[10.4.17]}$$

In a similar fashion, we obtain the polhode equations on the $b_2 b_3$ and $b_1 b_3$ planes as

$b_2 b_3$ plane: $\qquad I_2(I_1 - I_2)\rho_2^2 + I_3(I_1 - I_3)\rho_3^2 = I_1 - D \qquad \textbf{[10.4.18]}$

$b_1 b_3$ plane: $\qquad I_3(I_2 - I_3)\rho_3^2 + I_1(I_2 - I_1)\rho_1^2 = I_2 - D \qquad \textbf{[10.4.19]}$

All the coefficients of Eqs. [10.4.17] and [10.4.18] have the same sign, indicating that these two equations represent ellipses for all values of D. By contrast, the coefficients on the left side of Eq. [10.4.19] have differing signs, thus describing hyperbolas. Furthermore, the sign of $I_2 - D$ depends on the problem. The three projections are given in Figs. 10.6. The special case of $D = I_2$ has an interesting interpretation. The projection on the $b_1 b_3$ plane is in the form of straight lines that separate the hyperbolas. In analogy from stability theory, these lines are called separatrices (plural of separatrix), separating different regions of the motion that have different properties.

From the information above, we can construct the polhode curves in three dimensions. For a rigid body with $I_1 < I_2 < I_3$, the curves are given in Fig. 10.7. Note that when $D < I_2$, the polhode curves are closed around the b_1 axis, and when

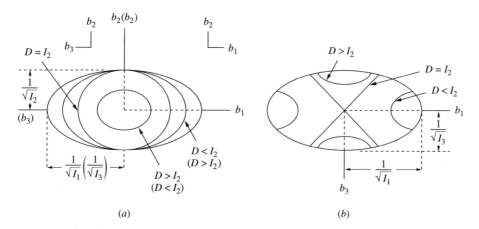

Figure 10.6 Polhode projections. (a) $b_1 b_2$ and $b_2 b_3$ planes, (b) $b_1 b_3$ plane.

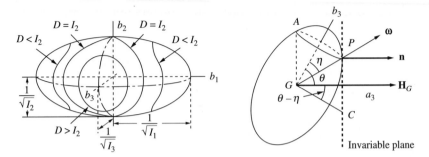

Figure 10.7   **Figure 10.8**

$D > I_2$, the polhode curves are closed around the b_3 axis. The separatrices, which are constructed for $D = I_2$, give the limiting cases. As D approaches I_1 or I_3, the polhode curves become smaller in size. The locus of the line joining the center of mass and the polhode curves defines the body cones when viewed from a body-fixed reference frame. However, unlike the symmetric body considered in Section 10.2, the cones have no symmetry.

We can analyze the motion of the body by examining the polhode curves. Consider the case where we give a large angular velocity in one direction and much smaller angular velocities in the other directions. According to the stability results in Example 8.6, initial motion that is predominantly about the largest or smallest axes of inertia will be stable, and initial motion about the intermediate axis of inertia (b_2 in our case) will be unstable. In this latter case, angular velocities about the other axes will grow with time, and all three angular velocities will become comparable in magnitude. We can demonstrate this using the polhode curves. The angular velocity of the body is along the line connecting the center of mass to the polhode curves. When the initial angular velocity is along the b_1 or b_3 axes, the value of D is slightly larger than I_1 or slightly smaller than I_3, respectively. In both cases, the polhode curves are small. Hence, the motion retains its characteristics. By contrast, if the bulk of the initial angular velocity is along the b_2 axis, then $D \approx I_2$ and the polhode curves are close to the separatrices. Hence, the polhode curves are large, which means that the magnitude and direction of the angular velocity varies a lot. This, in essence, is the explanation for the wobbly motion and also why it is difficult to visualize the motion of an arbitrary body.

Now consider the special case of axisymmetric bodies. Without loss of generality, we set $I_1 = I_2$ and $I_1 < I_3$. The inertia and momentum ellipsoids are described by

$$I_1(\rho_1^2 + \rho_2^2) + I_3\rho_3^2 = 1 \qquad I_1^2(\rho_1^2 + \rho_2^2) + I_3^2\rho_3^2 = D \qquad [10.4.20]$$

so that both ellipsoids are symmetric about the b_3 axis. It follows that the intersection of the ellipsoids defines circles about the b_3 axis, with the radius of the circles determined by the value of D. Indeed, introducing $I_1 = I_2$ into Eq. [10.4.17]

$$\rho_1^2 + \rho_2^2 = \frac{D - I_3}{I_1(I_1 - I_3)} \qquad [10.4.21]$$

In this case D varies between I_1 and I_3 and $D = I_2 = I_1$ no longer represents a separatrix. The body cone generated by the locus of the line joining the center of mass and the polhode curve is axisymmetric. To show that these body cones are indeed the same as the cones discussed in Section 10.2, consider Fig. 10.8, which shows the ellipsoid for a given value of D. This figure can also be considered as the projection of the ellipsoid to the plane generated by the intersection of b_3 and a_3. To show that the angles η and β (from the definition of body cone) are the same, we need to demonstrate that Eq. [10.2.25] holds, or

$$I_1 \tan \eta = I_3 \tan \theta \qquad [10.4.22]$$

The proof is left as an exercise. The projections of the polhode curves onto the $b_2 b_3$ or $b_1 b_3$ planes represent straight lines and they do not have a special physical interpretation. One can show that the cone described by the points G, P, and C is the space cone.

10.5 MOTION OF A SPINNING TOP

In the previous sections we considered moment-free bodies. We now switch our attention to bodies subjected to external moments. An interesting case is the motion of a spinning top, such as the one shown in Fig. 10.9. A top is basically described as a body that possesses inertial symmetry and that terminates at a sharp point along the symmetry axis. This point is called an *apex* or *vertex*. Tops are designed to be axisymmetric, as such a construction increases stability and minimizes the friction due to the air mass around the top. The motion of the top can be viewed as the balancing of the gravitational moment about the apex by the gyroscopic moment. The spin of the top gives it its stability.

We assume that the apex is in continuous contact with a plane in a way that it has no translation. (The translational motion of the apex, commonly referred to as *drift*, is due to the initial translational motion given to the top as well as to the unevenness and roughness of the plane on which the top moves.)

The equations of motion for a spinning top were derived in Example 8.10 using a 3-1-3 Euler angle transformation. Here, we will analyze the integrals of the motion

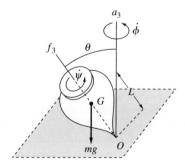

Figure 10.9

and qualitatively assess the behavior of the top. The top being an axisymmetric body, we can write the kinetic energy of the top as

$$T = T_{\text{rot}} = \frac{1}{2}i(I_1\omega_1^2 + I_2\omega_2^2 + I_3\omega_3^2) \quad [\mathbf{10.5.1}]$$

where the components of the mass moment of inertia are about point O. Introducing the values for the angular velocities for a 3-1-3 transformation, we obtain for the kinetic energy

$$T = \frac{1}{2}\left[I_1(\dot{\phi}^2 \sin^2\theta + \dot{\theta}^2) + I_3(\dot{\phi}\cos\theta + \dot{\psi})^2\right] \quad [\mathbf{10.5.2}]$$

The potential energy is

$$V = mgL\cos\theta \quad [\mathbf{10.5.3}]$$

The only other forces acting on the spinning top are those at the point of contact. They do no work, as they are being applied to a fixed point. Note that the moment about the apex O is due to the weight of the top.

The integrals of motion are the total energy and generalized momenta associated with precession and spin. The latter two integrals are recognized by noting that neither ϕ nor ψ are present in the Lagrangian. From Example 8.10, these integrals of the motion are

$$\pi_\phi = \frac{\partial T}{\partial \dot{\phi}} = I_1\dot{\phi}\sin^2\theta + I_3(\dot{\phi}\cos\theta + \dot{\psi})\cos\theta = \text{constant}$$

$$\pi_\psi = \frac{\partial T}{\partial \dot{\psi}} = I_3(\dot{\phi}\cos\theta + \dot{\psi}) = I_3\omega_3 = \text{constant} \quad [\mathbf{10.5.4}]$$

These integrals of the motion represent the components of the angular momentum along the a_3 and f_3 axes. The equations of motion associated with the precession and spin can be obtained by differentiating the above two equations. Using Lagrange's equations, the equation of motion for θ is

$$I_1\ddot{\theta} - (I_1 - I_3)\dot{\phi}^2\sin\theta\cos\theta + I_3\dot{\phi}\dot{\psi}\sin\theta - mgL\sin\theta = 0 \quad [\mathbf{10.5.5}]$$

From Eqs. [10.5.4] one can solve for the precession and spin rates as

$$\dot{\phi} = \frac{\pi_\phi - \pi_\psi\cos\theta}{I_1\sin^2\theta} \qquad \dot{\psi} = \pi_\psi\left(\frac{1}{I_3} + \frac{\cos^2\theta}{I_1\sin^2\theta}\right) - \pi_\phi\frac{\cos\theta}{I_1\sin^2\theta} \quad [\mathbf{10.5.6}]$$

and substitute into Eq. [10.5.5]. Once the nutation angle is solved for, the precession and spin rates can be calculated using Eqs. [10.5.6] and the initial conditions.

Another way of deriving the equation of motion for θ is to make use of Routh's method for ignorable coordinates, as ϕ and ψ are such coordinates. Defining the Routhian as

$$\mathcal{R} = L - \pi_\phi\dot{\phi} - \pi_\psi\dot{\psi} \quad [\mathbf{10.5.7}]$$

and introducing the expressions for $\dot{\phi}$ and $\dot{\psi}$ into it, we obtain

$$\mathcal{R} = \frac{1}{2}I_1\dot{\theta}^2 - \frac{(\pi_\phi - \pi_\psi \cos\theta)^2}{2I_1 \sin^2\theta} - \frac{\pi_\psi^2}{2I_3} - mgL\cos\theta \qquad [\mathbf{10.5.8}]$$

The Routhian can now be treated as the Lagrangian of a single degree of freedom natural system with kinetic and potential energies T' and V' of the form

$$T' = \frac{1}{2}I_1\dot{\theta}^2 \qquad V' = \frac{(\pi_\phi - \pi_\psi \cos\theta)^2}{2I_1 \sin^2\theta} + \frac{\pi_\psi^2}{2I_3} + mgL\cos\theta \qquad [\mathbf{10.5.9}]$$

Obviously, the energy integral is $E = T' + V'$. To analyze the energy integral in more detail, we introduce the constant quantities

$$\alpha = \frac{2}{I_1}\left(E - \frac{\pi_\psi^2}{2I_3}\right) \qquad \beta = \frac{2mgL}{I_1} \qquad a = \frac{\pi_\phi}{I_1} \qquad b = \frac{\pi_\psi}{I_1} \qquad [\mathbf{10.5.10}]$$

and the variable $u = \cos\theta$. Note that a and b have the units of angular velocity and that all of the above quantities are calculated from the initial conditions. Also, u is a nondimensional quantity describing the elevation of a point on the symmetry axis that is at a distance of unity from the apex. The energy integral can thus be written as a cubic function

$$\dot{u}^2 = f(u) = (\alpha - \beta u)(1 - u^2) - (a - bu)^2 \qquad [\mathbf{10.5.11}]$$

We will study the characteristics of the motion by qualitatively analyzing the function $f(u)$. Other methods of analysis include numerical integration of Eq. [10.5.11] or of Eq. [10.5.5], as well as a separation of variables. The latter leads to an elliptic integral for time in terms of u.

A very interesting special case is when $\dot{u} = 0$, as it corresponds to zero nutation rate. The values of u that lead to a zero nutation rate range can be obtained by solving $f(u) = 0$. To this end, we consider the characteristics of $f(u)$. Because u is defined as $u = \cos\theta$, we are interested in the roots of $f(u)$ in the range $-1 \leq u \leq 1$. Further, we are interested in the values of u that are larger than zero, as a negative value for u implies that the top is spinning below the platform it is on. We make the following observations regarding the roots of $f(u)$:

1. Both $f(1)$ and $f(-1)$ are less than zero.
2. As u becomes larger, $f(u) \approx \beta u^3 > 0$, since $\beta > 0$.

Hence, $f(u)$ has a root for $u > 1$ and $f(u) < 0$ for $u = -1$. This implies that in the range of $-1 \leq u \leq 1$ there either are no roots, or there are two roots. We discount the possibility that $f(u)$ has no roots in this range, as this would imply that there are no positive values for the nutation angle. We thus consider the case of two roots for $f(u)$ in the range $-1 \leq u \leq 1$. Fig. 10.10 shows a typical plot of $f(u)$. The two roots, denoted by u_1 and u_2, are in general both positive or both negative. Let us consider the case when they are both positive, corresponding to an upright top, and order them without any loss of generality as $u_1 \leq u_2$. This way, u_2 corresponds to a smaller nutation angle. The physical range

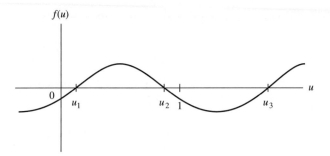

Figure 10.10

of the motion of the top is then between u_1 and u_2. The motion can be visualized as the change in the nutation angle from $\theta_2 = \cos^{-1}(u_2)$, corresponding to the highest elevation of the top, to $\theta_1 = \cos^{-1}(u_1)$, corresponding to the lowest elevation.

The motion of a point on the symmetry axis of the top at a distance of unity from the apex can be viewed as tracing a curve on the unit sphere, as in Fig. 10.11. The circles on the sphere that correspond to the locus of all points with $\theta = \cos^{-1}(u_1)$ and $\cos^{-1}(u_2)$ are called the *bounding circles,* and the f_3 axis basically moves between these circles. The precession rate, from Eq. [10.5.6] can be shown to be

$$\dot{\phi} = \frac{a - bu}{1 - u^2} \qquad [10.5.12]$$

so that the precession rate depends on the value of $a - bu$. Let us introduce the quantity u_0, based on the initial conditions, and defined as

$$u_0 = \frac{\pi_\phi}{\pi_\psi} = \frac{a}{b} = \frac{I_1 \sin^2 \theta_0 \dot{\phi}_0}{I_3(\cos \theta_0 \dot{\phi}_0 + \dot{\psi}_0)} + \cos \theta_0 \qquad [10.5.13]$$

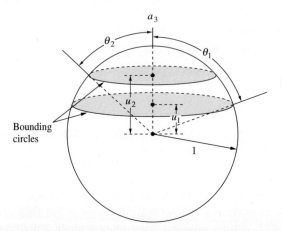

Figure 10.11   Bounding circles

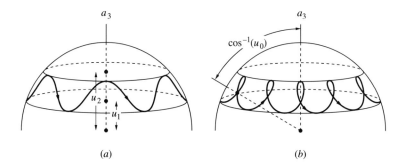

Figure 10.12 (a) Unidirectional and (b) looping precession

For the general case of motion, $u_0 \neq \cos\theta_0$. It turns out that characteristics of the motion of the top depend on the relationship of u_0 to u_1 and u_2. To find u_0 one begins with the initial conditions $\dot\phi_0$, θ_0, and $\dot\psi_0$, and the property that $\dot\theta_0 = 0$. The generalized momenta π_ϕ, π_ψ, and a and b are calculated using Eqs. [10.5.10] and [10.5.4]. The results are substituted into Eq. [10.5.13] to find u_0.

We identify the following different types of motion based on the initial conditions:

Unidirectional Precession In this case, $u_0 > u_2$ (or $u_0 < u_1$). This case can also be described as having the same precession direction at the bounding circles, so that $a > bu_2$ (or $a < bu_1$). The precession rate is not zero at both bounding circles. This kind of motion can be initiated by releasing a top from its highest elevation point $u = u_2$, with zero nutation rate and a precession rate high enough to make u_0 greater than u_2. After the top is released, it begins to fall as a result of the action of gravity. In the process, the top attains precessional motion. At $u = u_1$, when the maximum kinetic energy is reached, the nutation rate changes sign and the top begins to rise until $u = u_2$, and the process repeats itself. The resulting motion is a periodic motion of the nutation angle, like a sinusoid, as shown in Fig. 10.12a.

Looping Precession Here, $u_1 < u_0 < u_2$, and the direction of precession is different at the bounding circles (Fig. 10.12b). At the bounding circles, the nutation rate becomes zero. At the top bounding circle, $\theta = \theta_2$, the precession rate is negative; when $\theta = \theta_1$, the precession rate is positive. When u becomes equal to u_0 the precession rate becomes zero, which explains the looping nature of the motion of the symmetry axis. Such motion can be initiated by releasing the top at $\theta = \theta_2$, with a small precession rate and zero nutation rate so as to satisfy $u_1 < u_0 < u_2$. The top begins with its initial precession, and when the point corresponding to $\theta = \cos^{-1}(a/b) = \cos^{-1} u_0$ is reached, the precession rate changes sign. The motion continues in this fashion until the nutation angle reaches the lower bounding circle $u = u_1$. At that point, the elevation of the symmetry axis begins to increase. When $u = u_0$ is reached, the precession again changes sign. Fig. 10.13 gives a typical plot of the precession rate $\dot\phi$. In essence, the precession rate oscillates between its bounds at θ_1 and θ_2, changing sign at $\theta = \cos^{-1}(u_0)$.

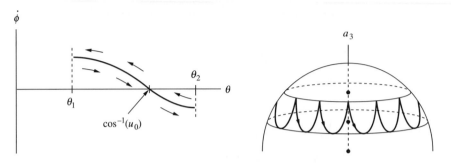

Figure 10.13 **Figure 10.14**

Cuspidal (or Cuspidial) Motion This is the limiting case between the two other types of precession. Here, $u_0 = u_2$, so that $a = bu_2$. From Eq. [10.5.12] we observe that the precession rate becomes zero at $u = u_2$ and is greater than zero at all other times. This type of motion can be initiated by releasing the top at $u = u_2$, with no initial precessional or nutational motion. The motion characteristics are similar to those in unidirectional precession, with the difference being a zero precession rate at $u = u_2$. The top is released at $u = u_2$ with zero precession. As the top begins to precess, its symmetry axis begins to fall, resulting in an increase of kinetic energy and reduction in potential energy. At $u = u_1$, the top begins to rise, and when it reaches $u = u_2$ the precession rate becomes zero. The symmetry axis traces cusps, as shown in Fig. 10.14. One cannot have cuspidal motion if the top is released with $u_0 = u_1$, because the tendency of the top is to fall.

To obtain more insight into the nature of the motion of tops, we analyze Eq. [10.5.11] further. Recall that all of the above results were obtained for $f(u) = \dot{u} = 0$, which corresponds to zero nutation rate as an initial condition. Because cuspidal motion represents a limiting case, consider an initial condition corresponding to this type of motion. Hence, we have an initial spin rate of $\dot{\psi}_0$ and zero initial precession, with an initial nutation angle of θ_0 and no initial nutation rate. In this special case, u_0 and θ_0 are related by Eq. [10.5.13] and that $u_0 = \cos\theta_0$. It follows from Eqs. [10.5.4] that $\pi_\psi = I_3\dot{\psi}_0 = I_3\omega_3$, and ω_3 is constant. We also have as initial conditions

$$a = bu_0 \quad \text{or} \quad u_0 = u_2 \qquad [10.5.14]$$

which, when substituted into Eq. [10.5.11] and $f(u)$ set to zero, yields $\alpha = \beta u_0$. Introduction of this into Eq. [10.5.11] yields

$$f(u) = (u_0 - u)[\beta(1 - u^2) - b^2(u_0 - u)] \qquad [10.5.15]$$

so that one root of $f(u)$, denoted by u_2, is $u_2 = u_0$. The remaining equation to solve for the other roots can be written as

$$u^2 - 2\mathcal{E}u + 2\mathcal{E}u_0 - 1 = 0 \qquad [10.5.16]$$

where

$$\mathcal{E} = \frac{b^2}{2\beta} = \frac{\pi_\psi^2}{4I_1 mgL} = \frac{I_3^2\dot{\psi}_0^2}{4I_1 mgL} \qquad [10.5.17]$$

From a physical perspective, $\mathscr{E}$ denotes the ratio of kinetic energy to the gravitational potential energy, multiplied by the ratio of the mass moments of inertia. The kinetic energy is associated with the gyroscopic moment and the potential energy is due to gravity. The other two roots are

$$u_1 = \mathscr{E} - \sqrt{\mathscr{E}^2 - 2u_0\mathscr{E} + 1} \qquad u_3 = \mathscr{E} + \sqrt{\mathscr{E}^2 - 2u_0\mathscr{E} + 1} \qquad [10.5.18]$$

Of these roots, u_3 is greater than one and without physical significance. To show this, recall that the magnitude of u_0 is smaller than 1, so that $\sqrt{\mathscr{E}^2 - 2u_0\mathscr{E} + 1} > \sqrt{\mathscr{E}^2 - 2\mathscr{E} + 1} = |\mathscr{E} - 1|$. By the same argument, one can show that $|u_1| < 1$. If the top is given a fast spin as an initial condition, then $\mathscr{E} \gg 1$, and u_1 can be approximated using a two-term Taylor series expansion as

$$u_1 = \mathscr{E} - \mathscr{E}\sqrt{1 - \frac{2u_0}{\mathscr{E}} + \frac{1}{\mathscr{E}^2}} \approx u_0 - \frac{1 - u_0^2}{2\mathscr{E}} \qquad [10.5.19]$$

which gives an indication of how the lower bounding circle varies as a function of $\mathscr{E}$. Because the difference between u_1 and u_0 is not large, one can introduce the perturbation parameter $\varepsilon = u_0 - u$ to Eq. [10.5.11] and obtain

$$\dot{\varepsilon}^2 = \varepsilon\beta[1 - (u_0 - \varepsilon)^2] - b^2\varepsilon^2 \qquad [10.5.20]$$

Neglecting terms in ε that are of higher order than 2, and noting that $b^2 \gg 2\beta u_0$ for fast spin, this equation can be simplified as

$$\dot{\varepsilon}^2 = \varepsilon\beta(1 - u_0^2) - b^2\varepsilon^2 \qquad [10.5.21]$$

This equation can be solved by converting it into an elliptic integral by taking the square roots of both sides and separating variables. A more physically meaningful solution can be obtained by differentiating it with respect to time and dividing the resulting equation by $\dot{\varepsilon}$, which yields

$$\ddot{\varepsilon} + b^2\varepsilon = \beta\frac{1 - u_0^2}{2} \qquad [10.5.22]$$

subject to the initial conditions $\varepsilon(0) = 0$, $\dot{\varepsilon}(0) = 0$. This basically is the equation of a simple sinusoid subjected to a constant excitation. The solution can be shown to be

$$\varepsilon(t) = \frac{\beta(1 - u_0^2)}{2b^2}(1 - \cos bt) = \frac{1 - u_0^2}{4\mathscr{E}}(1 - \cos bt) \qquad [10.5.23]$$

Differentiating the above equation and recalling the definition of ε, we obtain the rate of change of u as

$$\dot{u} = -\dot{\theta}\sin\theta = -\frac{b(1 - u_0^2)}{4\mathscr{E}}\sin bt \qquad [10.5.24]$$

and, because the value of the angle θ does not change too much during the motion, its sine can be approximated as $\sin\theta = \sqrt{1 - u^2} \approx \sqrt{1 - u_0^2}$. Hence, an approximate expression for $\dot{\theta}$ can be written as

$$\dot{\theta}(t) \approx \frac{b\sqrt{1-u_0^2}}{4\mathcal{E}} \sin bt = \frac{b \sin \theta_0}{4\mathcal{E}} \sin bt \qquad [\mathbf{10.5.25}]$$

In a similar fashion, we approximate the precession rate. Introducing Eqs. [10.5.23] into Eq. [10.5.12] and using the approximation $(1 - u^2) \approx (1 - u_0^2)$, we obtain for the precession rate

$$\dot{\phi}(t) \approx \frac{b}{4\mathcal{E}}(1 - \cos bt) \qquad [\mathbf{10.5.26}]$$

The motion can be interpreted as follows. For cuspidal motion the precession rate oscillates between $\dot{\phi} = 0$ and $\dot{\phi} = b/2\mathcal{E}$, so that its average value is

$$\dot{\phi}_{av} = \frac{b}{4\mathcal{E}} = \frac{\beta}{2b} = \frac{mgL}{I_3 \dot{\psi}_0} \qquad [\mathbf{10.5.27}]$$

The nutation rate oscillates like a sinusoidal of amplitude $b \sin \theta_0 / 4\mathcal{E}$. The frequency of oscillation is the same for both nutation and precession. At the point of zero nutation, $\sin bt = 0$ and $\cos bt = \pm 1$, so that the precession rate is either zero (which corresponds to the highest elevation of the top) or at its maximum value (which corresponds to the lowest elevation).

Steady Precession As the initial spin rate of the top is increased, the amplitudes of the precession and nutation become smaller, while the frequency of oscillation b becomes larger. As a result, u_1 and u_2 become very close to each other, the nutation and the oscillatory part of the precession become smaller and more difficult to observe visually, which gives the appearance that the top is precessing uniformly with no variation in the nutation angle. This case is known as *steady precession*.

To analyze steady precession in more detail, we set $\ddot{\theta} = 0$ in the equation of motion for the nutation angle. We rearrange Eq. [10.5.5] as a quadratic function in terms of $\dot{\phi}$ as

$$\dot{\phi}^2 - \frac{I_3 \omega_3}{I_1 \cos \theta} \dot{\phi} + \frac{mgL}{I_1 \cos \theta} = 0 \qquad [\mathbf{10.5.28}]$$

We solve this equation for $\dot{\phi}$ to yield

$$\dot{\phi}_{1,2} = \frac{I_3 \omega_3}{2 I_1 \cos \theta} \pm \frac{I_3 \omega_3}{2 I_1 \cos \theta} \sqrt{1 - \frac{4 I_1 mgL \cos \theta}{I_3^2 \omega_3^2}} \qquad [\mathbf{10.5.29}]$$

In order to have steady precession, these solutions must be real, which implies that the radical in the above equation must be positive. From this we obtain the critical angular velocity requirement for steady precession as

$$\omega_3^2 > \frac{4 I_1 mgL \cos \theta}{I_3^2} \qquad [\mathbf{10.5.30}]$$

Equation [10.5.30] describes the minimum angular velocity that must be imparted in the b_3 direction for steady precession. For large values of the spin, one can treat the term $4I_1 mgL \cos\theta/(I_3\omega_3)^2$ as small and approximate the precession rates in Eq. [10.5.29] as

$$\dot{\phi}_2 \approx \frac{mgL}{I_3\omega_3} = \frac{\beta}{2b} \qquad \dot{\phi}_1 \approx \frac{I_3\omega_3}{I_1 \cos\theta} = \frac{b}{\cos\theta} \qquad [10.5.31]$$

The two roots correspond to fast and slow precession. The smaller root, $\dot{\phi}_2$, corresponds to the slow precession rate; it is the same as the average precession rate derived for cuspidal motion in Eq. [10.5.27]. The larger root, $\dot{\phi}_1$, corresponds to fast precession; it is usually not attainable, because of the very high spin rate required to achieve this motion. Actually, there is an interesting interpretation of the fast precession rate. Expressing it as

$$I_1 \cos\theta \dot{\phi}_1 = I_3\omega_3 = I_3(\cos\theta \dot{\phi}_1 + \dot{\psi}) \qquad [10.5.32]$$

we can relate the precession and spin rates as

$$\dot{\phi}_1 = \frac{I_3 \dot{\psi}}{(I_1 - I_3)\cos\theta} \qquad [10.5.33]$$

This equation is the same as Eq. [10.2.24], which we derived for torque-free motion. The spin in this case is so high that the moment exerted on the top by gravity becomes insignificant.

A special case of steady precession is when the nutation angle θ is zero, which implies that the top is rotating in the upright position. To an observer the top appears motionless (the axis of the top and F frame are not moving) unless there are distinguishing marks on it. This type of motion is known as the *sleeping top*. The precession and spin rates are added linearly, $\omega_3 = \dot{\phi} + \dot{\psi}$, and from Eq. [10.5.30] the minimum value of the angular velocity required becomes

$$\omega_{cr}^2 = \frac{4I_1 mgL}{I_3^2} \qquad [10.5.34]$$

To have the sleeping top motion, the top must be released in an exactly upright position with a minimum rotation rate given in the above equation. This initial condition, of course, is hard to achieve. Also, there always is some amount of friction at the apex, which slows the top. Most tops, when released almost vertically and with a high spin, begin their motion as steady precession and with a very small nutation angle. As the effects of friction begin to build up over time, the spin rate decreases below the critical speed and nutational motion is observed. As the spin rate decreases further, the nutation angle gets bigger, until the top hits the ground. Table 10.1 summarizes the types of motion that can be achieved depending on the initial conditions. Here, we denote by θ_0 the initial angle with which the top is released (note that $u_0 = \cos\theta_0$ for this special case), and recognize from Eq. [10.5.17] that the ratio of the initial angular velocity $\omega_3 = \dot{\psi}_0$ to

Table 10.1 Summary of top motions

Initial Precession Rate $\dot{\phi}_0$	Initial Condition u_0	Initial Spin Rate $\dot{\psi}_0$	Type of Motion
0	1	$\dot{\psi}_0 = \omega_3$	Sleeping top
0	$u_0 < 1, u_0 = u_1 = u_2$	$\dot{\psi}_0 = \omega_3$	Steady precession
0	$u_0 < 1, u_0 = u_2$	$\dot{\psi}_0 = \omega_3$	Cuspidal motion
$\neq 0$	$u_0 > u_2$ or $u_0 < u_1$		Unidirectional precession
$\neq 0$	$u_1 < u_0 < u_2$		Looping precession

ω_{cr} can be expressed as

$$\left(\frac{\omega_3}{\omega_{cr}}\right)^2 = \frac{I_3^2 \dot{\psi}_0^2}{4 I_1 mgL} = \mathcal{E} \qquad [10.5.35]$$

Be aware that the values of u_1 and u_2 are dependent on the initial precession and spin rates, as well as on the nutation angle. When deciding on the type of initial condition to impart to the top, one has to conduct an analysis relating u_1 and u_2 to the initial values of the precession nutation and spin.

Up to now we hardly considered the conditions under which a top will no longer spin and fall to the ground. Confining our analysis to tops on horizontal platforms without elevated pivots, the instant the side of the top makes contact with the platform is when the nutation angle θ plus the half angle of the top becomes equal to 90 degrees (Fig. 10.15). Once this value of the nutation angle is found, the corresponding u_1 is calculated and the remaining conditions are then determined. A simple special case is cuspidal motion and setting $\theta = 90°$, so that $u_1 = 0$. Setting $u_1 = 0$ in Eq. [10.5.18] and solving for u_0 we obtain

$$u_0 = \cos\theta_0 = \frac{1}{2\mathcal{E}} = \frac{2 I_1 mgL}{I_3^2 \dot{\psi}_0^2} \qquad [10.5.36]$$

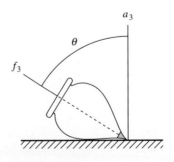

Figure 10.15

which can be solved for the initial spin rate as

$$\dot{\psi}_0^2 = \frac{2I_1 mgL}{I_3^2 \cos\theta_0} \qquad [10.5.37]$$

Equation [10.5.37] gives some interesting results. If the top is released from the upright position, $\cos\theta_0 = 1$, the minimum speed required to prevent the top from falling down is $\sqrt{2I_1 mgL/I_3^2} = \omega_{cr}/\sqrt{2}$. Any initial spin between $\omega_{cr}/\sqrt{2}$ and ω_{cr} results in a precessional motion. Also, as the initial nutation angle increases, the initial angular velocity required to keep the top spinning increases, an expected result.

10.6 ROLLING DISK

In Section 7.9 we explained the kinematics of a rolling disk, and in Chapters 8 and 9 we derived the equations of motion for the case of no slip, using a variety of approaches. Here, we qualitatively analyze the equations of motion for the case of no slip and investigate the range of validity of the no-slip assumption. As in the spinning top discussion, the general motion of a disk can be described as the contest between the moment due to gravity and the gyroscopic moment. Intuitively, we know that a disk rolled with a higher initial speed will fall down later than a disk rolled with a lower initial speed.

The disk configuration is shown in Fig. 10.16. Because of the no-slip assumption, the friction force at the contact point does no work, so there are no forces that are nonconservative. We hence conclude that energy is conserved. With regards to the generalized momenta, recall the kinetic energy expression from Eq. [e] in Example 8.13 as

$$T = \frac{1}{2}m(\dot{X}^2 + \dot{Y}^2 + R^2\dot{\theta}^2\cos^2\theta) + \frac{1}{2}\left(I_1(\dot{\phi}^2\sin^2\theta + \dot{\theta}^2) + I_3(\dot{\phi}\cos\theta + \dot{\psi})^2\right) \qquad [10.6.1]$$

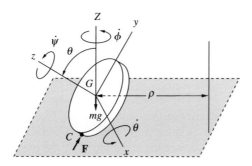

Figure 10.16

Substituting the values of the velocity components of the center of mass $\dot{X}$ and $\dot{Y}$ from Eqs. [b] and [c] in Example 8.13

$$\dot{X} = R\dot{\theta}\sin\phi\sin\theta - R(\dot{\phi}\cos\theta + \dot{\psi})\cos\phi$$
$$\dot{Y} = -R\dot{\theta}\cos\phi\sin\theta - R(\dot{\phi}\cos\theta + \dot{\psi})\sin\phi \qquad [\mathbf{10.6.2}]$$

we obtain for the kinetic energy

$$T = \frac{1}{2}\left((I_1 + mR^2)\dot{\theta}^2 + (I_3 + mR^2)(\dot{\phi}\cos\theta + \dot{\psi})^2 + I_1\dot{\phi}^2\sin^2\theta\right) \qquad [\mathbf{10.6.3}]$$

Even though the kinetic energy does not have contributions from ϕ and ψ, the generalized momenta associated with these generalized coordinates are not constant. This is because we are dealing with a constrained system, and Lagrange multipliers are present in the formulation.

The equations of motion were derived to be

$$(I_3 + mR^2)(\ddot{\phi}\cos\theta - \dot{\phi}\dot{\theta}\sin\theta + \ddot{\psi}) - mR^2\dot{\phi}\dot{\theta}\sin\theta = 0 \qquad [\mathbf{10.6.4}]$$
$$(I_1 + mR^2)\ddot{\theta} + (I_3 + mR^2)\dot{\phi}\sin\theta(\dot{\phi}\cos\theta + \dot{\psi})$$
$$-I_1\dot{\phi}^2\sin\theta\cos\theta + mgR\cos\theta = 0 \qquad [\mathbf{10.6.5}]$$
$$I_1\ddot{\phi}\sin\theta + 2I_1\dot{\phi}\dot{\theta}\cos\theta - I_3\dot{\theta}(\dot{\phi}\cos\theta + \dot{\psi}) = 0 \qquad [\mathbf{10.6.6}]$$

These are considerably more complicated than the equations for the spinning top. Consequently, it is more difficult to conduct a qualitative analysis. One special case is steady motion, with a constant nutation angle θ. To analyze the characteristics of this motion, we remove all derivatives of θ from Eqs. [10.6.4]–[10.6.6], with the result

$$(I_3 + mR^2)(\ddot{\phi}\cos\theta + \ddot{\psi}) = 0 \qquad [\mathbf{10.6.7}]$$
$$(I_3 + mR^2)\dot{\phi}\sin\theta(\dot{\phi}\cos\theta + \dot{\psi}) - I_1\dot{\phi}^2\sin\theta\cos\theta + mgR\cos\theta = 0 \qquad [\mathbf{10.6.8}]$$
$$I_1\ddot{\phi}\sin\theta = 0 \qquad [\mathbf{10.6.9}]$$

From Eq. [10.6.9] we conclude that $\ddot{\phi}$ is zero, corresponding to a constant precession rate $\dot{\phi}$. This is to be expected, as zero nutation results in steady precession in axisymmetric bodies. Introducing this to Eq. [10.6.7], we conclude that the spin rate $\dot{\psi}$ is also constant. We calculated the velocity of the center of mass for the general case in Section 7.9 as

$$\mathbf{v}_G = R\omega_x\mathbf{k} - R\omega_z\mathbf{i} = -R(\dot{\phi}\cos\theta + \dot{\psi})\mathbf{i} + R\dot{\theta}\mathbf{k} \qquad [\mathbf{10.6.10}]$$

For steady precession $\dot{\theta} = 0$, and because the precession and spin rates are constant, one concludes that the velocity of the center of mass has constant amplitude. The center of the disk follows a circular path. The radius of this path, denoted by ρ, is basically the speed of the center of mass v divided by the precession rate

$$\rho = \frac{v}{\dot{\phi}} \qquad [\mathbf{10.6.11}]$$

in which

$$v = -R(\dot\phi\cos\theta + \dot\psi) \qquad [\mathbf{10.6.12}]$$

is the constant speed of the center of mass. Combining Eqs. [10.6.11] and [10.6.12], we can express the spin rate in terms of the precession rate as

$$\dot\psi = -\dot\phi\left(\cos\theta + \frac{\rho}{R}\right) \qquad [\mathbf{10.6.13}]$$

The motion can be described in terms of two parameters, the precession and spin rates, or in terms of v and one of ρ or θ. We explore the relationship between v and these parameters by introducing Eqs. [10.6.11] and [10.6.12] into Eq. [10.6.8], with the result

$$-(I_3 + mR^2)\frac{v}{\rho}\sin\theta\frac{v}{R} - I_1\left(\frac{v}{\rho}\right)^2\sin\theta\cos\theta + mgR\cos\theta = 0 \qquad [\mathbf{10.6.14}]$$

which can be rearranged and solved for v^2 as

$$v^2 = \frac{mgR^2\rho^2\cot\theta}{(I_3 + mR^2)\rho + I_1 R\cos\theta} \qquad [\mathbf{10.6.15}]$$

For a thin disk, the mass moments of inertia can be written as $I_3 = m\kappa^2$, $I_1 = m\kappa^2/2$, where κ is the radius of gyration, so that the above equation can be expressed as

$$v^2 = \frac{2gR^2\rho^2\cot\theta}{2(\kappa^2 + R^2)\rho + \kappa^2 R\cos\theta} \qquad [\mathbf{10.6.16}]$$

For a uniformly flat disk, $\kappa^2 = R^2/2$. Hence, given an initial speed of v and initial nutation angle θ, with no nutation rate, one can determine the radius of curvature for steady precession from Eq. [10.6.11]. Of course, this relationship holds as long as the no-slip assumption is not violated.

A special case of Eq. [10.6.16] is when the disk is released in the upright position. The minimum velocity for this case can be obtained by making appropriate substitutions in Eq. [10.6.16] (θ approaches 90 degrees and ρ approaches ∞). Another way is to conduct a perturbation analysis. We expand the nutation angle and precession and spin rates about their nominal values by

$$\theta = \theta_0 + \varepsilon\theta_1 \qquad \dot\phi = \dot\phi_0 + \varepsilon\dot\phi_1 \qquad \dot\psi = \dot\psi_0 + \varepsilon\dot\psi_1 \qquad [\mathbf{10.6.17}]$$

in which ε is a small parameter. This is tantamount to assuming that the disk, rolling with steady precession, is acted upon by an impulsive force and as a result the precession, nutation, and spin values have slightly changed. Introducing these perturbed values into the equations of motion, Eqs. [10.6.4]–[10.6.6], and ignoring terms that are quadratic or higher order in ε, one gets a set of three linear equations in terms of θ_1, $\dot\phi_1$, and $\dot\psi_1$. The eigenvalues of these equations are then used to ascertain the conditions under which steady precession is possible.

For the special case of upright release there is no precession, $\dot{\phi}_0 = 0$, and the nutation angle is $\theta_0 = \pi/2$. We have

$$\cos\theta \approx -\varepsilon\theta_1 \qquad \sin\theta \approx 1 \qquad \dot{\phi} = \varepsilon\dot{\phi}_1 \qquad \dot{\theta} = \varepsilon\dot{\theta}_1 \qquad \dot{\psi} = \dot{\psi}_0 + \varepsilon\dot{\psi}_1 \quad \text{[10.6.18]}$$

Introducing these values into the equations of motion and neglecting higher-order terms, the linearized equations can be shown to be

$$(I_3 + mR^2)\ddot{\psi}_1 = 0 \qquad \text{[10.6.19]}$$

$$(I_1 + mR^2)\ddot{\theta}_1 + (I_3 + mR^2)\dot{\psi}_0\dot{\phi}_1 - mgR\theta_1 = 0 \qquad \text{[10.6.20]}$$

$$I_1\ddot{\phi}_1 - I_3\dot{\psi}_0\dot{\theta}_1 = 0 \qquad \text{[10.6.21]}$$

Equation [10.6.19] indicates that $\ddot{\psi}_1$ is zero, showing, as expected, the constancy of the spin rate. Integrating Eq. [10.6.21], we obtain

$$\dot{\phi}_1 = \frac{I_3\dot{\psi}_0}{I_1}\theta_1 \qquad \text{[10.6.22]}$$

which, when introduced into Eq. [10.6.20], yields

$$(I_1 + mR^2)\ddot{\theta}_1 + \left(\frac{(I_3 + mR^2)I_3\dot{\psi}_0^2}{I_1} - mgR\right)\theta_1 = 0 \qquad \text{[10.6.23]}$$

In order for the above equation to have a nondivergent solution, the coefficient of θ_1 must be positive. For this, the initial spin rate must satisfy the condition

$$\dot{\psi}_0^2 > \frac{mI_1gR}{(I_3 + mR^2)I_3} = \frac{gR}{2(\kappa^2 + R^2)} \qquad \text{[10.6.24]}$$

A disk rolled with an initial angular velocity $\dot{\psi}_0$ larger than the value given above will continue rolling upright. For a disk to roll in an exactly upright position it has to be released exactly upright. Otherwise, the disk will attain some precessional motion, as evidenced by Eq. [10.6.16]. In reality, because of friction and air resistance a disk rolled upright initially will eventually begin to tilt and precess. As this happens, the center of the disk no longer follows a circular path, but a path like a spiral. Ultimately either the no-slip condition is violated or the angular velocity of the disk becomes too small to sustain rolling.

To analyze the no-slip assumption in more detail, it is useful to look at the motion using the $X'Y'Z'$ frame, which is obtained after the first Euler angle rotation. Fig. 10.17 shows the free-body diagram for this case. The normal force is in the Z direction and the friction forces are shown along the X' and Y' axes. We see that if the disk slips, the slip can take place in two forms: One type of slip is along the X' direction, which is the line of nodes and basically denotes the tangent to the path of the point C'. Point C' is on the F frame and coincides with the contact point at all times. Recall from Eq. [f] in Example 9.5 that the velocity of C' is calculated as

$$\mathbf{v}_{C'} = -R\dot{\psi}\mathbf{f}_1 = -R\dot{\psi}\mathbf{I}' \qquad \text{[10.6.25]}$$

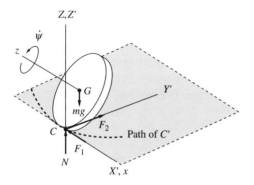

Figure 10.17

Hence, for slip that occurs along the line of nodes, the disk basically spins in place. We refer to this type of slip as *spin slip*. An example is someone moving a car from rest by depressing the accelerator fully or trying to move a car on an icy surface. If the friction between the car and road is not sufficient to prevent slip, the tires spin in place. Intuitively, we expect that in order for spin slip to occur, a large force or moment must be applied to the disk.

The second type of slip is along the Y' axis, and is very similar to the sliding of a body on a surface. We refer to this type of slip as *sliding slip*, and expect intuitively that for a freely rolling disk, if slipping occurs it will be of this type.

To examine whether slipping occurs or not, we write the translational equations of motion of a disk along the $X'Y'Z'$ axes and equate them to the external forces

$$m\mathbf{a}_G = \mathbf{F} = -F_1\mathbf{I}' + F_2\mathbf{J}' + (N - mg)\mathbf{K} \qquad [10.6.26]$$

in which N is the normal force and the expression for $\mathbf{a}_G$ is given in Eq. [7.9.26]. The relationship between the unit vectors in the F frame (xyz) and the $X'Y'Z'$ frame is given by

$$\mathbf{i} = \mathbf{I}' \qquad \mathbf{j} = \cos\theta \mathbf{J}' + \sin\theta \mathbf{K}' \qquad \mathbf{k} = -\sin\theta \mathbf{J}' + \cos\theta \mathbf{K}' \qquad [10.6.27]$$

so, introducing Eqs. [10.6.27] into Eq. [7.9.26], the components of Eq. [10.6.26] become

$$F_1 = -mR(-\ddot{\phi}c\theta + 2\dot{\phi}\dot{\theta}s\theta - \ddot{\psi}) = mR\frac{d}{dt}(\dot{\phi}c\theta + \dot{\psi}) - mR\dot{\phi}\dot{\theta}s\theta$$

$$F_2 = mR[-\dot{\phi}c\theta(\dot{\phi}c\theta + \dot{\psi}) - \dot{\theta}^2]c\theta - mR[\ddot{\theta} + \dot{\phi}s\theta(\dot{\phi}c\theta + \dot{\psi})]s\theta$$

$$= -mR\dot{\phi}(\dot{\phi}c\theta + \dot{\psi}) - mR\frac{d}{dt}(\dot{\theta}s\theta)$$

$$N - mg = mR[-\dot{\phi}c\theta(\dot{\phi}c\theta + \dot{\psi}) - \dot{\theta}^2]s\theta + [\ddot{\theta} + \dot{\phi}s\theta(\dot{\phi}c\theta + \dot{\psi})]c\theta$$

$$= mR\frac{d}{dt}(\dot{\theta}c\theta) \qquad [10.6.28]$$

The no-slip condition is given by

$$\sqrt{F_1^2 + F_2^2} \leq \mu N \qquad [10.6.29]$$

where μ is the coefficient of friction. After finding the values of F_1 and F_2, one should check if Eq. [10.6.29] holds. If analyzing the motion of the disk by numerical simulation, one should check Eq. [10.6.29] at every time step.

Let us examine the physical significance of Eqs. [10.6.28]. The Z component depends only on the change in nutation. This is logical, as only the change of elevation of the disk should affect the value of the normal force. The first term for F_1 describes the rate of change of ω_z, the angular velocity in the z direction. In the absence of large torques applied to the disk, this change will be small. The second term for F_1 is the product of the precession rate multiplied by the nutation rate, so that if the nutation rate is small, this term is small also. We conclude that unless large torques are applied along the z axis, F_1 is usually a small quantity, corroborating the earlier intuitive comments regarding spin slip. By contrast, the first term for F_2 has the precession rate multiplied by the spin rate, a much larger term than the precession rate multiplied by the nutation rate. Hence, in the absence of large torques about the z axis, the friction force in the Y' direction uses up most of the available friction force.

An interesting application of these results is for steady motion. In this case, the nutation angle and precession and spin rates are constant. Using Eqs. [10.6.11] and [10.6.12], Eqs. [10.6.28] reduce to

$$F_1 = 0 \qquad N - mg = 0 \qquad F_2 = m\frac{v}{\rho}v = \frac{mv^2}{\rho} \qquad [10.6.30]$$

Hence, there is no spin slip in this case and the friction force is required to only prevent sliding slip. The normal force is equal to the gravitational force. This is to be expected, because as the elevation of the disk does not change, the normal force remains the same. The last term in Eq. [10.6.30] is the same expression as the centripetal acceleration for constant speed. Indeed, examining Fig. 10.17 and considering steady precession we see that the X' axis corresponds to the tangential direction and the Y' axis to the normal direction. The binormal direction is the vertical.

The above results can be visualized by taking a coin and rolling it with different initial values of θ and different initial speeds of the center of the coin. Above a certain elevation of the coin, if the initial speed is large enough, the coin will roll. Below a certain elevation or if the initial speed is not large enough, the coin will slip and, depending on the initial speed, will exhibit a rolling motion with a very small radius of curvature or just fall flat down.

Example 10.2

A disk of radius 12 cm is released with a speed of its center of mass of 5 m/s and at a nutation angle $\theta_0 = 15°$ and zero nutation rate. Determine the radius of curvature tracked by the center of the disk and the minimum value of the coefficient of friction required to prevent slipping.

Solution

We assume the disk to be perfectly thin, so that the radius of gyration is related to the radius by $\kappa^2 = R^2/2$, and Eq. [10.6.16] becomes

$$v^2 = \frac{4g\rho^2 \cot\theta}{6\rho + R\cos\theta} \qquad [a]$$

which can be written as a quadratic expression in terms of the radius of curvature as

$$4g\cot\theta\,\rho^2 - 6v^2\rho - Rv^2\cos\theta = 0 \qquad \text{[b]}$$

Introducing the given information, we obtain

$$4g\cot\theta = 146.4 \text{ m/s}^2 \qquad 6v^2 = 150 \text{ m}^2/\text{s}^2 \qquad Rv^2\cos\theta = 2.898 \text{ m}^3/\text{s}^2 \qquad \text{[c]}$$

We solve the quadratic equation for ρ. The result is a positive and negative root. We discard the negative root, as it has no physical significance. The positive root is $\rho = 1.044$ m.

Because we have steady precession, the friction force is $F_1 = 0$, so that the total friction force is F_2. Also, the normal force is equal to the force of gravity. Introducing the values for v and ρ into the force balance, we obtain for the limiting case of friction

$$F_2 = \frac{mv^2}{\rho} = \mu N = \mu mg \qquad \text{[d]}$$

We can solve this equation for the friction coefficient μ to prevent slipping

$$\mu = \frac{v^2}{\rho g} = \frac{25}{1.044 \times 9.807} = 2.442 \qquad \text{[e]}$$

which indeed is a very high friction coefficient. This is to be expected, because the initial nutation angle was very low. To try a more realistic case, we set $\theta_0 = 60°$, which gives

$$4g\cot\theta = 22.65 \text{ m/s}^2 \qquad 6v^2 = 150 \text{ m}^2/\text{s}^2 \qquad Rv^2\cos\theta = 1.5 \text{ m}^3/\text{s}^2 \qquad \text{[f]}$$

and solve for ρ, with the result $\rho = 6.633$ m. It follows that the minimum coefficient of friction needed to sustain the no-slip condition is

$$\mu = \frac{v^2}{\rho g} = \frac{25}{6.633 \times 9.807} = 0.3843 \qquad \text{[g]}$$

An examination of Eqs. [c] and [f] shows that for this choice of disk radius (12 cm, indicating a small disk) and for other problems when R/ρ is small, the $Rv^2\cos\theta$ term will always be much smaller than the other terms. For such a case only, we can drop the last term in Eq. [b] and approximate the radius of curvature as

$$\rho \approx \frac{3v^2\tan\theta}{2g} \qquad \text{[h]}$$

so that the friction coefficient necessary to prevent slippage becomes

$$\mu = \frac{v^2}{\rho g} \approx \frac{2\cot\theta}{3} \qquad \text{[i]}$$

For example, when $\theta = 15°$, Eq. [i] gives the approximate value for $\mu = 2.49$, and when $\theta = 60°$, $\mu = 0.38$, which match the results above very closely. This indicates that for a small disk, the friction coefficient required to sustain no slip is dependent primarily on the nutation angle with which the disk is released. For a larger disk, the speed becomes more important as the factor to sustain the no-slip condition.

If we compare the roll of a coin to the free roll of an automobile tire (not attached to the automobile), we have two important distinctions. The radius of the coin is much smaller and the coefficient of friction between the tire and any platform is higher than that of the coin and the platform. Hence, given some initial motion, the tire keeps rolling for much longer distances, a fact known to people who have chased a tire or watched one roll off a racing car. The higher coefficient of friction and the size of the tire keeps the tire rolling more.

10.7 THE GYROSCOPE

Gyroscopes have traditionally been integral parts of vehicles as navigation systems and as devices that provide guidance by measuring directions and angular velocities. A gyroscope basically consists of a rotor spinning rapidly about its symmetry axis. The symmetry axis is allowed to rotate through a system of gimbals. When there are two gimbals, the outer gimbal permits precessional motion and the inner gimbal, to which the symmetry axis is attached, permits the nutational motion. In essence, the F frame is attached to the inner gimbal. The location of the rotor on the symmetry axis, the number, weight, and inertia properties of the gimbals, as well as weights, springs, and dashpots attached to the gimbals determine the nature of the motion.

As inertial guidance systems for vehicles, gyroscopes provide an inertial reference frame that moves with the vehicle. A gyrocompass is designed such that its precession rate is the same as that of the earth, so that it always points in the same direction. Gyroscopes that measure angular velocities are usually designed as single-axis gyroscopes.

Gyroscopes used for navigation are manufactured to very high precision standards. The analysis here can be considered as a gross simplification. From a historical standpoint, towards the end of the 20th century the importance of the classical gyroscopes—those with rotors and gimbals—somewhat diminished. This is because of the development of satellite-based navigational systems, as well as the manufacture of certain types of gyroscopes out of piezoelectric materials, with fewer moving parts.

In this section we study the free gyroscope, the gyrocompass, and the single-axis gyroscope. Except where noted, we ignore the mass and inertia properties of the gimbals. For a more complete analysis of gyroscopes, the reader is encouraged to consult the references at the end of this chapter.

10.7.1 FREE GYROSCOPE

A free gyroscope is illustrated in Fig. 10.18, with L denoting the distance of the center of the rotor from the center of the inner gimbal. When $L = 0$, the gyroscope is referred to as a *balanced gyroscope*. Such gyroscopes are used for inertial navigation systems. In this case, the rotor can rotate without any precession and nutation. The symmetry axis of the rotor is fixed in rotation, which provides the navigational requirement of a translating reference frame. The orientation of the vehicle is measured with respect to this reference frame. This measurement is also used to drive servomotors that maintain a platform inside the vehicle in a fixed orientation with respect to the earth. Accelerometers on the platform measure the translational motion of the platform with respect to the earth. It is obvious that the design and operation of such a device requires tremendous precision.

When $L \neq 0$, and the center of the rotor is above the center of the gyroscope, the gyroscope behaves like a spinning top. Unless the initial conditions are specified as those of a sleeping top, any motion of the gyroscope involves precession and nutation. When the center of the rotor is below the center of the gyroscope, we have a gyropendulum, such as the one considered in Example 8.5.

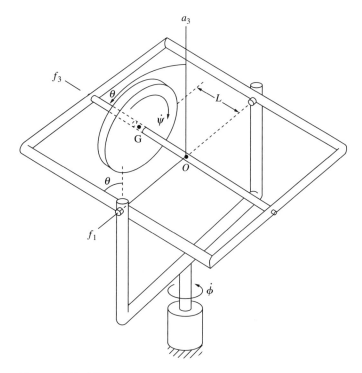

Figure 10.18

An interesting case is a gyroscope in steady precession when a disturbance is encountered. To analyze this case, we use the Routhian in Eq. [10.5.8] and derive the equation of motion in terms of the generalized momenta. The result can be shown to be

$$I_1\ddot{\theta} + I_1 \frac{ab(1 + \cos^2\theta) - (a^2 + b^2)\cos\theta}{\sin^3\theta} - mgL\sin\theta = 0 \quad \textbf{[10.7.1]}$$

where a and b are defined in Eq. [10.5.10]. Let us now take the case when the angle θ is disturbed from its steady value and analyze the ensuing motion. We can accomplish this in two ways. One is to take the expression for V in Eq. [10.5.9], differentiate it twice, and substitute the value of θ for steady precession. This approach is based on the developments in Chapter 5, for small motions about equilibrium. A second and simpler way is to replace θ in Eq. [10.7.1] with $\theta_s + \varepsilon$, where ε is a small parameter and θ_s is the nutation angle at steady state. We expand the transcendental functions in Eq. [10.7.1] and retain terms linear in ε, with the result

$$\cos\theta = \cos(\theta_s + \varepsilon) \approx \cos\theta_s - \varepsilon\sin\theta_s \qquad \sin\theta = \sin(\theta_s + \varepsilon) \approx \sin\theta_s + \varepsilon\cos\theta_s$$
$$\textbf{[10.7.2]}$$

We introduce Eqs. [10.7.2] into Eq. [10.7.1] and recognize that the terms not involving ε describe the steady precession condition, so that they vanish. After some manipulation, we obtain the linear equation

$$\ddot{\varepsilon} + \Omega^2\varepsilon = 0 \quad \textbf{[10.7.3]}$$

where

$$\Omega^2 = b^2 + \dot{\phi}_2^2 \sin^2 \theta_s - \frac{4mgL}{I_1} \cos \theta_s \qquad [10.7.4]$$

in which $\dot{\phi}_2$ is the value of the precession rate for slow steady precession, given in Eq. [10.5.31]. As the system is conservative, from the potential energy theorem, the ensuing motion will be stable if $\Omega^2 > 0$. We show that this is indeed so, by considering from Eq. [10.5.30], the critical value of the angular velocity for which steady precession is possible. Considering the definition of b as $b = I_3 \omega_3 / I_1$, we rewrite Eq. [10.5.30] as

$$b^2 - \frac{4mgL \cos \theta_s}{I_1} > 0 \qquad [10.7.5]$$

Comparing Eq. [10.7.5] with Eq. [10.7.4], once motion is initiated with steady precession, Ω^2 will be larger than zero. We conclude that a free gyroscope subjected to a small impulsive moment will oscillate about the steady precession position. In reality, as the friction and air resistance take effect, the small oscillation eventually dies out and the motion reverts to that of steady precession. We can actually visualize this by taking a spinning top and putting it into motion with steady precession. Tipping the top slightly will make the symmetry axis oscillate a bit, with the oscillation dying out. In this case, because the apex of the top is not attached, there is energy loss associated with the translation of the apex. This energy loss changes the nutation angle, as well as damping out the oscillation of the symmetry axis.

10.7.2 Gyrocompass

The gyrocompass is a device designed to determine the direction of true north (or south). Unlike a free gyroscope, whose symmetry axis is fixed in space, the gyrocompass is designed to always point to the north. This is accomplished by adding a counterweight under the inner gimbal, which results in precessional motion that can be adjusted to match the precession of the earth.

Fig. 10.19 shows a schematic of the gyrocompass. The rotor is placed at the middle of the inner gimbal and a pendulous mass m is attached to the bottom of the inner gimbal. We use a coordinate system $e_1 e_2 e_3$ attached to the earth, where e_1 is in the north, e_2 is toward the west, and e_3 is the local vertical. This system is the same as the xyz coordinate system attached to the earth that we used in Chapter 2. The angular velocity of this coordinate system is

$$^A\boldsymbol{\omega}^E = \omega_e (\cos \lambda \mathbf{e}_1 + \sin \lambda \mathbf{e}_3) \qquad [10.7.6]$$

where λ is the latitude and ω_e is the angular velocity of the earth, $\omega_e = 7.292(10^{-5})$ rad/s. As before, the inner gimbal—the F frame—is obtained by a 3-1 transformation, so that the angular velocity of the rotor with respect to the earth is

$$^E\boldsymbol{\omega}^B = \dot{\theta} \mathbf{f}_1 + \dot{\phi} \sin \theta \mathbf{f}_2 + (\dot{\phi} \cos \theta + \dot{\psi}) \mathbf{f}_3 \qquad [10.7.7]$$

10.7 THE GYROSCOPE

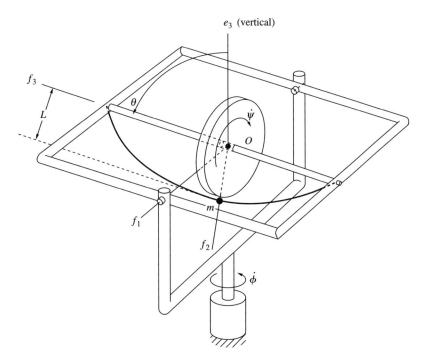

Figure 10.19

Using Table 7.1 at the end of Chapter 7, we find the unit vectors in the E and F frames are related by

$$\mathbf{e}_1 = \cos\phi \mathbf{f}_1 - \sin\phi \cos\theta \mathbf{f}_2 + \sin\phi \sin\theta \mathbf{f}_3 \qquad \mathbf{e}_3 = \sin\theta \mathbf{f}_2 + \cos\theta \mathbf{f}_3 \qquad [\mathbf{10.7.8}]$$

so we can write the angular velocity of the rotor as

$$^A\boldsymbol{\omega}^B = {}^A\boldsymbol{\omega}^E + {}^E\boldsymbol{\omega}^B = \omega_1 \mathbf{f}_1 + \omega_2 \mathbf{f}_2 + \omega_3 \mathbf{f}_3 \qquad [\mathbf{10.7.9}]$$

where

$$\omega_1 = \omega_e \cos\lambda \cos\phi + \dot{\theta} \qquad \omega_2 = -\omega_e \cos\lambda \sin\phi \cos\theta + \omega_e \sin\lambda \sin\theta + \dot{\phi}\sin\theta$$
$$\omega_3 = \omega_e \cos\lambda \sin\phi \sin\theta + \omega_e \sin\lambda \cos\theta + \dot{\phi}\cos\theta + \dot{\psi} \qquad [\mathbf{10.7.10}]$$

Because the precession and nutation rates are slow, we ignore the kinetic energy of the pendulous mass, so that the kinetic energy of the system is only due to the rotor. Further, we ignore terms quadratic in ω_e, as the angular velocity of the earth is very small. Recalling that our objective is to see if there is a set of precession and nutation angles for which the rotor axis and spin rate are constant, we consider the precession and nutation rates to be zero. Ignoring also the gimbal inertias, the kinetic energy becomes

$$T = \frac{1}{2}I_1(\omega_1^2 + \omega_2^2) + \frac{1}{2}I_3\omega_3^2 \approx \frac{1}{2}I_3\left[\dot{\psi}^2 + 2\dot{\psi}\omega_e(\cos\lambda \sin\phi \sin\theta + \sin\lambda \cos\theta)\right]$$
$$[\mathbf{10.7.11}]$$

The potential energy is due to the pendulous mass as

$$V = -mgL \sin \theta \qquad [10.7.12]$$

We next invoke Lagrange's equations for ϕ and θ, noting that the precession, nutation, and spin rates are constant. We obtain

for ϕ, $\dfrac{\partial L}{\partial \phi} = 0 \rightarrow \dot{\psi} \omega_e \cos \lambda \cos \phi \sin \theta = 0$

for θ, $\dfrac{\partial L}{\partial \theta} = 0 \rightarrow I_3 \dot{\psi} \omega_e (\cos \lambda \sin \phi \cos \theta - \sin \lambda \sin \theta) + mgL \cos \theta = 0$

$$[10.7.13]$$

For the spin angle, we note that the associated generalized momentum, π_ψ, is constant, which we can express as

$$\dot{\psi} + \omega_e(\cos \lambda \sin \phi \sin \theta + \sin \lambda \cos \theta) = \text{constant} \qquad [10.7.14]$$

We are interested in the values of ϕ and θ that make Eqs. [10.7.13] valid. For the first equation to hold, either $\cos \phi = 0$ or $\sin \theta = 0$. The case when $\sin \theta = 0$ is not meaningful, as it corresponds to the case where the inner gimbal has not moved. We then consider that $\cos \phi = 0$. This is possible for $\phi = \pm \pi/2$. Let us take the case when $\phi = \pi/2$, so that $\sin \phi = 1$. We introduce this value into the second of Eqs. [10.7.13] and solve for θ, with the result

$$\cot \theta = \frac{I_3 \dot{\psi} \omega_e \sin \lambda}{I_3 \dot{\psi} \omega_e \cos \lambda + mgL} \qquad [10.7.15]$$

so that if the gyrocompass is released from rest with the initial nutation angle given in Eq. [10.7.15], it will acquire a precession rate that is equal to the component of the angular velocity of the earth along the local horizontal.

We can simplify the expression for the nutation angle by noting that mgL is much larger than $I_3 \dot{\psi} \omega_e$, and that the mgL term dominates the above equation. Further, $\cot \theta$ is very small, and θ is very close to $\pi/2$. Introducing the expression

$$\theta = \frac{\pi}{2} - \varepsilon \qquad [10.7.16]$$

where ε is small, we can simplify Eq. [10.7.15] to

$$\varepsilon = \frac{I_3 \dot{\psi} \omega_e \sin \lambda}{mgL} \qquad [10.7.17]$$

One can show that upon the application of a small impulsive moment, the gyrocompass retains its stability. The translational motion of the gyrocompass is neglected from the above analysis. It turns out that translational motion of the gyrocompass, especially motion along a curved trajectory, reduces the accuracy of the gyrocompass even more. For this reason, gyrocompasses are more widely used in slowly moving vehicles, such as ships.

10.7.3 SINGLE-AXIS GYROSCOPE

A single-axis gyroscope is a useful tool for measuring the angular velocity of a vehicle in a particular direction. This gyroscope consists of a platform, which is attached to the vehicle, a single gimbal, and a rotor attached to the gimbal, as shown in Fig. 10.20. The symmetry axis of the rotor is normal to the axis of the gimbal. Furthermore, a torsional spring of constant k and torsional dashpot of constant c are attached to the gimbal, such that they resist the motion of the gimbal. We denote by XYZ the reference frame attached to the vehicle, with the F frame (xyz or $f_1 f_2 f_3$) attached to the gimbal, rotating about the X axis by an angle θ. The angular velocities of the vehicle, gimbal, and rotor are

$$\boldsymbol{\omega}_{\text{vehicle}} = \omega_X \mathbf{I} + \omega_Y \mathbf{J} + \omega_Z \mathbf{K}$$

$$\boldsymbol{\omega}_f = \boldsymbol{\omega}_{\text{vehicle}} + \dot{\theta}\mathbf{I}$$
$$= (\omega_X + \dot{\theta})\mathbf{i} + (\omega_Y \cos\theta + \omega_Z \sin\theta)\mathbf{j} + (-\omega_Y \sin\theta + \omega_Z \cos\theta)\mathbf{k}$$

$$\boldsymbol{\omega} = \boldsymbol{\omega}_f + \dot{\psi}\mathbf{k} = (\omega_X + \dot{\theta})\mathbf{i} + (\omega_Y \cos\theta + \omega_Z \sin\theta)\mathbf{j}$$
$$+ (-\omega_Y \sin\theta + \omega_Z \cos\theta + \dot{\psi})\mathbf{k} \qquad [\mathbf{10.7.18}]$$

We use a Lagrangian approach to analyze the system. To this end, we note that the angular velocities of the vehicle are given quantities and not generalized coordinates. Hence, we have a two degree of freedom system.

Ignoring the gimbal inertia and the effects due to the translational motion of the vehicle, the kinetic energy has the form

$$T = \frac{1}{2}I_1(\omega_1^2 + \omega_2^2) + \frac{1}{2}I_3\omega_3^2 = \frac{1}{2}I_1\left\{(\omega_X + \dot{\theta})^2 + (\omega_Y \cos\theta + \omega_Z \sin\theta)^2\right\}$$
$$+ \frac{1}{2}I_3(-\omega_Y \sin\theta + \omega_Z \cos\theta + \dot{\psi})^2 \qquad [\mathbf{10.7.19}]$$

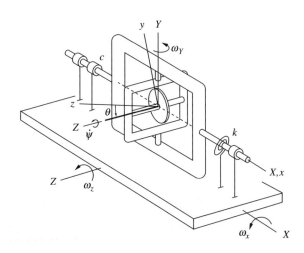

Figure 10.20

The potential energy and Rayleigh's dissipation function have the form

$$V = \frac{1}{2}k\theta^2 \qquad \mathcal{F} = \frac{1}{2}c\dot{\theta}^2 \qquad [10.7.20]$$

from which we notice that ψ is a cyclic coordinate and that the associated generalized momentum is constant

$$\pi_\psi = I_3(-\omega_Y \sin\theta + \omega_Z \cos\theta + \dot{\psi}) = \text{constant} \qquad [10.7.21]$$

Invoking Lagrange's equations, we obtain the equation of motion for θ as

$$I_1\ddot{\theta} + c\dot{\theta} + k\theta - I_1(\omega_Y \cos\theta + \omega_Z \sin\theta)(-\omega_Y \sin\theta + \omega_Z \cos\theta)$$
$$+ \pi_\psi(\omega_Y \cos\theta + \omega_Z \sin\theta) = -I_1\dot{\omega}_X \qquad [10.7.22]$$

Assuming that θ is small and that the spin of the rotor is much larger than the angular velocities of the platform, we can linearize the above equation to

$$I_1\ddot{\theta} + c\dot{\theta} + k\theta \approx \pi_\psi \omega_Y - I_1\dot{\omega}_X \qquad [10.7.23]$$

which represents the equation of motion of a damped oscillator. For steady motion of the vehicle in the X direction, or when $\dot{\omega}_X$ is either small or it takes place over a short period of time, its effect damps out. The transient effects also die out over time, and the steady-state solution of Eq. [10.7.23] becomes

$$\theta = \frac{\pi_\psi}{k}\omega_Y \qquad [10.7.24]$$

Thus, the gimbal of the rotor is tilted by an angle proportional to the angular velocity of the platform in the Y direction, that is, perpendicular to the plane of the platform. By placing three such gyroscopes on each of the XY, XZ, and YZ planes, one can measure the angular velocities of the vehicle about the X, Y, and Z axes.

Returning to the gyroscope in Fig. 10.20, if the spring is removed, then one can obtain a value of θ by integrating the angular velocity. Indeed, using either a convolution integral or a Laplace transform solution, the response of θ can be shown to be

$$\theta(t) = \frac{\pi_\psi}{c}\int_0^t \omega_Y(\sigma)\left[1 - \exp\frac{-c(t-\sigma)}{I_1}\right]d\sigma \qquad [10.7.25]$$

where we note that $\omega_Y(t)$ does not need to be constant. If the damping ratio is high, the second term on the right side of the above equation vanishes rapidly and we have

$$\theta(t) = \frac{\pi_\psi}{c}\int_0^t \omega_Y(\sigma)\,d\sigma \qquad [10.7.26]$$

so that the tilt of the gimbal is proportional to the integral of the angular velocity ω_Y over time. Such a gyroscope is called an *integrating* gyroscope.

REFERENCES

Arnold, R. N., and M. Maunder. *Gyrodynamics and Its Engineering Applications.* New York: Academic Press, 1961.
Ginsberg, J. H. *Advanced Engineering Dynamics.* New York: Harper & Row, 1988.
Goldstein, H. *Classical Mechanics.* 2nd ed. Reading, MA: Addison-Wesley, 1981.
Greenwood, D. T. *Classical Dynamics.* Englewood Cliffs, NJ: Prentice-Hall, 1977.
Junkins, J. L., and J. D. Turner. *Optimal Spacecraft Rotational Maneuvers.* Amsterdam-New York: Elsevier, 1986.
Meirovitch, L. *Methods of Analytical Dynamics.* New York: McGraw-Hill, 1970.

HOMEWORK EXERCISES

SECTION 10.2

1. A football (mass 400 g, radii of gyration 4.5 cm, 6 cm) is thrown and at the instant considered has an angular velocity about the longitudinal axis of 4 rad/s, and its center of mass has a speed of 100 km/hr. A player tips the football, and observers can see that the football begins to wobble with a nutation angle of 15°. Assuming that the tipping of the ball can be treated as an impulse, find (*a*) the angular velocity and velocity of the center of mass, (*b*) the angle β, and precession as well as spin rates, (*c*) if the impulse took 0.05 seconds, the magnitude of the applied force by the tip.

2. In Problem 8.39 sketch the body and space cones before and after impact.

3. An axisymmetric spacecraft, with $I_1/I_3 = 2.5$ has a precession rate of 0.2 rad/s and spin rate of 0.3 rad/s. It is desired to eliminate the precession by firing rockets attached to the satellite. The rockets exert impulsive moments on the spacecraft. Find the minimum value of the angular impulse that the rockets need to exert. What is the spin rate afterwards?

SECTION 10.3

4. Consider a rigid body with three distinct moments of inertia ($I_1 < I_2 < I_3$), and where $I_2 = I_1 + \varepsilon$. Solve for the precession, nutation, and spin rates in terms of ε, and find a value for ε such that the body can be treated as axisymmetric.

SECTION 10.4

5. Show that Eq. [10.4.22] holds.

SECTION 10.5

6. Write a computer program to simulate the motion of a top. Then, supply the program with the initial conditions derived in Section 10.5 and verify the results

in Section 10.5. When dealing with the sleeping top, the initial condition of θ should be taken as a very small number.

7. Use the modified Euler's equations and derive the relationship for ω_3 for steady precession.

8. Given a top with the following properties $m = 250$ g, $\kappa_1 = 4$ cm, $\kappa_3 = 2$ cm (both about center of mass), $L = 6$ cm, select the initial precession, nutation, and spin rates such that the top will have the following initial conditions:
    a. Unidirectional precession ($u_1 = 0.6$, $u_2 = 0.8$, $u_0 = 0.9$).
    b. Looping precession ($u_1 = 0.6$, $u_2 = 0.8$, $u_0 = 0.75$).
    c. Cuspidal motion ($u_1 = 0.6$, $u_2 = 0.8$, $u_0 = 0.8$).

9. Consider a spinning symmetric top. The apex of the top is on a rough surface, with coefficient of friction μ. Assuming that the normal force and friction forces are the only source for the reaction forces, derive a relationship for the minimum value of μ necessary to prevent the apex from slipping. Consider cuspidal motion.

SECTION 10.6

10. Consider a disk of weight 0.5 lb and radius $R = 4$ in. The disk is released with an initial speed of its center of 2 ft/sec, a zero nutation rate, and a finite nutation angle. Obtain the coefficient of friction required to prevent the disk from slipping as a function of the nutation angle.

11. Consider the rolling disk, rolling without slipping with a constant nutation angle of 75°. An impulsive force is applied to the disk in the f_3 direction at point A. Calculate the maximum value of the impulsive force, as a function of the precession and spin rates, as well as the coefficient of friction that will prevent the disk from slipping.

SECTION 10.7

12. Consider a free gyroscope and that a servomotor is used to keep the spin rate $\dot{\psi} = \Omega = $ constant. Derive the equations of motion, and identify the integrals of the motion. Then, given that the initial conditions are specified as steady precession, linearize the equation for θ for small disturbances, and comment on whether instability can result.

13. Consider the free gyroscope in Fig. 10.18. Now, consider that the mass moments of inertia of the gimbals are not negligible. The mass moments of inertia of the inner gimbal about the F frame are J_1, J_2, and J_3, and the mass moment of inertia of the outer gimbal about the inertial a_3 axis is J. Derive the Lagrangian, and the integrals of the motion. Use the motion integrals to obtain a single equation for θ.

14. Consider the gyrocompass in Fig. 10.19 and derive the equations of motion in the presence of gimbal inertia. Then, linearize the resulting equations about $\theta = \dfrac{\pi}{2}$ and show that Eq. [10.7.17] holds.

15. Consider the single-axis gyroscope in Fig. 10.20 and derive the equations considering the gimbal inertia. Discuss the effects of the gimbal inertia on the measurement of the angular velocity.

16. Consider the single-axis gyroscope in Fig. 10.20 and that a servomotor is used to keep the spin rate $\dot{\psi} = \Omega$ = constant. Derive the equation of motion, linearize it, and discuss the performance of the resulting angular velocity measurement system.

17. Consider the gyropendulum in Example 8.5 and find the condition for steady precession. The angular velocity of the shaft is $\dot{\phi}$. Evaluate the stability of steady precession.

chapter

11

DYNAMICS OF LIGHTLY FLEXIBLE BODIES

11.1 INTRODUCTION

Considering a body as rigid is an approximation whose validity needs to be checked at all times. Under many circumstances elastic effects have to be included in the mathematical model. Dynamical systems consisting of both rigid and elastic components have widespread applications. Examples include rotating shafts, spacecraft with flexible solar panels and antennae, and robots with both rigid and elastic links. Interactions and energy transfer between the rigid and elastic motions of a system is of utmost importance. We saw a significant example of this in 1958 in the field of spacecraft dynamics, when the Explorer satellite was launched. The vibration of the antennae resulted in energy transfer between the rigid and elastic motions of the satellite and led to nutational instabilities.

In this chapter we analyze dynamical systems that undergo large-angle rigid body motion as well as a small amount of elastic motion. The subject of motion of deformable bodies is commonly treated in depth in vibration books and there are several excellent texts available. We do not go into such an analysis in detail here and assume that the reader is familiar with the introductory theory of vibration of continuous systems. We use near linear elasticity theory and assume that the elastic deformation is small compared to the overall dimensions of the body. When the rigid body component of the motion is small, one can use linear theory and the principle of superposition for the entire motion. Small rigid body rotations generally correspond to 20 degrees or less. For larger-angle motions, linear superposition of the rigid and elastic motions loses its accuracy and interaction effects between the rigid and elastic motions become significant.

There are a number of ways to derive the equations of motion of a deformable body. One approach is Newtonian, and it is based on force and moment balances. It uses the geometry of the problem and stresses as a starting point. Another approach is analytical, and it is based on strain energy. Yet another analytical approach is the integral formulation. We use the strain energy approach in this chapter, because it is a natural extension of the analytical approaches we developed earlier. We will see a very elegant application of the extended Hamilton's principle, where both the equations of motion and the boundary conditions are derived simultaneously.

11.2 KINETIC AND POTENTIAL ENERGY: SMALL OR NO RIGID BODY MOTION

We first consider the case where the rigid body motion is small, so that it is possible to linearly superpose the rigid and elastic motions. The equations of motion can be derived the same way as when the motion is due to the elastic effects only. We first study the kinematics of the structure, and then develop expressions for the kinetic and potential energies.

11.2.1 KINEMATICS AND GEOMETRY

Consider an axially long structure, such as a straight beam of length L. We use an inertial xyz coordinate system, where the x axis is along the axis of the beam, going through the centroid of the cross section, as shown in Fig. 11.1. This axis is referred to as the *beam axis*. As a result of deformation, the beam axis moves and point A on the beam axis moves to point A^*, as shown in Fig. 11.2. The position of A^* can be expressed as

$$\mathbf{r}(x,t) = (x + u(x,t))\mathbf{i} + v(x,t)\mathbf{j} + w(x,t)\mathbf{k} \qquad [\mathbf{11.2.1}]$$

where the $u(x,t)$, $v(x,t)$, and $w(x,t)$ denote the components of the deformation in the x, y, and z directions, respectively. The distance along the beam axis to point A^* is denoted by s, similar to the definition in Chapter 1 of the distance traversed along a curve.

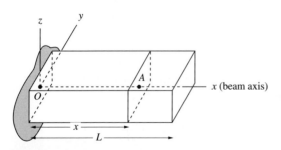

Figure 11.1

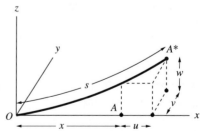

Figure 11.2 Deformation of beam axis

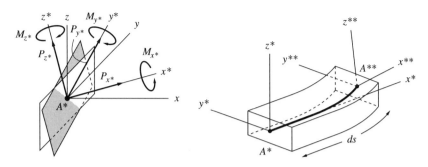

Figure 11.3   **Figure 11.4**

In addition to the deformation of the beam axis, the plane shown in Fig. 11.1 rotates, and a differential element around the beam axis assumes an orientation $x^*y^*z^*$, shown in Fig. 11.3.[1] We will quantify this rotation by the angles $\theta_x(x, t)$, $\theta_y(x, t)$, and $\theta_z(x, t)$.

We hence have six variables at each point x to describe the deformation of the beam, the three displacements and the three rotations. Fig. 11.3 shows the free-body diagram of the cross section, where there are three internal forces and three internal moments, shown along their components in the $x^*y^*z^*$ coordinate system. These are referred to as

P_{x^*}: Axial force P_{y^*}, P_{z^*}: Shear forces
M_{y^*}, M_{z^*}: Bending moments M_{x^*}: Twisting moment

We now introduce additional assumptions that simplify the problem: All deformations u, v, and w are small compared to the length of the beam, all angles θ_x, θ_y, and θ_z are small, and all shear deformation effects can be ignored. The first two assumptions are justified by limiting the analysis to beams that deform very little. The assumption of no shear deformation effects is justified analytically by taking a full-blown model of a beam and analyzing the effects of shear deformation. For slender beams, it can be shown that the shear deformation effects become negligible compared to the bending and torsion.

To examine the effects of these assumptions, consider an infinitesimal element of the beam that has an undeformed length of dx, whose deformed length becomes ds (Fig. 11.4). The coordinate axes $x^*y^*z^*$ and $x^{**}y^{**}z^{**}$ at the two ends of the differential element are obtained by the rotations $d\theta_x$, $d\theta_y$ and $d\theta_z$. Because these angles are very small, the rotation sequence to go from the $x^*y^*z^*$ to the $x^{**}y^{**}z^{**}$ coordinates is immaterial. Consider now the projections of the differential element onto the x^*y^* and x^*z^* planes, shown in Figs. 11.5 and 11.6. One can relate the rotation angles $d\theta_z$ and $d\theta_y$ to ds by ρ_y and ρ_z, the radii of curvature of the projections of the curve on the xy and xz planes, by

$$\frac{1}{\rho_y} = \frac{d\theta_z}{ds} \qquad \frac{1}{\rho_z} = -\frac{d\theta_y}{ds} \qquad [11.2.2]$$

---

[1] We are using starred coordinates $x^*y^*z^*$ to denote the coordinates associated with the rotated plane, as we use primes in this chapter to denote spatial derivatives.

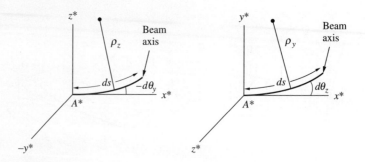

Figure 11.5 **Figure 11.6**

From Chapter 1 the radii of curvature along the xy and xz planes are given by

$$\frac{1}{\rho_y} = \frac{v''}{(1+v'^2)^{3/2}} \qquad \frac{1}{\rho_z} = \frac{w''}{(1+w'^2)^{3/2}} \qquad [11.2.3]$$

in which the prime denotes partial differentiation with respect to the spatial variable x, that is, $v'(x, t) = \partial v(x, t)/\partial x$. Because of the small angles and deflections, we approximate the expressions $1/\rho_y$ and $1/\rho_z$ in the above equation by v'' and w'', respectively, and we approximate differentiation with respect to s with differentiation with respect to x in Eq. [11.2.2]. Equating the different expressions for the radii of curvature and integrating both sides over x, we obtain

$$\theta_y = -\frac{\partial w(x, t)}{\partial x} = -w'(x, t) \qquad \theta_z = \frac{\partial v(x, t)}{\partial x} = v'(x, t) \qquad [11.2.4]$$

Hence, the six variables used for the description of the deformation are reduced to four: u, v, w, and θ_x. The relations from the force and moment balances corresponding to Eqs. [11.2.4] indicate that shear force is the derivative of the bending moment. We next assume that the torsional motion of the beam can be separated from the axial and bending motions. Axial and bending motions are coupled to torsional motions only through higher-order terms.

Having separated torsion from the rest of the motion, we now consider the axial stretch and bending in the y and z directions. Thus, the only component of the stress and strain that we consider are σ_{xx} and ε_{xx}. Effects that will be included in the formulation are the shortening of the projection of the deformed member, which results from the curvature of the beam, and rotatory inertia, which contributes to the kinetic energy as a result of rotational motion of the cross-section due to the curvature. A beam modeled this way is referred to as a *Rayleigh beam*. Table 11.1 summarizes the commonly used beam models and the assumptions they make.

The validity of the assumptions we use depends not only on the physical dimensions of the cross section of the beam, but also on the boundary conditions and the loading. The significance of the shortening of the projection actually depends not so much on the amplitude of the elastic motion, but on the presence of large axial loads or rapid rigid body motions. Rotational motion can be viewed as generating a centrifugal force, which can be treated as an axial load. The rotatory

11.2 KINETIC AND POTENTIAL ENERGY: SMALL OR NO RIGID BODY MOTION

Table 11.1 Comparison of beam models

	Effects Included			
Model	Rotatory Inertia	Bending Strain	Shear Strain	Shortening of Projection
Euler-Bernoulli	No	Yes	No	Yes or no
Rayleigh	Yes	Yes	No	Yes or no
Timoshenko	Yes	Yes	Yes	No
Shear	No	No	Yes	No

inertia is usually significant when there is rapid rotation about the x axis. We also ignore temperature effects.

To illustrate the shortening of the projection, consider a deformed beam. Fig. 11.7 shows the projection of the deformation of the beam axis on the xy plane. The slope of the deformation along the xy and xz planes is given by Eq. [11.2.4]. It follows that the slope of the beam axis at any point x is

$$\theta(x, t) = \sqrt{v'^2(x, t) + w'^2(x, t)} \qquad [11.2.5]$$

Because shear effects are ignored we neglect any other rotational deformation of the beam. We denote by

$$s(x, t) = x + e(x, t) = \text{Distance traversed along the beam axis from} \\ x = 0 \text{ to the deformed position } A^*$$

$$e(x, t) = \text{How much the beam axis has stretched}$$

$$\xi(x, t) = x + u(x, t) = \text{Projection of the beam axis position onto the } x \\ \text{axis}$$

Consider now a differential element of the beam, whose undeformed length is dx and whose deformed length is ds. Fig. 11.8 shows the projection of this differential element on the xy plane. The projection of ds onto the x axis is $d\xi$ and it can be expressed as

$$d\xi = ds \cos \theta = ds \cos \sqrt{\left(\frac{\partial v}{\partial x}\right)^2 + \left(\frac{\partial w}{\partial x}\right)^2} \qquad [11.2.6]$$

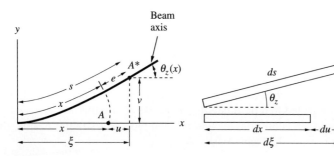

Figure 11.7   **Figure 11.8**

The difference between $d\xi$ and dx, $d\xi - dx = du$, is referred as the *shortening of the projection*. The nomenclature can be explained by noting that if there is no stretch, du will be a negative quantity. Because the deformations are small, we approximate $\cos\theta$ in the above equation by a Taylor series expansion, so that

$$d\xi \approx ds\left(1 - \frac{1}{2}\theta^2\right) = ds - \frac{1}{2}\left[\left(\frac{\partial v}{\partial x}\right)^2 + \left(\frac{\partial w}{\partial x}\right)^2\right]ds \quad [11.2.7]$$

Integrating Eq. [11.2.7], we obtain

$$\xi(x,t) = x + u(x,t) \approx s(x,t) - \frac{1}{2}\int_0^{s(x,t)}\left[\left(\frac{\partial v}{\partial \sigma}\right)^2 + \left(\frac{\partial w}{\partial \sigma}\right)^2\right]d\sigma \quad [11.2.8]$$

We introduce the definitions of s and ξ into this equation. Also, because the stretch $e(x,t)$ is small we replace the upper limit in the integral, $s = x + e$, by x. Hence, we have an expression relating the deformation in the x direction to the stretch and the curvature in the y and z directions as

$$u(x,t) \approx e(x,t) - \frac{1}{2}\int_0^x\left[\left(\frac{\partial v}{\partial \sigma}\right)^2 + \left(\frac{\partial w}{\partial \sigma}\right)^2\right]d\sigma \quad [11.2.9]$$

This relation can also be obtained by considering the nonlinear strain displacement relationships. The first and second-order terms in the expression for the strain in the x direction are

$$\varepsilon_{xx} = \frac{\partial u}{\partial x} + \frac{1}{2}\left[\left(\frac{\partial u}{\partial x}\right)^2 + \left(\frac{\partial v}{\partial x}\right)^2 + \left(\frac{\partial w}{\partial x}\right)^2\right] + \text{higher-order terms} \quad [11.2.10]$$

Assuming that the axial deformation is much smaller than the transverse deformations, we ignore the $\left(\frac{\partial u}{\partial x}\right)^2$ as well as any cubic or higher-order terms. Integration of the remaining terms in Eq. [11.2.10] over x leads to Eq. [11.2.9]. We also observe from Eqs. [11.2.8]–[11.2.10] that

$$\varepsilon_{xx} = e'(x,t) \quad [11.2.11]$$

indicating the well-known relation between the axial strain along the beam axis and the stretch.

To find the components of the strain at any point along the beam, consider Fig. 11.9, drawn for the case of deformation in the y direction only. We now invoke the assumption that plane sections remain plane after deformation. This permits one to express the deformation of a point on the beam in terms of the distance of that point from the beam axis. One can show that the beam axis coincides with the neutral axis. The *neutral axis* is defined as the axis at which there is no stretch or contraction when the beam is subjected to a pure bending load. Be aware that for a curved beam, the neutral axis and the centroidal axis do not coincide. For straight beams, the stretch of a point away from the beam axis is given by

$$s(x,y,t) = s(x,t) - yv'(x,t) \quad \text{or} \quad e(x,y,t) = e(x,t) - yv'(x,t) \quad [11.2.12]$$

11.2 KINETIC AND POTENTIAL ENERGY: SMALL OR NO RIGID BODY MOTION

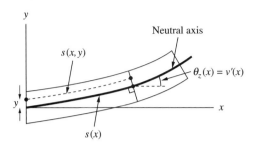

Figure 11.9

It follows that for the general case where there is transverse deformation in both the y and z directions, the expression for strain becomes

$$\varepsilon_{xx}(x, y, z, t) = e'(x, t) + zw''(x, t) - yv''(x, t) \qquad [11.2.13]$$

The stress is obtained by Hooke's law as

$$\sigma_{xx} = E\varepsilon_{xx} \qquad [11.2.14]$$

with all of the remaining components of the stress being zero. E is the *modulus of elasticity*, a material property.

11.2.2 KINETIC AND POTENTIAL ENERGY FOR BEAMS

We next calculate the kinetic and potential energies for a beam. Differentiating Eq. [11.2.1], the velocity of a point on the beam axis becomes

$$\dot{\mathbf{r}}(x, t) = \dot{u}(x, t)\mathbf{i} + \dot{v}(x, t)\mathbf{j} + \dot{w}(x, t)\mathbf{k} \qquad [11.2.15]$$

To find the angular velocity of a differential element, consider Figs. 11.5 and 11.6. We saw earlier that, as a result of the elastic deformation, the cross section rotates. The angular velocity of the cross section is due to this rotation. The rotation angles are $\theta_z = v'(x, t)$ about the z axis and $\theta_y = -w'(x, t)$ about the y axis. Because we assume the rotation angles to be small, the order of the rotation is not significant. We then approximate the angular velocity of the differential element as

$$\boldsymbol{\omega}(x, t) \approx -\dot{w}'(x, t)\mathbf{j} + \dot{v}'(x, t)\mathbf{k} \qquad [11.2.16]$$

We also ignore the effects of the angular velocity of the differential element on the velocity of a certain point on the differential element. Hence, the translational velocity of every point on the cross section is approximated as being the same. Defining the mass per unit length at x as $\mu(x)$, we can write the translational kinetic energy as

$$T_{\text{tran}} = \frac{1}{2}\int_0^L \mu(x)\dot{\mathbf{r}}(x, t) \cdot \dot{\mathbf{r}}(x, t)\, dx$$

$$= \frac{1}{2}\int_0^L \mu(x)\left(\dot{u}^2(x, t) + \dot{v}^2(x, t) + \dot{w}^2(x, t)\right)dx \qquad [11.2.17]$$

To find the rotational kinetic energy we note that the angular velocity has components along the y and z directions only and consider the mass moments of inertia of the differential elements as $dI_{xx} = \mathcal{I}_{xx}(x)\,dx$, $dI_{yy} = \mathcal{I}_{yy}(x)\,dx$, $dI_{zz} = \mathcal{I}_{zz}(x)\,dx$, $dI_{yz} = \mathcal{I}_{yz}(x)\,dx$, denoting mass moments of inertia per unit length. The rotational kinetic energy has the form

$$T_{\text{rot}} = \frac{1}{2}\int_0^L \{\omega\}^T [dI_G]\{\omega\}\,dx = \frac{1}{2}\int_0^L \left(\mathcal{I}_{yy}(x)\dot{w}'^2 + 2\mathcal{I}_{yz}(x)\dot{v}'\dot{w}' + \mathcal{I}_{zz}(x)\dot{v}'^2\right)dx$$

[11.2.18]

In general, the contribution of the rotatory inertia to the kinetic energy is very small and is ignored. The potential energy is due to the elastic deformation. Noting that, for a beam where the shear deformation is ignored and the beam is not subjected to any torsional loading, all of the stress components except σ_{xx} vanish, the potential energy can be shown to be

$$V = \frac{1}{2}\int \sigma_{xx}\varepsilon_{xx}\,d(\text{Vol}) = \frac{1}{2}\int E\left(e'(x,t) + zw''(x,t) - yv''(x,t)\right)^2 d(\text{Vol})$$

[11.2.19]

We can write the differential volume element as $d(\text{Vol}) = dA\,dx$, where $dA = dy\,dz$ is a differential area element of the cross section. Because the beam axis is also the neutral axis, the following area integrals hold

$$\int_A dA = A(x) \qquad \int_A y\,dA = \int_A z\,dA = 0$$

$$\int_A y^2\,dA = I_z(x) \qquad \int_A z^2\,dA = I_y(x) \qquad \int_A yz\,dA = I_{yz}(x) \quad [11.2.20]$$

where $A(x)$ is the cross-sectional area of the beam at point x, $I_y(x)$ and $I_z(x)$ are the area moments of inertia about the y and z axes, respectively, and $I_{yz}(x)$ is the area product of inertia. If the y and z axes are selected such that they are principal axes,[2] the product of inertia vanishes, $I_{yz} = 0$. The reader should not confuse area product of inertia and mass product of inertias.

Expansion of Eq. [11.2.19] and use of Eqs. [11.2.20] yields the following terms for the potential energy:

$$V = \frac{1}{2}\int_0^L \Big(EA(x)[e'(x,t)]^2 + EI_y(x)[w''(x,t)]^2 + EI_z(x)[v''(x,t)]^2$$

$$- 2EI_{yz}(x)v''(x,t)w''(x,t)\Big)dx \qquad [11.2.21]$$

The term $EA(x)$ is known as the *axial stiffness*, and the $EI_y(x)$, $EI_{yz}(x)$, and $EI_z(x)$ terms denote the *bending stiffness*. When the y and z axes are selected as principal axes, $I_{yz} = 0$ and one can show through a stress analysis that the internal bending moments and the internal axial forces are related to the deformations by

$$M_y(x,t) = EI_y(x)w''(x,t) \qquad M_z(x,t) = EI_z(x)v''(x,t)$$

$$P(x,t) = EA(x)e'(x) \qquad [11.2.22]$$

---

[2] See Chapter 6 for cases on how I_{yz} vanishes due to symmetry.

The virtual work is due to the external forces and moments acting on the beam. One can write the virtual work as

$$\delta W = \int_0^L \left(p_x(x,t)\delta u + p_y(x,t)\delta v + p_z(x,t)\delta w + m_z(x,t)\delta v' - m_y(x,t)\delta w' \right) dx \quad [11.2.23]$$

in which p_x, p_y, and p_z denote the distributed external forcing (force/length) in the x, y, and z directions, respectively, and m_y and m_z denote the distributed moments (moment/length) about the y and z axes.

11.2.3 KINETIC AND POTENTIAL ENERGY FOR TORSION

Next, we consider torsion. In linear elasticity theory, the axial and transverse motions are not coupled to torsion. Hence, we analyze torsional deformation by itself. Fig. 11.10 shows a slender bar subjected to a distributed torsional load $m_x(x,t)$. As a result, the cross section at a distance x from a fixed end twists by an angle $\theta_x(x,t)$ about the neutral axis. The internal moment at point x is denoted by $M_x(x)$. Using the semi-inverse method of St. Venant, we assume that the only stress and strain components are σ_{xy}, σ_{xz}, ε_{xy}, and ε_{xz}. The components of the stress can conveniently be studied using the stress function formulation. We will not pursue this analysis here, but concentrate on expressions for the strain. Neglecting nonlinear terms, the components of strain have the form

$$\varepsilon_{xy} = \frac{\partial v}{\partial x} + \frac{\partial u}{\partial y} \qquad \varepsilon_{xz} = \frac{\partial w}{\partial x} + \frac{\partial u}{\partial z} \quad [11.2.24]$$

and the stress and strain components are related by Hooke's law, $\sigma_{xy} = G\varepsilon_{xy}$, $\sigma_{xz} = G\varepsilon_{xz}$, in which G is the *shear modulus,* related to the modulus of elasticity E via the *Poisson's ratio* ν by

$$G = \frac{E}{2(1+\nu)} \quad [11.2.25]$$

A major difference between torsion and bending is that in torsion, the shape of the cross section affects the nature of the deformation. If the cross section is not circular, warping occurs, where plane sections no longer remain plane but warp out of their original planes. As hypothesized by St. Venant, for small deformations the projection of the warped cross section onto the yz plane coincides with the original shape of the cross section. Fig. 11.11 shows an arbitrary point B, at a distance r from the neutral axis and at an angle β from the y axis that deforms to point B'. The distance between B' and the neutral axis is still r. Approximating the arclength between B and B' as a straight line we have

$$v = -r\theta_x \sin\beta = -z\theta_x \qquad w = r\theta_x \cos\beta = y\theta_x \quad [11.2.26]$$

We further assume that the deformation in the x direction is proportional to the rate of change of $\theta_x(x,t)$ with respect to x. Introducing the angle of twist per unit length $\psi(x,t) = \partial \theta_x(x,t)/\partial x$, we write

$$u(x,t) = \psi(x,t) f(y,z) \quad [11.2.27]$$

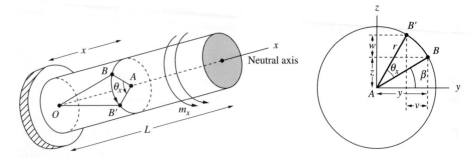

Figure 11.10   **Figure 11.11**

in which $f(y, z)$ is the *warping function* at point x. The warping function is a property of the cross section. One can show that the warping function satisfies the relation $\nabla^2 f = 0$, in which ∇^2 is the two-dimensional Laplace operator. Calculation of the warping function is, in general, a complicated task, and beyond the scope of this text. Introducing Eqs. [11.2.26] and [11.2.27] to Eqs. [11.2.24], we obtain for the strain

$$\varepsilon_{xy} = \psi\left(\frac{\partial f}{\partial y} - z\right) \qquad \varepsilon_{xz} = \psi\left(\frac{\partial f}{\partial z} + y\right) \qquad [11.2.28]$$

We can then write the potential energy as

$$V = \frac{1}{2}\int (\sigma_{xy}\varepsilon_{xy} + \sigma_{xz}\varepsilon_{xz})\,d(\text{Vol}) = \frac{1}{2}\int G(\varepsilon_{xy}^2 + \varepsilon_{xz}^2)\,dy\,dz\,dx$$

$$= \frac{1}{2}\int G\psi^2\left[\left(\frac{\partial f}{\partial y} - z\right)^2 + \left(\frac{\partial f}{\partial z} + y\right)^2\right]dy\,dz\,dx \qquad [11.2.29]$$

We define the *torsion constant* $J(x)$ as

$$J(x) = \int_A \left[\left(\frac{\partial f}{\partial y} - z\right)^2 + \left(\frac{\partial f}{\partial z} + y\right)^2\right]dy\,dz \qquad [11.2.30]$$

and can write the potential energy as

$$V = \frac{1}{2}\int_0^L GJ(x)[\theta_x'(x, t)]^2\,dx \qquad [11.2.31]$$

The term $GJ(x)$ is also called the *torsional stiffness*. It can be shown through a stress analysis that the torsion constant is related to the internal twisting moment M_x by

$$J(x) = \frac{M_x}{G\psi} \qquad [11.2.32]$$

For a circular cross section there is no warping, so that $f(y, z)$ vanishes, and considering Eqs. [11.2.30], the expression for $J(x)$ becomes

$$J(x) = \int_A (y^2 + z^2)\,dy\,dz = I_y(x) + I_z(x) \qquad [11.2.33]$$

and $J(x)$ is recognized as the *polar area moment of inertia*.

To calculate the kinetic energy we consider Fig. 11.11 and the velocity of point B, which we can write as

$$v_B(x, y, z, t) = r\dot{\theta}_x(x, t) \quad [11.2.34]$$

The differential mass element can be expressed as $dm = \rho\, dy\, dz\, dx$, in which ρ is the density, that is, mass per unit volume. The kinetic energy can then be written as

$$T = \frac{1}{2}\int_{\text{body}} v_B^2\, dm = \frac{1}{2}\int \rho r^2 \dot{\theta}_x^2(x,t)\, dy\, dz\, dx = \frac{1}{2}\int \rho(y^2 + z^2)\dot{\theta}_x^2(x,t)\, dy\, dz\, dx \quad [11.2.35]$$

We defined earlier by $\mathcal{I}_{xx}(x)$ the mass moment of inertia per unit length about the x axis. The explicit expression for $\mathcal{I}_{xx}$ has the form

$$\mathcal{I}_{xx}(x) = \int \rho(y^2 + z^2)\, dy\, dz \quad [11.2.36]$$

so that for circular rods with constant density $\mathcal{I}_{xx}(x) = \rho J(x)$. We can now write the kinetic energy expression as

$$T = \frac{1}{2}\int_0^L \mathcal{I}_{xx}(x)\dot{\theta}_x^2(x,t)\, dx \quad [11.2.37]$$

The virtual work is

$$\delta W = \int_0^L m_x(x,t)\delta\theta_x\, dx \quad [11.2.38]$$

To analyze the general case of coupled axial, transverse, and torsional motion one needs to consider higher-order terms in the bending and shear stresses. The interested reader is referred to the text by Rivello.

11.2.4 Operator Notation

The kinetic energy and quadratic parts of the potential energy can conveniently be expressed in operator notation. To this end, we introduce the displacement vector $\{\mathcal{U}(\mathcal{D})\}$, in which $\mathcal{D}$ is the spatial domain. For the beam considered above, $\{\mathcal{U}\} = [u \ v \ w \ \theta]^T$ and $\mathcal{D} = x$. Recall that the rotational deformation of the beam about the y and z axes was expressed by the slopes $w'(x, t)$ and $v'(x, t)$. Hence, we have four degrees of freedom. Note that what we refer to as a degree of freedom here is different from a degree of freedom for a rigid body. Here, the degree of freedom is for each point on the body, perhaps more properly described by the term *continuous degrees of freedom*.

The derivatives of $\{\mathcal{U}\}$ with respect to the spatial variables have the form

$$\{\mathcal{U}'\} = \frac{\partial \{\mathcal{U}\}}{\partial x} \qquad \{\mathcal{U}''\} = \frac{\partial^2 \{\mathcal{U}\}}{\partial x^2} \quad [11.2.39]$$

Here, we introduce the inner product notation between two vectors $\{X(\mathcal{D})\}$ and $\{Y(\mathcal{D})\}$ as

$$< \{X(\mathcal{D})\}, \{Y(\mathcal{D})\} > = \int_{\mathcal{D}} \{X(\mathcal{D})\}^T \{Y(\mathcal{D})\} d\mathcal{D} \qquad [11.2.40]$$

and we can express the kinetic energy as

$$T = \frac{1}{2} < \{\dot{u}\}, [M]\{\dot{u}\} > + \frac{1}{2} < \{\dot{u}'\}, [M_1]\{\dot{u}'\} > \qquad [11.2.41]$$

in which the matrices $[M]$ and $[M_1]$ have the form

$$[M] = \begin{bmatrix} \mu(x) & 0 & 0 & 0 \\ 0 & \mu(x) & 0 & 0 \\ 0 & 0 & \mu(x) & 0 \\ 0 & 0 & 0 & \mathcal{I}_{xx}(x) \end{bmatrix} \qquad [M_1] = \begin{bmatrix} 0 & 0 & 0 & 0 \\ 0 & \mathcal{I}_{zz}(x) & \mathcal{I}_{yz}(x) & 0 \\ 0 & \mathcal{I}_{yz}(x) & \mathcal{I}_{yy}(x) & 0 \\ 0 & 0 & 0 & 0 \end{bmatrix}$$

$$[11.2.42]$$

The $[M_1]$ matrix contains the rotational inertia terms. To describe the potential energy, we note from Eq. [11.2.19] that the potential energy is more conveniently expressed in terms of the stretch e, as Eq. [11.2.19] is in quadratic form. We introduce the vector $\{W(x)\} = [e \quad v \quad w \quad \theta]^T$ and the matrices $[S_0]$, $[S_1]$, and $[S_2]$. We can then write the quadratic part of the potential energy as

$$V = \frac{1}{2} < \{W\}, [S_0]\{W\} > + \frac{1}{2} < \{W'\}, [S_1]\{W'\} > + \frac{1}{2} < \{W''\}, [S_2]\{W''\} >$$

$$[11.2.43]$$

For the beam considered,

$$[S_2] = \begin{bmatrix} 0 & 0 & 0 & 0 \\ 0 & EI_z(x) & -EI_{yz}(x) & 0 \\ 0 & -EI_{yz}(x) & EI_y(x) & 0 \\ 0 & 0 & 0 & 0 \end{bmatrix} \quad [S_1] = \begin{bmatrix} EA(x) & 0 & 0 & 0 \\ 0 & 0 & 0 & 0 \\ 0 & 0 & 0 & 0 \\ 0 & 0 & 0 & GJ(x) \end{bmatrix}$$

$$[S_0] = [0] \qquad [11.2.44]$$

The matrix $[S_0]$ has nonzero entries in the presence of springs attached to the body. Also, in the presence of axial forces that can be expressed as part of the potential energy, $[S_1]$ has additional terms. Often, the notation for potential energy in Eq. [11.2.43] is shortened to the *energy inner product*

$$V = \frac{1}{2} [\![\{W\}, \{W\}]\!] \qquad [11.2.45]$$

The energy inner product is a quadratic term. Gravitational potential energy is not a quadratic function, thus it is not included in the energy inner product, but as an additional term.

Example 11.1

Write the kinetic and potential energies and the virtual work for the beam shown in Fig. 11.12, which is of uniform density and is allowed to deflect in the z direction only. Ignore rotatory inertia.

11.2 KINETIC AND POTENTIAL ENERGY: SMALL OR NO RIGID BODY MOTION

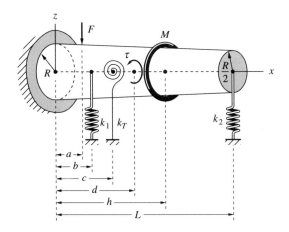

Figure 11.12

Solution

The positive z direction for the deflection of $w(x, t)$ is taken as upward. From Eq. [11.2.17] the kinetic energy has the form

$$T(t) = \frac{1}{2}\int_0^L \mu(x)\dot{w}^2(x, t)\,dx + \frac{1}{2}M\dot{w}^2(h, t) \qquad \text{[a]}$$

The potential energy is due to the elasticity, the concentrated springs, and the gravitational force. It has the form

$$V(t) = \frac{1}{2}\int_0^L \left(EI_y(x)[w''(x, t)]^2 + \mu(x)gw(x, t)\right)dx$$
$$+ \frac{1}{2}\left(k_1[w(b, t)]^2 + k_T[w'(c, t)]^2 + k_2[w(L, t)]^2\right) \qquad \text{[b]}$$

in which g is the gravitational constant. To calculate the mass density and area moments of inertia, we recall that the mass density is the mass per unit length. The expressions for $\mu(x)$ and $I_y(x)$ for a circular cross-secton are

$$\mu(x) = \frac{m(x)}{L} = \rho A(x) = \rho\pi r^2(x) \qquad I_y(x) = \frac{1}{4}\pi r^4(x) = \frac{1}{4}A(x)r^2(x) \qquad \text{[c]}$$

where r is the radius and ρ is the density. The radius is given by $r(x) = R(1 - x/2L)$, so that

$$\mu(x) = \rho\pi R^2\left(1 - \frac{x}{2L}\right)^2 \qquad I_y(x) = \frac{1}{4}\pi R^4\left(1 - \frac{x}{2L}\right)^4 \qquad \text{[d]}$$

The virtual work is due to the external force F and torque τ. We can write it as

$$\delta W = -F\delta w(a) + \tau\delta w'(d) \qquad \text{[e]}$$

This expression is obtained by substituting into Eq. [11.2.23] the expressions for the distributed force and moment

$$p_z(x, t) = -F\underline{\delta}(x - a) \qquad m_y(x, t) = \tau\underline{\delta}(x - d) \qquad \text{[f]}$$

in which $\underline{\delta}$ denotes the Dirac delta function.

11.3 EQUATIONS OF MOTION

In this section, we use the expressions for the kinetic and potential energy developed in the previous section and invoke the extended Hamilton's principle. The process yields the equations of motion, as well as the boundary condititons.

11.3.1 EXTENDED HAMILTON'S PRINCIPLE

Consider the axial and transverse deformations only and select the y and z axes as principal axes. Here we have two choices of variables to use: u, v, and w, or e, v, and w. Let us use u, v, and w. From Chapter 4, the extended Hamilton's principle states

$$\delta \int_{t_1}^{t_2} [T(t) - V(t)]\, dt + \int_{t_1}^{t_2} \delta W\, dt = 0 \qquad [11.3.1]$$

where t_1 and t_2 are any two time instances. The expressions for the kinetic and potential energies are given in Eqs. [11.2.17] and [11.2.21]. We ignore the contribution of the rotatory inertia. Taking the variation, we obtain

$$\int_{t_1}^{t_2} \int_0^L \left\{ \mu(x)(\dot{u}\delta\dot{u} + \dot{v}\delta\dot{v} + \dot{w}\delta\dot{w}) - EA(x)e'\left(\frac{\partial e'}{\partial u'}\delta u' + \frac{\partial e'}{\partial v'}\delta v' + \frac{\partial e'}{\partial w'}\delta w'\right) \right.$$
$$- EI_y(x)w''\delta w'' - EI_z(x)v''\delta v'' + p_x\delta u +$$
$$\left. p_y\delta v + p_z\delta w + m_z\delta v' - m_y\delta w' \right\} dx\, dt = 0 \qquad [11.3.2]$$

To write the extended Hamilton's principle in terms of δu, δv, and δw only, we eliminate the variations of the derivatives of u, v, and w by integration by parts. Recall that the order of differentiation and variation can be exchanged when the differentiation is with respect to an independent variable, in this case the time t and the spatial variable x. The procedure yields integrals as well as integrated terms. We observe that three types of terms result. Examples of each type are

$$\int_0^L m_z \delta v'\, dx = \int_0^L m_z \frac{d}{dx}(\delta v)\, dx$$
$$= \left. m_z \delta v \right|_0^L - \int_0^L \frac{d}{dx}(m_z)\, \delta v\, dx$$
$$= -\int_0^L \frac{d}{dx}(m_z)\, \delta v\, dx \qquad [11.3.3]$$

$$\int_{t_1}^{t_2} \mu(x)\dot{u}\,\delta\dot{u}\, dt = \int_{t_1}^{t_2} \mu(x)\dot{u}\frac{d}{dt}(\delta u)\, dt$$
$$= \left. \mu(x)\dot{u}\,\delta u \right|_{t_1}^{t_2} - \int_{t_1}^{t_2} \mu(x)\ddot{u}\,\delta u\, dt$$
$$= -\int_{t_1}^{t_2} \mu(x)\ddot{u}\,\delta u\, dt \qquad [11.3.4]$$

$$\int_0^L EI_y(x) w'' \delta w'' \, dx = EI_y \frac{\partial^2 w}{\partial x^2} \delta w' \bigg|_0^L - \int_0^L \frac{\partial}{\partial x} \left(EI_y \frac{\partial^2 w}{\partial x^2} \right) \delta w' \, dx$$

$$= \left[EI_y \frac{\partial^2 w}{\partial x^2} \delta w' - \frac{\partial}{\partial x} \left(EI_y \frac{\partial^2 w}{\partial x^2} \right) \delta w \right]_0^L + \int_0^L \frac{\partial^2}{\partial x^2} \left[EI_y \frac{\partial^2 w}{\partial x^2} \right] \delta w \, dx \quad \textbf{[11.3.5]}$$

The integrated term in Eq. [11.3.3] vanishes because $m_z(x)$ is a moment per unit length. The integrated term in Eq. [11.3.4] vanishes due to the definition of the variation. Recall that the varied paths for u, v, and w are selected such that they vanish at times t_1 and t_2. By contrast, there are two integrations by parts in Eq. [11.3.5], both involving spatial differentiation, and none of the integrated terms disappear automatically. Performing all the integrations and rearranging yields

$$\int_{t_1}^{t_2} \int_0^L \left(\left\{ -\mu(x)\ddot{u} + \frac{\partial}{\partial x} \left[EA(x) \frac{\partial u}{\partial x} + \frac{1}{2} EA(x) \left[\left(\frac{\partial v}{\partial x} \right)^2 + \left(\frac{\partial w}{\partial x} \right)^2 \right] \right] + p_x \right\} \delta u \right.$$

$$+ \left\{ -\mu(x)\ddot{v} - \frac{\partial^2}{\partial x^2} \left[EI_z \frac{\partial^2 v}{\partial x^2} \right] + \frac{\partial}{\partial x} \left[P(x,t) \frac{\partial v}{\partial x} \right] + p_y - \frac{\partial}{\partial x} m_z \right\} \delta v$$

$$+ \left\{ -\mu(x)\ddot{w} - \frac{\partial^2}{\partial x^2} \left[EI_y \frac{\partial^2 w}{\partial x^2} \right] + \frac{\partial}{\partial x} \left[P(x,t) \frac{\partial w}{\partial x} \right] + p_z + \frac{\partial}{\partial x} m_y \right\} \delta w \right) dx \, dt$$

$$+ \int_{t_1}^{t_2} \left(\left\{ EA(x) \frac{\partial u}{\partial x} + \frac{1}{2} EA(x) \left[\left(\frac{\partial v}{\partial x} \right)^2 + \left(\frac{\partial w}{\partial x} \right)^2 \right] \right\} \delta u \right.$$

$$- \left(EI_z \frac{\partial^2 v}{\partial x^2} \right) \delta v' + \left\{ \frac{\partial}{\partial x} \left(EI_z \frac{\partial^2 v}{\partial x^2} \right) - P(x,t) \frac{\partial v}{\partial x} \right\} \delta v$$

$$- \left(EI_y \frac{\partial^2 w}{\partial x^2} \right) \delta w' + \left\{ \frac{\partial}{\partial x} \left(\left(EI_y \frac{\partial^2 w}{\partial x^2} \right) - P(x,t) \frac{\partial w}{\partial x} \right) \right\} \delta w \right) \bigg|_0^L dt = 0 \quad \textbf{[11.3.6]}$$

in which we have used the notation

$$P(x,t) = EA(x) \frac{\partial u}{\partial x} + \frac{1}{2} EA(x) \left[\left(\frac{\partial v}{\partial x} \right)^2 + \left(\frac{\partial w}{\partial x} \right)^2 \right] = EA(x) e'(x,t) \quad \textbf{[11.3.7]}$$

Note that $P(x, t)$ denotes the internal force in the x direction. By virtue of the arbitrariness of the variations of u, v and w, in order for the integrals to be equal to zero, the integrands and integrated terms must vanish, and they must vanish individually. The three integrands in [11.3.6] lead to the equations of motion, and the integrated terms lead to the boundary expressions. We have, for the axial deformation

$$\mu(x)\ddot{u} - \frac{\partial}{\partial x} \left\{ EA(x) \frac{\partial u}{\partial x} + \frac{1}{2} EA(x) \left[\left(\frac{\partial v}{\partial x} \right)^2 + \left(\frac{\partial w}{\partial x} \right)^2 \right] \right\} = p_x$$

$$\left\{ EA(x) \frac{\partial u}{\partial x} + \frac{1}{2} EA(x) \left[\left(\frac{\partial v}{\partial x} \right)^2 + \left(\frac{\partial w}{\partial x} \right)^2 \right] \right\} \delta u \bigg|_0^L = 0 \quad \textbf{[11.3.8]}$$

For the transverse deformation in the y direction

$$\mu(x)\ddot{v} + \frac{\partial^2}{\partial x^2}\left[EI_z(x)\frac{\partial^2 v}{\partial x^2}\right] - \frac{\partial}{\partial x}\left[P(x,t)\frac{\partial v}{\partial x}\right] = p_y - \frac{\partial}{\partial x}m_z$$

$$\left[EI_z(x)\frac{\partial^2 v}{\partial x^2}\right]\delta v'\bigg|_0^L = 0$$

$$\left\{\frac{\partial}{\partial x}\left[EI_z(x)\frac{\partial^2 v}{\partial x^2}\right] - P(x,t)\frac{\partial v}{\partial x}\right\}\delta v\bigg|_0^L = 0 \qquad [11.3.9]$$

For the transverse deformation in the z direction

$$\mu(x)\ddot{w} + \frac{\partial^2}{\partial x^2}\left[EI_y(x)\frac{\partial^2 w}{\partial x^2}\right] - \frac{\partial}{\partial x}\left[P(x,t)\frac{\partial w}{\partial x}\right] = p_z + \frac{\partial}{\partial x}m_y$$

$$\left[EI_y(x)\frac{\partial^2 w}{\partial x^2}\right]\delta w'\bigg|_0^L = 0$$

$$\left\{\frac{\partial}{\partial x}\left[EI_y(x)\frac{\partial^2 w}{\partial x^2}\right] - P(x,t)\frac{\partial w}{\partial x}\right\}\delta w\bigg|_0^L = 0 \qquad [11.3.10]$$

In a similar fashion, we obtain the equation of motion and boundary expressions for the torsional vibration. The results can be shown to be

$$\mathcal{I}_{xx}(x)\ddot{\theta}_x - \frac{\partial}{\partial x}\left[GJ(x)\frac{\partial \theta_x}{\partial x}\right] = m_x \qquad \left[GJ(x)\frac{\partial \theta_x}{\partial x}\right]\delta\theta_x\bigg|_0^L = 0 \qquad [11.3.11]$$

11.3.2 BOUNDARY CONDITIONS

The boundary conditions are ascertained from the boundary expressions by examining the geometry at the boundaries. Fig. 11.13 shows commonly encountered boundary conditions. In each boundary expression there are two terms. One term is a variation of a deformation (or variation of a slope), and the other term is an internal force (or moment). By examining the geometry, one determines which one of the terms is zero, and identifies the boundary condition. This procedure is basically an application of a fundamental principle from structural mechanics:

> At any point on the body, if the displacement (slope) is known, then the internal force (moment) is not known at that point; and if the loading force (moment) is known, then the displacement (slope) is not known at that point, until the problem is solved.

This principle is applicable at the boundaries as well as in the interior. Confining the analysis to the boundaries, if the beam is restrained to move in a certain direction at a boundary, then the associated boundary condition is geometric, as it is determined from the geometry of the deformation. For example, if the beam is fixed at $x = 0$,

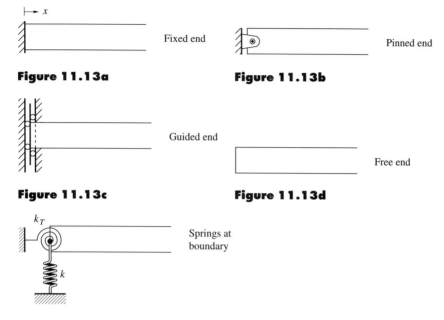

Figure 11.13a Fixed end
Figure 11.13b Pinned end
Figure 11.13c Guided end
Figure 11.13d Free end
Figure 11.13e Springs at boundary

the boundary conditions become

$$u(0, t) = 0 \quad v(0, t) = 0 \quad w(0, t) = 0$$
$$v'(0, t) = 0 \quad w'(0, t) = 0 \quad \theta_x(0, t) = 0 \quad \text{[11.3.12]}$$

Such boundary conditions are known as *essential*, or *geometric*, or *Dirichlet type* boundary conditions, or *boundary conditions of the first kind*. When the boundary conditions are essential, it follows that the force and moment balance terms, which are the coefficients of δu, δv, δw, $\delta v'$, $\delta w'$, and $\delta \theta_x$, are unknown. Furthermore, the magnitudes of these internal forces and moments do not vanish; if a point is restrained to move in a certain direction, there must be a force or moment at that point preventing the structure from moving.

The magnitudes of the internal forces and moments can only be determined after the system differential equations are solved, as stated in the fundamental principle above. We will refer to these internal reactions (the coefficients of δu, δv, δw, $\delta v'$, $\delta w'$, and $\delta \theta_x$) as *complementary boundary conditions* (CBC). The reason for defining these terms is that, when constructing approximate solutions by series expansions, trial functions can be selected that violate the force and moment balances at the boundaries. Unless the CBC are considered, the approximate solution may incorrectly set the internal forces and moments to zero at boundaries and interfaces. The net effect is use of an incomplete set of trial functions and slow or lack of convergence. Table 11.2 gives the geometric boundary conditions and associated CBCs.

When the boundaries are unrestrained, the variations of the deformations and their slopes do not vanish. It follows that the coefficients of the variations must be zero. For example, if the beam is free to move in the y direction at $x = L$, the

Table 11.2 Essential boundary conditions and complementary boundary conditions

Type of Boundary	Geometric Boundary Condition (Known)	CBC (Unknown Quantity)	Geometric Boundary Condition	CBC
Transverse deformation (v)				
Fixed end	$v = 0$	$\frac{\partial}{\partial x}\left(EI_z \frac{\partial^2 v}{\partial x^2}\right) - P\frac{\partial v}{\partial x}$	$v' = 0$	$EI_z \frac{\partial^2 v}{\partial x^2}$
Pinned end	$v = 0$	$\frac{\partial}{\partial x}\left(EI_z \frac{\partial^2 v}{\partial x^2}\right) - P\frac{\partial v}{\partial x}$		
Guided end	$v' = 0$	$EI_z \frac{\partial^2 v}{\partial x^2}$		
Axial deformation (u)				
Fixed end	$u = 0$	$EA(x)\left\{\frac{\partial u}{\partial x} + \frac{1}{2}\left[\left(\frac{\partial v}{\partial x}\right)^2 + \left(\frac{\partial w}{\partial x}\right)^2\right]\right\}$		
Torsional deformation (θ_x)				
Fixed end	$\theta_x = 0$	$GJ(x)\frac{\partial \theta_x}{\partial x}$		

associated boundary conditions become

$$EI_z \frac{\partial^2 v}{\partial x^2}\bigg|_{x=L} = 0 \qquad \left[\frac{\partial}{\partial x}\left(EI_z \frac{\partial^2 v}{\partial x^2}\right) - P(x,t)\frac{\partial v}{\partial x}\right]\bigg|_{x=L} = 0 \qquad \textbf{[11.3.13a,b]}$$

These boundary conditions are known as *dynamic* or *natural*. Equation [11.3.13a] shows a type also called the *Neumann* type, or a *boundary condition of the second kind*. Equation [11.3.13b] is also known as a *boundary condition of the third kind*, as it involves not only the shear force but also the transverse force due to the axial motion. Such boundary conditions are also encountered in the presence of springs at the boundaries.

Natural boundary conditions indicate that the force and moment balances are zero at the boundaries. This is intuitively expected, because, if the structure is free to translate (rotate) at a boundary, there should be no force (moment) at that point to restrain it from moving.

11.3.3 SIMPLIFICATION

Equations [11.3.8]–[11.3.10] can be simplified in many cases by ignoring the deformation in the axial direction, setting the axial stretch $e(x, t)$ and all its derivatives to zero. The simplification is valid when the axial internal force $P(x, t)$ is not large or not a function of time (or it is a slowly varying function of time). Typical examples include the helicopter blade problem and the buckling of columns. Note that once the axial deformation is ignored, $u(x, t)$ is no longer an independent variable. From

Eq. [11.2.9], setting $e(x,t) = 0$ we obtain

$$u(x,t) = -\frac{1}{2}\int_0^x \left[\left(\frac{\partial v}{\partial \sigma}\right)^2 + \left(\frac{\partial w}{\partial \sigma}\right)^2\right] d\sigma \qquad [11.3.14]$$

The time derivative of $u(x,t)$ is also usually ignored. The force $P(x)$ now becomes the internal axial force that balances the external axial force $p_x(x)$. A simple force balance in the axial direction (or a reexamination of Eq. [11.3.8]) indicates that

$$\frac{\partial}{\partial x}P(x) = -p_x(x) \qquad [11.3.15]$$

which basically states that the external force (per unit length) is balanced by the derivative of the internal force. The equations of motion for the transverse deflection become

$$\mu(x)\ddot{v} + \frac{\partial^2}{\partial x^2}\left[EI_z(x)\frac{\partial^2 v}{\partial x^2}\right] - \frac{\partial}{\partial x}\left[P(x)\frac{\partial v}{\partial x}\right] = p_y - \frac{\partial}{\partial x}m_z \qquad [11.3.16a]$$

$$\mu(x)\ddot{w} + \frac{\partial^2}{\partial x^2}\left[EI_y(x)\frac{\partial^2 w}{\partial x^2}\right] - \frac{\partial}{\partial x}\left[P(x)\frac{\partial w}{\partial x}\right] = p_z + \frac{\partial}{\partial x}m_y \qquad [11.3.16b]$$

When one ignores the deformations due to axial elasticity from the beginning, the equations of motion can be derived more simply. One can treat the effect of the internal axial force $P(x)$ in two ways: as part of the virtual work or as added potential energy. Consider the virtual work expression due to the external axial load $p_x(x)$ as

$$\delta W = \int_0^L p_x(x)\delta u \, dx \qquad [11.3.17]$$

Introduce Eq. [11.3.15] to this equation and integrate by parts to obtain

$$\delta W = -\int_0^L \frac{d}{dx}P(x)\delta u \, dx = -P(x)\delta u\Big|_0^L + \int_0^L P(x)\delta u' \, dx = \int_0^L P(x)\delta u' \, dx \qquad [11.3.18]$$

The integrated term disappears because at a fixed end u vanishes, and at a free end the internal force P is zero. Differentiating Eq. [11.3.14], we get the expression for $u'(x,t)$ as

$$u'(x,t) = -\frac{1}{2}\left[\left(\frac{\partial v}{\partial x}\right)^2 + \left(\frac{\partial w}{\partial x}\right)^2\right] \qquad [11.3.19]$$

so the virtual work expression becomes

$$\delta W = -\frac{1}{2}\int_0^L P(x)\delta\left[\left(\frac{\partial v}{\partial x}\right)^2 + \left(\frac{\partial w}{\partial x}\right)^2\right] dx = -\int_0^L P(x)[v'\delta v' + w'\delta w'] \, dx \qquad [11.3.20]$$

The effect of an axial force can also be expressed as a potential energy contribution due to the internal axial force $P(x)$ acting through the shortening of the projection. Recall that $P(x)$ is taken as positive along the x direction and the shortening of the projection $dx - d\xi = -du$ is positive along the negative x direction. We hence write the additional potential energy as

$$V_{\text{additional}} = \int_0^L P(x)(dx - d\xi) \qquad [11.3.21]$$

Introducing the expression for du from Eq. [11.3.19] into the above equation yields

$$V_{\text{additional}} = \frac{1}{2}\int_0^L P(x)\left\{\left(\frac{\partial v}{\partial x}\right)^2 + \left(\frac{\partial w}{\partial x}\right)^2\right\} dx \qquad [11.3.22]$$

Introduction of Eq. [11.3.22] or Eq. [11.3.20] into the extended Hamilton's principle yields Eqs. [11.3.16]. When the axial loading is not large or is not present, the $P(x)$ term is eliminated and one arrives at the well-known *Euler-Bernoulli* equations.

If $P(x)$ is tensile, then according to Eq. [11.3.22] the additional potential energy is a positive quantity. Considering that the potential energy is strain energy, representing the stiffness or resistance to deformation, a tensile force can be considered as one that adds to the stiffness. This increase in stiffness has a stabilizing effect, a phenomenon readily observable, such as in the helicopter blade. By contrast, a compressive axial force reduces the potential energy, thereby making the component less stiff. A classic example of this is the buckling of columns. Columns become unstable and collapse when the applied compressive load becomes too high.

The above derivations are in terms of an internal axial force along the x axis. Sometimes it is more convenient to describe the internal axial force along the beam axis (x^* direction). Consider an external force which generates an internal force $F(x)$ along the beam axis, as depicted in Fig. 11.14 for a two-dimensional system. We resolve the internal force into its components along the x and y axes as

$$\mathbf{F}(x) = F(x)\cos\theta_z\mathbf{i} + F(x)\sin\theta_z\mathbf{j} \approx F(x)\mathbf{i} + F(x)v'(x, t)\mathbf{j} \qquad [11.3.23]$$

The contribution of the component in the x direction is the same as in Eq. [11.3.16], with F replacing P. The contribution of the component in the y direction can be obtained by noting from Eq. [11.3.15] that $F(x)$ is related to the distributed external load by $p_x(x) = -F'(x)$. Hence, the component in the y direction can be expressed

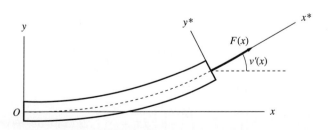

Figure 11.14

as an additional distributed external force as

$$p_y(x) = -F'(x)v'(x) \quad [11.3.24]$$

From Eq. [11.3.16a], the total contribution of $F(x)$ becomes

$$\frac{\partial}{\partial x}\left[F(x)\frac{\partial v}{\partial x}\right] + p_y(x) = F(x)\frac{\partial^2 v}{\partial x^2} + F'(x)v'(x) - F'(x)v'(x) = F(x)\frac{\partial^2 v}{\partial x^2} \quad [11.3.25]$$

In a similar fashion, we obtain the contribution of $F(x)$ in the z direction. Consequently, the equations of motion, Eqs. [11.3.16], have the form

$$\mu(x)\ddot{v} + \frac{\partial^2}{\partial x^2}\left[EI_z(x)\frac{\partial^2 v}{\partial x^2}\right] - F(x)\frac{\partial^2 v}{\partial x^2} = p_y - \frac{\partial}{\partial x}m_z$$

$$\mu(x)\ddot{w} + \frac{\partial^2}{\partial x^2}\left[EI_y(x)\frac{\partial^2 w}{\partial x^2}\right] - F(x)\frac{\partial^2 w}{\partial x^2} = p_z + \frac{\partial}{\partial x}m_y \quad [11.3.26]$$

11.3.4 OPERATOR NOTATION

It is convenient to express the equations of motion of a flexible body in terms of mass and stiffness operators, especially when the equations are linearized. Let us consider an elastic foundation under the beam, modeled as a distributed spring $k_y(x)$. The added potential energy due to the spring is

$$V_{spring} = \frac{1}{2}\int_0^L k_y(x)v^2(x,t)\,dx \quad [11.3.27]$$

and the effect on the equation of motion in the y direction is an additional $ak_y(x)v(x,t)$ term on the left side of Eq. [11.3.16a].

Recall the displacement vector $\{\mathcal{U}(\mathcal{D})\}$ in which $\mathcal{D}$ is the spatial domain. For the simplified beam model considered above $\{\mathcal{U}\} = [v\ w]^T$ and $\mathcal{D} = x$. Equations [11.3.16] (with the added term due to the elastic foundation) can then be expressed in operator form as

$$\mathcal{L}\{\mathcal{U}(\mathcal{D})\} + \mathcal{M}\{\ddot{\mathcal{U}}(\mathcal{D})\} = \{F\} \quad [11.3.28]$$

in which $\mathcal{L}$ is the matrix stiffness operator and $\mathcal{M}$ is the matrix mass operator in the form

$$\mathcal{L} = \begin{bmatrix} \frac{\partial^2}{\partial x^2}\left[EI_z(x)\frac{\partial^2}{\partial x^2}\right] - \frac{\partial}{\partial x}\left[P(x)\frac{\partial}{\partial x}\right] + k_y(x) & 0 \\ 0 & \frac{\partial^2}{\partial x^2}\left[EI_y(x)\frac{\partial^2}{\partial x^2}\right] - \frac{\partial}{\partial x}\left[P(x)\frac{\partial}{\partial x}\right] \end{bmatrix}$$

$$\mathcal{M} = \begin{bmatrix} \mu(x) & 0 \\ 0 & \mu(x) \end{bmatrix} \quad [11.3.29]$$

and $\{F\}$ is the forcing vector in the form $\{F\} = [p_y - \frac{\partial}{\partial x}m_z \quad p_z + \frac{\partial}{\partial x}m_y]^T$. The operators are diagonal because the y and z axes are selected as principal axes. The equations of motion in the z and y directions are uncoupled, unless $P(x)$ is a function of both v and w.

We observe from the above equation that the mass operator is nothing but the mass matrix in Eq. [11.2.42], and the stiffness operator can be expressed in terms of the $[S_i]$ ($i = 0, 1, 2$) matrices as

$$\mathscr{L} = \frac{\partial^2}{\partial x^2}\left([S_2]\frac{\partial^2}{\partial x^2}\right) - \frac{\partial}{\partial x}\left([S_1]\frac{\partial}{\partial x}\right) + [S_0] \qquad [11.3.30]$$

This expression can be derived from the kinetic and potential energies. And the boundary conditions can also be expressed in terms of the matrices $[S_i]$ ($i = 0, 1, 2$). For conservative systems, the operators $\mathscr{L}$ and $\mathscr{M}$ have the property of being *self-adjoint*, which is defined as follows. Consider two functions $\{\Theta(\mathscr{D})\}$ and $\{\Gamma(\mathscr{D})\}$, that satisfy all boundary conditions and are as many times differentiable as the highest-order derivative in the stiffness operator. Such functions are called *comparison functions*. The operators $\mathscr{L}$ and $\mathscr{M}$ are called self-adjoint if the following relationships hold:

$$<\{\Theta(\mathscr{D})\}, \mathscr{L}\{\Gamma(\mathscr{D})\}> = <\{\Gamma(\mathscr{D})\}, \mathscr{L}\{\Theta(\mathscr{D})\}> \qquad [11.3.31]$$

$$<\{\Theta(\mathscr{D})\}, \mathscr{M}\{\Gamma(\mathscr{D})\}> = <\{\Gamma(\mathscr{D})\}, \mathscr{M}\{\Theta(\mathscr{D})\}> \qquad [11.3.32]$$

Also, for two comparison functions the following relationship holds:

$$<\{\Theta(\mathscr{D})\}, \mathscr{L}\{\Gamma(\mathscr{D})\}> = [\![\{\Theta(\mathscr{D})\}, \{\Gamma(\mathscr{D})\}]\!] \qquad [11.3.33]$$

Another important group of functions is *admissible functions*. These satisfy only the geometric boundary conditions and are half as many times differentiable as the highest-order derivative in the stiffness operator. The relationship [11.3.31] cannot be written for admissible functions. However, Eq. [11.3.32] holds. Also, one can write the energy inner product for two admissible functions

$$[\![\{\Theta(\mathscr{D})\}, \{\Gamma(\mathscr{D})\}]\!] = [\![\{\Gamma(\mathscr{D})\}, \{\Theta(\mathscr{D})\}]\!] \qquad [11.3.34]$$

Example 11.2

Consider the beam in Example 11.1; derive the equation of motion and identify the boundary conditions.

Solution

We will take the kinetic and potential energies from Example 11.1, as well as the virtual work, obtain their variations, and perform the necessary integrations by parts. We can write the kinetic energy as

$$T = \frac{1}{2}\mathscr{M}\dot{w}^2(x, t) \qquad \text{[a]}$$

in which the mass operator $\mathscr{M}$ has the form

$$\mathscr{M} = \mu(x) + M\underline{\delta}(x - h) \qquad \text{[b]}$$

The stiffness terms can also be expressed as

$$S_2 = EI_y(x) \qquad S_1 = k_T \underline{\delta}(x - c) \qquad S_0 = k_1 \underline{\delta}(x - b) + k_2 \underline{\delta}(x - L) \qquad [\mathbf{c}]$$

Note that the contribution due to the weight of the beam is not in quadratic form. It is more convenient to treat the weight as part of the external forces, in the virtual work. The potential energy is

$$V = \frac{1}{2} \int_0^L \{EI_y(x)[w''(x,t)]^2 + k_1 \underline{\delta}(x-b) w^2(x,t)$$
$$+ k_T \underline{\delta}(x-c) w'^2(x,t) + k_2 \underline{\delta}(x-L) w^2(x,t)\} dx \qquad [\mathbf{d}]$$

We can express the virtual work as

$$\delta W = \int_0^L [-F\underline{\delta}(x-a)\delta w + \tau \underline{\delta}(x-d)\delta w' - \mu(x) g \delta w] dx \qquad [\mathbf{e}]$$

The variation of the kinetic energy is straightforward, as all integrated terms drop out due to the definition of the variation. In the virtual work expression, the only term we must integrate by parts is due to the torque. Integrating by parts gives

$$\int_0^L \tau \underline{\delta}(x-d) \delta w' \, dx = -\int_0^L \tau \underline{\delta}'(x-d) \delta w \, dx \qquad [\mathbf{f}]$$

Taking the variation of the potential energy and performing the necessary integrations by parts, the individual terms become

$$\delta \left[\frac{1}{2} \int_0^L EI_y(x) w''^2 \, dx \right] = \left[EI_y \frac{\partial^2 w}{\partial x^2} \delta w' - \frac{\partial}{\partial x}\left(EI_y \frac{\partial^2 w}{\partial x^2}\right) \delta w \right]\Big|_0^L + \int_0^L \frac{\partial^2}{\partial x^2}\left(EI_y \frac{\partial^2 w}{\partial x^2}\right) \delta w \, dx$$

$$\delta \left[\frac{1}{2} \int_0^L k_1 \underline{\delta}(x-b) w^2(x,t) \, dx \right] = \int_0^L k_1 \underline{\delta}(x-b) w \delta w \, dx$$

$$\delta \left[\frac{1}{2} \int_0^L k_T \underline{\delta}(x-c) w'^2(x,t) \, dx \right] = \int_0^L k_T \underline{\delta}(x-c) w' \delta w' \, dx = -\int_0^L \frac{\partial}{\partial x}\left[k_T \underline{\delta}(x-c) \frac{\partial w}{\partial x}\right] \delta w \, dx$$

$$\delta \left[\frac{1}{2} \int_0^L k_2 \underline{\delta}(x-L) w^2(x,t) \, dx \right] = \delta \left[\frac{1}{2} k_2 w^2(L,t) \right] = k_2 w(L,t) \delta w(L) \qquad [\mathbf{g}]$$

Note that we treated the spring at the end differently than the springs in the middle. This is because the variation of the displacement at the end $x = L$ enters the boundary terms. Performing an integration by parts on the virtual work in Eq. [e], we obtain

$$\delta W = \int_0^L \left\{ -F\underline{\delta}(x-a) - \frac{\partial}{\partial x}[\tau \underline{\delta}(x-d)] - \mu g \right\} \delta w \, dx \qquad [\mathbf{h}]$$

We next invoke the extended Hamilton's principle. For the equation of motion, which is the sum of all the integrands, we obtain

$$\{\mu(x) + M\underline{\delta}(x-h)\}\ddot{w} + \frac{\partial^2}{\partial x^2}\left[EI_y \frac{\partial^2 w}{\partial x^2}\right] + k_1 \underline{\delta}(x-b) w - \frac{\partial}{\partial x}\left[k_T \underline{\delta}(x-c) \frac{\partial w}{\partial x}\right]$$
$$= -F\underline{\delta}(x-a) - \frac{\partial}{\partial x}[\tau \underline{\delta}(x-d)] - \mu g \qquad 0 < x < L \qquad [\mathbf{i}]$$

and the boundary terms become

$$\left[EI_y \frac{\partial^2 w}{\partial x^2} \delta w' - \frac{\partial}{\partial x}\left(EI_y \frac{\partial^2 w}{\partial x^2}\right)\delta w \right]\Bigg|_0^L + k_2 w(L,t)\delta w(L) = 0 \qquad \text{[j]}$$

At the end $x = 0$, the beam is fixed, so that the boundary conditions are geometric and they are

$$w(0,t) = 0 \qquad w'(0,t) = 0 \qquad \text{[k]}$$

At the end $x = L$, the displacement and slope are not fixed, hence their variations are not zero. The coefficients of δw and $\delta w'$ must vanish. The boundary conditions are recognized as

$$EI_y \frac{\partial^2 w}{\partial x^2}\bigg|_{x=L} = 0 \qquad \frac{\partial}{\partial x}\left(EI_y \frac{\partial^2 w}{\partial x^2}\right)\bigg|_{x=L} = k_2 w(L,t) \qquad \text{[l, m]}$$

These are natural boundary conditions, with Eq. [m] being a boundary condition of the third kind. This boundary condition represents the force balance at $x = L$, where the spring force counteracts the shear force.

Example 11.3

Given the uniform rotating blade in Fig. 11.15 spinning with the constant angular velocity Ω, find the equations of motion.

Solution

There are several ways of solving this problem. One way is to consider that the rotation of the blade causes a centrifugal force. Another is to consider a relative frame to which the base of the blade is attached. Using the first way, the external forces acting on a differential element are shown in Fig. 11.16. Note that the x axis is the original undeformed axis, and the $x*$ axis is along the elastic curve. The external forces acting on the differential element are the force of gravity $p_y\, dx = -\mu g\, dx$ and the centrifugal force $p_x\, dx = \mu \Omega^2 x\, dx$. The total centrifugal force acting at any point x can be obtained by integrating the centrifugal force on

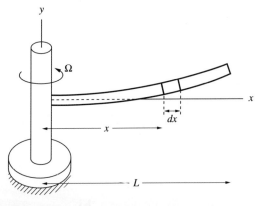

Figure 11.15

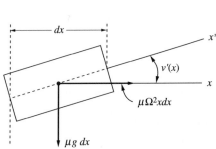

Figure 11.16

the differential element (from the free end to point x)

$$P(x) = -\int_L^x p_x \, dx = \int_x^L \mu\Omega^2 x \, dx = \frac{\mu\Omega^2}{2}(L^2 - x^2) \qquad \textbf{[a]}$$

Introducing Eq. [a] and the force of gravity into Eq. [11.3.16a], we obtain the equation of motion

$$\mu\ddot{v} + EI_z \frac{\partial^4 v}{\partial x^4} - \frac{\mu\Omega^2}{2} \frac{\partial}{\partial x}\left[(L^2 - x^2)\frac{\partial v}{\partial x}\right] = -\mu g \qquad \textbf{[b]}$$

We next derive the equation of motion for the rotating blade using a rotating reference frame approach. We attach a reference frame to the undeformed position of the beam, with the z-axis along the vertical. The position of a point can be expressed as

$$\mathbf{r}(x, t) = [x + u(x, t)]\mathbf{i} + v(x, t)\mathbf{j} \qquad \textbf{[c]}$$

The xyz frame is rotating with constant angular velocity $\Omega\mathbf{j}$, so that the velocity of a point can be expressed as

$$\dot{\mathbf{r}}(x, t) = \dot{u}(x, t)\mathbf{i} + \dot{v}(x, t)\mathbf{j} + \Omega\mathbf{j} \times \mathbf{r}(x, t) = \dot{u}(x, t)\mathbf{i} + \dot{v}(x, t)\mathbf{j} - \Omega[x + u(x, t)]\mathbf{k} \qquad \textbf{[d]}$$

leading to the kinetic energy expression

$$T = \frac{1}{2}\int_0^L \mu \dot{\mathbf{r}}(x, t) \cdot \dot{\mathbf{r}}(x, t) = \frac{1}{2}\int_0^L \mu\left(\dot{u}^2(x, t) + \dot{v}^2(x, t) + \Omega^2[x + u(x, t)]^2\right)dx$$

$$= \frac{1}{2}\int_0^L \mu\left(\dot{u}^2(x, t) + \dot{v}^2(x, t) + \Omega^2 x^2 + 2\Omega^2 x u(x, t) + \Omega^2 u^2(x, t)\right)dx \qquad \textbf{[e]}$$

Because the angular velocity of the blade is large, the term $\dot{u}^2(x, t)$ is much smaller than any term containing Ω^2, and we ignore it. The term $\Omega^2 u^2(x, t)$ is quadratic in u and can be ignored as it is much smaller than the $2\Omega^2 x u(x, t)$ term. Ignoring the axial elasticity, we write $u(x, t)$ as

$$u(x, t) = -\frac{1}{2}\int_0^x \left(\frac{\partial v}{\partial \sigma}\right)^2 d\sigma \qquad \textbf{[f]}$$

It follows that the kinetic energy has one added term, $2\Omega^2 x u(x, t)$, which, in light of the above equation, has the form

$$T_{\text{additional}} = -\frac{1}{2}\mu\Omega^2 \int_0^L x \int_0^x \left(\frac{\partial v}{\partial \sigma}\right)^2 d\sigma \, dx \qquad \textbf{[g]}$$

Note that the $\Omega^2 x^2$ term in Eq. [e] does not affect the equations of motion, as its derivatives with respect to the variables u and v are zero. By applying some integration trickery, Eq. [g] can be expressed as

$$T_{\text{additional}} = -\frac{1}{2}\mu\Omega^2 \int_0^L x \int_0^x \left(\frac{\partial v}{\partial \sigma}\right)^2 d\sigma \, dx = -\frac{1}{2}\mu\Omega^2 \int_0^L \left(\frac{\partial v}{\partial x}\right)^2 \int_x^L \sigma \, d\sigma \, dx$$

$$= -\frac{1}{4}\mu\Omega^2 \int_0^L \left(\frac{\partial v}{\partial x}\right)^2 (L^2 - x^2) \, dx \qquad \textbf{[h]}$$

The potential energy is

$$V = \frac{1}{2}\int_0^L EI_z(x)\left(\frac{\partial^2 v}{\partial x^2}\right)^2 dx \qquad [\text{i}]$$

Integrating by parts gives the variation of the additional kinetic energy the form

$$\delta T_{\text{additional}} = -\frac{1}{2}\mu\Omega^2 \int_0^L (L^2 - x^2)\left(\frac{\partial v}{\partial x}\right)\delta\left(\frac{\partial v}{\partial x}\right) dx$$

$$= -\frac{1}{2}\mu\Omega^2(L^2 - x^2)\left(\frac{\partial v}{\partial x}\right)\delta v\bigg|_0^L + \frac{1}{2}\mu\Omega^2 \int_0^L \frac{\partial}{\partial x}\left[(L^2 - x^2)\left(\frac{\partial v}{\partial x}\right)\right]\delta v\, dx \qquad [\text{j}]$$

The boundary terms drop out because at $x = 0$, $v(0, t) = 0$, and at $x = L$, the coefficient $L^2 - x^2$ vanishes. Invoking the definition of $P(x)$ from Eq. [a] we can express Eq. [j] as

$$\delta T_{\text{additional}} = \int_0^L \frac{\partial}{\partial x}\left[P(x)\frac{\partial v}{\partial x}\right]\delta v\, dx \qquad [\text{k}]$$

Yet another way of obtaining the equations of motion is to use Eq. [11.3.22] and treat the axial force $P(x)$ in conjunction with the shortening of the projection. In this case, we have the additional term in the potential energy,

$$V_{\text{additional}} = \frac{1}{2}\int_0^L P(x)\left(\frac{\partial v}{\partial x}\right)^2 dx = \frac{1}{4}\mu\Omega^2 \int_0^L (L^2 - x^2)\left(\frac{\partial v}{\partial x}\right)^2 dx \qquad [\text{l}]$$

$T_{\text{additional}}$ in Eq. [h] is the negative of $V_{\text{additional}}$ in Eq. [l], which is to be expected.

11.4 EIGENSOLUTION AND RESPONSE

In this section, we summarize the solution of the eigenvalue problem associated with linear, self-adjoint continuous systems and obtain a general form for the response by modal analysis. Consider the simply supported beam in Fig. 11.17 and assume that the axial deformation and transverse deformation in the z direction are negligible, and that the axial force $P(x)$ is zero. The equation of motion becomes

$$\mu(x)\ddot{v} + \frac{\partial^2}{\partial x^2}\left[EI_z(x)\frac{\partial^2 v}{\partial x^2}\right] = p_y - \frac{\partial}{\partial x}m_z \qquad 0 < x < L \qquad [\mathbf{11.4.1}]$$

The boundary conditions are due to the deflection and moment balance being equal

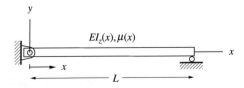

Figure 11.17

to zero at both ends, thus

$$v(x, t) = 0 \qquad EI_z(x)\frac{\partial^2 v(x, t)}{\partial x^2} = 0 \qquad \text{at } x = 0, x = L \qquad [11.4.2]$$

We will seek a solution by modal analysis. The procedure that we will use is very similar to the eigensolution of a multidegree of freedom vibrating system in Section 5.5. The reader is urged to compare every step here with the procedure in Section 5.5.

We first consider the homogeneous problem (no external excitation and no explicitly time-dependent boundary conditions) and use separation of variables to express the deformation $v(x, t)$ as

$$v(x, t) = \phi(x)e^{\lambda t} \qquad [11.4.3]$$

where $\phi(x)$ is the amplitude function and $e^{\lambda t}$ denotes the time dependence. Introducing Eq. [11.4.3] into Eq. [11.4.1] and setting $p_y = 0$, $m_z = 0$, we obtain

$$\left(\lambda^2 \mu(x)\phi(x) + \frac{d^2}{dx^2}\left[EI_z(x)\frac{d^2\phi}{dx^2}\right]\right)e^{\lambda t} = 0 \qquad [11.4.4]$$

Introducing Eq. [11.4.3] to the boundary expressions in Eq. [11.4.2], we obtain the boundary conditions that $\phi(x)$ has to satisfy as

$$\phi(x) = 0 \qquad EI_z(x)\frac{d^2\phi}{dx^2} = 0 \qquad \text{at } x = 0, x = L \qquad [11.4.5]$$

In order for Eq. [11.4.4] to have a nontrivial solution, we must have

$$\lambda^2 \mu(x)\phi(x) + \frac{d^2}{dx^2}\left[EI_z(x)\frac{d^2\phi}{dx^2}\right] = 0 \qquad [11.4.6]$$

which is a variable coefficient, ordinary differential equation of order four. Equation [11.4.6], subject to the boundary conditions [11.4.5], constitutes the *boundary value problem*, where one needs to find the values of λ that lead to nontrivial solutions $\phi(x)$ of Eq. [11.4.6]. The solutions λ are called *eigenvalues*, and the corresponding functions $\phi(x)$ are called the *eigenfunctions*.

To demonstrate the solution procedure, we further simplify the problem by assuming that the beam has a uniform cross-section, setting $\mu(x) = \mu = $ constant, $EI_z(x) = EI = $ constant. Expecting λ to be pure imaginary, we write Eq. [11.4.6] as

$$\frac{d^4\phi}{dx^4} - \beta^4\phi(x) = 0 \qquad [11.4.7]$$

where $\beta^4 = -\lambda^2\mu/EI$. The general solution of Eq. [11.4.7] is

$$\phi(x) = c_1 \sin \beta x + c_2 \cos \beta x + c_3 \sinh \beta x + c_4 \cosh \beta x \qquad [11.4.8]$$

where the coefficients $c_i (i = 1, 2, 3, 4)$ are found by substituting in the boundary conditions. Note that because the right side of Eq. [11.4.7] is zero, $\phi(x)$ can only be determined to within a multiplicative constant.

First, take $\phi(x)$ and evaluate at $x = 0$. This gives $c_2 + c_4 = 0$. Then, take $\phi''(x)$ and evaluate it at $x = 0$. This gives $-c_2 + c_4 = 0$, from which we conclude that $c_2 = c_4 = 0$. We repeat the same procedure for the boundary $x = L$. Doing this, we end up with $c_3 = 0$, and the relation that remains is

$$c_1 \sin \beta L = 0 \qquad [11.4.9]$$

For Eq. [11.4.9] to have a nontrivial solution we must have

$$\sin \beta L = 0 \qquad [11.4.10]$$

which is known as the *characteristic equation*. The characteristic equation has an infinite number of solutions. For the problem at hand, the solutions are $\beta_r L = r\pi$ ($r = 1, 2, \ldots$). Considering the definition of β, all the eigenvalues λ_r are pure imaginary. From Eq. [11.4.3] we conclude that the motion is oscillatory. Defining the *natural frequency* $\omega_r = -i\lambda_r$ ($i^2 = -1$), we can write

$$\lambda_r = i(r\pi)^2 \sqrt{\frac{EI}{\mu L^4}} \qquad \omega_r = (r\pi)^2 \sqrt{\frac{EI}{\mu L^4}} \qquad r = 1, 2, \ldots \qquad [11.4.11]$$

Since $c_2 = c_3 = c_4 = 0$, the eigenfunction corresponding to each eigenvalue (or natural frequency) is $\phi_r(x)$, and

$$\phi_r(x) = A_r \sin \beta_r x = A_r \sin \frac{r\pi x}{L} \qquad [11.4.12]$$

where A_r is an arbitrary constant. The solution to the eigenvalue problem yields the shape of the eigenfunctions uniquely, but not their amplitudes. It is convenient to normalize the eigenfunctions. A commonly used normalization scheme is with respect to the mass distribution, thus

$$\int_0^L \mu(x)\phi_r^2(x)\,dx = 1 \qquad [11.4.13]$$

Using Eq. [11.4.13], one can show that for the pinned-pinned uniform beam, $A_r = \sqrt{2/\mu L}$. The first three eigenfunctions are plotted in Fig. 11.18. We note

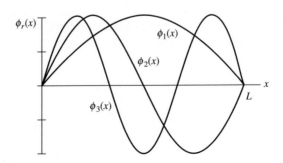

Figure 11.18 First three eigenfunctions of a simply-supported uniform beam

that except for the first, the eigenfunctions cross the x axis and the number of zero crossings increases with the mode number. The points of zero crossing are called *nodes*. It can be shown that the rth mode has $r - 1$ nodes.

Another very interesting and useful property of the eigenfunctions is that they are orthogonal to each other. The orthogonality results from the self-adjointness of the mass and stiffness operators. For the beam problem at hand, the orthogonality relations can be expressed as

$$\int_0^L \mu(x)\phi_r(x)\phi_s(x)\,dx = 0 \quad \text{when } r \neq s, \quad r, s = 1, 2, \ldots$$

$$\int_0^L \phi_r(x)\frac{d^2}{dx^2}\left(EI_z\frac{d^2\phi_s}{dx^2}\right)dx = 0 \quad \text{when } r \neq s \quad [11.4.14]$$

Considering the normalization scheme in Eq. [11.4.13], one can combine Eqs. [11.4.13] and [11.4.14] to obtain the *orthonormality relations*

$$\int_0^L \mu(x)\phi_r(x)\phi_s(x)\,dx = \underline{\delta}_{rs} \qquad \int_0^L \phi_r(x)\frac{d^2}{dx^2}[EI_z\frac{d^2\phi_s}{dx^2}]\,dx = \omega_r^2\underline{\delta}_{rs}$$
$$[11.4.15a,b]$$

where $\underline{\delta}_{rs}$ was defined earlier as the Kronecker delta.

The eigenfunctions $\phi_r(x)$ constitute a *complete set*. That is, any comparison function $Y(x)$ can be expanded as a uniformly convergent series of the eigenfunctions as

$$Y(x) = \sum_{r=1}^{\infty} a_r\phi_r(x) \quad [11.4.16]$$

The coefficients a_r can be found by multiplying $Y(x)$ by $\mu(x)\phi_s(x)$, integrating along the domain and invoking the orthogonality relation [11.4.15a]. We then have

$$\int_0^L \mu(x)\phi_s(x)Y(x)\,dx = \int_0^L \mu(x)\phi_s(x)\sum_{r=1}^{\infty} a_r\phi_r(x)\,dx$$

$$= \sum_{r=1}^{\infty} a_r \int_0^L \mu(x)\phi_s(x)\phi_r(x)\,dx = \sum_{r=1}^{\infty} a_r\underline{\delta}_{rs} = a_s \quad [11.4.17]$$

so that we can write

$$a_r = \int_0^L \mu(x)\phi_r(x)Y(x)\,dx \quad [11.4.18]$$

Equations [11.4.16] and [11.4.18] describe the *expansion theorem*. Note the similarity of the expansion theorem to Fourier series expansions or other orthogonal series expansions. One very important application of the expansion theorem is in derivation of the modal equations of motion. Here, the function to be expanded is the displacement profile $v(x, t)$

$$v(x, t) = \sum_{r=1}^{\infty} \eta_r(t)\phi_r(x) \qquad [11.4.19]$$

The coefficients of the eigenfunctions, $\eta_r(t)$, are time-dependent terms. We refer to $\eta_r(t)$ as the *modal coordinates*. Introducing this expansion into the equation of motion, Eq. [11.4.1], multiplying with $\phi_s(x)$, integrating along the domain, and invoking the orthogonality conditions, we obtain

$$\int_0^L \phi_s(x)\left\{\mu(x)\ddot{v} + \frac{\partial^2}{\partial x^2}\left(EI_z\frac{\partial^2 v}{\partial x^2}\right) - \left[p_y(x,t) - \frac{\partial}{\partial x}m_z(x,t)\right]\right\} dx$$

$$= \sum_{r=1}^{\infty} \ddot{\eta}_r(t)\int_0^L \mu(x)\phi_s(x)\phi_r(x)\,dx + \sum_{r=1}^{\infty} \eta_r(t)\int_0^L \phi_s(x)\frac{\partial^2}{\partial x^2}\left[EI_z\frac{\partial^2 \phi_r}{\partial x^2}\right] dx$$

$$- \int_0^L \phi_s(x)\left[p_y(x,t) - \frac{\partial}{\partial x}m_z(x,t)\right] dx$$

$$= \sum_{r=1}^{\infty} \ddot{\eta}_r(t)\underline{\delta}_{rs} + \sum_{r=1}^{\infty} \eta_r(t)\omega_r^2\underline{\delta}_{rs} - \mathcal{N}_s(t) \qquad [11.4.20]$$

Where

$$\mathcal{N}_s(t) = \int_0^L \phi_s(x)\left[p_y(x,t) - \frac{\partial}{\partial x}m_z(x,t)\right] dx \qquad [11.4.21]$$

is called the *modal force*. Hence, we obtain the modal equations of motion as

$$\ddot{\eta}_s(t) + \omega_s^2\eta_s(t) = \mathcal{N}_s(t) \qquad s = 1, 2, \ldots \qquad [11.4.22]$$

The initial conditions associated with each mode can be found by applying the expansion theorem to the initial conditions $v(x, 0)$ as

$$v(x,0) = \sum_{r=1}^{\infty} \eta_r(0)\phi_r(x) \qquad \dot{v}(x,0) = \sum_{r=1}^{\infty} \dot{\eta}_r(0)\phi_r(x) \qquad [11.4.23]$$

Multiplying the above equations by $\mu(x)\phi_s(x)$, integrating along the domain, and invoking the orthogonality conditions, one obtains the modal initial conditions

$$\eta_s(0) = \int_0^L \mu(x)\phi_s(x)v(x,0)\,dx \qquad \dot{\eta}_s(0) = \int_0^L \mu(x)\phi_s(x)\dot{v}(x,0)\,dx$$

$$[11.4.24]$$

Once the modal equations [11.4.22] are solved, the response can be obtained by substitution of the modal responses into Eq. [11.4.19]. In a real situation, one can integrate and sum only a finite number of modal equations. This raises issues about the number of modal equations of motion to retain in a specific problem and the participation of each mode to the motion. We begin analyzing this issue qualitatively, by making use of the asymptotic behavior of the eigensolution.

An extension of a theorem from Courant and Hilbert states that as the index $r \to \infty$, the asymptotic behavior of the natural frequencies ω_r and eigenfunctions $\phi_r(x)$ are governed by

$$\omega_r \sim Cr^p \qquad \phi_r(x) \sim f(Drx) \qquad r = 1, 2, \ldots \qquad [11.4.25]$$

in which $2p$ is the highest-order spatial derivative associated with the elastic motion. The constants C and D depend on the mass and stiffness properties. When $p = 1$, such as in axial or torsional deformation, the natural frequencies increase with arithmetic progression, and $f(Drx) = A_r \sin(Drx) + B_r \cos(Drx)$. When $p = 2$, such as in the beam equation, the natural frequencies increase in geometric progression, and $f(Drx) = A_r \sin(Drx) + B_r \cos(Drx) + A'_r \sinh(Drx) + B'_r \cosh(Drx)$. In the simply supported (pinned-pinned) beam we discussed, $p = 2$ and $\omega_r = (r\pi)^2 \sqrt{EI/\mu L^4}$; all of the natural frequencies increase with geometric progression.

When studying beam vibrations, retaining the first four or five modes is usually sufficient for an accurate mathematical model, except for cases when the initial conditions on certain modes are large and the external forces and moments excite higher modes more than the others.

By contrast, when studying axial and torsional vibrations, because the asymptotic behavior of the rth mode is governed by r, the amplitude of each mode falls off much slower. Hence, one must retain a larger number of modes for an accurate representation of the motion. A similar statement can be made about plates and shells, where the modes are much more closely spaced than in beams.

The solution of Eq. [11.4.22] subject to the initial conditions [11.4.24] can be accomplished in many ways, as discussed in Section 5.7. Equation [5.7.7] gives the general form of the response.

The preceding developments can be generalized by making use of the operator notation. Using the equations of motion in the form of Eq. [11.3.28], we assume a solution in the form of

$$\{\mathcal{U}(\mathcal{D}, t)\} = \{\Phi(\mathcal{D})\}e^{\lambda t} \qquad [11.4.26]$$

where $\{\Phi(\mathcal{D})\}$ is the amplitude vector and $e^{\lambda t}$ denotes the time dependency. Introducing this into the homogeneous part of Eq. [11.3.28] and using the same argument as before, we obtain the eigenvalue problem

$$\mathcal{L}\{\Phi\} + \lambda^2 \mathcal{M}\{\Phi\} = \{0\} \qquad [11.4.27]$$

The solution yields an infinite number of eigenvalues $\lambda_r (r = 1, 2, \ldots)$ and corresponding vector eigenfunctions $\{\Phi_r(\mathcal{D})\}$. When the mass and stiffness operators are self-adjoint, the eigenfunctions can be normalized as

$$< \{\Phi_r(\mathcal{D})\}, \mathcal{M}\{\Phi_s(\mathcal{D})\} > = \delta_{rs}$$

$$< \{\Phi_r(\mathcal{D})\}, \mathcal{L}\{\Phi_s(\mathcal{D})\} > = \omega_r^2 \delta_{rs} \qquad r, s = 1, 2, \ldots \qquad [11.4.28]$$

The inner product notation between two vectors was defined in Eq. [11.2.40].

One can then invoke the expansion theorem and expand any comparison function $\{\Pi(\mathcal{D})\}$ as

$$\{\Pi(\mathcal{D})\} = \sum_{r=1}^{\infty} a_r \{\Phi_r(\mathcal{D})\} \qquad [11.4.29]$$

and the coefficients a_r can be shown to be

$$a_r = <\{\Pi(\mathcal{D})\}, \mathcal{M}\{\Phi_r(\mathcal{D})\}> \qquad [11.4.30]$$

To find the response, we expand the deformation $\{\mathcal{U}(\mathcal{D}, t)\}$ as

$$\{\mathcal{U}(\mathcal{D}, t)\} = \sum_{r=1}^{\infty} \eta_r(t)\{\Phi_r(\mathcal{D})\} \qquad [11.4.31]$$

Introducing this expression into the equations of motion, left-multiplying with $\{\Phi_s(\mathcal{D})\}^T$, and integrating over the domain yields the modal equations [11.4.22], with the modal excitation $\mathcal{N}_s(t)$ and initial conditions $\eta_s(0)$ and $\dot{\eta}_s(0)$ being

$$\mathcal{N}_s(t) = <\{F(\mathcal{D}, t)\}, \{\Phi_s(\mathcal{D})\}>$$
$$\eta_s(0) = <\{\mathcal{U}(\mathcal{D}, 0), \mathcal{M}\{\Phi_s(\mathcal{D})\}>$$
$$\dot{\eta}_s(0) = <\{\dot{\mathcal{U}}(\mathcal{D}, 0), \mathcal{M}\{\Phi_s(\mathcal{D})\}> \qquad [11.4.32]$$

Example 11.4

Find the eigensolution of the uniform pinned-free beam shown in Fig. 11.19.

Solution

We consider Eq. [11.4.1] as the equation of motion. The boundary conditions are that the displacement and moment vanish at the pinned end, and that the moment and shear vanish at the free end; thus

$$v(0, t) = 0 \qquad EI\frac{\partial^2 v(x, t)}{\partial x^2} = 0 \quad \text{at } x = 0 \text{ and } x = L$$

$$EI\frac{\partial^3 v(x, t)}{\partial x^3} = 0 \qquad \text{at } x = L \qquad [a]$$

Introducing Eq. [11.4.3] to Eq. [11.4.1], we obtain the differential eigenvalue problem

$$\lambda^2 \mu \phi(x) + EI\frac{d^4\phi}{dx^4} = 0 \qquad [b]$$

and associated boundary conditions

$$\phi(0) = 0 \qquad \phi''(0) = 0 \qquad \phi''(L) = 0 \qquad \phi'''(L) = 0 \qquad [c]$$

Using the variable $\beta^4 = -\lambda^2 \mu / EI$, we write Eq. [b] as

$$\frac{d^4\phi}{dx^4} - \beta^4 \phi(x) = 0 \qquad [d]$$

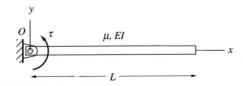

Figure 11.19

which has the general solution

$$\phi(x) = c_1 \sin\beta x + c_2 \cos\beta x + c_3 \sinh\beta x + c_4 \cosh\beta x \qquad [e]$$

To find the constants c_i ($i = 1, 2, 3, 4$) we invoke the boundary conditions. At $x = 0$, $\phi(0) = 0$ yields

$$\phi(0) = c_2 + c_4 = 0 \qquad [f]$$

Differentiating Eq. [e] twice gives

$$\phi''(x) = \beta^2(-c_1 \sin\beta x - c_2 \cos\beta x + c_3 \sinh\beta x + c_4 \cosh\beta x) \qquad [g]$$

At $x = 0$ we have

$$\phi''(0) = \beta^2(-c_2 + c_4) = 0 \qquad [h]$$

A comparison of Eqs. [f] and [h] indicates that $c_2 = c_4 = 0$. The boundary conditions at $x = L$ yield

$$\phi''(L) = \beta^2(-c_1 \sin\beta L + c_3 \sinh\beta L) = 0 \qquad [i]$$

$$\phi'''(L) = \beta^3(-c_1 \cos\beta L + c_3 \cosh\beta L) = 0 \qquad [j]$$

From Eq. [i] we obtain $c_1 = c_3 \sinh\beta L / \sin\beta L$, which, when introduced into Eq. [j] yields the characteristic equation

$$\cos\beta L \sinh\beta L = \sin\beta L \cosh\beta L \qquad [k]$$

Unlike the pinned-pinned uniform beam, solution of the characteristic equation here cannot be obtained in closed form, and one has to resort to numerical techniques to solve Eq. [k]. The first five roots of the characteristic equation can be found to be

$$\beta_1 L = 0 \qquad \beta_2 L = 3.9266 \qquad \beta_3 L = 7.0686 \qquad \beta_4 L = 10.210 \qquad \beta_5 L = 13.352$$

The first root is zero, which implies that the first natural frequency of the pinned-free beam is zero. This mode is known as the rigid body mode. To find the corresponding eigenfunction, we take Eq. [d] and set $\beta = 0$, leading to the equation $\phi_1''''(x) = 0$, which has the solution

$$\phi_1(x) = A_0 + A_1 x + A_2 x^2 + A_3 x^3 \qquad [l]$$

Introducing the boundary conditions, we obtain

$$\phi(0) = 0 \rightarrow A_0 = 0 \qquad \phi''(0) = 0 \rightarrow A_2 = 0 \qquad \phi'''(L) = 0 \rightarrow A_3 = 0 \qquad [m]$$

The only term that survives is A_1. We conclude that the rigid body mode has the form

$$\phi_1(x) = A_1 x \qquad [n]$$

where A_1 is the amplitude. Using the normalization procedure in Eq. [11.4.13], one obtains $A_1 = \sqrt{3/\mu L^3}$. We observe that, because the mass moment of inertia of a rigid uniform beam about the pinned end is $I_O = mL^2/3 = \mu L^3/3$, the orthogonality constant can be expressed as[3]

$$A_1 = \frac{1}{\sqrt{I_O}} \qquad [o]$$

---

[3]Equation [o] is valid for pinned-free beams with nonuniform cross sections as well. To verify, introduce Eq. [n] into Eq. [11.4.13] and invoke the definition of the mass moment of inertia, $I_O = \int \mu(x) x^2 \, dx$.

Let us investigate the rigid body eigenfunction in more detail. With the rigid body mode, the beam does not deform elastically, but moves at an angle with respect to the original undeformed axis. That is, the first mode represents the motion of the beam as if it had no elasticity. Because in the rigid body mode there usually is no elasticity, there is no elastic restoring force.[4] Hence, one can physically explain that there is no natural frequency associated with this mode, or that the natural frequency is zero.

The remaining modes describe the elastic deformation of the beam. From Eqs. [e], [i], and [j], we can show that the eigenfunctions associated with the elastic motion have the form

$$\phi_r(x) = A_r(\sinh \beta_r L \sin \beta_r x + \sin \beta_r L \sinh \beta_r x) \qquad r = 2, 3, \ldots \qquad \textbf{[p]}$$

where the amplitudes A_r can be found by introducing Eq. [p] into Eq. [11.4.13]. The total motion of the pinned-free beam is a linear superposition of the rigid and elastic motions. The expansion of the motion can be written as

$$v(x, t) = \sum_{r=1}^{\infty} \phi_r(x)\eta_r(t) = v_{\text{rigid}}(x, t) + v_{\text{elastic}}(x, t) = \phi_1(x)\eta_1(t) + \sum_{r=2}^{\infty} \phi_r(x)\eta_r(t) \qquad \textbf{[q]}$$

Note that in the same way the rigid body mode contains no contributions from the elastic motion, the elastic motion has no contributions from the rigid motion. We can show this by invoking the orthogonality properties of the modes. Multiplying the rigid and elastic motions and integrating over the domain, we obtain

$$\int_0^L \mu v_{\text{rigid}}(x, t) v_{\text{elastic}}(x, t)\, dx = \int_0^L \mu A_1 x \eta_1(t) \left[\sum_{r=2}^{\infty} \phi_r(x)\eta_r(t)\right] dx$$

$$= \eta_1(t) \sum_{r=2}^{\infty} \eta_r(t) \int_0^L \mu \phi_1(x)\phi_r(x)\, dx$$

$$= \eta_1(t) \sum_{r=2}^{\infty} \eta_r(t) \underline{\delta}_{1r} = 0 \qquad \textbf{[r]}$$

The first three eigenfunctions are plotted in Fig. 11.20. In order for the linear superposition of the rigid and elastic motions to be possible, the rigid body motion must be small. In general, rigid body angles less than 20 degrees can be analyzed using this approach.

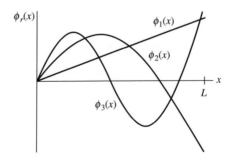

Figure 11.20 First three eigenfunctions of a pinned free beam

---

[4] It is possible to find cases where a rigid body mode will have a nonzero natural frequency.

Consider the pinned-free uniform link in Example 11.4 and obtain the modal equations of motion considering that a moment $\tau(t)$ is applied at the pinned end, as shown in Fig. 11.19. Then, analyze the contribution of each mode to the response.

Example 11.5

Solution

The external excitation can be expressed as

$$p_y(x, t) = 0 \qquad m_z(x, t) = \tau(t)\underline{\delta}(x) \qquad \text{[a]}$$

Substituting Eqs. [a] into Eq. [11.4.21] and using integration by parts yields

$$\mathcal{N}_r(t) = -\int_0^L \phi_r(x) \frac{\partial}{\partial x} [\tau(t)\underline{\delta}(x)] \, dx =$$

$$-\tau(t)\int_0^L \phi_r(x)\underline{\delta}'(x) \, dx = \tau(t)\phi_r'(0) \qquad r = 1, 2, \ldots \qquad \text{[b]}$$

For the first mode, the rigid body mode, the excitation can be written as

$$\mathcal{N}_1(t) = \phi_1'(0)\tau(t) = A_1\tau(t) = \sqrt{\frac{3}{\mu L^3}}\tau(t) \qquad \text{[c]}$$

which, when substituted into the modal equations of motion, gives the equation of motion for the first mode as

$$\ddot{\eta}_1(t) = \sqrt{\frac{3}{\mu L^3}}\tau(t) \qquad \text{[d]}$$

The rigid body modal coordinate, $\eta_1(t)$, can be related to the rigid body angle as follows. From Eq. [q] in Example 11.4 we have

$$v_{\text{rigid}}(x, t) = \phi_1(x)\eta_1(t) \qquad \text{[e]}$$

Since we consider the angle θ to be small, we can write for any point x

$$\theta \approx \sin\theta \approx \tan\theta = \frac{v_{\text{rigid}}(x, t)}{x} = \frac{\phi_1(x)\eta_1(t)}{x} = \sqrt{\frac{3}{\mu L^3}}\eta_1(t) \qquad \text{[f]}$$

Substituting this into Eq. [d], and noting that the mass moment of inertia about the pin joint is given by $I_0 = \mu L^3/3$, we obtain for the rigid body mode

$$I_0\ddot{\theta}(t) = \tau(t) \qquad \text{[g]}$$

which is recognized as the rotational equation of motion of a uniform rigid pinned-free link of length L.

From Eq. [11.4.22], the modal equations associated with the elastic coordinates have the form

$$\ddot{\eta}_r(t) + \omega_r^2 \eta_r(t) = \tau(t)\phi_r'(0) \qquad r = 2, 3, \ldots \qquad \text{[h]}$$

Let us next investigate the mode participation. Consider zero initial conditions. If we introduce the excitation from Eq. [h] into the response equation [5.7.7] we obtain

$$\eta_r(t) = \phi_r'(0)\frac{1}{\omega_r}\int_0^t \tau(t - \sigma)\sin\omega_r\sigma \, d\sigma \qquad \text{[i]}$$

From Eqs. [11.4.25] we estimate the orders of magnitudes of the terms in this equation as

$$\phi'_r(0) = O(r) \qquad \frac{1}{\omega_r} = O\left(\frac{1}{r^2}\right) \qquad [\text{j}]$$

and the order of the integral depending on the nature of $\tau(t)$. Let us denote the order of the integral in Eq. [j] by $O(I)$. It follows that the modal coordinates have orders of magnitude of

$$O(\eta_r(t)) = O\left(\frac{1}{r}\right)O(I) \qquad [\text{k}]$$

For the case when $\tau = 1$, the integral has the form

$$\int_0^t \tau(t-\sigma)\sin\omega_r\sigma\,d\sigma = \frac{1}{\omega_r}(1-\cos\omega_r t) = O\left(\frac{1}{r^2}\right) \qquad [\text{l}]$$

so that the response is of order

$$O(\eta_r(t)) = O\left(\frac{1}{r^3}\right) \qquad [\text{m}]$$

As the mode number increases, the amplitudes of the individual modes become smaller by the cube of the mode number. The second mode has an amplitude of about 1/8 the first mode, the third mode 1/27, and the fourth mode, 1/64. Hence, the participation of the higher modes rapidly falls, which is typical for beams.

11.5 APPROXIMATE SOLUTIONS: THE ASSUMED MODES METHOD

The pinned-pinned uniform beam considered in Section 11.4 and the pinned-free beam in Example 11.4 lend themselves to closed-form solutions for the eigenfunctions. These two examples are more of an exception than the general case. In most cases, and even for simple geometries, it is not possible to find a closed-form solution to the eigenvalue problem. Members with nonuniform cross sections, interconnected bodies, and almost any structure with a complex geometry fall in this category. When a closed-form solution cannot be found, one resorts to approximate techniques to solve for the eigenvalues.

All approximations to continuous systems are in essence discretization procedures. The eigenvalue problem associated with Eq. [11.4.6] is in differential form, which implies that the solution is of infinite order. By discretizing the eigenvalue problem, one converts the differential eigenvalue problem into an algebraic one that is of finite order. Algebraic eigenvalue problems are generally easier to solve than differential eigenvalue problems, as they permit the use of matrix algebra.

11.5.1 APPROXIMATION TECHNIQUES AND TRIAL FUNCTIONS

Approximation techniques can be classified into two broad categories: lumping and series expansions. When using the lumping approach, one lumps parts of the system into discrete elements. The stiffness parameters become the connections between the lumped masses, and they are modeled as springs. This approach is not analytical. It is crude but easy to use and leads to discretized models of relatively low order. The approach was more popular before the advent of fast digital computers that made it possible to formulate and solve discretized problems of very large order.

When using series expansions, one expands the deformation in a finite series of known functions, multiplied by undetermined coefficients such as

$$v(x, t) = \sum_{r=1}^{n} \psi_r(x) q_r(t) \qquad [11.5.1]$$

where $\psi_r(x)$ $(r = 1, 2, \ldots, n)$ are the known *trial functions*, $q_r(t)$ are their amplitudes, and n is the order of the expansion. The coefficients $q_r(t)$ are in essence a set of generalized coordinates. The trial functions must be from a complete set and must satisfy certain conditions, depending on the discretization method. In general, the trial functions can be classified into two groups:

1. *Comparison functions.* As discussed earlier, these functions satisfy all the boundary conditions and are as many times differentiable as the highest-order spatial derivative in the equations of motion. In order to be consistent with the force and moment balances, the comparison functions should not violate the complementary boundary conditions.

2. *Admissible functions.* These functions satisfy only the geometric boundary conditions and are half as many times differentiable as the highest-order spatial derivative in the equations of motion. In order to be a consistent set, they should not violate the complementary boundary conditions, and they should not prevent the natural boundary conditions from being satisfied.

We introduce two new categories of trial functions: *consistent admissible functions* and *consistent comparison functions*. The properties of these and traditional admissible and comparison functions are summarized in Table 11.3. As before, the highest-order spatial derivative is denoted by $2p$.

Table 11.3 Classification of trial functions

Trial Function	Properties			
	Complete Set	Differentiability	Boundary Conditions to Be Satisfied	Complementary Boundary Conditions
Admissible functions	Yes	p times	All geometric boundary conditions	No requirement
Comparison functions	Yes	$2p$ times	All boundary conditions	No requirement
Consistent admissible functions	Yes	p times	All geometric boundary conditions do not prevent dynamic boundary conditions from being satisfied	Do not violate
Consistent comparison functions	Yes	$2p$ times	All boundary conditions	Do not violate

The added requirements regarding consistent trial functions are for physical completeness. Trial functions that violate the CBC and do not make it possible to satisfy the natural boundary conditions lead to slow convergence.

It is usually more desirable to deal with admissible functions than comparison functions, as it is easier to generate a set of admissible functions. Whether one can use admissible functions depends on the discretization method. Methods that permit use of admissible functions are more widely used.

Having selected a set of trial functions, one performs an operation to the series expansion in Eq. [11.5.1], which leads to an algebraic eigenvalue problem. The nature of the eigenvalue problem depends on the type of expansion and discretization. Commonly used discretization procedures are

a. Those based on *variational principles,* such as the Rayleigh-Ritz, assumed modes, and finite-element methods.

b. Those based on *weighted residuals,* such as the Galerkin and colocation methods.

When the discretization method is based on variational principles, the series expansion is plugged into the variational principle. Stationary values of the variational principle lead to an eigenvalue problem. For example, with the Rayleigh-Ritz method, one seeks stationary values of the Rayleigh quotient. For the method of assumed modes, the variational principle used is the extended Hamilton's principle.

When using a weighted residual approach, one seeks minimization of the difference between the actual solution and the series expansion in Eq. [11.5.1]. The difference, referred to as the *residual,* is minimized by the use of weighting functions.

Another consideration when selecting the trial functions is the domain of the trial functions. We distinguish between *global* trial functions, defined over the entire body, and *local* trial functions, defined over a portion of the domain. The finite element method basically makes use of local trial functions, a very suitable approach for problems with complex geometry. In this chapter, we consider global trial functions.

11.5.2 Method of Assumed Modes

We outline the *assumed modes* method, as it is convenient to use and is applicable to nonlinear systems, as will be the case when we consider combined large-angle rigid and elastic motion. The trial functions $\psi_r(x)$ can be admissible functions. We first take the expansion in Eq. [11.5.1] and substitute it into the expressions for the kinetic and potential energies. The method requires use of the energy inner product expression for the potential energy. Otherwise, one cannot use admissible functions as the expansion functions.

We demonstrate the method with an illustration. For a beam that deforms only in the y direction and is subjected to an internal axial load $P(x)$ and where the axial elasticity is ignored, the kinetic and potential energies can be written as

$$T = \frac{1}{2}\int_0^L \mu(x)\dot{v}^2(x,t)\,dx$$

$$V = \frac{1}{2}\int_0^L \left\{ EI_z(x)\left[\frac{\partial^2 v(x,t)}{\partial x^2}\right]^2 + P(x)\left[\frac{\partial v(x,t)}{\partial x}\right]^2 \right\} dx \quad \textbf{[11.5.2]}$$

Substituting the expansion [11.5.1] into these equations yields

$$T = \frac{1}{2}\int_0^L \mu(x)\left[\sum_{r=1}^n \psi_r(x)\dot{q}_r(t)\right]\left[\sum_{s=1}^n \psi_s(x)\dot{q}_s(t)\right] dx$$

$$= \frac{1}{2}\sum_{r=1}^n \sum_{s=1}^n \dot{q}_r(t)\dot{q}_s(t) \int_0^L \mu(x)\psi_r(x)\psi_s(x)\,dx$$

$$V = \frac{1}{2}\int_0^L \left\{ EI_z(x)\left[\sum_{r=1}^n \psi_r''(x)q_r(t)\right]\left[\sum_{s=1}^n \psi_s''(x)q_s(t)\right] \right.$$

$$\left. + P(x)\left[\sum_{r=1}^n \psi_r'(x)q_r(t)\right]\left[\sum_{s=1}^n \psi_s'(x)q_s(t)\right] \right\} dx$$

$$= \frac{1}{2}\sum_{r=1}^n \sum_{s=1}^n q_r(t)q_s(t)\int_0^L \left(EI_z(x)\psi_r''(x)\psi_s''(x) + P(x)\psi_r'(x)\psi_s'(x)\right) dx \quad \textbf{[11.5.3]}$$

Introducing the notation

$$m_{rs} = \int_0^L \mu(x)\psi_r(x)\psi_s(x)\,dx \quad \textbf{[11.5.4]}$$

$$k_{rs} = \int_0^L \left(EI_z(x)\psi_r''(x)\psi_s''(x) + P(x)\psi_r'(x)\psi_s'(x)\right) dx \quad r,s = 1,2,\ldots,n \quad \textbf{[11.5.5]}$$

where m_{rs} are the *mass coefficients* and k_{rs} are the *stiffness coefficients*, we express the kinetic and potential energies as

$$T = \frac{1}{2}\sum_{r=1}^n \sum_{s=1}^n m_{rs}\dot{q}_r(t)\dot{q}_s(t) \qquad V = \frac{1}{2}\sum_{r=1}^n \sum_{s=1}^n k_{rs}q_r(t)q_s(t) \quad \textbf{[11.5.6]}$$

or, in matrix form

$$T = \frac{1}{2}\{\dot{q}(t)\}^T[M]\{\dot{q}(t)\} \qquad V = \frac{1}{2}\{q(t)\}^T[K]\{q(t)\} \quad \textbf{[11.5.7]}$$

in which $\{q(t)\}$ is a column vector containing the coordinates $q_r(t)$ ($r = 1, 2, \ldots n$) in the form $\{q(t)\} = [q_1(t) \; q_2(t) \; \ldots \; q_n(t)]^T$, and $[M]$ and $[K]$ are the mass and stiffness matrices. The mass matrix is always symmetric and positive definite. For the system described by Eq. [11.5.2], the stiffness matrix is symmetric. The sign-definiteness of $[K]$ depends on the internal axial force $P(x)$ as well as on the boundary conditions. For example, for a beam admitting rigid body motions, $[K]$ is positive semidefinite. If $P(x)$ is tensile, as in a helicopter blade, $[K]$ is positive definite. When $P(x)$ is compressive, as in columns, the sign of $[K]$ depends on the value of $P(x)$.

In the presence of an external force $p_y(x, t)$, the virtual work can be expressed in terms of the expansion [11.5.1] as

$$\delta W = \int_0^L p_y(x,t)\,\delta v(x,t)\,dx = \int_0^L p_y(x,t)\,\delta\left[\sum_{r=1}^n \psi_r(x)q_r(t)\right]dx$$

$$= \sum_{r=1}^n \delta q_r \int_0^L p_y(x,t)\psi_r(x)\,dx \qquad [11.5.8]$$

Introducing the notation

$$Q_r(t) = \int_0^L p_y(x,t)\psi_r(x)\,dx \qquad r = 1, 2, \ldots, n \qquad [11.5.9]$$

and substituting it into [11.5.8], we can express the virtual work as

$$\delta W = \sum_{r=1}^n Q_r \delta q_r = \{Q\}^T \delta\{q\} \qquad [11.5.10]$$

where $\{Q(t)\} = [Q_1(t)\,Q_2(t) \ldots Q_n(t)]^T$. We recognize $Q_r(t)$ to be the generalized force associated with the generalized coordinate $q_r(t)$.

We have now expressed the kinetic and potential energies in terms of n coordinates, thus approximating the system as an n-dimensional discrete one. We then use these discretized forms of the kinetic and potential energies together with the extended Hamilton's principle. Because the problem is now in terms of a finite set of generalized coordinates, one can obtain the equations of motion directly by using Lagrange's equations.

Next, let us discuss the solution when the mass and stiffness coefficients are independent of time, which is the case for the linear problem considered here. From Section 5.3, we can directly write the equations of motion of the approximate system as

$$[M]\{\ddot{q}(t)\} + [K]\{q(t)\} = \{Q(t)\} \qquad [11.5.11]$$

To find the response of Eq. [11.5.11], we use the same approach as in Sections 5.5–5.7. We first solve the eigenvalue problem. Introducing

$$\{q(t)\} = \{a\}e^{\lambda t} \qquad [11.5.12]$$

where $\{a\}$ is a time-invariant amplitude vector, and $e^{\lambda t}$ denotes the time dependency, into Eq. [11.5.11] yields

$$(\lambda^2[M] + [K])\{a\}e^{\lambda t} = \{0\} \qquad [11.5.13]$$

which leads to the characteristic equation

$$\det(\lambda^2[M] + [K]) = 0 \qquad [11.5.14]$$

11.5 APPROXIMATE SOLUTIONS: THE ASSUMED MODES METHOD

The roots of the *characteristic equation* are the eigenvalues of the discretized system. As discussed in Chapter 5, when the stiffness matrix is positive definite, all the eigenvalues λ_r ($r = 1, 2, \ldots, n$) are pure imaginary. We introduce the notation $\lambda_r = -i\hat{\omega}_r$, where $\hat{\omega}_r$ are the natural frequencies of the approximate system.

Associated with each eigenvalue λ_r, there is a corresponding eigenvector $\{a_r\}$ such that

$$(\lambda_r^2[M] + [K])\{a_r\} = \{0\} \qquad r = 1, 2, \ldots, n \qquad [\mathbf{11.5.15}]$$

The eigenvalues are approximations to the first n eigenvalues of the actual system. The eigenvectors can be used to approximate the first n eigenfunctions by

$$\hat{\phi}_r(x) = \{a_r\}^T \{\psi(x)\} \qquad [\mathbf{11.5.16}]$$

where $\hat{\phi}_r(x)$ are the approximate eigenfunctions and $\{\psi(x)\}$ is a column vector containing the trial functions in the form $\{\psi(x)\} = [\psi_1(x)\psi_2(x)\ldots\psi_n(x)]^T$.

The eigenvectors $\{a_r\}$ can be shown to be orthogonal with respect to the mass and stiffness matrices. They can be normalized to yield

$$\{a_r\}^T[M]\{a_s\} = \underline{\delta}_{rs} \qquad \{a_r\}^T[K]\{a_s\} = \hat{\omega}_r^2 \underline{\delta}_{rs} \qquad [\mathbf{11.5.17}]$$

The approximate eigenfunctions are orthogonal with respect to the original mass distribution and stiffness, expressed in terms of the energy inner product as

$$\int_0^L \mu(x)\hat{\phi}_r(x)\hat{\phi}_s(x)\,dx = \underline{\delta}_{rs} \qquad \int_0^L \left(EI_z(x)\hat{\phi}_r''(x)\hat{\phi}_s''(x) + P(x)\hat{\phi}_r'(x)\hat{\phi}_s'(x)\right)dx$$

$$= \hat{\omega}_r^2 \underline{\delta}_{rs} \qquad r, s = 1, 2, \ldots, n \qquad [\mathbf{11.5.18}]$$

To prove this, we introduce Eq. [11.5.16] into Eqs. [11.5.18], which yields

$$\int_0^L \mu(x)\hat{\phi}_r(x)\hat{\phi}_s(x)\,dx = \{a_r\}^T \left[\int_0^L \mu(x)\{\psi(x)\}\{\psi(x)\}^T\,dx\right]\{a_s\} \qquad [\mathbf{11.5.19}]$$

$$\int_0^L \left(EI_z(x)\hat{\phi}_r''(x)\hat{\phi}_s''(x) + P(x)\hat{\phi}_r'(x)\hat{\phi}_s'(x)\right)dx$$

$$= \{a_r\}^T \int_0^L \left(EI_z(x)\{\psi''(x)\}\{\psi''(x)\}^T + P(x)\{\psi'(x)\}\{\psi'(x)\}^T\right)dx\{a_s\} \qquad [\mathbf{11.5.20}]$$

A closer examination of the term inside the square brackets indicates that

$$\int_0^L \mu(x)\{\psi(x)\}\{\psi(x)\}^T\,dx = [M]$$

$$\int_0^L \left(EI_z(x)\{\psi''(x)\}\{\psi''(x)\}^T + P(x)\{\psi'(x)\}\{\psi'(x)\}^T\right)dx = [K] \qquad [\mathbf{11.5.21}]$$

Thus, Eqs. [11.5.19]–[11.5.20] become identical to Eq. [11.5.17]. Note that the orthogonality with respect to stiffness is demonstrated using the energy inner product only. To have orthogonality with respect to the stiffness operator, one must use comparison functions as trial functions.

11.5.3 Convergence Issues

It can be shown that the approximate eigenvalues are higher in magnitude than the actual eigenvalues. This is because using a finite series of trial functions is mathematically equivalent to placing constraints on a system. An exact representation of a flexible system requires expansion by an infinite series. Constraints make a system stiffer and raise its natural frequencies. This property is common to all discretization procedures that are based on series expansions and variational principles. Increasing the order of approximation in essence relaxes some of the constraints. Hence, the eigenvalues become lower in value. As the model order is increased, the eigenvalues approach their actual values from above. By contrast, when the discretization is in the form of lumping, the approximate eigenvalues usually approach their actual values from below.

The property that the approximate eigenvalues approach the actual eigenvalues from above can be demonstrated by the *inclusion principle,* which can be stated as follows: Let $\hat{\omega}_r$ $(r = 1, 2, \ldots, n)$ denote the natural frequencies associated with an nth-order approximation. Now, consider an approximation of order $n + 1$, where the first n trial functions are the same as before, and the next term in the set of admissible functions is the $(n + 1)$th trial function.

Denoting the eigenvalues of the order $n + 1$ system by $\hat{\Omega}_r$ $(r = 1, 2, \ldots, n + 1)$, the inclusion principle states that

$$\hat{\Omega}_1 \leq \hat{\omega}_1 \leq \hat{\Omega}_2 \leq \hat{\omega}_2 \leq \ldots \leq \hat{\Omega}_n \leq \hat{\omega}_n \leq \hat{\Omega}_{n+1} \quad [\mathbf{11.5.22}]$$

Proof of the inclusion principle is based on the minimax theorem; it can be found in advanced vibration texts.

We next consider the issue of convergence. Convergence of the approximate solution to the actual solution depends on the type and number of trial functions used, as well as on the mass and stiffness properties and the loading. For example, if the system has discrete (also called *concentrated,* to distinguish from continuous) springs acting on it, these concentrated springs cause a discontinuity in the shear profile. This discontinuity becomes hard to approximate by continuous trial functions. Also, if the trial functions are selected such that the complementary boundary conditions are violated or it is not possible to satisfy the natural boundary conditions, convergence becomes slower. The sensitivity of the trial functions to various parameters in them are also important. For example, dealing with hyperbolic sines and cosines often presents problems, as these functions are very sensitive to their arguments.

In general, if one uses an nth order model, one can assume the first $n/2$ eigenvalues to be accurate. The exception is when simple polynomials are used as trial functions, where usually fewer than $n/2$ eigenvalues are estimated accurately. Also, if the trial functions are selected as functions resembling the eigenfunctions, convergence is much faster. For example, to find the eigensolution of a beam with a nonuniform cross section, a good choice of admissible functions is the eigenfunctions of a uniform beam with the same boundary conditions. Convergence issues associated with trial function expansions of solutions constitute a very broad subject, one that is beyond the scope of this text.

The assumed modes method can be described in terms of the general operator notation. We expand the elastic deformation as

$$\{\mathcal{U}(\mathcal{D},t)\} = \sum_{r=1}^{n}\{\psi_r(\mathcal{D})\}q_r(t) \qquad [11.5.23]$$

in which $\{\psi_r(\mathcal{D})\}$ $(r = 1, 2, \ldots, n)$ are a set of admissible functions. The entries of the mass and stiffness matrices and the generalized forces become

$$m_{rs} = \langle\{\psi_r(\mathcal{D})\}, \mathcal{M}\{\psi_s(\mathcal{D})\}\rangle \quad k_{rs} = \langle\{\psi_r(\mathcal{D})\}, \{\psi_s(\mathcal{D})\}\rangle$$

$$Q_r = \langle\{\psi_r(\mathcal{D})\}, \{F\}\rangle \quad r, s = 1, 2, \ldots, n \qquad [11.5.24]$$

in which $\{F\}$ is the external excitation vector. One then invokes Lagrange's equations and obtains the equations of motion, which have the form of Eq. [11.5.11] when $[M]$ and $[K]$ are constant coefficient matrices.

Example 11.6

Consider the rotating blade problem in Example 11.3, which is modeled as a uniform fixed-free beam of length L, rotating with angular velocity Ω. Compare the natural frequencies of the beam for the rotation speeds $\Omega^2/(EI/\mu L^4) = 0, 0.1, 1,$ and 10, and verify that the asymptotic behavior of the eigenvalues is governed by Eq. [11.4.25].

Solution

We use the method of assumed modes. The trial functions must satisfy the geometric boundary conditions of

$$\psi_r(0) = 0 \quad \psi'_r(0) = 0 \quad r = 1, 2, \ldots, n \qquad [a]$$

The trial functions must also not violate the CBC, such that at least one of the trial functions $\psi_r(x)$ must have a nonvanishing second derivative at $x = 0$ and one must have a nonvanishing third derivative at $x = 0$. Also, the trial functions should not prevent the natural boundary conditions to be satisfied at the free end.

A suitable set of consistent admissible functions is the eigensolution of a uniform, fixed-free, nonrotating beam. Another choice is polynomials in the form

$$\psi_r(x) = \left(\frac{x}{L}\right)^{r+1} \quad r = 1, 2, \ldots, n \qquad [b]$$

We saw in Example 11.3 that the effects of the rotation can be modeled in a number of ways. Here, let us consider it as an added potential energy. From Eq. [a] of Example 11.3, the internal axial force has the form

$$P(x) = \frac{\mu\Omega^2(L^2 - x^2)}{2} \qquad [c]$$

Table 11.4 Natural frequencies of helicopter blade as a function of the rotor speed

	R				Ratio of $\frac{\lambda_{R=10}}{\lambda_{R=0}}$
Mode no.	0	0.1	1	10	
1	3.516	3.533	3.681	4.918	1.399
2	22.03	22.05	22.18	23.46	1.065
3	61.72	61.73	61.86	63.15	1.024
4	128.4	128.4	128.5	129.7	1.010

The entries of the mass and stiffness matrices are found using Eqs. [11.5.4] and [11.5.5], to have the form

$$m_{rs} = \frac{1}{L^{r+s+2}} \int_0^L \mu x^{r+1} x^{s+1}\, dx = \frac{\mu L}{r+s+3}$$

$$k_{rs} = \frac{(r+1)r(s+1)s}{L^{r+s+2}} \int_0^L EI x^{r-1} x^{s-1}\, dx + \frac{(r+1)(s+1)}{L^{r+s+2}} \int_0^L P(x) x^r x^s\, dx$$

$$= \frac{EI}{L^3} \frac{(r+1)(s+1)rs}{r+s-1} + \mu\Omega^2 L \frac{(r+1)(s+1)}{2}\left(\frac{1}{r+s+1} - \frac{1}{r+s+3}\right) \quad [d]$$

Let us divide both the mass and stiffness coefficients by μL and introduce the ratio $R = \Omega^2 \mu L^4 / EI$. As R is increased, the blade rotates faster. We can then write the mass and stiffness coefficients as

$$m_{rs} = \frac{1}{r+s+3}$$

$$k_{rs} = \frac{EI}{\mu L^4}(r+1)(s+1)\left[\frac{rs}{r+s-1} + \frac{R}{2}\left(\frac{1}{r+s+1} - \frac{1}{r+s+3}\right)\right] \quad [e]$$

Letting $EI/\mu L^4 = 1$, we solve the eigenvalue problem for various values of R. We use a discretized model of order $n = 6$. The first four eigenvalues are given in Table 11.4, together with the ratio of the eigenvalues for $R = 10$ and $R = 0$.

We observe that as R increases and the rotation speed gets bigger, all of the eigenvalues become larger. As expected from the asymptotic analysis, the increase in the natural frequencies is more pronounced in the lower modes.

Example 11.7

As an illustration of the importance of satisfying the complementary boundary conditions and associated convergence issues, we consider a uniform bar of length L in axial vibration, fixed at one end ($x = 0$), and attached to a spring of constant k at the other end ($x = L$), as shown in Fig. 11.21. The kinetic and potential energies have the form

$$T = \frac{1}{2} \int_0^L \mu \dot{u}^2(x,t)\, dx \qquad V = \frac{1}{2} \int_0^L EA[u'(x,t)]^2\, dx + \frac{1}{2} k u^2(L,t) \quad [a]$$

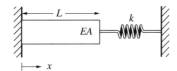

Figure 11.21

For simplicity we select unit length and unit mass and stiffness distributions, $L = 1$, $\mu = 1$, $EA = 1$. The boundary conditions can be shown to be

$$u(0, t) = 0 \quad EAu'(L, t) + ku(L, t) = 0 \qquad \textbf{[b]}$$

The boundary condition at $x = 0$ is geometric, and the associated complementary boundary condition is that $EAu'(0, t)$ is unspecified. At $x = L$, we have a boundary condition of the third kind. To find the eigensolution we consider three sets of admissible functions $\psi_{1r}(x)$, $\psi_{2r}(x)$, $\psi_{3r}(x)$, each from a complete set:

Set 1: $\psi_{1r}(x) = x^r$ **[c]**

Set 2: $\psi_{2r}(x) = \sin \dfrac{(2r-1)\pi x}{2L}$ **[d]**

Set 3: $\psi_{3r}(x) = (\sin \beta_r L - \sinh \beta_r L)(\sin \beta_r x - \sinh \beta_r x) + (\cos \beta_r L + \cosh \beta_r L)(\cos \beta_r x - \cosh \beta_r x)$ **[e]**

The mass and stiffness matrices have the entries

$$m_{rs} = \int_0^L \mu \psi_{ir}(x) \psi_{is}(x)\, dx$$

$$k_{rs} = \int_0^L EA\psi'_{ir}(x)\psi'_{is}(x)\, dx + k\psi_{ir}(L)\psi_{is}(L) \qquad i = 1,2,3 \qquad \textbf{[f]}$$

The first set is simple polynomials, which do not violate the CBC or prevent the natural boundary condition from being satisfied. Hence, they qualify as a set of consistent admissible functions. The second set is the eigenfunctions of a fixed-free bar in axial motion. This set is the exact eigensolution when there is no spring, i.e., $k = 0$. In the presence of a spring, this set of functions cannot satisfy the natural boundary condition at $x = L$, as all trial functions have vanishing first derivatives at $x = L$. The third set, Eq. [e], is the eigenfunctions of a fixed-free beam in bending, where $\beta_r L$ are the solutions of the associated characteristic equation. For this set, both the displacement and slope are zero at $x = 0$, thus violating the complementary boundary condition at $x = 0$. Hence, the second and third sets do not qualify as consistent admissible functions.

We compare the accuracy of the solution obtained by using all three sets for varying orders of approximation and for different values of the spring constant k. Table 11.5 compares the first three eigenvalues for the different sets of admissible functions, and for varying values of the spring constant, using approximation orders of five and seven. Even though simple polynomials constitute a relatively poor choice of admissible functions, the difference in accuracy and rates of convergence between simple polynomials and the other two sets is obvious.

Analyzing the eigenvalues we make the following observations:

1. As expected, in all cases the eigenvalues become smaller as the order of approximation increases and the inclusion principle holds.

Table 11.5   Comparison of eigenvalues

	$k = 0$		$k = 100$		$k = 1000$	
	$n = 5$	$n = 7$	$n = 5$	$n = 7$	$n = 5$	$n = 7$
Set 1: Polynomials as consistent admissible functions						
ω_1	1.5708	1.5708	3.1105	3.1105	3.1385	3.1385
ω_2	4.7132	4.7124	6.2227	6.2211	6.2787	6.2769
ω_3	12.174	7.8550	9.9508	9.3479	10.091	9.4154
Set 2: Eigenfunctions of fixed-free uniform bar as trial functions						
ω_1	1.5708	1.5708	3.2417	3.2027	3.2717	3.2348
ω_2	4.7124	4.7124	6.4938	6.4091	6.5515	6.4726
ω_3	7.8540	7.8540	9.7727	9.6239	9.8524	9.7157
Set 3: Eigenfunctions of fixed-free uniform beam as trial functions						
ω_1	1.6221	1.6187	3.2090	3.2043	3.2387	3.2341
ω_2	4.8963	4.8816	6.4390	6.4322	6.4965	6.4903
ω_3	8.1518	8.0818	9.5568	9.5163	9.6339	9.5978

2. When $k = 0$, which implies a free end at $x = L$, the eigenfunctions of the fixed-free bar give the exact natural frequencies. This is because the eigenfunctions of the fixed-free bar become the closed-form eigensolution for this case. Comparing polynomials and set 3, for $n = 7$, polynomials exactly match the first two natural frequencies, with the third being off by about 0.01 percent. However, using the eigenfunctions of the fixed-free beam gives errors in all eigenvalues, with the error in the first natural frequency being about 2.5 percent. Interestingly, the error in the third natural frequency is also about 2.5 percent off the actual natural frequency. This is an indication of problems encountered when the CBC are violated, because one expects the lower natural frequencies to be more accurate than the higher ones.

3. For nonzero values of k, polynomials again give the best results, except for ω_3 for $n = 5$. This is expected, because the polynomials are a set of consistent admissible functions and the other two sets are not. The natural frequencies again have an error of 2.5 to 3.0 percent for sets 2 and 3. The lower frequencies are not estimated more accurately than the higher frequencies, clear indications of the problems encountered when the CBC are violated and the natural boundary condition cannot be satisfied.

4. For nonzero values of k, while the accuracy obtained from sets 2 and 3 is comparable for ω_1, eigenfunctions of the fixed-free beam give better results for ω_3. For ω_2, set 2 generally gives more accurate results.

5. Also, for nonzero values of k, the eigenvalues obtained from sets 2 and 3 never get very close to the actual values. For example, for $k = 100$, the solution for ω_1 does not improve at all for set 1 as we go from $n = 5$ to $n = 7$. This is because the solution has almost converged. However, for set 2 there is quite a change as the order of the approximation is increased. For set 3, there seems to be convergence, but not to the correct natural frequencies.

6. When polynomials are used as trial functions, one obtains very accurate results for the lower modes but the higher modes are inaccurate. Usually, for an approximation of order n, fewer than $n/2$ modes give accurate results. This can be observed by looking at ω_3 for $n = 5$.

7. Finally, one can observe the asymptotic behavior of the eigenvalues by looking at the results for different values of k. Consider the results for polynomials and $n = 7$, so that we can treat the first three natural frequencies as exact. For $k = 100$ and $k = 1000$, the first natural frequency is almost doubled than ω_1 for $k = 0$, while ω_2 increases by 33 percent and ω_3 increases by about 20 percent. As in Example 11.6, addition of an extra source of stiffness affects the lower modes much more than the higher modes.

11.6 KINEMATICS OF COMBINED ELASTIC AND LARGE ANGLE RIGID BODY MOTION

An important case in vibrations is the combined elastic and large-angle rigid motion of a body. Consider, for example, the beam in Fig. 11.22. Analyzing the position of a point compared with an initial, undeformed position, the rigid body and elastic motions can no longer be linearly superposed, as the large-angle motion changes the projection of the beam axis onto the undeformed axis substantially.

When the elasticity of the body is small, it is convenient to view the elastic motion from a set of reference axes, selected such that the motion observed from these axes is small and linear vibration theory can be applied. When the elastic deformation becomes large compared with the dimensions of the body, one needs to consider theories other than those presented in this chapter.

Consider Figs. 11.22–11.24, describing beams undergoing both elastic and large-angle rigid body motion. The two ends of the beam are denoted by c and d. These beams represent commonly encountered situations when dealing with the combined elastic and large-angle rigid body motion. Figure 11.22, depicting a pinned-free beam, is representative of a robotic arm; Figure 11.23 describes a free-free beam, as in a satellite with antennae; and Fig. 11.24 is representative of an intermediate link of a mechanical system.

We will observe the motion of these beams from a relative frame such as the one shown in Fig. 11.25. The frame has an origin O and a certain orientation, and is called the *shadow frame* or the *tracking frame*. We will refer to the motion of this frame as the *primary motion*, and the motion observed from the frame as the *secondary motion*. The reason for describing the motion in two parts is to select origin and

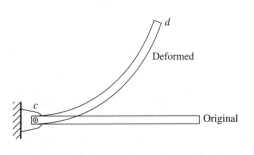

Figure 11.22   Deformed pinned-free beam

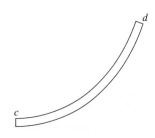

Figure 11.23   Deformed free-free beam

Figure 11.24 Deformed intermediate link

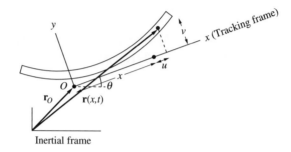

Figure 11.25

orientation of the tracking frame such that one can treat the secondary motion using the near linear theory discussed earlier in this chapter. With this in mind, we write the displacement of a point on the beam as

$$\mathbf{r}(x, t) = \mathbf{r}_O + (x + u(x, t))\mathbf{i} + v(x, t)\mathbf{j} + w(x, t)\mathbf{k} \quad [11.6.1]$$

in which $\mathbf{r}_O + x\mathbf{i}$ denotes the primary motion and $u(x, t)\mathbf{i} + v(x, t)\mathbf{j} + w(x, t)\mathbf{k}$ denotes the secondary motion.

We note that $u(x, t)$, $v(x, t)$, and $w(x, t)$ are not necessarily the same as their counterparts in the previous sections of this chapter and that the primary and secondary motions do not necessarily correspond to the rigid and elastic motions. The values $u(x, t)$, $v(x, t)$, or $w(x, t)$ depend on the choice of the tracking reference frame and they may contain components from the rigid body motion.

The angular velocity of the reference frame is written in terms of its components along the reference frame as

$$\boldsymbol{\omega} = \omega_x \mathbf{i} + \omega_y \mathbf{j} + \omega_z \mathbf{k} \quad [11.6.2]$$

The velocity of point x becomes

$$\begin{aligned}
\dot{\mathbf{r}}(x, t) &= \dot{\mathbf{r}}_O + \dot{u}\mathbf{i} + \dot{v}\mathbf{j} + \dot{w}\mathbf{k} + \boldsymbol{\omega} \times [(x + u)\mathbf{i} + v\mathbf{j} + w\mathbf{k}] \\
&= \dot{\mathbf{r}}_O + (\dot{u} - \omega_z v + \omega_y w)\mathbf{i} + (\dot{v} + \omega_z x + \omega_z u - \omega_x w)\mathbf{j} \\
&\quad + (\dot{w} + \omega_x v - \omega_y x - \omega_y u)\mathbf{k}
\end{aligned} \quad [11.6.3]$$

We will discuss the following issues associated with the modeling of lightly flexible bodies undergoing combined rigid and elastic motions:

1. Selection of the origin and orientation of the tracking reference frame.
2. The type of trial functions to use to expand the secondary motion.
3. Analysis of the equations of motion.

11.6 KINEMATICS OF COMBINED ELASTIC AND LARGE ANGLE RIGID BODY MOTION

One can expect the nature of the reference frame to dictate the nature of the equations of motion. The primary criterion is to enable one to use the near linear theory and to describe the secondary motion in a series expansion. As we will see in the next section, in many cases it is necessary to keep the $u(x, t)$ in the formulation, because for fast rigid body motions the centrifugal force is quite large.

The expansions for $v(x, t)$, $w(x, t)$, and $e(x, t)$ have the form

$$v(x, t) = \sum_{r=1}^{n} \psi_r(x) q_r(t) \qquad w(x, t) = \sum_{r=1}^{n^*} \gamma_r(x) q_{n+r}(t)$$

$$e(x, t) = \sum_{r=1}^{m} \beta_r(x) q_{n+n^*+r}(t) \qquad [11.6.4]$$

in which $\psi_r(x)$, $\gamma_r(x)$, and $\beta_r(x)$ are sets of suitable admissible functions and $q_r(t)$ are generalized coordinates, with n, n^*, and m denoting the order of the expansion. It follows from Eq. [11.2.6] that $u(x, t)$ can be expressed in terms of the generalized coordinates as

$$u(x, t) = e(x, t) - \frac{1}{2} \int_0^x \left[\left(\frac{\partial v}{\partial \sigma} \right)^2 + \left(\frac{\partial w}{\partial \sigma} \right)^2 \right] d\sigma = \sum_{r=1}^{m} \beta_r(x) q_{n+n^*+r}(t)$$

$$- \frac{1}{2} \sum_{r=1}^{n} \sum_{s=1}^{n} q_r(t) q_s(t) \int_0^x \psi_r'(\sigma) \psi_s'(\sigma) \, d\sigma$$

$$- \frac{1}{2} \sum_{r=1}^{n^*} \sum_{s=1}^{n^*} q_{n+r}(t) q_{n+s}(t) \int_0^x \gamma_r'(\sigma) \gamma_s'(\sigma) \, d\sigma \qquad [11.6.5]$$

There are two primary choices for the selection of the relative frame:

a. *To select the relative frame such that its motion is known a priori*. This choice is relevant when one has a good idea as to what the rigid component of the motion is going to look like. An example is the maneuver of robots or spacecraft with little flexibility. A set of inputs is applied to move the body along a desired trajectory. For a variety of reasons, the actual trajectory of the arm will differ from the desired one. The secondary motion makes up for this difference. With this model, the motion is measured from a tracking frame whose motion is known.

Note that in this case both $\mathbf{r}_O$ and θ are treated as known quantities, so that when deriving the equations of motion their variation is zero. Hence, in this formulation the primary motion is independent of the secondary motion. The boundary conditions on the secondary motion remain the same as in the initial description of the beam.

b. *To select the relative frame such that its location and motion depend on the orientation of the structure*. In this formulation one takes into consideration the current position and orientation of the body. Hence, the primary motion is not selected independently and it is related to the secondary motion. The quantities describing the primary motion are variables, because their values depend on the location of the body. This implies that we may be adding redundant degrees of freedom. Such redundancy necessitates imposition of constraints on the secondary motion. The imposition of constraints can change the boundary conditions of the secondary motion.

One can view the geometric boundary conditions of a body as configuration constraints imposed on an otherwise free-free member. From this perspective, one can view the constraints imposed on the secondary motion as changing the boundary conditions. However, one should keep in mind that the imposition of constraints on the secondary motion is at a mathematical level, so that the existing internal forces and moments at the boundaries, hence the CBC, do not change. One must be cautious in describing the secondary motion such that the newly generated geometric boundary conditions are satisfied and the already existing CBC are not violated.

As a consequence of using constrained coordinates, the resulting equations of motion will be in the form of constrained equations. One way to circumvent dealing with a constrained set of equations is to select the generalized coordinates that describe the secondary motion in a way such that the imposed constraints are automatically satisfied. We will adopt this approach. In essence, by selecting the trial functions as consistent admissible functions for the secondary motion one can satisfy the constraints automatically.

There are several ways of defining the primary motion. Selection of the relative frame is, in general, complicated for the case of general three-dimensional motion. Furthermore, for three-dimensional problems, physical representation of the choice of the relative frame becomes difficult, as well as any attempt to analyze the nature of the resulting equations of motion. On the other hand, when the angular velocity has certain properties, some interesting descriptions of the motion result. Here, we will analyze cases where the angular velocity of the reference frame is in one direction only, and the secondary motion is on a plane. We consider the undeformed beam axis to be the x direction, one of $w(x, t)$ or $v(x, t)$ to be zero, and the angular velocity to be along one of the x, y, or z directions. It turns out that there are three distinct types of motion possible depending on the directions of the secondary motion and angular velocity.

Example 11.3 showed the rotating blade problem, where the angular velocity was in the y direction and the secondary motion was on the xy plane, with $w(x, t) = 0$. When the angular velocity is in the z direction and $v(x, t)$ is set to zero we get the same type of behavior. We refer to such problems as *Case 1*. In this type of problem, there are no constraints imposed on the secondary motion. The angular velocity of the reference frame is along the plane of the secondary motion; thus the additional rotational variable due to the rotation of the tracking frame is an added degree of freedom and not a redundant one. For Case 1 problems, the angular velocity of the tracking frame has a stiffening effect on the secondary motion, as we saw in Example 11.3. This case is the most straightforward of the three types of motion.

When the angular velocity is in the x direction only, the motion is similar to that of a rotating shaft. We will analyze this case, referred to as *Case 3,* in Section 11.9, together with the effects of the rotation on the secondary motion. As in Case 1, selection of the tracking frame does not impose a constraint on the secondary motion, for the same reasons.

By far the most interesting case is *Case 2,* when the secondary motion is in the xy plane and the angular velocity is in the z direction (or secondary motion in the xz plane and angular velocity in the y direction). The angular velocity is perpendicular

11.6 KINEMATICS OF COMBINED ELASTIC AND LARGE ANGLE RIGID BODY MOTION

Table 11.6   Types of large-angle rigid motion

Type of Motion	Plane of Secondary Motion	Angular Velocity Direction
Case 1 (rotating blade, stiffening effect)	xy	y
	xz	z
Case 2 (robotic arm, spacecraft)	xy	z
	xz	y
Case 3 (rotating shaft, whirling)	yz	x

to the plane of the secondary motion. Hence, selection of the reference frame imposes a constraint on the secondary motion. Table 11.6 outlines the three cases.

For the case of angular velocity in all three directions, general conclusions cannot be drawn, because of the mixed terms that enter the formulation. The exception is when the angular velocities are small; then, one can rely on the results from this chapter to describe the stiffening and softening effects.

In the remainder of this section we consider Case 2 and outline three approaches to select the origin and orientation of the reference frame, together with the constraints they impose on the secondary motion. Without loss of generality, in all descriptions we consider the secondary motion to be in the xy plane and the angular velocity in the z direction.

The Rigid Body Constraint   In this approach, one selects the moving coordinate frame such that there is no rigid body component in the secondary motion. That is, all of the rigid component of the motion is described by the angle θ and by $\mathbf{r}_O$. Figures 11.26 to 11.28 show this constraint for the three beam configurations considered earlier.

Let us consider the selection of the origin of the reference frame. For the pinned-free beam, the origin O is selected as the pin joint. For the free-free and interconnected beams the origin is selected as the center of mass, so that

$$\int_c^d \mathbf{r}(x,t)\, dm = m\mathbf{r}_O \qquad [11.6.6]$$

where $dm = \mu(x)\, dx$. This leads to the constraint for the center of mass

$$\int_c^d \mu(x)[(x + u(x,t))\mathbf{i} + v(x,t)\mathbf{j}]\, dx = \mathbf{0} \qquad [11.6.7]$$

Note the component of this equation in the y direction indicates orthogonality with the translational rigid body mode. The component in the x direction is basically the lateral motion. This motion is usually very small, as it is the integral of $u(x, t)$ over the beam. For the intermediate link in Fig. 11.28, one can select the origin at one of the joints, but this complicates the physical interpretation of the motion.

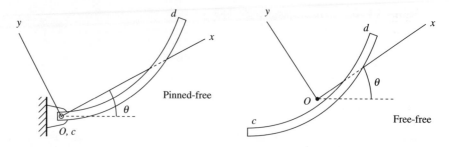

Figure 11.26   Pinned-free

Figure 11.27   Free-free

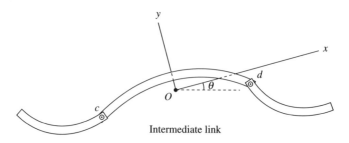

Figure 11.28   Rigid body constraint

To determine the orientation of the tracking frame, a number of approaches can be followed. One choice is to have the components of the motions in the x and y directions be orthogonal to each other. This leads to the *principal axes constraint*

$$\int_c^d \mu(x)[x + u(x,t)]v(x,t)\,dx = 0 \qquad [11.6.8]$$

Another choice is to select the orientation such that $v(x, t)$ contains no rigid body components. That is, $v(x, t)$ is orthogonal to the rotational rigid body mode. For both a pinned-free and a free-free beam, the rotational rigid body eigenfunction is (note that x is measured from point O)

$$\phi_1(x) = \frac{1}{\sqrt{I_O}} x \qquad [11.6.9]$$

The constraint can then be expressed as

$$\int_c^d \mu(x)\phi_1(x)v(x,t)\,dx = 0 \qquad [11.6.10]$$

For the interconnected beam in Fig. 11.28 it is much harder to distinguish between the rigid and elastic motions, as both ends of the beam are not free, but one can use the above equation as the constraint.

Another choice for the orientation constraint is known as the *modified Tisserand constraint*, and it has the form

$$\int_c^d \mu(x)[x + u(x,t)]\mathbf{i} \times \dot{v}(x,t)\mathbf{j}\,dx = 0 \qquad [11.6.11]$$

For the types of motion of interest here, the deformation $u(x, t)$ is usually small, so that any product of $u(x, t)$ and $v(x, t)$ and their derivatives can be ignored. Using this assumption, both the constraints [11.6.8] and [11.6.11] become equivalent to Eq. [11.6.10]. We will use Eq. [11.6.10] as the rigid body constraint from now on.

Let us next investigate the boundary conditions on the secondary motion. For the pinned-free and free-free beams the boundary conditions do not change after the primary frame is introduced. Hence, to expand $u(x, t)$ and $v(x, t)$, one can use any set of trial functions used in the analysis of pinned-free or free-free beams. For the pinned-free and free-free beams, use of the rigid body mode constraint has the appearance of being a natural choice for the shadow frame.

For the intermediate link, the situation is different. Introduction of the reference frame changes the boundary conditions of the secondary motion at c and d to basically free ends. The internal force at a free end is zero. However, for the actual beam, at that point there is an internal force due to the joint. So one ends up with a situation where the displacement and slope are unspecified at the boundary, while the internal forces and moments are not zero.

A variant of the rigid body mode constraint is to not use the mass density term in the constraint equations. Rather, one has

$$\int_0^L v(x,t)\,dx = 0 \qquad \int_0^L \left(1 - \frac{x}{L}\right)v(x,t)\,dx = 0 \qquad [11.6.12]$$

This constraint makes it easier to generate a set of trial functions.

The rigid body constraint has the advantage that it provides a physical explanation for the selection of the reference frame. Also, for the pinned-free and free-free beams, it separates the rigid and elastic components of the overall motion into the primary and secondary motions. On the other hand, in an actual measurement or control application, where one needs to have measurements of the origin and orientation of the reference frame, it is difficult to deal with this constraint. Because the reference axes are not attached to the body, the location and orientation of the reference frame cannot be measured directly and they must be calculated from other measurements.

The Zero Slope Constraint   The origin of the reference frame is attached to a point on the body, and the orientation of the reference frame is selected such that the secondary motion has zero slope at the origin, as shown in Figs. 11.29 to 11.31. In general, if there is a hub on the body the hub location is selected as the origin, as it is easier to attach a sensor or actuator there for measurement, navigation, or control purposes. For the interconnected beam, the reference frame is usually placed at one of the joints. The constraint on the secondary motion is expressed as

$$v'(0, t) = 0 \qquad [11.6.13]$$

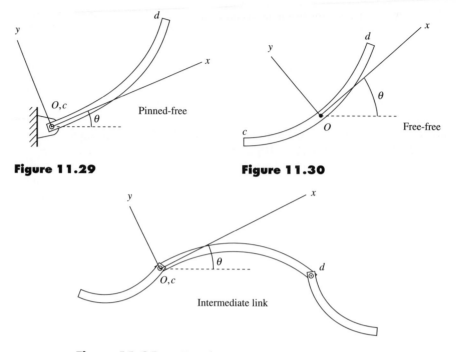

Figure 11.29

Figure 11.30

Figure 11.31   Zero slope constraint

The resulting geometric boundary conditions on the secondary motion at the origin become

$$v(0, t) = 0 \qquad v'(0, t) = 0 \qquad [11.6.14]$$

in essence giving the secondary motion fixed-free end conditions. This choice of the reference frame changes the boundary conditions on the secondary motion of all three types of beams considered in this section. For the beams in Figs. 11.30 and 11.31, if the origin O is placed on a point in the interior of the beam, one can view the beam as consisting of two beams, one from O to one end c, and the other from O to the other end d.

As in the rigid body mode constraint, the fixed-free end conditions specify that the internal force is zero at end d. For the interconnected beam, there is the problem of having the boundary conditions of a free end and the CBC of a pinned end when modeling the secondary motion.

The Zero Tip Deformation Constraint   The orientation of the reference frame is obtained by drawing a straight line between the two ends of the beam, as shown in Figs. 11.32 to 11.34. The origin is taken as one of the ends. As a result, the constraints on the secondary motion become

$$v(0, t) = 0 \qquad v(L, t) = 0 \qquad [11.6.15]$$

where L is the original length of the beam. These constraint equations are the geometric boundary conditions associated with the secondary motion, those of a

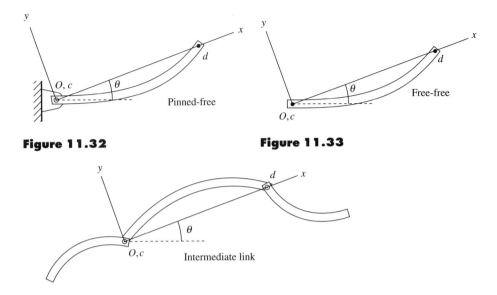

Figure 11.32 Pinned-free

Figure 11.33 Free-free

Figure 11.34 Zero tip deformation constraint — Intermediate link

pinned-pinned beam. Note that the distance between the two ends is not necessarily equal to L, due to the shortening of the projection. Intuitively, this choice of the reference frame seems suitable for the intermediate link, Fig. 11.34.

With this choice of the tracking reference frame, the boundary conditions as well as complementary boundary conditions at ends c and d are changed for the pinned-free and free-free beams. At the free end, the internal force and internal moment are zero. Once the constraint is imposed, the free end becomes a pinned end. However, because the internal force and moment are zero, there is no restriction on the types of functions one can use to expand the secondary motion other than the geometric boundary conditions. Hence, the imposition of the zero tip deformation constraint does not represent an inconsistency from a physical standpoint.

11.7 DYNAMICS OF COMBINED ELASTIC AND LARGE ANGLE RIGID BODY MOTION

We will consider elastic deformation along the beam axis and along the y direction and use an analytical approach to derive the equations of motion. We will write the kinetic and potential energies, as well as the virtual work, and then invoke the extended Hamilton's principle. We first introduce the displacements $u(x, t)$ and $v(x, t)$ in their general form to the kinetic and potential energies. As a result, the equations of motion will have the form of partial differential equations for $u(x, t)$ and $v(x, t)$, and ordinary differential equations for $\mathbf{r}_O(t)$ and $\theta(t)$. Then, we discretize the displacement, $u(x, t)$ and $v(x, t)$ and derive the equations of motion in discretized form.

Consider a differential element of the beam, and Case 2 discussed in the previous section. The angular velocity of the differential element can be expressed as

$\boldsymbol{\omega} = [\dot{\theta}(t) + \dot{v}'(x, t)]\mathbf{k}$. Hence, the kinetic and potential energies are

$$T = T_{\text{tran}} + T_{\text{rot}}$$
$$= \frac{1}{2}\int_c^d \mu(x)\dot{\mathbf{r}}(x,t) \cdot \dot{\mathbf{r}}(x,t)\,dx + \frac{1}{2}\int_c^d \mathcal{I}_{zz}(x)[\dot{\theta}(t) + \dot{v}'(x,t)]^2\,dx$$
[11.7.1a]

$$V = \frac{1}{2}\int_c^d \left(EI_z(x)v''^2(x,t) + P(x)v'^2(x,t) + EA(x)e'^2(x,t)\right)dx$$
[11.7.1b]

where we have included in the potential energy an axial force in the x direction. The virtual work has the form

$$\delta W = \int_c^d \left(\mathbf{f}(x,t) \cdot \delta\mathbf{r} + m_z(x,t)\delta[\theta + v'(x)]\right)dx$$
[11.7.2]

in which $\mathbf{f}(x, t) = p_x(x, t)\mathbf{i} + p_y(x, t)\mathbf{j}$ denotes the distributed external force. The distributed moment acts through the total rotation of point x, which is $\theta(t) + v'(x, t)$. Note that when the reference frame is selected a priori, $\theta(t)$ becomes a known quantity, instead of a variable.

11.7.1 EQUATIONS OF MOTION IN HYBRID FORM

Derivation of the equations of motion in the form of partial differential equations is quite cumbersome, even for the case considered here. For this task, we simplify the model further by ignoring the stretch associated with the axial deformation, $e(x, t) \approx 0$. Further, we consider a pinned-free beam, so that $\mathbf{r}_O = \mathbf{0}$. We also ignore the rotational kinetic energy T_{rot}, which is due to the rotatory inertia. As an external input, we consider a concentrated moment at the pin joint of magnitude τ. The kinetic energy then has the form

$$T = \frac{1}{2}\int_0^L \mu(x)\Big([\dot{u}(x,t) - \dot{\theta}v(x,t)]^2 + [\dot{v}(x,t) + \dot{\theta}x + \dot{\theta}u(x,t)]^2\Big)dx$$
$$= \frac{1}{2}\int_0^L \mu(x)\Big(\dot{u}^2 + \dot{\theta}^2 v^2 - 2\dot{\theta}\dot{u}v + \dot{v}^2$$
$$+ \dot{\theta}^2 x^2 + \dot{\theta}^2 u^2 + 2\dot{\theta}\dot{v}x + 2\dot{\theta}\dot{v}u + 2\dot{\theta}^2 ux\Big)dx$$
[11.7.3]

We simplify the expression further by noting that $u(x, t)$ is much smaller than $v(x, t)$. Any term quadratic in u or involving $\dot{u}$, as well as terms involving products of u and v, are ignored. We are then left with

$$T \approx \frac{1}{2}\int_0^L \mu(x)\Big(\dot{\theta}^2 v^2 + \dot{v}^2 + \dot{\theta}^2 x^2 + 2\dot{\theta}\dot{v}x + 2\dot{\theta}^2 ux\Big)dx$$
[11.7.4]

11.7 DYNAMICS OF COMBINED ELASTIC AND LARGE ANGLE RIGID BODY MOTION

When the mass density is constant the last term in the kinetic energy can be written in terms of v, using Eq. [h] in Example 11.3, as

$$\frac{1}{2}\int_0^L \mu 2\dot{\theta}^2 ux\, dx = -\frac{1}{2}\int_0^L \dot{\theta}^2 x\mu \int_0^x \left(\frac{\partial v}{\partial \sigma}\right)^2 d\sigma\, dx$$

$$= -\frac{1}{4}\dot{\theta}^2 \int_0^L \mu(L^2 - x^2)\left(\frac{\partial v}{\partial x}\right)^2 dx \qquad [11.7.5]$$

The potential energy has the same form as in Eq. [11.7.1b]. The virtual work is

$$\delta W = \tau\delta[\theta + v'(0)] \qquad [11.7.6]$$

Because one variable—the body angle θ—is added to the system description, one corresponding constraint equation needs to be specified. For the time being, describe the constraint in the general form

$$C = \int_0^L c(x)v(x, t)\, dx = 0 \qquad [11.7.7]$$

where $c(x)$ is the constraint operator. We continue with the case of constant mass density. We add Eq. [11.7.7] to the Lagrangian by using the Lagrange multiplier λ, which yields $L = T - V + \lambda C$, and we invoke the extended Hamilton's principle. Taking the variation of the Lagrangian with respect to the two variables $\theta(t)$ and $v(x, t)$ and their derivatives, we write

$$\delta L = \frac{\partial L}{\partial \dot{\theta}}\delta\dot{\theta} + \frac{\partial L}{\partial v}\delta v + \frac{\partial L}{\partial \dot{v}}\delta\dot{v} + \frac{\partial L}{\partial v'}\delta v' + \frac{\partial L}{\partial v''}\delta v''$$

$$= \int_0^L \mu[\dot{\theta}v^2 + x\dot{v} + \dot{\theta}x^2]\delta\dot{\theta}\, dx + \int_0^L \mu\dot{\theta}^2 v\, \delta v\, dx + \int_0^L \mu(\ddot{v} + \dot{\theta}x)\delta\dot{v}\, dx$$

$$- \frac{1}{2}\dot{\theta}\,\delta\dot{\theta} \int_0^L \mu(L^2 - x^2)\left(\frac{\partial v}{\partial x}\right)^2 dx - \frac{1}{2}\dot{\theta}^2 \int_0^L \mu(L^2 - x^2)\left(\frac{\partial v}{\partial x}\right)\delta v'\, dx$$

$$- \int_0^L EI_z(x)\left(\frac{\partial^2 v}{\partial x^2}\right)\delta v''\, dx - \int_0^L P(x)\left(\frac{\partial v}{\partial x}\right)\delta v'\, dx \qquad [11.7.8]$$

The applied load can be expressed as $m_z(x) = \tau\underline{\delta}(x)$, and the virtual work can be expressed as

$$\delta W = \int_0^L \tau\underline{\delta}(x)\delta\theta\, dx + \int_0^L \tau\underline{\delta}(x)\delta v'\, dx = \int_o^L \tau\underline{\delta}(x)\delta\theta\, dx - \int_0^L \tau\underline{\delta}(x)\delta v\, dx$$

$$[11.7.9]$$

Performing the integrations by parts in the variation of the Lagrangian and collecting coefficients of $\delta\theta$ and δv, we obtain two equations of motion. For the primary motion we have

$$\mu\left\{x^2\ddot{\theta} + x\ddot{v} + \ddot{\theta}\left[v^2 - \frac{1}{2}(L^2 - x^2)v'^2\right] + 2\dot{\theta}\left(v\dot{v} - \frac{1}{2}(L^2 - x^2)v'\dot{v}'\right)\right\} = \tau\underline{\delta}(x)$$

$$[11.7.10]$$

and for the secondary motion

$$\mu\left\{x\ddot{\theta} + \ddot{v} - \dot{\theta}^2 v - \frac{1}{2}\frac{\partial}{\partial x}\left[\dot{\theta}^2(L^2 - x^2)\frac{\partial v}{\partial x}\right]\right\}$$
$$+ \frac{\partial^2}{\partial x^2}\left[EI_z(x)\frac{\partial^2 v}{\partial x^2}\right] - \frac{\partial}{\partial x}\left[P(x)\frac{\partial v}{\partial x}\right] = -\tau\underline{\delta}'(x) - \lambda c(x) \quad [\mathbf{11.7.11}]$$

When the mass density is not constant, the above equations have a more complicated form. A few observations about these equations of motion are in order. The equation for the primary motion can be expressed as

$$\mu\frac{d}{dt}\left[x^2\dot{\theta} + x\dot{v} + v^2\dot{\theta} + \frac{1}{2}\dot{\theta}(L^2 - x^2)v'^2\right]$$
$$= \mu\frac{d}{dt}[x^2\dot{\theta} + x\dot{v} + v^2\dot{\theta} + 2xu\dot{\theta}] = \tau\underline{\delta}(x) \quad [\mathbf{11.7.12}]$$

To understand this equation better, it is useful to look at the expressions for the linear and angular momentum. The linear momentum is expressed as

$$\mathbf{p} = \int_0^L \mu(x)\dot{\mathbf{r}}(x, t)\,dx = \int_0^L \mu(x)[(\dot{u} - \dot{\theta}v)\mathbf{i} + (\dot{v} + \dot{\theta}x + \dot{\theta}u)\mathbf{j}]\,dx \quad [\mathbf{11.7.13}]$$

The angular momentum about O is

$$\mathbf{H}_O = \int_0^L \mu(x)\mathbf{r}(x, t) \times \dot{\mathbf{r}}(x, t)\,dx$$
$$= \int_0^L \mu(x)[(x + u)\mathbf{i} + v\mathbf{j}]$$
$$\times [(\dot{u} - \dot{\theta}v)\mathbf{i} + (\dot{v} + \dot{\theta}x + \dot{\theta}u)\mathbf{j}]\,dx$$
$$= \int_0^L \mu(x)[(x + u)(\dot{v} + \dot{\theta}x + \dot{\theta}u) - v(\dot{u} - v\dot{\theta})]\mathbf{k}\,dx$$
$$\approx \int_0^L \mu(x)(x\dot{v} + x^2\dot{\theta} + 2xu\dot{\theta} + v^2\dot{\theta})\mathbf{k}\,dx \quad [\mathbf{11.7.14}]$$

in which we ignored terms quadratic in u, $\dot{u}$ and products of u and v. Comparing Eqs. [11.7.12] and [11.7.14], we observe that Eq. [11.7.12] represents the rate of change of angular momentum per unit length about O. This is expected, because we were considering a pinned-free beam whose equation of motion for the primary motion is a moment balance. One can show that for the free-free beam, the translational equations of motion represent force balances.

When we use the approach of selecting θ and its derivatives a priori, then θ and its derivatives become known quantities and they have no variations. Equation [11.7.11] remains the sole equation of motion, as there is no longer an equation of motion for the primary motion.

11.7.2 Equations of Motion in Discretized Form

The equations of motion derived above are in the form of hybrid equations, a combination of ordinary and partial differential equations. One way of dealing with these equations is to discretize Eqs. [11.7.10] and [11.7.11]. To this end, one can make use of several discretization methods. An alternative is to begin with a discretized description of the secondary motion and introduce this description to the extended Hamilton's principle. We illustrate this approach next and make an assumed modes expansion of the secondary motion of order n as

$$v(x, t) = \sum_{r=1}^{n} \psi_r(x) q_r(t) \qquad [11.7.15]$$

where $\psi_r(x)$ are admissible functions and $q_r(t)$ are generalized coordinates.

For the pinned-free beam under consideration, introducing Eq. [11.7.15] to the kinetic energy and potential energy in Eqs. [11.7.4] and [11.7.1b], we obtain

$$\frac{1}{2} \int_0^L \mu(x) \dot{\theta}^2 v^2(x, t) \, dx = \frac{1}{2} \dot{\theta}^2 \sum_{r=1}^{n} \sum_{s=1}^{n} m_{rs} q_r q_s = \frac{1}{2} \{q\}^T [M] \{q\} \dot{\theta}^2$$

$$\frac{1}{2} \int_0^L \mu(x) \dot{v}^2(x, t) \, dx = \frac{1}{2} \sum_{r=1}^{n} \sum_{s=1}^{n} m_{rs} \dot{q}_r \dot{q}_s = \frac{1}{2} \{\dot{q}\}^T [M] \{\dot{q}\}$$

$$\frac{1}{2} \int_0^L 2x \mu(x) \dot{\theta} \dot{v}(x, t) \, dx = \dot{\theta} \sum_{r=1}^{n} a_r \dot{q}_r = \{a\}^T \{\dot{q}\} \dot{\theta}$$

$$\frac{1}{2} \int_0^L \mu(x) 2 \dot{\theta}^2 u x \, dx = -\frac{1}{2} \sum_{r=1}^{n} \sum_{s=1}^{n} h_{rs} q_r q_s \dot{\theta}^2 = -\frac{1}{2} \{q\}^T [h] \{q\} \dot{\theta}^2$$

$$\frac{1}{2} \int_0^L E I_z(x) v''^2(x, t) \, dx + \frac{1}{2} \int_0^L P(x) v'^2(x, t) \, dx$$

$$= \frac{1}{2} \sum_{r=1}^{n} \sum_{s=1}^{n} k_{rs} q_r q_s = \frac{1}{2} \{q\}^T [K] \{q\} \qquad [11.7.16]$$

in which m_{rs} and k_{rs} are defined in Eqs. [11.5.4] and [11.5.5], and

$$a_r = \int_0^L x \mu(x) \psi_r(x) \, dx \qquad h_{rs} = \int_0^L x \mu(x) \int_0^x \psi_r'(\sigma) \psi_s'(\sigma) \, d\sigma \qquad [11.7.17a,b]$$

When the mass density is constant Eq. [11.7.17b] becomes

$$h_{rs} = \frac{1}{2} \int_0^L \mu(L^2 - x^2) \psi_r'(x) \psi_s'(x) \, dx \qquad [11.7.18]$$

From Eqs. [11.7.6] and [11.7.9], the virtual work is expressed as

$$\delta W = \tau [\delta \theta + \delta v'(0)] = \tau \delta \theta + \sum_{r=1}^{n} \tau \psi_r'(0) \delta q_r \qquad [11.7.19]$$

and the variation of the constraint becomes

$$\delta C = \lambda \sum_{r=1}^{n} c_r \delta q_r \qquad c_r = \int_0^L c(x)\psi_r(x)\,dx \qquad [11.7.20]$$

From now on, we consider that the admissible functions have been selected such that $c_r = 0\,(r = 1, 2, \ldots, n)$. Next, we invoke Lagrange's equations. For the coordinate θ, the integral of the $\mu(x)x^2\dot{\theta}^2$ term in the kinetic energy gives $I_0\dot{\theta}^2$. Carrying out the algebra, we obtain the equation of motion for the primary motion as

$$I_0\ddot{\theta} + \sum_{r=1}^{n} a_r \ddot{q}_r + \ddot{\theta}\sum_{r=1}^{n}\sum_{s=1}^{n}(m_{rs} - h_{rs})q_r q_s$$

$$+ 2\dot{\theta}\sum_{r=1}^{n}\sum_{s=1}^{n}(m_{rs} - h_{rs})\dot{q}_r q_s = \tau \qquad [11.7.21]$$

or in column vector form as

$$I_0\ddot{\theta} + \{a\}^T\{\ddot{q}\} + \{q\}^T([M] - [h])\{q\}\ddot{\theta} + 2\{\dot{q}\}^T([M] - [h])\{q\}\dot{\theta} = \tau \qquad [11.7.22]$$

This equation can also be written as

$$\frac{d}{dt}\left(I_0\dot{\theta} + \{a\}^T\{\dot{q}\} + \{q\}^T([M] - [h])\{q\}\dot{\theta}\right) = \tau \qquad [11.7.23]$$

and the term in the big brackets is recognized as the spatially discretized form of the angular momentum about point O.

For the secondary motion, we follow a similar procedure and obtain the equations of motion as

$$a_r\ddot{\theta} + \sum_{s=1}^{n} m_{rs}\ddot{q}_s + \sum_{s=1}^{n}[\dot{\theta}^2(h_{rs} - m_{rs}) + k_{rs}]q_s = \tau\psi'_r(0) \qquad r = 1, 2, \ldots, n$$

$$[11.7.24]$$

or, in column vector format as

$$\{a\}\ddot{\theta} + [M]\{\ddot{q}\} + (([h] - [M])\dot{\theta}^2 + [K])\{q\} = \tau\{\psi'(0)\} \qquad [11.7.25]$$

Both the primary and secondary motion equations are nonlinear. The nonlinearity and a source of coupling between the two motions is through the $[M] - [h]$ term and $\dot{\theta}^2$. Had we not included the shortening of the projection to the formulation, the $[h]$ matrix would be absent from the equations of motion and the coupling would be influenced only by $[M]$. For Case 2 problems, it turns out that $[h]$, $[M]$, and $[h] - [M]$ are all positive definite (see article by Smith and Baruh), so that neglecting the shortening of the projection changes the sign of the coupling and leads to an erroneous mathematical model. The contribution of $[h]\dot{\theta}^2$ in Eq. [11.7.25] is known as the *centrifugal stiffening* and the contribution of $[M]\dot{\theta}^2$ is known as the *centrifugal softening*. For Case 2 problems, the centrifugal stiffening is always larger than the centrifugal softening. When the primary motion has three rotational components, centrifugal stiffening does not always dominate the centrifugal softening.

When the angular velocity of the tracking frame is slow, one can ignore the term $([h] - [M])\dot{\theta}^2$ from the secondary motion and $\{q\}^T([M] - [h])\{q\}\dot{\theta}$ from the primary motion. As a result, one obtains a set of linearized equations that describe the overall motion in the form

$$I_O \ddot{\theta} + \{a\}^T \{\ddot{q}\} = \tau$$
$$\{a\}\ddot{\theta} + [M]\{\ddot{q}\} + [K]\{q\} = \tau\{\psi'(0)\} \quad [11.7.26]$$

These two equations can be combined into a single matrix equation of order $n + 1$

$$[m^*]\{\ddot{q}^*\} + [k^*]\{q^*\} = \{F^*\} \quad [11.7.27]$$

where $\{q^*\} = [\theta \ q_1 \ q_2 \ \ldots \ q_n]^T$, $\{F^*\} = \tau[1 \ \psi_1'(0) \ \psi_2'(0) \ \ldots \ \psi_n'(0)]^T$, and the matrices $[m^*]$ and $[k^*]$ are of order $n + 1$ and have the form

$$[m^*] = \begin{bmatrix} I_O & a_1 & a_2 & \ldots & a_n \\ a_1 & m_{11} & m_{12} & \ldots & m_{1n} \\ a_2 & m_{21} & m_{22} & \ldots & m_{2n} \\ \ldots & \ldots & \ldots & \ldots & \ldots \\ a_n & m_{n1} & m_{n2} & \ldots & m_{nn} \end{bmatrix} \quad [k^*] = \begin{bmatrix} 0 & 0 & 0 & \ldots & 0 \\ 0 & k_{11} & k_{12} & \ldots & k_{1n} \\ 0 & k_{21} & k_{22} & \ldots & k_{2n} \\ \ldots & \ldots & \ldots & \ldots & \ldots \\ 0 & k_{n1} & k_{n2} & \ldots & k_{nn} \end{bmatrix}$$

[11.7.28]

with m_{rs} and k_{rs} ($r,s = 1, 2, \ldots, n$) denoting the entries of $[M]$ and $[K]$.

It should also be noted from Eq. [11.7.25] that the location of the external moment affects the equations of motion for the secondary motion, but not that of the primary motion, corroborating the statements in Chapter 8 that the location of a concentrated moment does not affect the rigid body motion but that it affects the elastic motion.

11.8 ANALYSIS OF THE EQUATIONS OF MOTION

In this section we examine the nature of the equations of motion, the individual terms in the equations of motion for the various constraints discussed in Section 11.6, and suitable trial functions to be used as admissible functions. We primarily consider the pinned-free beam from the previous section, so that $\mathbf{r}_O = \mathbf{0}$. We also discuss issues associated with the free-free and interconnected beams.

The Rigid Body Constraint In this approach one selects the body axes such that there is no rigid body component in $v(x, t)$. All of the rigid component of the motion is described by the body angle θ. The constraint equation is given by Eq. [11.6.10], from which the constraint function $c(x)$ is recognized as

$$c(x) = \mu(x)x \quad [11.8.1]$$

We now investigate trial functions $\psi_r(x)$ ($r = 1, 2, \ldots, n$) to use when expanding the secondary motion. We are looking for trial functions that automatically satisfy this constraint. One suitable set consists of the eigenfunctions of a pinned-free beam with the same mass and stiffness characteristics except for the rigid body eigenfunction. We then have

$$\psi_r(x) = \phi_{r+1}(x) \qquad [11.8.2]$$

which, of course, satisfies the relationship

$$\int_0^L c(x)\psi_r(x)\,dx = \int_0^L \mu(x)x\phi_{r+1}(x)\,dx = 0 \qquad [11.8.3]$$

by virtue of the orthogonality of the eigenfunctions to the rigid body mode $\phi_1(x) = x$. Another suitable set of trial functions is polynomials orthogonalized using a Gram-Schmidt process. Such polynomials can be normalized to satisfy Eq. [11.8.3].

When using the eigenfunctions of the pinned-free beam as trial functions, we obtain the following values for the coefficients discussed in the previous section

$$m_{rs} = \delta_{rs} \quad k_{rs} = \omega_r^2 \underline{\delta}_{rs} \quad a_r = 0 \quad c_r = 0 \qquad [11.8.4]$$

where ω_r are the natural frequencies associated with the elastic motion of a pinned-free beam. When orthogonalized polynomials are used, the entries of k_{rs} are no longer natural frequencies of the corresponding beam. Also, $[K]$ is fully populated, as opposed to being diagonal.

When the eigenfunctions of the pinned-free beam are used as admissible functions the equation of motion for the primary motion becomes

$$I_O \ddot{\theta}(t) + \ddot{\theta} \sum_{r=1}^n \sum_{s=1}^n (\delta_{rs} - h_{rs})q_r q_s + 2\dot{\theta} \sum_{r=1}^n \sum_{s=1}^n (\delta_{rs} - h_{rs})\dot{q}_r q_s = \tau(t) \qquad [11.8.5]$$

As discussed earlier, this equation can be expressed as an angular momentum balance

$$\frac{d}{dt}\left([I_O + \{q\}^T([M] - [h])\{q\}]\dot{\theta}\right) = \tau(t) \qquad [11.8.6]$$

where the terms inside the square brackets

$$I_O + \{q\}^T([M] - [h])\{q\} \qquad [11.8.7]$$

are recognized as the mass moment of inertia of the deformed beam about the z axis. To see this, we consider the definition of the mass moment of inertia about the z axis,

$$I_O = \int \mathbf{r} \cdot \mathbf{r}\,dm = \int [(x+u)\mathbf{i} + v\mathbf{j}] \cdot [(x+u)\mathbf{i} + v\mathbf{j}]\,dm$$

$$= \int (x^2 + u^2 + 2xu + v^2)\,dm \qquad [11.8.8]$$

Ignoring the u^2 term, introducing the discretization [11.7.15] and the definition of u, gives Eq. [11.8.7]. In essence, the elastic deformation changes the magnitude of the mass moment of inertia. This change is dependent on the sign of $\delta_{rs} - h_{rs}$. As discussed earlier, $[h]-[M]$ is positive definite for the type of motion considered here, so that while these terms increase the stiffness in the secondary motion equations, they reduce the mass moment of inertia. When $\{q\}^T([M] - [h])\{q\}$ becomes very large, one may reach a very small or even negative value for the inertia term in Eq. [11.8.7], which is not possible physically. This, of course, takes place when either one or both of $\dot{\theta}$ and the secondary motion amplitudes are high. These parameters basically define the accuracy range for the assumption used in viewing the motion as a primary plus a secondary motion.

For the secondary motion we obtain

$$\ddot{q}_r(t) + \sum_{s=1}^{n}[\dot{\theta}^2(h_{rs} - \underline{\delta}_{rs}) + k_{rs}]q_s(t) = \tau\phi'_{r+1}(0) \quad r = 1, 2, \ldots, n \quad \textbf{[11.8.9]}$$

An interesting property when using the rigid body constraint is that the equation of motion for the primary motion contains no acceleration terms from the secondary motion, and vice versa. The inertia matrix for the combined system is diagonal, thus

$$[m^*] = \text{diag}(I_O, 1, 1, \ldots, 1) \quad \textbf{[11.8.10]}$$

This feature makes it easier to integrate or manipulate the equations of motion numerically. In addition, the rigid body constraint gives one a better physical interpretation of the terms involved. While these features are very desirable for analysis and simulation of the system behavior, the rigid body constraint loses its appeal when one needs to take measurements or conduct experiments. This is because of the additional effort needed in locating the relative frame from actual measurements.

For the free-free beam, one can select as trial functions the eigenfunctions of that beam, or orthogonalized polynomials, as in the pinned-free case. For the interconnected beam, the situation is again different. Here, if one selects the eigenfunctions of a free-free beam as the admissible functions, the CBC at the ends will be violated, because these trial functions describe a system with zero shear force, and there are nonzero shear forces at the ends of interconnected links. Hence, pinned-free or free-free eigenfunctions do not constitute a set of consistent admissible functions for the interconnected beam. Orthogonalized polynomials do not have this problem. However, the problem they do have is regarding convergence and numerical accuracy of the solution. Polynomials of order higher than seven or eight often lead to singularity. Also, as discussed in Section 11.5, a larger number of terms are required from polynomials than from other types of trial functions, for the same desired accuracy levels.

The Zero Slope Constraint This constraint selects the rigid body angle θ such that the slope of the secondary motion is zero at the pinned end, or

$$v'(0, t) = 0 \quad \textbf{[11.8.11]}$$

The constraint function $c(x)$ is expressed as

$$c(x) = \underline{\delta}'(x) \quad \textbf{[11.8.12]}$$

Substituting the constraint in Eq. [11.8.11] into Eq. [11.7.20] we obtain

$$c_r = \int_0^L c(x)\psi_r(x)\,dx = \int_0^L \underline{\delta}'(x)\psi_r(x)\,dx = -\psi'_r(0) = 0 \quad r = 1, 2, \ldots, n$$

$$\textbf{[11.8.13]}$$

The trial functions $\psi_r(x)$ that are used in the expansion of the secondary motion are selected such that they have zero slopes at $x = 0$. It follows that the essential boundary conditions these trial functions should satisfy are

$$\psi_r(0) = 0 \quad \psi_r'(0) = 0 \qquad [11.8.14]$$

There are several choices when selecting the trial functions. One reason why there are more choices here than for the rigid body mode constraint is that the trial functions need not satisfy any orthogonality relations. Two functions that immediately come to mind are the eigenfunctions of a fixed-free beam and simple polynomials, where $\psi_r(x) = x^{r+1}$. Using the trial functions that satisfy Eqs. [11.8.14] together with the assumed modes method, we obtain the equation of motion for the primary motion of the pinned-free beam as

$$I_0 \ddot{\theta} + \sum_{r=1}^{n} a_r \ddot{q}_r + \ddot{\theta} \sum_{r=1}^{n} \sum_{s=1}^{n} (m_{rs} - h_{rs}) q_r q_s + 2\dot{\theta} \sum_{r=1}^{n} \sum_{s=1}^{n} (m_{rs} - h_{rs}) \dot{q}_r q_s = \tau$$

[11.8.15]

and for the secondary motion as

$$a_r \ddot{\theta} + \sum_{s=1}^{n} m_{rs} \ddot{q}_s + \sum_{s=1}^{n} [\dot{\theta}^2 (h_{rs} - m_{rs}) + k_{rs}] q_s = \tau \psi_r'(0) \quad r = 1, 2, \ldots, n$$

[11.8.16]

If the trial functions $\psi_r(x)$ are selected as the eigenfunctions of a fixed-free beam with the same cross section and stiffness properties as the beam at hand, then $m_{rs} = \delta_{rs}$ and $k_{rs} = \omega_r^2 \delta_{rs}$ $(r,s = 1,2,\ldots,n)$. However, the a_r terms, which lead to a nondiagonal inertia matrix, do not vanish, regardless of the choice of trial functions. The discussion in the preceding subsection with regards to the dominance of the centrifugal stiffening over the softening terms is valid when the zero slope constraint is used. However, one can show that the magnitudes of the $h_{rs} - m_{rs}$ terms are much larger when the zero slope constraint is used, as compared with the rigid body constraint. This leads to a less accurate mathematical model, whose range of applicability for $\dot{\theta}$ is much smaller than the model obtained using the rigid body constraint.[5]

For a free-free beam, the formulation and the types of trial functions one can use are the same as for the pinned-free beam. The origin can be selected as a point in the interior of the beam, or as one of the ends. When the interior is selected one needs to use two sets of functions, from O to the ends.

For an interconnected beam, one can select the origin as any point on the beam, including the end points. However, the eigenfunctions of the fixed-free beam do not constitute a set of consistent trial functions, as they violate the CBC at the interconnected ends. The same situation is encountered when using eigenfunctions of a pinned-free beam with the rigid body constraint.

The primary advantage of this constraint is that it makes it very easy to measure the location and orientation of the reference frame.

---

[5] See L. Yu, master's thesis, Rutgers University, 1995.

The Zero Tip Deformation Constraint This constraint leads to the following geometric boundary conditions for the admissible functions:

$$\psi_r(0) = 0 \quad \psi_r(L) = 0 \quad r = 1, 2, \ldots, n \qquad [11.8.17]$$

where L is the undeformed length of the beam. The equations of motion for the primary and secondary motions have the same form as Eqs. [11.8.15] and [11.8.16]. Again, if the eigenfunctions of a pinned-pinned beam with the same measurements and mass and stiffness properties of the beam considered are used as trial functions, then $m_{rs} = \delta_{rs}$ and $k_{rs} = \omega_r^2 \delta_{rs}$ $(r,s = 1,2,\ldots,n)$. Also, $a_r \neq 0$, regardless of the choice of trial functions. Another set of suitable trial functions are simple sinusoidals, which become the actual eigenfunctions if the beam is uniform. Sinusoidals are easier to deal with than hyperbolic sines and cosines or polynomials. Further, they have better convergence characteristics and very few singularity problems.

Another advantage of the zero tip deformation constraint is that the trial functions do not violate internal force and moment balances for all of the pinned-free, free-free, and interconnected beams. This is especially important for the interconnected beam, as in the other two constraints we discussed, the only set of trial functions that did not violate force and moment balances at the boundaries were polynomials. Also, from an implementation perspective, the location and orientation of the tracking reference frame can be measured directly by taking measurements at the joints.

Comparison of the Constraints and Convergence Issues When modeling Case 2 problems, the type of constraint to select and the types of trial functions to use depend on a variety of factors. The rigid body constraint appears more logical, especially for pinned-free and free-free beams, as it permits use of the exact eigensolution of the beam as trial functions. The boundary conditions of the secondary motion are the same as the original beam. The discretized equations of motion are likely to have fewer terms, making it easier to numerically integrate these equations. Plus $a_r = 0$ $(r = 1, 2, \ldots, n)$, another desirable feature for numerical integration. On the other hand, the choice of trial functions is limited. And, there is the problem of locating the origin and measuring the orientation angle θ when seeking measurements.

With the zero slope constraint, the discretized equations are more complicated, but there are more and easier choices for selecting the trial functions. The trial functions do not need to satisfy any orthogonality conditions. A disadvantage is that the assumption of viewing the motion as a superposition of rigid and elastic motions loses its accuracy much faster than the rigid body mode constraint. In addition, polynomials have much slower convergence properties than sines or hyperbolic sines. On the other hand, the origin and orientation of the reference frame can be easily measured.

The zero tip deformation constraint emerges as an alternative somewhere in the middle. For pinned-free and free-free beams it does not have the nice properties of the rigid body mode constraint, but it does not have the convergence problems associated with the zero slope constraint. For the interconnected beam, it definitely is the better alternative, as a set of consistent admissible functions is very easy to find.

Table 11.7 summarizes the different types of trial functions discussed in this section and whether they qualify as consistent admissible functions or not. When we refer to eigenfunctions of pinned-free, fixed-free, or free-free beams, we mean eigenfunctions of a beam with the same dimensions, material properties; and cross-section as the beam being considered.

The accuracy of the eigenfrequencies of the linearized problem is one indicator of the accuracy to be expected of the overall motion, but it is not the only indicator. Another indicator of accuracy is the numerical complexities associated with the manipulation of trial functions. Especially when one deals with hyperbolic sines and cosines, as well as with polynomials, the trial functions are difficult to handle numerically. Hyperbolic sines and cosines are extremely sensitive to the coefficients involved. For instance, if the model parameters are not known accurately, the trial functions constructed using these erroneous values will not be a complete set and, hence, they will give incorrect results. The function $e^{\beta L(1+\varepsilon)} = e^{\beta L} e^{\beta L \varepsilon}$, and even if ε is small, $\beta L \varepsilon$ may not be small, especially for the higher modes. This can reduce the accuracy of the mathematical model substantially. With polynomials, one should not use too many trial functions, as this leads to singularity problems. Also

Table 11.7 Comparison of certain trial functions

Type of Constraint	Type of Beam			m_{rs}, k_{rs} Diagonal	$a_r = 0$
	Pinned-free Consistent Admissible Function	Free-free Consistent Admissible Function	Interconnected Consistent Admissible Function		
Rigid body					
Eigenfunctions of pinned-free beam	Yes	Yes	No	Yes	Yes
Eigenfunctions of free-free beam	Yes	Yes	No	Yes	Yes
Orthogonalized polynomials	Yes	No	Yes	No	No
Zero slope					
Eigenfunctions of fixed-free beam	Yes	Yes	No	Yes	No
Eigenfunctions of fixed-free uniform beam	Yes	Yes	No	No	No
Simple polynomials	Yes	No	Yes	No	No
Zero tip deformation					
Simple sinusoids	Yes	Yes	Yes	No	No
Eigenfunctions of pinned-pinned beam	Yes	No	Yes	Yes	No

with polynomials, only very few of the approximate eigenvalues are close to their actual values.

Another interesting comparison of the different models involves a comparison of the strain. While the magnitude of the secondary motion is different for the different constraints used, the strains are comparable, as they are derivatives of the secondary motion.

We summarize the issues to be considered when selecting the model to describe the secondary motion and the trial functions:

1. The objective in the mathematical modeling. Is it analysis, measurement, or navigation and control?
2. Are the trial functions a set of consistent admissible functions?
3. Are the coefficient matrices diagonal and are $a_r = 0$?
4. How sensitive are the trial functions to numerical as well as other types of errors?
5. What are the convergence characteristics of the trial functions and how many trial functions are needed to have a certain number of accurate modes?

Example 11.8

A pinned-free uniform link is made of steel and has the following dimensions properties: cross section 3/8 × 6 in, length = 12 ft, $E = 27.5(10)^6$ psi, unit weight = 0.28 lb/in^3. Find the eigensolution of the beam for small motions, and calculate the h_{rs} terms and compare them with the natural frequencies using the rigid body mode constraint.

Solution

The normalized eigenfunctions of a pinned-free beam are used as trial functions. These functions were calculated in Example 11.4. For the linear elastic model, $m_{rs} = \delta_{rs}$, and $k_{rs} = \omega_r^2 \delta_{rs}$, where ω_r are the natural frequencies, by virtue of orthonormality of the trial functions. These diagonal elements, along with the centrifugal stiffening coefficients h_{rs} calculated using Eq. [11.7.18], are given in Table 11.8.

As one can see, the h_{rs} terms are much larger than the m_{rr} terms. Further, comparing the elements k_{rr} we find them to be substantially larger than the $h_{rs} - m_{rs}$ terms, so that these terms have an effect on the secondary motion only when the primary motion is quite rapid. However, their effect on the primary motion is more important (see article by Baruh and Tadikonda).

Table 11.8 Comparison of coefficients in equations of motion

	h_{rs}					
r/s	1	2	3	4	m_{rr}	k_{rr}
1	6.397	1.861	−0.366	0.121	1.000	250.30
2	1.861	17.90	6.195	−1.480	1.000	2629.1
3	−0.366	6.195	35.99	12.64	1.000	11,445
4	0.121	−1.480	12.64	60.67	1.000	33,467

Example 11.9

Consider the pinned-free beam in the previous example and compare the eigenvalues using the three approaches.

Solution

The linearized equations are given by Eq. [11.7.27]. Expressing the configuration vector as $\{q^*(t)\} = \{z\}e^{\lambda t}$, the associated eigenvalue problem has the form

$$(\lambda^2[m^*] + [k^*])\{z\} = \{0\} \qquad [a]$$

We use the following trial functions in each problem:

Rigid body constraint: (a) Eigenfunctions of a pinned-free uniform beam with the same dimensions as the beam in the previous example. (b) Orthogonalized polynomials.

Zero slope constraint: (a) Eigenfunctions of a fixed-free uniform beam of the same geometry as in the previous example. (b) Simple polynomials, x^r ($r = 1, 2, \ldots$).

Zero tip deformation constraint: Simple sinusoids, $\sin r\pi x/L$, where $L = 12$ ft, the beam length.

All the trial functions qualify as consistent admissible functions. The polynomial trial functions give the same results for both constraints they are used with, even though the $[m^*]$ and $[k^*]$ matrices will be different for the different constraints used. The nonzero natural frequencies are given in Table 11.9.

As expected, the pinned-free eigenfunctions give the best results, because they are the exact solution for the linearized problem. The results obtained using the fixed-free eigenfunctions are very close to the actual values, except for the last natural frequency. The frequencies obtained using simple sinusoids are slightly less accurate than the fixed-free eigenfunction results, except for the last frequency, when they are much better than the fixed-free results. Polynomials give relatively good results only for the first four frequencies, after which their accuracy diminishes substantially.

Note that this problem can also be viewed as an example for Section 11.5, where the accuracy of different trial functions is compared for the linear vibration of a uniform pinned-free beam. The rigid body mode constraint results are basically the exact solution.

Table 11.9 Comparison of nonzero natural frequencies for pinned-free beam

Mode No.	Rigid Body Constraint	Zero Slope Constraint	Zero Tip Deformation Constraint	Polynomials (for Both Constraints)
2	15.827	15.827	15.827	15.827
3	51.288	51.289	51.293	51.292
4	107.01	107.02	107.05	107.15
5	182.99	183.07	183.21	194.72
6	279.24	279.72	280.08	325.45
7	395.74	398.32	398.82	1030.6
8	532.51	895.21	549.96	1943.4

11.9 CASE 3, ROTATING SHAFTS

When the angular velocity of the reference frame is along the x axis, the motion is similar to that of a rotating shaft. Because in such problems the cross section is circular, we will consider both $v(x, t)$ and $w(x, t)$ in the formulation.

Consider a rotating shaft, shown in Fig. 11.35. The reference frame is attached to one end of the shaft and the x axis is along the neutral axis. The velocity of a point on the center line of the shaft can be obtained by setting $\mathbf{r}_O$, ω_y, and ω_z to zero in Eq. [11.6.3] and using the notation $\omega_x = \dot{\theta}$, which yields

$$\dot{\mathbf{r}}(x, t) = \dot{u}\mathbf{i} + \dot{v}\mathbf{j} + \dot{w}\mathbf{k} + \omega_x \mathbf{i} \times [(x+u)\mathbf{i} + v\mathbf{j} + w\mathbf{k}]$$
$$= \dot{u}\mathbf{i} + (\dot{v} - \dot{\theta}w)\mathbf{j} + (\dot{w} + \dot{\theta}v)\mathbf{k} \quad [11.9.1]$$

Assuming that $\dot{u}$ is negligible compared to the other terms, the velocity expression further reduces to

$$\dot{\mathbf{r}}(x, t) = (\dot{v} - \dot{\theta}w)\mathbf{j} + (\dot{w} + \dot{\theta}v)\mathbf{k} \quad [11.9.2]$$

The expression for velocity gives the indication that we are dealing with a gyroscopic system. The kinetic energy associated with the translational motion of the shaft then becomes

$$T_{\text{tran}} = \frac{1}{2} \int_0^L \mu(x) \dot{\mathbf{r}}(x, t) \cdot \dot{\mathbf{r}}(x, t) \, dx$$
$$= \frac{1}{2} \int_0^L \mu(x) \left(\dot{v}^2 + \dot{w}^2 + \dot{\theta}^2(v^2 + w^2) + 2\dot{\theta}(v\dot{w} - w\dot{v}) \right) dx \quad [11.9.3]$$

To find the rotational kinetic energy, we need to calculate the angular velocity of a differential element on the shaft. Consider the formulation in Section 11.2. The difference here is that the xyz frame is rotating with $\dot{\theta}$ while in Section 11.2 the

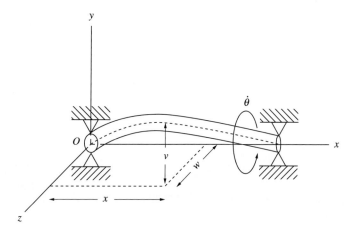

Figure 11.35   A rotating flexible shaft

xyz frame was fixed. Here, the rotations are by $v'(x, t)$ about the z axis and then by $-w'(x, t)$ about the rotated y axis (or first by $-w'$ and then by v'). This is basically a 1-3-2 (or 1-2-3) Euler angle transformation. Making a small-angles assumption and recalling from Fig. 11.3 that $x^*y^*z^*$ is the rotated frame, that of the beam axis, we approximate the angular velocity of the differential element as

$$\boldsymbol{\omega} = \dot{\theta}\mathbf{i} + \dot{v}'\mathbf{k} - \dot{w}'\mathbf{j}^* = \dot{\theta}(cv'cw'\mathbf{i}^* - sv'\mathbf{j}^* - cv'sw'\mathbf{k}^*) + \dot{v}'(sw'\mathbf{i}^* + cw'\mathbf{k}^*) - \dot{w}'\mathbf{j}^*$$
$$\approx (\dot{\theta} + \dot{v}'w')\mathbf{i}^* + (-\dot{\theta}v' - \dot{w}')\mathbf{j}^* + (-\dot{\theta}w' + \dot{v}')\mathbf{k}^* \quad [11.9.4]$$

In view of the fact that $\dot{\theta}$ is often much larger than the secondary motion, the angular velocity can be simplified to

$$\boldsymbol{\omega} = \dot{\theta}\mathbf{i}^* - \dot{\theta}v'\mathbf{j}^* - \dot{\theta}w'\mathbf{k}^* \quad [11.9.5]$$

The differential element has mass moments of inertia of $dI_{xx} = \mathcal{I}_{xx}(x)\,dx$, $dI_{yy} = \mathcal{I}_{yy}(x)\,dx$, $dI_{zz} = \mathcal{I}_{zz}(x)\,dx$. Assuming that the xyz axes are selected as principal axes, so are $x^*y^*z^*$, and the rotational kinetic energy can be approximated as

$$T_{\text{rot}} = \frac{1}{2}\int_0^L \left(\mathcal{I}_{xx}(x)\dot{\theta}^2 + \mathcal{I}_{yy}(x)(\dot{\theta}v')^2 + \mathcal{I}_{zz}(x)(\dot{\theta}w')^2\right)dx \quad [11.9.6]$$

Often, T_{rot} is small enough to be ignored, or, as in Eq. [11.9.6], only very few terms are retained in its description. T_{rot} usually becomes significant in the presence of disks attached to the shaft, and it is calculated only for the disks. The potential energy has the form

$$V = \frac{1}{2}\int_0^L \left(EI_y(x)w''^2 + EI_z(x)v''^2\right]dx \quad [11.9.7]$$

Invoking the extended Hamilton's principle, we obtain the equations of motion. Note that when evaluating the variation of a term like $\dot{w}'$, one uses two integrations by parts, one with respect to time and the other with respect to the spatial variable. Omitting the details of the integration by parts, the equations of motion can be shown to have the form

$$\mathcal{I}_{xx}(x)\ddot{\theta} + \mathcal{I}_{yy}(x)(\ddot{\theta}v'^2 + 2\dot{\theta}\dot{v}'v') + \mathcal{I}_{zz}(x)(\ddot{\theta}w'^2 + 2\dot{\theta}\dot{w}'w')$$
$$+ \mu(x)\ddot{\theta}(v^2 + w^2) + \mu(x)\dot{\theta}(2v\dot{v} + 2w\dot{w}) + \mu(x)(v\ddot{w} - w\ddot{v}) = 0$$
$$0 \leq x \leq L$$

$$\mu(x)\ddot{v} - \mu(x)\ddot{\theta}w - \dot{\theta}^2\mu(x)\dot{\theta}\dot{w} - \mu(x)\dot{\theta}^2 v + \dot{\theta}^2[\mathcal{I}_{yy}(x)v']' + (EI_z(x)v'')'' = 0$$
$$0 \leq x \leq L$$

$$\mu(x)\ddot{w} + \mu(x)\ddot{\theta}v + 2\mu(x)\dot{\theta}\dot{v} - \mu(x)\dot{\theta}^2 w + \dot{\theta}^2[\mathcal{I}_{zz}(x)w']' + (EI_y(x)w'')'' = 0$$
$$0 \leq x \leq L \quad [11.9.8]$$

The equations of motion associated with the secondary motion contain Coriolis as well as centrifugal terms. The centrifugal terms arise from the rotatory inertia and from the centripetal acceleration. They have a softening effect, while the Coriolis terms contribute to added stiffness.

11.9 CASE 3, ROTATING SHAFTS

Obviously, the equations of motion are quite complicated, so that one usually seeks an approximate solution. We demonstrate the use of the assumed modes method in Example 11.10.

Example 11.10

Consider the uniform circular shaft of length L in Fig. 11.36. On the shaft there is a rigid disk of mass M and moments of inertia I and J, with I being about the shaft axis, located at a distance a from end O. Derive the equations of motion using the assumed modes method, by assuming that the shaft speed is kept constant by a servomotor at $\dot{\theta} = \Omega$ and by ignoring the kinetic energy of the shaft.

Solution

The secondary motion has the same boundary conditions as that of a pinned-pinned beam in both transverse directions. We will use a one term expansion for both $v(x, t)$ and $w(x, t)$ in the form

$$v(x, t) = \psi(x) q_1(t) \qquad w(x, t) = \psi(x) q_2(t) \qquad \textbf{[a]}$$

in which $\psi(x) = \sin \pi x / L$ is the first eigenfunction of a pinned-pinned uniform beam. There is no problem with using the same trial function for both $v(x, t)$ and $w(x, t)$. The justification for ignoring the kinetic energy of the shaft is based on the fact that the mass moment of inertia of the disk is usually much larger than that of the shaft. The rotational kinetic energy has the form

$$T_{\text{rot}} = \frac{1}{2}\left(I\Omega^2 + J\Omega^2 [v'^2(a, t) + w'^2(a, t)] \right) \qquad \textbf{[b]}$$

From Eq. [11.9.3] the translational kinetic energy has the form

$$T_{\text{tran}} = \frac{1}{2} M \left(\dot{v}^2(a, t) + \dot{w}^2(a, t) + \Omega^2 [v^2(a, t) + w^2(a, t)] + 2\Omega [v(a, t)\dot{w}(a, t) - w(a, t)\dot{v}(a, t)] \right) \qquad \textbf{[c]}$$

with the potential energy defined by Eq. [11.9.7].

We next introduce the assumed modes expansion of $v(x, t)$ and $w(x, t)$ into Eqs. [b], [c] and [11.9.7], with the results

$$T_{\text{rot}} = \frac{1}{2}[I\Omega^2 + JZ^2\Omega^2(q_1^2 + q_2^2)] = \frac{1}{2}I\Omega^2 + \frac{1}{2}JZ^2\Omega^2\{q\}^T\{q\}$$

$$T_{\text{tran}} = \frac{1}{2} MY^2[(\dot{q}_1^2 + \dot{q}_2^2) + \Omega^2(q_1^2 + q_2^2) + 2\Omega(q_1\dot{q}_2 - q_2\dot{q}_1)]$$

$$= \frac{1}{2} MY^2\{\dot{q}\}^T\{\dot{q}\} + \frac{1}{2} MY^2\Omega^2\{q\}^T\{q\} + MY^2\Omega\{\dot{q}\}^T[g]\{q\}$$

$$V = \frac{1}{2} B^2\omega^2(q_1^2 + q_2^2) = \frac{1}{2} B^2\omega^2\{q\}^T\{q\} \qquad \textbf{[d]}$$

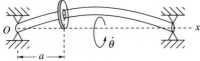

Figure 11.36

in which $Y = \psi(a)$, $Z = \psi'(a)$, ω is the first natural frequency of a pinned-pinned bar with the same geometry and material properties, and $[g]$ is a skew-symmetric matrix in the form

$$[g] = \begin{bmatrix} 0 & -1 \\ 1 & 0 \end{bmatrix} \qquad [\text{e}]$$

The constant B relates the first normalized eigen funtions $\phi(x)$ to $\psi(x)$ by $\psi(x) = B\phi_1(x)$, for a uniform beam $B = \mu L/2$. The terms in Eqs. [d] can be combined as

$$T = \frac{1}{2}\{\dot{q}\}^T[M_2]\{\dot{q}\} + \frac{1}{2}\Omega\{\dot{q}\}^T[G]\{q\} + \frac{1}{2}\Omega^2\{q\}^T[M_0]\{q\} \quad V = \frac{1}{2}\{q\}^T[K]\{q\} \qquad [\text{f}]$$

in which

$$[M_2] = MY^2 \begin{bmatrix} 1 & 0 \\ 0 & 1 \end{bmatrix} \quad [G] = 2MY^2[g] \quad [M_0] =$$
$$(JZ^2 + MY^2)\begin{bmatrix} 1 & 0 \\ 0 & 1 \end{bmatrix} \quad [K] = B^2\omega^2\begin{bmatrix} 1 & 0 \\ 0 & 1 \end{bmatrix} \qquad [\text{g}]$$

Because all coefficient matrices are time invariant, we make use of the developments in Section 5.3 and directly write the equations of motion as

$$[M_2]\{\ddot{q}\} + \Omega[G]\{\dot{q}\} + ([K] - \Omega^2[M_0])\{q\} = \{0\} \qquad [\text{h}]$$

Both the terms that have Ω^2 in them (JZ^2 and MY^2) contribute to the softening effect. The centrifugal component MY^2 is from T_{tran}, and it can be visualized easily. The contribution due to the rotatory inertia, JZ^2, may seem puzzling at first, but can be explained by considering that rotatory inertia adds to the kinetic energy that has to be balanced by the potential energy. The kinetic energy becomes higher when one includes the rotatory inertia term, so that in essence this corresponds to a lower relative stiffness. Be aware that Eq. [h] represents a gyroscopic system so that the reduction in stiffness due to increasing Ω does not describe the entire dynamics.

When $a = L/2$, that is, when the disk is in the middle of the shaft, the rotatory inertia terms have no contribution for the order of expansion considered, because $\psi'(L/2) = 0$. In this case, to capture the rotatory inertia contribution and to have a more accurate model one should use more than one trial function in the expansion of $v(x, t)$ and $w(x, t)$.

Let us next introduce the inertia of the shaft to the problem. In this case, we can safely assume that the contribution of the shaft to T_{rot} will be negligible and the contribution to T_{tran} is given by Eq. [11.9.3]. Introducing the assumed mode expansion in Eqs. [a] to Eq. [11.9.3], we obtain

$$T_{\text{tran (additional)}} = \frac{1}{2}M^*\left\{(\dot{q}_1^2 + \dot{q}_2^2) + \Omega^2(q_1^2 + q_2^2) + 2\Omega(q_1\dot{q}_2 - q_2\dot{q}_1)\right\} \qquad [\text{i}]$$

in which

$$M^* = \int_0^L \mu(x)\psi^2(x)\,dx \qquad [\text{j}]$$

This expression is then added to the total energy. It follows that the equation of motion retains the same form as Eq. [h]. The elements of the coefficient matrices change as we replace MY^2 with $MY^2 + M^*$ in $[M_2]$, $[G]$ and $[M_0]$. As before, when seeking a two or higher-order term expansion of the secondary motion, the additional T_{tran} will be more complicated.

A very important phenomenon in the motion of shafts is that of whirling. At certain rotational speeds the shaft undergoes large amplitude vibrations. Such vibrations are often

11.10 HYBRID PROBLEMS

In previous sections we saw derivation of the equations of motion and analysis of these equations for continuous beams. The angular velocity was always in one direction. More complex models of flexible systems undergoing rigid body motion include angular velocities in more than one direction and bodies comprised of rigid as well as elastic components. In such a system the kinetic energy will have contributions from both the rigid and elastic parts. However, the potential energy will be due to the elastic part alone. This makes it difficult for a single continuous function to approximate the entire body.

Consider, as an example, the model of a satellite with a hub of radius R and two antennae, as shown in Fig. 11.37. A suitable placement for the tracking reference frame is to attach its origin to the center of the hub and to align it along the undeformed position of the antennae. The position of a point on the antenna can be written as

Antenna 1: $\mathbf{r}_1(x_1, t) = [R + x_1 + u_1(x_1, t)]\mathbf{i} + v_1(x_1, t)\mathbf{j} + w_1(x_1, t)\mathbf{k}$

Antenna 2: $\mathbf{r}_2(x_2, t) = [R + x_2 + u_2(x_2, t)]\mathbf{i} + v_2(x_2, t)\mathbf{j} + w_2(x_2, t)\mathbf{k}$ **[11.10.1]**

with x_1 and x_2 denoting the spatial variables. The secondary motion of the antennas are then obtained using separate expansions for each antenna. One can expand v_i and w_i ($i = 1, 2$) using the zero slope constraint and use fixed-free eigenfunctions or simple polynomials as trial functions. One can also invoke the other constraints we have discussed, but these constraints do not have as good a physical interpretation as the zero slope constraint for this problem. For example, if the zero tip deformation constraint is used one draws a line from one end of the antenna to the other end.

If one wishes to use a single set of admissible functions to describe the motion (from the tip of $\mathbf{r}_1$ to the tip of $\mathbf{r}_2$), the functions used should be able to model the zero elastic deformation in the hub. This may be a viable approach if the hub is small compared to the rest of the structure, but not so if the hub dimensions are anywhere close to the size of the antennae.

For interconnected bodies, such as the one shown in Fig. 11.38, the zero tip deformation constraint appears more suitable than the other two, because of the attractiveness of using simple sinusoids as consistent admissible functions. The deformations on the individual links can be expressed as

Link 1: $\mathbf{r}_1(x_1, t) = [x_1 + u_1(x_1, t)]\mathbf{i}_1 + v_1(x_1, t)\mathbf{j}_1 + w_1(x_1, t)\mathbf{k}_1$

Link 2: $\mathbf{r}_2(x_2, t) = \mathbf{r}_1(L_1, t) + [x_2 + u_2(x_2, t)]\mathbf{i}_2 + v_2(x_2, t)\mathbf{j}_2 + w_2(x_2, t)\mathbf{k}_2$

[11.10.2]

where x_1 and x_2 are the spatial variables and L_1 is the length of the first link. Note that there is a separate set of unit vectors and orientation angles (θ_1 and θ_2) for each link.

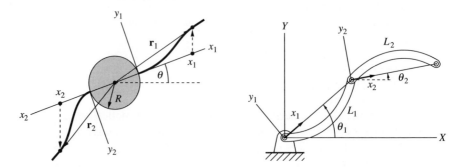

Figure 11.37 **Figure 11.38**

References

Baruh, H., and S. S. K. Tadikonda. "Issues in the Dynamics and Control of Flexible Robot Manipulators." *Journal of Guidance, Control, and Dynamics* 12, no. 5, 1989, pp. 659–71.

Courant, R., and D. Hilbert. *Methods of Mathematical Physics*. 1. New York: Wiley Interscience, 1953.

Gottlieb, D., and S. A. Orszag. *Numerical Analysis of Spectral Methods: Theory and Applications*. Society for Industrial and Applied Mathematics, J.W. Arrowsmith, 1977.

Laskin, R. A., P. W. Likins, and R. W. Longman. "Dynamical Equations of a Free-Free Beam Subject to Large Overall Motions," *Journal of the Astronautical Sciences* 31, 1983, pp. 507–27.

Meirovitch, L. *Elements of Vibration Analysis*. 2nd ed. New York: McGraw-Hill, 1986.

———. *Computational Methods in Structural Dynamics*. The Netherlands: Sijthoff Noordhoff, 1980.

Rivello, R. M. *Theory and Analysis of Flight Structures*. New York: McGraw-Hill, 1969.

Smith, C., and H. Baruh. "The Dominance of Stiffening Effects in Rotating Flexible Beams." *Journal of Guidance, Control, and Dynamics* 14, no. 5, 1991, pp. 1072–74.

Tadikonda, S. S. K., and H. Baruh. "Gibbs Phenomenon in Structural Mechanics." *AIAA Journal* 29, no. 9, 1991, pp. 1488–97.

Yu, L. "Modeling and Control of Flexible Multilink Manipulators," master's thesis, Rutgers University, May 1995.

Homework Exercises

Section 11.2

1. Write the kinetic and potential energies, as well as the virtual work, for the beam shown in Fig. 11.39. The cross section varies as follows: b is constant, while $h(x) = h_0(1 - x/2L)$.

2. Find $\mathcal{I}_{xx}$, $\mathcal{I}_{yy}$, $\mathcal{I}_{zz}$ for the beam in Example 11.1.

3. Find the kinetic and potential energies for the bar undergoing torsion in Fig. 11.40.

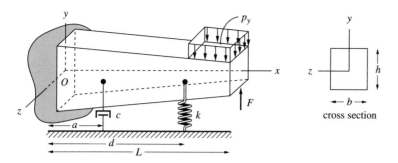

Figure 11.39

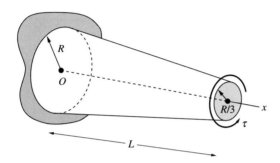

Figure 11.40

SECTION 11.3

4. Find the equation of motion and boundary conditions of the tapered bar in axial vibration shown in Fig. 11.41. The cross section varies as follows: b is constant, while $h(x) = h_0(1 - x/2L)$.
5. Find the equation of motion and boundary conditions of the beam shown in Fig. 11.39.
6. Find the equation of motion and boundary conditions of the bar undergoing torsion in Fig. 11.40.
7. Find the eigensolution and first five eigenvalues of a free-free uniform beam of length L, stiffness EI, and mass density μ.

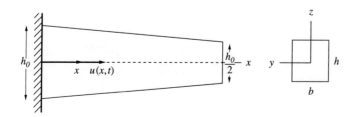

Figure 11.41

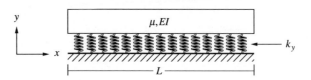

Figure 11.42

8. Find the eigensolution for the free-free uniform beam resting on an elastic foundation, shown in Fig. 11.42. The elastic foundation is modeled as a continuous spring of constant $k_y(x) = 0.4EI/L^4$. Comment on the rigid body mode.

9. Derive an expression for the response of a pinned-pinned uniform beam to an impulsive force $p_y(x, t) = \hat{F}\underline{\delta}(t)\underline{\delta}(x - a)$, in which a denotes the point of application of the force. Analyze the mode participation, and draw conclusions for arbitrary values of a. Then, let $a = L/2$ and reevaluate your results.

SECTION 11.5

10. Using the assumed modes method, find the approximate eigensolution of the tapered fixed-free beam shown in Fig. 11.39. Ignore the dashpot, let $p_y = 0$, $a = L/2, k = 2EI/L^3$. Use approximation orders of 3, 5, and 7, and use as trial functions the eigenfunctions of a fixed free beam. Verify numerically that the approximate eigenfunctions obtained from the discretization are indeed orthogonal with respect to the mass distribution.

11. For a pinned-pinned uniform steel beam of length $L = 2$ m and circular cross section with radius $R = L/10$, analyze the contribution of the rotatory inertia. Consider deformation in both the y and z directions, and obtain the eigensolution with and without the rotatory inertia terms. Then compare the two sets of eigenvalues. Use as trial functions the eigenfunctions of a pinned-pinned beam and three terms in the expansion.

12. Given a uniform column subjected to an axial load P as shown in Fig. 11.43, find the load P that makes the first eigenvalue zero using the assumed modes method and three trial functions. Then, compare this value for P with the critical load required for buckling, $P_{cr} = \pi^2 EI/L^2$. Plot the first natural frequency versus P. *Note:* Finding P requires some trial and error calculations.

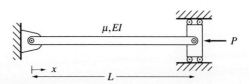

Figure 11.43

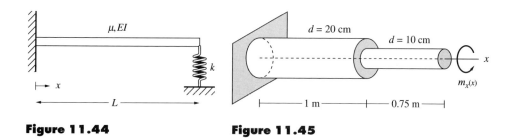

Figure 11.44

Figure 11.45

13. Using the assumed modes method, find the eigensolution of the beam shown in Fig. 11.44. Use simple polynomials as trial functions and an approximation order of 5. Compare the results for the following values of k: 0.1 EI/L^3, 0.5 EI/L^3, 3 EI/L^3.

14. Consider the shaft undergoing torsion in Fig. 11.45. The shaft manufactured of steel ($G = 79$ GPa, $\rho = 7860$ kg/m^3) by connecting two shafts of different diameter. Calculate the eigensolution of the shaft and list the first three frequencies. Use as trial functions the eigensolution of a uniform fixed-free shaft.

Sections 11.7 and 11.8

15. Consider Example 11.8 and calculate the values of h_{rs} when the zero slope constraint is used and as trial functions simple polynomials (x^2, x^3, ...) are used. Normalize the trial functions by setting $m_{ii} = 1$ ($i = 1, 2, \ldots, 4$), and compare h_{ij} with the results in Example 11.8.

16. Write a computer program to model the beam in Example 11.8. Use the rigid body constraint and the trial functions in Example 11.8. Simulate the response for a period of 15 seconds for an external moment of $M(t) = 1.2 \sin 20t$ N·m at the pin joint. Do the simulation for $n = 1$, $n = 2$, and $n = 3$. Plot the values for $\theta(t)$ and $q_r(t)$ for each case. Compare the results as the number of terms that describe the secondary motion are varied.

Section 11.9

17. Derive the equations of motion of the rotating shaft considered in Example 11.10. This time use two trial functions each to expand $v(x, t)$ and $w(x, t)$. Calculate the elements of the coefficient matrices. Include gravitational potential energy in your formulation.

18. Consider the previous problem and a shaft made of steel of length $L = 2$ m and radius 3 cm. The rigid disk is of mass 20 kg and radius 10 cm. Conduct a parametric study and plot the first two eigenvalues of the approximate model as a function of the driving frequency Ω, in the range of $0 \leq \Omega \leq 2000$ rpm.

Section 11.10

19. Write the kinetic and potential energies for the double-link robot arm shown in Fig. 11.38. Consider that both links are flexible only in the plane of motion (XY).

20. Find the equations of motion for the robot arm shown in Fig. 11.38. Assume that link 1 is rigid and link 2 is flexible. Use one trial function to describe the secondary motion of link 2. Justify your choice for the reference frame.

21. Figure 11.46 describes a uniform robotic arm of length L attached to a pin joint. The beam has elasticity in both transverse directions. Find the equation of motion of the beam, using the assumed modes method. Discuss your choice of trial functions for each transverse direction. Neglect axial stretch.

22. Find the equations of motion for the robot arm in Problem 20 by considering that both arms are flexible. Use one trial function for each arm and justify the motion constraints that you are using for each arm.

23. Consider the model of a satellite in Fig. 11.37. The satellite consists of a hub of mass M, radius R, and mass moment of inertia J. To the hub are attached two antennae, each of length $L = 3R$, mass density $\mu = M/5L$, and stiffness EI. In the undeformed position the two antennae are along the same line. Write the kinetic and potential energies of the hub.

24. Using the assumed modes method, derive the equations of motion for the hub in the previous problem. Which constraint should you use? *Note:* Each antenna needs to be described as a separate body.

25. Consider the equations of motion of the satellite in Problem 23. Linearize these equations and solve the resulting eigenvalue problem. Let $I = 0.3MR^2$. Use two trial functions for each antenna. Plot the eigenfunctions.

26. Consider the satellite model in Fig. 11.47. The satellite has the same dimensions as the satellite considered in Problem 23, but now there are four identical antennae, whose undeformed positions are perpendicular to each other. Obtain the equations of motion using the assumed mode, method, and one trial function per antenna.

27. Derive the linearized equations of motion for the satellite in the previous problem and solve for the eigensolution. Use one trial function per antenna.

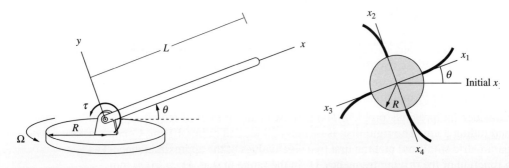

Figure 11.46　　　　　　　　　　　**Figure 11.47**

APPENDIX

A BRIEF HISTORY OF DYNAMICS

A.1 INTRODUCTION

This appendix presents a historical overview of the field of analytical mechanics, with emphasis on the history of rigid body dynamics. We also discuss the lives and achievements of the people who made key contributions. We present the accomplishments of each individual, and explore the interactions that took place among these brilliant minds.

One of the greatest challenges when writing a historical document is that of perspective. Many of the scientists who contributed to this field also made significant contributions in other fields. For example, when studying the life of Isaac Newton, dynamicists tend to think of Newton's laws of motion as his greatest contribution, while mathematicians view Newton's greatest contribution as his development of the calculus. There are several publications about the history of mechanics, each one emphasizing different accomplishments. Even two major books on the subject (Truesdell and Dugas) at times disagree on the level of contribution.

Although a well-balanced history of dynamics may be hard to attain, the process is extremely fascinating and rewarding. In this author's view, we get a much deeper understanding of mechanics when we know the historical background. It is my recommendation to every student of mechanics (or any other branch of science, mathematics, or engineering) to learn its history and benefit from it. The interested reader should consult not only this appendix and the references cited, but also other texts on the history of science and engineering.

The sections that follow give a historical survey of the evolution of dynamics and the life stories of key contributors. The technical community is much indebted to these people. We should also acknowledge those who thought to study the history of science and technology. In mechanics, Pierre Duhem (1861–1916) deserves the most credit; he is regarded by many as the founder of the field of history of science. Another individual who contributed to this subject significantly is Ernst Mach (1838–1916).

A.2 HISTORICAL EVOLUTION OF DYNAMICS

That the motion of moving bodies has intrigued people throughout history should not come as a surprise. People have always been interested not only in moving things from one place to another, but also in the motions of animals, clouds, and celestial bodies such as the earth, moon, sun, and stars. For moving and lifting objects, we have used two mechanisms since earliest times: levers and pulleys. Expectedly, some of the first significant contributions to mechanics were for levers and pulleys, as well as for weaponry. And the ancient study of celestial mechanics was carried out for scientific as well as religious purposes.

The seminal contribution to mechanics is attributed to the Greek scientist Aristotle (384–322 B.C.E.). He is credited with the first written work, *Problems of Mechanics*. Aristotle derived the law of equilibrium for a lever. He stated that when a lever rotates, the velocities of the weights at the ends of the lever will be proportional to the lengths of the arms of the lever. Interesting to note is that his starting point was *power*, as Aristotle stated the principle that "forces balance each other if they are inversely proportional to the velocities."

Because the principle Aristotle stated involves velocities, and balance of a lever involves equilibrium and small deviations from equilibrium, Aristotle is also credited as having introduced the concept of a virtual displacement. Aristotle's other research, which includes general motion analysis, celestial mechanics, and the shape of the earth, is not as well posed as his results on levers. Aristotle believed that wherever there is a motion, there must be a force causing it, and he tried to relate velocity to force. His work on celestial mechanics was primarily intuitive and based on his observations. He thought that the only natural motion was circular and that the sun and planets moved around the earth in circular paths. Aristotle's theory on celestial mechanics was refined slightly by Ptolemy, but it was not challenged for over 1500 years, until the time of Copernicus and Galileo. Similarly, his work on levers went unrefined for a very long period of time.

The next significant contribution to mechanics is attributed to Archimedes (287–212 B.C.E.), another Greek scientist. Archimedes built on Aristotle's work, refined Aristotle's equilibrium laws, and proposed some of his own. These laws were published in his treatise, *On the Equilibrium of Planes or on the Centers of Gravity of Planes*. Unlike Aristotle, who attempted to solve a wide variety of static and dynamic problems, Archimedes concentrated primarily on bodies at rest. The fame of Archimedes is primarily due to his contributions for static fluid mechanics and floating bodies (Archimedes' principle). He also made contributions to geometry and the measurement of circles and of the number π.

Although the famous Egyptian mathematician Euclid (365–~300 B.C.E.) is not usually credited with a major contribution to mechanics, many Arab authors refer to him in their writings as having had some contribution, particularly with regards to the equilibrium of levers. However, there seems to be no historical evidence that Euclid's work on mechanics was known to Aristotle or Archimedes. We should note that Euclidean geometry is ideally suited for the study of Newtonian mechanics; thus, Euclid's contributions to mathematics paved the way for the advances in Newtonian mechanics.

In Egypt, other mathematicians, Hero of Alexandria (second century C.E.) and Pappus (fourth century C.E.), made contributions to mechanics. Hero did work on pulleys and levers and considered levers of nonuniform cross sections and density. He knew about Aristotle's results and introduced the concept of a moment, albeit implicitly. Pappus is more known for his contributions to mathematics but he explored equilibrium problems on inclined planes, as well as the center of gravity.

Between the time of Aristotle and Galileo there were minor contributions to mechanics, which we briefly discuss here. In the first millennium of the common era, primary contributions came from Alexandria and the Arab Peninsula. As mentioned, Hero of Alexandria contributed through his discussions of machines, like levers, pulleys, and screws. Thabit ibn Kurrah (836–901), also of Alexandria, did work on geometry and on levers. There is uncertainty whether the book *Liber Charistonis* is the Latin translation of his book or was written by the second-century mathematician Charistion.

In the 13th century, the equilibrium of levers and of bodies on inclined surfaces was the focus of attention. The German scientist Jordanus (1225–1260) developed a theory where bodies would be heavier and lighter depending on the type of motion they were undergoing. Hence, he realized the concept of inertia, but without considering accelerations. One of the reasons mechanics did not evolve as fast as many other subjects (mathematics, for example) is that the study focused primarily on position and velocity. This is understandable, as it is much easier to visualize displacement and velocity than to visualize accelerations.

The 14th and 15th centuries were the precursors to the scientific revolution. During this period, three major concepts were analyzed more extensively, and certain Aristotelian concepts began to be challenged:

1. *The shape and motion of the earth.* Even though the correct shape of the earth was known to many Greek and Egyptian astronomers and physicists, the flat or cylindrical shape theory persisted for centuries, perhaps because of the influence of the church and the widely accepted Ptolemaic theory, which placed the earth at the center of the universe. Albert of Saxony (1316–1390) theorized that the earth was round in the north–south direction.

2. *The concept of impetus.* This related the tendency of a body to continue moving. Impetus was a concept close to linear momentum, but it was not defined as such. Impetus was a combination of momentum and a bit of energy, and it made way for the development of momentum and energy as concepts that indicate why motion continues. Impetus contradicts the Aristotelian theory. The person credited with first discussing impetus is Buridan (~1300–~1357) of France. Albert of Saxony and his French contemporary Oresme (1323–1387) are also credited with development of impetus as a property of motion.

3. *The use of acceleration as a variable to describe motion.* Albert of Saxony and Oresme are credited also with contributions in this area. Truesdell indicates that scholars at Merton College in Oxford, England, between the years 1328–1350 considered accelerations to describe motion. However, it was only in the 15th and 16th centuries that acceleration began to be applied more widely.

The Italian scientists Parma, Nicolas of Cues, Leonardo Da Vinci (1452–1519), and Dominic De Soto (1494–1560) are credited with blending these concepts and furthering new thoughts in mechanics. Da Vinci is the most famous, though primarily for his paintings and contributions to geometry. Leonardo did work on inclined planes, resolution of forces (his results were incorrect), free-falling objects (again, with incorrect results, even though he considered accelerations), shape of the earth, energy and momentum, and weaponry. De Soto is credited for work on the effects of gravity. The discussions initiated by the Italian school continued in other European institutions, especially Italian and French universities.

At this point, let us consider a brief history of the evolution of orbital theory, as this evolution was instrumental in bringing about the scientific revolution and the developments in dynamics.

Copernicus was the first to propose a heliocentric system for the motion of planets. He theorized a system where the planets and the earth were in circular orbit around the sun. (One should not be overly critical of his error in theorizing that the orbits were circular. The orbits of the earth and Mars around the sun have very small eccentricities, of 0.0167 and 0.093, respectively.) His theory, refined during the years 1513–1543, was widely spurned, as it went against the teachings of the church that the earth was the center of the universe. However, scientists began to notice, among them Galileo, Tycho Brahe of Denmark and later on Prague, and Kepler. Astronomical observations soon began to show that the Copernican theory was on the right track but not very accurate.

In 1600, Kepler (1571–1630) became assistant to Tycho Brahe (1546–1601), who was making accurate observations of the planets. But Tycho Brahe's thinking was deeply influenced by his religious beliefs and he did not accept the Copernican theory.

After Brahe died in 1601, Kepler continued his work. The level of accuracy of Tycho Brahe's and Kepler's observations is very impressive, especially Kepler's in his resolve to explain the smallest discrepancies in the motion of planets. Kepler showed that a planet moves around the sun in an elliptical orbit that has the sun in one of its two foci. He also showed that a line joining the planet to the sun sweeps out equal areas in equal times as the planet describes its orbit. Both these laws were first formulated for the motion of Mars around the sun, and published in 1609. Kepler's third law appeared in 1619. It is ironic that while Kepler's third law was widely accepted from the beginning, his first two laws were not warmly received for over 75 years. It was the work of Newton, as well as of Hooke and Halley, that finally confirmed Kepler's laws. Newton and Hooke both take credit for the inverse square law that describes gravitational attraction. As a result of discussions and exchange of information between Halley, Hooke, Wren, and Newton, both Hooke and Newton showed that the inverse square law of gravitational attraction leads to elliptic orbits. These developments led to the publication of the *Principia* in 1687.

Returning to the history of dynamics, one 16th-century scientist to make an important contribution was Stevin (1548–1620), who lived in Belgium and Holland. Stevin refined the theory of statics and considered equilibrium of masses on inclined planes, and of pulleys. He used concepts of energy and, informally, the principle of virtual work. His work was followed by the contributions of Galileo, who confirmed Stevin's results and showed equilibrium directly from virtual work. Galileo (1564–1642) recognized that, contrary to Aristotelian theory, an applied force is not necessary to maintain motion with no change in velocity. This observation is Galileo's *law of inertia,* and it lays the foundation of what constitutes an inertial reference frame. Galileo also deduced from his measurements of pendulums that inertial mass and gravitational mass are identical.

Galileo worked on the projectile problem and obtained accurate results, primarily because he used acceleration in analyzing motion. Interestingly, Galileo's correct results on the projectile problem came without knowledge of Newton's second law. Galileo is one of the first to discuss that in the free-fall problem, displacement is proportional to the square of time. Toricelli (1608–1647), who was a student—more a disciple—of Galileo, put Galileo's work on projectile motion into a systematic form. Of course, Toricelli is most famous for his work on pressure and the barometer. Galileo is also credited with an accurate analysis of tides, which included the effects of both the sun and the moon.

The more rapid dissemination of information and theories from one institution of learning to the next had begun. Works of scientists and astronomers were being translated, and the influence of the church was beginning to decline. These occurrences were important catalysts in the rapid developments in dynamics that began with the second half of the 17th century.

One scientist who made no significant contributions himself, but who was responsible for initiating discussion and correspondence and spreading information, is Father Marinne Mersenne (1588–1648). Mersenne lived most of his life in Paris. He was in continuous communication with his French colleagues Descartes (1596–1650), Pascal (1623–1662), and Roberval (1602–1675), as well as with Galileo in Italy and Huygens (1629–1695) in Holland. Mersenne defended Descartes' and Galileo's work to others.

Contributions around the middle of the 17th-century also came from Roberval, who is credited with the parallelogram law of forces; Huygens, who invented the pendulum clock and made it possible to measure time more accurately; and Descartes, whose contributions in algebra and geometry are far more significant. Huygens also worked on the geometry of curves, oscillations, and centers of oscillations. Huygens is credited with the first published treatise of relative motion. Descartes contributed the concepts of work and linear momentum; however, his free-fall theory was incorrect, as he (like many before him) assumed that velocity was proportional to distance traversed.

Another major development in the middle of the 17th century was the increased interest in impact of bodies. The problem attracted the interest of the Royal Society in London, which promoted research on the subject. The fact that bodies with different levels of elasticity behave differently in a collision was known quite early, and a distinction was made between *soft impact* and *hard impact,* precursors to the coefficient of restitution. It is interesting to note that the initiative of the Royal Society is concurrent with Newton's discoveries—but before Newton published his discoveries. Actually, Huygens in 1700 used Newton's first law as one of his hypotheses to describe the equations governing impact. Other contributors to the developments in impact include the British scientists Wallis (1616–1703) and Wren (1632–1723), and Mariotte (1620–1684) of France.

We next come to Isaac Newton. Newton (1642–1727) began his contributions by laying the foundation for differential and integral calculus, several years before its independent discovery by Leibniz (1646–1716) of Germany. Newton considered differentiation as the basic operation of calculus and thus was able to develop solution methods that unified many separate techniques such as finding areas, tangents, the lengths of curves, and the maxima and minima of functions.

Newton's laws of motion are dated to 1666, at which time he had already formulated their early versions. Newton made use of the concepts of momentum and force as the fundamental quantities. He defined quantities as mass, impressed force (*vis impressa*), and inertia force (*vis vista,* or innate force of matter). He showed that the period of a pendulum is independent of its mass.

Newton also made significant contributions to celestial mechanics. The Copernican theory and Kepler's laws had been proposed for the motion of planets. Newton proposed that the motion of the moon around the earth was governed by the same laws that described the motion of planets. Newton's laws of motion and the universal gravitational law were refined in part due to his communication (often unpleasant—see Newton's biography at the end of this appendix) with his fellow British scientists Hooke (1635–1702) and Halley (1656–1742).

Newton published the *Philosophiae naturalis principia mathematica,* or *Principia,* in 1687, which included his laws of motion and his contributions to celestial mechanics. Newton also discovered the law giving the centrifugal force on a body moving uniformly in a circular path. However, he did not have a correct understanding of the mechanics of circular motion. In addition, Newton studied the precession of the vernal equinox and the shape of the earth.

Newton's laws provide a firm basis for describing the motion of particles, and his law of universal gravitation unifies the works of Copernicus, Kepler, and Galileo. It should be noted that Newton stated his laws of motion for rigid bodies but did not consider rotational motion. This may be attributed to his incomplete understanding of circular motion.

What we know as Newtonian analysis, which is based on force and moment balances, continued with Euler. Euler (1707–1783) built on Newton's work, and he developed descriptive equations for systems of particles in 1740. He then proposed the rigid body rotational equations of motion in 1750, while he was working on the motion of ships. In that same paper, Euler restated Newton's second law as a fundamental law of motion applicable to all bodies.

In dynamics, Euler is remembered primarily for his contributions to the calculus of variations and rigid body dynamics, as well as for his book *Theoria motus corporum solidorum seu rigidorum* (*Theory of the Motions of Rigid Bodies*), published in 1765. He is considered to be the first to use the components of the angular velocity as kinematic variables. Euler's equations use a set of body-fixed coordinates, a novel concept for its time. They also involve mass moments of inertia and principal axes, all of which were developed by Euler.

Euler is also credited with considering angular momentum as a fundamental quantity. By the early 1730s Euler, Daniel Bernoulli, and Johann Bernoulli had recognized that Newton's laws of motion could not explain rigid body motion, and they began to look for other quantities to describe rigid body rotations. Huygens had already developed the correct equations for a pendulum by this time. Euler and Daniel Bernoulli began to look into rigid body motion and Euler continued the work after Daniel Bernoulli gave up on this subject. While developing the rigid body equations of motion, Euler began to consider angular momentum as a fundamental quantity. His results were also influenced by his work on deformable bodies, where the angular momentum leads to the symmetry of the stress tensor. Finally, in 1775, Euler formally stated $\mathbf{F} = \dot{\mathbf{p}}$ and $\mathbf{M} = \dot{\mathbf{H}}$ to be the two fundamental laws of mechanics. These two equations are known as *Euler's laws of mechanics,* and they have given rise to the term *Newton-Euler formulation.*

The work of Newton and Euler completes the analysis of rigid bodies and particles by force and moment balances. At the same time as the Newtonian mechanics (or Newton-Euler formulation) was being refined, the analytical approach to mechanics was being developed. Actually, both Newton and Euler contributed to analytical mechanics, Newton for the development of calculus, and Euler for the calculus of variations and the principle of least action.

Leibniz (1646–1716) took a different approach to describe motion; in contrast to Newton's approach of using change in linear momentum, Leibniz used energy (he defined a quantity called *vis viva,* which is equal to twice the kinetic energy). He stated that motion could be described by equating the change in kinetic energy to the work done by the acting forces, hence the work-energy theorem. Leibniz is also credited for inventing calculus independently of Newton's findings, as well as for his work on impact of bodies.

The next major development in analytical mechanics came from Johann Bernoulli (1667–1748) and D'Alembert (1717–1783). Making use of Newton's and Leibniz's contributions to calculus and variational calculus, in 1717, Johann Bernoulli formally stated the principle of virtual work as a general principle, from which all problems of statics could be solved. Many consider him to be the most important contributor to this subject. D'Alembert added to this by considering moving bodies and the inertia force. He viewed the inertia force as a force that produces equilibrium. By including the inertia force in the formulation, D'Alembert showed that the principle of virtual work can be used to analyze systems that are in motion. This statement is dated to 1743. Actually, there is debate on the scope of what D'Alembert actually stated. In mechanics, two equations are referred to as D'Alembert's principle, as discussed in Chapter 4. D'Alembert also worked on impact problems, as well as the living force concept and the center of oscillation.

The concept of a differential equation of motion was put forward by D'Alembert and Euler in 1743, in two separate articles that each published that year.

Variational calculus and the principle of virtual work continued to attract interest. Johann II and Daniel Bernoulli made contributions to the subject, and Euler worked maximum and minimum problems. These developments eventually led to the formulation of the principle of least action. Lagrange is usually credited with formally stating this principle, but the principle was initially conceived by Fermat (1601–1665) of France, who is credited with stating the first minimum principle that is not trivial, and Mauperitus (1698–1759), who lived in France and Switzerland. Mauperitus theorized that there is a fundamental quantity called *action,* and that its minimization described motion. However, he was not able to establish exactly what action was. As a result, his contributions were not appreciated by many.

The next major contribution came from Lagrange (1736–1813). In addition to his contributions to celestial mechanics and mathematics, Lagrange is credited for formulating the principle of least action and for taking advantage of generalized coordinates. This made it possible to extend the treatment of dynamical systems by different coordinate systems and different variables, and gave one the power to select the set of variables best suited to solve a problem. Lagrange's equations were introduced in the *Méchanique Analitique* in 1788. In the introduction, Lagrange states that one needs only kinetic and potential energy and a set of generalized coordinates to solve for the equations of motion. In their original derivation, these equations were written for conservative systems only. Lagrange is also credited with formally stating the generalized principle of D'Alembert in the form that we study it today. Lagrange was influenced by both Euler's and D'Alembert's work. What we call Lagrange's equations are also referred to as the Euler-Lagrange equations (see Appendix B).

Another of Lagrange's contributions to mechanics was in the treatment of systems subjected to constraints. He introduced the Lagrange multipliers, which we use to formulate and solve constrained systems. He also considered equilibrium problems, developing the theory of small motions around equilibrium, and stability theorems based on potential energy.

The contributions of Lagrange put the field of analytical mechanics into a structured form. The next significant contributions to the field came from Hamilton (1805–1865), along with his developments in relativistic mechanics. In between the two, a number of notables made minor contributions to mechanics—minor in the sense that, although significant, they were not nearly as significant as Lagrange's or Hamilton's contributions. All these scientists made more significant contributions to other branches of science. We discuss their contributions to mechanics next.

Coulomb (1736–1806) developed the model of dry friction in 1781, as the winning essay of a competition sponsored by the French Academy of Sciences. Coulomb's primary contributions to science were on electricity and magnetism. Carnot (1753–1823) developed the impulse-momentum theorem. He also worked on impacts. His work was similar to that of Lagrange. Carnot was certainly familiar with D'Alembert's work. How much he was aware of Lagrange's work when he published his essays in 1783 and 1803 is not accurately known.

Laplace (1749–1827), who was a contemporary of Lagrange, contributed to celestial mechanics and the analysis of the solar system. He discovered the invariability of planetary mean motions. These results appear in his *Traité du Mécanique Céleste* published in five volumes over 26 years (1799–1825). Laplace also introduced the potential function and Laplace coefficients. Fourier (1768–1830) worked with Lagrange and Laplace on mathematics, with emphasis on analysis. He succeeded Lagrange in his position at the École Polytechnique. His contributions to mechanics include work on virtual displacements and constrained motion.

Gauss (1777–1855) developed the principle of least constraint in 1829, which is a precursor to the Gibbs-Appell equations. He regarded Lagrange's work as brilliant, but did not consider the principle of virtual work as very intuitive and thus worked on developing a more intuitive principle. Gauss's principle does not involve time integration and gives the true minimum rather than a stationary value. However, the principle makes use of accelerations, which are harder to deal with than velocities. Poisson (1781–1840), who was a student of Lagrange and Laplace, considered generalized momentum (without giving it that name) and generated the Poisson brackets. His work was included in the second edition of the *Méchanique Analitique,* which was published in 1811. Poisson also contributed to the theory of impact, by developing the Poisson's hypothesis, and to the theory of elasticity, by developing the Poisson ratio.

In the 18th and 19th centuries, while the advances were being made in dynamics, a number of scientists (primarily French scientists at the École Polythechnique) refined the concepts in kinematics, rotations and relative motion: Clairaut (1713–1765) obtained a more accurate but still incorrect form for relative acceleration. He also worked on celestial mechanics problems, such as analyzing the motions of the moon and of Halley's comet. He studied the shape of the earth and verified Newton's result that the earth is an oblate spheroid. Coriolis (1792–1843) analyzed the dynamics of motion viewed from a moving reference frame and developed the concept of Coriolis force. It is interesting to note that while his is regarded as a contribution to kinematics, Coriolis always considered the dynamics of relative motion, hence the name Coriolis force. Foucault (1819–1868) conducted experiments on the effects of the rotation of the earth on bodies. He was inspired by Reich's experimental work on free-fall in mine shafts (1833). One of his earlier experiments was with a sphere of 5 kg suspended from a steel wire of length 2 m. He later used a gyroscope attached to the pendulum and demonstrated that the symmetry axis of the gyroscope did not change orientation as the pendulum moved. He related the rotation of the swing plane of the pendulum to the latitude of the location of the experiment. Chasles (1793–1880), whose primary work was in geometry and conic sections, is known for Chasles' theorem, an extension of Euler's theorem.

The next most important contributions to mechanics came from Hamilton and the German scientist Jacobi (1804–1851). Hamilton is credited with unifying the field of analytical mechanics. He developed the most general form of the principle of least action (called in this text the extended Hamilton's principle). He showed that the fundamental principles of mechanics and optics are very similar. Hamilton also succeeded in developing the Hamiltonian, from which the equations of motion can be written as first-order differential equations that are in state form. Jacobi called these equations *Hamilton's canonical equations.* Jacobi developed the transformation theory of canonical equations and showed that these equations have applications in many other fields. He gave a new formulation to the principle of least action, eliminating the time integral. He is also credited for the Jacobean, which becomes the most general form of the energy integral. Hamilton's other significant contribution to mechanics includes quaternions.

Hamilton's contributions consist of the fundamental principles that govern the motion of bodies. The contributions that came after Hamilton deal with the analysis of Lagrange's and Hamilton's equations, as well as certain special cases. We outline these developments next. By the time of Hamilton, communication among scientists had become more frequent and information was being disseminated more rapidly. This led to substantial collaboration. This, in many cases, makes it hard to credit a certain development to a particular individual.

One of the offshoots of the Lagrangian and Hamiltonian methods was the examination of the types of variables that we can use as generalized coordinates. Advances in this subject came mostly in the second half of the 19th century and the first few years of the 20th century. The terminology *generalized coordinates* was introduced by Thomson (1824–1907) and Tait (1831–1901) of Scotland, in their 1867 text *Treatise on Natural Philosophy.* Routh (1831–1907), Helmholtz (1821–1894) of Germany, and Thomson recognized the importance of cyclic variables, i.e., ignorable coordinates, and the way they could be eliminated from the formulation. Such variables are mostly encountered in rotating systems. Kelvin, Thomson, and Tait are also credited with work on gyroscopic systems. Helmholtz analyzed energy conservation. Hertz (1857–1894) looked at how mechanics can be described as a whole using the four concepts of space, mass, force, and motion. Lord Rayleigh (1842–1919) introduced the Rayleigh's quotient, a way of treating dissipative effects in Lagrangian mechanics.

Nonholonomic systems began to attract more interest toward the end of the 19th century. The German mechanician Heun (1859–1929) and his student Hamel (1877–1954) are recognized as the initial contributors. Appell (1855–1930) considered nonholonomic constraints and the types of approaches one can use when dealing with them. He used quasi-velocities with nonholonomic systems. He laid the foundations for the Gibbs-Appell equations, called by many the Appell equations.

Jourdain (1879–1921) developed Jourdain's variational principle, which closes the gap between D'Alembert's principle and Gauss's principle of least constraint, 80 years after Gauss's principle was stated. Gauss's principle was extended for inequality constraints in 1879 by Gibbs. An interpretation of the various theories was written by Duhem in 1903, in his book *Évolution de la Mécanique.* Klein (1849–1925) and Cayley (1821–1895) did work on description of rotations, Cayley's work based on Hamilton's research on quaternions. More recent contributors to mechanics, such as Lanczos (1893–1974) and Pars (1896–1985), did work on both classical mechanics as well as relativistic mechanics.

Another extension of the Lagrangian and Hamiltonian methods has been the qualitative analysis of dynamics problems, especially regarding stability theory. Lie (1842–1899) introduced the group theory to canonical transformations. He also considered infinitesimal transformations.

Significant contributions to the analysis of motion and stability theory came from Poincaré (1854–1912), who looked into stability issues, integral invariants, perturbation theory, differential geometry, and topology. The index of Poincaré and analysis of the three-body problem by perturbation techniques are some of his contributions to mechanics.

Table A.1 gives the dates of major contributions to the field of analytical mechanics. Be aware that many of the dates are approximate, and there are differences of opinion regarding the contributor and levels of contribution. The dates used here are mostly the ones given by Papastavridis and Dugas. Also, we list contributions up to 1910, as after that date there was substantially more cooperation among dynamicists.

Table A.1 Key dates in the history of dynamics

Year	Contributor(s)	Contribution
1513	Copernicus	Copernican theory introduced
1609	Kepler	First two of Kepler's laws
1638	Galileo	Particle motion in one dimension, projectile motion
1687	Newton	*Principia* published, laws of motion and law of gravitational attraction
1717	Johann Bernoulli	Principle of virtual work
1740	Euler	Systems of particles
1743	D'Alembert	Inertia force, principle of D'Alembert
1750	Euler	Rigid body rotational equations
1760	Lagrange	Principle of least action
1765	Euler	Theory of the motions of rigid bodies published
1775	Euler	Declares force and moment balances as the two fundamental equations of mechanics
1780	Lagrange	Lagrange's equations
1788	Lagrange	*Méchanique Analitique* published
1811	Lagrange	Lagrangian derivation of rigid body Euler's equations, 2nd edition of *Méchanique Analitique*
1829	Gauss	Principle of least constraint
~1830	Hamilton	Canonical equations of motion, Hamilton's principle
~1840	Jacobi	Transformation theory
~1860	Kelvin, Thomson, Tait	Gyroscopic systems
~1870	Routh, Helmholtz	Cyclic (ignorable) coordinates, stability
1879	Gibbs	Gauss's principle for inequality constraints
1870–1910	Heun, Hamel, Appell	Nonholonomic systems, quasi-coordinates, velocity constraints, Gibbs-Appell equations
1909	Jourdain	Jourdain's variational principle

A.3 Biographies of Key Contributors to Dynamics

This section presents brief life histories of key contributors to mechanics.

Albert of Saxony (1316–1390). Few details are known about the life of Albert of Saxony, except that he studied in the Sorbonne and was rector of the University of Paris. He became rector of the University of Vienna in 1365 and Bishop of Halberstadt from 1366 until his death.

Albert did work on the concept of impetus, and began to use acceleration as a kinematic quantity to describe motion. He is credited with work on locating the center of gravity. He also did work on projectile motion, with incorrect results. Albert is credited more for dissemination of scientific information than for developing new theories himself.

Appell, Paul Emile (1855–1930). Appell was born in Strasbourg, France. In 1885 he was appointed to the Chair of Mechanics at the Sorbonne. He served as rector of the University of Paris from 1920 to 1925.

Appell's first paper (1876) was based on projective geometry, continuing the work of Chasles. He then continued research on algebraic functions, differential equations, and complex analysis. He was a good friend of Poincaré. His paper "Notice sur lea travaux scientifique," *Acta Mathematica* 45 (1925), lists his 140 works in analysis, 30 works in geometry, 87 works in mechanics, and many textbooks, addresses, and lectures on the history of mathematics and on mathematical education. His primary contributions in mechanics were the treatment of nonholonomic systems and the development of the Appell equations.

Appell took high government positions in addition to scientific posts, which exposed him to war, espionage, and political controversy. He was involved with the Dreyfus affair, and served on the commission that exonerated Dreyfus. During World War I, Appell founded the Secours National, a semi-official organization, which gave help to civilian victims of the war. After the war, he served as secretary-general for the French Association to the League of Nations.

Despite his many contributions to science, politics, and government, Appell is not nearly as well known as his peers. This is mostly because he was a problem solver, not a developer of general theories. It may be a fitting conclusion to his life story that the importance of Appell's contributions began to be appreciated 50 years after his death.

Aristotle (384–322 B.C.E.). Born in northern Greece, Aristotle made contributions to a broad range of fields, including mechanics, logic, anatomy, and nature. In 367 B.C.E., Aristotle came as a student to Plato's academy in Athens, where he soon became a teacher. In 335 B.C.E., he formed his own school, the Lyceum in Athens.

In 343 B.C.E., he became tutor to Alexander the Great. He died a year after Alexander the Great died. One of the most remarkable of Aristotle's attributes was his tremendous capacity to observe phenomena and then deduce laws that describe those phenomena. His contributions to mechanics primarily are intuitive deductions from his observation of motion.

Archimedes (287–212 B.C.E.). Archimedes lived almost his entire life in Syracuse, Sicily, with a short stay in Egypt. His primary contributions were in the areas of geometry, statics, and floating bodies. His contributions to geometry are viewed as a precursor to integral calculus. He did calculations on the area of a circle and the volume of a sphere. He narrowed down the value of π to between $3\frac{10}{71}$ and $3\frac{1}{7}$ (3.1408 and 3.1429). In mechanics, Archimedes worked on the laws of equilibrium for levers and centers of gravity. He used his results in centers of gravity to study buoyancy and developed the famous Archimedes' principle. While he was in Egypt, he developed a machine called Archimedes' screw, which is used as a pump even to this day.

Bernoulli family. This was a large Swiss family who for three generations dominated the field of mathematics. Several members of the family occupied the chair in mathematics in the University of Basel. The first major contributors, Johann—also known as Jean—(1667–1748) and Jacob—also known as Jacques—(1654–1705) were the children of Nicolaus (1623–1708). Their other brother Nicolaus (1662–1716) was not as prominent, but his son Nicolaus I (1687–1759) was. Jacob Bernoulli studied the center of oscillation problem, which became a precursor to D'Alembert's principle.

Johann took his brother Jacob's chair at Basel after Jacob died in 1705. Based on Newton's and Leibniz's work, he formulated the principle of virtual work and put the field of calculus of variations on a solid foundation. He was the first to recognize that the principle of virtual work was a general principle from which static equilibrium problems could be solved.

Johann Bernoulli had three sons who become prominent mathematicians: Johann II (1710–1790), Daniel (1700–1782), who worked closely with his brother Johann II, and Nicolaus II (1695–1726), whose contributions appear to be mainly in mathematics and who died very young. Johann II took his father's chair when his father died in 1748. He was 38 years old at the time.

Johann II and Daniel did a lot of collaboration, with Johann's work primarily concentrating on heat and light, and Daniel's on hydrodynamics. Johann II initially studied law, and he pursued mathematics independently of his father. He won the Prize of the Paris Academy four times by himself. Johann II is credited for extending the works of Newton and Leibniz into a broader spectrum of problems. Johann Bernoulli II introduced the brachistochrone problem.

Daniel is credited with Bernoulli's principle. In 1725, Daniel and his brother Nicolaus were invited to work at the St. Petersburg Academy of Sciences. There he collaborated with Euler (who came in as his student). One of the areas of collaboration was ship dynamics, based on which Euler developed the rigid body equations of motion. Daniel also had a wider range of interests, such as anatomy, statistics, life insurance, and kinetic theory of gases.

Johann II had three sons, Johann III (1744–1807), Daniel II (1751–1834), and Jacob II (1759–1789). Of these three children, only Johann III became prominent in mathematics (probability theory, decimals, and theory of equations).

Copernicus, Nicolaus (1473–1543). Copernicus was born and died in Poland. He studied in Krakow (then the capital of Poland) and also in Bologna, Padua, and Ferrara, Italy, and he graduated in 1503 with a degree in canon law. Shortly after, he returned to Poland and settled in Frauenberg (Frombok). In addition to his religious duties, he practiced medicine and wrote articles on monetary reform. While doing all that, he began to work on the motion of the earth and planets. He was the first to challenge the views of Aristotle and Ptolemy and to formally state that the sun is the center of the solar system. In his theory, the earth and the other planets revolve around the sun in circular orbits. He further theorized that the earth rotates about its own axis.

Copernicus first wrote his theory, called *Commentoriolus,* and known to this date as the *Copernican theory,* in 1513. His theory was widely rebuffed, as it went against what was accepted as fact for well over 1000 years. He was spurned by many as a result of his writings. For 30 years he worked on refining his theory and his book *On the Revolutions of the Heavenly Spheres* (in Latin) was published in 1543, a few days before he died. A couple of generations after his death, Kepler and Galileo came to his defense (Galileo was under house arrest for eight years for doing this) and their use of the Copernican theory was the basis for the development of the laws of planetary motion. Many historians regard the Copernican theory as the beginning of the scientific revolution.

Coriolis, Gustave Gaspard de (1792–1843). Coriolis was born and died in Paris, and lived there for most of his life. He studied mechanics and engineering mathematics at the École Polytechnique, Paris, then taught mathematics there. He also worked as a military engineer. His major contribution is what is known as the Coriolis force and the laws of motion in a rotating reference frame. He showed that to write the equations of motion in a rotating frame of reference, the Coriolis force has to be included. Coriolis also did work on solid mechanics and on collisions, the latter based on collisions of billiards. In 1835 he published his *Théorie Mathématique des Effets du Jeu de Billiard.*

Coulomb, Charles Augustin de (1736–1806). While this French physicist and engineer is primarily known for his contributions to static electricity and magnetism (a Coulomb is a unit of electrical charge), Coulomb also contributed to mechanics. Guided by the interest of the French Academy of Sciences in friction and ropes, as well as their military applications, Coulomb began experimental work on friction, extending the results of Amontons. His award-winning essay "Theory of Simple Machines" discusses the relation between forces (pressure) exerted by contacting bodies and friction, and develops Coulomb's law (of proportionality) to describe friction. To this day, dry friction is referred to as Coulomb friction. Coulomb also worked on torsional vibrations and inclined surfaces.

D'Alembert, Jean Le Rond (1717–1783). D'Alembert lived all his life in Paris. He turned down many lucrative offers, such as an offer from Frederick II to go to Prussia as president of the Berlin Academy, and an invitation from Catherine II of Russia to tutor her son.

In 1741 at the age of 24, D'Alembert was admitted to the Paris Academy of Science, where he worked for the rest of his life. He was a contemporary and friend of Voltaire. His primary work was the study of differential equations and their applications, as well as theory of functions, derivatives, and limits. In mechanics, he is best known for the D'Alembert's principle, which appeared in his *Traité de Dynamique* (1742). He is credited with developing the concept of an inertia force ($-m\mathbf{a}$) which makes it possible to apply the principle of virtual work to systems in motion. A year later he applied his results to the equilibrium and motion of fluids.

D'Alembert's other major work is on the concept of equations of motion as differential equations. He first published an article on this in 1743. His more famous article, entitled *Différentiel,* was published in 1754.

Euler, Leonhard (1707–1783). Born in Basel, Euler joined the University of Basel as a theology student, following in the footsteps of his father. There, he began to like mathematics and, at the insistence of Johann Bernoulli II, he became Bernoulli's student. After graduating, Euler went to St. Petersburg Academy of Sciences in 1727. He began working with and living in the home of Daniel Bernoulli. In 1741, Euler joined the Berlin Academy of Science at the invitation of Frederick the Great. He stayed there for 25 years, returning to St. Petersburg in 1766. Euler lost his vision in one eye at age 31 and became blind in 1765, but he nevertheless continued his research for many years.

Euler's book *Theory of the Motions of Rigid Bodies* was published in 1765. His contributions also include logarithms, the Euler number, function theory, and the function notation. He is credited, too, with contributions to the solutions of differential equations, geometry, and topology. In addition, Euler worked on the mathematical theory of music, as well as on deformable bodies.

Euler is regarded by many as the most prolific writer of mathematics of all time. Some view him as the great architect of mechanics, at a level higher than Newton and Lagrange, who traditionally receive more credit for their contributions than Euler did for his. Euler's complete works include over 800 books and papers, which continued to be published by the St. Petersburg Academy for 50 years after his death.

Despite his tremendous achievements, Euler remained a very modest person, always praising other people's work and refraining from criticism. In an incident involving himself, Lagrange, and Mauperitus (1698–1759), Euler went to great lengths and fought hard to defend Mauperitus from charges of plagiarism and to get Mauperitus the credit due for his work on the principle of least action. His conduct should serve as a guide and example to all.

Galilei, Galileo (1564–1642). Galileo was born and lived in Italy. His father was a professional musician. Galileo was a professor of mathematics in Pisa and Padua, where he also taught astronomy. In addition to his contributions in astronomy and natural philosophy, Galileo contributed to the analysis of free-falling objects and also to the theory of vibrations through his research on the simple pendulum strings. He is known as the father of the experimental method. He is credited with developments in hydrostatics. While teaching astronomy at Padua, Galileo began to develop an interest in the Copernican theory and Kepler's observations. In 1609, after seeing a new spyglass in Venice, Galileo made his own telescope, which was the most powerful of its time. He is reported to have seen mountains on the moon, and the satellites of Jupiter, and to have discerned the nature of the Milky Way. He received an open letter from Kepler in 1610, praising him for his measurements and endorsing his discoveries.

In 1613, while he was Mathematician and Philosopher to the Grand Duke of Tuscany, Galileo in his observations discovered that Venus orbits the Sun. His rejection of the theory that the earth was the center of the universe and adoption of the Copernican theory earned him many enemies. In 1616 he was warned by the Italian government not to defend the Copernican theory. He did not heed and his *Dialogue Concerning the Two Greatest World Systems* was published in Florence in 1632. Galileo was summoned to Rome, found to be suspected of heresy, and condemned to house arrest, for life, at his villa. Under house arrest, he continued working with his followers and wrote a book on the strength of materials, which was smuggled out of Italy and published in the Netherlands in 1638. He died under house arrest.

Gauss, Carl Friedrich (1777–1855). Born in the Duchy of Brunswick, Gauss showed himself to be a genius at a very early age. Gauss entered the academy Brunswick Collegium Carolinum in 1792; he left, returned, and obtained his degree in 1799. He received financial support from the Duke of Brunswick. He published the book *Disquisitiones Arithmeticae* in 1801. His second book, *Theoria motus corporum coelestium in sectionibus conicis Solem ambientium,* was published in 1809, a major two-volume treatise on the motion of celestial bodies. In 1822, he won the Copenhagen University Prize. Gauss worked in a wide variety of fields in both mathematics and physics, including number theory, analysis, differential geometry, geodesy, magnetism, astronomy, and optics. His primary contribution to mechanics is the principle of least constraint. Some of Gauss's principal contributions include Bode's law, the binomial theorem, the arithmetic-geometric mean, prime numbers, construction of the 17-gon by ruler and compass, least squares approximation, orbital measurements, Gaussian curvature, and potential theory.

Gauss's life was full of personal tragedies. Soon after his first marriage, the Duke of Brunswick died, leaving him without financial support. He was forced to move to Gottingen. A year later his wife died after childbirth, soon to be followed by the death of the baby. He remarried and then his second wife died a few years later.

Gibbs, Josiah Willard (1839–1903). Gibbs was born, raised, worked, and died in Connecticut. He was the son of a professor of literature at Yale. He received his doctorate in engineering from Yale in 1863, the first engineering doctorate given in the United States. He studied until 1869 in Paris, Berlin, and Heidelberg, and became professor of mathematical physics at Yale in 1871.

Gibbs's major publications began in 1873, on thermodynamics and equilibrium. His subsequent work included electromagnetic theory of light, statistical mechanics, astrodynamics, and vector analysis. In mechanics he is recognized for his work on inequality constraints and contributions toward the Gibbs-Appell equations. His contributions were appreciated more in Europe than they were in the United States. Gibbs was known as a true gentleman, kind and unassuming in his manners. He lived a quiet and dignified life.

Hamilton, Sir William Rowan (1805–1865). Born in Dublin, Ireland, Hamilton gave evidence of genius as a young child. By age five he knew four languages. He was very interested in mathematics and, by the time he was 15, was studying the works of Newton and Laplace. At age 17, Hamilton found an error in Laplace's *Méchanique céleste*.

Hamilton entered Trinity College in Dublin at the age of 18. In 1827, at age 22, even though he was an undergraduate, he was appointed astronomer royal at Dunsink Observatory and professor of astronomy at Trinity College. He had not applied for the post, for which there were applicants, but rather, was invited to accept it.

In 1833, Hamilton published a study of vectors as ordered pairs, as well his first work on dynamics, *The Problem of Three Bodies by my Characteristic Function*. He used algebra in treating dynamics in *On a General Method in Dynamics* in 1834. This work, in essence, describes his most important contribution to mechanics. In 1843, he began to formulate the theory of quaternions (Euler parameters). He devoted the rest of his life to quaternions, and wrote two books on the subject, but neither book was well received. The second book was published posthumously, with the final chapter incomplete.

Hamilton was the first foreign member elected to the U.S. National Academy of Sciences. The latter years of his life were unhappy ones, and he suffered from alcoholism.

Jourdain, Philip Edward Bertrand (1879–1921). This British scientist lived a short life that was full of medical problems. He was already severely handicapped when he began his studies at Cambridge in 1898. His undergraduate years were very difficult. He did poorly in his degree. However, he was awarded the Allen studentship for research in 1904.

In mechanics, Jourdain is credited with Jourdain's variational principle (1909), which closes the gap between D'Alembert's principle and Gauss's principle. His most famous paper on this subject is *On some Points in the Foundation of Mathematical Physics* (1908).

Jourdain also worked in mathematical logic and set theory. He wrote a number of articles on this subject between 1906 and 1913 under the title *Development of the Theory of Transfinite Numbers*. In 1913 he proposed the card paradox. This was a card which on one side said: "The sentence on the other side of this card is TRUE." On the other side of the card the sentence read: "The sentence on the other side of this card is FALSE."

Kepler, Johannes (1571–1630). Kepler, born in Germany, is known primarily for formulating the three mathematical laws of planetary motion. He is also credited for work on mathematics, primarily on logarithms, as well as optics.

Kepler attended the University of Tübingen and took a teaching position in mathematics at Graz. At Tübingen he was exposed to both the Ptolemaic and Copernican theories, and he came to accept the Copernican theory. In 1596, Kepler published his *Mysterium Cosmographicum*, which argues for the truth of the Copernican theory.

In 1600, Kepler went to Prague, as assistant to Tycho Brahe. Brahe died in 1601, but Kepler went on to use his observations to calculate planetary orbits. The level of accuracy of his measurements is astounding. His first two laws were published in *Astronomia Nova* in 1609. His third law appeared in *Harmonice mundi* in 1619. Kepler's other books are *Epitome Astronomiae Copernicanae* (1618 to 1621), and *Rudolphine Tables* (1627), based on Tycho Brahe's observations and Kepler's laws.

Lagrange, Joseph Louis (1736–1813). Born in Turin, Italy, Lagrange's interest in mathematics began at a very early age when he read a book by Halley.

Lagrange served as professor of geometry at the Royal Artillery School in Turin from 1755 to 1766 and helped to found the Royal Academy of Science there in 1757. In 1764 he was awarded his first prize of many, when the Paris Academy awarded a prize for his essay on the libation of the moon. When Euler left the Berlin Academy of Science, at his request, Lagrange succeeded him as director of mathematics (1766). Development of what we know as Lagrange's equations was accomplished during this period. These developments were made possible by the advances that Lagrange introduced to the calculus of variations. He also contributed to hydrodynamics.

In 1787 Lagrange left Berlin to become a member of the Paris Academy of Science, where he remained for the rest of his career. He published his *Méchanique Analitique* in 1788, which includes Lagrange's equations. This work summarized all the work done in the field of mechanics since the time of Newton; it is notable for its use of the theory of differential equations. In it, Lagrange transformed mechanics into a branch of mathematical analysis. He survived the French revolution. During the 1790s he worked on the metric system and advocated a decimal base. He also taught at the École Polytechnique, which he helped to found. In 1797 he published the first theory of functions of a real variable. Napoleon named him to the Legion of Honor and Count of the Empire in 1808. In 1811 he published the second edition of *Méchanique Analitique*. Lagrange also excelled in all fields of analysis and number theory, as well as in celestial mechanics.

Hamilton referred to Lagrange as the "Shakespeare of mathematics," in appreciation of the mathematical depth and elegance of Lagrange's contributions. Lagrange also worked on the history of science.

Newton, Sir Isaac (1642–1727). Regarded by many as the most important contributor to the field of dynamics, Isaac Newton was born in Lincolnshire to a family of farmers. In his teenage years his academic performance was very weak. His mother removed him from grammar school, and at the urging of an uncle, in 1661 Newton entered Trinity College, Cambridge, to study law.

At Cambridge, Newton was attracted to philosophy, but also to algebra and analytical geometry. He was very impressed with the Copernican theory. The plague in 1665—which closed Cambridge University for two years—gave him the opportunity to go home to Lincolnshire and work on his own. He made several significant contributions in mathematics, optics, physics, and astronomy during that time. His first contribution was in calculus and unification of the previously scattered developments in this subject. His *De Methodis Serierum et Fluxionum* was written in 1671 (he failed to get it published and it did not appear until 1736). After returning to Cambridge in 1669, he was appointed to the Lucasian chair. He was 27 years old at the time, and his genius was well known. Newton held this chair until 1687. Newton was elected a fellow of the Royal Society in 1672. Also in 1672 he published his first scientific paper on light and color in the *Philosophical Transactions of the Royal Society*. While well received in general, his paper was objected to by Hooke and Huygens. This objection resulted in further strained relations among them.[1] He published his *Opticks* in 1704, a year after the death of Hooke.

Newton's laws of mechanics were published after debates with Hooke and Halley. In 1684, Halley asked Newton what orbit a body would follow under an inverse square force, and Newton replied that it would be an ellipse and that he had solved this problem five years earlier. Halley, being very impressed with this answer, urged Newton to publish his law of planetary motion, which he did in 1687 in the third part of his *Principia*.

Newton suffered a nervous breakdown in 1693, after which he lost interest in science. He retired from research to take up a government position in the Royal Mint in London. In 1703 he was elected president of the Royal Society and was re-elected each year until his death. He was knighted in 1708. The falling apple episode seems to be a myth, as it is reported that this incident occurred in 1666, whereas by 1666 Newton had already formulated his laws of motion.

---

[1] Like many before and after him, Newton did not take kindly to scientific criticism. In 1679, Hooke wrote to Newton proposing a central force that governs the motion of planets. Newton's response to Hooke discussed an inverse square law, but it also had an error, which Newton had to admit grudgingly. As a result, both Hooke and Newton worked independently on the law of planetary motion, and both claimed credit for the inverse square law.

Poincaré, Jules Henri (1854–1912). Poincaré was an applied mathematician who is credited with several advances in algebraic topology, stability theory, and the theory of analytic functions. Poincaré studied in the École Polytechnique and at the École des Mines. In 1881 he was appointed to a chair of mathematical physics at the Sorbonne. He studied optics, electricity, telegraphy, elasticity, thermodynamics, potential theory, quantum theory, theory of relativity, and cosmology. He also studied the three-body problem, and theories of light and electromagnetic waves. He is acknowledged as a codiscoverer, with Einstein and Lorentz, of the special theory of relativity.

His major works include *Analysis situs* (1895), an early systematic treatment of topology, *Les Méthods nouvelle de la méchanique celeste* (1892–1899), and *Leçons de mecanique celeste* (1905). He also wrote many popular scientific articles. In addition to his work on manifolds and what is known in dynamics as the *index of Poincaré*, Poincaré was the first to consider the possibility of chaos in a deterministic system. Little interest was shown in this work until the modern study of chaotic dynamics began in 1963.

Poincaré was critical of the way mechanics was taught by the British. He thought the British treated mechanics as an experimental science, whereas in continental Europe mechanics was taught as a deductive science.

Poisson, Siméon Denis (1781–1840). Originally forced to study medicine, Poisson began to study mathematics in 1798 at the École Polytechnique. His teachers were Laplace and Lagrange. He also made contact with Legendre, as a result of a memoir on finite differences, written when he was 18 years old. Poisson taught at École Polytechnique from 1802 until 1808 when he became an astronomer at Bureau des Longitudes. In 1809 he was appointed to the chair of pure mathematics in the Faculté des Sciences. Poisson's most important works were in mathematics, mechanics, astronomy, electricity, and magnetism. In mathematics, he advanced the theory of definite integrals and Fourier series, as well as Poisson's integral, Poisson's equation and Poisson's brackets. His contribution to probability theory is best remembered by the Poisson distribution. The Poisson distribution is widely used to model several phenomena (such as the famous gas station attendant problem: arrival of cars is modeled by the Poisson distribution while service time is modeled using the exponential distribution). In electricity, he is known for the Poisson's constant.

Poisson was a prolific author; he wrote over 300 articles. His *Traité de mécanique,* published in 1811 and again in 1833, was considered the standard text on mechanics for many years.

Ptolemy, Claudius (85–165 C.E.). Ptolemy was born in Egypt and lived most of his life in Alexandria, where he died. He developed the geocentric theory, with the motion of the planets and the sun as epicycles (combination of circles). His astronomical observations and his theory on the motion and location of the planets were considered valid for over a millennium, until they were disputed by Copernicus.

Routh, Edward John (1831–1907). Born in Canada, Routh went to England in 1842. He entered Peterhouse College at the same time as Maxwell, but Maxwell transferred to Trinity College. Much of his life was spent in competition with Maxwell. In 1854, the Smith Prize was divided between Routh and Maxwell (the first time the prize had been awarded jointly).

Routh published famous advanced treatises which became standard applied mathematics texts, such as *A Treatise on Dynamics of Rigid Bodies* (1860), *A Treatise on Analytic Statistics* (1891), and *A Treatise on Dynamics of a Particle* (1898).

Routh was elected to the Royal Society in 1872. His contributions to mechanics include advances to stability theory, gyroscopic systems, cyclic coordinates, and Routh's method for ignorable coordinates. In 1877, Routh was awarded the Adams Prize for work on dynamic stability. He was revered as an outstanding teacher and coach. He died in Cambridge.

REFERENCES

Dugas, René. *A History of Mechanics*. New York: Dover Publications, 1988.
Lanczos, C. *The Variational Principles of Mechanics*. 4th ed. New York: Dover Publications, 1970.
O'Connor, John J., and Edmund F. Robertson. *History of Mathematics Archive*. (World Wide Web site) URL: http://www-groups.dcs.st-andrews.ac.uk/~history/index.html
Papastavridis, J. G. "A Panoramic Overview of the Principles and Equations of Motion of Advanced Engineering Dynamics." *Applied Mechanics Reviews* 51, no. 4 (1998), pp. 239–65.
Truesdell, C. *Essays in the History of Mechanics*. New York: Springer-Verlag, 1960.

APPENDIX B

CONCEPTS FROM THE CALCULUS OF VARIATIONS

B.1 INTRODUCTION

In dynamics one frequently uses concepts from the calculus of variations. Variational principles such as D'Alembert's principle and Hamilton's principle are considered as foundations of analytical mechanics.

One of the earliest problems in the calculus was the determination of the minimum and maximum values of a function. The first major developments in this subject came in the second half of the 17th century. Newton, in 1671, and Leibnitz, in 1684, made the first original contributions, by noting that the rate of change of a continuous function must be zero for a minimum or maximum. Newton also looked into the problem of minimization of a functional. Johann Bernoulli and his son Daniel extended the works of Newton and Leibnitz into a broader spectrum of problems. Johann Bernoulli introduced the brachistochrone problem. Finally, Euler, who was a student of Johann's sons Johann II and Daniel Bernoulli, built on the Bernoulli family's work and developed the calculus of variations to its current form. Today the calculus of variations is used not only in analytical mechanics but also in several other problems from mathematics, physics, and engineering, such as Sturm-Liouville problems and optimal control theory, as well as in generating methods—such as the Rayleigh-Ritz and finite element methods—for finding approximate solutions to boundary value problems.

An introduction to the variation of a function is also given in Chapter 4, which develops variational principles used in dynamics. One can follow the developments of the text by just reading Chapter 4, without consulting this appendix. Here, the reader will find a source of additional details and more in-depth developments.

B.2 STATIONARY VALUES OF A FUNCTION

Consider a function of a single variable $f(x)$ over an interval $\mathcal{I}$, where x is the variable and $\mathcal{I} = (a, b)$ is the interval. The function $f(x)$ has an *extremum,* a local minimum or maximum, in the interior of this interval if there exists a point x^* such that $f(x) \geq f(x^*)$ or $f(x) \leq f(x^*)$ for all values of x. For $f(x)$ to have an extremum in the interior of the interval, $f'(x) = df(x)/dx$ has to vanish at $x = x^*$, or it has to not exist. Otherwise, the minimum and maximum values of $f(x)$ will be at the boundaries, at $f(a)$ or at $f(b)$.

One determines whether the extremum point $x = x^*$ represents a local minimum or maximum by examining the second derivative of $f(x)$. If $f''(x^*) > 0$, x^* represents a local minimum; if $f''(x^*) < 0$, x^* represents a local maximum. When $f''(x^*) = 0$, no conclusion can be drawn. A simple example of this is to compare the functions $f(x) = x^3$ and $f(x) = x^4$.

The first derivative of a function determines the slope, and it is a *necessary* condition for the existence of a minimum or maximum. The second derivative describes the shape of the function, and it constitutes a *sufficient* condition for the extremum to be a local maximum or minimum. When its second derivative is positive, $f(x)$ is convex upwards, as in a bowl. When the second derivative is negative, the curve is convex downwards (also called concave), as in an inverted bowl. A point where the shape of the curve changes from convex to concave or vice versa is called an *inflection* point. At an inflection point, $f''(x) = 0$.

Next, consider a function of n independent variables, $f(x_1, x_2, \ldots, x_n)$, in which $x_1, x_2, \ldots, x_n$ are the variables. We denote by $X(x_1, x_2, \ldots, x_n)$ a point in the interval $\mathcal{I}$ ($a_k \le x_k \le b_k$, $k = 1, 2, \ldots, n$). Let $f(X)$ be continuously differentiable in all its variables. The necessary condition for $f(X)$ to have a local minimum or a maximum in the interior of $\mathcal{I}$ is for all of its first derivatives with respect to $x_1, x_2, \ldots, x_n$ to vanish. That is, the point $X^* = (x_1^*, x_2^*, \ldots, x_n^*)$ is a local minimum or maximum if

$$\frac{\partial f}{\partial x_k} = 0 \quad \text{at} \quad x_1 = x_1^*, x_2 = x_2^*, \ldots, x_n = x_n^* \quad k = 1, 2, \ldots, n \quad \text{[B.1]}$$

The point X^*, where all the first derivatives of f with respect to x_k vanish, is referred to as a *stationary value* of the function $f(X)$. One arrives at Eq. [B.1] also by considering the differential of $f(x_1, x_2, \ldots, x_n)$, which has the form

$$df = \frac{\partial f}{\partial x_1} dx_1 + \frac{\partial f}{\partial x_2} dx_2 + \cdots + \frac{\partial f}{\partial x_n} dx_n \quad \text{[B.2]}$$

Because the variables x_k ($k = 1, 2, \ldots, n$) are independent, their differentials dx_k ($k = 1, 2, \ldots, n$) are also independent. The differential of f vanishes at a stationary point, $df = 0$. For df to vanish, all of the coefficients of dx_k ($k = 1, 2, \ldots, n$) must vanish independently, and hence the result in Eq. [B.1].

To determine whether the stationary value X^* represents a minimum or maximum, one needs to investigate the second derivatives of $f(X)$. Define the Hessian matrix $[H]$ by

$$[H] = \begin{bmatrix} \frac{\partial^2 f}{\partial x_1^2} & \frac{\partial^2 f}{\partial x_1 \partial x_2} & \cdots & \frac{\partial^2 f}{\partial x_1 \partial x_n} \\ \frac{\partial^2 f}{\partial x_2 \partial x_1} & \frac{\partial^2 f}{\partial x_2^2} & \cdots & \frac{\partial^2 f}{\partial x_2 \partial x_n} \\ \vdots & \vdots & \vdots & \vdots \\ \frac{\partial^2 f}{\partial x_n \partial x_1} & \frac{\partial^2 f}{\partial x_n \partial x_2} & \cdots & \frac{\partial^2 f}{\partial x_n^2} \end{bmatrix} \quad \text{[B.3]}$$

whose entries are $H_{ij} = \partial^2 f / \partial x_i \partial x_j$ ($i, j = 1, 2, \ldots, n$) evaluated at $X = X^*$. Note that $[H]$ is square and symmetric. A theorem from the calculus of variations states that if the Hessian matrix is positive (negative) definite at the stationary point X^*, then the stationary point X^* is a local minimum (maximum). For a symmetric matrix to be positive definite, it must satisfy Sylvester's criterion, which states that the following must hold:[1]

---

[1] For a symmetric matrix to be negative definite, all the diagonal elements must be negative, and the sign of the principal minor determinants must alternate.

1. All diagonal elements must be greater than 0, and
2. All principal minor determinants must be greater than 0.

If the determinant of $[H] = 0$, no conclusion can be drawn. If the determinant of $[H]$ is not zero and $[H]$ is indefinite, the stationary point is neither a maximum nor a minimum and is referred to as a *saddle point*. The Hessian matrix is also denoted by $[H] = \nabla^2 f$.

Consider now the case when the variables $x_1, x_2, \ldots, x_n$ are not independent and are related to each other by the equality constraints

$$c_j(x_1, x_2, \ldots, x_n) = 0 \qquad j = 1, 2, \ldots, m \qquad \text{[B.4]}$$

where m is the number of constraints. To find the stationary values of $f(x_1, x_2, \ldots, x_n)$ subject to the constraints $c_j(x_1, x_2, \ldots, x_n)$ a number of approaches can be adopted. One approach is based on rewriting the constraint equations such that m of the variables are expressed in terms of the remaining $n - m$ variables. Theoretically, at least, it is possible to do so. Then the function is described by $n - m$ independent variables and the above developments apply. For instance, say x_l is the variable we wish to eliminate from the formulation. We seek to express x_l in terms of the other $n - 1$ variables as

$$x_l = g(x_1, x_2, \ldots, x_{l-1}, x_{l+1}, \ldots, x_n) \qquad \text{[B.5]}$$

and to substitute Eq. [B.5] into $f(X)$. If this can be done for all the constraint relations, one can effectively express $f(X)$ in terms of $n - m$ independent variables and use the procedure outlined earlier to find the stationary values.

Many times, it is difficult to find a relation of the form [B.5] for all the constraint relations. Even if such relations are found, they may complicate the expression for $f(X)$ immensely. Also, certain constraints are described in the differential form

$$a_{j1} dx_1 + a_{j2} dx_2 + \cdots + a_{jn} dx_n = 0 \qquad \text{[B.6]}$$

and it is not possible to integrate Eq. [B.6] to the form of Eq. [B.4]. For these cases, there is another procedure for finding the stationary values of $f(X)$. This method makes use of Lagrange multipliers. We construct an augmented function, denoted by $\phi(X) = \phi(x_1, x_2, \ldots, x_n)$ and referred to as the *Lagrangian*, which has the same value as $f(X)$, but has the form

$$\phi(x_1, x_2, \ldots, x_n) = f(x_1, x_2, \ldots, x_n) + \sum_{j=1}^{m} \lambda_j c_j(x_1, x_2, \ldots, x_n) \qquad \text{[B.7]}$$

where λ_j ($j = 1, 2, \ldots, m$) are called the *Lagrange multipliers*. Consider the differential of the augmented function ϕ. To this end, we first take the differentials of the constraint relations

$$dc_1 = \frac{\partial c_1}{\partial x_1} dx_1 + \frac{\partial c_1}{\partial x_2} dx_2 + \cdots + \frac{\partial c_1}{\partial x_n} dx_n = 0$$

$$dc_2 = \frac{\partial c_2}{\partial x_1} dx_1 + \frac{\partial c_2}{\partial x_2} dx_2 + \cdots + \frac{\partial c_2}{\partial x_n} dx_n = 0$$

$$\cdot \qquad \cdot \qquad \cdot \qquad \cdot$$

$$dc_m = \frac{\partial c_m}{\partial x_1} dx_1 + \frac{\partial c_m}{\partial x_2} dx_2 + \cdots + \frac{\partial c_m}{\partial x_n} dx_n = 0 \qquad \text{[B.8]}$$

Multiplying each one of Eqs. [B.8] by the Lagrange multipliers λ_j, and adding these equations to Eqs. [B.2], one obtains $d\phi$, the differential of the augmented function, as

$$d\phi = df + \sum_{j=1}^{m} \lambda_j dc_j = \left[\frac{\partial f}{\partial x_1} + \frac{\partial c_1}{\partial x_1}\lambda_1 + \frac{\partial c_2}{\partial x_1}\lambda_2 + \cdots + \frac{\partial c_m}{\partial x_1}\lambda_m\right] dx_1$$

$$+ \left[\frac{\partial f}{\partial x_2} + \frac{\partial c_1}{\partial x_2}\lambda_1 + \frac{\partial c_2}{\partial x_2}\lambda_2 + \cdots + \frac{\partial c_m}{\partial x_2}\lambda_m\right] dx_2 + \cdots$$

$$+ \left[\frac{\partial f}{\partial x_n} + \frac{\partial c_1}{\partial x_n}\lambda_1 + \frac{\partial c_2}{\partial x_n}\lambda_2 + \cdots + \frac{\partial c_m}{\partial x_n}\lambda_m\right] dx_n = 0 \qquad \text{[B.9]}$$

When the constraint is in the form of Eq. [B.6], we multiply each constraint equation by λ_j and add it to df to obtain $d\phi$. In essence, we replace $\partial c_j/\partial x_k$ by a_{jk} in Eq. [B.9]. To solve Eq. [B.9] one selects the Lagrange multipliers λ_j ($j = 1, 2, \ldots, m$) such that each of the expressions in the square brackets, that is, the coefficients of dx_k ($k = 1, 2, \ldots, n$), vanish individually. This leads to a total of n equations in Eq. [B.9] and m equations in Eq. [B.4] or Eq. [B.6], which can be solved for the $n + m$ unknowns $x_1, x_2, \ldots, x_n, \lambda_1, \lambda_2, \ldots, \lambda_m$.

The determination of whether the stationary values of constrained systems constitute minima or maxima is a more complicated subject. This determination requires use of developments from nonlinear optimization theory and Kuhn-Tucker conditions. We will not go into this subject in detail, but briefly mention the theorem from Avriel (see the text by Pike), which establishes a set of sufficient conditions to establish minima or maxima.

After finding the stationary values of a constrained function using the approach outlined above, one constructs the matrices $[H_p]$ ($p = m + 1, m + 2, \ldots, n$), such that

$$[H_p] = \begin{bmatrix} [H^*] & [c'] \\ [c']^T & [0] \end{bmatrix} \qquad \text{[B.10]}$$

The matrices $[H^*]$ and $[c']$ are defined as

$$H_{ij}^* = \frac{\partial^2 \phi}{\partial x_i \partial x_j} \quad i, j = 1, 2, \ldots, p \qquad c'_{sr} = \frac{\partial c_r}{\partial x_s} \quad r = 1, 2, \ldots, m; s = 1, 2, \ldots, p$$

[B.11]

where all the derivatives are evaluated at the stationary points. Avriel's theorem states that

If $(-1)^m \det[H_p] > 0$ for all $p = m + 1, m + 2, \ldots, n$, the stationary point is a local minimum.

If $(-1)^p \det[H_p] > 0$ for all $p = m + 1, m + 2, \ldots, n$, the stationary point is a local maximum.

If neither one of the above conditions holds, no conclusions can be drawn. The interested reader is referred to texts on optimization theory for further details on equality constraints, as well as for inequality constraints.

Example B.1

Find the stationary values of the function $f(x, y) = y^2/2 + x^3/24 + xy + x + 1$ in the interval $-4 \leq x \leq 8, -8 \leq y \leq 4$.

Solution

The stationary value of f is given by $df = 0$. Taking the differential of f we obtain

$$df = \frac{\partial f}{\partial x} dx + \frac{\partial f}{\partial y} dy = 0 \qquad \text{[a]}$$

which holds only if both $\frac{\partial f}{\partial x} = 0$ and $\frac{\partial f}{\partial y} = 0$. Setting these first derivatives equal to zero we obtain

$$\frac{\partial f}{\partial x} = \frac{x^2}{8} + y + 1 = 0 \qquad \frac{\partial f}{\partial y} = y + x = 0 \qquad \text{[b]}$$

Substituting the second of the above equations into the first we obtain

$$\frac{x^2}{8} - x + 1 = 0 \qquad \text{[c]}$$

the solution of which yields two stationary values: $x_1^* = 4 + 2\sqrt{2}$ and $x_2^* = 4 - 2\sqrt{2}$. Both stationary values are inside the interval given above. It follows from the second of Eqs. [b] that $y_1^* = -x_1^* = -4 - 2\sqrt{2}$ and $y_2^* = -x_2^* = -4 + 2\sqrt{2}$.

To check whether the stationary points are local minima, maxima, or neither, we analyze the Hessian. The second derivatives of f have the form

$$\frac{\partial^2 f}{\partial x^2} = \frac{x}{4} \qquad \frac{\partial^2 f}{\partial y^2} = 1 \qquad \frac{\partial^2 f}{\partial x \partial y} = 1 \qquad \text{[d]}$$

so that the Hessian becomes

for $X_1^* = (x_1^*, y_1^*) = (4 + 2\sqrt{2}, -4 - 2\sqrt{2}) \qquad [H] = \nabla^2 f = \begin{bmatrix} 1 + \sqrt{1/2} & 1 \\ 1 & 1 \end{bmatrix}$

for $X_2^* = (x_2^*, y_2^*) = (4 - 2\sqrt{2}, -4 + 2\sqrt{2}) \qquad [H] = \nabla^2 f = \begin{bmatrix} 1 - \sqrt{1/2} & 1 \\ 1 & 1 \end{bmatrix} \qquad \text{[e]}$

For the stationary point (x_1^*, y_1^*) the Hessian is positive definite. Hence, the stationary point (x_1^*, y_1^*) is a local minimum. The Hessian for the stationary point (x_2^*, y_2^*) is neither positive nor negative definite. It is indefinite. It follows that this second stationary point is neither a minimum nor a maximum, but a saddle point. To check if X_1^* constitutes a global minimum in the interval, one must evaluate the values of f at the boundaries of the interval as well.

Example B.2

Find the stationary values of the function $f(x, y, z) = x^3 + y^2 + z^2$ subject to the constraint $c(x, y, z) = y - x^2 = 0$.

Solution

We can solve this problem either by introducing the constraint into $f(x, y, z)$ and expressing f in terms of two variables, or by using the Lagrange multipliers. For illustrative purposes, we select the latter approach. We form the augmented function $\phi(x, y, z) = f(x, y, z) + \lambda c(x, y, z)$ and take partial derivatives of ϕ, which yields

$$\frac{\partial \phi}{\partial x} = 3x^2 - 2\lambda x \qquad \frac{\partial \phi}{\partial y} = 2y + \lambda \qquad \frac{\partial \phi}{\partial z} = 2z \qquad \text{[a]}$$

Selecting λ such that all of the first derivatives of ϕ with respect to x, y, and z vanish, and considering the constraint equation, we obtain the four equations

$$3x^2 - 2\lambda x = 0 \qquad 2y + \lambda = 0 \qquad 2z = 0 \qquad y - x^2 = 0 \qquad \text{[b]}$$

to be solved for the unknowns x, y, z, and the Lagrange multiplier λ. The third equation yields $z = 0$. From the last equation, $y = x^2$, and from the second equation, $\lambda = -2y$. Combining these equations we obtain $\lambda = -2x^2$, which, when substituted into the first equation gives

$$3x^2 - 2\lambda x = 3x^2 + 4x^3 = 0 \qquad \text{[c]}$$

Equation [c] can be factored to yield $3x^2(1 + \frac{4}{3}x) = 0$. This equation has three solutions, a double root at $x = 0$ and $x = -3/4$. Combining with Eqs. [b], we identify the stationary points X_1^* and X_2^* to be

$$X_1^* \to x_1^* = 0 \; y_1^* = 0 \; z_1^* = 0 \qquad X_2^* \to x_2^* = -\frac{3}{4} \; y_2^* = \frac{9}{16} \; z_2^* = 0 \qquad \text{[d]}$$

To calculate whether the stationary points constitute local minima or maxima, we generate the $[H_p]$ matrices. For this problem we have $n = 3$, $m = 1$, and two values for p, $p = 2, 3$. There are two $[H_p]$ matrices to check. The $[H^*]$ matrix is diagonal with the entries

$$H_{11}^* = \frac{\partial^2 \phi}{\partial x^2} = 6x - 2\lambda \qquad H_{22}^* = \frac{\partial^2 \phi}{\partial y^2} = 2 \qquad H_{33}^* = \frac{\partial^2 \phi}{\partial z^2} = 2 \qquad \text{[e]}$$

All other entries of $[H^*]$ are zero. Substituting the value of the Lagrange multiplier into H_{11}^*, we obtain

$$H_{11}^* = 6x - 2\lambda = 6x - 2(-2x^2) = 6x + 4x^2 \qquad \text{[f]}$$

For the first stationary point $x_1^* = 0$, so that $H_{11}^* = 0$. For the second stationary point $x_2^* = -0.75$ and $H_{11}^* = -2.25$. Consider the first stationary point. For the case when $p = 2$

$$[H^*] = \begin{bmatrix} 0 & 0 \\ 0 & 2 \end{bmatrix} \qquad [c'] = \begin{bmatrix} -2x \\ 1 \end{bmatrix} = \begin{bmatrix} 0 \\ 1 \end{bmatrix} \qquad \text{[g]}$$

This results in the $[H_2]$ matrix

$$[H_2] = \begin{bmatrix} 0 & 0 & 0 \\ 0 & 2 & 1 \\ 0 & 1 & 0 \end{bmatrix} \qquad \text{[h]}$$

whose determinant is zero. The first stationary point is neither a minimum nor a maximum, but a saddle point. We dispense with the calculation of $[H_3]$ for this stationary point and consider the next stationary point. For x_2^* the $[H_2]$ matrix has the form

$$[H_2] = \begin{bmatrix} -2.25 & 0 & 1.5 \\ 0 & 2 & 1 \\ 1.5 & 1 & 0 \end{bmatrix} \qquad \text{[i]}$$

and $(-1)^m \det[H_2] = 2.25$.

To determine $[H_3]$, we form the $[H^*]$ and $[c']$ matrices as

$$[H^*] = \begin{bmatrix} -2.25 & 0 & 0 \\ 0 & 2 & 0 \\ 0 & 0 & 2 \end{bmatrix} \qquad [c'] = \begin{bmatrix} 1.5 \\ 1 \\ 0 \end{bmatrix} \qquad \text{[j]}$$

resulting in

$$[H_3] = \begin{bmatrix} -2.25 & 0 & 0 & 1.5 \\ 0 & 2 & 0 & 1 \\ 0 & 0 & 2 & 0 \\ 1.5 & 1 & 0 & 0 \end{bmatrix} \qquad \text{[k]}$$

It can be shown that $(-1)^m \det[H_3] = 4.50$. As both quantities are larger than zero, the second equilibrium position constitutes a local minimum.

B.3 STATIONARY VALUES OF A DEFINITE INTEGRAL

A large number of problems in mechanics and in mathematics can be posed as the determination of the stationary values of a definite integral. A simple definition of the problem is as follows: Find the function $y(x)$ which is continuously differentiable in the interval (a, b) which yields the stationary values of the integral

$$I = \int_a^b F\left(x, y, \frac{dy}{dx}\right) dx \qquad [\text{B.12}]$$

In this equation, I is referred to as the *functional* and $y(x)$ satisfies certain boundary conditions. For $y(x)$ to be the solution, any function other than $y(x)$ should not lead to a stationary value. We introduce a continuous function $\eta(x)$, which is arbitrary, together with the constant ε, and we construct the varied function $y(x) + \varepsilon\eta(x)$, as shown in Fig. B.1. When dealing with constrained systems $\eta(x)$ must be consistent with the constraints. The values of $\eta(x)$ at the boundaries depend on the boundary conditions of $y(x)$. Consider initially the case where $y(a)$ and $y(b)$ are specified. Such boundary conditions are known as *geometric boundary conditions* or *boundary conditions of the first kind*. In this case, the varied function vanishes at the boundaries, $\eta(a) = \eta(b) = 0$. If $y(x)$ has fixed values at certain points then the varied function cannot be arbitrary at those points. Introducing the varied function $y(x) + \varepsilon\eta(x)$ into the integral whose stationary values are sought, we obtain

$$I(\varepsilon) = \int_a^b F(x, y + \varepsilon\eta, y' + \varepsilon\eta') dx \qquad [\text{B.13}]$$

with primes denoting differentiation with respect to x. For the integral I to have a stationary value, it must satisfy

$$\frac{dI}{d\varepsilon} = 0 \quad \text{when} \quad \varepsilon = 0 \qquad [\text{B.14}]$$

Differentiating Eq. [B.13] using the chain rule and evaluating at $\varepsilon = 0$ we obtain

$$\left.\frac{dI}{d\varepsilon}\right|_{\varepsilon=0} = \int_a^b \left(\frac{\partial F}{\partial(y + \varepsilon\eta)} \frac{\partial(y + \varepsilon\eta)}{\partial \varepsilon} + \frac{\partial F}{\partial(y' + \varepsilon\eta')} \frac{\partial(y' + \varepsilon\eta')}{\partial \varepsilon}\right) dx \bigg|_{\varepsilon=0}$$

$$= \int_a^b \left(\frac{\partial F}{\partial y}\eta + \frac{\partial F}{\partial y'}\eta'\right) dx = 0 \qquad [\text{B.15}]$$

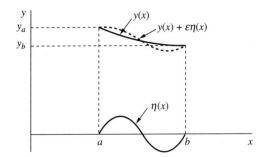

Figure B.1

We integrate the second term on the right hand side of the above equation by parts to obtain[2]

$$\int_a^b \frac{\partial F}{\partial y'} \eta' \, dx = \frac{\partial F}{\partial y'} \eta \bigg|_a^b - \int_a^b \frac{d}{dx}\left(\frac{\partial F}{\partial y'}\right) \eta \, dx \qquad \text{[B.16]}$$

The objective of this integration by parts is to have $\eta(x)$ as a coefficient of all the terms in the integrand. Introducing Eq. [B.16] into Eq. [B.15] we obtain

$$\int_a^b \left[\frac{\partial F}{\partial y} - \frac{d}{dx}\left(\frac{\partial F}{\partial y'}\right)\right] \eta(x) \, dx + \frac{\partial F}{\partial y'} \eta \bigg|_a^b = 0 \qquad \text{[B.17]}$$

The integral and boundary terms in this equation vanish individually. Because $\eta(x)$ is arbitrary in the interval (a, b), in order for the integral to be equal to zero the integrand must vanish at every point in the interval, so that

$$\frac{\partial F}{\partial y} - \frac{d}{dx}\left(\frac{\partial F}{\partial y'}\right) = 0 \qquad a < x < b \qquad \text{[B.18]}$$

Equation [B.18] is known as *Euler's equation* or the *Euler-Lagrange equation*. In addition, each boundary term in Eq. [B.17] is zero by itself, leading to

$$\frac{\partial F}{\partial y'} \eta \bigg|_{\eta=a} = 0 \qquad \frac{\partial F}{\partial y'} \eta \bigg|_{\eta=b} = 0 \qquad \text{[B.19a,b]}$$

Equation [B.18] and the end conditions constitute the necessary condition for the establishment of the stationary value of the definite integral. Previously we considered geometric boundary conditions, so that η vanishes at $x = a$ and $x = b$. As a result, Eqs. [B.19a] and [B.19b] are automatically satisfied.

Consider now the case when $y(x)$ is not specified at one or both of $x = a$ and $x = b$. Hence $\eta(x)$ does not vanish at those boundaries and has arbitrary values, in addition to the interior. To drive the boundary terms in Eqs. [B.19a] and [B.19b] to zero, the coefficient of $\eta(x)$ must vanish, or

If $y(a)$ is not specified, then $\quad \dfrac{\partial F}{\partial y'} = 0 \quad$ at $\quad x = a$

If $y(b)$ is not specified, then $\quad \dfrac{\partial F}{\partial y'} = 0 \quad$ at $\quad x = b$

The conditions above are the boundary conditions. Note that these boundary conditions are related to the function. They are called *natural* boundary conditions, or *boundary conditions of the second kind*. In mechanics they are also referred to as *dynamic boundary conditions*. Sometimes, the boundary conditions have the form $g(y(a), y'(a)) = 0$, $h(y(b), y'(b)) = 0$. Such boundary conditions are known as *boundary conditions of the third kind*.

An interesting special case is encountered when the function F is not an explicit function of x, that is, $F = F(y, y')$ only. We can take the derivative of F with respect to x and obtain

$$\frac{dF}{dx} = \left(\frac{\partial F}{\partial y}\right)\frac{dy}{dx} + \left(\frac{\partial F}{\partial y'}\right)\frac{dy'}{dx} = \left(\frac{\partial F}{\partial y}\right)\frac{dy}{dx} + \left(\frac{\partial F}{\partial y'}\right)\frac{d^2y}{dx^2} \qquad \text{[B.20]}$$

---

[2] Note that we are assuming that F is twice differentiable with respect to x, y, and y'. See text by Hildebrand for cases when the second derivatives of F have discontinuities.

From Eq. [B.18] we have

$$\frac{\partial F}{\partial y} = \frac{d}{dx}\left(\frac{\partial F}{\partial y'}\right) \quad \text{[B.21]}$$

Introducing Eq. [B.21] into Eq. [B.20] we obtain

$$\frac{dF}{dx} = \frac{d}{dx}\left(\frac{\partial F}{\partial y'}\right)\frac{dy}{dx} + \left(\frac{\partial F}{\partial y'}\right)\frac{d^2 y}{dx^2} = \frac{d}{dx}\left[y'\left(\frac{\partial F}{\partial y'}\right)\right],$$

$$\text{or} \quad \frac{d}{dx}\left[F - y'\left(\frac{\partial F}{\partial y'}\right)\right] = 0 \quad \text{[B.22]}$$

leading to the result

$$F - y'\left(\frac{\partial F}{\partial y'}\right) = \text{constant} \quad \text{[B.23]}$$

so that when F is not an explicit function of x, then the expression $F - y'(\partial F/\partial y')$ represents a *first integral*. The subject of first integrals for dynamical systems is discussed in Section 1.12.

The above relations describe a very useful property which lets one recognize the existence of first integrals. In Chapters 4 and 5 we make use of the above properties and observe the relation between the function F and the Lagrangian.

Example B.3

For the functional $F(x, y, y')$ given by $F = y'^2 - 4y^2 - 6y^4$, find Euler's equation and the associated first integral.

Solution

Taking the partial derivatives of F, we obtain

$$\frac{\partial F}{\partial y} = -8y - 24y^3 \qquad \frac{\partial F}{\partial y'} = 2y' \qquad \frac{d}{dx}\left(\frac{\partial F}{\partial y'}\right) = 2y'' \quad \text{[a]}$$

which, when introduced into Euler's equation, gives

$$-2y'' - 8y - 24y^3 = 0 = 2y'' + 8y + 24y^3 \quad \text{[b]}$$

To find the first integral we note that x is absent from F. From Eq. [a] we write

$$y'\left(\frac{\partial F}{\partial y'}\right) = 2y'^2 \quad \text{[c]}$$

so that the first integral becomes

$$F - y'\left(\frac{\partial F}{\partial y'}\right) = y'^2 - 4y^2 - 6y^4 - 2y'^2 = \text{constant} \quad \text{[d]}$$

which can be rewritten as

$$y'^2 + 4y^2 + 6y^4 = \text{constant} \quad \text{[e]}$$

We can draw an analogy between this example and the stiffening spring considered in Chapter 1. Indeed, if we think of x as time, y as the spring deflection, then F represents the difference between twice the kinetic and potential energies, and $\partial F/\partial y$ represents the spring force. The integral of the motion is, of course, the total energy.

B.4 THE VARIATIONAL NOTATION

The variational notation streamlines the formulation of variational mechanics problems, and it demonstrates the difference between differential calculus and variational calculus more clearly. Consider the function $F(x, y, y')$ and stationary values of its integral. In the previous section, we defined the varied function as $y(x) + \varepsilon\eta(x)$. Let ε be a small quantity and define the variation of $y(x)$ as the change $\varepsilon\eta(x)$, and denote it by δy, so that

$$\delta y = \varepsilon\eta(x) \qquad [\text{B.24}]$$

The change in F due to the variation in y, for a fixed value of x, can be expressed as

$$\Delta F = F(x, y + \varepsilon\eta, y' + \varepsilon\eta') - F(x, y, y') \qquad [\text{B.25}]$$

which can be expanded in a Taylor series

$$\Delta F = \frac{\partial F}{\partial y}\varepsilon\eta + \frac{\partial F}{\partial y'}\varepsilon\eta' + \text{(higher-order terms in } \varepsilon\text{)} \qquad [\text{B.26}]$$

The *variation* of F, denoted by δF, is defined as the terms in the above equation linear in ε, so that

$$\delta F = \frac{\partial F}{\partial y}\varepsilon\eta + \frac{\partial F}{\partial y'}\varepsilon\eta' = \frac{\partial F}{\partial y}\delta y + \frac{\partial F}{\partial y'}\delta y' \qquad [\text{B.27}]$$

where $\delta y'$ denotes the variation of y', $\delta y' = \varepsilon\eta'(x)$. δF is also referred to as the *first variation* of F.

The above relationship is similar to the exact differential of F, but it is different. The exact differential of F has the form

$$dF = \frac{\partial F}{\partial x}dx + \frac{\partial F}{\partial y}dy + \frac{\partial F}{\partial y'}dy' \qquad [\text{B.28}]$$

and it has one extra term due to the derivative of F with respect to x. The difference between the exact differential and variation of F arises from the way the variation operation treats the variables x and y. For the function F, x is the independent variable and $y(x)$ is the dependent variable, dependent in the sense that it is a function of x, and that it is not constrained in any form. When we obtain the variation of F, we hold the independent variable fixed and vary the dependent variable $y(x)$ and its derivative $y'(x)$ to get $y(x) + \varepsilon\eta(x)$ and $y'(x) + \varepsilon\eta'(x)$. The variation is from one curve (or path) to another while x is held fixed, so that there is no variation in x. Hence, $\delta x = 0$. By contrast, the differential of a function describes the change in that function along a particular curve (or path), and the independent variable x varies along that path as well. This point highlights the difference between differential calculus and variational calculus: The variation of a function includes changes of only the dependent variables, whereas the differential of a function considers changes of both the dependent and independent variables.

The rules associated with mathematical operations on variation are the same as the rules of differentiation. The variation operation and the differentiation operation commute when the differentiation is with respect to an independent variable. We have, for example

$$\frac{d}{dx}(\delta y) = \frac{d}{dx}(\varepsilon\eta) = \varepsilon\eta' = \delta y' \qquad [\text{B.29}]$$

The variation operation is applicable to scalar functions as well as to vectors. Also, when we refer to the variation of a function $f(x_1, x_2, \ldots, x_n)$, where all the variables x_k ($k = 1, 2, \ldots, n$) are independent of each other, the variation is exactly analogous to differentiation and has the form

$$\delta f = \frac{\partial f}{\partial x_1} \delta x_1 + \frac{\partial f}{\partial x_2} \delta x_2 + \cdots + \frac{\partial f}{\partial x_n} \delta x_n \qquad \text{[B.30]}$$

One should compare Eq. [B.30] with Eq. [B.2].

We next derive the Euler-Lagrange equation by using the variation of the integral in Eq. [B.12]. Taking its variation, we obtain

$$\delta I = \delta \int_a^b F(x, y, y')\,dx = \int_a^b \left\{ \frac{\partial F}{\partial y} \delta y + \frac{\partial F}{\partial y'} \delta y' \right\} dx = 0 \qquad \text{[B.31]}$$

Integration of the second term in the integrand by parts yields

$$\int_a^b \left\{ \frac{\partial F}{\partial y} \delta y + \frac{\partial F}{\partial y'} \delta y' \right\} dx = \int_a^b \left\{ \frac{\partial F}{\partial y} - \frac{d}{dx}\left(\frac{\partial F}{\partial y'}\right) \right\} \delta y\,dx + \frac{\partial F}{\partial y'} \delta y \bigg|_a^b = 0 \qquad \text{[B.32]}$$

Noting that the integral and the boundary expressions vanish individually, we obtain

$$\int_a^b \left\{ \frac{\partial F}{\partial y} - \frac{d}{dx}\left(\frac{\partial F}{\partial y'}\right) \right\} \delta y\,dx = 0 \qquad \frac{\partial F}{\partial y'} \delta y \bigg|_a^b = 0 \qquad \text{[B.33a,b]}$$

By virtue of the arbitrariness of δy, Eq. [B.33a] leads to the Euler equation in Eq. [B.18]. Equation [B.33b] leads to the boundary conditions.

The variational formulation can also be used to find the stationary values of an integral subject to constraints. The general formulation of the constrained problem will not be derived here. We consider this problem in the study of analytical mechanics in Chapter 4, within the context of applications.

Example B.4

A very important application of variational calculus is the *Sturm-Liouville problem*, which can be derived by seeking the values for $y(x)$ which render stationary values of the ratio of two integrals I_1 and I_2, defined as

$$\gamma = \frac{I_1}{I_2} = \frac{\int_a^b [p(x)y'^2(x) - q(x)y^2(x)]\,dx}{\int_a^b r(x)y^2(x)\,dx} \qquad \text{[a]}$$

where γ is the ratio of the integrals, $p(x)$, $q(x)$, and $r(x)$ are known functions of x, and a and b are the boundaries. Moreover, $p(x)$ and $r(x)$ are positive. Note that x is the independent variable. Invoking the product rule, we obtain for the variation of γ

$$\delta \gamma = \frac{I_2 \delta I_1 - I_1 \delta I_2}{I_2^2} = 0 \qquad \text{[b]}$$

Because $I_2 \neq 0$, we can multiply the above equation with it, which yields

$$\delta I_1 - \frac{I_1}{I_2} \delta I_2 = \delta I_1 - \gamma \delta I_2 = 0 \qquad \text{[c]}$$

The variations of the individual integrals are

$$\delta I_1 = 2\int_a^b [p(x)y'\,\delta y' - q(x)y\,\delta y]\,dx \qquad \delta I_2 = 2\int_a^b r(x)y\,\delta y\,dx \qquad \text{[d]}$$

We integrate the first of Eq. [d] by parts and obtain

$$\delta I_1 = 2\int_a^b \left\{ -\frac{d}{dx}\left(p(x)\frac{d}{dx}\right) - q(x)y \right\} \delta y \, dx + p(x)y'\,\delta y \Big|_a^b \qquad \text{[e]}$$

Introduction of the second of Eqs. [d] and [e] into Eq. [c] and division by 2 yields

$$\int_a^b \left\{ -\frac{d}{dx}\left[p(x)\frac{dy}{dx}\right] - q(x)y - \gamma r(x)y \right\} \delta y \, dx + p(x)y'\,\delta y \Big|_a^b = 0 \qquad \text{[f]}$$

The integral in Eq. [f] is evaluated over the interval (a, b) and the integrated term is evaluated at the boundaries. By definition of the variation of y, the two expressions are independent of each other and, for Eq. [f] to hold, they must both vanish individually. For the integral to vanish for all possible values of δy, the integrand must be zero, leading to the Euler-Lagrange equation

$$\frac{d}{dx}[p(x)y'] + q(x)y + \gamma r(x)y = 0 \qquad a < x < b \qquad \text{[g]}$$

The second term in Eq. [f] contains the boundary expressions, from which the boundary conditions are ascertained. Consider the boundary $x = a$. If $y(x)$ is specified at $x = a$, say $y(a) = y_a$, then the variation of y at that point is zero. This type of boundary condition is an *essential boundary condition,* or a *boundary condition of the first kind.* If, on the other hand, $y(x)$ is not specified at $x = a$, its variation δy does not vanish at that point. It follows that, to render the boundary term zero, the coefficient of δy, $p(x)y'$, must vanish. The relation

$$p(x)y'(x) = 0 \quad \text{at} \quad x = a \quad \text{or} \quad p(a)y'(a) = 0 \qquad \text{[h]}$$

becomes the boundary condition. This type of boundary condition is referred to as a *natural boundary condition,* or a *boundary condition of the second kind.* The Euler-Lagrange equation and the boundary conditions constitute the so-called *boundary value problem.*

Examples of Sturm-Liouville problems from mechanics include the transverse vibration of strings, axial vibration of bars, and torsional vibration of rods. Table B.1 gives the corresponding quantities for $y(x)$, $p(x)$, $q(x)$, and $r(x)$ for these problems.

Table B.1 Sturm-Liouville Problems

Type of Problem	$y(x)$	$p(x)$	$q(x)$	$r(x)$
String vibrations	Transverse deformation $y(x)$	Tension $-T(x)$	0	Mass/unit length $\mu(x)$
Axial vibration of bars	Axial deformation $u(x)$	Stiffness $-EA(x)$	0	Mass/unit length $\mu(x)$
Torsional vibration of rods	Torsional deformation $\theta(x)$	Torsional stiffness $-GJ(x)$	0	per unit length $\mathcal{I}(x)$

In all three cases the expression for $p(x)$ is a negative quantity. Derivation of the equations of motion of deformable bodies is considered in Chapter 11. Note that even though the form of the equations looks similar, the transverse vibration of beams is not a Sturm-Liouville problem. The highest order spatial derivative is four, as opposed to two in Sturm-Liouville problems. However, the way the equations of motion are derived is the same. For the simple cases of vibration considered here, $q(x) = 0$. This expression becomes nonzero in more complex vibration problems.

B.5 APPLICATION OF THE VARIATIONAL NOTATION TO DYNAMICS PROBLEMS

The variational notation is ideally suited for dynamics problems because it makes the formulation concise and it has a meaningful physical interpretation.

Consider, for example, the motion of a particle. When the motion is described in terms of Cartesian coordinates, we use the variables x, y, and z to describe position and the variable t to describe the time dependence. The vector $\mathbf{r}(x(t), y(t), z(t), t)$ describes the position. It follows that x, y, and z are the dependent variables and t is the independent variable. There is no variation of the position vector with respect to time, and

$$\delta \mathbf{r} = \frac{\partial \mathbf{r}}{\partial x} \delta x + \frac{\partial \mathbf{r}}{\partial y} \delta y + \frac{\partial \mathbf{r}}{\partial z} \delta z \qquad [\text{B.34}]$$

The variations of the spatial coordinates are known as *virtual displacements*. Because there is no variation in time, the virtual displacements can be considered as occurring instantaneously. This leads to a very interesting physical interpretation. A virtual displacement can be thought of as imagining the dynamical system in a different position while holding time fixed. The arbitrary position is consistent with the system constraints and it is within the admissible paths the particle can follow. This concept is illustrated in Fig. B.2 for the x coordinate.

The variation of the position vector is intimately related to the velocity. Indeed, the velocity of the particle considered above can be written as

$$\mathbf{v}(x, y, z, t) = \frac{d\mathbf{r}}{dt} = \frac{\partial \mathbf{r}}{\partial x} \dot{x} + \frac{\partial \mathbf{r}}{\partial y} \dot{y} + \frac{\partial \mathbf{r}}{\partial z} \dot{z} + \frac{\partial \mathbf{r}}{\partial t} \qquad [\text{B.35}]$$

This brings about the possibility of calculating a virtual displacement from the expression for velocity, instead of taking derivatives of the position vector. We demonstrate this in Chapter 4.

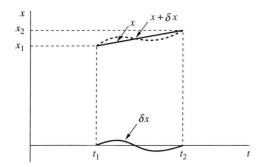

Figure B.2

REFERENCES

Greenberg, M. D. *Foundations of Applied Mathematics.* Englewood Cliffs, NJ: Prentice-Hall, 1978.
Hildebrand, F. B. *Methods of Applied Mathematics.* Englewood Cliffs, NJ: Prentice-Hall, 1965.
Meirovitch, L. *Methods of Analytical Dynamics.* New York: McGraw-Hill, 1970.
Pike, R. W. *Optimization for Engineering Systems.* New York: Van Nostrand Reinhold, 1970.

HOMEWORK EXERCISES

SECTION B.2

1. Find the points on the curve $xy = 1$ that are the closest to the origin.
2. Find the points on the ellipsoid $4x^2 - xy + 4y^2 - yz + 3z^2 = 1$ that are the closest to the origin. Note: This problem can be posed as an eigenvalue problem.
3. Find the stationary values of the function $f(x) = 3x_1^2 + x_2^2 + 2x_1x_2 + 6x_1 + 2x_2$ subject to the constraint $2x_1 - x_2 = 4$. Check for minima and maxima.

SECTIONS B.3 AND B.4

4. Given two points on the xy plane (x_1, y_1) and (x_2, y_2), make use of Eq. [1.3.33] to show that one must minimize the integral

$$I = \int_{x_1}^{x_2} \sqrt{1 + y'^2}\, dx \quad \text{with} \quad y(x_1) = y_1, y(x_2) = y_2$$

in order to find the arc of minimum length that passes through two given end points. Show that, given no restrictions on the path from (x_1, y_1) to (x_2, y_2), the solution is a straight line.

5. Obtain the Euler-Lagrange equation associated with the functional $F(x, y, y') = xy'^2 - yy' + y$.

6. In a similar way to Problem 4, show that the minimal surface of revolution about the x axis passing through the two points (x_1, y_1) and (x_2, y_2) can be found by minimizing the integral

$$I = \int_{x_1}^{x_2} y\sqrt{1 + y'^2}\, dx \quad \text{with} \quad y(x_1) = y_1, y(x_2) = y_2$$

Then, find the resulting Euler's equation, and the solution.

7. Find the first integral associated with the function $F(x, y, y') = y'^2 + 5y^3 - 2y^4$.

8. Consider the problem of axial deformation of a circular rod of length L and uniform density ρ, fixed at $x = 0$ and free at $x = L$, as shown in Fig. B.3. The radius of the cross section of the rod $R(x)$ varies with the relation $R(x) = R_0(1 - x/3L)$. Find Euler's equation and boundary conditions.

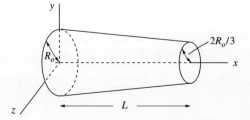

Figure B.3

APPENDIX C

COMMON INERTIA PROPERTIES

Note: xyz always denote centroidal coordinates.

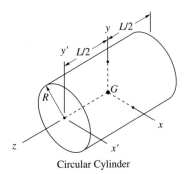

Circular Cylinder

Volume $= \pi R^2 L$

$I_{zz} = \dfrac{1}{2} m R^2$

$I_{xx} = I_{yy} = \dfrac{1}{4} m R^2 + \dfrac{1}{12} m L^2$

$I_{x'x'} = I_{y'y'} = \dfrac{1}{4} m R^2 + \dfrac{1}{3} m L^2$

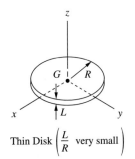

Thin Disk $\left(\dfrac{L}{R} \text{ very small} \right)$

Volume $= \pi R^2 L$

$I_{zz} = \dfrac{1}{2} m R^2$

$I_{xx} = I_{yy} = \dfrac{1}{4} m R^2$

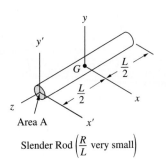

Slender Rod $\left(\dfrac{R}{L} \text{ very small} \right)$

Volume $= AL$

$I_{zz} \approx 0$

$I_{xx} = I_{yy} = \dfrac{1}{12} m L^2$

$I_{x'x'} = I_{y'y'} = \dfrac{1}{3} m L^2$

APPENDIX C • Common Inertia Properties

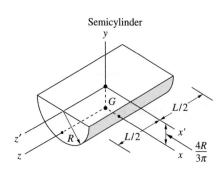

Semicylinder

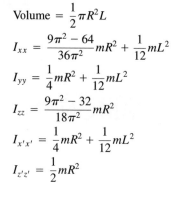

Volume $= \dfrac{1}{2}\pi R^2 L$

$I_{xx} = \dfrac{9\pi^2 - 64}{36\pi^2}mR^2 + \dfrac{1}{12}mL^2$

$I_{yy} = \dfrac{1}{4}mR^2 + \dfrac{1}{12}mL^2$

$I_{zz} = \dfrac{9\pi^2 - 32}{18\pi^2}mR^2$

$I_{x'x'} = \dfrac{1}{4}mR^2 + \dfrac{1}{12}mL^2$

$I_{z'z'} = \dfrac{1}{2}mR^2$

Rectangular Prism

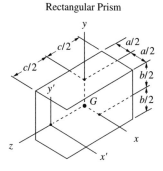

Volume $= abc$

$I_{xx} = \dfrac{1}{12}m(b^2 + c^2)$

$I_{yy} = \dfrac{1}{12}m(a^2 + c^2)$

$I_{zz} = \dfrac{1}{12}m(a^2 + b^2)$

$I_{x'x'} = \dfrac{1}{12}m(b^2 + 4c^2)$

$I_{y'y'} = \dfrac{1}{12}m(a^2 + 4c^2)$

Thin Plate (One of a, b, or c small. Say c)

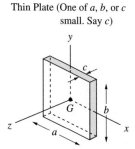

Volume $= abc$

$I_{xx} = \dfrac{1}{12}mb^2$

$I_{yy} = \dfrac{1}{12}ma^2$

$I_{zz} = \dfrac{1}{12}m(a^2 + b^2)$

Sphere

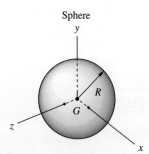

Volume $= \dfrac{4}{3}\pi R^3$

$I_{xx} = I_{yy} = I_{zz} = \dfrac{2}{5}mR^2$

Hemisphere

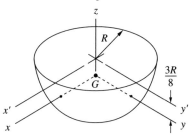

Volume $= \dfrac{2}{3}\pi R^3$

$I_{xx} = I_{yy} = \dfrac{83}{320} mR^2$

$I_{zz} = \dfrac{2}{5} mR^2$

$I_{x'x'} = I_{y'y'} = \dfrac{2}{5} mR^2$

Right Circular Cone

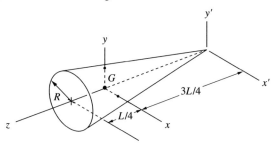

Volume $= \dfrac{1}{3}\pi R^2 L$

$I_{xx} = I_{yy} = \dfrac{3}{20} mR^2 + \dfrac{3}{80} mL^2$

$I_{zz} = \dfrac{3}{10} mR^2$

$I_{x'x'} = I_{y'y'} = \dfrac{3}{20} mR^2 + \dfrac{3}{5} mL^2$

Half Cone

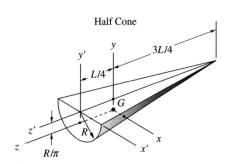

Volume $= \dfrac{1}{6}\pi R^2 L$

$I_{xx} = \left(\dfrac{3}{20} - \dfrac{1}{\pi^2}\right) mR^2 = \dfrac{3}{80} mL^2$

$I_{yy} = \dfrac{3}{20} mR^2 + \dfrac{3}{80} mL^2$

$I_{zz} = \left(\dfrac{3}{10} - \dfrac{1}{\pi^2}\right) mR^2$

$I_{yz} = -\dfrac{1}{20\pi} mRL \qquad I_{xy} = I_{xz} = 0$

$I_{x'x'} = I_{y'y'} = \dfrac{3}{20} mR^2 + \dfrac{1}{10} mL^2$

$I_{z'z'} = \dfrac{3}{10} mR^2$

Ellipsoid

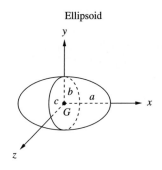

Volume $= \dfrac{4}{3}\pi abc$

$I_{xx} = \dfrac{1}{5} m(b^2 + c^2)$

$I_{yy} = \dfrac{1}{5} m(a^2 + c^2)$

$I_{zz} = \dfrac{1}{5} m(a^2 + b^2)$

Surface defined by $\dfrac{x^2}{a^2} + \dfrac{y^2}{b^2} + \dfrac{z^2}{c^2} = 1$

Right Triangular Prism

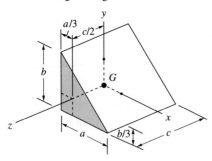

$$\text{Volume} = \frac{1}{2}abc$$

$$I_{xx} = \frac{1}{18}mb^2 + \frac{1}{12}mc^2$$

$$I_{yy} = \frac{1}{18}ma^2 + \frac{1}{12}mc^2$$

$$I_{zz} = \frac{1}{18}m(a^2 + b^2)$$

$$I_{xy} = -\frac{1}{36}mab$$

$$I_{xz} = I_{yz} = 0$$

Triangular Plate ($c \ll a$, $c \ll b$)

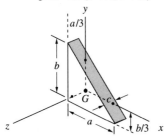

$$\text{Volume} = \frac{1}{2}abc$$

$$I_{xx} = \frac{1}{18}mb^2$$

$$I_{yy} = \frac{1}{18}ma^2$$

$$I_{zz} = \frac{1}{18}m(a^2 + b^2)$$

Rectangular Tetrahedron

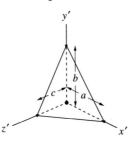

$$\text{Volume} = \frac{1}{6}abc$$

$$I_{x'x'} = \frac{1}{10}m(b^2 + c^2)$$

$$I_{y'y'} = \frac{1}{10}m(a^2 + c^2)$$

$$I_{z'z'} = \frac{1}{10}m(a^2 + b^2)$$

Note: Centroidal coordinates xyz not shown.

Coordinates of G: $\left(\dfrac{a}{4}, \dfrac{b}{4}, \dfrac{c}{4}\right)$

$$I_{xx} = \frac{3}{80}m(b^2 + c^2)$$

$$I_{yy} = \frac{3}{80}m(a^2 + c^2)$$

$$I_{zz} = \frac{3}{80}m(a^2 + b^2)$$

$$I_{xy} = -\frac{1}{80}mab$$

$$I_{xz} = -\frac{1}{80}mac$$

$$I_{yz} = -\frac{1}{80}mbc$$

APPENDIX C COMMON INERTIA PROPERTIES

Half Torus

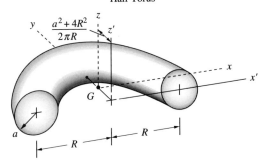

$$\text{Volume} = \pi a^2 R$$
$$I_{x'x'} = I_{y'y'} = \frac{1}{2}mR^2 + \frac{5}{8}ma^2$$
$$I_{z'z'} = mR^2 + \frac{3}{4}ma^2$$

Half Circular Rod $\left(\frac{a}{R} \text{ small}\right)$

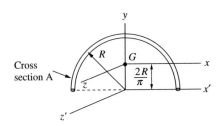

$$\text{Volume} = \pi R A$$
$$I_{xx} = m\left(\frac{1}{2} - \frac{4}{\pi^2}\right)R^2$$
$$I_{yy} = \frac{mR^2}{2}$$
$$I_{zz} = m\left(1 - \frac{4}{\pi^2}\right)R^2$$
$$I_{x'x'} = \frac{mR^2}{2}$$
$$I_{z'z'} = mR^2$$

Conical Shell

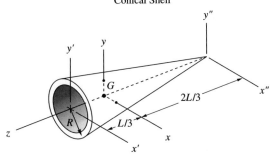

$$I_{xx} = I_{yy} = \frac{1}{4}mR^2 + \frac{1}{18}mL^2$$
$$I_{zz} = \frac{1}{2}mR^2$$
$$I_{x'x'} = I_{y'y'} = \frac{1}{4}mR^2 + \frac{1}{6}mL^2$$
$$I_{x''x''} = I_{y''y''} = \frac{1}{4}mR^2 + \frac{1}{2}mL^2$$

Circular Cylindrical Shell

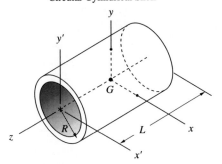

$$I_{zz} = mR^2$$
$$I_{xx} = I_{yy} = \frac{1}{2}mR^2 + \frac{1}{12}mL^2$$
$$I_{x'x'} = I_{y'y'} = \frac{1}{2}mR^2 + \frac{1}{3}mL^2$$

Half Cylindrical Shell

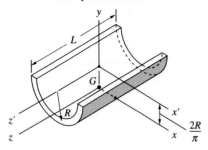

$$I_{zz} = \left(1 - \frac{4}{\pi^2}\right)mR^2$$

$$I_{z'z'} = mR^2$$

$$I_{x'x'} = I_{y'y'} = \frac{1}{2}mR^2 + \frac{1}{12}mL^2$$

$$I_{xx} = \left(\frac{1}{2} - \frac{4}{\pi^2}\right)mR^2 + \frac{1}{12}mL^2$$

Spherical Shell

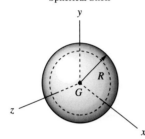

$$I_{xx} = I_{yy} = I_{zz} = \frac{2}{3}mR^2$$

Hemispherical Shell

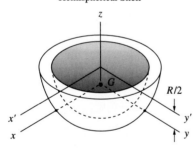

$$I_{yy} = I_{xx} = \frac{5}{12}mR^2$$

$$I_{zz} = \frac{2}{3}mR^2$$

$$I_{x'x'} = I_{y'y'} = \frac{2}{3}mR^2$$

INDEX

Index

A

Acceleration
  centripetal, 11, 19, 125
  Coriolis, 19, 23, 125
  normal, 11
  plane motion, 125
  relative, 124
  tangential, 11
Action integral, 251
Albert of Saxony, 671, 678
Angle
  attitude, 374
  azimuthal, 21
  bank, 374
  heading, 374
  phase, 65
  polar, 21
Angle of twist, 599
Angular acceleration, 117
  apparent, 158
  axisymmetric bodies, 372
  conservation, 40, 449
  in terms of Euler angles, 371
Angular momentun
  particles, 39, 158
  rigid bodies, 96, 423, 652
  transformation properties, 425
Angular velocity
  critical, 570
  definition of, 109, 111, 389
  of differential element, 597
  matrix, 110, 365
  mean, 188
  partial, 379, 507
  as quasi-velocity, 378
  simple, 111, 357
  in terms of Euler angles, 369
  in terms of Euler parameters, 385
  vector, 109, 389
Anomaly
  eccentric, 187
  mean, 189
  true, 187
Apex, 563

Apocenter, 179
Apogee, 180
Appell, P.E., 517, 676, 678
Appell equations, 518
Archimedes, 670, 678
Areal rate, 181
Argument of pericenter, 189
Aristotle, 669, 678
Assumed modes method, 628, 649
Asymptotic behavior, 621
Axial force, 593
Axis of rotation, 359, 360
Axis of symmetry, 328, 372, 565
Axisymmetric body, 341, 372, 499, 548

B

Baruh, H., 650
Beam
  axis, 592
  Rayleigh, 594
  theory, 592
Beat phenomenon, 143
Bending moment, 593
Bending stiffness, 598
Bernoulli, D., 679, 687
Bernoulli, Johann, 242, 674, 679
Bernoulli, Johann, II, 679, 687
Body-fixed
  axes, 356, 370
  coordinates, 371, 435
  rotation, 100, 105
Boltzmann-Hamel equations, 499
Boundary condition
  complementary, 607
  dynamic, 608
  essential, 607
  of the first kind, 607, 693
  geometric, 607, 693
  natural, 608, 694
  of the second kind, 608, 694
  of the third kind, 608, 694
Boundary value problem, 617

Bounding circles, 567
Brahe, T., 671

C

Center of curvature, 9
Center of mass, 154, 324
   composite bodies, 325
Centrifugal softening, 650
Centrifugal stiffening, 650
Characteristic equation, 56, 59, 289,
   343, 618, 630
Characteristic value, 290
Chasles's theorem, 363
Coefficient of restitution, 166, 193,
   478
Complete set, 619
Compression period, 165, 478
Cone
   body, 359, 550
   space, 359, 550
Configuration space, 217
Conic section, 179
Constraint, 217
   configuration, 219
   equation, 219
   force, 219
   holonomic, 220, 404
   modified Tisserand, 643
   nonholonomic, 220, 405
   number of, 35
   in Pfaffian form, 220
   rigid body, 641, 651
   velocity, 220
   workless, 223
   zero slope, 643, 653
   zero tip deformation, 643, 655
Constraint relaxation method, 262,
   464
Convolution integral, 69
Coordinates, body-fixed, 356
Coordinates, Cartesian, 5
   constrained, 218
   curvilinear, 8
   cyclic, 300

   cylindrical, 17
   extrinsic, 5
   generalized, 216
   ignorable, 300, 303
   independent, 218
   intrinsic, 8
   normal-tangential, 6
   number of, 35
   polar, 18
   rectilinear, 5
   right-handed system, 4
   spherical, 21
Copernicus, N., 671, 679
Coriolis, G. G., 675, 679
Coriolis acceleration, 19, 23,
   125
Coriolis force, 136
Coulomb, C. A., 675, 680
Coulomb friction, 29
Courant, R., 621
Critical angular velocity, 570
Curvilinear coordinates, 8
Cuspidal motion, 568
Cycloid, 402
Cylindrical pair, 396

D

D'Alembert, J. L. R., 680
D'Alembert's principle, 246
   direct application of, 247, 311
   Lagrange's form, 254
   rigid bodies, 247, 474, 526
Da Vinci, L., 671
Damped linearized systems,
   287
Damped natural frequency, 61
Damping
   coefficient, 59
   factor, 59
   matrix, 287
   proportional, 297
Datum, 46, 47
Degree of freedom, 34, 218, 601
Dependent variable, 231

Derivative
  in different reference frame, 121
  global, 88
  local, 88, 117
Differential-algebraic equations, 261, 313
Dirac delta function, 37
Direction
  azimuthal, 21
  binormal, 11
  normal, 9
  polar, 21
  radial, 18, 21
  tangential, 9
  transverse, 18
Direction angle, 98
Direction cosine, 98, 343
  in terms of Euler parameters, 386
Dissipation function, 286
Distance traversed, 8, 595
Dry friction, 29
Dual spin satellite, 494
Dugas, A., 669
Duhem, P., 669
Dynamic balancing, 444
Dynamic potential, 276

E

Earth
  mass of, 176
  as a moving reference frame, 133, 582
Eccentricity, 179
Eigenfunction, 617
Eigenvalue, 290, 343, 361, 617, 631
  repeated, 344
Eigenvector, 290, 343, 361, 631
Ellipse
  apsis, 179
  semimajor axis, 179
  semiminor axis, 179
Energy
  principle of conservation of, 49
  total, 49, 453

Energy inner product, 602
Energy of acceleration, 517
Equation of motion, 31
  damped systems, 255
  differential-algebraic, 261
  Euler's, 437
  Gibbs-Appell, 518
  Kane's, 525
  Lagrange's, 255
  linearized, 55, 282, 287
  in matrix form, 287
  rigid body, 431
  in state form, 76, 312, 447
Equilibrium
  critically stable, 55
  dynamic, 245
  nonnatural systems, 276
  small motions about, 279
  stable, 54
  static, 54, 241
  unstable, 55
Escape velocity, 26, 183
Euclid, 670
Euler, L., 26, 360, 673, 680, 687
Euler angles
  for aircraft problems, 373, 419
  angular acceleration, 371
  angular velocity, 369
  properties of, 419
  sequence of, 368, 460
Euler parameters, 381, 463
  in terms of direction cosines, 387
Euler's equations, 435, 437
  modified, 436, 490
Euler-Bernoulli equation, 610
Euler-Lagrange equation, 255
Expansion, 138, 441, 575, 581
Expansion theorem, 293, 619
Extremum, 687

F

F frame, 372
First integral, 72, 301, 695

First variation, 696
Flat body, 554
Flight path angle, 180
Force
   central, 41, 175
   centrifugal, 136
   circulatory, 287
   conservative, 44
   constraint, 136, 155
   Coriolis, 136
   definition of, 29
   effective, 135
   external, 155
   friction, 30
   generalized, 239, 508
   impulsive, 37, 164, 305
   inertia, 245, 431
   internal, 155
   nonconservative, 49
   normal, 30
Foucault, 675
Foucault's pendulum, 141, 283
Free body diagram, 31, 255
Frenet's formulas, 13
Frequency
   average, 143
   beat, 143
   natural, 59, 290
   response, 63
Friction
   coefficient of, 29
   Coulomb, 29
   dry, 29
   treatment in analytical mechanics, 260, 444
Function
   admissible, 612, 627
   comparison, 612, 619, 627
   consistent admissible, 627
   consistent comparison, 627
Functional, 693
Fundamental equations, 527
   Gibbs-Appell form, 527
   Kane's form, 527
Fundamental function, 518

G

Galileo, G., 672, 681
Gauss, C.F., 675, 681
Gauss's variational principle, 315
Generalized coordinate, 216
   constrained, 218, 261
   ignorable, 300
   independent, 218
Generalized force, 239, 457
   in assumed modes method, 630
   inertia, 525
   for partial velocity, 507, 513
Generalized impulse
   for generalized coordinates, 305
   for generalized speeds, 539
Generalized inertia force, 525
Generalized momentum, 300, 459
Generalized speed, 378, 504
   constrained, 504
   independent, 504, 513
   surplus, 504, 513
Generalized velocity, 232
Generatrix, 359
Geosynchronous orbit, 182
Gibbs, J.W., 517, 681
Gibbs function, 517
Gibbs-Appell equations, 518
Gibbs-Appell function, 517
   for a rigid body, 520
Gimbal lock, 394
Ginsberg, J.H., 24
Gradient operation, 92
Gravitational constant, 3, 28
   universal, 28, 173
Guide bars, 397
Gyrocompass, 582
Gyropendulum, 377, 438
Gyroscope
   balanced, 581
   free, 581
   integrating, 586
   single-axis, 585
Gyroscopic effect, 358
Gyroscopic matrix, 283

Gyroscopic moment, 434
Gyroscopic motion, 547
Gyroscopic system, 142
Gyroscopic vector, 275
Gyrostat, 494

H

Hamilton, W.R., 249, 681
Hamilton's equations, 251
Hamilton's principle
   for continuous systems, 604
   extended, 251
   of varying action, 250
Hamiltonian, 308
Herpolhode, 559
Hessian matrix, 688
Hilbert, D., 621
Hohmann transfer, 185
Holonomic
   constraint, 220, 405
   system, 220, 226
Hooke's law, 48, 597, 599
Hybrid systems, 663
Hyperbolic excess speed, 184

I

Ignorable coordinates, 300
Impact
   elastic, 167
   line of, 164, 477
   particle, 164
   plastic, 167
   rigid body, 477
Impetus, 671
Impulse, 36
Impulse response, 69
Impulse-momentum theorem
   angular, 40, 159, 204, 449
   linear, 36, 157, 204, 449
Impulsive force, 37
Inclusion principle, 632

Inertia ellipsoid, 347, 558
   normal to, 559
Inertia matrix, 275, 328
Inertia properties
   axisymmetric bodies, 344
   parallel axis theorem, 336
   rotation of coordinates, 337
   translation of coordinates, 335
Inertia torque, 525
Inertial symmetry, 372, 548
Inflection point, 688
Inner product, 602
Instantaneous center zero velocity, 200
Integral of the motion, 72, 301, 695
Internal properties, 329
International Gravity Formula, 29
Invariable plane, 559

J

Jacobi integral, 301, 309
Joint
   ball and socket, 392
   Cardan, 393
   constant velocity, 394
   prismatic, 130, 395
   revolute, 130, 391
   universal, 393
Jordanus, 670
Jourdain, P.E.B., 678, 682
Jourdain's variational principle, 314
Junkins, J.L., 371

K

Kane's equations, 525
Kepler, J., 181, 671, 682
Kepler's laws of planetary motion, 181
Kepler's equation, 189
Kinematic differential equations, 370, 446, 505

Kinematics
  infinitesimal rotation, 107
  joints, 391
  plane motion rotation, 195
  rigid body, 355
  rolling, 200, 400
Kinetic energy
  beams, 597
  linearized form, 281
  particles, 43, 162
  rigid bodies, 203, 451
  rotational, 162, 452
  torsion, 599
  translational, 162, 453, 597
  two body problem, 175
Kronecker delta, 293

L

Lagrange, J. L., 255, 674, 683
Lagrange multipliers, 260, 464, 689
Lagrange's equations, 255
  for constrained systems, 260, 464
  for quasi-coordinates, 534
  for rigid bodies, 456
  in terms of Euler angles, 462
  in terms of Euler parameters, 463
Lagrangian, 251, 303
Line of impact, 164, 477
Line of nodes, 190, 369, 577
Linear momentum
  conservation of, 37
  particles, 26, 155
  rigid bodies, 422
Linear range, 47
Linearization
  damped systems, 287
  about equilibrium, 283
  of free gyroscope, 581
  nonnatural systems, 283
  of rolling motion, 576
  Taylor series expansion, 55, 278
Longitude of ascending node, 191
Longitude of descending node, 191

M

Mach, E., 669
Magnification factor, 64
Mass, 2
Mass flow rate, 172
Matrix
  circulatory, 287
  identity, 99
  inertia, 275, 328
  mass, 275
  orthogonal, 343
  positive-definite, 280, 290, 328, 688
  rotation, 101
  stiffness, 280
  symmetric, 280, 328, 688
  unitary, 99
Meirovitch, L., 108, 298
Mixed descriptions, 24
Modal analysis, 296, 617
Modal coordinates, 296
Modal equations, 296
Modal force, 296, 620
Modal vector, 291
Modified Euler equations, 436, 490
Modulus of elasticity, 48, 597
Moment equations
  arbitrary bodies, 434
  about an arbitrary point, 498
  axisymmetric bodies, 436, 490
  classification, 503
Moment of inertia
  area, 330
  area polar, 330
  mass, 195, 327, 423, 598, 652
  principal, 329, 343
Momentum
  angular, 39, 196, 423
  generalized, 300, 459
  linear, 26, 36, 196, 422
Momentum wheel, 112
Motion
  primary, 637
  relative to earth, 134
  secondary, 637

N

Natural frequency, 59, 290
Natural mode, 291
Newton, I., 26, 166, 478, 673, 683
   law of gravitation of, 28
   laws of, 27, 434
   Newton-Euler formulation, 432
Node, 619
Nonholonomic, 109
   angular velocity as, 109
   constraint, 220
   system, 223
North polar axis, 189
Nutation, 369

O

Orbit
   circular, 179
   elliptic, 179
   hyperbolic, 183
   parabolic, 183
Orbital inclination, 190
Orbital parameters, 187
Orthogonality, 293, 631
Orthonormality, 293
Osculating plane, 11

P

Parallel axis theorem, 197, 336
Partial velocity, 506
   matrix, 506
Particle, definition of, 26
Particle assumption, 26
Passarello-Huston equations, 533
Path variables, 8
Pericenter, 179
Perigee, 180
Perturbation, 138, 441, 575
Pfaffian form, 220
Phase portrait, 72
Pitch rate, 375
Plane motion, 193
Plane of contact, 400
Plane of symmetry, 328
Poincari, J.H., 677, 684
Poinsot's construction, 557
Poisson, S.D., 165, 478, 675, 684
Poisson's hypothesis, 165, 478
Poisson's ratio, 599
Polhode, 559
Position vector
   cylindrical coordinates, 18
   rectilinear coordinates, 5
   spherical coordinates, 21
Potential energy, 44, 239
   beams, 598
   elastic, 48, 598, 600
   gravitational, 44, 163
   modified, 276
   springs, 46
   theorem, 56, 281
   torsion, 600
Power, 50, 454
Precession, 369
   direct, 553
   looping, 567
   retrograde, 554
   steady, 569, 574
   unidirectional, 567
Principal axes, 343, 348
   axisymmetric bodies, 343, 348
Principal coordinates, 296
Principal line, 360
Principal plane, 345
Principia, 26
Principle of least action, 251
Principle of virtual work, 243
Product of inertia
   area, 598
   mass, 327
Ptolemy, C., 684
Pure rotation, 356

Q

Quasi-coordinate, 505
  variation of, 508
Quasi-velocity, 378, 505
  constrained, 378, 505

R

Radian, 1
Radius of curvature, 9, 400, 593
Radius of gyration, 197, 332
Rayleigh's dissipation function, 286
Reference frame
  for axisymmetric bodies, 372
  body-fixed, 356
  fixed, 88
  fixed to earth, 133
  inertial, 26
  moving, 88
  relative, 133, 630
Relative acceleration, 125
Relative velocity, 124
Resonance, 64
Response
  damped systems, 61, 298
  free, 58, 298
  general, 68
  to harmonic excitation, 65
  in phase, 65
  out of phase, 65
Restitution period, 166, 478
Resultant, 196, 429
Rigid body
  assumption, 323, 591
  equations of motion, 431
  kinematics, 355
Rigid body mode, 291, 624
Rivello, R.M., 601
Rodriques parameters, 415
Roll rate, 375
Rolling
  of inertia ellipsoid, 559
  kinematics, 400
  with slip, 200, 401
  without slip, 201, 401, 573
Rotating shaft, 659
Rotation, 102
  of earth, 133
  finite, 103
  about a fixed axis, 356, 443
  about a fixed point, 358, 563
  infinitesimal, 107
  about principal line, 360
Rotation sequence, 107
Routh, E.J., 684
Routhian, 303

S

Saddle point, 689
Sampling period, 75
Semilatus rectum, 179
Semimajor axis, 179
Semiminor axis, 179
Separatrix, 72, 561
Servomotor, 440
Shadow frame, 637
Shear force, 593
Shear modulus, 599
Shortening of projection, 595
Sleeping top, 571
Slender body, 553
Slider, 395
Sliding in rolling, 577
Smith, C., 650
Space-fixed rotation, 100
Spherical pair, 392
Spherical pendulum, 23
Spin, 369
Spin slip in rolling, 577
Spring
  axial, 46
  linear range, 47
  softening, 47
  stiffening, 47
  torsional, 47
Spring constant, 46
St. Venant's hypothesis, 599

Stability
  asymptotic, 55
  critical, 55
  free motion, 441, 562
  neutral, 55
  theorem, 56, 281
State form, 76, 308, 312
Stationary value
  of constrained systems, 690
  of a definite integral, 693
  of a function, 688
Steady motion, 574
Steady precession, 574, 582, 584
Stiffness
  axial, 598
  bending, 598
  torsional, 600
Stiffness coefficients, 280
Strain
  axial, 48
  shear, 599
Stress
  axial, 48
  shear, 599
Sturm-Liouville problem, 697
Sylvester's criterion, 328, 688
Synchronous motion, 289
System
  conservative, 56
  holonomic, 220, 226
  natural, 275
  nonnatural, 275
  rheonomic, 220
  scleronomic, 220
  undamped, 59
  underdamped, 61

T

Tadikonda, S.S.K., 654
Taylor series expansion, 55, 279
Time of pericenter passage, 189
Torque-free motion, 548, 556

Torsion of curve, 12
Torsional stiffness, 600
Tracking frame, 637
Transfer orbit, 185
Transmissibility, 67
Transport theorem, 117, 121
Trial function, 628
Truesdell, C.T., 432, 669
Turner, J., 371
Twisting moment, 593
Two body problem, 173

U

Unit step function, 69
Unit vector, 4
  cylindrical coordinates, 18
  rectilinear coordinates, 4
Units
  SI, 2
  U.S. Customary, 2

V

Variable mass, 170
Variation of latitude, 552
Variation of quasi-coordinates, 508
Variational notation, 230, 696
Vector
  algebraic, 91
  column, 91
  geometric, 91
  gyroscopic, 275
  spatial, 91
Vehicle dynamics, 224
Velocity
  base, 124
  partial, 379, 459, 506
  relative, 124
  transport, 125
Vernal equinox, 189
Vertex, 563

Virtual displacements, 230, 699
   analytical approach, 232
   kinematical approach, 232
Virtual work, 233
   constraint force, 233
   independent quasi-velocities, 512

W

Warping function, 598
Weighted residual, 628
Whirling, 662
Wobbling motion, 552

Work
   done by a conservative force, 48
   done by a constraint force, 223
   done by a nonconservative force, 49
   incremental, 43
   virtual, 230, 233, 457
Work-energy theorem, 43, 163, 455

Y

Yaw rate, 375
Yu, L., 654